Nucleic Acid Detection

Zhengping Li
Editor

Nucleic Acid Detection

Springer

Chemical Industry Press Co., Ltd.

Editor
Zhengping Li
School of Chemistry and Biological
Engineering
University of Science and Technology
Beijing
Beijing, China

ISBN 978-981-95-7294-6 ISBN 978-981-95-7295-3 (eBook)
https://doi.org/10.1007/978-981-95-7295-3

This Springer imprint is published by the registered company Springer Nature Singapore Pte Ltd.
The registered company address is: 152 Beach Road, #21-01/04 Gateway East, Singapore 189721, Singapore

Preface

The Human Genome Project has spurred rapid advancements in high-throughput nucleic acid sequencing technologies, enabling us to acquire an ever-increasing amount of nucleic acid sequence information and uncover the correlations between nucleic acid sequences and biological processes. As a result, nucleic acid analysis has become increasingly crucial in life sciences, precision medicine, and the diagnosis of major diseases. With the rapid development of society and science, new challenges and higher demands are continuously being placed on nucleic acid analysis, necessitating the establishment of novel analytical methods to meet the significant needs of scientific research and human health. However, until now, there has been no systematic professional book dedicated to nucleic acid analysis. This book aims to fill that gap.

The preparation and completion of this book has received strong support of my colleagues and the enthusiastic encouragement of Editor Li from the publishing house. Although I have been engaged in nucleic acid analysis research for over two decades and have long intended to systematically summarize the field along with some of my research insights, limitations in energy and expertise had previously deterred me from doing so. The outbreak of COVID-19 has underscored the importance of nucleic acid analysis for human health, prompting us to invite many young scholars to contribute to the relevant chapters of this book, building on the foundation of our professional course on nucleic acid analysis.

This book is divided into two parts: the Fundamental Section and the Application Section, comprising a total of 12 chapters. The Fundamental Section (Chaps. 1, 2, 3, 4, 5, 6, and 7) introduces the principles of various nucleic acid amplification reactions and the detection methods for amplification products, including Polymerase Chain Reaction (PCR), Ligase Chain Reaction (LCR), Loop-mediated Isothermal Amplification (LAMP), Rolling Circle Amplification (RCA), Exponential Amplification Reaction (EXPAR), Helicase-dependent Amplification (HAD), Recombinase Polymerase Amplification (RPA), and nucleic acid detection based on the gene editing system (CRISPR/Cas). The Application Section (Chaps. 8, 9, 10, 11, and 12) discusses the biological functions and analytical methods of various nucleic acid biomarkers, including DNA Single Nucleotide Polymorphisms (SNPs), DNA methylation, microRNA, messenger RNA splice variants and fusion gene transcripts, messenger RNA methylation, and novel coronavirus RNA. Every chapter and the authors are listed as follows: Chap. 1 was written by Dr. Pengbo Zhang, Chap. 2 was written by Dr. Fengxia Su and Yuting Jia, Chap. 3 was written by Dr. Hui Wang, Chap. 4 was written by Dr. Fengxia Su and Gaoting Wang, Chap. 5 was written by Dr. Honghong Wang, Chap. 6 was written by Dr. Hui Wang and Weixiang Hong, Chap. 7 was written by Dr. Zhian Hu, Song Wang, Mai Zhang and Yanfei Che, Chap. 8 was written by Dr. Pengbo Zhang, Chap. 9 was written by Dr. Fengxia Su and Weiliang Liu, Chap. 10 was written by Dr. Honghong Wang and Kejian Gao, Chap. 11 was written by Dr. Hui Wang and Weiliang Liu, Chap. 12 was written by Dr. Yang Sun and Weixiang Hong.

This book strives to embody systematicity, scientific rigor, and accuracy, balancing foundational knowledge with cutting-edge research findings. It is designed to serve both as an introductory text for those new to nucleic acid analysis and as a reference for scientific research. The writing of this book has been greatly supported and assisted by the leadership of the School of Chemistry and Biological Engineering, University of Science and Technology Beijing, the Academic Affairs Office, the Graduate School, and the Chemical Industry Press. Graduate students Pian Hongru, Zhao Yuecan, Liu Kai, Zhang Baoshui, Wang Hao, Luo Zhiheng, Wang Chunfang, Wang Song, Li Xiang, Wen Jingshuai, and Zhao Meng contributed to the illustration and proofreading of this book. We extend our heartfelt thanks to all of them.

This book is intended for researchers and practitioners in the fields of chemistry, biology, medicine, and nucleic acid analysis, and can also be used as a textbook or reference for graduate and advanced undergraduate students in higher education institutions.

Finally, due to the broad scope of the book and the limitations of the author's expertise, there may be inevitable inaccuracies, overlaps, or omissions in the content. We sincerely welcome readers' criticisms and suggestions for improvement.

Beijing, China
March 2025

Zhengping Li

Contents

About the Editor

Zhengping Li is a Professor and Doctoral Supervisor at the University of Science and Technology Beijing, and the Dean of the School of Chemical and Biological Engineering. He also serves as the Director of the Beijing Key Laboratory of Bioengineering and Sensing Technology. He obtained his Ph.D. in Science from the College of Chemistry and Molecular Engineering at Peking University in 1999. From 2000 to 2002, he conducted postdoctoral research at the Central Research Laboratory of Hitachi, Japan. He was recognized as a National Excellent Teacher in 2007 and received the National Science Fund for Distinguished Young Scholars in 2009. In 2011, he was awarded funding from the Ministry of Education's Innovation Team and received continued support. In 2016, he was selected as an expert receiving special government allowances from the State Council. His primary research focuses on biochemical analysis and molecular diagnostics, including high-sensitivity analysis of nucleic acid and protein biomarkers and disease diagnosis, biosensing analysis at micro-nano interfaces, single-cell analysis, and high-throughput nucleic acid sequencing. He was awarded the first and second prizes of the Hebei Province Natural Science Award in 2010 and 2018, respectively, and the second prize of the Shaanxi Province Natural Science Award in 2021 (as the second contributor). He serves as an editorial board member of *Chinese Journal of Analytical Chemistry* and is a council member of the 31st Chinese Chemical Society.

Part I
Fundamental Section

Chapter 1
Polymerase Chain Reaction (PCR)

Pengbo Zhang

1.1 Brief History of PCR

With the in-depth study of the physical and chemical properties of DNA, scientists gained deeper understanding of DNA denaturation/refolding processes and semi-conservative replication mechanisms, laying the foundation for modern molecular biology. In the late 1960s and early 1970s, researchers shifted their research focus to developing in vitro gene isolation techniques. However, the low gene content limited the in vitro gene isolation to some extent. In 1971, Khorana first proposed the concept of in vitro nucleic acid amplification: "After denaturation, appropriate primers can hybridize with denatured DNA and extend under the catalysis of DNA polymerase. This repetitive process can yield large quantities of tRNA gene fragments." By 1983, Kary Mullis developed the prototype of PCR technology using Klenow DNA polymerase fragment isolated from *Escherichia coli* to establish the in vitro DNA amplification. After 2 years of dedicated research, Mullis and colleagues published the first academic paper about PCR in Science in 1985 [1], and secured the first related patent, earning him the 1993 Nobel Prize in Chemistry. However, the initial Klenow DNA polymerase lacked thermostability and was prone to inactivation during thermal cycles, requiring frequent enzyme addition after the denaturation of each cycle. This complexity led to operational difficulties and frequent errors, ultimately limiting the technology's widespread adoption. In early 1988, Keohanog replaced Klenow DNA polymerase with T4 DNA polymerase for PCR amplification, achieving significant improvements in results. After PCR amplification, the amplified products showed uniform fragments with detectable DNA segments of expected molecular weights. However, T4 DNA polymerase required a 37 °C optimal reaction temperature and still needed to be supplemented after each

P. Zhang (✉)
University of Science and Technology Beijing, Beijing, China
e-mail: zhangpengbo@ustb.edu.cn

Z. Li (ed.), *Nucleic Acid Detection*, https://doi.org/10.1007/978-981-95-7295-3_1

thermal denaturation cycle, making the process cumbersome. Scientists hypothesized that a heat-resistant DNA polymerase could enable PCR to operate spontaneously by adjusting temperatures. In 1988, Saiki et al. [2] isolated a thermostable DNA polymerase from *Thermus aquaticus* (a thermophilic bacterium found in hot springs), naming it Taq DNA polymerase. This enzyme maintained activity during high-temperature denaturation of PCR without requiring additional enzyme supplementation, simplifying procedures and enhancing amplification efficiency. The discovery and commercialization of Taq DNA polymerase drove the invention of fully automated thermal cyclers (PCR instruments), enabling PCR automation and significantly accelerating the adoption of this technology.

Since its inception, PCR has undergone continuous evolution. From using DNA as template for amplification to directly using RNA as the template, that is reverse transcription PCR. Expanding from amplifying DNA fragments between two known sequences to detecting unknown regions on both sides of target sequences (reverse PCR). Evolving from qualitative detection of target genes to quantitative analysis (quantitative PCR), and even to achieve absolute quantification of the original template (digital PCR). From originally amplifying only a few kb of genes to handling DNA fragments spanning up to tens of kb. With its rapid development, PCR has been widely used in molecular biology, genetics, microbiology, and even life science research, significantly advancing the field's development.

1.2 Basic Principles of PCR

PCR is an in vitro nucleic acid amplification technique that mimics the natural DNA replication process. While its DNA synthesis mechanism resembles cellular replication, PCR uniquely controls reaction progression through temperature cycling [3, 4]. As shown in Fig. 1.1a, using the DNA molecules as the template, the amplification process begins with high-temperature denaturation. When the

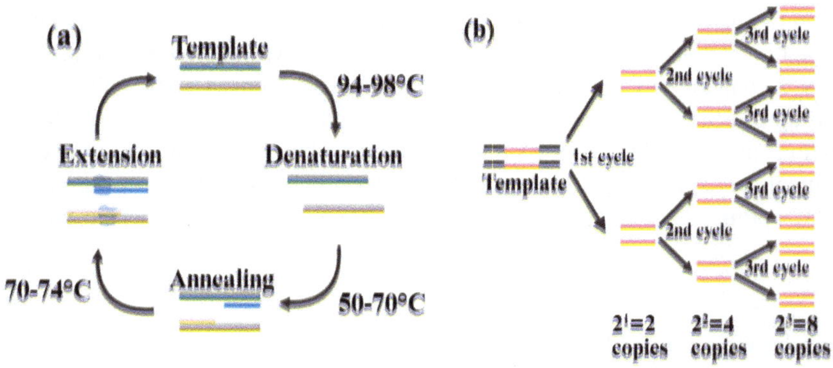

Fig. 1.1 (**a**) Schematic diagram of PCR. (**b**) The exponential amplification of PCR

temperature reaches 94–98 °C, hydrogen bonds between the double-stranded DNA are broken, forming two single-stranded DNA. The second step is low temperature annealing, where the temperature decreases to 50–70 °C, a pair of oligonucleotide fragments (primers) hybridize to the 3′ ends of the two single-stranded templates, respectively. The third step is optimal temperature extension. When the temperature rises to the optimal temperature of Taq DNA polymerase (70–74 °C), the primers are extended along the template by DNA polymerase until the formation of the two complete DNA duplexes. Repeating this process continuously allows target amplification. As shown in Fig. 1.1b, PCR is exponentially amplified in the form of 2^n (n is the number of cycles) theoretically. The final products are specific double-stranded DNA fragments with primers as termini. Throughout PCR, the denaturation of template DNA, primer annealing, and synthesis of new DNA strands are precisely controlled by temperature regulation, enabling continuous replication cycles through precise temperature management. Each newly synthesized DNA strand can serve as a template for DNA replication in the next cycle, so one of the newly synthesized DNA strands comes from the product of the previous cycle amplification and the other is produced by primer-mediated generation. In this respect, the PCR amplification process still follows the semi-conservative replication mechanism.

1.3 System Design of PCR

1.3.1 Reaction System of PCR

The PCR reaction system includes the following basic components: template, primers, DNA polymerase, deoxynucleotide triphosphates (dNTPs), divalent cations, and buffer.

(1) Template

Template DNA refers to DNA containing the desired amplification fragment. The quantity and purity of this DNA serve as critical determinants for PCR success. Almost all DNA types, such as genomic DNA, phage DNA, plasmid DNA, pre-amplified DNA, and single-stranded DNA, can be used as templates for PCR. While PCR reactions have relatively low requirements for template purity, certain components in samples such as proteases, nucleases, DNA polymerase inhibitors, and DNA-binding proteins may affect reaction efficiency. In addition to purified DNA, PCR can also directly use DNA from cell lysates as templates for amplification.

(2) Primers

A primer is an artificially synthesized oligonucleotide fragment that specifically hybridizes to the 3′ end of the target amplification region, serving as the starting point for DNA replication. Primers are categorized into forward primers (FP) and reverse primers (RP). They are crucial for PCR specificity and directly influence amplification efficiency. The primer concentration typically ranges from

0.1 to 0.5 μmol/L. Insufficient concentration may reduce yield, while excessive concentration could lead to mismatches, nonspecific amplification, and increased formation of primer dimers.

The most important principle in primer design is to maximize the amplification efficiency and specificity while minimizing the nonspecific amplification [5]. The following principles are generally followed when designing PCR primers: (I) Primers should be sufficiently specific. For amplifying specific DNA sequences from complex templates like genomes, consider the conservation of target DNA sequences to avoid nonspecific regions in selected primer designs. If unavoidable, the homology between the primers and the nonspecific sequences should be kept below 70% or there should not be 8 consecutive complementary bases. (II) Avoid template regions prone to secondary structure formation, including intramolecular complementary sequences, hairpin structures, or intermolecular complementary sequences. Secondary structures complicate primer-template hybridization and amplification, so these areas should be minimized. Template secondary structure prediction software can be used before primer design. (III) Primers typically range between 18 and 30 bases, with 20-base length being common. Primer length correlates positively with reaction specificity and melting temperature (T_m). Shorter primers may reduce amplification specificity, while longer ones risk forming secondary structures or dimerization. (IV) The T_m difference between the two primers should be less than 2–3 °C. Significant T_m differences can directly lead to reduced amplification efficiency or failure. Generally, the optimal annealing temperature forDiagram illustrating a molecular process PCR should be approximately 5 °C lower than the T_m value of primer. (V) Primer base composition also affects T_m values, with GC content typically maintained between 40% and 60%. Insufficient GC content leads to low primer T_m values, causing primers' dissociation from templates and poor amplification efficiency. Excessive GC content results in high primer T_m values, increasing hybridization with nonspecific sequences and leading to nonspecific amplification. (VI) Primers should ideally have random distribution of four bases, avoiding repeated polypurines or polypyrimidines sequences (more than four repetitions). Notably, the 3′ end should not contain three consecutive G or C bases, as this may cause mismatches in GC-rich regions and reduce amplification specificity. (VII) Avoid complementary sequences within or between primers. Continuous complementary bases should not exceed three per primer and four between primers, particularly avoiding overlapping at the 3′ end. (VIII) The 5′ end of the primers can be modified, while the 3′ end cannot. The 5′ end of the primers determines the length of the PCR products and allows modifications such as adding restriction enzyme sites, biotin labeling, fluorescent groups, or introducing point mutations—none of which affect amplification specificity. Since extension starts from the 3′ end, no modifications are permitted there. (IX) The 3′ end of the primer cannot be A-terminated. Different bases at the 3′ end significantly influence mismatch initiation efficiency: A-terminal bases may initiate chain synthesis even in the case of mismatch, whereas T-terminal bases greatly reduce mismatch initiation efficiency. G and C mismatches show intermediate efficiency compared to A and T, making T the preferred 3′ end choice. (X) The 3′ end should avoid codon third

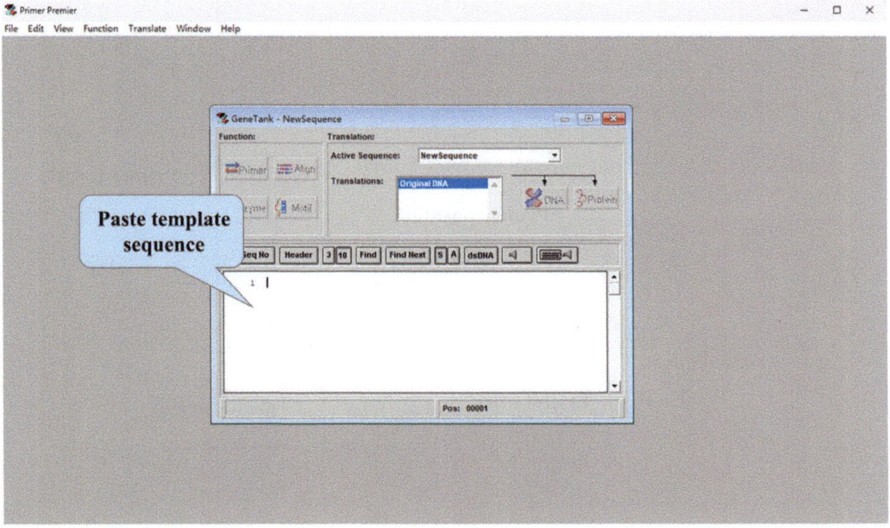

Fig. 1.2 Primer Premier running interface

positions. If the amplification sequence is located in the gene coding region, the 3′ end of the primers should not be chosen at the third position of the terminate codon, because the third position of the codon is prone to degeneracy, which will affect the specificity and efficiency of amplification.

When designing PCR primers, it's challenging to fully adhere to the aforementioned principles. To save time and minimize issues during amplification, primers can be designed, selected, and optimized using software. Commonly used primer design software includes Primer Premier, Oligo, DNAMAN, IDT, and Primer-Blast. Taking Primer Premier as an example, open the software, click "File-New-DNA sequence" in the top left corner. In the blank area of the dialog box, either manually enter the template DNA sequence or directly paste it into the designated space as shown in Fig. 1.2.

After pasting, there are four options in the pop-up dialog box. "As is" means using the input DNA sequence as the template directly; "Reversed" means the reverse sequence of the pasted sequence will be used as the template; "Complemented" means using the complementary sequence of the pasted sequence as the template; "Reverse Complemented" means using the reverse complementary sequence of the pasted sequence as the template. After selecting the template DNA sequence, click the Primer button in the top-left corner to open the primer design dialog box. Click the Search button to display various primer design parameters, as shown in Fig. 1.3.

Enter the expected position, length and PCR product size for the forward and reverse primers, respectively, and click the OK button. The dialog box that appears lists key parameters to consider during primer design, such as T_m value, GC content, error triggering, etc. After selecting relevant options, click OK again. The software-

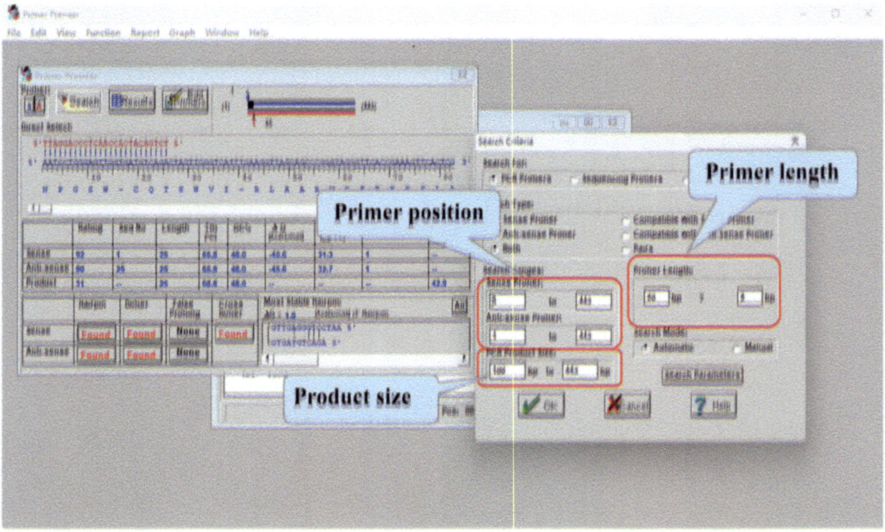

Fig. 1.3 Parameter settings for designing primers using Primer Premier

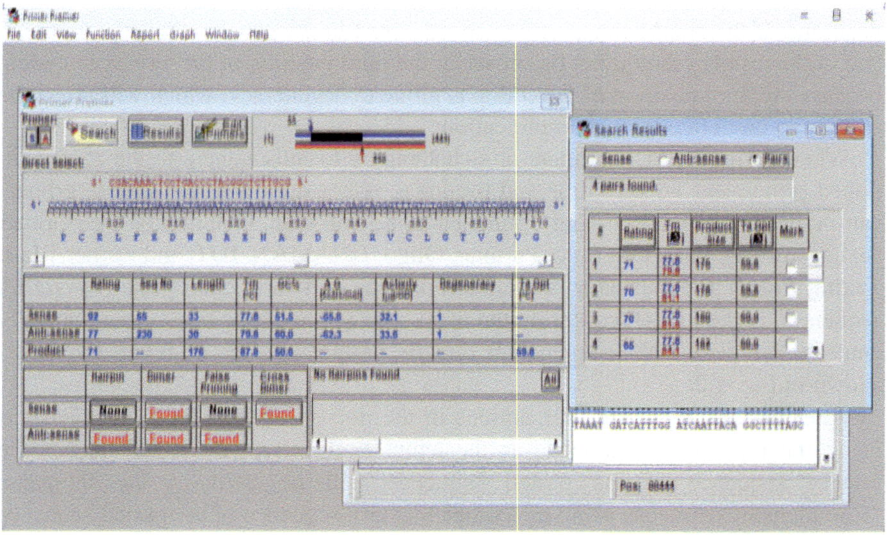

Fig. 1.4 Primer information of Primer Premier design

generated PCR primers are then displayed in a sorted list by evaluation scores. When selecting a desired primer, the Primer dialog box on the left will show details including length, T_m value, GC content, hairpin structure, dimer formation, false priming, etc., as shown in Fig. 1.4. If the designed primers do not meet the requirements, click the Edit Primers button to modify and optimize the primers until satisfactory results are achieved.

(3) DNA Polymerase

Thermostable DNA polymerase is the key to achieving automation in PCR. Currently, the most commonly used PCR polymerase is Taq DNA polymerase, which is isolated from the thermophilic aquatic bacterium *Thermus aquaticus* YT-1 strain found in hot springs. Taq DNA polymerase contains a full-length sequence of 2499 base pairs that translates into a protein with a molecular weight of 94 kDa and 832 amino acids. The main characteristics of Taq DNA polymerase are as follows.

High thermal stability: Taq DNA polymerase has half-lives of 130 min, 40 min, and 5–6 min at 92.5 °C, 95 °C, and 97.5 °C, respectively. The denaturation time of DNA in PCR reactions is usually 30 s. If a total of 30 cycles are performed, the cumulative denaturation time is about 15 min. At this time, Taq DNA polymerase still maintains a fairly high activity to meet the efficiency of PCR amplification.

High catalytic activity: Taq DNA polymerase has the highest catalytic DNA synthesis activity of all the thermostable DNA polymerases found so far, reaching 200,000 U/mg. It has been demonstrated that the activity of Taq DNA polymerase is temperature-dependent, and its optimal temperature for catalyzing DNA synthesis is 75–80 °C, at which point the elongation rate is 150–300 nucleotides/(s*molecule). The elongation rate is 60 nucleotides/(sec-enzyme molecule) at 70 °C; 22 nucleotides/(s*enzyme molecule) at 55 °C; 1.5 nucleotides/(s*enzyme molecule) at 37 °C and 0.25 nucleotides/(s*enzyme molecule) at 22 °C, respectively. When the reaction temperature exceeds 80 °C, Taq DNA polymerase could barely catalyze the synthesis of DNA, which could also be caused by the unstable binding of primers to the template at high temperature.

Reaction fidelity: Taq DNA polymerase has 5′–3′ exonuclease activity, i.e., excision and repair function, which can hydrolyze the DNA strand in front of the DNA extension strand from the 5′–3′ direction. Taq DNA polymerase does not have 3′–5′ exonuclease activity and proofreading function. Therefore, it is prone to cause base mismatches during catalytic DNA extension.

During PCR, even a short incubation at temperatures significantly below the T_m can lead to the formation of nonspecific amplification products such as primer dimers. To avoid such problems, hot-start DNA polymerase can be used. As shown in Fig. 1.5, hot-start Platinum Taq DNA polymerase is a recombinant Taq

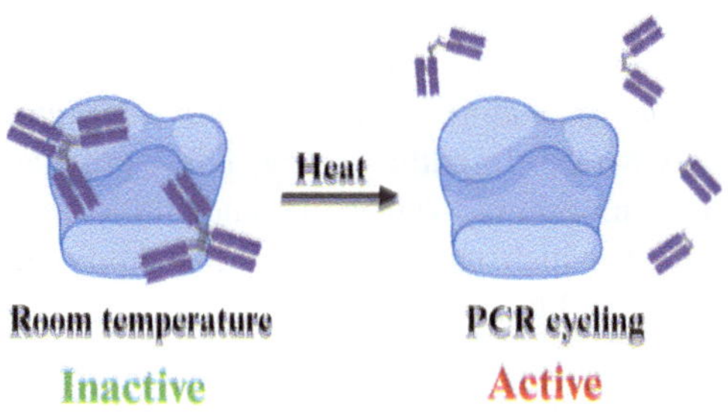

Fig. 1.5 Schematic diagram of hot-start Taq DNA polymerase

DNA polymerase compounded with an anti-Taq DNA polymerase monoclonal antibody. At room temperature, the anti-Taq DNA polymerase monoclonal antibody specifically binds to Taq DNA polymerase, at which point the active site of Taq DNA polymerase is blocked and cannot induce primer extension. Incubation at 94 °C for 2 min is sufficient to inactivate the antibody and restore 90% of the Taq DNA polymerase activity. Therefore, before performing PCR amplification, the reaction system is incubated at 94 °C for 2 min to destroy the anti-Taq DNA polymerase monoclonal antibody, and the active site of Taq DNA polymerase is re-exposed and can catalyze primer extension. The presence of hot-start Taq DNA polymerase greatly reduces the nonspecific amplification during PCR and improves the specificity of the assay.

(4) dNTPs

The dNTPs used in standard PCR system contain four equal molar concentrations of dATP, dCTP, dGTP, and dTTP. In PCR reaction, the concentration of dNTPs is generally 100–250 μmol/L. Too low concentration will reduce the PCR yield, and too high concentration will quench Mg^{2+}, resulting in a lower concentration of free Mg^{2+} and thus inhibiting the PCR reaction.

(5) Divalent Cation

The activity of DNA polymerase requires divalent cations as activator, and Mg^{2+} is commonly used. Buffers containing Mn^{2+} can also make some polymerases catalytically active, but their efficiency is much lower than Mg^{2+}. In contrast, Ca^{2+} cannot activate DNA polymerase. The concentration of Mg^{2+} is generally between 1.5 and 2.0 mmol/L. The concentration of free Mg^{2+} in the reaction system has an important effect on the PCR reaction. A low concentration of Mg^{2+} will cause the activity loss of Taq DNA polymerase and decrease the amount of PCR products. A high concentration of Mg^{2+} will affect the reaction specificity. The concentration of Mg^{2+} also affects the annealing of primers, the *Tm* of the template and PCR products, and the formation of primer dimer. Since Mg^{2+} can bind to negative ions, the concentration of dNTPs and EDTA in the reaction system affects the concentration of free Mg^{2+} in the reaction.

(6) Buffer

During PCR amplification, the reaction buffer, such as pH and salt concentration, can provide the necessary acidity and certain ionic strength for the reaction. In the actual PCR reaction system, the pH of the buffer is between 6.8 and 7.8. The commonly used buffer nowadays is Tris-HCl buffer, which has a pH of about 7.2 at 72 °C.

The reaction buffer also contains KCl. KCl below 50 mmol/L is beneficial for primer annealing, while KCl above 50 mmol/L inhibits the activity of Taq DNA polymerase. In addition, Taq DNA polymerase protectors, such as bovine serum protein, gelatin, Tween-20, and dithiothreitol, need to be added to the reaction buffer, which can improve the stability of Taq DNA polymerase and facilitate PCR amplification.

1.3.2 Basic Steps of PCR

PCR is a DNA synthesis process consisting of three basic steps, denaturation, annealing, and extension, which constitute one cycle. The newly synthesized DNA molecules can be used as template for the next cycle of DNA synthesis, so the DNA fragment can be efficiently amplified after several cycles.

(1) Denaturation
Under high temperature conditions, the hydrogen bonds of the DNA double helix are broken. At this point, the double-stranded DNA is denatured to single-stranded DNA, which becomes the active template for primer binding. The denaturation temperature of the template is determined by its GC content. The higher the GC content, the higher the melting temperature. The time of denaturation is determined by the length of the template. The longer the DNA strand, the longer it takes for the two strands to completely separate. If the denaturation temperature is too low or the time is too short, only some regions in the template DNA are denatured, and as the temperature decreases, the template DNA will compound into a double-stranded structure and cannot bind to the primer. During PCR, the denaturation temperature is usually between 94 and 98 °C and the denaturation time is usually 30–50 s.

(2) Annealing
When the temperature of the system is lowered to a specific temperature, the denatured template binds specifically to the primers in a base-complementary pairing manner, forming a primer-template complex. The annealing temperature is an important factor affecting the specificity of PCR and depends on the length of the primer, base composition, concentration, and the length of the target sequence, which is generally selected near the T_m value of the primer. If the annealing temperature is too high, the primers cannot bind well to the template, resulting in the lowered amplification efficiency. Too low annealing temperature makes primers prone to produce nonspecific binding, leading to the formation of nonspecific amplification products. The annealing temperature is usually between 50 and 70 °C, and the annealing time is usually 30–90 s.

(3) Extension
At the optimal temperature of Taq DNA polymerase, the primer-template complex is recognized by Taq DNA polymerase and catalyzed to synthesize a new DNA strand complementary to the template DNA using the four dNTPs as raw materials and single-stranded DNA as the template, resulting in the formation of a complete double-stranded DNA. The extension temperature is usually 70–74 °C and the extension time is usually 60–120 s.

(4) Cycle Numbers
The cycle numbers required for PCR amplification depend on the number of starting templates in the reaction system and the efficiency of primer extension and amplification. Once the PCR enters an exponential growth phase, the reaction proceeds automatically until a component is exhausted and becomes the limiting factor, which

brings the reaction to a plateau. For Taq DNA polymerase, 30 cycles are generally sufficient.

1.3.3 The Nonspecific Amplification of PCR

The nonspecific amplification of PCR can be roughly divided into two categories according to the mechanism. The first type is caused by the nonspecific binding of primer and template, and the second type is caused by the hybridization and extension between primers, known as primer dimer. The nonspecific amplification products can be detected by electrophoresis and appear as bands or large diffuse bands that do not match the length of the target fragments. The formation process of primer dimer is shown in Fig. 1.6. If there are matched bases in the forward primer or reverse primer itself, or there are matched bases between the forward primer and reverse primer, especially when the matched bases are located at the 3′ end of the primer, then the matched part of the primers can extend to form a primer dimer under the catalysis of Taq DNA polymerase. In consideration of the short length of primers and the high amplification efficiency, once formed, the primer dimer will take

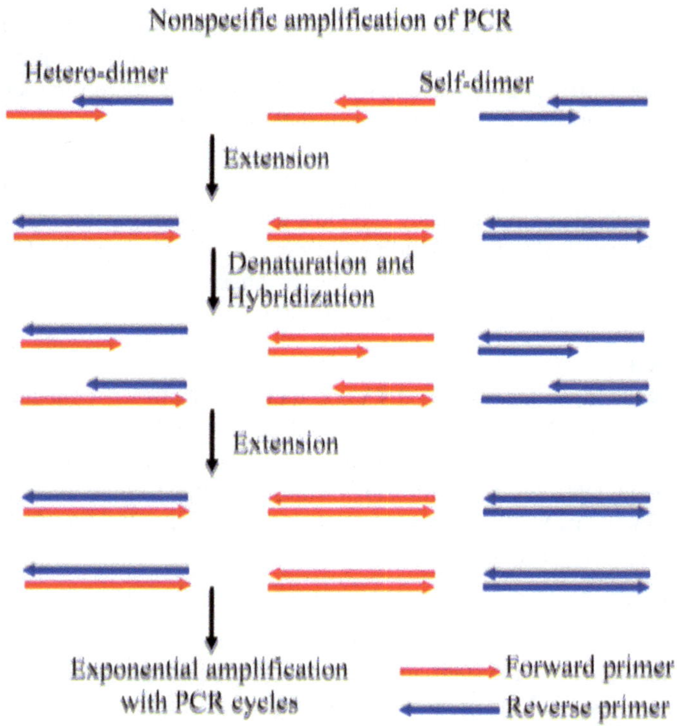

Fig. 1.6 The formation process of primer dimer during PCR

precedence over the specific products for efficient amplification in the subsequent PCR, resulting in the formation of large amounts of nonspecific amplification products. Similarly, the final extension products are not the target fragments when there are matched bases between the primer and the template in the nontarget region. These nonspecific amplification products will be amplified exponentially in the subsequent PCR and eventually produce a large number of nonspecific amplification products. In addition to the low specificity of primers, the causes of nonspecific amplification include too high concentration of template or primer, too much enzyme, too high concentration of Mg^{2+}, low annealing temperature, too many cycles, and so on. These factors may cause nonspecific amplification alone or through interaction.

To eliminate these nonspecific amplification products, researchers continuously optimize PCR reaction conditions. These methods include redesigning primers, appropriately reducing template or primer concentration, decreasing enzyme dosage, lowering Mg^{2+} concentration, raising annealing temperature, reducing cycle numbers, or employing nested PCR. Such strategies aim to minimize interference from nonspecific amplification and enhance PCR amplification efficiency.

1.4 Product Detection of PCR

After PCR amplification, it is necessary to further check whether the results are as expected and whether the products are specifically amplified. Therefore, specific identification and analysis of the amplification products must be performed. The most commonly used methods are based on electrophoresis and fluorescence detection, where electrophoresis mainly includes agarose gel electrophoresis and polyacrylamide gel electrophoresis, and fluorescence detection mainly uses dyes that can bind to double-stranded DNA. Electrophoresis is the easiest and most common method to detect PCR products and is also commonly used to identify, isolate and purify nucleic acids. The electrophoresis detection gives direct information on whether PCR amplification is performed, whether there is nonspecific amplification, and the size of the amplification products.

1.4.1 Agarose Gel Electrophoresis

Agarose gel electrophoresis is a zonal electrophoresis using agarose gel as a support to differentiate nucleic acid molecules by their different migration rates in the gel. Agarose is a linear polysaccharide polymer extracted from algal sugar with a melting temperature between 62 and 65 °C. It is generally dissolved in water when heated to above 90 °C and forms a semi-solid gel with certain pores and a uniform three-dimensional mesh structure when the temperature drops to about 30 °C. When running electrophoresis, alkaline electrophoresis buffer needs to be added. Since

DNA molecules are negatively charged, they will migrate through the pores in the agarose gel from negative electrode to positive electrode under electric field force. DNA molecules of different sizes are subjected to different resistance when they pass through the pores of the gel. At a specific voltage, the migration rate of DNA molecules in a certain concentration of agarose gel electrophoresis is inversely proportional to the logarithm of their molecular weight, i.e., the larger the DNA molecular weight, the greater the resistance to the migration process and the slower the migration. The migration rate of DNA molecules of the same molecular weight in different concentrations of gels is also different. The greater the concentration of agarose gel, the smaller the pores formed, the harder for DNA molecules to pass, and the slower the migration. The migration rate of DNA molecules accelerates with the increase of voltage. The larger the molecular weight of DNA, the greater the magnitude of the migration rate elevates by the increase of electric field strength.

Electrophoresis must be performed in a certain buffer, which provides a certain alkaline environment and buffer conditions to facilitate DNA migration. Commonly used electrophoresis buffers are Tris-acetic acid-EDTA (TAE) buffer and Tris-boric acid-EDTA (TBE) buffer. TAE is commonly used in agarose gel electrophoresis due to its low buffer capacity, and TBE has a high buffer capacity and is commonly used in polyacrylamide gel electrophoresis. The buffer is usually prepared as a concentrated solution and used after dilution. The EDTA in the buffer can chelate divalent metal ions, thus inhibiting the enzyme activity and preventing DNA degradation, and it can also terminate the reaction to avoid the reaction during electrophoresis, thus refraining from inaccurate results.

To ensure that the samples can be completely added to the loading wells of gel electrophoresis, the samples need to be mixed with the loading buffer. The loading buffer is composed of bromophenol blue, sucrose, glycerol, or polysucrose 400. Bromophenol blue is an electrophoresis indicator, which is violet-blue in the electrophoresis buffer. After mixing with the sample, bromophenol blue makes the sample colorful, facilitating visual observation and making the sample adding operation more convenient. Besides, the loading buffer can also increase the specific gravity of the sample to ensure that the sample can sink into the wells evenly, and it can also form a naked-eye visible indicator band to predict the speed and position of nucleic acid electrophoresis.

Fluorescent dye staining is commonly used to detect the results of agarose gel electrophoresis. Commonly used nucleic acid dyes, such as ethidium bromide (EB), Goldview, SYBR, and Genefinder, are intercalated into double-stranded DNA and emit fluorescence under UV light, allowing DNA that cannot be observed with the naked eye to be seen under UV light. EB is a fluorescent dye that shows red color under UV light. In the early days, when EB was used as a fluorescent dye, agarose gels were usually immersed in a certain concentration of EB after electrophoresis, left for 10–15 min, and then observed in a UV light box for taking photograph. However, EB is somewhat carcinogenic, gloves are required when using it, and the waste solution of EB has to be treated before it can be discarded. As more and more dyes have been discovered, EB has been replaced by other dyes. Goldview is less toxic and can be added during the preparation of agarose gel, i.e., the prepared

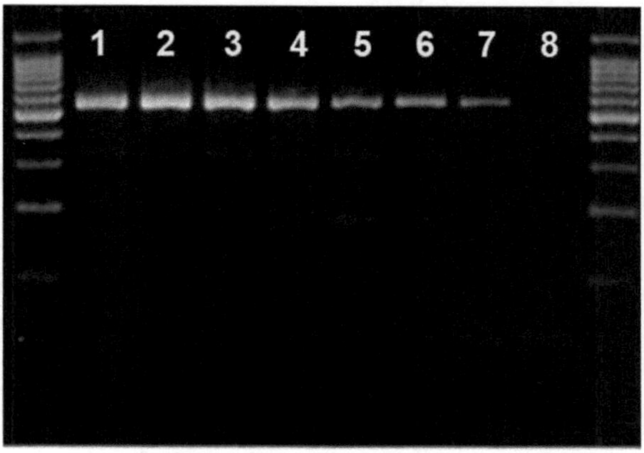

Fig. 1.7 The electrophoretic results of PCR products

agarose gel contains the dye, eliminating the need to soak and stain after electrophoresis and shortening the detection time. Under UV light, Glodview shows green color after binding to double-stranded DNA and red color after binding to single-stranded DNA, so Goldview can better distinguish between double-stranded and single-stranded DNA.

Figure 1.7 shows the electrophoretic results of PCR products, in which lane 8 represents for blank, and there are no amplification bands in this lane, indicating that PCR does not produce any nonspecific amplification products. Lanes 1–7 represent for samples. These samples show uniform bands without any amplification bands, and the length of PCR products can be obtained by comparing with marker. In addition, the gray scale of the bands decreases from lane 1 to 7, indicating that the initial concentration of sample 1 is the highest and that of sample 7 is the lowest. The results of gel electrophoresis can be used to initially determine the length, purity, and other information of PCR products.

In order to objectively compare the relative expression of target gene in different samples and avoid some systematic and human errors during RNA extraction, spiking, measurement, cDNA synthesis, and PCR reaction, it is necessary to select appropriate internal reference genes, also known as housekeeping genes. The expression of the internal reference genes in cell or the copy number of the internal reference genes in the genome is constant, and it is little affected by environmental factors. During the operation process, the internal reference genes always undergo the same process as the sample to be tested to achieve the purpose of correction. Commonly used internal reference genes are *GAPDH*, *β-actin*, *18S rRNA*, *28S rRNA*, and so on. After electrophoresis, the gray values of the amplification products generated by sample gene and the reference gene are calculated first. By comparing the target gene and internal reference gene of one sample, this relative value is then compared with other parallel groups, generally there are more than three samples per group. The selection of internal reference gene depends on the specific circumstances of the target gene and the experiment due to the differed life and physiological processes of different organisms. To ensure the credibility of the experimental results, internal references must be used for each experiment.

1.4.2 Polyacrylamide Gel Electrophoresis

Polyacrylamide gel electrophoresis (PAGE), also known as vertical electrophoresis, has the same principle as agarose gel electrophoresis, except that PAGE is a three-dimensional mesh-like gel formed by the polymerization of monomeric acrylamide and methylenebisacrylamide. The principle of PAGE is that monomer acrylamide polymerizes to long chains under the catalysis and induction of ammonium persulfate (APS) and tetramethylenediamine (TEMED). When methylenediacrylamide is involved in the polymerization, the formed long chains will cross-link to form a three-dimensional network structure. The pore size of PAGE is determined by the length of the chain formed by the cross-linking of acrylamide and its cross-linking degree. Compared to agarose gel electrophoresis, PAGE has a high resolution and can distinguish differences down to a few bases. Moreover, the loading capacity of PAGE is much larger than agarose gel electrophoresis, and up to 10 μg of DNA can be added to a single sample well.

For the detection of amplification products, the gels are usually immersed in the liquid containing the dye for staining after PAGE is completed and then placed under UV light for observation and photography. It is important to note that acrylamide has neurotoxic side effects and can be absorbed through the skin and accumulate in the body. Masks and gloves are required for the preparation and use of acrylamide. Although the consensus is that polyacrylamide is nontoxic, it still needs to be handled with care due to the possibility of containing small amounts of unpolymerized acrylamide.

1.5 Real-Time Fluorescence Quantitative PCR

Electrophoretic detection of PCR products can only qualitatively and semiquantitatively analyze the endpoint products of PCR, and cannot accurately quantify the starting amount of the template or monitor the amplification process in real time. Applied Biosystems introduced real-time fluorescence quantitative PCR technology for the first time in 1996, which enabled PCR to make a leap from qualitative to quantitative detection. Real-time quantitative fluorescence PCR adds a dye or fluorescent probe to conventional PCR and uses the change of fluorescence signal to detect the amount of amplification products in each PCR cycle in real time. Combining with computer software to analyze the obtained fluorescence signal, the initial concentration of DNA or RNA in the sample can be calculated finally. Real-time fluorescence quantitative PCR has the advantages of high sensitivity, high specificity, good reproducibility, fast reaction speed, fully closed-tube reaction, and no post-processing.

1.5.1 *Principle of Real-Time Fluorescence Quantitative PCR*

Real-time fluorescence quantitative PCR is a molecular biology technique that integrates PCR amplification with simultaneous product detection. By incorporating fluorescent dyes or fluorescent probes into conventional PCR, the PCR process can be monitored in real time. As the PCR reaction proceeds, the accumulating amplification products lead to a corresponding increase in fluorescent signal. This fluorescence signal is collected at each cycle, allowing continuous observation throughout the entire amplification process. By plotting the cycle number on the horizontal axis and the fluorescence signal on the vertical axis, a characteristic fluorescence amplification curve is generated, as shown in Fig. 1.8. The amplification curve of real-time fluorescence quantitative PCR typically exhibits four distinct phases: background phase, exponential growth phase, the linear growth phase, and plateau phase. During the background phase, the fluorescence signal generated by amplification products remains below the detection threshold, masked by background noise, making it difficult to discern changes in product quantity. As cycles proceed and products accumulate, the PCR reaction enters the exponential growth phase, amplifying with an efficiency of $(1 + E)^n$, where E represents the PCR amplification efficiency. At 100% efficiency, PCR follows a 2^n trend. As the reaction continues, key PCR components, such as Taq DNA polymerase, dNTPs, and primers, are progressively consumed, causing amplification efficiency to decline and the rate of product growth to slow, at which point PCR amplification enters the linear growth phase. Once any essential component is exhausted, PCR enters the plateau phase. Since PCR is sensitive to environmental variables, the timing of entering the plateau phase and the height of the plateau phase vary greatly and are not precisely controllable. Even under identical experimental conditions, replicate reactions often yield different final product amounts and plateau levels. Therefore, the endpoint quantification of PCR products is inherently unreliable.

Traditional PCR quantification methods typically measure the endpoint products rather than the initial DNA concentration. Since there is no direct proportional relationship between the amount of PCR products and the starting template, quantitative analysis is made possible by introducing two key concepts derived from the PCR amplification curve: the threshold and the Ct value (cycle threshold), where the Ct value allows for the estimation of the initial amount of DNA. In practice, the

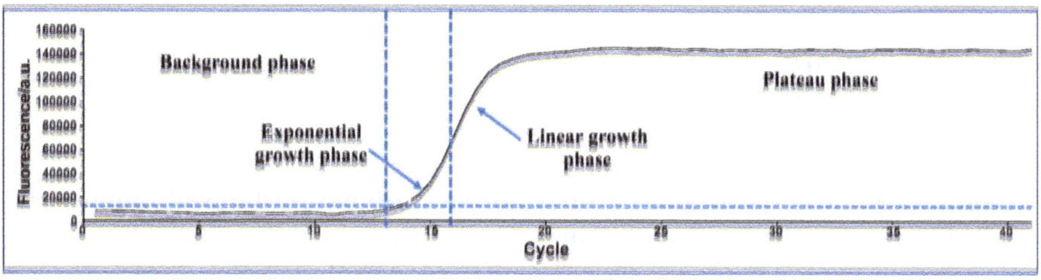

Fig. 1.8 Fluorescence amplification curve of real-time fluorescence quantitative PCR

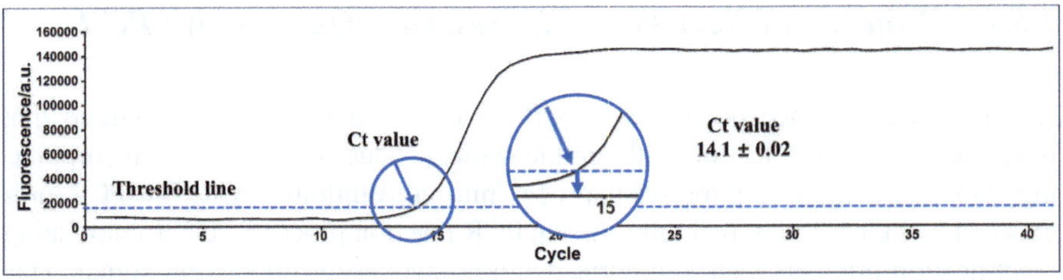

Fig. 1.9 Determination of the threshold and Ct values

threshold is commonly set at 10 times the standard deviation of the fluorescence signal from cycles 3–15 and is positioned within the exponential growth phase of the amplification curve. For a given fluorescence curve, the threshold remains constant. The Ct value is defined as the cycle number at which the fluorescence signal reaches the threshold, as shown in Fig. 1.9. The Ct value is also considered to be the cycle number corresponding to the inflection point at which the fluorescence signal transitions from the background phase into the exponential growth phase. A higher initial DNA concentration results in a lower Ct value, while a lower initial DNA concentration yields a higher Ct value. When the initial concentration is identical, the corresponding Ct value remains consistent even if the final amount of amplification products vary significantly during the plateau phase. Thus, the Ct value depends solely on the initial sample concentration and is unaffected by later reaction variability.

So how does the Ct value relate to the initial concentration of the sample? The ideal PCR amplifies in the form of 2^n, setting the starting template amount to X_0, and the amount of products after the nth cycle of amplification to X_n, so that $X_n = X_0 \times 2^n$. For a nonideal PCR reaction with amplification efficiency E, $X_n = X_0 \times (1 + E)^n$.

At the time when the fluorescence signal reaches the threshold, the corresponding cycle number at this point is Ct. Since the threshold of the fluorescence curve is a known constant set to M, $X_{\mathrm{Ct}} = X_0 \times (1 + E)^{\mathrm{Ct}} = M$.

Take the logarithm of both sides of the above equation simultaneously yields $\log_2 M = \log_2 X_0 + \log_2 (1 + E)^{\mathrm{Ct}} = \log_2 X_0 + \mathrm{Ct}\log_2 (1 + E)$. Thus, $\log_2 X_0 = \log_2 M - \mathrm{Ct} \log_2 (1 + E)$. Provided that the fluorescence curve is known, $\log_2 \mathrm{M}$ and $\log_2 (1 + E)$ are constants; therefore, the Ct value is proportional to the negative logarithm of the starting concentration of the sample, which can be calculated from the Ct value.

1.5.2 Principle of Fluorescence Signal Generation in Real-Time Fluorescence Quantitative PCR

At present, the fluorescent groups added to the real-time fluorescence quantitative PCR systems are divided into two major categories: fluorescent dyes that can be embedded in the double helix structure of DNA and fluorescent probes that can specifically bind to PCR products. Embedded fluorescent dyes are mainly used, such as SG and PicoGreen (PG). However, these dyes are not specific and can only measure the total amount of double-stranded DNA. Fluorescent probes mainly include TaqMan probe and molecular beacon probe, which add a target recognition step and improve the specificity of the assay. Both have advantages and disadvantages.

1.5.2.1 Fluorescent Dye-Based Method

SG is a dye with a green excitation wavelength that binds to all double-stranded DNA within the minor groove region of the double-helix structure. The maximum absorption wavelength of SG is approximately 497 nm, and the maximum emission wavelength is approximately 520 nm. SG does not emit fluorescence when bound to single-stranded DNA, but the fluorescence can be enhanced by thousands of times when bound to double-stranded DNA. This is determined by the molecular structure of SG and PG, whose chemical structure is shown in Fig. 1.10 [6]. Although SG and PG have multiple positive charges, only the positive charges inside the red circle in the figure can bind to the phosphate backbone of DNA by electrostatic interactions. The rest of the positive charges are embedded between the DNA double strands, which limits the rotation of the SG and PG molecules, increasing the rigidity of the molecules and enhancing the fluorescence signal. As the fluorescence intensity of SG and PG is related to the amount of double-stranded DNA, the amount of double-stranded DNA present in the PCR system can be deduced from the fluorescence intensity of SG or PG. Although SG and PG have similar sensitivity for detecting double-stranded DNA, PG binds more strongly to double-stranded DNA because PG

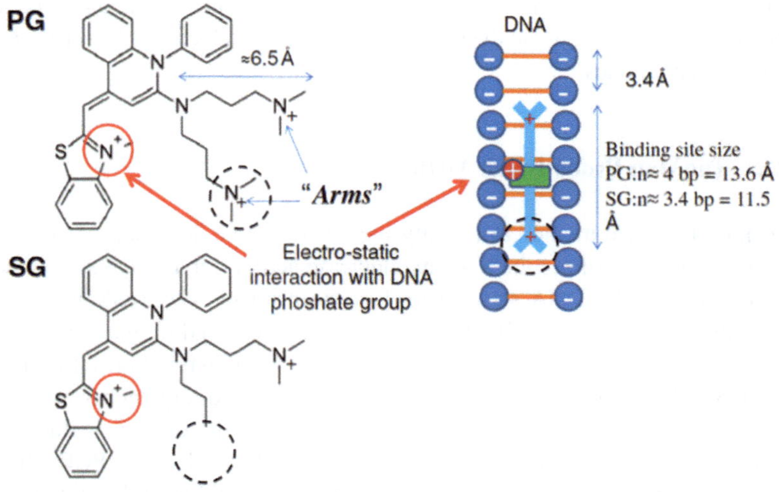

Fig. 1.10 Chemical structure and working principle of SG and PG. (Revised and reprinted with permission from Ref. [6])

has one more positive charge than SG. Therefore, PG may inhibit the amplification reaction, resulting in prolonged amplification reaction time. In addition, PG is more expensive, so SG is more commonly used as a fluorescent dye for detection in amplification reactions.

In view of the specific interaction of SG with double-stranded DNA, the PCR process can be monitored in real time by detecting the fluorescence intensity of SG as follows: during the denaturation step, all DNA exists in a single-stranded state and SG shows negligible fluorescence. As the temperature decreases, the primers anneal to their complementary template and extend in the presence of dNTPs and DNA polymerase, ultimately forming the complete double-stranded DNA. At this point, SG binds to the amplification products and emits fluorescence. In the subsequent denaturation step, SG dissociates from the double-stranded DNA, causing a sharp drop in fluorescence. Therefore, fluorescence data are typically collected during the extension step of PCR cycle, enabling real-time detection. The SG-based method is highly versatile, cost-effective, and requires no customized probes. However, because SG binds indiscriminately to any dsDNA and fluoresces regardless of sequence specificity, the method lacks discrimination between target amplicons, non-specific amplification products, and primer dimers. This compromises the accuracy of quantitative detection. To address this limitation, melting curve analysis can be performed to verify the specificity of the amplification products. After PCR amplification, the temperature is gradually increased, while the fluorescence is monitored. When the melting temperature of the double-stranded DNA is reached, SG dissociates from the double-stranded DNA and returns to its free state, resulting in a decrease in fluorescence signal. Plotting the fluorescence signal against temperature yields a melting curve, as shown in Fig. 1.11a. The negative derivative of this curve (–dF/dT) reveals characteristic peaks (Fig. 1.11b), with each peak's temperature corresponding to the T_m of a double-stranded DNA species. Since non-specific products such as primer dimers typically have different T_m values from target amplicons, melting curve analysis allows specific products to be distinguished from non-specific ones.

1.5.2.2 TaqMan Probe-Based Method

TaqMan probe-based method is a highly specific real-time fluorescence quantitative PCR technique that utilizes the 5′–3′ exonuclease activity of Taq DNA polymerase, which hydrolyzes the DNA strand ahead of the primers. The TaqMan probe is a single-stranded DNA, generally 20–24 nt in length, designed to bind specifically to the target sequences between the two PCR primers. As shown in Fig. 1.12, the probe is dual-labeled with a reporter fluorophore at one end and a quencher at the other. When intact, the proximity of the quencher suppresses the reporter's fluorescence through fluorescence resonance energy transfer (FRET). During PCR, after the denaturation step separates the DNA strands, both the TaqMan probe and the primer anneal to the single-stranded template during the annealing phase, and the 3′ end of the primer extends along the template under the catalysis of Taq DNA polymerase.

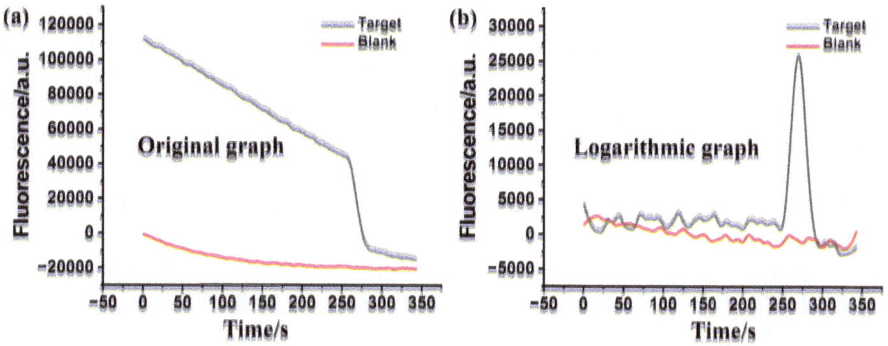

Fig. 1.11 Melting curve of PCR amplification products

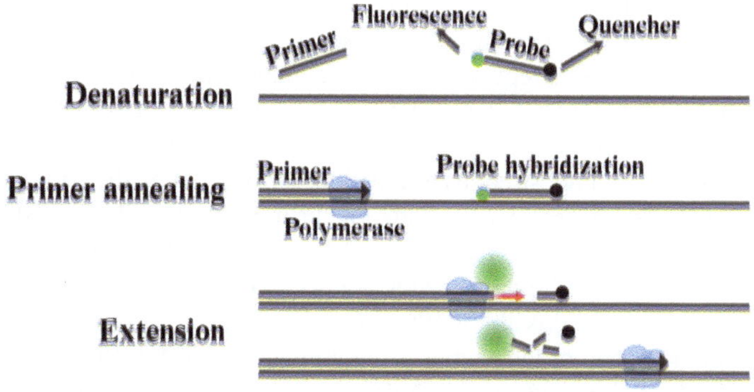

Fig. 1.12 The working principle of TaqMan probes

The 5′–3′ exonuclease activity of Taq DNA polymerase cleaves the bound TaqMan probe once synthesis reaches it, separating the fluorescent group from the quencher, resulting in a detectable fluorescent signal. Since one probe is hydrolyzed for each newly synthesized target amplicon, the accumulation of fluorescence during the extension step of each cycle directly correlates with the amount of amplified products, generating a quantifiable fluorescence curve. Consequently, the TaqMan probe-based method significantly enhances the specificity and accuracy of PCR by ensuring that fluorescence is generated only from the intended target sequence. However, the requirement for custom-synthesized, dual-labeled probes makes this technique more expensive than dye-based methods. Furthermore, each probe is sequence-specific, which limits its broad application to predetermined targets.

To improve the ability of TaqMan probe-based method to discriminate single-base variants, the TaqMan probe is modified with the nonfluorescent quencher (NFQ) at the 3′ end. Unlike conventional quenchers, the NFQ does not emit fluorescence of its own, thereby substantially reducing background signal. In addition, the 3′ end of the probe is also modified with a minor groove binder (MGB) moiety. The MGB stabilizes the probe–target duplex, raising the probe's T_m by approximately 10 °C. To obtain the same T_m value, the MGB probe can be designed shorter than a conventional TaqMan probe. This shorter length increases the ΔT_m between perfectly matched and mismatched probe–target hybrids, making it easier to distinguish single-nucleotide variants. Consequently, the TaqMan MGB probe not only improves the accuracy of mutation detection, but also reduces costs through the use of shorter oligonucleotides.

The design principles of TaqMan probes are as follows: (I) the T_m value of TaqMan probes should be about 10 °C higher than the primer to ensure it is fully hybridized before primer extension. An ideal T_m value typically ranges from 65 to 72 °C. At this point, the Taq DNA enzyme exhibits optimal 5′–3′l exonuclease activity. (II) Avoid guanine (G) at the 5′ end of the probe. Even after cleavage, a 5′-terminal G can quench the reporter fluorophore, thereby reducing signal intensity. (III) The 3′ end of the TaqMan probe must be blocked to prevent it from extending as a primer during PCR. This is commonly achieved by attaching a quencher (e.g., TAMRA), an amino linker, or another appropriate modifier. (IV) Maintain a GC content of 40–60%. Avoid repetition of single nucleotide sequences, especially G, and ensure the number of G does not exceed the number of C within the probe. (V) Optimal probe length is about 20–24 nucleotides (nt). This balances binding specificity and avoid undesired hybridization or overlap with primers. (VI) To amplify AT-rich target sequences, both the primers and probes must be longer enough to reach the required T_m value. The probe should generally not exceed 40 nt, otherwise the quenching efficiency is low. (VII) The probes should be placed as close to the primer as possible without overlapping, and a minimum spacing of one base from the 3′ end of the primer is recommended. (VIII) When discriminating single-nucleotide polymorphisms (SNPs) or mutations, position the mismatch (es) centrally within the probe rather than at the ends. Using a shorter probe enhances the ΔT_m between matched and mismatched duplexes, improving specificity. (IX) When using TaqMan probe for mRNA expression analysis, the probe sequence should include exon/exon boundaries whenever possible.

1.5.2.3 Molecular Beacon-Based Method

Molecular beacon (MB) is a single-stranded stem-loop DNA probe with a hairpin structure. The two ends of MB are inversely complementary, and the middle loop is designed to be complementary to a specific target sequence. The two ends of MB are labeled with a fluorescent group and a quencher group, respectively. In the absence of a target, the stem keeps the fluorophore and quencher in close proximity, enabling FRET and thus quenching the fluorescence. When the target sequence is present, the loop part of MB hybridizes to the target to form a double-stranded hybrid, forcing the stem to open. This separation of the fluorophore and quencher reduces FRET efficiency, leading to the restoration of fluorescence, as shown in Fig. 1.13. As PCR amplification proceeds, the amplicons become more and more numerous, the number of bound MB also increases, resulting in enhanced fluorescence signal.

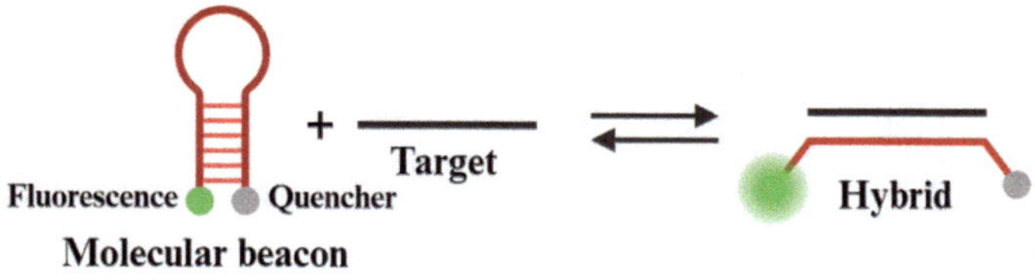

Fig. 1.13 Working principle of MB

The design principles of MB are as follows: (I) The total probe length should be 15–33 nt. The loop region, which is complementary to the target sequence, confers specificity for target recognition. (II) The calculated T_m value of the loop region should be 7–10 °C higher than the annealing temperature. When calculating the T_m value of the probe sequence, only the loop sequence should be considered. (III) The probe sequence must be designed to hybridize to an accessible region of the target. It should avoid complementarity to segments prone to forming stable secondary structures within the target itself. (IV) The MB should bind near the center of the target sequence, and a minimum spacing of >6 nt should be maintained between the 3′ end in the target upstream and the 5′ end of the probe. (V) The stem region of the MB should be 5–8 bp with a GC content of 70–80%. An excessively long or stable stem can slow hybridization kinetics and reduce the efficiency of target binding. (VI) Since G residues can act as static quenchers, the nucleotide immediately adjacent to the fluorophore (usually at the 5′ end) should be avoided. A cytosine (C) is preferred at this position. (VII) Complementarity between MB and PCR primers should be avoided; otherwise, it will cause an increase in background signal due to probe- primer hybridization. (VIII) The choice of fluorescent dye and quencher pairs is critical for performance. Common fluorophores include Texas Red and fluorescein derivatives, and DABCYL is selected as the quenching group.

1.5.3 Quantitative Methods for Real-Time Fluorescence Quantitative PCR

The quantitative analysis of fluorescent quantitative PCR is based on the Ct value of the fluorescence curve for each sample. The commonly used quantification methods are divided into absolute and relative quantification. The difference lies in whether the concentration of the target gene used to make the standard curve is known. If the concentration of the DNA standard is known, a standard curve can be plotted using Ct value as the vertical axis and the negative logarithm of the standard concentration as the horizontal axis. The concentration of the sample can be calculated by putting the Ct value of the sample to the standard curve, and this is the absolute quantification, as shown in Fig. 1.14. If the concentration of the DNA standard is unknown, the multiple of the concentration difference between the sample and the standard can be

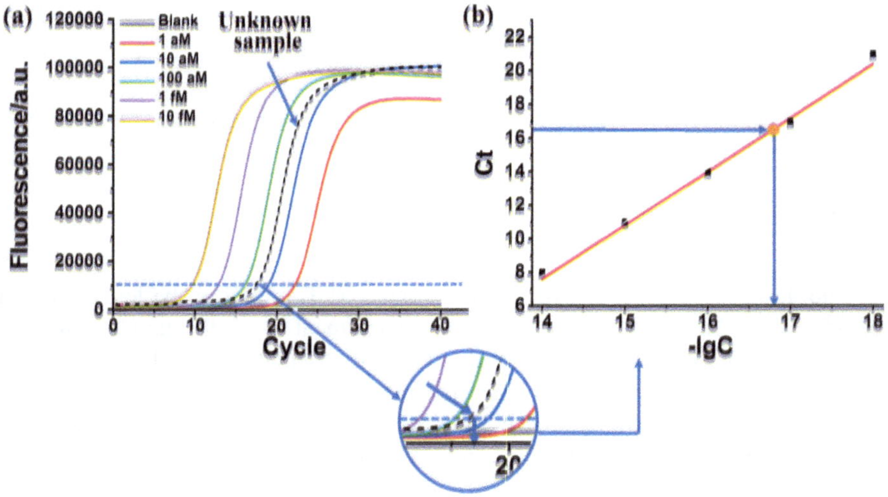

Fig. 1.14 Absolute quantification of real-time quantitative PCR

calculated based on the difference between the Ct value of the standard and the sample, and this is the relative quantification.

After introducing the internal reference gene, the relative quantification of real-time quantitative PCR can be divided into two types: the double standard curve method and the $2^{-\Delta\Delta Ct}$ method. The double standard curve method quantifies the target gene and the internal reference gene in control sample and sample to be tested by the standard curve. Then the relative expression is obtained according to the calculation formula (1.1). The double standard curve method is suitable for cases where the amplification efficiency of the target gene and the amplification efficiency of the internal reference have large differences. This method is simple in analysis and experimental optimization, but the standard curve must be done for each gene and each round of experiment, which increases the operation steps. $2^{-\Delta\Delta Ct}$ method can only calculate the relative expression of the target gene and the internal reference gene in control and sample. Equation (1.2) can calculate the relative concentration multiplication of the sample relative to the control. This method is suitable for cases where the amplification efficiency of the target gene and the internal reference gene are very close to each other. There is no need to make a standard curve, and only the Ct values of the sample and the control are needed. However, the amplification efficiencies of the target and reference genes are not always the same in every experiment, so the experimental conditions need to be strictly optimized.

$$\text{Corrected value} = \frac{\text{The results of target gene}}{\text{The results of internal reference gene}}$$

$$\text{Relative value} = \frac{\text{The corrected value of sample to be tested}}{\text{The corrected value of control sample}} \tag{1.1}$$

$$\text{The relative value} = 2^{-\left(\begin{matrix}\text{The mean Ct value}\\ \text{of target gene in}\\ \text{sample to be tested}\end{matrix} - \begin{matrix}\text{The mean Ct value of}\\ \text{internal reference gene}\\ \text{in sample to be tested}\end{matrix}\right) - \left(\begin{matrix}\text{The mean Ct value}\\ \text{of target gene in}\\ \text{control sample}\end{matrix} - \begin{matrix}\text{The mean Ct value of}\\ \text{internal reference gene}\\ \text{in control sample}\end{matrix}\right)} \tag{1.2}$$

1.6 Reverse Transcription PCR (RT-PCR)

Conventional PCR uses DNA as the template. However, when the target is RNA, direct amplification becomes technically challenging. Although RNA can be extracted directly from cells, it is prone to degradation, difficult to store long-term, and typically present in low quantities, making it impractical to obtain in sufficient amounts for reliable PCR. In contrast, DNA is far more stable and can be preserved for extended periods [7]. So RNA is first reverse transcribed into complementary DNA (cDNA), and the RNA content of a gene can be obtained indirectly by detecting the amount of cDNA. This approach led to the development of reverse transcription PCR (RT-PCR).

1.6.1 Principle of RT-PCR

RT-PCR is a technique that combines reverse transcription of RNA with PCR amplification of the resulting cDNA. The process consists of two sequential steps. The first step is reverse transcription, RNA serves as the template for the synthesis of cDNA, catalyzed by reverse transcriptase. In the subsequent PCR step, the cDNA is first dissociated from the RNA strand by heating, and subsequently annealed to PCR primers. DNA polymerase extends these primers to generate double-stranded DNA, which then serves as the template for PCR amplification.

1.6.2 RT Primers

RT primers are commonly categorized into three types: random hexamers, oligo (dT) primers, and gene-specific primers, each suited for different experimental purposes. The random hexamers are short, 6 nt primers in which each position is randomly occupied by one of the four deoxyribonucleotides, and thus they present a mixture of 4^6 possible sequences. They can prime reverse transcription from any RNA template, and approximately 96% of the synthesized cDNA comes from ribosomal RNA (rRNA) and transfer RNA (tRNA). Therefore, if mRNA is analyzed, it is better to remove tRNA and rRNA from total RNA first. Oligo (dT) primers are generally 18–24 nt. They can anneal to the poly(A) tail at the 3′ end of mRNA, enabling selective reverse transcription of mRNA. However, for very long mRNA transcripts, reverse transcriptase may not synthesize full-length cDNA due to processivity limitations. Gene-specific primers are designed to be complementary to a particular target RNA and produce only cDNA from that gene, which is more useful when working with low-abundance RNA targets. Therefore, in terms of target specificity, the order of RT primers is gene-specific primers > oligo(dT) primers > random hexamers.

1.6.3 Reverse Transcriptase

Reverse transcriptases are RNA-mediated DNA polymerases that synthesize cDNA using RNA as the template and dNTPs as substrates. Most reverse transcriptases possess three key enzymatic activities: First, RNA-directed DNA polymerase activity, which catalyzes the synthesis of a DNA strand complementary to the RNA template. Second, RNase H activity. This activity degrades the RNA strand in an RNA-DNA hybrid. RNase H does not cleave single- or double-stranded DNA nor does it digest single-stranded RNA. Third, DNA-directed DNA polymerase activity. The first single-stranded DNA synthesized by reverse transcription is used as the template for the synthesis of a second DNA molecule using dNTPs as a substrate.

The currently available reverse transcriptases for synthesizing cDNA strands are AMV reverse transcriptase, M-MuLV reverse transcriptase, ProtoScript II reverse transcriptase, WarmStart RTx reverse transcriptase, and Superscript reverse transcriptase.

AMV reverse transcriptase derives from Avian myelomonocytoma virus and has an optimal reaction temperature of 42 °C. The enzyme uses RNA or single-stranded DNA as template and RNA or DNA as primer for cDNA synthesis. AMV reverse transcriptase has high RNase H activity, but lacks 3′–5′ exonuclease activity. Therefore, AMV reverse transcriptase can hydrolyze the RNA strand in the DNA-RNA hybrid strand formed during reverse transcription to release cDNA for subsequent PCR. However, this property also limits the length of cDNA synthesis.

M-MuLV reverse transcriptase derives from Moloney murine leukemia virus and has an optimal reaction temperature of 37 °C. Compared to AMV reverse transcriptase, M-MuLV reverse transcriptase has relatively weak RNase H activity and is therefore more suitable for the synthesis of long-strand cDNA.

ProtoScript II reverse transcriptase is a recombinant M-MuLV reverse transcriptase with reduced RNase H activity and increased thermal stability. ProtoScript II reverse transcriptase has an optimal reaction temperature of 42 °C and an active temperature of up to 50 °C. It can synthesize cDNA at higher temperatures and has high specificity, long synthetic cDNA length, and high yield.

WarmStart RTx reverse transcriptase derives from recombinant *E. coli* and has an optimal temperature of 50–55 °C with intact RNase H activity. The activity is inhibited below 40 °C and elevating the temperature activates WarmStart RTx reverse transcriptase to further improve the specificity of the reaction. Thus, WarmStart RTx reverse transcriptase can react at high temperatures and is particularly suitable for loop-mediated isothermal amplification reaction (LAMP).

Superscript reverse transcriptase is an M-MuLV reverse transcriptase mutant that lacks RNase H activity and features more efficient reverse transcription and longer synthetic cDNA length.

1.6.4 One-Step RT-PCR

The RT-PCR described above consists of an independent RT and PCR. One-step PCR, as the name implies, condenses RT and PCR into one step, which requires the addition of both reverse transcriptase and hot-start DNA polymerase to the reaction system. Therefore, the buffer for the reaction poses a challenge and needs to be optimized to screen for the optimal buffer system. One-step PCR is first carried out at a lower temperature for the reverse transcription reaction, when the thermostable DNA polymerase is inactivated due to its activity being blocked by the antibody, and only the reverse transcriptase is active in the system. After the reverse transcription reaction is completed, the reaction temperature is increased to 95° C, and the reverse transcriptase is inactivated while the hot-start DNA polymerase is used for subsequent PCR amplification.

Compared with two-step PCR, one-step PCR simplifies the experimental procedures while minimizing sample contamination by eliminating the need for lid opening. The simultaneous reaction of RT and PCR in a single tube provides enhanced sensitivity for low-abundance RNA detection. However, since both enzymes share the same buffer solution, neither operates under optimal conditions. This necessitates increased enzyme dosage, resulting in resource waste.

1.6.5 Application of RT-PCR

RT-PCR is widely used in various applications, including analyzing gene transcripts, detecting RNA viral in cells, synthesizing cDNA probes, and directly cloning specific gene sequences. In clinical settings, it plays a vital role in genetic disease diagnosis, cancer screening, and identifying RNA viruses such as SARS-CoV-2, HAV, and HIV in patient specimens. Li et al. [8] used RT-PCR to detect the expression of MUC1 and GST-π mRNA in 108 HE-stained negative lymph nodes from 20 breast cancer patients, in which the positive detection rates of MUC1 and

GST-π in HE-stained negative lymph nodes were 66% and 39%, respectively. In breast cancer tissues, HE-positive lymph nodes, and normal breast tissues, the positive detection rates of MUC1 were all 100%, and the positive rates of GST-π were 88%, 71%, and 0%, respectively. The results indicated that MUC1 positivity in HE-stained negative lymph nodes suggested the presence of tumor micrometastases, whereas GST-π, which is overexpressed during tumor development, serves as a crucial biomarker for breast cancer. Wang et al. used a combination of serum total antibody and RT-PCR to detect SARS-CoV-2 virus [9]. The detection limit was 2.0 copies/μL with 98.6% sensitivity and 98.7% specificity. Corman et al. developed a novel real-time RT-PCR method for the detection of SARS-CoV-2 virus with 95% specificity and up to 100% sensitivity [10], In addition, the detection limits were 3.9 and 3.6 copies for the E gene and RdRp gene, respectively. Burchard et al. investigated a dye-embedded RT-PCR for the detection of HIV-1 in human plasma [11]. The detection time was less than 2.5 h and the limit of detection was 50 copies/mL in plasma, which was rapid and sensitive. Ye et al. used RT-PCR to identify a novel gene transcript in a variety of tumor cells [12]. The transcript was a fusion product of lipid synthase (FAS) and estrogen receptor (ER), which could not be detected by RNase protection assay due to its extremely low expression levels. The majority of transcripts (86%) were expressed at less than 5 copies per cell, accounting for 25% of total cellular transcript expression, while the remaining transcripts (14%) were expressed at significantly higher quantities over 5 copies per cell, accounting for 75% of the total cellular transcript expression.

1.7 Multiplex PCR

The PCR mentioned above typically uses a single pair of primers to amplify one specific target, yielding a single amplicon. When several different targets need to be analyzed simultaneously in an experiment, multiple pairs of primers can be designed and added to a single PCR system. This allows for the concurrent amplification of multiple different target sequences, a technique known as multiplex PCR. In 1988, Chamberlain et al. pioneered this approach by designing six pairs of exon-specific primers and applied multiplex PCR for the first time to detect exon deletions in the Duchenne muscular dystrophy (DMD) gene [13]. Multiplex PCR retains the specificity and sensitivity of conventional PCR, but reduces the operational steps and reagent consumption, greatly improving the detection efficiency while saving human, material, and financial resources. At present, multiplex PCR has become a PCR technique, which is widely applied in clinical and scientific research fields, such as pathogen identification, transgenic identification and genetic disease diagnosis.

1.7.1 Principle of Multiplex PCR

Multiplex PCR is an advanced PCR amplification technique developed from conventional PCR. While sharing identical reaction principles, reagents, and procedures with standard PCR, it uniquely combines two or more primer pairs within a single reaction system to amplify distinct target DNA fragments, as shown in Fig. 1.15. For electrophoretic detection, each primer pair should produce amplification products of varying lengths to enable simultaneous detection.

Performing multiplex PCR requires systematic optimization. The process begins by testing each primer pair individually to confirm its ability to amplify the intended target. The reaction conditions of each primer pair are optimized independently to determine the optimal primer concentrations. Once established, primer pairs are added sequentially into a single reaction, and the reaction conditions are continuously adjusted until all primer pairs can amplify the targets under the same condition. Multiple primer pairs may increase the possibility of 3′-end complementation and thus nonspecific amplification such as primer dimers and other nonspecific products. Thus, the primer design for multiplex PCR becomes critical. The primers should be highly specific to prevent nonspecific amplification in nontarget regions. Primers should have comparable T_m values to function effectively at a shared annealing temperature. If the nonspecific amplification is serious during the reaction, test each primer pair individually to determine the source of the problem, and then the primers responsible can be redesigned. If no specific pair is identified, consider the possibility of inter-primer mismatch-induced dimer formation. Perform multiplex PCR by sequentially adjusting primer pairs, using a process of elimination to isolate the problematic pair.

1.7.2 Application of Multiplex PCR

Multiplex PCR can be applied to several fields of biological research, such as pathogen identification, gender screening, hereditary disease diagnosis, forensic medicine research, and analysis of gene deletions, mutations, and polymorphisms. Multiplex PCR has three distinctive features: (I) high efficiency, i.e., multiple genes and pathogenic microorganisms can be detected simultaneously in the same PCR

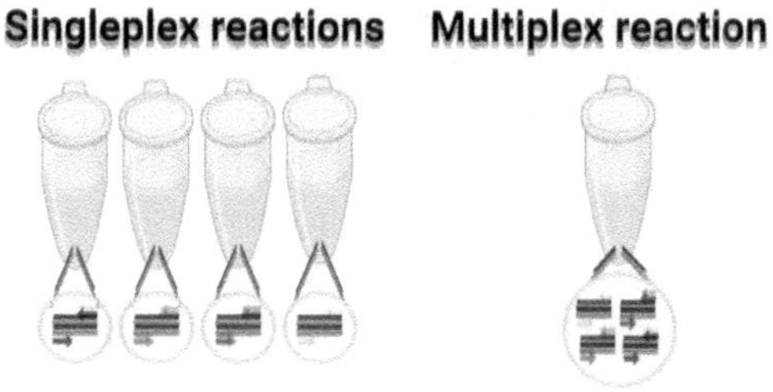

Fig. 1.15 Principle of multiplex PCR

reaction tube, or multiple types of target genes can be genetyped, especially multiple genes can be detected with a very small amount of specimen material, which greatly saves sample usage. (II) Systematic. Multiplex PCR is very suitable for the detection of groups of pathogens, such as hepatitis virus, intestinal pathogenic bacteria, venereal diseases, non-budding anaerobic bacteria, war wound infection bacteria and bacterial warfare agents at the same time. (III) Economic simplicity. Multiple genes are detected simultaneously in the same reaction tube, which greatly saves detection time, reagent consumption, spiking work intensity, and cost, and can provide more diagnostic information to the clinic more quickly. Wu et al. developed a multiplex PCR and CRISPR/Cas12a-based assay system to simultaneously visual detection of the CaMV35S promoter and coagulant in transgenic soybean flour and lectin gene [14].

1.7.3 Multiplex Ligation-Dependent Probe Amplification (MLPA)

The above-mentioned multiplex PCR often requires the careful optimization of primer concentrations. Because different primer pairs can exhibit varying amplification efficiencies, it is common to reduce the concentration of highly efficient primers and increase that of less efficient ones to achieve balanced yields of all PCR products. To address the challenge of different amplification efficiencies among multiple primer pairs, Schouten et al. developed multiplex ligation-dependent probe amplification (MLPA) reaction in 2002 [15], which enables simultaneous detection of 40–50 nucleic acid targets and has been widely applied for multiplex detection of mutations, deletions, copy number variations, and DNA methylation.

The schematic diagram of MLPA is shown in Fig. 1.16. A pair of probes is designed according to the target sequence. Each probe includes a target-specific sequence (pink part in the figure) and a universal primer-specific sequence (black part in the figure), where one of the probes has a stuffer sequence of 19–370 nt between the two regions to regulate the length of the final PCR products for simultaneous detection. For different targets, the universal primer-specific sequences are identical for all probe pairs. Upon addition of the target sequence, the two probes bind adjacently to their complementary regions to form a double-stranded structure with a nick. Under the catalysis of DNA ligase, the two probes are ligated into a long single-stranded DNA. For different targets, a series of ligation products with length intervals of 6–9 bases can be formed. A single pair of universal primers is then used to amplify all ligated products simultaneously via PCR, yielding amplicons ranging from approximately 130 to 480 bp. These products are finally separated and detected by capillary electrophoresis. The technique relies on the high specificity of the ligation step: only when both probes perfectly match the target sequence can they be ligated to long-stranded DNA to trigger subsequent PCR. By strategically placing

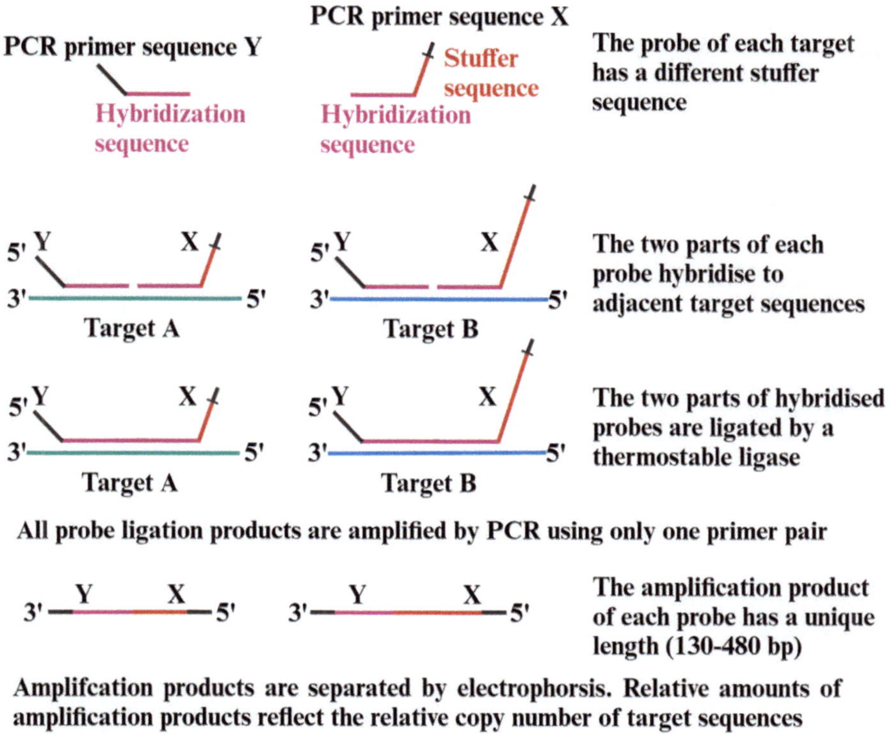

Fig. 1.16 Principle of MLPA reaction. (Revised and reprinted with permission from Ref. [15])

the mutant base at the ligation junction, even a single mismatch can prevent proper ligation, thereby blocking the PCR reaction and enabling MLPA to specifically identify single-base mutations. Because all probes share the same universal primer sequences, every target is amplified with identical efficiency, which solves the problem of different amplification efficiency caused by different primer sequences in conventional multiplex PCR, and also reduces the probability of nonspecific amplification.

The MLPA method has been successfully applied to the exon deletion detection of the breast cancer-related gene BRCA1. Probes were designed separately for each of the 24 exons of the BRCA1 gene to detect multiple exon deletions simultaneously [16], as shown in Fig. 1.17. Moreover, probes were designed separately for each of the 16 exons of the MSH2 gene to detect the deletion of exons 1–6 in the MSH2 gene simultaneously [17]. The MLPA technique was improved by Sørensen et al. through applying this method to microsamples, such as dried blood samples [18]. Results from six patients with congenital adrenocortical hyperplasia (CAH) showed 99.9% accuracy, with detection of target from samples as low as 1.6 ng DNA.

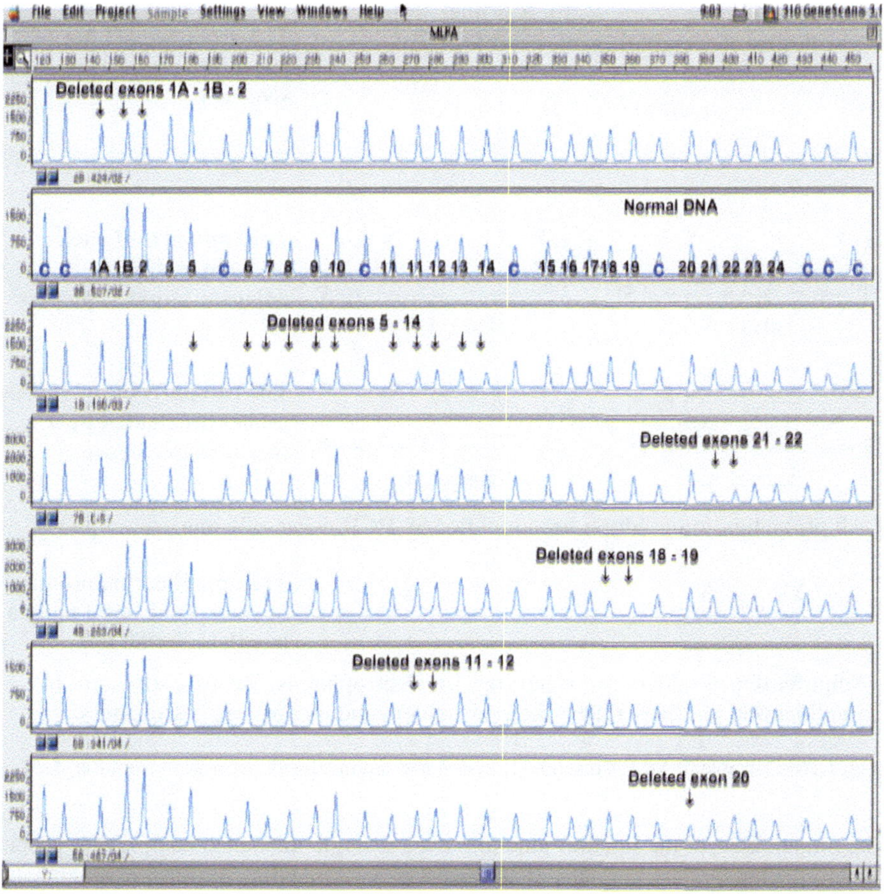

Fig. 1.17 MLPA to detect BRCA1 exon deletion. (Revised and reprinted with permission from Ref. [16])

1.7.4 MeltArray

Fluorescent quantitative PCR is the most widely used nucleic acid detection technology in clinical settings. However, a fundamental limitation is its low capacity to simultaneously detect a variety of different targets in a single reaction. This is primarily due to the limited differentiation capacity of PCR instruments for different fluorescent colors, typically offering only 4–6 channels, which constraint the multiplex detection. Professor Qingge Li of Xiamen University innovatively developed a multiplex detection method called "MeltArray" [19], which is a highly multiplex PCR method. MeltArray is achieved by integrating the 5′-flap endonuclease (FEN) activity of the Taq DNA polymerase with multiple annealing sites of the MB reporters to realize the breakthrough of simultaneously identify as many as

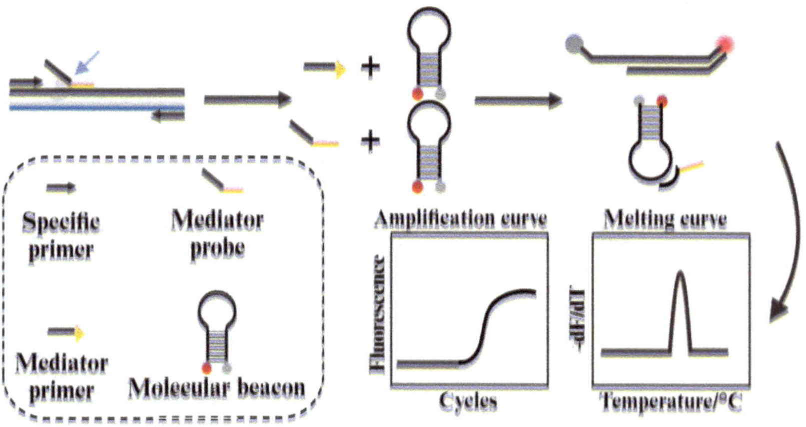

Fig. 1.18 The principle of MeltArray. (Revised and reprinted with permission from Ref. [19])

72 different target sequences in a single closed-tube. Compared to current real-time PCR, this approach demonstrates over 10 times higher throughput while eliminating the need for additional enzymes or separate reactions.

MeltArray uses a mediator probe to avoid the nonspecific amplification caused by multiple universal and specific primers of traditional multiplex PCR. As shown in Fig. 1.18, the mediator probe contains two regions: a target-specific region (as shown in yellow) and a universal mediator region (as shown in black) with the 3′ end of the probe blocked. After hybridizing with target, the mediator probes can be cleaved to mediator primers leveraging the FEN activity of Taq DNA polymerase in the presence of PCR primers. Then the mediator primers can bind and extend along the MB reporter to generate fluorescent double-stranded DNA. The uncleaved mediator probe remains intact and cannot prime extension after hybridizing with the MB reporter probe, resulting in no fluorescent signal generated. Finally, each target sequence is identified by a unique combination of fluorescence color and T_m using multicolor melting curve analysis (MMCA).

In order to realize multiple detection, the loop region of traditional MB probe used in MeltArray was designed to be a long sequence, capable of binding multiple mediator primers simultaneously. As shown in Fig. 1.19a, the MeltArray assay consists of 12 different mediator probes, all of which contain a unique mediator primer region that complementary to different predefined regions within the loop of MB reporter. Therefore, when the mediator primers of these 12 different sequences bind to and extend along the loop of the MB probe under the catalysis of Taq DNA polymerase, the stem-loop structure of the MB probe can be opened to produce partial double-stranded structures with different lengths. After melting curve analysis, different T_m values can be generated in the range of 56–91 °C with intervals of 3 °C. Then a set of 12 distinguishable T_m values can be obtained to achieve the simultaneous detection of 12 targets. Building upon multiplex PCR, MeltArray employs a unique design approach. First, a 20-nt-long "tag sequence," which is designed not to be complementary to any target sequence, is added to the 5′-end of

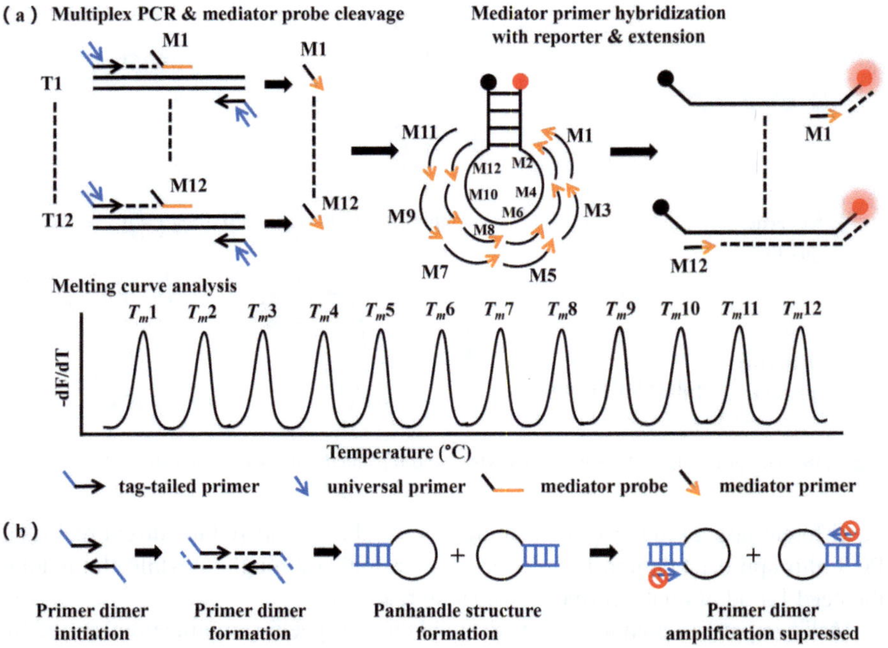

Fig. 1.19 Detection of target sequences in MeltArray assay. (**a**) A 12-plex MeltArray assay utilizing one fluorophore. (**b**) Elimination of primer-dimer exponential amplification. (Revised and reprinted with permission from Ref. [19])

each target-specific primer. After several rounds of amplification, each of the obtained amplification products has a complete primer sequence (including the tag sequence) at its 5′ end and a complementary sequence of the primer (including the complementary sequence of the tag sequence) at the 3′ end. There is also a single universal primer with high concentration in the detection mixture, whose sequence is the same as the tag sequence. Therefore, these abundant universal primers hybridize with the complementary sequences at the 3′ end of the amplification products, which can effectively initiate subsequent exponential amplification of the amplification products generated by all the target sequences in the samples. After that, the universal primers corresponding to the tag sequence complete subsequent PCR, which balances the amount of different amplification products, so that multiple targets are detected without any omission.

To achieve efficient amplification of targets, secondary structures such as primer dimers should be avoided. Here, the Hom-Tag Assisted Non-Dimer System, named as HANDS, is utilized (Fig. 1.19b). By designing the same tag sequence of primers, the two single-stranded DNA extended by primers can spontaneously fold into a stem-loop structure with tag sequence as the stem region. The T_{m} value of the stem-loop structure is higher than that of the primer dimer, which can avoid the false positive signal generated by the primer dimer. In addition, a universal primer is added to the 5′ end of the primer, which turns the multiplex PCR into a singleplex PCR, reducing the interference between multiple primers. This design ensures that the primers remain functional for subsequent extension cycles after high-temperature denaturation, while simultaneously preventing nonspecific dimer formation. Notably, even when primers have complementary sequences to the stem region of the stem loop structure, their extension remains inhibited.

Through the optimization of amplification and detection techniques, MeltArray can simultaneously detect up to 12 targets using a single fluorescent channel. Using six fluorescence channels, such as Atto 425 (450 nm), FAM, HEX (565 nm), ROX, Cy5, and Quasar 705 (705 nm), it is possible to simultaneously detect 72 targets in a single tube, which is undoubtedly a major breakthrough, realizing the transformation of fluorescence quantitative PCR to DNA array.

The authors further use MeltArray to achieve quantitative detection of human Y chromosome, *Escherichia coli,* and respiratory pathogens, which covers the mainstream items of IVD detection, namely, genetic disease detection, infectious disease detection, and tumor detection, reflecting the typical characteristics of MeltArray as a mature nucleic acid detection. MeltArray combines methods such as MMCA, mediator probes, universal primers, and "Hands," and flexibly uses the MB probe, which can be hybridized with the target. This combination finally forms the exquisite detection method. However, although the peak height of the melting curve is proportional to the target concentration within a certain range, MeltArray method has limited quantitative detection capability as an endpoint detection mode. Therefore, MeltArray is mainly used for screening analysis, rather than quantitative detection, and this limitation needs further research and improvement. In general, MeltArray, a highly multiplex PCR method, has realized a major breakthrough in 72-plex detection without requiring additional enzymes or separate reactions. This innovation fills the technical gap between low-order PCR and sequencing technologies for a long time, demonstrating promising application prospects.

1.7.5 Sequence-Specific Nanoparticle Barcode Method

Recently, Ningshao Xia, Shengxiang Ge, and Shiyin Zhang group developed a sequence-specific nanoparticle barcode method (SSNB) for simultaneous detection of 10 HEV types [20]. As shown in Fig. 1.20, in SSNB method, the target-specific primers are conjugated covalently to distinct nanoparticles with different fluorescence properties. The amplification products of different targets are separated into relatively independent physical spaces by particle unit, which effectively reducing the mismatches between primers, primers and templates in conventional multiple detection, thus maintaining the exceptionally high specificity and sensitivity of multiple amplification. In addition, the nanoparticles used in SSNB can realize up to 500 fluorescence codes, and each fluorescence can be a unique code for the amplification of target on the surface of nanoparticles. By decoding the fluorescence signal of nanoparticles, the corresponding detection target properties can be

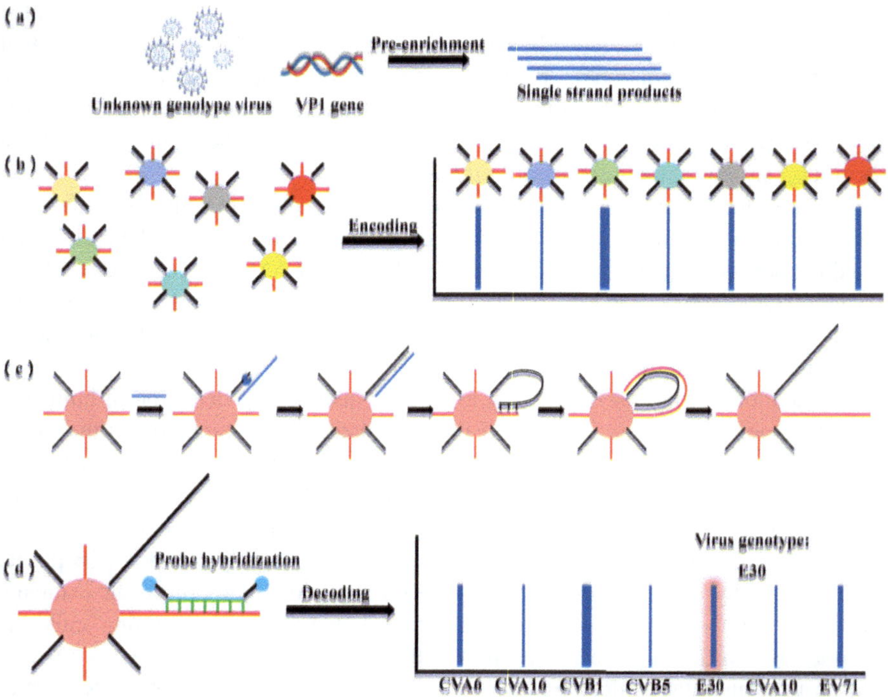

Fig. 1.20 Schematic overview of SSNB for HEV genotyping. (Revised and reprinted with permission from Ref. [20])

determined, and the fluorescence signal of the amplified products on the surface of nanoparticles can indicate negative or positive result. Therefore, SSNB not only solves the problem of insufficient sensitivity and specificity of traditional multiple amplification, but also breaks the limitation of detection channel of fluorescence quantitative PCR instruments, indicating that SSNB can theoretically achieve the detection of up to 500 samples.

The SSNB method demonstrates a 100% specificity in accurately distinguishing between 10 different HEV types and other prevalent clinical viruses, and shows 100% consistency with sequencing results in type identification accuracy. It also has the ability to identify multitype co-infections. While providing super-multiple analysis, SSNB presents a significant increase in detection sensitivity, which can be improved by $10–10^6$ times compared with established multiplex hybridization genotyping techniques. In clinical specimen analysis, the sensitivity of SSNB is comparable to that of PCR for typing single sample. The detection time is greatly shortened, and the identification workload is reduced. As a clinical testing tool, SSNB demonstrates ideal repeatability with accurate and reliable results. More importantly, each reaction in SSNB can form an independent amplification microenvironment, which is not affected by the amplification of other targets. When identifying new pathogens, the only need is to add nanoparticles encoded by new primers in the system without the requirement for major readjustments of amplification and detection systems, which have better target scalability and flexibility. The successful application of SSNB in HEV typing has verified its ideal sensitivity and specificity, super-multiplex detection capabilities, the ability to expand new targets, and the convenience brought by the exclusion of professional bioinformatics analysis. In the future, this technology can also be applied to the identification of syndromic pathogens, the detection of polygenetic diseases, the detection of tumor high-frequency mutations, and the analysis of multi-drug resistance sites in different nucleic acid super-multiple target analysis fields, indicating that SSNB is emerging as a new powerful tool in multiplex detection of nucleic acids.

1.8 Touchdown PCR

The PCR amplification process is influenced by multiple factors, such as annealing temperature, buffer composition, primer concentration, etc. Each PCR reaction requires an optimal reaction condition determined through empirical optimization. The ideal PCR is that a target produces only a single band in electrophoresis; however, weaker bands, nonspecific amplification bands are often present. Increasing the annealing temperature suppresses nonspecific amplification to some extent, but at the same time decreases the yield of the target. Decreasing the annealing temperature increases the yield of the target, but nonspecific amplification is also elevated. To address low PCR efficiency caused by inappropriate annealing temperature, touchdown PCR is developed, which improves the sensitivity and specificity of PCR by optimizing only the annealing temperature without changing the buffer composition, primer concentration, and other conditions.

In 1991, Don et al. first invented touchdown PCR, which is based on the principle of first selecting a high annealing temperature, then gradually lowering it by 1 °C every one or several cycles until a lower annealing temperature is reached, and finally performing about 10 cycles at this annealing temperature [21]. The initial annealing temperature is generally set about 15 °C above the T_m of primer, and in the following cycles, the annealing temperature is gradually reduced by 1–2 °C per cycle until it drops 5 °C below the T_m of primer. Compared with conventional PCR, the initial annealing temperature of touchdown PCR is set above the T_m of primer. At this elevated temperature, only perfectly matched primer-templates can hybridize, ensuring specific amplification while preventing primer dimer formation and nonspecific products. As the annealing temperature decreases, the number of specific amplification products gradually increase. Even though the final annealing temperature drops to the T_m for nonspecific hybridization, the exponential amplification of specific amplification products has already begun. During subsequent reactions, the specific products will compete to bind to DNA polymerase and dNTPs, gaining priority for amplification. This mechanism ensures dominant amplification of single-target products while minimizing nonspecific amplification. Through this approach,

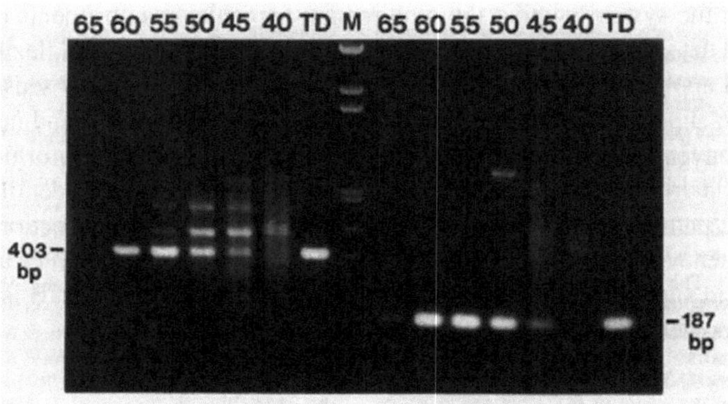

Fig. 1.21 Comparison of conventional PCR and touchdown (TD) PCR. (Revised and reprinted with permission from Ref. [22])

the desired PCR products are selectively amplified during the process, with minimal occurrence of nonspecific amplification.

Compared with conventional PCR, touchdown PCR only differs in the setting of annealing temperature, and the rest of the procedures and precautions are the same. However, touchdown PCR provides a good solution for nonspecific amplification during PCR, which not only effectively improves the specificity of the reaction, but also increases the yield of the target. Hecker et al. [22] experimentally compared the difference between conventional PCR and touchdown PCR with different annealing temperatures. The conventional PCR employed six annealing temperatures: 40, 45, 50, 55, 60, and 65 °C, and the annealing temperature of touchdown PCR was decreased from 60 to 46 °C by 1 °C every 2 cycles, and additionally 10 cycles were added at 50 °C. Two DNA fragments of 403 and 187 bp in length were amplified using conventional PCR and touchdown PCR, respectively, and both PCRs were amplified for 40 cycles for comparison. As shown in Fig. 1.21, conventional PCR exhibited significant nonspecific amplification at a low annealing temperature (40 °C), which competitively inhibited the generation of specific products. As the annealing temperature increased from 40 to 60 °C, the nonspecific amplification bands gradually disappeared, while the yield of specific amplification products gradually increased. When the annealing temperature was increased to 65 °C, no product was generated due to the high temperature. In contrast, touchdown PCR produced a single distinct band with deep intensity (TD line, Fig. 1.21). Although about 30 of the 40 cycles of touchdown PCR were performed below 55 °C, only specific products appeared. This contrasts sharply with conventional PCR, which generates substantial nonspecific amplification at this temperature, leading to false-positive results. These findings clearly demonstrate the advantages of drop PCR.

When simultaneously detecting CymMV and ORSV in orchid plants, Seoh et al. initially employed conventional PCR with a annealing temperature of 55 °C [23]. The results showed distinct banding patterns: CymMV produced a deep amplification product, while ORSV barely showed any detectable bands, indicating significantly higher amplification efficiency of CymMV. This phenomenon was attributed to the higher GC content of CymMV, which facilitates optimal annealing. However, lowering the annealing temperature increased ORSV amplification efficiency, but also generated nonspecific amplification products. To enhance ORSV amplification efficiency without compromising PCR accuracy, the authors implemented touchdown PCR with stepwise annealing: starting at 50 °C, 48 °C and 46 °C, respectively, and gradually decreasing the temperature by 0.5 °C per cycle for 30 cycles in total. Electrophoretic analysis revealed that when the initial annealing temperature was set at 48 °C, both gene amplification products exhibited comparable band intensities.

To expand the application scope of touchdown PCR, it can be combined with other methods, such as real-time quantitative-touchdown PCR, multiplex-touchdown PCR, etc. In the combined PCR system, the required materials and the detection methods of PCR products remain identical to those in real-time quantitative PCR and multiplex PCR, except that the annealing temperature in the PCR program is set to gradually decrease in touchdown PCR. Wang et al. used multiplex-touchdown PCR to simultaneously detect mecA, bla_{SHV}, $bla_{CTX\text{-}M}$, bla_{TEM}, and bla_{OXA} genes with 100% specificity and as low as 100 CFU/mL can be sensitively detected [24]. Larsen et al. used real-time quantitative-touchdown PCR to detect *Pneumocystis carinii* in 12 samples with 100% specificity, while the specificity was only 58% when conventional PCR was used [25].

1.9 Nested PCR

The advent of PCR technology has significantly enhanced the detection of trace substances. However, in certain scenarios where target concentrations are too low or specific components in the reaction are inhibited, products generated through single-round amplification using a pair of primers may fail to meet detection requirements. Nested PCR effectively addresses this limitation by improving amplification sensitivity.

Nested PCR is an enhanced version of conventional PCR that requires two primer pairs for two rounds of amplification. As illustrated in Fig. 1.22, the amplification process begins with one primer pair for amplification. A portion of first-round products are then used as template for a second round PCR using the second primer pair. The resulting products from this second round serve as the final products. Typically, since the second primer pair (inner primers) is positioned within the first primer pair (outer primers), the second-round products are shorter in length compared to those from the first round.

The primer design principles for nested PCR are similar to conventional PCR, with no standardized requirements for the spacer length between inner and outer primers, as specific conditions vary experimentally. As shown in Fig. 1.22, nested PCR undergoes two rounds of amplification. The first round follows a standard 15–30 cycle amplification protocol, while the second round targets the products from

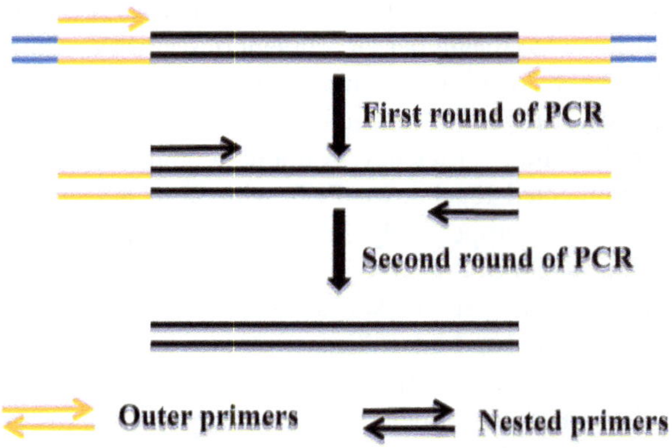

Fig. 1.22 Principle of nested PCR

the first round, conducting another 15–30 cycles. This design significantly enhances reaction sensitivity, enabling efficient amplification of sufficient product from limited starting DNA. Additionally, during PCR amplification, even if the outer primer generates nonspecific products, the probability of inner primers in the second round to pair with mismatched products and the subsequent amplification during the second round of PCR are reduced. Therefore, the specificity of PCR is enhanced by the second primer pair.

Nested PCR is predominantly employed in reactions with low template concentrations, where a single PCR cycle often fails to yield satisfactory results. Through two rounds of amplification, nested PCR can generate substantial target products. Massung et al. utilized nested PCR to detect the 16S rRNA gene associated with human granulocytic eminence [26]. The first primer pair (ge3a and ge10r) produced an amplification product of 932 bp, while the second pair (ge9f and ge2) generated a 546 bp product. Figure 1.23 demonstrates the electrophoretic patterns of DNA extracted from serum amplified by nested PCR, with Fig. 1.23a showing the first-round results and Fig. 1.23b presenting the second-round results. The data reveal that only samples 1–4 exhibited faint target bands following initial amplification, whereas all samples 1–8 displayed distinct single target band after the second round of PCR. These results indicate that nested PCR significantly enhances both sensitivity and specificity.

1.10 Inverse PCR

Conventional PCR amplifies the sequence between two known primers. Even when the internal sequence remains unknown, sequencing can be performed after PCR amplification to obtain complete DNA sequence information. However, it cannot reveal information about the sequences outside the primers, such as which part of a missing gene was lost or what the promoter sequence of a cDNA is. On this basis, Ochman and Triglia et al. established the inverse PCR based on conventional PCR,

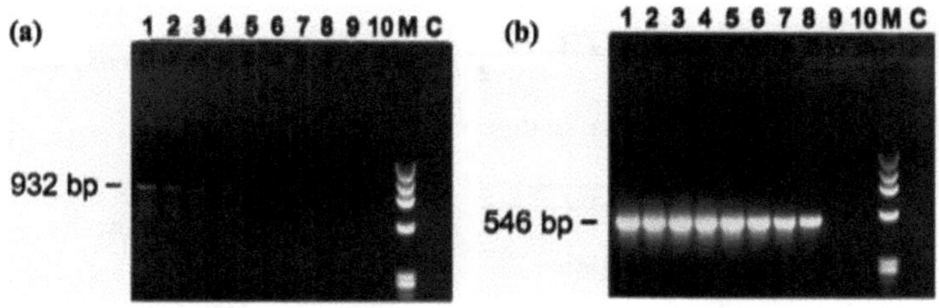

Fig. 1.23 Electrophoresis images of nested PCR products. (**a**) The electrophoresis image of conventional PCR. (**b**) The electrophoresis image of nested PCR. (Revised and reprinted with permission from Ref. [26])

which uses reverse primers to amplify and obtain the sequence information on the outer side of the primers quickly [27, 28].

Inverse PCR uses a pair of primers that match the target, but amplify in the opposite direction. Unlike conventional PCR, this method cannot achieve exponential amplification and only produces linearly amplified products at both ends of the primers. To perform inverse PCR, the target strand must first be cleaved with a restriction enzyme to create sticky ends (as shown in Fig. 1.24). The cleaved strand is then ligated using a ligase to form circular DNA. When paired with inverse primers, the circular DNA serves as the template for inverse PCR. The resulting products can be directly analyzed or sequenced to obtain detailed sequence information.

The primer design principle of inverse PCR is the same as that of conventional PCR, with the only difference being the opposite direction of the primers. Inverse PCR is often used to determine the sequence of adjacent unknown regions, which is helpful for studying the promoter sequence of a gene, oncogenic chromosomal rearrangements, such as gene fusion and viral gene integration.

Originally designed to isolate naturally occurring transposons such as the Tn5 flank sequence [29], inverse PCR has evolved through molecular biology advancements to enable the extraction of DNA fragments containing unknown flanking sequences within known regions. Key applications include: amplifying flanking sequence of transposon insertion sites (ISL) in natural *E. coli* isolates; identifying gene encoding surface antigen precursors on Plasmodium falciparum's main schizonts; obtaining flanking regions at the 5′ or 3′ ends of transcribed genes; and analyzing cellular DNA flanks containing exogenous viral DNA integrated into mouse cells for studying viral genome integration sites. Inverse PCR also facilitates research on gene promoters, cloning continuous genomic DNA fragments, and performing site-directed mutagenesis in DNA [30].

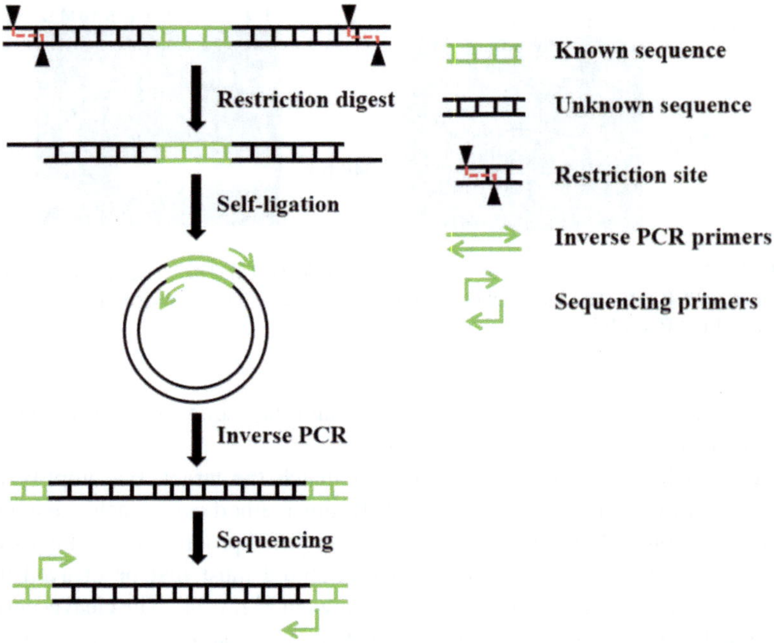

Fig. 1.24 Principle of inverse PCR

1.11 Asymmetric PCR

Conventional PCR amplifies DNA using two equal amount of primers through multiple cycles of denaturation, annealing, and extension, resulting in exponential growth of double-stranded DNA products. However, for subsequent hybridization reactions, researchers prefer single-stranded DNA products. To address this challenge, Gyllensten et al. [31] developed asymmetric PCR in 1988—a novel amplification method that produces large quantities of uniformly length of single-stranded DNA [31].

Asymmetric PCR is a method that uses a pair of primers with different concentrations to amplify DNA, producing large quantities of single-stranded DNA. The lower concentration primer is called the restriction primer, while the higher concentration one is termed the nonrestrictive primer. The amount of restriction primers is crucial because only when they are completely consumed can the remaining nonrestrictive primers be used for single-stranded DNA synthesis. Unlike conventional PCR where nonrestrictive primers remain in excess throughout the process, asymmetric PCR maintains a significant concentration difference between the two primers, with the optimal ratio typically being 1:50 or 1:100. During the initial amplification phase (typically the first 15 cycles), both primers can synthesize DNA under DNA polymerase catalysis, resulting in double-stranded DNA products. As the concentration of restricted primers decreases, the reaction transitions from exponential to linear growth until complete consumption of the restriction primers. At this stage, only nonrestrictive primers continue to synthesize single-stranded DNA products. All subsequent amplifications use these single-stranded DNA products as templates, undergoing linear amplification guided by the nonrestrictive primers.

In asymmetric PCR, the quantity of primers serves as a critical factor for ensuring the reaction proceeds. This involves both the absolute amount and relative quantity of primers, with the absolute amount of restriction primers being the most crucial. The template quantity is another key factor determining whether asymmetric PCR can proceed. When the initial template concentration is low, increasing the amount of restriction primers can enhance the number of cycles for double-stranded DNA formation and improve the efficiency of single-stranded DNA synthesis.

With the advancement of DNA sequencing technologies, various methods have been developed to directly sequence PCR amplification products. However, current sequencing approaches typically require purification of PCR products to remove residual primers, a process that is not only cumbersome and time-consuming but also prone to DNA loss. In asymmetric PCR, where the two primers differ in quantity, restriction primers are first consumed, resulting in predominantly single-stranded DNA during subsequent amplification. This eliminates the need for purification steps, allowing direct sequencing [32]. Conventional PCR products are double-stranded DNA, which, even after high-temperature denaturation, tends to reanneal into double-stranded structures upon brief single-strand formation. With the rapid development of nucleic acid hybridization techniques in molecular biology, efficient probe hybridization has become crucial for successful reactions. During hybridization, single-stranded DNA demonstrates higher efficiency than double-stranded DNA as it can directly hybridize with probes without denaturation. Therefore, the single-stranded DNA products generated by asymmetric PCR are particularly suitable for in situ hybridization detection.

Liu et al. developed multiplex asymmetric PCR for detecting various pathogens [33]. The process begins with extracting genomic DNA from pathogens, followed by amplification using pathogen-specific primers with universal primers at the 5′ end. This asymmetric PCR step produces single-stranded DNA products featuring identical universal sequences at the 5′ end and pathogen-specific sequences at the 3′ end. Targeted detection is achieved by immobilizing specific probes for different pathogens in distinct test strip regions. Since the 5′ ends of PCR products share identical universal primers, fluorescently labeled primers targeting these universal sequences can be immobilized on the test strips. Fluorescence detection across different regions enables simultaneous identification of multiple pathogens. Poddar compared conventional PCR and asymmetric PCR for adenovirus detection using gel electrophoresis and molecular beacon technology [34]. Gel electrophoresis results showed significantly higher yields of amplified products from conventional PCR compared to asymmetric PCR. However, molecular beacon technology, which requires competing reactions with double-stranded DNA from conventional PCR amplification, produced stronger fluorescence signals than conventional PCR.

1.12 In Situ PCR

With the rapid advancement of PCR technology, this method has been widely applied across various fields of life sciences, including pathogen detection, early cancer diagnosis, and genetic testing. Traditional PCR amplification in homogeneous solutions releases products throughout the system, making it impossible to perform localized analysis on initial samples. In situ hybridization (ISH) detects specific DNA or RNA sequences within tissues or cells by designing probes that bind specifically to target molecules. However, this approach exhibits low sensitivity, failing to detect low-abundance targets in cells or tissues. To address the challenge of detecting low-expression or even single-copy DNA/RNA sequences in situ, Haase et al. pioneered in situ PCR in 1990 [35]. This innovative approach integrates PCR amplification with in situ detection, enabling enhanced sensitivity for low-copy or even single-copy target DNA/RNA analysis within tissues or cells.

In situ PCR begins with immobilizing target tissues or cells using appropriate fixatives (typically formaldehyde or glutaraldehyde) to preserve their structural integrity. The process then involves pretreatment with protease to enhance cell membrane permeability. Following this, in situ PCR amplification is performed on DNA or RNA within the cells, followed by product analysis. Depending on the specific labeling agents used during amplification, in situ PCR can be classified into two methods: direct in situ PCR and indirect in situ PCR.

In the direct in situ PCR, the sample is fixed and treated with protease, followed by PCR amplification using labeled primers or dNTPs, directly labeling the products. Commonly used labels include isotopes, fluorescent molecules, biotin, and digoxin, with detection based on their properties. Isotope labeling is detected via autoradiography; fluorescence labeling through fluorescence microscopy; biotin labeling using avidin-based reactions; and digoxin labeling via antibody assays. Since amplified products contain labels, this method eliminates probe hybridization requirements. The direct detection capability allows simple operation with streamlined procedures. However, the high sensitivity of PCR amplification may lead to nonspecific amplification, potentially causing false-positive results.

In the indirect in situ PCR, after fixing and treating test samples with protease, unlabeled primers and dNTPs are directly used for PCR amplification. After the amplification, labeled probes are added to hybridize with the amplification products in situ, forming labeled hybrid molecules. Detection is then performed based on different markers. Although this method enhances detection specificity through in situ hybridization of amplified products, it also increases the testing process and time required.

For RNA detection in cells or tissues, in situ reverse transcription PCR can be employed. This method combines reverse transcription, PCR amplification, and in situ detection to identify low-abundance mRNA within cells or tissues. After fixing the sample and treating it with protease, an additional DNA digestion step is required to remove cellular DNA. Subsequently, reverse transcription converts mRNA into cDNA, which serves as the template for PCR amplification. Finally, the amplified products are analyzed for detection.

Since its initial publication, in situ PCR technology has garnered sustained attention from scholars worldwide and has been extensively applied in fields such as pathology, microbiology, and genetics. This technique can detect exogenous genes (e.g., viral or bacterial infections) and transplanted genes (e.g., those introduced through genetic cloning). It also serves as a diagnostic tool for detecting genetic mutations. Yin et al. demonstrated this capability by using a defective herpes simplex virus vector to deliver functional genes into the mammalian central nervous system [36]. In situ PCR enabled detection of single-copy DNA in brain neurons—a method unachievable with conventional PCR or in situ molecular hybridization. By directly analyzing the introduced gene itself, this approach allows precise tracking and localization of transplanted genetic material, effectively overcoming the limitations of marker-based methods. Tokusashi et al. utilized in situ PCR to distinguish between normal and mutated albumin genes in mouse liver tissue sections [37]. Furthermore, this technology can identify specific DNA sequence alterations acquired or inherited by individual cell types.

1.13 Immuno-PCR

When detecting a certain antigen or antibody, immunological detection methods based on antigen-antibody reaction are usually used. This method has high specificity, but the conventional enzyme-linked detection method has low sensitivity and cannot detect trace antigens or antibodies, so it cannot be used for the detection of low-content protein markers and early diagnosis of diseases.

In 1992, Sano et al. first reported immuno-PCR, a novel method combining immune reactions with PCR amplification [38]. This approach builds upon the enzyme-linked immunosorbent assay (ELISA) while integrating the specificity of antigen-antibody interactions with the sensitivity of PCR amplification. As shown in Fig. 1.25, the principle involves labeling a DNA fragment with a known sequence onto either an antigen or antibody that matches the target analyte. Through specific antigen-antibody binding, the DNA is immobilized on the target analyte, followed by removal of free antibodies or antigens with DNA modification. Using this DNA as template for PCR amplification, quantitative detection of PCR products determines the presence and initial concentration of the target antigen. While immuno-PCR shares similarities with ELISA in methodology, their final detection systems differ: ELISA uses enzyme-labeled antibodies to detect antigens through substrate chromogenic reactions, whereas immuno-PCR employs antibodies labeled with DNA fragments to quantify target products via PCR amplification. Compared to enzyme-based amplification methods, immuno-PCR demonstrates 10–1000-fold higher detection sensitivity [39].

Compared to conventional antigen detection methods, immuno-PCR has achieved a sensitivity improvement of several orders of magnitude. In recent

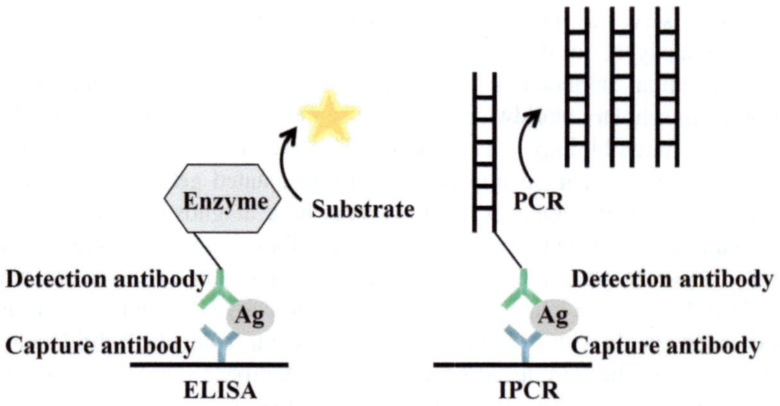

Fig. 1.25 Schematic representation of ELISA and immuno-PCR. (Revised and reprinted with permission from Ref. [39])

years, this technology has seen rapid development and is now widely applied in research areas such as pharmaceuticals, cell biology, microbiology, and disease diagnosis. Zhou et al. demonstrated that immuno-PCR can detect human transcription factor ETS1 with a limit of detection 100,000 times higher than conventional ELISA [40]. Abnormal conformation of prion protein (PrPC) can cause transmissible spongiform encephalopathy in mammals, including bovine spongiform encephalopathy, sheep scab, and human kuru and Creutzfeldt-Jakob diseases. Barletta et al. utilized immuno-PCR to detect PrPC in brain tissue homogenates of scab-infected hamsters at concentrations as low as 70–700 fg/mL, achieving a sensitivity one million times greater than ELISA [41].

1.14 Digital PCR

Over the three decades since the first invention of PCR in 1985, the technology has evolved through three generations, depending on detection methods. The first-generation PCR used gel electrophoresis for qualitative and semi-quantitative analysis of products. The second-generation real-time fluorescence quantitative PCR monitors the entire amplification process by collecting fluorescence signals during extension, using Ct values to calculate target gene expression levels. While achieving nucleic acid quantification, this method relied on Ct values and standard curves, resulting in relative quantification. Its sensitivity and accuracy for low-copy-target genes still required validation. This led to the emergence of third-generation digital PCR (dPCR). As an absolute quantification method for nucleic acids, dPCR demonstrates significant advantages over second-generation PCR in detecting low-abundance samples and rare mutations under complex backgrounds. It has been widely applied in gene expression analysis, miRNA research, cancer biomarker identification, and single-cell analysis.

1.14.1 Principle of Digital PCR

In 1999, Vogelstein and Kinzler introduced the concept of digital PCR, which enabled quantitative analysis of K-ras mutations—an oncogene associated with colon cancer [42]. This technology distributes a fluorescent quantitative PCR into numerous microreactors, each containing either no target molecules or one to several molecules. Each microreactor undergoes independent single-molecule template PCR amplification. After amplification, the endpoint fluorescence signal in each microreactor is detected: positive fluorescence signals indicate a reaction unit, while negative signals indicate a negative unit. By counting positive reaction units, the copy number of target in the sample can be calculated. Unlike traditional methods that rely on fluorescence curves or Ct values, digital PCR uses fluorescence presence as the sole criterion for quantification. This approach eliminates the need for control samples or standard curves for precise analysis. The distributed reaction system significantly reduces background signal from competing sequences, making it particularly effective for detecting rare mutations in complex genomic environments (Fig. 1.26a).

The reaction was conducted in 96-well and 384-well plates, where the target DNA molecules were diluted to an average concentration of 1 copy per two wells. PCR amplification was performed under optimized conditions, with amplified products detected using MB probes (Fig. 1.26b). As shown in Fig. 1.26c, two fluorescently labeled molecular beacons were introduced to distinguish between wild-type and mutant DNA. One MB probe specifically binds to sequences outside the mutation site, enabling both wild-type and mutant DNA amplifications to produce fluorescence signals. The other MB probe only hybridizes with wild-type DNA amplifications, ensuring fluorescence emission exclusively from wild-type DNA. By detecting fluorescence signals in each well, the relative abundance and ratio of wild-type to mutant DNA within the same sample can be determined.

1.14.2 Droplet Digital PCR

The principle of droplet digital PCR (ddPCR) is illustrated in Fig. 1.27. The method involves microdroplet processing of prepared PCR systems, where the target molecule-containing reaction mixture is encapsulated within nanoliter to picoliter-sized microdroplets formed by oil–water phases. These microemulsion droplets function as independent reaction units, while the rest of the process follows standard digital PCR protocols. Each droplet contains 0, 1, or a few target molecules. After PCR amplification, microdroplets with target molecules emit fluorescence signals

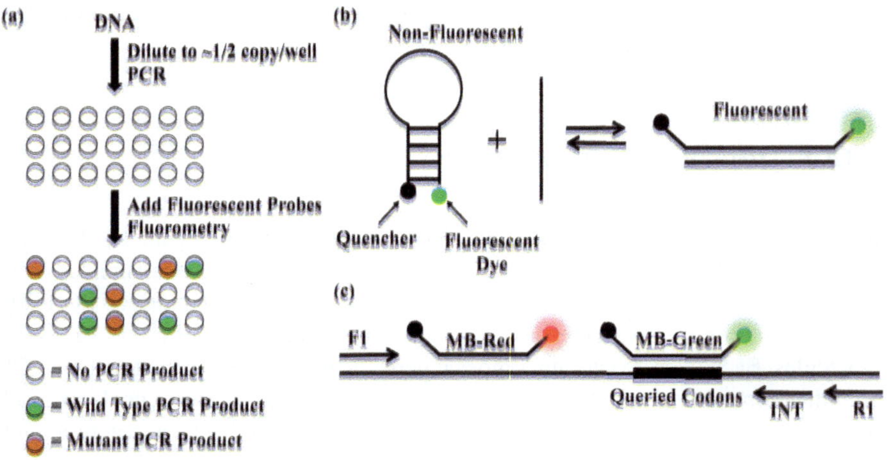

Fig. 1.26 Principles of digital PCR. (**a**) The process of digital PCR. (**b**) Principles of digital PCR detected by MB-based method. (**c**) Detection of PCR products generated by wild-type and mutant targets by MB-based method. (Revised and reprinted with permission from Ref. [42])

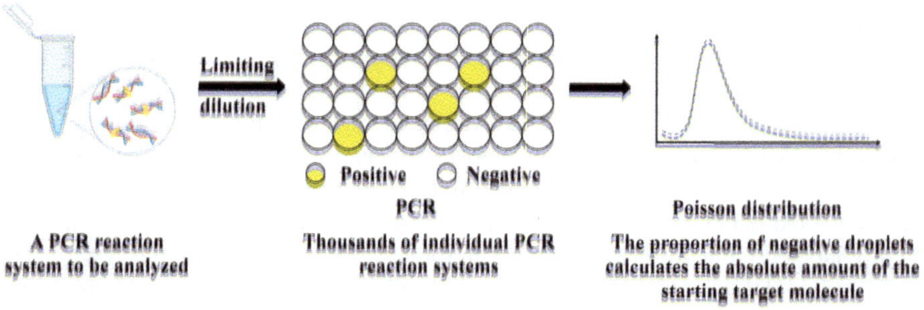

Fig. 1.27 Schematic representation of ddPCR

are interpreted as "1," while those without show no fluorescence are interpreted as "0." By applying Poisson distribution principles and analyzing the number and proportion of positive droplets, the initial copy number or concentration of the target molecule can be determined.

How does digital PCR enable detection of rare mutations in complex backgrounds? Taking a 20 μL reaction volume as an example, the system contains 40,000 wild-type targets (background nucleic acids) and 40 mutant targets (target genes), as shown in Fig. 1.28. Conventional PCR would detect only 0.1% of the mutant targets, requiring highly sensitive and specific methods. In contrast, ddPCR works by dividing the 20 μL PCR mixture into 20,000 nL microdroplets. Of these, 19,960 contain two wild-type targets, while 40 contain both two wild-type targets and one mutant target. After amplification, the 19,960 microdroplets containing wild-type targets show no fluorescence. Among the 40 fluorescent microdroplets,

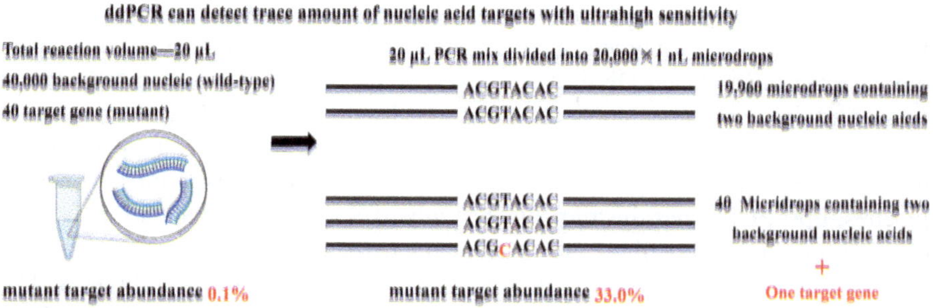

Fig. 1.28 Schematic representation of ddPCR for detection of low-abundance mutations

the mutant targets account for 33%, making them easily detectable. Hindson et al. demonstrated that ddPCR can detect rare mutations in the BRAF V600 gene with low background signal, achieving detection ratios as low as 0.001% [43]. This approach reduces background noise, increases target molecule concentration, and enables detection of trace amounts of biomarkers.

1.15 PCR Contamination and Control

PCR amplifies large quantities of fixed-length DNA fragments from trace amounts of initial DNA. However, its exceptional amplification efficiency makes it highly susceptible to contamination, particularly in detection methods that rely on the presence of amplification products as diagnostic criteria, where PCR contamination can severely impact results. Current contamination sources primarily include: contamination from exogenous genomic DNA, contamination from experimental materials, and contamination from PCR products of the same target molecule [44]. Among these, contamination from PCR products is the most severe and challenging to control. Due to the high copy numbers of amplified products, repeated operations like opening sample tubes and pipetting can easily disperse amplified products into the air, forming aerosol contamination. As early as 1989, scientists identified aerosol contamination as a serious issue [45, 46]. Even trace amounts of aerosol contamination can lead to false-positive PCR results, making it essential to design negative controls for detecting contamination. Commonly used methods include using non-template DNA or adding primers that are not complementary to the target DNA. Theoretically, such systems should not produce amplified products. If amplified products still emerge under these conditions, it indicates contamination, requiring further investigation to identify the source.

Ultraviolet irradiation can partially eliminate DNA contamination, a method described as "rapidly destroying all residual DNA on exposed surfaces" [47, 48]. UV-induced DNA damage typically occurs through the formation of cyclohexane rings between adjacent bases [49]. These cyclobutane rings create intra-strand pyrimidine dimers, thereby inhibiting polymerase-mediated chain extension. The formation of dimers is reversible, establishing a homeostatic equilibrium that favors monomers over dimers. Consequently, approximately <10% dimers exist in irradiated DNA at any given time. Since inclined surfaces weaken light intensity, laboratory applications of UV irradiation for DNA contamination control require perpendicular alignment of irradiated surfaces with light sources to maximize efficacy. Potential contamination sources during experiments include pipettes, tips, PCR tubes, instruments, test tube racks, and door handles. However, only partial surfaces of these items maintain perpendicular alignment with light sources when exposed, resulting in ineffective UV disinfection. Additionally, UV radiation proves more effective for longer DNA molecules, while shorter PCR products containing fewer adjacent pyrimidine bases show reduced sensitivity to UV exposure. Although UV irradiation offers some assistance in DNA contamination removal, further studies indicate that prolonged exposure duration and sequence-dependent characteristics limit its efficiency in achieving complete DNA decontamination.

The most widely used method for eliminating PCR contamination is the dUTP/UDG enzymatic method. UDG enzymes catalyze the hydrolysis of uracil bases and N-glycosidic bonds in the DNA backbone, creating pyrimidine-deficient sites that release free uracil and cleave PCR products, preventing them from serving as templates for further amplification [44, 50, 51]. The procedure involves replacing dTTP with dUTP during PCR preparation to introduce uracil-containing products. Before initiating a new PCR cycle, adding UDG enzymes for 10 min at 25 °C pre-reaction incubation removes uracil from PCR products, generating dozens or hundreds of pyrimidine-deficient sites. During subsequent thermal denaturation in the new cycle, the UDG enzyme becomes inactive, while high temperatures facilitate hydrolysis and cleavage of phosphate bonds at these sites, effectively eliminating contamination from uracil-containing products. Native DNA without modifications contains no uracil and remains unaffected by UDG, allowing continued PCR amplification. This method is highly effective in removing residual PCR products before thermal cycling. However, since PCR products shorter than 100 bp contain fewer uracil bases, complete elimination cannot be guaranteed using UDG. Consequently, longer PCR products demonstrate higher UDG-based contamination elimination efficiency. Currently, the dUTP/UDG enzymatic method has been widely adopted to eliminate aerosol contamination in various nucleic acid amplification reactions [52, 53].

1.16 PCR Applications and Perspectives

Currently, researchers can use real-time fluorescence quantitative PCR to conduct research on miRNA, DNA methylation, single-nucleotide polymorphism, and exogenous gene copy number in transgenic plants, providing an effective technical approach for the development of genetics. The details will be described in detail in subsequent chapters.

With the rapid development of PCR, in addition to its applications in basic biology research, food and environmental testing, PCR can also be applied to medical testing and diagnosis, especially cancer diagnosis and treatment. The mechanisms of cancer development are very complex, and the molecular mechanisms and histopathological pathways leading to cancer development vary for different cancers and even for patients with the same cancer. As the researches continue at the gene-molecular level, more and more tumor cell signaling pathways have been discovered, and a large number of clinical studies have shown that the amplification, mutation, and expression status of specific genes in the pathways are closely related to the efficacy of targeted therapies and chemotherapeutic drugs. Therefore, the detection of amplification, mutation, and expression of specific genes in clinical practice allows for personalized treatment plans tailored to each patient's needs, thus maximizing treatment effectiveness, reducing drug toxicity and side effects, and avoiding delayed treatment due to improper medication use.

Currently, the overall effectiveness of chemotherapy ranges from 30% to 40%, while genetic screening for patients with potential benefits can increase to 80%. The primary technology in cancer genetic testing is real-time fluorescence quantitative PCR, which boasts high accuracy, strong specificity, high sensitivity, ease of operation, broad applicability, and high-throughput capabilities. In clinical practice, PCR is mainly used for: (I) Detecting chromosomal translocations in hematologic malignancies, particularly for detecting microscopic residual lesions to aid early diagnosis; (II) Rapidly analyzing tumor prognosis through detection of activated oncogenes; (III) Identifying and classifying tumor-associated viruses linked to human cancers to guide treatment strategies; (IV) Detecting alterations in tumor suppressor genes to analyze pathogenesis and predict prognosis; (V) Analyzing drug resistance genes to optimize chemotherapy regimens; (VI) Detecting metastasis-inducing and metastasis-suppressing genes to determine tumor spread status and inform surgical decisions.

References

1. Saiki RK, Scharf S, Faloona F, Mullis KB, Horn GT, Erlich HA, Arnheim N. Enzymatic amplification of beta-globin genomic sequences and restriction site analysis for diagnosis of sickle cell anemia. Science. 1985;230:1350–4.
2. Saiki RK, Gelfand DH, Stoffel S, Scharf SJ, Higuchi R, Horn GT, Mullis KB, Erlich HA. Primer-directed enzymatic amplification of DNA with a thermostable DNA polymerase. Science. 1988;239:487–91.
3. Mullis K, Faloona F, Scharf S, Saiki R, Horn G, Erlich H. Specific enzymatic amplification of DNA in vitro: the polymerase chain reaction. In: Cold Spring Harbor symposia on quantitative biology, vol. 51. New York: Cold Spring Harbor Laboratory Press; 1986. p. 263–73.
4. Pelt-Verkuil EV, Witt RT. Principles of PCR. In: Molecular diagnostics. Singapore: Springer Press; 2019. p. 131–215.
5. Hyndman DL, Mitsuhashi M. PCR primer design. In: PCR protocols. Totowa: Humana Press; 2003. p. 81–8.

6. Dragan AI, Casas-Finet JR, Bishop ES, Strouse RJ, Schenerman MA, Geddes CD. Characterization of PicoGreen interaction with dsDNA and the origin of its fluorescence enhancement upon binding. Biophys J. 2010;99:3010–9.
7. Kawasaki ES. Amplification of RNA. In: PCR protocols: a guide to methods and applications. San Diego: Academic Press; 1990. p. 21–7.
8. Li JM, Li CH, Wang QQ, Tang RZ, Sun LY, Hao P, Jiang P. Detection of MUC1and GST-π gene by means of RT-PCR in axillary lymph nodes taken from breast cancer patients. Tumor Control Res. 2001;28:351–3.
9. Wang P. Combination of serological total antibody and RT-PCR test for detection of SARS-CoV-2 infections. J Virol Methods. 2020;283:113919.
10. Corman VM, Landt O, Kaiser M, Molenkamp R, Meijer A, Chu DK, Bleicker T, Brunink S, Schneider J, Schmidt ML, Mulders DG, Haagmans BL, Veer BVD, Brink SVD, Wijsman L, Goderski G, Romette JL, Ellis J, Zambon M, Peiris M, Goossens H, Reusken C, Koopmans M, Drosten C. Detection of 2019 novel coronavirus (2019-nCoV) by real-time RT-PCR. Eurosurveillance. 2020;25:2000045.
11. Burchard PR, Abou Tayoun AN, Scherer A, Tsongalis GJ. A rapid RT-PCR assay for the detection of HIV-1 in human plasma specimens. Exp Mol Pathol. 2014;97:111–5.
12. Ye Q, Chung LW, Li S, Zhau HE. Identification of a novel FAS/ER-α fusion transcript expressed in human cancer cells. Biochim Biophys Acta Gene Struct Expression. 2000;1493: 373–7.
13. Chamberlain JS, Gibbs RA, Rainer JE, Nguyen PN, Thomas C. Deletion screening of the Duchenne muscular dystrophy locus via multiplex DNA amplification. Nucleic Acids Res. 1988;16:11141–56.
14. Wu H, Qian C, Wu C, Wang Z, Wang D, Ye Z, Ping J, Wu J, Ji F. End-point dual specific detection of nucleic acids using CRISPR/Cas12a based End-point dual specific detection of nucleic acids using CRISPR/Cas12a based portable biosensor. Biosens Bioelectron. 2020;157: 112153.
15. Schouten JP, McElgunn CJ, Waaijer R, Zwijnenburg D, Diepvens F, Pals G. Relative quantification of 40 nucleic acid sequences by multiplex ligation-dependent probe amplification. Nucleic Acids Res. 2002;30:e57.
16. Vasickova P, Machackova E, Lukesova M, Damborsky J, Horky O, Pavlu H, Kuklova J, Kosinova V, Navratilova M, Foretova L. High occurrence of BRCA1 intragenic rearrangements in hereditary breast and ovarian cancer syndrome in The Czech Republic. Occurrence of BRCA1 intragenic rearrangements in hereditary breast and ovarian cancer syndrome in The Czech Republic. BMC Med Genet. 2007;8:1–10.
17. Taylor CF, Charlton RS, Burn J, Sheridan E, Taylor GR. Genomic deletions in MSH2 or MLH1 are a frequent cause of hereditary nonpolyposis colorectal cancer: identification of novel and recurrent deletions by MLPA. Hum Mutat. 2003;22:428–33.
18. Sørensen KM, Andersen PS, Larsen LA, Schwartz M, Schouten JP, Nygren AO. Multiplex ligation-dependent probe amplification technique for copy number analysis on small amounts of DNA material. Anal Chem. 2008;80:9363–8.
19. Huang Q, Chen D, Du C, Liu Q, Lin S, Liang L, Xu Y, Liao Y, Li Q. Highly multiplex PCR assays by coupling the 5′-flap endonuclease activity of Taq DNA polymerase and molecular beacon reporters. Proc Natl Acad Sci USA. 2022;119:e2110672119.
20. Zhong Z, Su X, Yang K, Huang W, Wang J, Zhuo Z, Xiang J, Lin L, He S, Li T, Zhang J, Ge S, Zhang S, Xia N. Sequence-specific nanoparticle barcode strategy for multiplex human enterovirus typing. Nat Commun. 2024;15:6478.
21. Don RH, Cox PT, Wainwright BJ, Baker K, Mattick JS. 'Touchdown' PCR to circumvent spurious priming during gene amplification. Nucleic Acids Res. 1991;19:4008.
22. Hecker KH, Roux KH. High and low annealing temperatures increase both specificity and yield in touchdown and stepdown PCR. Biotechniques. 1996;20:478–85.

23. Seoh ML, Wong SM, Zhang L. Simultaneous TD/RT-PCR detection of cymbidium mosaic potexvirus and odontoglossum ringspot tobamovirus with a single pair of primers. J Virol Methods. 1998;72:197–204.
24. Wang MY, Geng JL, Chen YJ, Song Y, Sun M, Liu HZ, Hu CJ. Direct detection of mecA, bla SHV, bla CTX-M, bla TEM and bla OXA genes from positive blood culture bottles by multiplex-touchdown PCR assay. Lett Appl Microbiol. 2017;64:138–43.
25. Larsen HH, Masur H, Kovacs JA, Gill VJ, Silcott VA, Kogulan P, Maenza J, Smith M, Lucey DR, Fischer SH. Development and evaluation of a quantitative, touch-down, real-time PCR assay for diagnosing Pneumocystis carinii pneumonia. J Clin Microbiol. 2002;40:490–4.
26. Massung RF, Slater K, Owens JH, Nicholson WL, Mather TN, Solberg VB, Olson JG. Nested PCR assay for detection of granulocytic ehrlichiae. J Clin Microbiol. 1998;36:1090–5.
27. Ochman H, Gerber AS, Hartl DL. Genetic applications of an inverse polymerase chain reaction. Genetics. 1988;120:621–3.
28. Triglia T, Peterson MG, Kemp DJ. A procedure for in vitro amplification of DNA segments that lie outside the boundaries of known sequences. Nucleic Acids Res. 1988;16:8186.
29. Huang G, Zhang L, Birch RG. Rapid amplification and cloning of Tn5 flanking fragments by inverse PCR. Lett Appl Microbiol. 2000;31:149–53.
30. Sentry JW, Kaiser K. Application of inverse PCR to site-selected mutagenesis of Drosophila. Nucleic Acids Res. 1994;22:3429.
31. Gyllensten UB, Erlich HA. Generation of single-stranded DNA by the polymerase chain reaction and its application to direct sequencing of the HLA- DQA locus. Proc Natl Acad Sci USA. 1988;85:7652–6.
32. Liu YG, Mitsukawa N, Whittier RF. Rapid sequencing of unpurified PCR products by thermal asymmetric PCR cycle sequencing using unlabeled sequencing primers. Nucleic Acids Res. 1993;21:3333.
33. Liu F, Liu H, Liao Y, Wei J, Zhou X, Xing D. Multiplex detection and genotyping of pathogenic bacteria on paper-based biosensor with a novel universal primer mediated asymmetric PCR. Biosens Bioelectron. 2015;74:778–85.
34. Poddar SK. Symmetric vs asymmetric PCR and molecular beacon probe in the detection of a target gene of adenovirus. Mol Cell Probes. 2000;14:25–32.
35. Haase AT, Retzel EF, Staskus KA. Amplification and detection of lentiviral DNA inside cells. Proc Natl Acad Sci USA. 1990;87:4971–5.
36. Yin J, Kaplitt MG, Pfaff DW. In situ PCR and in vivo detection of foreign gene expression in rat brain. Cell Vis. 1994;1:58–9.
37. Tokusashi Y, Nishikawa Y, Ogawa K. Differentiation of the normal and mutant rat albumin genes on hepatic tissue sections by in situ PCR. Nucleic Acids Res. 1995;23:3790–1.
38. Sano T, Smith CL, Cantor CR. Immuno-PCR: very sensitive antigen detection by means of specific antibody-DNA conjugates. Science. 1992;258:120–2.
39. Niemeyer CM, Adler M, Wacker R. Immuno-PCR: high sensitivity detection of proteins by nucleic acid amplification. Trends Biotechnol. 2005;23:208–16.
40. Zhou H, Fisher RJ, Papas TS. Universal immuno-PCR for ultra-sensitive target protein detection. Nucleic Acids Res. 1993;21:6038.
41. Barletta JM, Edelman DC, Highsmith WE, Constantine NT. Detection of ultra-low levels of pathologic prion protein in scrapie infected hamster brain homogenates using real-time immuno-PCR. Hamster brain homogenates using real-time immuno-PCR. J Virol Methods. 2005;12:154–64.
42. Vogelstein B, Kinzler KW. Digital PCR. Proc Natl Acad Sci USA. 1999;96:9236–41.
43. Hindson BJ, Ness KD, Masquelier DA, Belgrader P, Heredia NJ, Makarewicz AJ, Bright IJ,. Lucero MY, Hiddessen AL, Legler TC, Kitano TK, Hodel MR, Petersen JF, Wyatt PW, Steenblock ER, Shah PH, Bousse LJ, Troup CB, Mellen JC, Wittmann DK, Erndt NG, Cauley TH, Koehler RT, So AP, Dube S, Rose KA, Montesclaros L, Wang S, Stumbo DP, Hodges SP, Romine S, Milanovich FP, White HE, Regan JF, Karlin-Neumann GA, Hindson GM, Saxonov S, Colston BW High-throughput droplet digital PCR system for absolute quantitation of DNA copy number. Anal Chem., 2011;**83**:8604–8610.

44. Hartley JL, Rashtchian A. Dealing with contamination: enzymatic control of carryover contamination in PCR. Genome Res. 1993;3:S10–4.
45. Kwok SA, Higuchi R. Avoiding false positives with PCR. Nature. 1989;339:237–8.
46. Gibbs RA, Chamberlain JS. The polymerase chain reaction: a meeting report. Genes Dev. 1989;5:1095–8.
47. Sarkar G, Sommer SS. Shedding light on PCR contamination. Nature. 1990;343:27.
48. Ou CY, Moore JL, Schochetman G. Use of UV irradiation to reduce false positivity in polymerase chain reaction. BioTechniques. 1991;10:442–5.
49. Borst A, Box ATA, Fluit AC. False-positive results and contamination in nucleic acid amplification assays: suggestions for a prevent and destroy strategy. Eur J Clin Microbiol Infect Dis. 2004;23:289–99.
50. Longo MC, Berninger MS, Hartley JL. Use of uracil DNA glycosylase to control carry-over contamination in polymerase chain reactions. Gene. 1990;93:125–8.
51. Mifflin TE. Setting up a PCR laboratory, vol. 7. Cold Spring Harbor Protoc; 2007;5–14.
52. Hsieh K, Mage PL, Csordas AT, Eisenstein M, Soh HT. Simultaneous elimination of carryover contamination and detection of DNA with uracil- DNA-glycosylase-supplemented loop-mediated isothermal amplification (UDG-LAMP). Chem Commun. 2014;50:3747–9.
53. Tang Y, Chen H, Diao Y. Advanced uracil DNA glycosylase-supplemented real-time reverse transcription loop-mediated isothermal amplification (UDG-rRT-LAMP) method for universal and specific detection of Tembusu virus. Sci Rep. 2016;6:1–12.

Chapter 2
Ligase Chain Reaction (LCR)

Fengxia Su and Yuting Jia

Since its introduction in 1985, polymerase chain reaction (PCR) has greatly promoted the development of molecular biology research and disease diagnosis and also significantly promoted the identification of novel genes and other biomedical advances. However, for the detection of single-base mutations, there is still an urgent need for nucleic acid amplification and detection methods that accurately identify single-base differences. In 1988, Landegren demonstrated that a ligase-based reaction could distinguish the single-base mutated β-globin gene (β^S) in patients with sickle-cell anemia from the normal β-globin gene (β^A) [1]. Similar to PCR, which requires amplification of DNA molecules through multiple temperature cycles, ligase-catalyzed ligation reaction can also be cycled through high-temperature denaturation, annealing, and ligation steps to achieve signal or target amplification. In 1989, Wu further designed an amplification method for multiple rounds of ligation reaction, which has the ability of exponential amplification of DNA and realizes single-base differentiation of β-globin gene [2]. However, the T4 DNA ligase used at that time was not heat-resistant, resulting in the enzyme losing activity at the thermal denaturation stage of each cycle, and the ligase had to be frequently replenished to maintain the reaction. In order to overcome this problem, in 1991, Barany applied heat-resistant DNA ligase in ligase chain reaction (LCR) for the detection of single-base mutation of β-globin gene [3], avoiding the tedious steps of constantly replenishing enzymes during the thermal cycle and promoting the wide application of LCR.

F. Su (✉) · Y. Jia
University of Science and Technology Beijing, Beijing, China
e-mail: sufengxia@ustb.edu.cn

Z. Li (ed.), *Nucleic Acid Detection*, https://doi.org/10.1007/978-981-95-7295-3_2

2.1 The Principle of LCR

LCR relies on the cyclic ligation of two pairs of complementary DNA probes, where the 5′ end of the probes at the ligation site are modified with the phosphate group, respectively, to enable the subsequent ligation reaction. The reaction system undergoes multiple ligation reactions through multiple thermal cycles to complete exponential amplification of target DNA. The basic principle of LCR is shown in Fig. 2.1. First, four probes including A, A′, B, and B′ are designed based on the target sequence. Among the four probes, A and B can complement one of the target DNA strands and hybridize in adjacent positions, and A′ and B′ can complement another target DNA strand and hybridize in adjacent positions. According to the principle of complementary base pairing, A is complementary to A′, B is complementary to B′, and the 5′ end of probe A and probe B′ are modified with the phosphate group, respectively, to promote ligation. Step 1, high temperature denaturation causes the target DNA to form two single-stranded DNAs (ssDNAs). As the temperature is reduced to the annealing temperature (a temperature close to the probe melting temperature T_m value), probes A and B and probes A′ and B′ are hybridized with two strands of the target DNA, respectively. Step 2, under the catalysis of thermal stable ligase (such as *Taq* DNA ligase), probe A and probe B are ligated to form phosphodiester bonds to generate ligation product AB. And probe A′ and probe B′ are similarly ligated to generate ligation product A′B′. Step 3, high temperature denatures the double-stranded DNA (dsDNA), forming 4 kinds of ssDNA (2 target DNA strands, ligation product AB and A′B′). With the temperature again decreases to the annealing temperature, the 4 kinds of ssDNA mentioned above can serve as templates to hybridize again with probe A and B, probe A′ and B′. Step 4, under the catalysis of thermal stable ligase, the probes continue to ligate to form the products AB and A′B′. By continuously repeating the thermal cycles of steps 3 and 4 (generally 20–30 cycles), exponential amplification is formed, and a large number of LCR products are obtained. If the target has a single base mismatch with DNA probe at the ligation site, this mismatch can be recognized by the DNA ligase, which prevents it from catalyzing the ligation and prevents the production of LCR products. This

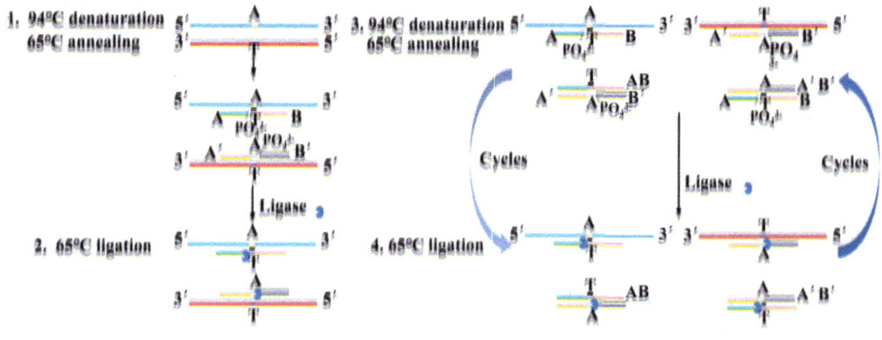

Fig. 2.1 Basic principle of LCR

process, in turn, enables the precise differentiation of single-base mutations in the target DNA. As an exponential amplification reaction, the specificity and sensitivity of LCR are affected by many factors, among which the target sequence, probe design, DNA ligase, and reaction conditions are the key factors.

2.1.1 Effect of Target Sequence on Ligation Efficiency

LCR is a cyclic ligation reaction catalyzed by DNA ligase. Because the ligase reaction has high specificity for distinguishing single-base mismatch, LCR is also endowed with the ability to distinguish single-base mutations. However, the ability of LCR to distinguish different nucleotide mismatches varies. The linear ligation using the *Taq* DNA ligase can distinguish all possible mismatch types, but for G-T and T-T mismatch, the efficiency is low, reaching up to 1.5% of the matching nucleotide, while for other types of mismatches, the efficiency is lower than 0.4% of the matching nucleotide [3]. Landegren designed different types of mismatch at the 3′ end of the probe at the ligation site by catalyzing the reaction with T4 DNA ligase [1]. The ligation efficiency of these probes with mismatched bases were tested, and the results showed that only T-G or G-T mismatch produced the ligation products at spermidine, high salt, and very low ligase concentrations. A study has shown that *Tth* DNA ligase has low fidelity for T-G and G-T mismatches at the 3′ end of the ligation site [4]. Whether *Tth* or T4 DNA ligase is used, T-G, G-T, A-C, A-A, C-A, G-A, or T-T mismatches were more difficult to distinguish at the 5′ end of the junction than at the 3′ end. The DNA ligase from *Saccharomyces cerevisiae* can recognize the mismatch A-G and T-G mismatches at 3′ end, but the effect of A-C, T-C, C-A, and G-A mismatches at the 5′ end on the ligation efficiency is very small [5]. Mammalian DNA ligase I and III show different efficiency in ligating C-T, G-T, and T-G mismatches [6]. It is easy for vaccinia virus DNA ligase to ligate C-T, G-T, T-T、, A-C, T-C, and C-C mismatches at 5′ end and C-A, C-T, G-T, T-T, and T-G mismatches at 3′ end [7].

In addition, the ligation efficiency of the ligase varies for different nucleotides. The properties of the 3′ end nucleotide of the probe also affect its ligation efficiency. The ligation efficiency of G or C at the 3′ end is higher than that of A or T. It may be that the GC base pairs, paired by three hydrogen bonds, which bind the template more closely, thus enabling the ligase to recognize the nick more effectively and accelerate the catalytic ligation reaction.

2.1.2 Probe Design

(1) Position and Length of the Probes

The LCR probes should be designed in the conserved sequence region of the target gene to increase the specificity of LCR amplification. Ensure that the mutation site is located exactly in the ligation site of two adjacent probes to achieve the highest degree of differentiation of mismatched bases and minimize nonspecific connections [8]. The specificity is improved significantly when the 3′ end of the probe is hybridized with the tested site, while the sensitivity is lower when the 5′ end of the probe is hybridized with the tested site. The artificial introduction of a mismatch base into the third base at the 3′ end of the ligation can increase the ability to distinguish the mismatch base. In addition, the use of nucleic acid analogs on the second or the third base at the 3′ end of the ligation can also increase its ability to distinguish single base mismatch, but it will reduce the efficiency of the ligation and detection sensitivity to a certain extent [4].

In addition, the length of the LCR probes should be moderate, preferably between 18 and 30 bases [8] to ensure that the probe can hybridize specifically with the target DNA and avoid nonspecific ligation. If the probe length is too short, the specificity of hybridization between the probe and the target sequence will become worse, resulting in nonspecific ligation. If the probe length is too long, it will easily form secondary structure and dimer between the probes, which will affect the subsequent ligation.

(2) The T_m Value of the Probes

The T_m values of the four probes should be close to each other, avoiding probes with high GC content, easy formation of secondary structures, or primer dimers as much as possible. The GC content of the probes is generally optimal between 40% and 60%. When the GC content is too low, the T_m value of the probe is too low, which makes the hybridization between the probe and the target unstable, and it is not easy to make the specific ligation of the probes. When the content of GC is too high, the T_m value of the probes is also increased, and the probes are easy to hybridize with the target, which may cause the nonspecific ligation of the probes. The T_m value of the DNA probes can be selected according to the optimal temperature of the DNA ligase, so that it is slightly higher than the temperature of the ligation reaction to obtain higher specificity and binding efficiency.

2.1.3 DNA Ligase

DNA ligase plays an important role in the process of DNA replication. DNA ligase ligates the nick in dsDNA, that is, catalyzes the formation of phosphodiester bond between the 5′ phosphate group and 3′ hydroxyl between adjacent probes [9–12]. Generally, DNA ligases can be divided into two types according to the different energy sources required [13–17], adenosine triphosphate (ATP) dependent and nicotinamide adenine dinucleotide (NAD^+) dependent. ATP-dependent ligases exist widely in eukaryotes, archaea, eubacteria, and viruses. NAD^+ dependent ligases are mainly present in archaea, eubacteria, and viruses. Common DNA ligases

1. **E + MgATP —— E·MgATP —— E·Mg_2ATP —— E-AMPMg + MgP_2O_7**
2. **E-AMPMg + ndsDNA —— E-AMPMg·ndsDNA —— E·n-MgAMP-dsDNA —— E + n-MgAMP-dsDNA**
3. **E·n-MgAMP-dsDNA —— E·dsDNA·MgAMP —— E + dsDNA + MgAMP**

Fig. 2.2 The catalysis principal mechanism of ligase. E stands for DNA ligase. ndsDNA indicates that dsDNA contains a 5′-phosphorylation nick. n-MgAMP-dsDNA indicates that the dsDNA is adenylated at the nicked 5′ phosphate. (Revised and reprinted with permission from Ref. [18])

include DNA ligase from *E. coli*, T4 DNA ligase from phage T4, and *Taq* DNA ligase from *Thermus aquaticus*.

Taking T4 DNA ligase as an example, the mechanism of catalysis of ligase includes three steps [18], enzyme adenylation, substrate adenylation, and nick closure. A complete ligation cycle is shown in Fig. 2.2. In the first step, DNA ligase (E) and cofactor (MgATP) transfer 5′-adenosine monophosphate (AMP) group to lysine residue through phosphoamide bond to obtain E-AMPMg; In the second step, the 5′-phosphate group of the dsDNA break attacks the E-AMPMg complex, and AMP is transferred to the 5′-phosphate group of the DNA strand, generating a DNA-AMP intermediate (n-MgAMP-dsDNA). In the third step, a phosphodiester bond is formed between the 3′-hydroxyl and 5′-phosphate group at the nick and AMP is released.

Although T4 DNA ligase can efficiently ligate dsDNA nicks, it is not resistant to high temperature and needs to be added frequently in the LCR thermal cycles, which is a complicated step. It was not until the discovery of the heat-resistant *Taq* DNA ligase that LCR technology became widely available. The ligase in LCR needs to have the following properties. (1) Thermal stability that ensures the activity of ligase in the thermal cycles. *Taq* DNA ligase is the first ligase applied to high-temperature reactions, which can withstand more than 30 high-temperature cycles [3, 19]. Other common DNA ligases include Hi-T4 thermal stable DNA ligase (a genetically engineered T4 DNA ligase that has significantly improved its thermal stability compared with T4 DNA ligase), HiFi *Taq* DNA ligase, Pfu DNA ligase (isolated and purified from *Escherichia coli* cloned with *Pyrococcus furiosis* DNA polymerase gene), etc. (2) High catalytic activity of the ligase. *Taq* DNA ligase commonly used in LCR is the first discovered thermostable ligase. 1 unit (40,000 units/mL) of *Taq* DNA ligase in a 50 μL reaction system can ligate 50% of 1 μg of BstEII digested λ DNA fragment (12 bp sticky end) in 15 min at 45 °C. (3) High-fidelity of the DNA ligase. *Taq* high-fidelity DNA ligase purified from recombinant *E. coli* carrying cloned thermophilic bacteria has been widely used in research. This enzyme not only inherits the heat-resistant property of *Taq* DNA ligase, but also ligates DNA probes with high fidelity and efficiency.

2.1.4 The Composition of LCR System

The main components of the LCR amplification reaction system are target DNA, four ligation probes, DNA ligase (such as *Taq* DNA ligase), and its reaction buffer solution (*Taq* DNA ligase reaction buffer: 20 mM Tris-HCl (pH 7.6 @ 25 °C), 25 mM KAc, 10 mM $MgAc_2$, 10 mM DTT, 1 mM NAD and 1% Triton X-100). Xiao studied the effect of additives (DMSO, Tween-20, glycerol, formamide and polyethylene glycol-6000 (PEG-6000)) on the efficiency and specificity of LCR [20]. The results showed that additives such as DMSO, glycerol, and formamide increased the specificity but decreased the ligation efficiency. At low concentration, Tween-20 can significantly improve the specificity of LCR, but too high concentration of Tween-20 can rapidly and completely inhibit the reaction. Among the five additives tested, PEG-6000 was proven to be the most effective one, with an optimal concentration of 5%. In addition, adding salmon sperm DNA to the system can also increase its ligation specificity, possibly because the large steric hindrance effect of salmon sperm DNA can effectively reduce template-independent ligation.

The standard procedure of LCR consists of the following steps. First, the mixture is denatured at a high temperature for a duration of 5 min. Subsequently, 20–30 thermal cycles are performed, each consisting of high temperature denaturation for 20–60 s, annealing and ligation at 60–65 °C for 1–5 min (the temperature and time are determined by the T_m value of the probes, which should ideally be close to the T_m value of the probes to ensure optimal reaction specificity). After the LCR amplification, the products are analyzed by electrophoresis, fluorescence, colorimetry, or electrochemistry.

2.1.5 Influence of Nonspecific Ligation in LCR

DNA ligase in LCR can trigger template-independent ligation to produce nonspecific ligation products with the same length as the specific product, and generate blank signal, which is the main factor affecting the sensitivity and specificity of LCR. In order to effectively reduce this nonspecific ligation, the primary strategy is to employ a highly thermal stable DNA ligase. In addition, the following measures can be taken to further optimize the LCR to reduce the generation of nonspecific products [21]: (1) Adding salmon sperm DNA to the reaction system; (2) Only the 5′ end of probes at the ligation site are modified by phosphate groups; (3) As shown in Fig. 2.3, the probes are designed to have a sticky end rather than a blunt end structure at the junction, that is, one base at the 3′ end of the junction (circle) is complementary to the base to be tested, which helps to improve the specificity of the reaction. In addition, adding a noncomplementary sequence as the tail of the complementary probe pair to increase the distance between probes can effectively reduce the interaction between the probes and reduce the nonspecific ligation. (4) Select a temperature close to the T_m value of DNA probes as the

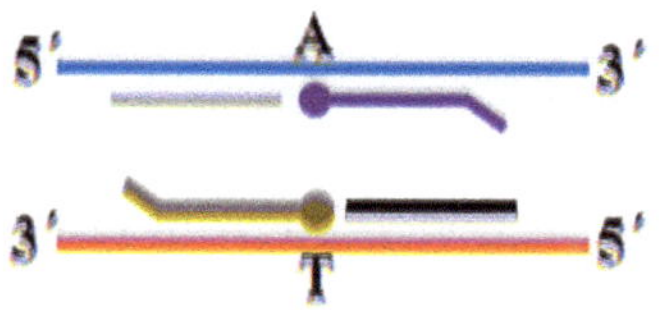

Fig. 2.3 Probe design diagram

temperature condition for the ligation step. Through the comprehensive application of the above strategies, the nonspecific ligation products in LCR can be effectively reduced, and the specificity and sensitivity of the reaction can be improved.

2.2 The Improvement of LCR Technology

LCR relies on the repeated ligation of four probes in the thermal cycles to achieve exponential amplification of initial DNA. However, these probes may produce nonspecific ligation independent of the template during hybridization and ligation process, which will interfere the experimental results. Therefore, it is necessary to improve the LCR technique to reduce the nonspecific ligation reaction and improve the specificity of LCR.

2.2.1 LDR

When Barany applied heat-resistant DNA ligase to LCR, he also performed a cyclic ligation reaction using only two adjacent probes, that is, ligase detection reaction (LDR). The principle of LDR is shown in Fig. 2.4. Two complementary DNA probes A and B are designed for the target DNA with single-base mutation. Probe A is a universal downstream probe for both mutant and wild-type targets, and its 5′ end is modified with the phosphate group. Probe B is a mutant target-specific upstream probe, and its 3′-terminal base is complementary to the base of the mutation site to be measured. As shown in Fig. 2.4a, high temperature denatures dsDNA to ssDNA. The temperature then drops to the annealing temperature, and probes A and B hybridize with the mutant DNA. Under the catalysis of heat-resistant DNA ligase, probes A and B are ligated to the AB. With the temperature cycles, denaturation, hybridization, and ligation are repeated, and LDR products can be obtained after 20–30 cycles of linear amplification. When there is only wild-type DNA, the 3′-end of probe B does not match the wild-type DNA, and probes A and B cannot be ligated at the annealing temperature, and no amplification occurs (Fig. 2.4b). In this way, the differentiation between mutant and wild-type targets is achieved.

The results of LDR and LCR for detecting single-base mutated β^S in patients with sickle-cell anemia and normal β^A are shown in Fig. 2.5. When 1 fmol of target DNA is present, LDR using perfectly matched probes can produce a significant ligation

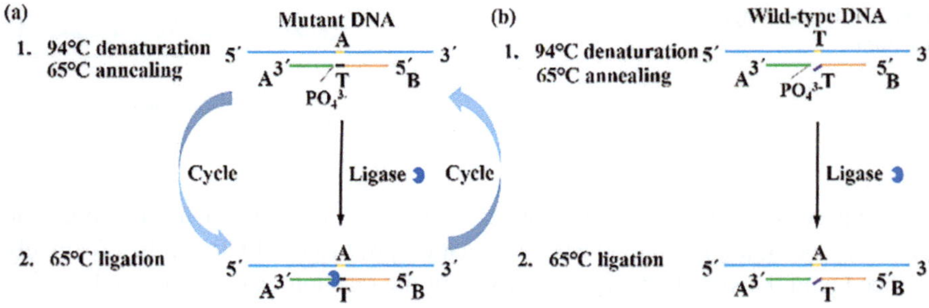

Fig. 2.4 Schematic diagram of LDR

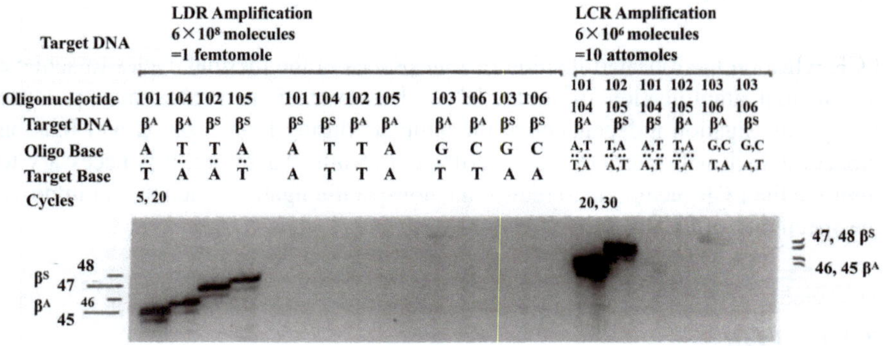

Fig. 2.5 Autoradiogram of single-base mutation analysis using LDR and LCR. (Revised and reprinted with permission from Ref. [3])

product, while LDR using mismatched probes can produce a ligation efficiency of 0.6% to <0.3% of that using perfectly matched probes. When amplify by LCR with perfectly matched probes, only 10 amol of target DNA can produce significant product bands. This result fully proves that exponentially amplified LCR has higher sensitivity than linearly amplified LDR during the analysis, but the relative ligation efficiency of LCR for mismatched probes ranges from 1.7% to <0.3%. The reason for this difference may be that in the LCR process, the nonspecific ligation products produced in the initial stage can be used as the target to trigger the ligation in the subsequent cycles, leading to the exponential amplification of the nonspecific ligation products, and ultimately affecting the detection specificity.

Compared with LCR, LDR only uses two probes for linear amplification of target DNA, and its amplification efficiency is lower than that of LCR. However, precisely because LDR omits a pair of complementary probes, it effectively avoids nonspecific ligation and provides better specificity than LCR. Compared with PCR amplification, LDR can achieve highly specific detection of single base mutation, but the sensitivity is generally not high. Therefore, LDR can be used in combination

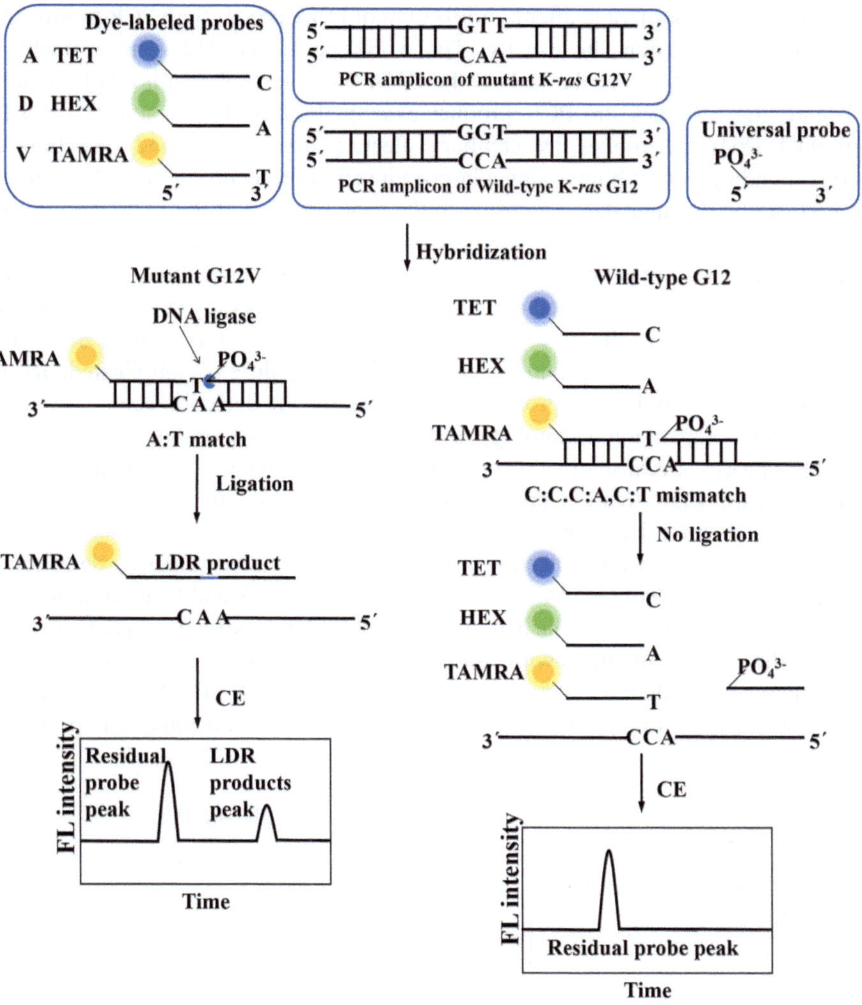

Fig. 2.6 Schematic diagram of single-nucleotide mutation detection by LDR and PCR coupling. (Revised and reprinted with permission from Ref. [22])

with PCR to achieve a highly sensitive and highly specific detection of single-base mutations.

In 2014, Hamada combined LDR with PCR to develop a method for detecting low abundance single-base mutation of K-*ras* oncogene codon 12 [22]. The detection principle is shown in Fig. 2.6. A series of adjacent DNA probes are designed with the mutation site at the ligation site, including probe A marked by TET at the 5′end and a C at the 3′end, probe D marked by HEX at the 5′ end and an A at the 3′ end, probe V marked by TAMRA at the 5′ end and a T at the 3′ end, and a universal probe modified the phosphate group at the 5′ end. After amplification of mutant K-*ras* G12V gene and wild-type K-*ras* G12 gene by PCR, LDR product can be formed by ligation only after hybridization of DNA probes with PCR product that matches the mutation site exactly. The wild-type K-*ras* G12 gene is unable to perform the LDR due to the lack of the DNA probe (a probe with a G base at the 3′ end) that matches the PCR product exactly. Finally, by separating the LDR product from the probes by capillary electrophoresis (CE) and detecting the corresponding fluorophores using a laser-induced fluorescence system, the mutation target can be distinguished according to the peak position of the LDR product, and rare mutations (about one mutation sequence in 100 normal sequences) can be detected in the excessive wild-type target. The method uses three probes labeled with different fluorophores at the same time, which can not only detect the mutations but also directly give the type of mutation base. However, this method uses CE to separate and detect LDR products, the operation procedure is complicated, and the use of fluorescently labeled probes greatly increases the detection cost.

Magnetic separation technology is often used as a separation method in the analysis process because of its simple operation and high speed. Watanabe modified biotin on the LDR ligation probe, used streptavidin-modified magnetic beads to separate and enrich LDR products, and then analyzed LDR products on the surface of magnetic beads with molecular beacon (MB) to realize sensitive detection of single-base mutation [23]. As shown in Fig. 2.7, probes A and B are first designed according to mutant DNA. The 5′ terminal of probe A is modified with the phosphate group, the 3′ end is modified with biotin, and the 3′-terminal nucleotide of probe B is complementary with the mutation site. Wild-type and mutant DNA are amplified by PCR respectively to obtain a large number of wild-type and mutant amplification

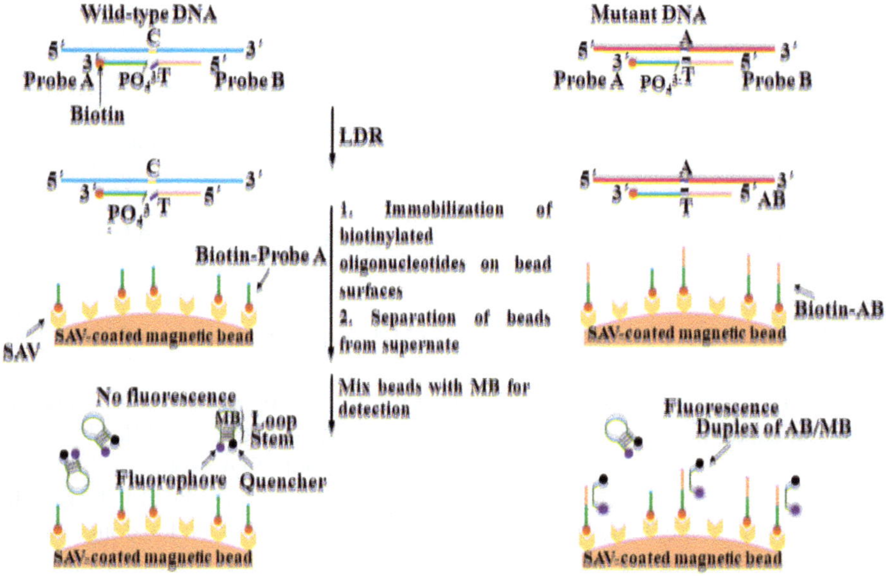

Fig. 2.7 Schematic diagram of microbead-based platform for LDR and PCR coupling to detect single-base mutation. (Revised and reprinted with permission from Ref. [23])

products. Probes A and B are completely complementary to the mutant amplification products (see the right side of Fig. 2.7) and can perform LDR under the action of ligase. The LDR product is fixed on the magnetic bead by the interaction of the biotin carried by itself with streptavidin on the surface of the magnetic beads. The LDR products are then analyzed by MB with modified fluorophore at 5′ end and quenching group at 3′ end, and the loop sequence of MB is complementary to the LDR product. Therefore, LDR products on the surface of magnetic beads can specifically trap MB and open its stem-loop structure to produce fluorescence signals. The magnetic beads are then gathered in a capillary tube by a magnet, and the mutation can be detected by scanning the encapsulated magnetic bead with a fluorescence scanning imager. If the initial target contains the mutation site, there are complementary LDR products on the magnetic bead surface and the stem-loop structures open, then fluorophore and the quenching group are separated to produce a fluorescence signal. In contrast, if the initial solution contains only the wild-type DNA, and there is a single-base mismatch between the 3′ end of probe B and the target sequence, LDR cannot be performed, the molecular beacon cannot be turned on, and therefore no fluorescence signal is generated. This method can successfully detect a single mutant target in a mixture containing 400 wild-type targets. In addition, this method takes 20 min for the detection of LDR products, and can effectively distinguish wild-type and mutant targets, saving operation time and improving the convenience of detection.

In summary, LDR has the ability to detect single-base mutations, but its wide application is limited by its relatively low detection sensitivity. In order to realize the full potential of LDR, it is often necessary to use it in combination with PCR. This combined method requires multiple amplification and detection, and the operation steps are relatively cumbersome, resulting in the experimental process complicated and time-consuming.

2.2.2 Nested LCR

The sensitivity of conventional LCR is higher than that of LDR, but its specificity is slightly less than that of LDR. Therefore, in order to further improve the specificity of LCR, scientists have proposed the concept of nested LCR. Nested LCR is an improved LCR amplification method proposed by Barany in 1991, also known as two-tube LCR technique, which includes two rounds of LCR reaction [21]. The principle is shown in Fig. 2.8. Taking the β^A as an example, the target site is the A-T base pair. In the first tube reaction, probes A and B are designed upstream of the "A" in β^A-1 sequence (probe B matches one base ahead of the "A"), and probes C and D are designed downstream of the "T" in β^A-1′ sequence (probe D has an "A" at the 3′ end and is complementary to the tested site "T"). Probes A and B are hybridized with β^A-1 and ligated to AB, while probes C and D are hybridized with β^A-1′ and ligated to CD under the action of DNA ligase. AB and CD are used as templates to ligate probes C and D and A, B, respectively, in subsequent cycles. After 20–30 cycles, a

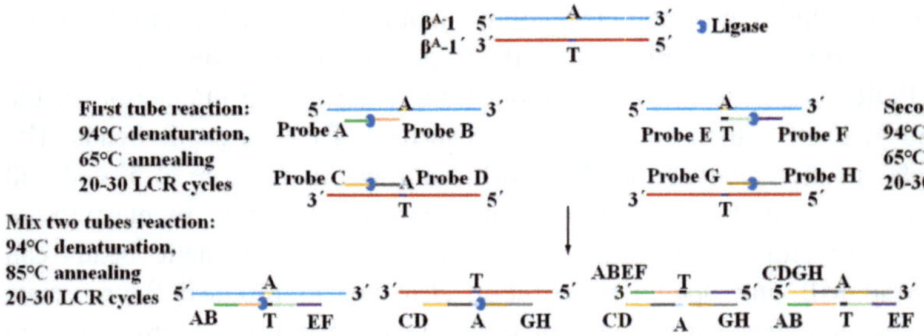

Fig. 2.8 Schematic diagram of the nested LCR. (Revised and reprinted with permission from Ref. [21])

large number of LCR products AB and CD are obtained. In the second tube reaction, probes E and F are designed according to the β^A-1 sequence (the 3′ end of probe E is a "T", complementary to the tested site "A"), and probes G and H are designed according to the β^A-1′ (the 5′ end of probe G matches a base to the right of the "T"). After denatured at 94 °C and annealed at 65 °C, probes E and F are hybridized with β^A-1, and then ligated to EF under the action of DNA ligase. Probes G and H are hybridized with β^A-1′, and GH is obtained after ligation. Similarly, EF and GH can be used as templates to ligate probes G, H, and E, F in subsequent cycles. The reaction is performed for 20–30 LCR cycles, and a large number of LCR products are obtained. After LCR is performed on both reaction systems, the products are mixed for a new round of LCR. The new LCR is amplified with AB, CD, EF, and GH as four probes, and the full-length LCR product (60 nt) is obtained. Thus, the specificity of LCR is significantly improved after the two-tube reaction.

Compared with conventional LCR, nested LCR achieves higher specificity. This is because, in the first tube LCR, product AB and CD are generated only when the target DNA perfectly matches probes A, B, C and D. Similarly, EF and GH are produced only when the target DNA matches with probes E, F, G and H. After the two-tube solutions are mixed, the AB, CD, EF, and GH generated above are used as new probes for LCR amplification. At this time, due to the hybridization and ligation temperature rising to 85 °C, the short probes in the original system could not undergo ligation, and only the correct ligation products in the previous step could be ligated. If the wrong ligation product is produced in the first amplification, the probability that the probe can be paired and amplified on the wrong fragment in the second amplification is extremely low. Therefore, the amplification specificity of nested LCR has been significantly improved.

2.2.3 *Gap-LCR*

In order to improve the sensitivity of the detection, the LCR process usually requires the addition of a large number of probes at the beginning of the reaction. However, an excess of probes may be present in double-stranded form and ligated in a target-independent manner by DNA ligase, that is, blunt-end ligation. The product of this blunt-end ligation can also be used as a template to lead background interference in subsequent cycles. In order to reduce this nonspecific ligation, Abravaya proposed gap-LCR to detect single-base mutation [24, 25], as shown in Fig. 2.9. In gap-LCR, four probes are designed. When the probes are hybridized with the target sequence, a gap of one to several nucleotides is formed between the two adjacent probes. The DNA extension reaction on the 3′ end of the probe is catalyzed by heat-resistant DNA polymerase (without 5′–3′ exonuclease activity), and the gap between two adjacent probes is filled. Subsequently, the two adjacent probes with a nick are ligated under the action of DNA ligase. The above extension and ligation reactions go through 20–30 cycles to obtain gap-LCR products.

Gap-LCR has been widely used because of its high sensitivity and specificity. Conventional LCR cannot distinguish mutant DNA from wild-type DNA with a ratio exceeding 100-fold [24, 26]. However, gap-LCR is able to detect 1 mutant DNA from 10^4 copies of wild-type DNA. Zhou developed a method for analyzing gap-LCR products based on DNAzyme to detect single-base mutations [27]. As shown in Fig. 2.10, four probes L, L*, R, and R* are designed according to the mutant DNA of β-globin gene. The 5′ end of the probe L* and R are modified with the phosphate group, respectively, and the 3′ end of probe R is modified with DNAzyme sequence. After the DNA is mixed with the 4 probes, the first step is

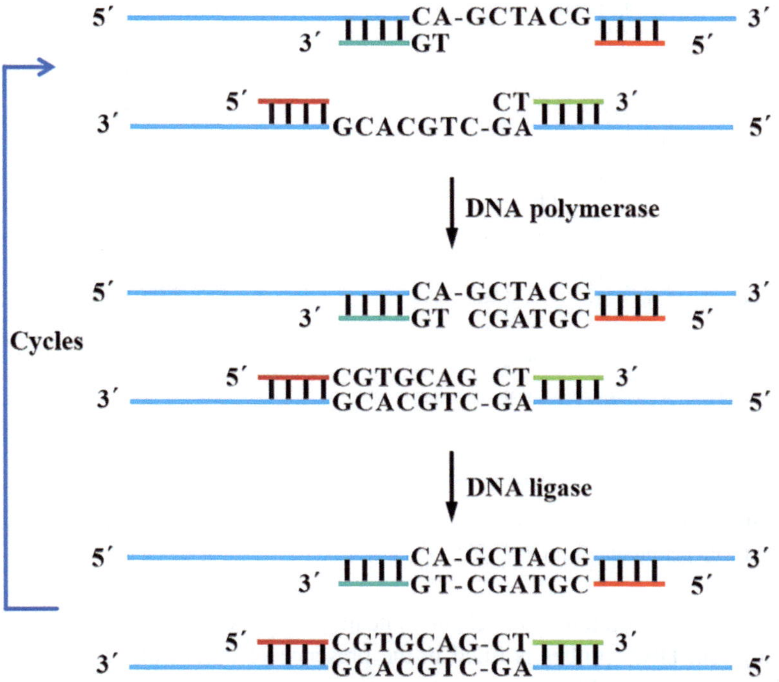

Fig. 2.9 Schematic diagram of gap-LCR. (Revised and reprinted with permission from Ref. [25])

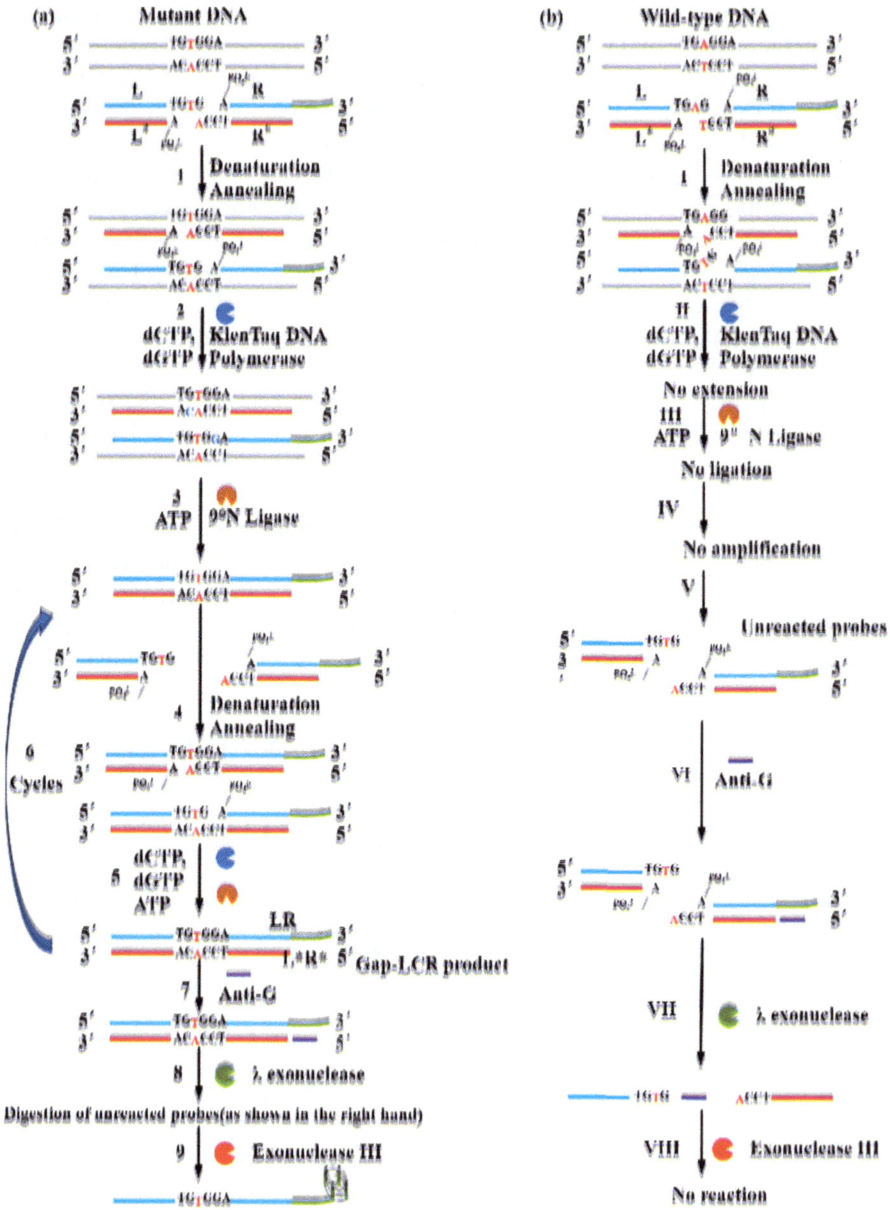

Fig. 2.10 Principal diagram of gap-LCR detection of single nucleotide mutation. (Revised and reprinted with permission from Ref. [27])

carried out: denaturation at 94 °C and annealing at 65 °C. Step 2: The gap between two adjacent DNA probes is filled in the presence of dCTP, dGTP, and KlenTaq DNA polymerase (which does not have 5′–3′ exonuclease activity). Step 3: 9°N DNA ligase catalyzes the ligation. After 26 cycles, a large number of gap-LCR products (LR/L*R*) are obtained. Subsequently, λ exonuclease and DNAzyme complementary probe Anti-G are added to the solution. λ exonuclease can efficiently degrade the 5′-end phosphorylated DNA strand in dsDNA to degrade the unreacted phosphorylated probe L* and R. In order to promote the degradation of DNAzyme sequence in probe R, probe Anti-G is added to complete its double-stranded structure. In addition, the 5′ non-phosphorylated dsDNA cannot be degraded by λ exonuclease, such as the existence of LCR products in double-stranded state, affecting the release and detection of DNAzyme sequence. Therefore, exonuclease III is then added to degrade the 3′ blunt-ended DNA (L*R*) in dsDNA, and only the single-strand product LR remained in the system. DNAzyme transforms into a single-stranded state, spontaneously forms a G-quadruplex structure, and exhibits peroxidase activity. After hybridizing the designed probe Anti-G with the DNAzyme sequence, the DNAzyme sequence protrudes 4 bases to ensure that the product LR will not be cut during exonuclease III digestion.

The DNAzyme sequence introduced on probe has the following properties: ① In the presence of Hemin, DNAzyme can stably bind Hemin and exist in the form of G-quadruplex. This complex can catalyze H_2O_2 to oxidize colorless ABTS to green, and this signal can be detected by naked eye or UV-visible spectrophotometer. ② In the presence of Hemin, the complex formed by the combination of DNAzyme and Hemin can catalyze H_2O_2 oxidation of nonfluorescent tyramine hydrochloride to produce fluorescent diphenol compounds. This method can detect mutant target specifically when the content of wild-type target is 10^5 times higher than that of mutant target.

2.2.4 Asymmetric Gap-LCR

In gap-LCR, the probes are hybridized with the target sequence to form a double-stranded structure with a gap by shortening the probe length. This gap must be filled before the ligation, which helps to reduce the background signal and improve the specificity. When the target is RNA, the efficiency of DNA ligase ligating DNA probe is low, resulting in the need to reverse transcribe RNA into cDNA. However, the introduction of dNTPs in the cDNA synthesis system may cause the LCR probe to fill the gap out of control and interfere with the progress of gap-LCR. Therefore, dNTPs must be completely removed from cDNA products before gap-LCR, which is a cumbersome process. Based on this, Marshall proposed asymmetric gap-LCR [28], allowing the direct use of cDNA products as the templates for gap-LCR. As shown in Fig. 2.11, a DNA primer and only three deoxynucleotides (dGTP, dCTP, and dATP) were used for reverse transcription of the RNA template. Since dTTP is not provided in the system, reverse transcription is terminated when A appears on the RNA template. After denaturation, the generated cDNA is hybridized with three DNA probes (probe 1, probe 2, and probe 3) to form a double-stranded structure with a gap. The gap between probe 1 and probe 3 is then filled with DNA polymerase and three dNTPs left in reverse transcription, and the adjacent probes are then ligated

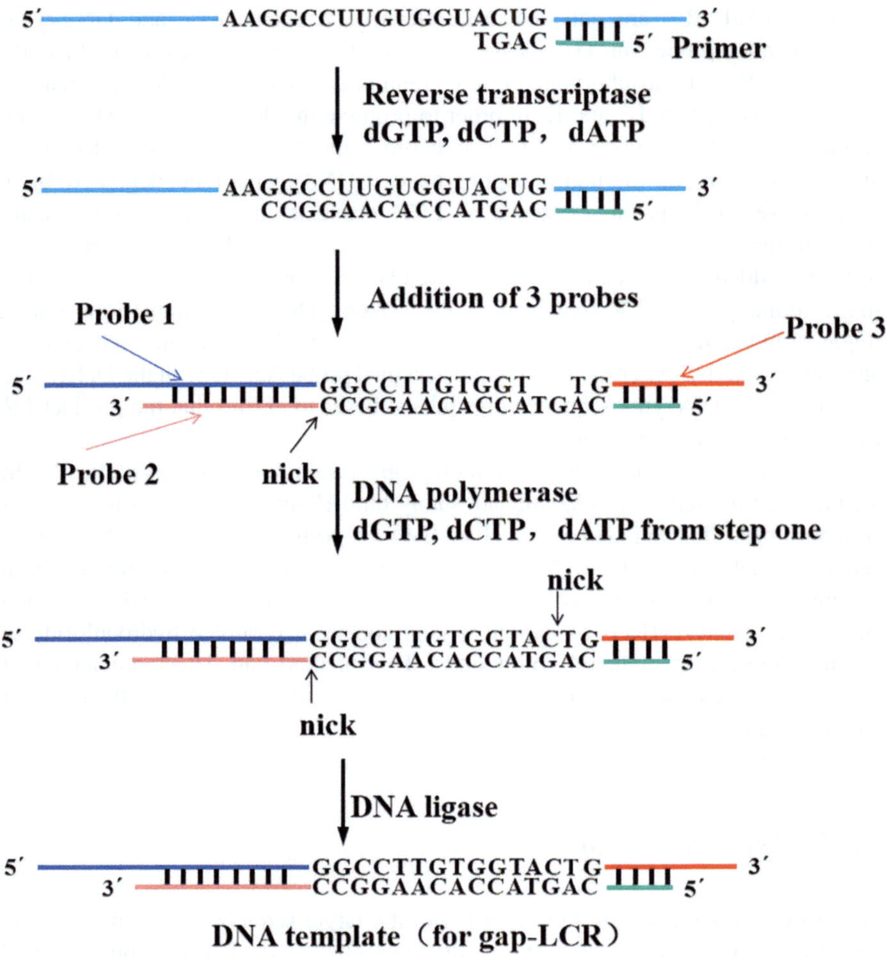

Fig. 2.11 Schematic of asymmetric gap-LCR. (Revised and reprinted with permission from Ref. [28])

with DNA ligase. The generated dsDNA is used as a template for gap-LCR. The method can detect as few as 20 copies of the recombinant RNA template. Asymmetric gap-LCR can also be used to analyze DNA, which is 10 times more sensitive than RNA.

Asymmetric gap-LCR realizes the automatic termination of reverse transcription by reducing the types of dNTP of reverse transcription, effectively avoiding the influence of the remaining dNTPs on the gap filling reaction, and avoiding the nonspecific amplification caused by blunt-end ligation, thus improving the detection sensitivity. It showed significant advantages in detecting RNA base mutations.

2.2.5 Multiple LCR

Usually, a disease is not only associated with a single-gene mutation, but the result of multiple gene changes. Therefore, for disease diagnosis and related research, the simultaneous detection of multiple gene mutations using multiple LCR is of great significance.

(1) Multiplex Gap-LCR for the Detection of Single-Nucleotide Polymorphism
Song performed fluorescence coding on LCR primers to achieve multiple detection of K-*ras* single-nucleotide polymorphism (SNP) [29]. As shown in Fig. 2.12a, first, four probes are designed according to the mutant target, namely a, b, c, and d. The 3′ end of probe c modifies biotin, and the 5′ end of probe d modifies fluorophore (Alexa488). In the presence of mutant target, these four probes can specifically hybridize with it. Under the action of DNA polymerase, probes a and d carry out elongation reaction, respectively, to fill the gap between adjacent probes. Subsequently, under the action of DNA ligase, the adjacent probes are ligated to form the products a + b and c + d. These products can be used as templates for the next rounds of amplification to achieve exponential amplification, and eventually produce a large number of gap-LCR products. Biotin on gap-LCR products can be combined with streptavidin on quantum dots (QDs) to form DNA-QD nanocomplexes, which can be directly analyzed using single-molecule spectroscopy (SMS) without separation. The principle of multiple detection is shown in Fig. 2.12b. Different fluorophore groups (Alexa488 and Alexa546) are modified on the 5′ end of probe d and probe e, respectively, corresponding to mutant and wild-type target. After gap-LCR,

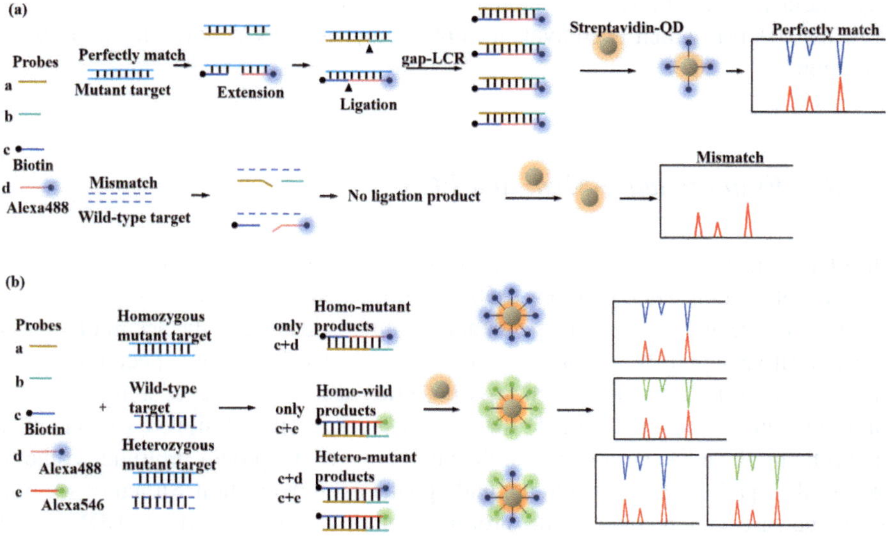

Fig. 2.12 Schematic diagram of fluorescence-coded gap-LCR probes combined with QD for multiple analysis of SNP. (Revised and reprinted with permission from Ref. [29])

multi-color SMS analysis of different amplification products can realize simultaneous detection of mutant and wild-type in complex samples. This method is able to detect target DNA as low as 170 zmol, and the detection specificity for K-*ras*12 single base mutation is near 100%.

(2) Multiplex LCR for the Detection of DNA Methylation
DNA methylation is an indispensable epigenetic modification in eukaryotic genomes and occurs specifically on cytosine of CpG dinucleotides. Research has shown that DNA methylation is closely associated with a variety of diseases, especially cancer. Given that a cancer may involve multiple CpG methylation sites, it is important to perform multiple detection for these sites. In 2015, Su treated methylated and unmethylated DNA with sodium bisulfite, amplified them with multiple LCR, and then analyzed LCR products with CE, finally realizing simultaneous detection of multiple methylation sites [30]. As shown in Fig. 2.13, genomic DNA is first treated with $NaHSO_3$ to convert unmethylated C into U, while methylated C (mC) remains unchanged. Four probes A_M, B_M, A'_M, and B'_M are designed according to the methylated target. The 3′ end of probe B_M is G, which is complementary to the C of the methylated target treated with $NaHSO_3$. The 5′ end of A'_M modifies a fluorophore, and the 5′ end of A and B'_M modifies the phosphate group, respectively. The four probes can perform LCR using methylated DNA as a template. However, when only unmethylated DNA is present, the LCR cannot be performed due to the mismatch between the probe and the template at the junction site. For different methylation sites, different amounts of adenine A are added to the 3′ end of probe B'_M for length coding, so that the amplified products corresponding to different target sites have different lengths and different migration rates in electrophoresis. The 5′ end of A'_M is fluorescently labeled, so the amplified product can be detected by a laser-induced fluorescence system. The method can detect methylated DNA as low as 0.01 fM, which is equivalent to 60 methylated DNA molecules in a 10 μL solution.

2.2.6 Endonuclease-Mediated LCR

In LCR, four highly concentrated oligonucleotide probes are usually used, and the contact of ligating sites between dsDNAs formed by complementary probes may trigger nonspecific ligation independent of template, leading to the accumulation of false positive signal. Based on this, Demchinskaya used structure-specific endonuclease to assist LCR reaction to improve detection accuracy while reducing template-independent ligation [31]. First, the probes are specially designed so that after hybridizing the adjacent probe with the target, one of the probes protrudes at the 5′ end of the ligation site and could not undergo the ligation reaction. Structure-specific endonuclease can recognize the protruding part of the 5′ end of dsDNA and hydrolyze the phosphodiester bond behind the first matching base to release the single-stranded part of dsDNA, forming the dsDNA structure with a nick, and

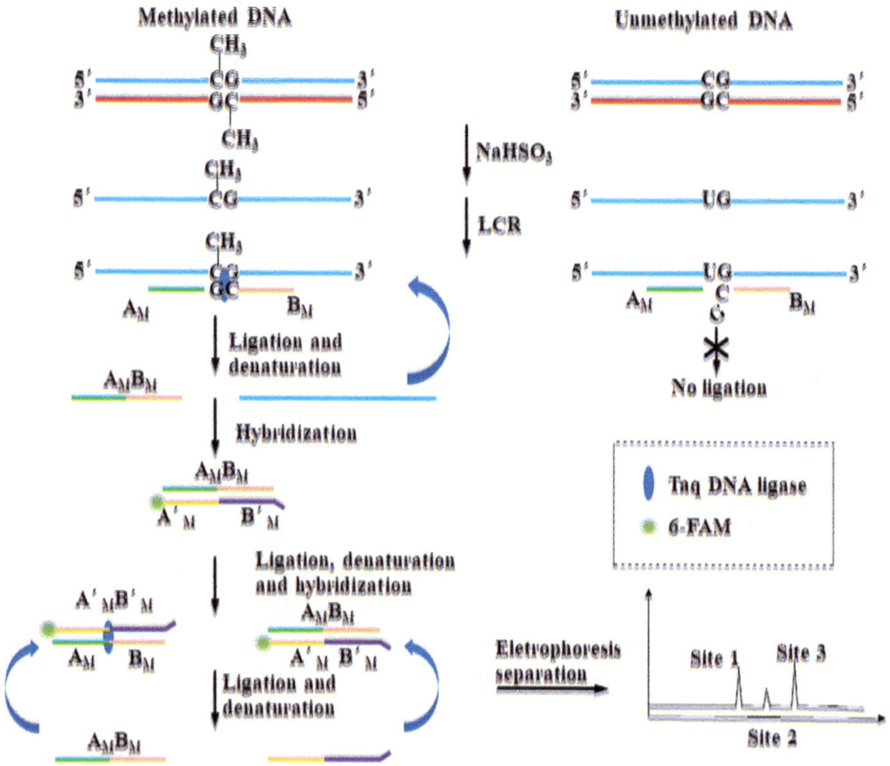

Fig. 2.13 Principal diagram of multiplex LCR detection of DNA methylation. (Revised and reprinted with permission from Ref. [30])

triggering LCR amplification. As shown in Fig. 2.14, after the hybridization of four probes with the target sequence, probes OL1-5B and OL2-5UP cannot be completely bind to the template sequence, and their 5′ ends contain a protruding sequence, respectively. The 3′ terminal bases of OL3-1 and OL4-1 do not fully hybridize with the template sequence due to the obstruction of adjacent probes. The structure is recognized by the structure-specific endonuclease, and the 5′ terminal nucleotides of probes OL1-5B and OL2-5UP are accurately removed, followed by hybridization of the 3′ terminal nucleotide of probes OL3-1 and OL4-1 with the target sequence to achieve complete complementation of the 4 probes with the target, and then LCR amplification occurs. In this work, enzyme digestion and subsequent LCR can only be performed after the probe has been annealed with the target sequence, which can avoid the nonspecific ligation independent of the template.

Since the 3′ end of probe OL1-5B is modified with biotin, the biotin-modified LCR products bind to streptavidin on the microwell plate and then are fixed to the bottom of the plate. The 3′ terminal UP region of the probe OL2-5UP can complement the DNA carrying alkaline phosphatase to oxidize NADPH (reduced coenzyme II) to NADH (reduced Coenzyme I). The amplifier solution is added to the system to

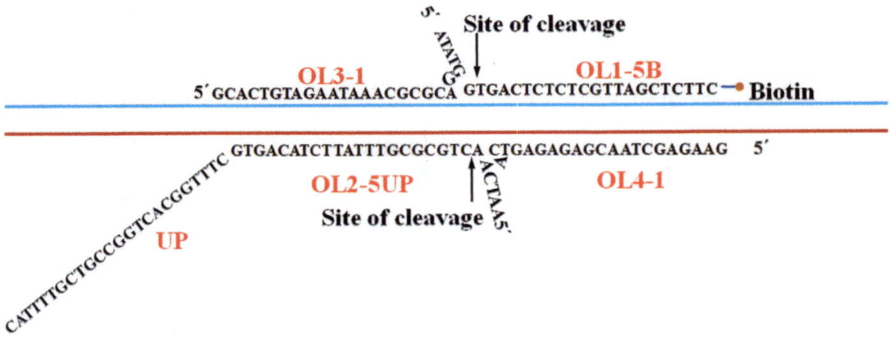

Fig. 2.14 Schematic diagram of structure-specific endonuclease-mediated LCR. (Revised and reprinted with permission from Ref. [31])

activate the NADH-driven secondary enzyme system, which converts NADH into NAD^+. Subsequently, NAD^+ is converted to NADH, where a REDOX cycle can be formed to produce colored substances, enabling colorimetric detection of target DNA as low as 30 aM. Due to the special design of the probe, no template-dependent ligation products are produced in LCR, and the specific recognition and cutting of structure-specific endonuclease improves the accuracy of the ligation reaction and further reduces the non-specificity of the reaction.

2.2.7 LCR in Combination with Other Amplification Methods

As an amplification method, LCR has good specificity in detecting SNP, but it is not very prominent from the point of view of sensitivity. In order to further improve the sensitivity, the researchers proposed the combination of LCR with other amplification methods.

(1) LCR in Combination with PCR

The method of detecting single-base mutation by combining LDR with PCR has been introduced in Sect. 2.2.1, which significantly increases the sensitivity of LDR. Here we focus on the combination of LCR and PCR, using LCR to distinguish single-base mutation, using PCR to further improve the sensitivity. Wang combined PCR with LCR to detect DNA methylation by surface-enhanced Raman scattering (SERS) [32]. The experimental principle is shown in Fig. 2.15. First, genomic DNA is treated with $NaHSO_3$, mC remains unchanged, unmethylated C is transformed into U, and then transformed into T after PCR amplification. Four LCR probes are designed according to the target sequences after $NaHSO_3$ transformation and PCR. In the presence of the target (methylated or unmethylated PCR products), the probes perform LCR to produce the corresponding products of methylated DNA and unmethylated DNA, respectively. The LCR probes are designed so that both LCR products have a universal sequence at one end, which can be hybridized with the

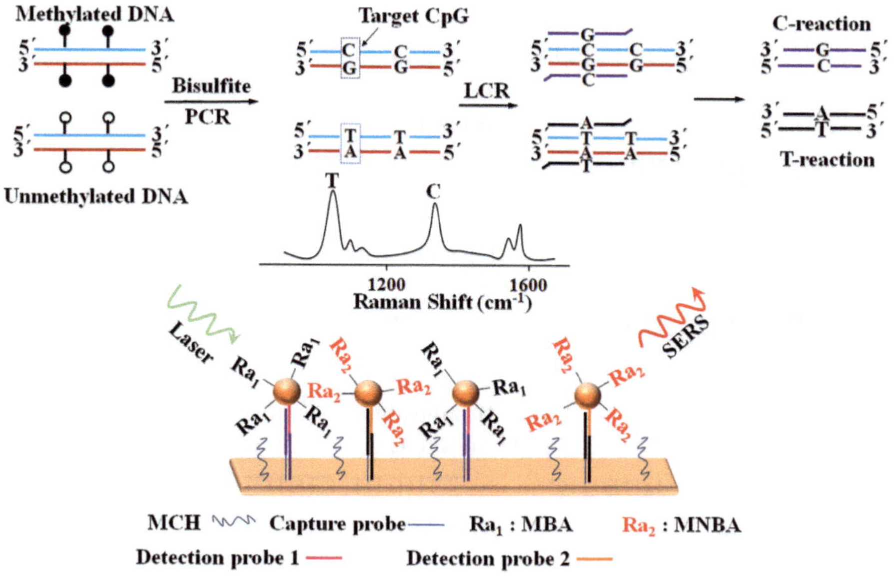

Fig. 2.15 Schematic diagram of LCR coupled with PCR for the detection of DNA methylation. (Revised and reprinted with permission from Ref. [32])

capture probe. The other end of the methylated LCR product is complementary to the detection probe 1, and the other end of the unmethylated LCR product is complementary to the detection probe 2, which are then labeled by different Raman signaling molecules. The LCR products are then fixed on SERS array by binding the universal sequence to the capture probe. The LCR products of methylated DNA target the detection probe 1 on the surface of gold nanoparticles, which is simultaneously modified to generate 4-mercaptobenzoic acid (MBA, Ra1) for SERS signal. The LCR products of unmethylated DNA target the detection probe 2 on the surface of another gold nanoparticle, which is simultaneously modified with 4-mercapto-3-nitro benzoic acid (MNBA, Ra2). Specific detection of methylated DNA and unmethylated DNA is realized through differential SERS signals. This method can detect as low as 10% methylated DNA in the presence of large excess of unmethylated DNA; however, its detection sensitivity is limited to 500 fM, which is insufficient for the sensitive detection of low-abundance methylated DNA. Further optimization of the amplification system is therefore required.

(2) LCR in Combination with Rolling Circle Amplification (RCA)

Cheng combined LCR with RCA to achieve high specificity and sensitivity analysis of SNP [33]. As shown in Fig. 2.16, three probes are designed according to the wild-type target, including probes X and Y for the upper strand and a padlock probe (PLP) for the lower strand. The 5′ end of probe X is modified with the phosphate group, and the T base at the 3′ end of probe Y is complementary to the A in the upper strand DNA. The 5′ end of PLP is modified with the phosphate group, and the A base at the

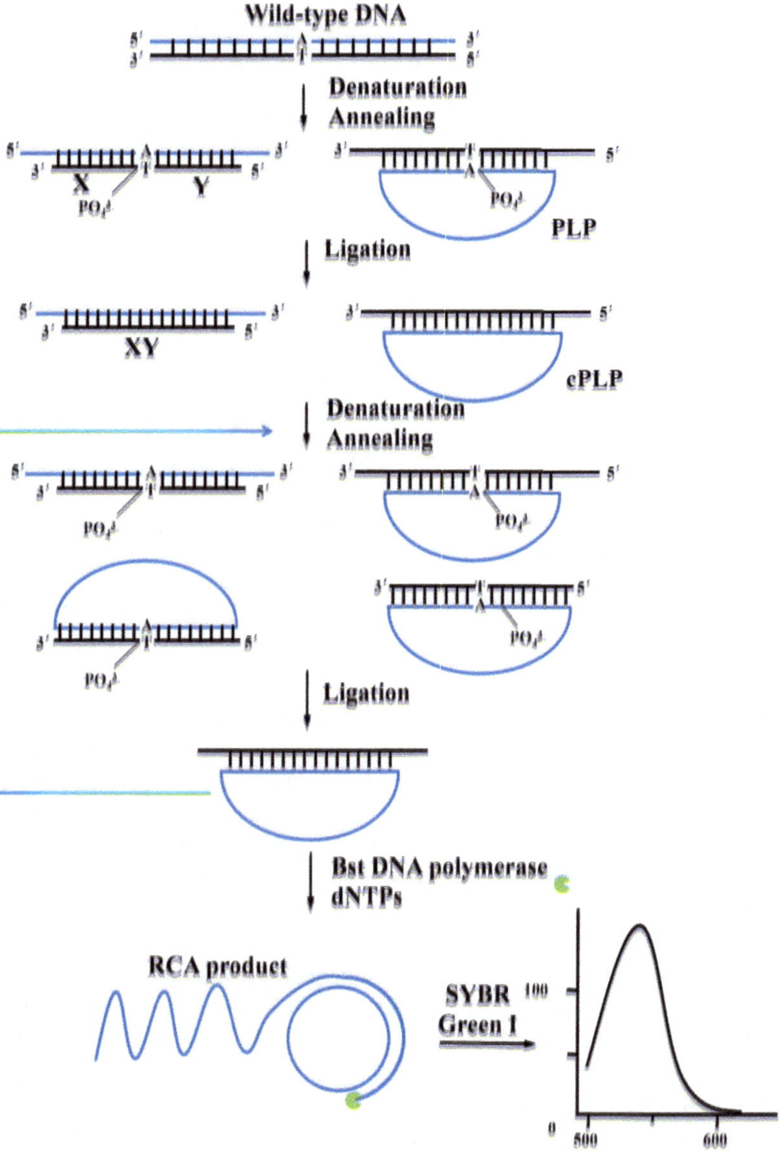

Fig. 2.16 Schematic diagram of LCR coupled with RCA for SNP detection. (Revised and reprinted with permission from Ref. [33])

3′ end is complementary to the T in the lower strand DNA. In the presence of wild-type DNA, probes X and Y hybridize adjacent positions of the upper strand in denaturation and annealing steps, and form XY under the action of DNA ligase. At the same time, the PLP is hybridized with the lower strand of the wild-type DNA and ligated to form a cyclized probe (cPLP). In the next cycle, XY can be used as a new template for the PLP ligation, again forming a cPLP. Similarly, the cPLP also acts as a template for probes X and Y, producing new XY. A large number of XY and cPLP can be generated by repeated LCR cycles. Subsequently, RCA is initiated with cPLP as the template and XY as the primer to produce a large number of long, repetitive ssDNA products. The fluorescence intensity of SYBR Green I embedded in the products is detected for the analysis of SNP. This method combines LCR and RCA, holding sufficient sensitivity and specificity to detect 1 fM of target DNA and as low as 1.0% of wild-type DNA in a large number of mutant DNA.

(3) LCR in Combination with Hyperbranched RCA (HRCA)

The BRAF gene V600E mutation is one of the most common pathogenic mutations in human cancer; therefore, it can be used as an important biomarker for tumor diagnosis, monitoring, and targeted therapy. In order to accurately identify low-abundance BRAF V600E mutations in clinical specimens, Zhang combined gap-LCR with HRCA to establish a highly sensitive method and applied it to clinical sample analysis [34]. The principle is shown in Fig. 2.17. First, a pair of padlock probes 1 and 2 are designed according to the mutant target sequence. When the mutant DNA is present, padlock probe 1 hybridizes with the upper strand, and padlock probe 2 hybridizes with the lower strand. By adding specific dNTPs to the reaction system, the padlock probes undergo extension reaction to fill the gap under the action of *Taq* DNA polymerase. With *Taq* DNA ligase, the padlock probes are ligated into two cyclized structures, respectively. In the next cycle, the mutant target can serve as a new template for the two padlock probes, forming two cyclized structures. Similarly, the newly formed cyclized structures can also be used as the templates of padlock probes 1 and 2, forming new cyclized structures again, and then cycling in turn, forming a large number of circular gap-LCR products. In the following, add proteinase K to inactivate *Taq* DNA polymerase and *Taq* DNA ligase to eliminate their impact on the subsequent HRCA. Numerous circular probes trigger

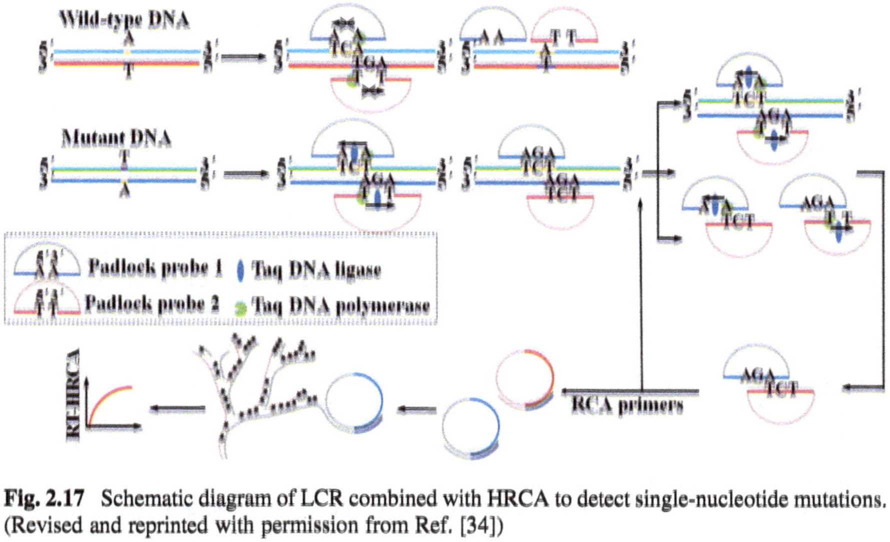

Fig. 2.17 Schematic diagram of LCR combined with HRCA to detect single-nucleotide mutations. (Revised and reprinted with permission from Ref. [34])

HRCA, generating large quantities of double-stranded amplification products. Real-time HRCA monitoring is performed with SYBR Green I. With a detection limit of 200 zmol, this method can identify BRAF V600E mutation as low as 0.01% and has been successfully applied to the actual clinical samples, demonstrating the application potential of detection of low-abundance somatic mutation in early cancer diagnosis and treatment.

2.3 LCR Product Detection

2.3.1 Gel Electrophoresis

(1) Polyacrylamide Gel Electrophoresis
Polyacrylamide gel electrophoresis (PAGE) is a common electrophoresis technique that uses the molecular sieve effects of mesh polyacrylamide gels to separate and detect amplified products. In non-denatured PAGE, DNA swims in a double-stranded state, and its mobility is affected by base composition and sequence. PAGE has a relatively high discrimination rate and can separate DNA molecules with a difference of 0.1% in length (that is, a difference of 1 bp in 1000 bp). The sequence length of LCR products is generally twice the length of its probes, so it can be separated and analyzed by PAGE. In 1991, Barany used PAGE to separate and autoradiographed LCR products [3], which could detect target molecules as low as 0.001 amol, and the color of the electrophoretic bands darkens gradually as the amount of target molecules gradually increases (Fig. 2.18).

(2) Agarose Gel Electrophoresis
Agarose gel electrophoresis is a technique that uses agarose gel as a support medium to separate oligonucleotides based on differences in the migration rates of nucleic acid molecules within the gel. Yan separated LCR products by agarose gel electrophoresis and quantified microRNA (miRNA) based on the pixel intensities of stained bands [35]. Figure 2.19 shows the analysis results of microRNA with a series of concentrations. This method achieves the analysis of target miRNA as low as 2 pM.

(3) Capillary Electrophoresis (CE)
CE is a liquid-phase separation technique that uses capillaries as separation channels and high-voltage direct current electric fields as driving forces to separate LCR products from probes. By labeling the fluorophores on the LCR probe, the products have fluorescence characteristics. The fluorescent products are then detected by laser induced fluorescence system. Lou used CE to detect LCR products to analyze a specific SNP site in genomic DNA [36]. As shown in Fig. 2.20, first, PCR is used to amplify the genomic DNA segment containing the SNP site. Specific LCR probes are designed for wild-type DNA, and the PCR product of wild-type DNA is used as the template for LCR amplification. Finally, CE is used to detect the LCR products, enabling qualitative and quantitative analysis of target based on peak position and peak area. For the mutant target, the probe contains a single-base

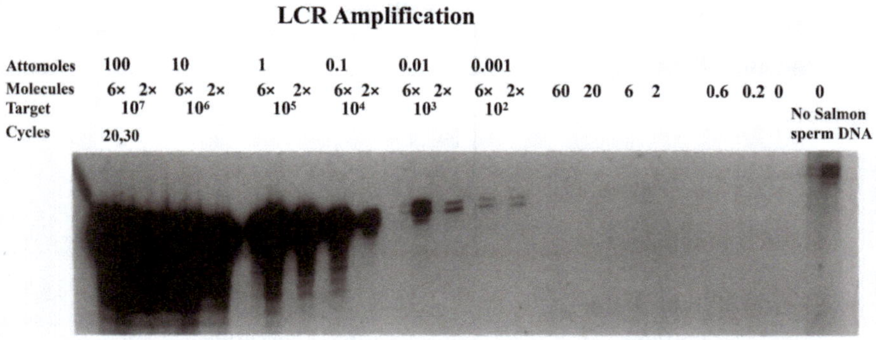

Fig. 2.18 The results of LCR products separated by PAGE. (Revised and reprinted with permission from Ref. [3])

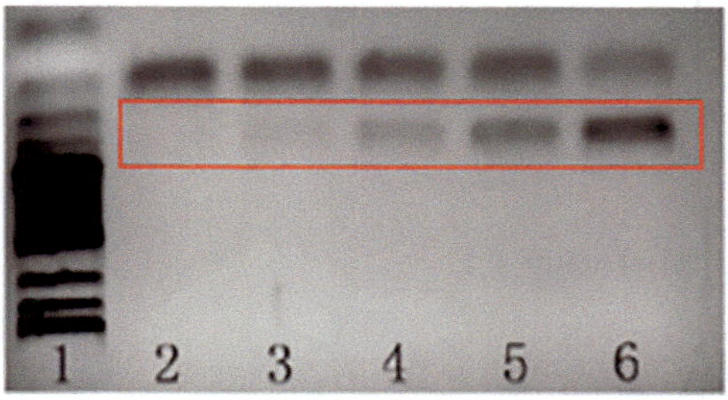

Fig. 2.19 Schematic diagram of miRNA detection using LCR and agarose gel electrophoresis. Lane 1: DNA marker; lane 2: blank control (no miRNA); lane 3 to lane 6: 2 pM, 20 pM, 200 pM, 2 nM miRNA. The red box indicates the product band. (Revised and reprinted with permission from Ref. [35])

mismatch at the junction and cannot undergo a ligation reaction, so no LCR products are generated.

Zhang realized multiple miRNA detection by encoding the length of LCR probes and differentiating LCR products of different lengths by CE [37]. As shown in Fig. 2.21, probes A-M and B are designed for the target miRNA, and the 3′ end of probe A-M modifies two RNA bases to improve the efficiency of ligating two DNA probes using RNA as a template. When the target miRNA is present, probe A-M and probe B are ligated to generate AMB using miRNA as the template under the action of T4 RNA ligase 2. Then, using AMB as a template and catalyzed by *Taq* DNA ligase, probe A′ and B′ are ligated to form A′B′. A′B′ can be used as a template to ligate probes A and B, and thus perform multiple LCR cycles. The probe A′ is labeled with fluorophore, and the probe B′ for different target miRNAs have different length coding sequences. Different miRNAs trigger corresponding probes to perform LCR, generating fluorescent LCR products of different lengths. After separation by CE, the LCR products are detected by laser-induced fluorescence system, enabling simultaneous analysis of let-7a, mir-92a, and mir-143. This method

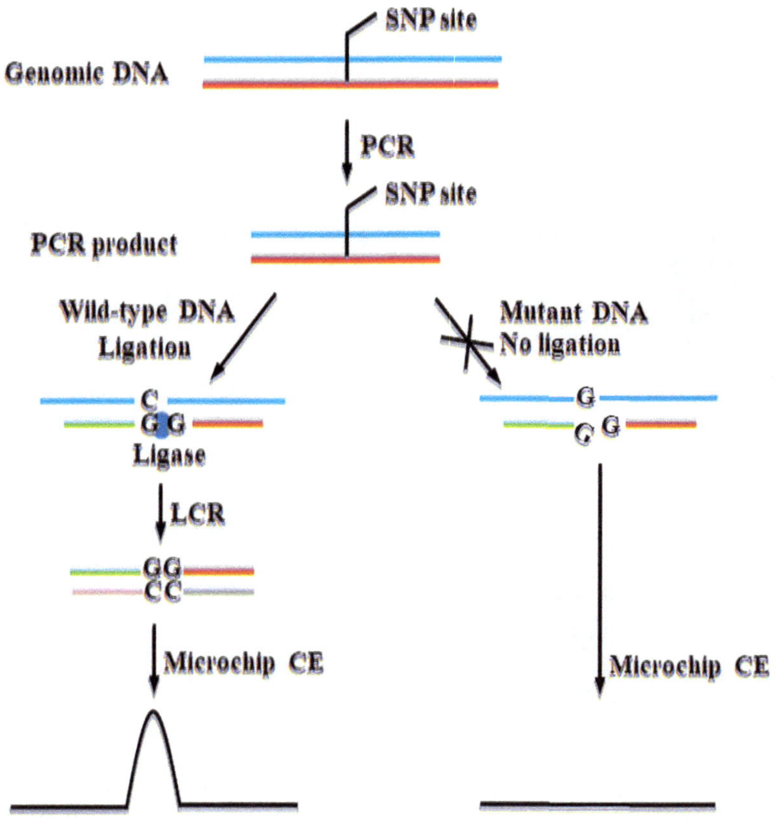

Fig. 2.20 Schematic diagram of using CE to detect LCR products for SNP analysis. (Revised and reprinted with permission from Ref. [36])

can accurately detect miRNAs as low as 0.2 fM and discriminate single-nucleotide differences between miRNA sequences with high specificity. In addition, the LCR probe can be fluorescently encoded, such that each target generates amplification products labeled with a distinct fluorophore, thereby enabling multiplex analysis.

2.3.2 Fluorescence Method

Fluorescence-based detection methods for LCR products are generally divided into three categories: real-time monitoring using embedded fluorescent dyes, endpoint detection in which fluorescently labeled probes generate fluorescent LCR products followed by separation and detection, and methods based on fluorescence resonance energy transfer (FRET).

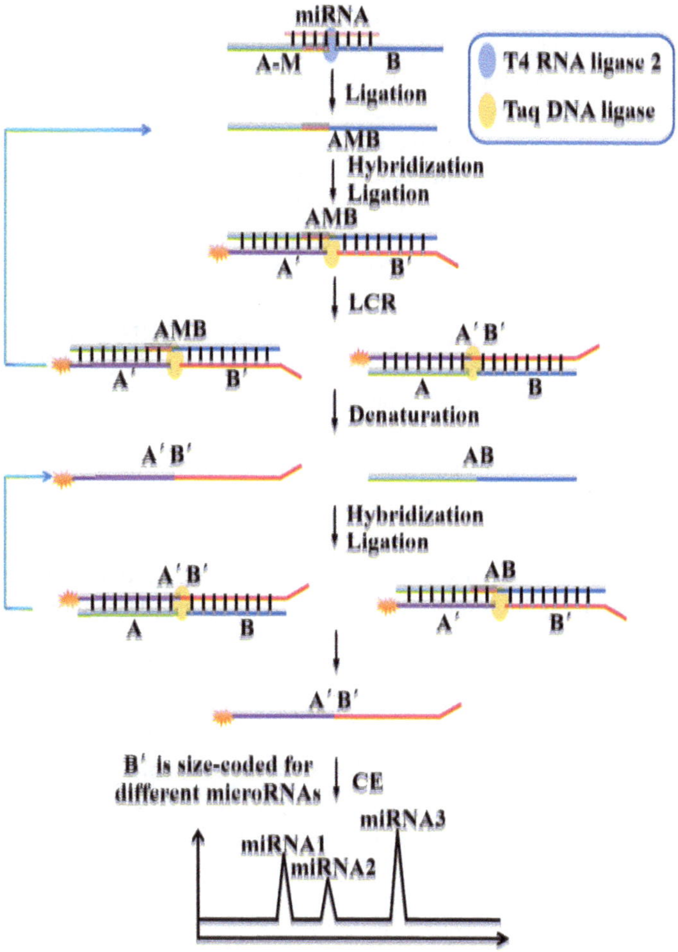

Fig. 2.21 Principal diagram of using CE to detect LCR product for multiplex detection of miRNAs. (Revised and reprinted with permission from Ref. [37])

(1) Embedded Fluorescent Dyes

EvaGreen is a nucleic acid dye that intercalates into dsDNA. It exhibits minimal fluorescence when unbound, but emits strong green fluorescence upon dsDNA intercalation, with fluorescence intensity positively correlated with dsDNA concentration. By measuring the fluorescence intensity of EvaGreen, the dsDNA concentration can be accurately quantified. SYBR Green I is a similar intercalating dye commonly used for dsDNA detection.

During LCR, unligated DNA probes can also form dsDNA during each hybridization and ligation step of thermal cycling. As a result, double-strand-specific dyes are unable to distinguish between ligated products and unligated double-stranded probes, posing a challenge for real-time monitoring of LCR. To address this, Psifidi analyzed the melt curves of both LCR products and probes to determine their respective T_m value, revealing that the T_m value of unligated probes is lower than that of ligated products [38]. A temperature between the two is then selected as the detection temperature, in which only the ligated products are in the double-stranded state, while the unligated probes are in the denatured state. Real-time monitoring of LCR products is achieved by detecting the fluorescence intensity of EvaGreen at this temperature (Fig. 2.22). In order to prevent interference from the probe dimer, design probes so that the ligation products have a high T_m (about 80 °C), while the probe T_m is at least 10 °C lower. Hu used RNA base-modified DNA probes for ligation reaction, and used SYBR Green I as an embedded dye for real-time fluorescent LCR to analyze the single-base changes of mRNA [39]. This method enables the detection of mRNA as low as 100 aM and the differentiation of single-base mutation in mRNA; however, the use of RNA base-modified probes increases the detection cost. Then, Su employed SplintR ligase to directly ligate DNA probes with RNA as a template for real-time fluorescent LCR, enabling quantitative analysis of different fusion gene transcripts in leukemia cells [40].

(2) Fluorescently Labeled Probes

Meng utilized streptavidin coated microbeads to capture biotin-labeled LCR products, thereby enriching fluorescently labeled LCR products [41]. Fluorescence on the microbeads was then visualized by fluorescence microscope, enabling both the detection of LCR products and the analysis of single-base mutation. In this approach, probes with different fluorescent modifications are designed for each target, and both targets are co-amplified by LCR. Mutant DNA generates amplification products with biotin at one end and ROX label at the other, while wild-type DNA produces products with biotin and a FAM label. The biotin-labeled LCR products are fixed on the microbead surface via streptavidin–biotin interaction. By detecting the

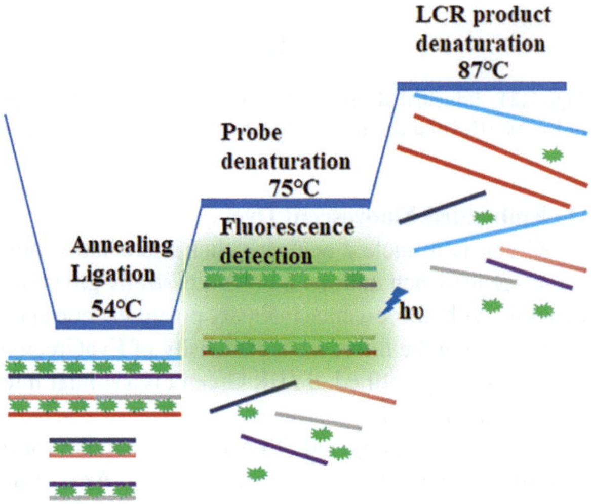

Fig. 2.22 Schematic diagram of real-time fluorescent LCR based on double-strand embedded dyes. (Revised and reprinted with permission from Ref. [38])

fluorescence intensity of ROX and FAM on the microbeads, wild-type and mutant DNA can be quantitatively detected simultaneously, achieving a detection limit of 1 fM.

Yuan used FRET between fluorescent molecule FAM and cationic conjugated polymer (CCP) to detect LCR products and analyzed miRNA [42]. As shown in Fig. 2.23, five DNA probes X′, X, Y, Xr, and Yr are designed according to the target miRNA. The 3′ end of probe X′ modifies two RNA bases, and the sequence of probe X is the same as that of X′, except for the RNA base modification. The 5′ end of probe Y modifies the phosphate group, and the 3′ end modifies FAM. Probe Xr complements probe X or X′, and modifies the phosphate group at its 5′ end and FAM at its 3′ end. Probe Yr and probe Y complement each other. Under the action of T4 RNA ligase 2, probe X′ and Y are ligated to X′Y using miRNA as the template. After denaturation and annealing, the probes Xr and Yr hybridize with X′Y and are ligated to form XrYr under the action of Ampligase. In the following cycles, single-stranded X′Y and XrYr are used as templates to hybridize with probes Xr and Yr and probes X and Y, respectively, and perform LCR to form a large number of double-stranded XY-XrYr. Probes Xr, X, Yr, and Y that are not involved in the ligation reaction are hybridized to form two double-stranded forms X-Xr and Y-Yr. Since λ exonuclease has a specific digestion effect on the 5′ end phosphorylated DNA strand in dsDNA, λ exonuclease is then added to digest the probes Xr and Y. The 5′ end of LCR product XY-XrYr lacks a phosphate group and is therefore resistant to digestion, allowing the fluorophores remain in the dsDNA. Given the partial overlap between the emission spectrum of CCP with the excitation spectrum of FAM, FRET can occur when the two fluorophores are brought into close proximity. Through electrostatic interaction, LCR products associate with CCP, shortening the distance between FAM and CCP, and thereby inducing FRET. In the absence of target miRNA, only free FAM mononucleotides are present, which exhibit negligible electrostatic interaction with CCP and fail to trigger FRET. Therefore, the homogeneous detection of LCR products can be realized. This method can detect as low as 0.1 fM miRNA and can discriminate single-base differences with high specificity.

The above methods achieve sensitive miRNA analysis through FRET detection of LCR products, but the endpoint detection requires multi-step operations. In order to realize real-time monitoring of LCR, Sun modified fluorophore and quenching groups on two adjacent LCR probes respectively, and constructed real-time LCR using FRET of fluorophores for detecting SNP [43], the principle of which is shown in Fig. 2.24. First, four DNA probes A, B, A′, and B′ are designed according to the mutant DNA. The four probes are hybridized with the target and are ligated to form AB and A′B′ under the catalysis of ligase. The middle position of probe A′ modifies FAM, and the middle position of probe B′ modifies BHQ1. In the process of LCR, probes A′ and B′ gradually ligate to form A′B′, bringing FAM and BHQ1 into proximity. This triggers FERT, quenching FAM fluorescence and resulting in a decrease in fluorescence intensity in the solution. SNP can be detected by real-time monitoring the decrease of fluorescence signal. If the probes do not match the DNA to be tested, no LCR amplification product is produced, and FRET cannot occur. The

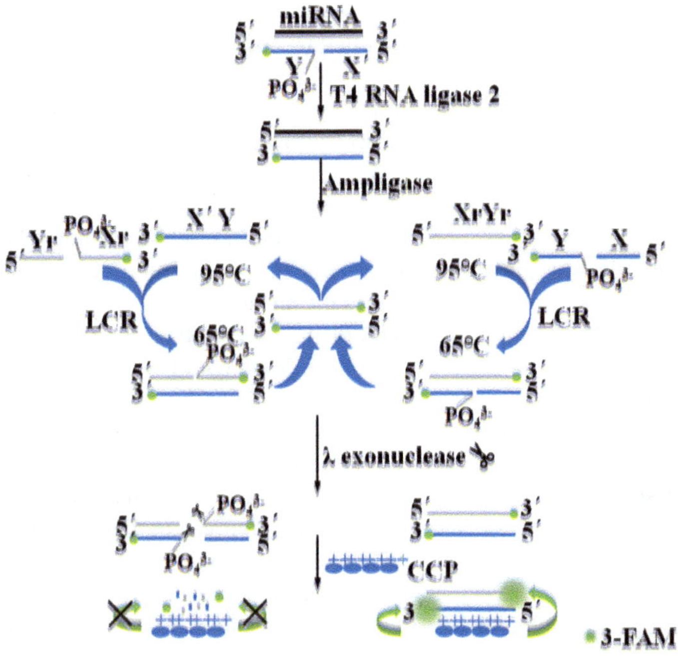

Fig. 2.23 Principal diagram of using FRET to detect LCR product for the detection of miRNA. (Revised and reprinted with permission from Ref. [42])

detection limit of this method is 10 aM, and 0.1% mutant DNA can be detected in the presence of a large amount of wild-type DNA.

2.3.3 *Colorimetric Method*

Shen used ferrofluidic nanoparticulate probes (FNPs) to develop a visual method to detect LCR products for analyzing SNP [44], as shown in Fig. 2.25. First, two probes are designed with the SNP site as the ligation site and modified on the surface of two FNPs to form FNP-1 and FNP-2. In addition, another two probes are designed complementary to the two probes above, respectively. In the presence of the target, the four probes perform LCR in thermal cycles. FNP-1 and FNP-2 hybridize adjacent to the target in LCR and undergo a ligation reaction, resulting in aggregation of FNPs. The amplified products are aggregated and separated by the magnetic properties of magnets, while the unreacted FNPs are still suspended in solution. Finally, the remaining dispersed FNPs catalyze the oxidation of 3,3′,5,5-′-tetramethylbenzidine (TMB) to produce colored substances. Through measuring the absorbance of the solution, LCR products can be detected and SNP can be analyzed. This method overcomes the difficulty of distinguishing probes from

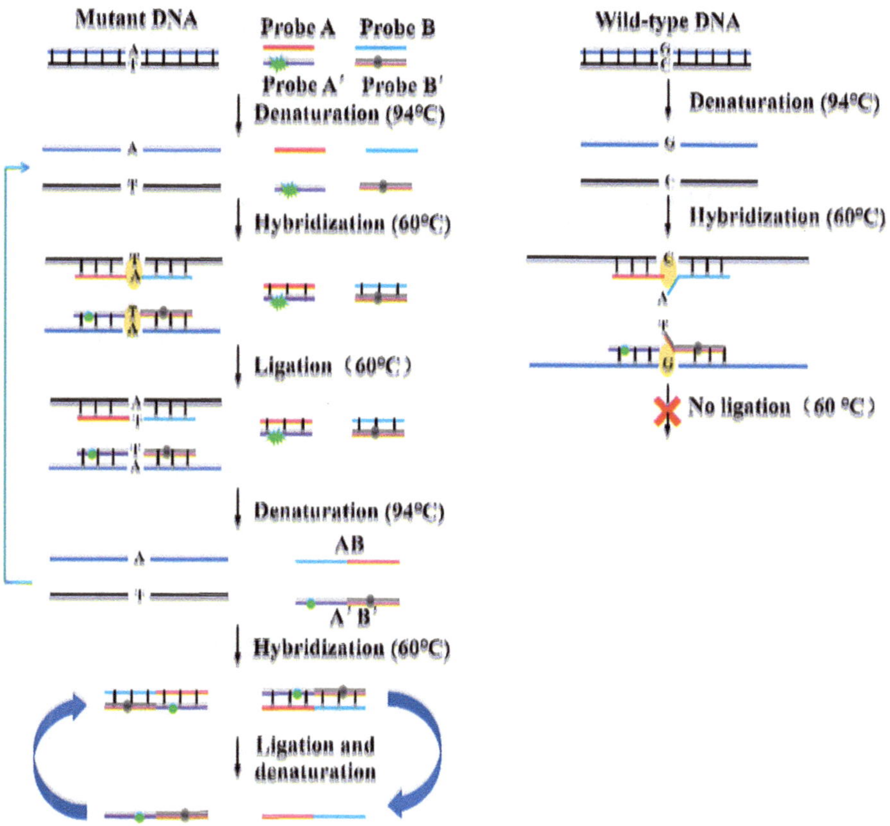

Fig. 2.24 Schematic diagram of real-time LCR detection of SNP based on FRET. (Revised and reprinted with permission from Ref. [43])

products inherent to homogeneous LCR, achieving fM- level sensitivity, and enabling the detection of 0.1% gene mutation.

In the state of uniform dispersion, colloidal gold appears red; As they gather, the surface plasmon resonance frequency changes, causing the color to change from red to purple or blue. This color change can be observed directly with the naked eye or quantitatively analyzed using UV-VIS spectroscopy. Shen used the color change principle of colloidal gold to analyze LCR products, enabling highly sensitive detection of SNP of K-*ras* gene [45]. As shown in Fig. 2.26, four LCR probes are designed for wild-type DNA, in which two adjacent probes are modified on the surface of gold nanoparticles, respectively, and the other two probes are complementary to the two above, respectively. In the presence of target DNA, the four probes are cyclically ligated for LCR amplification, which promoted the proximity of gold nanoparticles to form networked cross-linked aggregates, and the color of the solution gradually changed from red to blue. Real-time monitoring of absorbance changes during LCR enables LCR product detection and SNP analysis. This method achieves a limit of 20 aM, and can identify wild-type DNA at a frequency as low

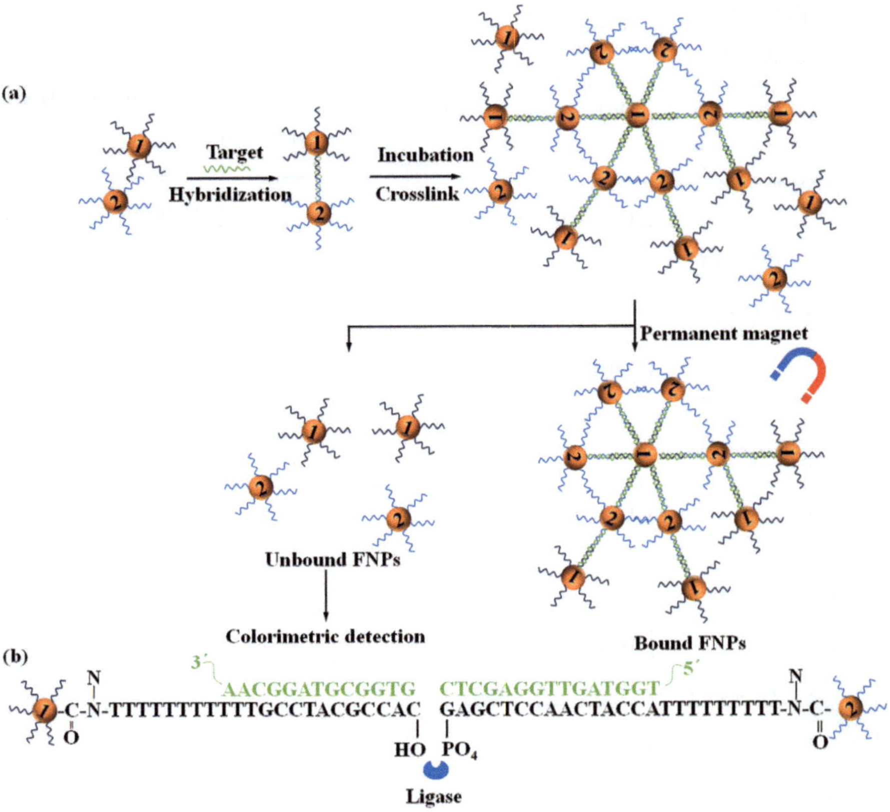

Fig. 2.25 Schematic diagram of LCR product detection based on FNPs. (Revised and reprinted with permission from Ref. [44])

as in 1500 mutant DNA. Separately, Su applied colloidal gold colorimetry for end point detection of LCR products and analysis of genomic DNA methylation level [46]. (See Fig. 9.15, Chap. 9 for details.)

2.3.4 Electrochemical Method

Chen detected LCR products based on the quenching effect of electrochemiluminescence of QD by hemin/G-quadruplex DNAzyme, and realized high sensitivity analysis of SNP [47], as shown in Fig. 2.27. First, four probes CP, DP, A, and B are designed according to the target sequence, wherein the 5′ end of probe CP and B modify the phosphate groups respectively and the 3′ end of probe CP modifies a amino group. Poly(diallyldimethyl ammonium chloride)@carboxylated multi-wall carbon nanotubes/quantum dots/chitosan/glass carbon electrode (PDDA@MWCNTs/QD/chitosan/GCE) is constructed by modifying the composite

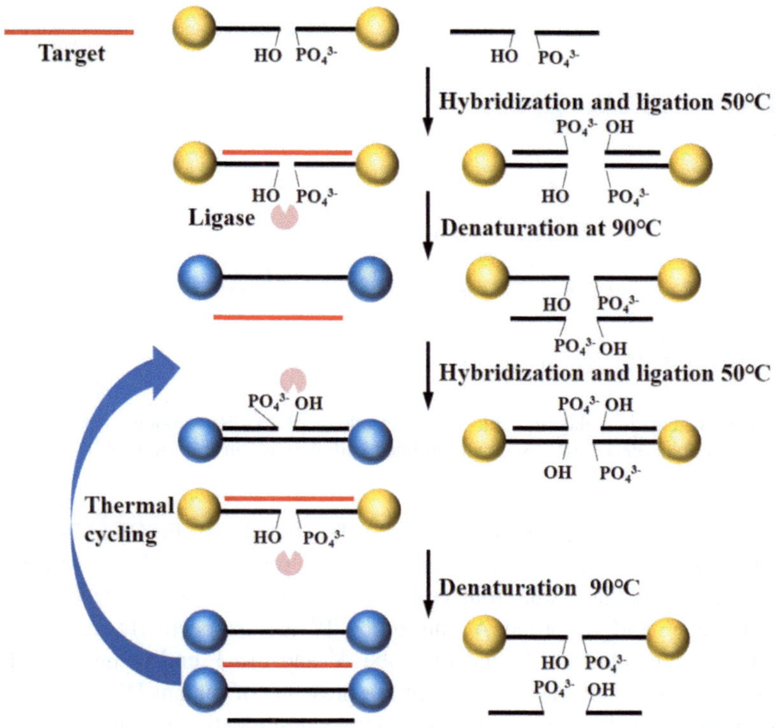

Fig. 2.26 Principal diagram of detecting LCR products based on colloidal gold. (Revised and reprinted with permission from Ref. [45])

on the surface of a GCE. Then glutaraldehyde is used as a crosslinking agent to form a covalent bond between the amino group of probe CP and chitosan, and CP is bound to the membrane of GCE. Subsequently, four probes are used for LCR in the presence of the target. Since the probe CP is fixed on the electrode surface, the LCR product CPDP is also fixed on the electrode surface. The hybridization of probe AP with the ligation product exposes the G-rich sequence to form a hemin/G-quadruplex DNAzyme. In the process of cathodic scanning, due to the reduction reaction of dissolved oxygen to produce H_2O_2, the CdTe QD on the surface of the GCE reacts with H_2O_2 to produce electrochemical luminescence. However, hemin/G-quadruplex DNAzyme can catalyze the reduction of O_2 to OH^-, thereby preventing the H_2O_2 reaction, resulting in a decrease in the intensity of electroluminescence. In the absence of the target, no ligation reaction occurs on the electrode surface, preventing the formation of G-quadruplex structures. As a result, the electrochemical luminescence intensity of QD remains unchanged, enabling SNP detection. This method achieved a detection limit of 45 fM for the mutant K-*ras* gene, and can detect as little as 0.1% mutant DNA in a background of wild-type sequences.

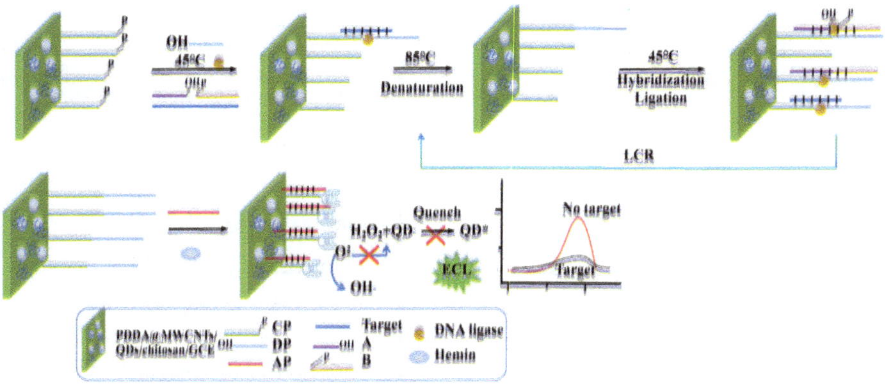

Fig. 2.27 Schematic diagram of the detection of LCR products based on QD electrochemiluminescence. (Revised and reprinted with permission from Ref. [47])

Zhang used the different electron transfer speeds of $Fe(CN)_6^{3-}/Fe(CN)_6^{4-}$ between different lengths of DNA to detect LCR products, enabling SNP analysis [48]. As shown in Fig. 2.28a, four probes are designed for mutant DNA sequence, wherein the 5′ end of probe A and probe B′ are, respectively, modified with phosphate groups, and the 5′ end of probe A′ and the 3′ end of probe B′ have a noncomplementary tail, respectively. In the presence of mutant DNA, LCR amplification is performed on four probes to generate amplification products with ssDNA at both ends. As shown in Fig. 2.28c, an amine-modified ssDNA capture probe (SP-DNA) is assembled via an amide bond to the surface of a GCE modified with MWCNTs. The LCR products are fixed on the electrode surface through the hybridization of SP-DNA with the ssDNA at one end. Subsequently, stem-loop DNA is added and hybridized with the ssDNA at the other end of the LCR product to increase the double-strand length. Due to the electrostatic repulsion between $Fe(CN)_6^{3-}/Fe(CN)_6^{4-}$ and the DNA skeleton phosphate groups on the electrode surface, the electron transfer and diffusion of $Fe(CN)_6^{3-}/Fe(CN)_6^{4-}$ are hindered, decreasing the electrical signal. In contrast, wild-type DNA could not trigger LCR in the presence of the above four probes (Fig. 2.28b), and the capture probe could only hybridize with one of the probes, resulting in shorter dsDNA and higher electrical signals. The detection limit of mutant and wild-type DNA is 1.18 aM and 0.35 aM, respectively, and 0.1% of SNP could be detected.

Hu built an electrochemical sensor based on ferrocene (Fc) to detect LCR products and analyze SNP [49]. First, an amino-modified ssDNA capture probe is assembled to the cMWCNT-NF/GCE surface via an amide bond (Fig. 2.29a). As shown in Fig. 2.29b, four probes are designed according to the mutant DNA sequence, namely, A, A′, B, and B′. The 5′ end of probe A modifies the phosphate group, and the 5′ end of probe A′ contains an unrelated sequence. The 3′ end of probe B′ modifies Fc, and the 5′ end modifies the phosphate group. In the presence of mutant DNA, four probes are used for LCR amplification. The LCR product hybridizes with the capture probe on the electrode through its 5′ end ssDNA and is

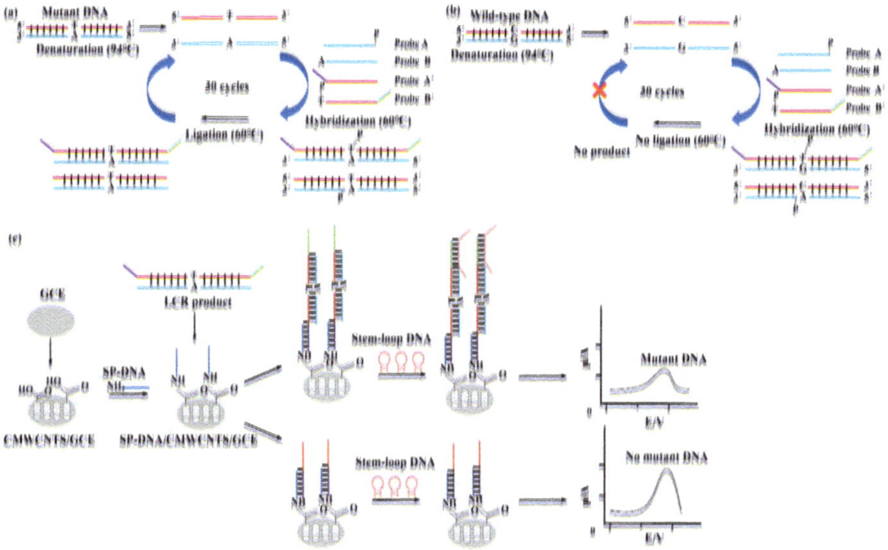

Fig. 2.28 Schematic diagram for LCR product detection based on $Fe(CN)_6^{3-}/Fe(CN)_6^{4-}$ electrochemical sensing. (Revised and reprinted with permission from Ref. [48])

fixed on the electrode. Since the 3′ end of A′B′ is marked with Fc, the proximity of Fc to the electrode enables a significant electrical signal (Fig. 2.29d). In contrast, in the presence of wild-type DNA alone, LCR fails to proceed, yielding only minimal mismatched dsDNA (Fig. 2.29c) and consequently a weak current signal. The detection limit of this test is 0.75 aM, and it can detect 0.01% of the mutant gene.

Electrochemical detection techniques show high sensitivity in LCR product analysis, with the ability to detect DNA as low as 1 aM. In addition, the instrument used in this technology is simple in design, relatively low in-cost, and requires minimal hands-on operation. However, the nonspecific adsorption phenomenon on the electrode surface may compromise the signal-to-noise ratio. Moreover, even minor perturbation in base pairing can lead to significant changes in charge transfer efficiency within dsDNA, resulting in corresponding fluctuations in electrochemical signals. As such, the stability of electrochemical detection warrants further improved.

2.4 Outlook of LCR

As a reaction that relies on thermal stable ligase to achieve exponential amplification by repeatedly ligating adjacent probes in thermal cycles, LCR exhibits comparable amplification capacity as PCR. Compared with primer extension reaction, ligation reaction offer superior specificity in discriminating single-base variations. Therefore,

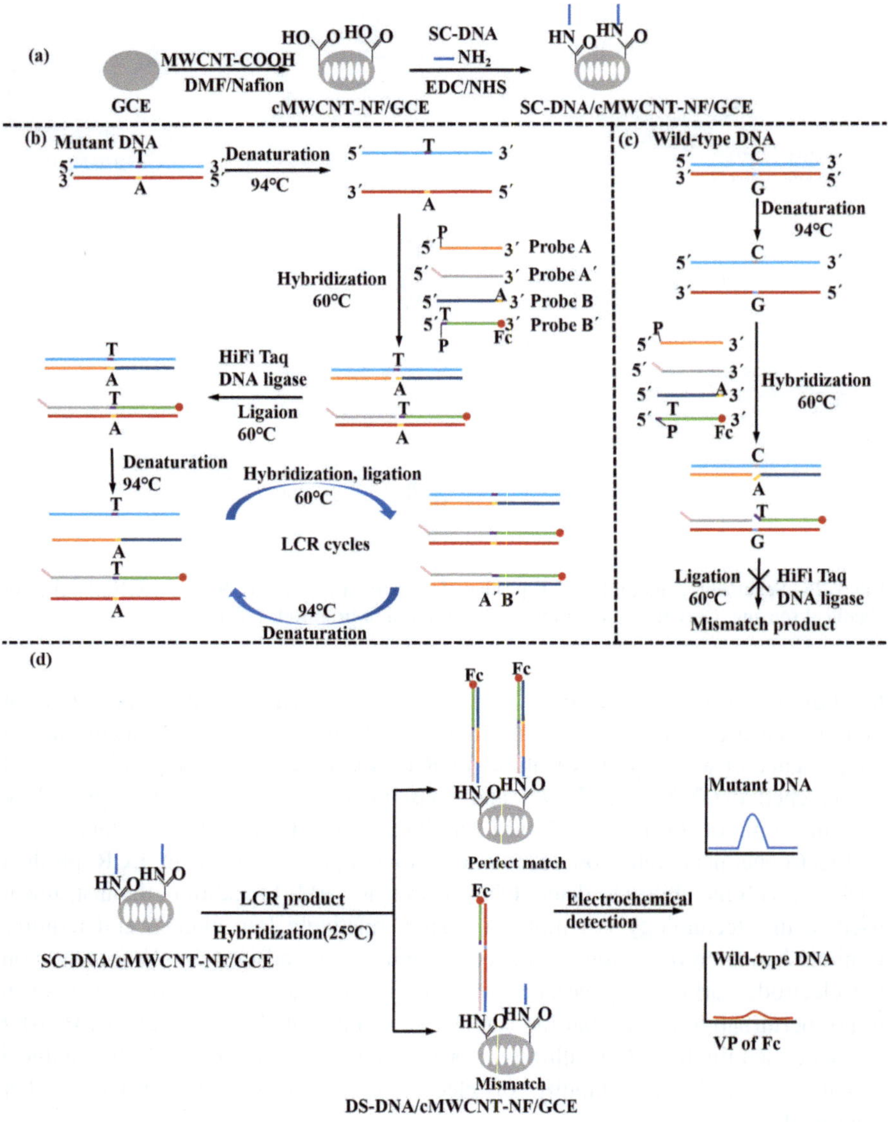

Fig. 2.29 Schematic diagram for detection of LCR products by electrochemical sensor based on ferrocene. (Revised and reprinted with permission from Ref. [49])

LCR shows higher specificity than PCR in detecting SNP, DNA methylation, and single-base difference in RNA, positioning it as a promising tool the diagnosis of infectious diseases, genetic disorders, viral infections, and cancers [50, 51].

However, LCR is susceptible to nonspecific ligation independent of template, which can compromise both the sensitivity and specificity of detection. To address this limitation, combining LCR with other amplification techniques, such as PCR or RCA, enables accurate identification of single-base changes while enhancing overall detection sensitivity. Moreover, such hybrid approaches are well-suited for multiplex analysis of relevant nucleic acid targets and hold significant potential for further development and application.

References

1. Landegren U, Kaiser R, Sanders J, Hood L. A ligase-mediated gene detection technique. Science. 1988;241:1077–80.
2. Wu DY, Wallace RB. The ligation amplification reaction (LAR)—amplification of specific DNA sequences using sequential rounds of template-dependent ligation. Genomics. 1989;4: 560–9.
3. Barany F. Genetic disease detection and DNA amplification using cloned thermostable ligase. Proc Natl Acad Sci USA. 1991;88:189–93.
4. Luo J, Bergstrom DE, Barany F. Improving the fidelity of Thermus thermophilus DNA ligase. Nucleic Acids Res. 1996;24:3071–8.
5. Tomkinson AE, Tappe NJ, Friedberg EC. DNA ligase I from Saccharomyces cerevisiae: physical and biochemical characterization of the CDC9 gene product. Biochemistry. 1992;31: 11762–71.
6. Husain I, Tomkinson AE, Burkhart WA, Moyer MB, Ramos W, Mackey ZB, Besterman JM, Chen J. Purification and characterization of DNA ligase III from bovine testes: homology with DNA ligase II and vaccinia DNA ligase. J Biol Chem. 1995;270:9683–90.
7. Shuman S. Vaccinia virus DNA ligase: specificity, fidelity, and inhibition. Biochemistry. 1995;34:16138–47.
8. Benjamin WH, Smith KR, Waites KB. Ligase chain reaction. Methods Mol Biol. 2003;226: 135–49. Humana Press
9. Wilson TE, Topper LM, Palmbos PL. Non-homologous end-joining: bacteria join the chromosome breakdance. Trends Biochem Sci. 2003;28:62–6.
10. Tomkinson AE, Chen L, Dong Z, Leppard JB, Levin DS, Mackey ZB, Motycka TA. Completion of base excision repair by mammalian DNA ligases. Prog Nucleic Acid Res Mol Biol. 2001;68:151–64.
11. Timson DJ, Singleton MR, Wigley DB. DNA ligases in the repair and replication of DNA. Mutat Res. 2000;460:301–18.
12. Lieber MR. The biochemistry and biological significance of nonhomologous DNA end joining: an essential repair process in multicellular eukaryotes. Genes Cells. 1999;4:77–85.
13. Martin IV, MacNeill SA. ATP-dependent DNA ligases. Genome Biol. 2002;3: reviews3005.1–3005.7.
14. Wilkinson A, Day J, Bowater R. Bacterial DNA ligases. Mol Microbiol. 2001;40:1241–8.
15. Doherty AJ, Suh SW. Structural and mechanistic conservation in DNA ligases. Nucleic Acids Res. 2000;28:4051–8.
16. Cao W. DNA ligases: structure, function and mechanism. Curr Org Chem. 2002;6:827–39.
17. Cao W. Recent developments in ligase-mediated amplification and detection. Trends Biotechnol. 2004;22:38–44.
18. Cherepanov AV, de Vries S. Dynamic mechanism of nick recognition by DNA ligase. Eur J Biochem. 2002;269:5993–9.
19. Lohman GJ, Tabor S, Nichols NM. DNA ligases. Curr Protoc Mol Biol. 2011;94:3.14.1–7.
20. Xiao Z-X, Cao H-M, Luan X-H, Zhao J-L, Wei D-Z, Xiao J-H. Effects of additives on efficiency and specificity of ligase detection reaction. Mol Biotechnol. 2007;35:129–33.
21. Barany F. The ligase chain reaction in a PCR world. PCR Methods Appl. 1991;1:5–16.

22. Hamada M, Shimase K, Tsukagoshi K, Hashimoto M. Discriminative detection of low-abundance point mutations using a PCR/ligase detection reaction/capillary gel electrophoresis method and fluorescence dual-channel monitoring. Electrophoresis. 2014;35:1204–10.
23. Watanabe S, Hagihara K, Tsukagoshi K, Hashimoto M. Microbead-based ligase detection reaction assay using a molecular beacon probe for the detection of low-abundance point mutations. Anal Chem. 2014;86:900–6.
24. Abravaya K, Carrino JJ, Muldoon S, Lee HH. Detection of point mutations with a modified ligase chain reaction (gap-LCR). Nucleic Acids Res. 1995;23:675–82.
25. Gibriel AA, Adel O. Advances in ligase chain reaction and ligation-based amplifications for genotyping assays: detection and applications. Mutat Res Rev Mutat Res. 2017;773:66–90.
26. Kälin I, Shephard S, Candrian U. Evaluation of the ligase chain reaction (LCR) for the detection of points mutation. Mutat Res. 1992;283:119–23.
27. Zhou L, Du F, Zhao Y, Yameen A, Chen H, Tang Z. DNAzyme based gap-LCR detection of single-nucleotide polymorphism. Biosens Bioelectron. 2013;45:141–7.
28. Marshall RL, Laffler TG, Cerney MB, Sustachek JC, Kratochvil JD, Morgan RL. Detection of HCV RNA by the asymmetric gap ligase chain reaction. Genome Res. 1994;4:80–4.
29. Song Y, Zhang Y, Wang TH. Single quantum dot analysis enables multiplexed point mutation detection by gap ligase chain reaction. Small. 2013;9:1096–105.
30. Su F, Wang L, Sun Y, Liu C, Duan X, Li Z. Highly sensitive and multiplexed analysis of CpG methylation at single-base resolution with ligation-based exponential amplification. Chem Sci. 2015;6:1866–72.
31. Demchinskaya AV, Shilov IA, Karyagina AS, Lunin VG, Sergienko OV, Voronina OL, Leiser M, Plobner L. A new approach for point mutation detection based on a ligase chain reaction. J Biochem Biophys Methods. 2001;50:79–89.
32. Wang Y, Wee EJ, Trau M. Highly sensitive DNA methylation analysis at CpG resolution by surface-enhanced Raman scattering via ligase chain reaction. Chem Commun. 2015;51:10953–6.
33. Cheng Y, Zhao J, Jia H, Yuan Z, Li Z. Ligase chain reaction coupled with rolling circle amplification for high sensitivity detection of single nucleotide polymorphisms. Analyst. 2013;138:2958–63.
34. Zhang L, Zhang Y, Huang L, Zhang Y, Li Y, Ding S, Cheng W. Ultrasrasensitive biosensing of low abundance BRAF V600E mutation in real samples by coupling dual padlock-gap-ligase chain reaction with hyperbranched rolling circle amplification. Sens Actuators B Chem. 2019;287:111–7.
35. Yan J, Li Z, Liu C, Cheng Y. Simple and sensitive detection of microRNAs with ligase chain reaction. Chem Commun. 2010;46:2432–4.
36. Lou XJ, Panaro NJ, Wilding P, Fortina P, Kricka LJ. Mutation detection using ligase chain reaction in passivated silicon-glass microchips and microchip capillary electrophoresis. BioTechniques. 2004;37:392–8.
37. Zhang P, Zhang J, Wang C, Liu C, Wang H, Li Z. Highly sensitive and specific multiplexed microRNA quantification using size-coded ligation chain reaction. Anal Chem. 2014;86:1076–82.
38. Psifidi A, Dovas C, Banos G. Novel quantitative real-time LCR for the sensitive detection of SNP frequencies in pooled DNA: method development, evaluation and application. PLoS One. 2011;6:e14560.
39. Hu Y, Jia H, Wang Y, Cheng Y, Li Z. Sensitive quantification of messenger RNA with a real-time ligase chain reaction by using a ribonucleotide-modified DNA probe. Chem Commun. 2014;50:13093–5.
40. Su F, Ji J, Zhang P, Wang F, Li Z. Real-time quantification of fusion transcripts with ligase chain reaction by direct ligation of adjacent DNA probes at fusion junction. Analyst. 2020;145: 3977–82.

41. Meng X, Yang X, Wang K, Guo Q, Tan Y, Mo Q, Xu X. Direct fluorescence detection of point mutations in human genomic DNA using microbead-based ligase chain reaction. Talanta. 2010;80:1725–9.
42. Yuan Z, Zhou Y, Gao S, Cheng Y, Li Z. Homogeneous and sensitive detection of microRNA with ligase chain reaction and Lambda exonuclease-assisted cationic conjugated polymer biosensing. ACS Appl Mater Interfaces. 2014;6:6181–5.
43. Sun Y, Lu X, Su F, Wang L, Liu C, Duan X, Li Z. Real-time fluorescence ligase chain reaction for sensitive detection of single nucleotide polymorphism based on fluorescence resonance energy transfer. Biosens Bioelectron. 2015;74:705–10.
44. Shen W, Lim CL, Gao Z. A ferrofluid-based homogeneous assay for highly sensitive and selective detection of single-nucleotide polymorphisms. Chem Commun. 2013;49:8114–6.
45. Shen W, Deng H, Gao Z. Gold nanoparticle-enabled real-time ligation chain reaction for ultrasrasensitive detection of DNA. J Am Chem Soc. 2012;134:14678–81.
46. Su F, Wang L, Sun Y, Liu C, Duan X, Li Z. Highly sensitive detection of CpG methylation in genomic DNA by AuNP-based colorimetric assay with ligase chain reaction. Chem Commun. 2015;51:3371–4.
47. Chen Y, Yang M, Xiang Y, Yuan R, Chai Y. Ligase chain reaction amplification for sensitive electrochemiluminescent detection of single nucleotide polymorphisms. Anal Chim Acta. 2013;796:1–6.
48. Zhang W, Hu F, Zhang X, Meng W, Zhang Y, Song Y, Wang H, Wang P, Gu Y. Ligase chain reaction-based electrochemical biosensor for the ultrasrasensitive and specific detection of single nucleotide polymorphisms. New J Chem. 2019;43:14327–35.
49. Hu F, Zhang W, Meng W, Ma Y, Zhang X, Xu Y, Wang P, Gu Y. Ferrocene-labeled and purification-free electrochemical biosensor based on ligase chain reaction for ultrasrasensitive single nucleotide polymorphism detection. Anal Chim Acta. 2020;1109:9–18.
50. Mano J, Oguchi T, Akiyama H, Teshima R, Hino A, Furui S, Kitta K. Simultaneous detection of recombinant DNA segments introduced into genetically modified crops with multiplex ligase chain reaction coupled with multiplex polymerasrase chain reaction. J Agric Food Chem. 2009;57:2640–6.
51. Blocker ME, Krysiak RG, Behets F, Cohen MS, Hobbs MM. Quantification of chlamydia trachomatis elementary bodies in urine by ligase chain reaction. J Clin Microbiol. 2002;40: 3631–4.

Chapter 3
Loop-Mediated Isothermal Amplification (LAMP)

Hui Wang

Loop-mediated isothermal amplification (LAMP) is an in vitro nucleic acid amplification technique invented by Notomi et al. in 2000 [1]. LAMP can detect DNA with high specificity and efficiency under constant temperature conditions. By using a set of specially designed primers and a DNA polymerase with high strand displacement activity, LAMP avoids the thermal cycling process and can amplify a few copies of DNA to 10^9 within 1 h. LAMP is highly specific for the target sequence assay, which is attributable to the special recognition of six regions on the target sequence by four primers. LAMP amplification products are a mixture of stem-loop with various stem lengths and cauliflower-like structural DNA fragments with multiple loops. LAMP amplification products can be detected by real-time fluorescence and real-time turbidimetry, as well as by electrophoresis, turbidity, fluorescence, electrochemistry, and lateral flow assay with an endpoint detection manner. Due to its constant temperature, specificity, high efficiency, rapidity, and flexible detection manners, LAMP amplification technology has been successfully applied to the detection of pathogenic microorganisms, nucleic acid biomarker analysis, genetically modified food identification, telomerase activity detection, etc., which is the most widely used isothermal nucleic acid amplification technology [2–10].

H. Wang (✉)
University of Science and Technology Beijing, Beijing, China
e-mail: winscavin@ustb.edu.cn

Z. Li (ed.), *Nucleic Acid Detection*, https://doi.org/10.1007/978-981-95-7295-3_3

3.1 Principle of LAMP

3.1.1 Primers and Amplification Mechanism of LAMP

As shown in Fig. 3.1, the classical LAMP amplification consists of four specially designed primers corresponding to six regions of the target gene. The four primers include two inner primers (FIP and BIP) and two outer primers (F3 and B3). The FIP consisted of two-part sequences (5′-F1c-F2-3′), in which the F1c is complementary to the sequence F1. The BIP also consisted of two-part sequences (5′-B1c-B2-3′), in which the B1c is complementary to the sequence B1. The outer primers F3 and B3 were complementary to the sequences F3c and B3c, respectively. In some specific cases, filler sequences can be designed between F1c and F2, or between B1c and B2. Additionally, based on the classical LAMP primer design, a pair of primers targeting the loop structures of amplification products can be introduced, known as loop primers (FLP and BLP). The FLP binding region is located between F1 and F2, while the BLP binding region is between B1 and B2. FLP and BLP primers are only used for the accelerated LAMP amplification reaction [10].

The LAMP amplification contains three steps by DNA polymerase with strand displacement activity [1]: the starting material producing stage (Fig. 3.2a), the cycling amplification stage (Fig. 3.2b), and the elongation and recycling stage (Fig. 3.2c). In the starting material-producing stage, the T_m value of F2 in inner primer FIP is higher than that of F3, and the concentration of F3 is lower than that of FIP. So the F2 in FIP preferentially hybridizes with the target DNA (Fig. 3.2a, Structure 1) and initiates the complementary strand synthesis by DNA polymerase (Fig. 3.2a, Structure 2). Next, the outer primer F3 hybridizes with the complementary region F3c on the target DNA and initiates the synthesis of another complementary strand (Fig. 3.2a, Structure 3). As the newly synthesized strand extends, it displaces the previously synthesized strand from the FIP extension, releasing a complementary strand that forms a loop structure at its 5′-end due to the complementary regions F1c and F1 (Fig. 3.2a, Structure 4). Simultaneously, the inner primer BIP hybridizes with the complementary region B2c on the released DNA strand. This hybridization guides the synthesis of a complementary strand, which

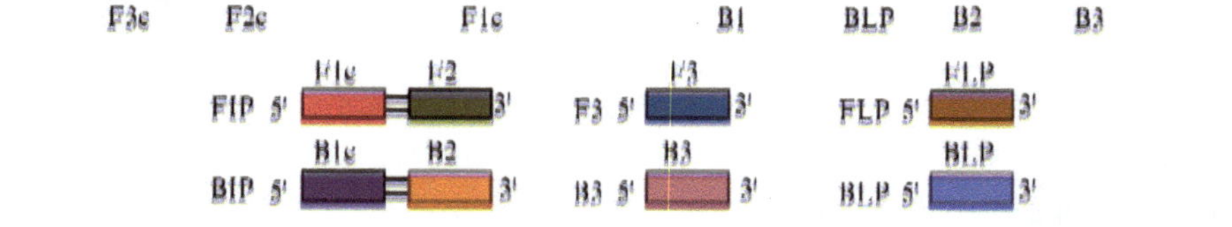

Fig. 3.1 Schematic diagram of primer design of LAMP [8, 10]. (Revised and reprinted with permission from Refs. [8, 10])

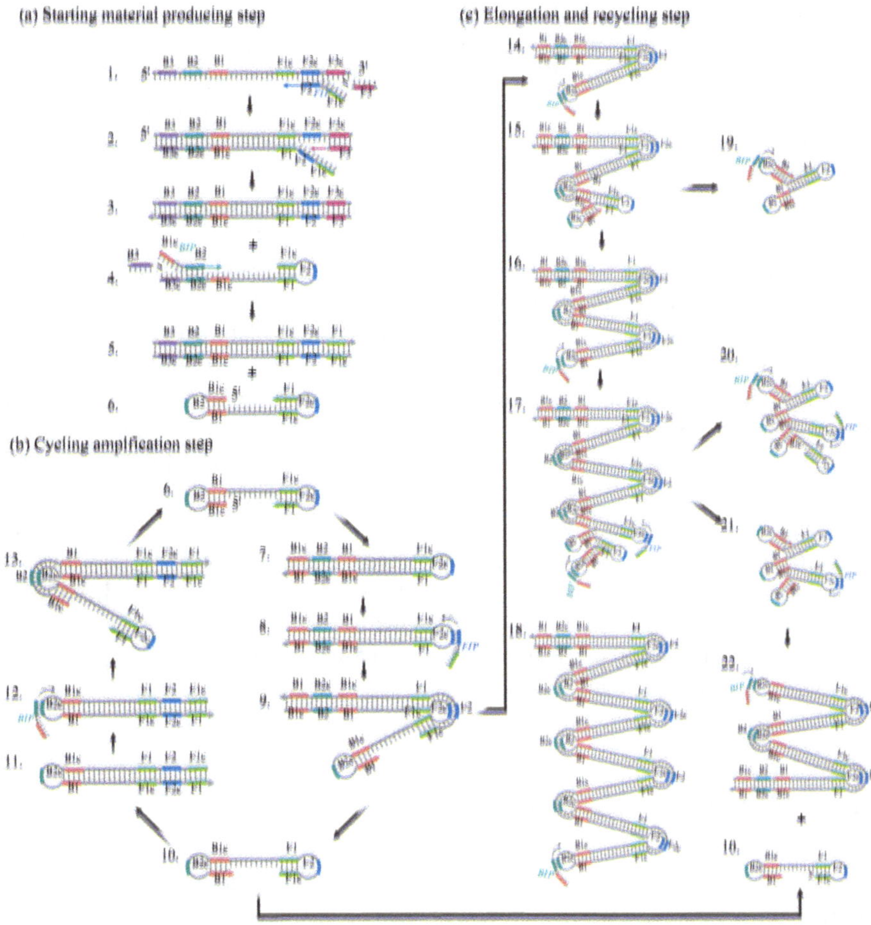

Fig. 3.2 Principle of LAMP [1]. (**a**) Starting material producing step. (**b**) Cycling amplification step. (**c**) Elongation and recycling step. (Revised and reprinted with permission from Ref. [1])

opens the loop structure formed by F1c and F1. The outer primer B3 then hybridizes with the complementary region B3c and extends to form a new complementary strand (Fig. 3.2a, Structure 5). This extension displaces the previously synthesized strand, releasing a complementary strand that has complementary regions B1c-B2-B1 and F1-F2c-F1c at both ends (Fig. 3.2a, Structure 6). These complementary regions allow the strand to self-anneal, forming a dumbbell DNA structure, which serves as the starting template for subsequent LAMP amplification.

In the cyclic amplification stage, the dumbbell DNA structure (Structure 6) self-primes at its 3′ end (F1), extending to form a single stem-loop DNA structure (Fig. 3.2b, Structure 7). Meanwhile, the F2 region in the inner primer FIP anneals to the looped F2c region on the single stem-loop DNA (Fig. 3.2b, Structure 8), leading to the synthesis of a new double-stranded DNA. This synthesis displaces a strand identical to the original template, and the displaced single strand's 3′ end (containing B1c and its complementary sequence) spontaneously forms another stem-loop structure (Fig. 3.2b, Structure 9). The 3′ end of Structure 9 then undergoes self-extension, producing a stem-loop structure with an elongated stem (Fig. 3.2c, Structure 14) while releasing a double-loop dumbbell DNA structure complementary to the previously formed strand (Fig. 3.2b, Structure 10). Structure 10 can, like Structure 6, undergo self-extension to form another stem-loop DNA structure (Fig. 3.2b, Structure 11). At this point, the B2 region of BIP can hybridize to B2c in Structure 11 and initiate another round of extension, displacing a strand (Fig. 3.2b, Structure 12). The displaced strand's 3′ end (containing F1c and its complementary sequence) again forms a stem-loop structure (Fig. 3.2b, Structure 13). The 3′ end of Structure 13 serves as a primer for self-extension, producing another stem-loop structure with a longer stem and releasing a double-loop dumbbell DNA structure complementary to Structure 10 (Fig. 3.2b, Structure 6). Meanwhile, the loop regions F2c in Structure 9 and B2c in Structure 10 continue to hybridize with FIP and BIP, maintaining the chain displacement and extension reactions.

During the elongation and recycling stage, the inner primers FIP and BIP continuously drive strand displacement reactions, steadily increasing the number of stem-loop structures. Ultimately, the final LAMP products are a collection of stem-loop DNA fragments of various lengths, all composed of repeated target sequences and multiple loop structures resembling cauliflower-like formations (Fig. 3.2c, Structures 15–22).

3.1.2 Reverse Transcription LAMP (RT-LAMP)

By adding the reverse transcriptase in the classical LAMP system, reverse transcription LAMP (RT-LAMP) was developed to detect RNA targets [1, 3, 9]. RT-LAMP uses reverse transcriptase to produce RNA target-complementary cDNA, and then the cDNA is amplified by LAMP. For RT-LAMP-based assay, all reaction reagents can be incubated in one reaction tube under constant temperature conditions, and the RNA target can be detected in one step, which is widely used in RNA-related assays.

3.1.3 Accelerated LAMP

Based on the classical four-primer LAMP, Nagamine et al. [10] introduced a pair of primers targeting the loop structures of LAMP products, called loop primers (LFP and LBP). The FLP binding region is located between F1 and F2, while the BLP binding region is between B1 and B2 (Fig. 3.1). The addition of the loop primers not only does not affect the binding of the original inner primers but also can bind to the region where the internal primer cannot bind, triggering the synthesis of more strand-substituted DNA, thereby greatly improving the speed of the LAMP reaction. The introduction of the loop primer shortens the original LAMP amplification reaction time by half and significantly increases the sensitivity [11]. However, the use of multiple primers increases the risk of primer–primer hybridization, resulting in no template amplification and false-positive results [12].

3.1.4 Selection of Target Sequence and Design of Primers

The target sequences of LAMP amplification are generally selected as conserved sequences of specific genes, and the length of target sequences (from F2 to B2) should not be too long (<300 bp). The optimal length of the target sequence is 130–200 bp, and more than 500 bp will significantly reduce LAMP amplification efficiency [1].

The design of LAMP amplification primers is relatively complex, but it is the core of LAMP amplification technology. Eiken Chemical Co., Ltd. (Tokyo, Japan) has developed the Primer Explorer [13], an online software for LAMP primer design, which can be used to design LAMP primers according to LAMP primer design rules. The LAMP primers, including the outer primers F3 and B3, inner primers FIP and BIP, and loop primers FLP and BLP, along with their corresponding binding regions and sequence lengths, are shown in Fig. 3.3. The key factors in the design of LAMP primers include primer length and inter-primer distance, primer T_m value, end stability, GC content, and secondary structure [1, 4, 8, 10, 14].

1. Primer length and inter-primer distance: the length of primers is determined by the GC content of the sequence, generally F2, F1, B1c, and B2c are between 18 and 24 bases in length, while F3, B3c, and the loop primer regions (FLPc and BLPc) are between 17 and 21 bases in length. The 5′ end of F2 to the 5′ end of B2 is the LAMP amplification target sequence, and the best distance is between 130 and 200 bases. The distance from the 5′ end of F1 to the 5′ end of F2 (or from the 5′ end of B2 to the 5′ end of B1) determines the size of the loop formation, and

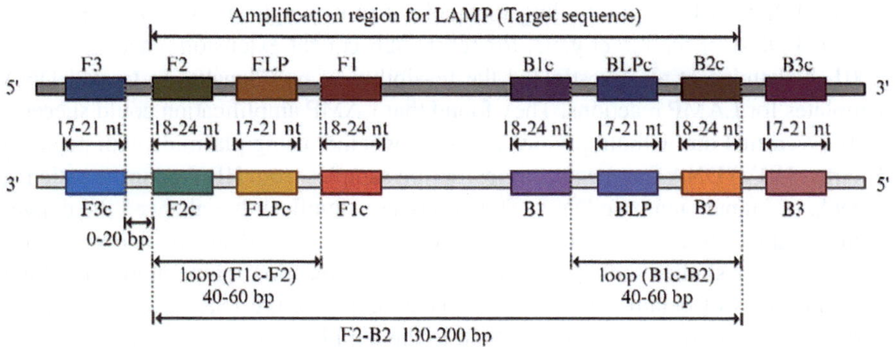

Fig. 3.3 Schematic diagram of LAMP primer binding region and size [8, 10]. (Revised and reprinted with permission from Refs. [8, 10])

the optimal distance between them is between 40 and 60 bases. The distance between F2 to F3 and B2 to B3 is between 0 and 60 bases.

2. T_m of Primers: the T_m value of each primer was controlled within the optimal reaction temperature range of *Bst* DNA polymerase, between 60 and 65 °C [1]. The T_m values of F1c and B1c are about 65 °C (64–66 °C), the T_m values of F2, B2, F3, and B3 are about 60 °C (59–61 °C), and the T_m values of loop primers are about 65 °C (64–66 °C). Generally, the experimental conditions, such as high salt concentration and primer concentration, affect the T_m values of primers. Primer Explorer software designs LAMP primers using default parameters: 0.1 μM primer concentration, 50 mM Na^+, and 4 mM Mg^{2+}. The software employed the nearest-neighbor method to estimate the T_m values of the primers, and this method is currently considered to be the approximation method that gives the closest actual values.
3. Stability of primer ends: the ends of primers are the starting point of DNA synthesis and thus must be stable. The 3′-end 6 bases of F2, B2, FLP, and BLP as well as the 5′-end 6 bases of F1c/B1c are designed with free energy (G) less than −4 kcal/mol. The hybridization reaction between primer and target sequence is an equilibrium reaction. The lower the G value, the easier the reaction will be.
4. GC content of primers: the GC content of LAMP primers is designed between 40% and 65%, and primers with GC content between 50% and 60% are relatively good primers.
5. Primer secondary structure: the primers should be designed to avoid secondary structure to prevent forming primer dimers. In addition, high AT content in the 3′ end of primers and complementary pairing with other sequences should be avoided.

3.1.5 Use of Non-denaturing Templates

Although LAMP amplification can be carried out under isothermal conditions, the classical LAMP protocol requires heated-denaturing the DNA template to 95 °C for 5 min, followed by rapid cooling to facilitate primer hybridization to the single-stranded DNA of the target gene, triggering subsequent extension reactions [1]. In 2001, Nagamine et al. investigated the feasibility of non-denaturing treated DNA templates for LAMP reactions. They found that LAMP amplification could successfully be carried out without pre-denaturation, whether using plasmids with hepatitis B virus (HBV) DNA fragments, clinical serum samples from HBV-positive patients, λDNA, or human genomic DNA. This discovery significantly enhanced the convenience and feasibility of LAMP technology for nucleic acid detection across various fields. Further studies indicated that, under higher reaction temperatures and high concentrations of betaine, the DNA double-strand structure may become unstable, and high primer concentrations can bind to the template DNA, initiating the amplification reaction. However, the specific molecular mechanisms underlying successful LAMP amplification without thermal denaturation require further investigation [15].

3.1.6 Ligation-LAMP

LAMP is currently one of the most sensitive methods for the analysis of DNA and RNA, which can detect DNA or RNA molecules down to a few copies, and the reaction is performed under isothermal conditions, thus eliminating the need for the thermal cycling process of amplification techniques such as PCR. However, there are still some shortcomings, mainly including (1) Challenging of target sequence: to ensure the specificity of LAMP, 4–6 primers need to be designed on the conserved sequence regions of target genes; however, it is challenging to choose correct and proper amplification sequence of conserved regions or specific target loci. (2) Length limitation of amplification sequence: the optimal length of the target sequence is 130–200 bp; therefore, short DNA/RNA sequence (such as miRNA) can not design LAMP amplification primers. (3) Complexity of primer design: Even though the Primer Explorer can be used to help with primer design, however, for certain circumstances, some of the interesting target sites might not be selected and, thus, would still need to design primers manually and perform systematic optimization and screening, complicating the primer design [3]. Moreover, for different target sequences, the different primer sequences will cause different amplification efficiencies of LAMP, which makes the quantitative analysis more difficult. (4) Lack of single-base resolution: Both classical LAMP, RT-LAMP, and accelerated LAMP are based on primer extension reactions, which cannot discriminate single-base variants in target sequences.

To address the above-mentioned issues and further expand the application of LAMP technology, Wang et al. first designed the two stem-loop probes and developed the general and ligation reaction-dependent LAMP (Ligation-LAMP) strategy for highly sensitive analysis of genetic biomarkers at single-base resolution [16]. The design principle of ligation-LAMP is shown in Fig. 3.4. Two stem-loop probes, P1 and P2, are first designed, in which each probe contains a stem-loop structure, an anti-target sequence (I and II) complementary to the target DNA/RNA, and a few thymine (T) bases as the spacer sequence between the stem and the anti-target sequence to avoid steric hindrance.

To detect single-base mutations in target DNA/RNA, it is sufficient to design the base at the 3′ end of P1 to be complementary to the mutated base in the target DNA/RNA. When the target DNA/RNA is present, it hybridizes with the specific recognition sequences in the P1 and P2 probes, bringing the 3′ end of P1 into proximity with the 5′ end of P2. Under the catalysis of DNA/RNA ligase, the P1 and P2 probes are ligated using the target DNA/RNA as a template, forming a LAMP double-loop dumbbell DNA template that initiates the LAMP amplification reaction. This strategy not only successfully extends LAMP amplification technology to high-sensitivity, high-specificity quantitative analysis of short nucleic acid sequences, but also introduces the ligase-specific ligation reaction, enabling LAMP to achieve high single-base resolution. For any target DNA/RNA molecule, universal stem-loop structure sequences and general LAMP primers can be designed, significantly reducing the complexity of primer design and eliminating amplification

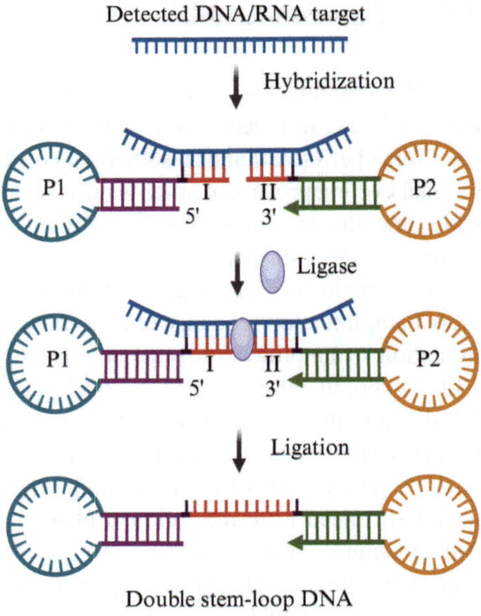

Fig. 3.4 Schematic diagram of ligation-LAMP [16]. (Revised and reprinted with permission from Ref. [16])

bias caused by different primers. Ligation-based LAMP is a highly sensitive, highly specific, and universally applicable new tool for DNA/RNA quantification and single-base mutation analysis. It has been successfully applied in single-nucleotide polymorphism (SNP) quantification, DNA methylation analysis, microRNA, and fusion gene detection [16–20].

3.2 Detection Technologies for LAMP

LAMP amplification product detection methods include turbidimetry, electrophoresis, visual colorimetry, fluorescence analysis, electrochemistry, lateral flow assay (LFA), etc.

3.2.1 Turbidity

During the LAMP amplification, the synthesis of new DNA will produce a large number of pyrophosphate ions ($P_2O_7^{2-}$, PPi) and hydrogen ions (H^+) from dNTPs (Fig. 3.5). Then, the releasing pyrophosphate ions react with magnesium ions (Mg^{2+}) in the reaction solution and produce a white precipitate of magnesium pyrophosphate ($Mg_2P_2O_7$). The LAMP amplification can be determined by visual observation of whether or not a white precipitate is produced at the bottom of the

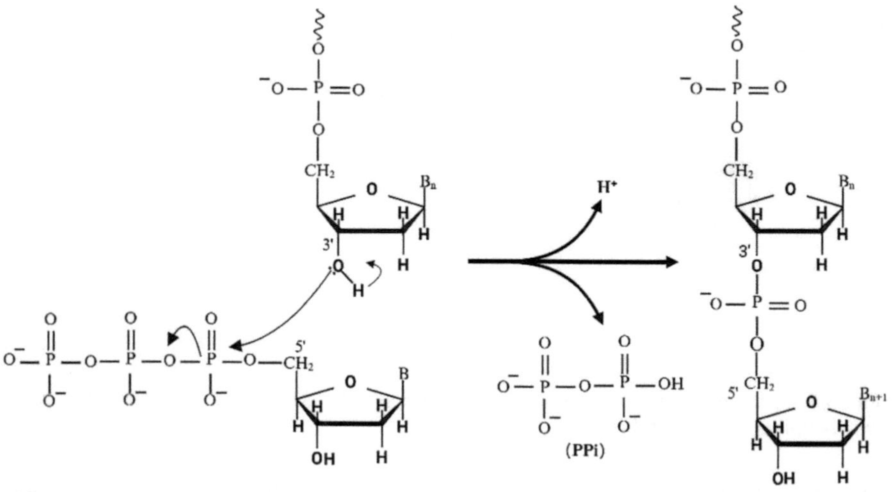

Fig. 3.5 Schematic diagram of the DNA extension process

reaction tube after brief centrifugation or standing [21, 22]. Eiken Chemical Co., Ltd. (Tokyo, Japan) has developed the LA series (LA-200, LA-320, and LA-500) real-time turbidimeter for LAMP product detection, which enables real-time monitoring of the turbidity during the LAMP amplification process and targets nucleic acid can be quantified by this measurement mode [23–25].

3.2.2 Gel Electrophoresis

Gel electrophoresis is a method for the separation and analysis of biomolecules (DNA, RNA, and proteins) based on their size and charge. It is a routine method for directly monitoring the LAMP amplification products [1, 26, 27]. The final amplification product of LAMP is a mixture of DNA fragments of varying sizes, which are separated by agarose electrophoresis or polyacrylamide gel electrophoresis. After staining with ethidium bromide (EB) or SYBR Green I, the positive LAMP reaction products can generate a ladder-like electrophoretic band on the gel. However, the technique requires a long time for electrophoretic separation and the open-cap operation will cause cross-contamination [22, 28].

3.2.3 Visualization Detection Based on Metal Ion Indicators

Metal ion indicators are used for visual colorimetric detection of LAMP amplification products, such as calcein and hydroxy naphthol blue (HNB). As shown in Fig. 3.6, before LAMP amplification, calcein molecules combine with manganese ions (Mn^{2+}), quenching calcein fluorescence. As the LAMP reaction processes in the

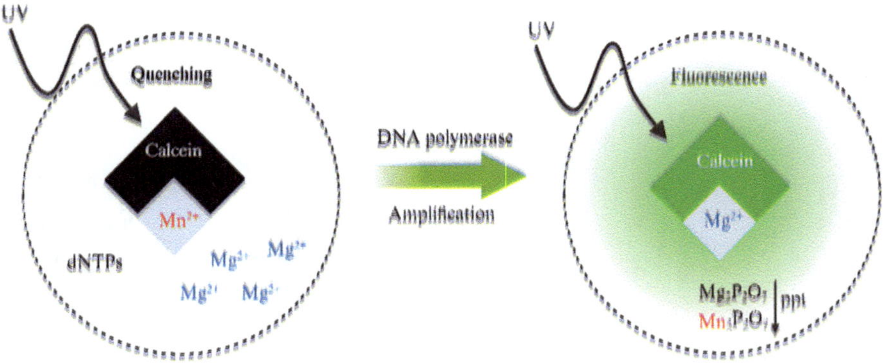

Fig. 3.6 Schematic diagram of LAMP product detection by calcein [8]. (Revised and reprinted with permission from Ref. [8])

presence of target DNA, the by-product of the LAMP amplification reaction, pyrophosphate ion ($P_2O_7^{4-}$), can seize Mn^{2+} bound to calcein and form an insoluble salt complex (manganese pyrophosphate), thus restoring the fluorescence of calcein. In addition, the free calcein molecules bind to magnesium ions (Mg^{2+}) in the reaction mixture, which further enhances the fluorescence of calcein. Under UV light (365 nm), the green fluorescence from calcein can be observed in the reaction tube with positive LAMP amplification. Furthermore, the reaction between calcein-$Mn^{2+}$and calcein-Mg^{2+} also causes the color change of the solution from orange to green, and the results of LAMP amplification can be judged by the naked eye under natural light, with green indicating a positive LAMP reaction and orange indicating a negative LAMP reaction [8, 22, 29]. The calcein-based end-point detection is the most commonly used method due to its highly sensitive and notable color change for naked-eye observation [30]. However, this method requires the addition of Mn^{2+} ions to the LAMP system, which has an impact on LAMP amplification efficiency and reduces the sensitivity of the LAMP reaction by an order of magnitude [31].

HNB, an Mg^{2+} ion indicator, appears sky blue and changes to violet while binding to Mg^{2+}. At the beginning of the LAMP reaction, HNB combines with Mg^{2+} in the reaction system showing a violet color. As the LAMP reaction processes, Mg^{2+} reacted with pyrophosphate ion, a by-product of LAMP reaction, to form magnesium pyrophosphate precipitate, and HNB lost Mg^{2+} to show sky blue color, while the target negative LAMP reaction still maintained violet color, which can be judged LAMP reaction results [32, 33]. Compared with the calcein-based end-point detection method, this method does not affect the sensitivity of the LAMP reaction [31].

3.2.4 Nucleic Acid Fluorescent Dyes-Based Detection

Nucleic acid-embedded fluorescent dyes can be used more conveniently and sensitively for LAMP reaction detection, such as SYBR Green I. SYBR Green I is a dye with a green excitation wavelength that can bind to the minor groove region of the DNA double helix. In the free state, SYBR Green I emits a weak fluorescence, and, after binding to double-stranded DNA, its fluorescence is enhanced by about 800–1000-fold. Therefore, the intensity of the fluorescence signal of SYBR Green I correlates with the amount of double-stranded DNA, and the amount of double-stranded DNA can be detected based on the fluorescence signal. After LAMP amplification, a high concentration of SYBR Green I is added to the reaction tube, the color in the positive LAMP amplification reaction tube appears yellow-green, while the color in the negative reaction tube appears orange-red. Under UV light, strong green fluorescence appears in the positive LAMP reaction tube, while there is no fluorescence in the negative reaction tube [11, 34–36]. However, the high concentration of LAMP amplification products easily causes aerosol contamination during the opening of the cap to avoid contamination. The most direct method is to add fluorescent dye before the LAMP reaction, but the visualization assay requires SYBR Green I of very high concentration, which will seriously inhibit the LAMP reaction. To solve this issue, SYBR Green I was sealed with microcrystalline wax at the bottom of the reaction tube. After the LAMP reaction, the microcrystalline wax was heated to melt and release the fluorescent dye. This strategy achieved a highly sensitive visual colorimetric detection of LAMP product fluorescence and solved the problem of aerosol contamination caused by opening the cap [37–40]. In addition, SYBR Green I can also be used to monitor the LAMP reaction process with real-time fluorescence [16, 19, 20, 41, 42]. Some other fluorescent dyes are also used for the detection of LAMP amplification products, including Malachite Green [43–45], Picogreen [46], Gelred [47, 48], SYTO [12, 49, 50], Evagreen [51, 52], and GoldView II [53, 54]. Berberine, a dye known as a natural plant alkaloid, has also been used for the detection of LAMP products. Studies have shown that berberine dyes have similar specificity and sensitivity to existing DNA insertion dyes and are inexpensive, so this dye can be used as a LAMP fluorescent dye for large-scale field analysis [30].

3.2.5 Visualization Detection Based on pH Indicators

During LAMP amplification, when DNA polymerase incorporates dNTP into the 3′ end of newly synthesized DNA, pyrophosphate ions and hydrogen ions (H^+) are released as by-products. The released H^+ can cause a small change in the pH of the reaction solution. Based on this, Jonathan Rothberg integrated an ion-sensitive field-effect transistor (ISFET) sensor, a temperature-controlled sensor, an RTD heating element, and a signal conversion and control element to build a pH-sensitive microwell array sensor and developed the Ion Torrent semiconductor high-throughput sequencing system. The Ion Torrent system can identify the four bases of GCAT by recognizing the different changes in pH caused by the release of H^+ from DNA synthesis in the microwell [55]. The pH sensor was also successfully used to detect DNA amplification in LAMP and PCR [56]. These studies have shown that DNA polymerase has good DNA synthesis activity in both weakly buffered and unbuffered systems. The New England Biolabs (NEB) reduced the concentration of Tris-HCl in the classical LAMP reaction system from 10 mM to 25–400 μM and subsequently investigated the activities of the DNA polymerases (*Bst* DNA polymerase large fragment and *Bst* 2.0 DNA polymerase) for LAMP amplification under weak buffer conditions. Experimental results suggested that these DNA polymerases have good pH tolerance and LAMP reaction can occur in the pH range of 6.0–10.0. The pH of the solution before and after the LAMP reaction decreased significantly from about 8.8 at the beginning to a final value between 6.0 and 6.5 [57]. The good pH tolerance of the LAMP reaction and the >2 units pH reduction set the basis for monitoring LAMP amplification with pH indicator dyes. Therefore, several commercial pH indicator dyes such as phenol red (8.2 ~ 6.8), cresyl red (8.8 ~ 7.2), neutral red (8.8 ~ 7.3), and cresyl violet (9.0 ~ 7.4) have been successfully applied for naked eye visualization of LAMP amplification [17, 58–61]. NEB has developed a commercial WarmStart Colorimetric LAMP Master Mix, which contains WarmStart *Bst* 2.0 DNA polymerase, RTx reverse transcriptase, dNTPs, pH indicator, and optimizing weak buffer solution. The large amount of H^+ generated during LAMP amplification results in a pH drop (>2 units), which can cause the color changing of the reaction solution from pink to yellow and enable visualization of LAMP amplification with the naked eye. The Master mixture can be used for any LAMP or RT-LAMP reaction, and the results of LAMP amplification can be visualized within 15–40 min, providing a user-friendly manner for rapid and one-step DNA and RNA detection [60–62]. In addition to the visualization of LAMP amplification results with the color change of the pH indicator, the pH of the reaction solution in LAMP amplification can also be measured quantitatively using a pH meter [63].

3.2.6 Electrochemical Methods

Compared to optical methods, electrochemical methods offer advantages such as fast response, low cost, and ease of miniaturization [64]. These significant benefits make electrochemical methods particularly suitable for the analysis of DNA amplification products [65]. Research in this area primarily focuses on using electrochemical sensors and chips to detect LAMP amplification products. Electrochemical active substances, such as Hoechst 33258, methylene blue (MB), and ruthenium hexamine, can bind to dsDNA and cause changes in the electrochemical signal. These substances are used as electrochemical signal indicators and have led to advancements in the development of electrochemical biosensors and electrochemical sensing chips. They have been successfully applied for endpoint detection and real-time quantitative analysis of LAMP amplification products [66–72].

3.2.7 Lateral Flow Assay (LFA)

Lateral flow assay (LFA) is an immunochromatographic technique consisting of a chromatographic separation and an immunochemical reaction. Usually, a nitrocellulose membrane (NC) is used as the chromatographic separation system for LFA. The immunochemical reaction is usually the interaction of antibody and antigen or DNA-specific interactions and specific hybridization between DNA probes [73, 74]. As exhibited in Fig. 3.7, the basic structure of a standard lateral flow assay strip consists of a support card and four components [74]. (1) A sample pad, an area on which the sample is dropped, is usually made of cellulose acetate/or glass fibers because they have a low affinity for proteins, which facilitates the rapid flow of proteins from the sample pad to the binding conjugate pad. (2) A binding conjugate pad, consisting of labeled biomolecule conjugate, on which labeled antibodies are used for protein targets, labeled nucleic acid capture probes are used for nucleic acid targets, and gold nanoparticles (AuNPs) [74–79], Quantum dots (QDs) [80–82], and other labels are used to generate color change or other detectable signals. (3) A reaction membrane (usually a nitrocellulose membrane), including a test line and a control line for target DNA-probe DNA hybridization or antigen-antibody interactions. The test line region is usually covalently coupled to an antibody that can react with the target protein or a DNA probe that specifically hybridizes with the nucleic acid targets. The control line region is usually covalently coupled to an antibody that can react with the labeled antibody or another DNA probe that can specifically hybridize with the labeled DNA capture probes [73]. (4) An absorption pad ensures the flow of the sample across the reaction membrane while storing the waste solution. When the sample is dropped onto the pad, the sample moves toward the end of the cellulose membrane by capillary action [83]. The sample pad transports the sample to the absorption pad, where the target molecules in the sample are specifically recognized and bound to the labeled recognition element and subsequently flow at a certain rate to the reaction membrane, where test and control lines are formed by antigen-antibody interactions or nucleic acid intermolecular hybridization, enabling rapid detection by color or simple readout devices. Compared with

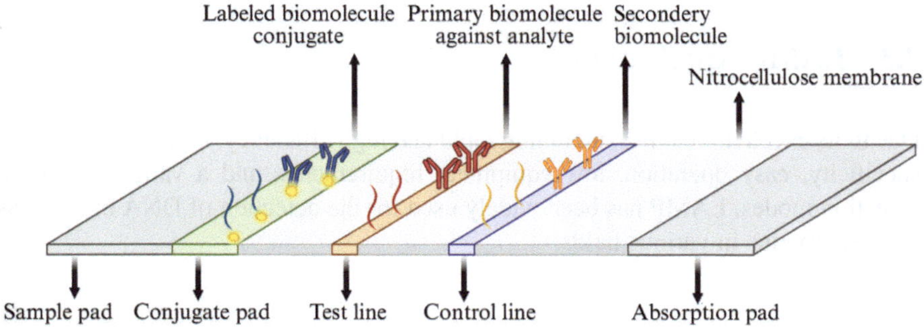

Fig. 3.7 Schematic diagram of the basic structure of LFA

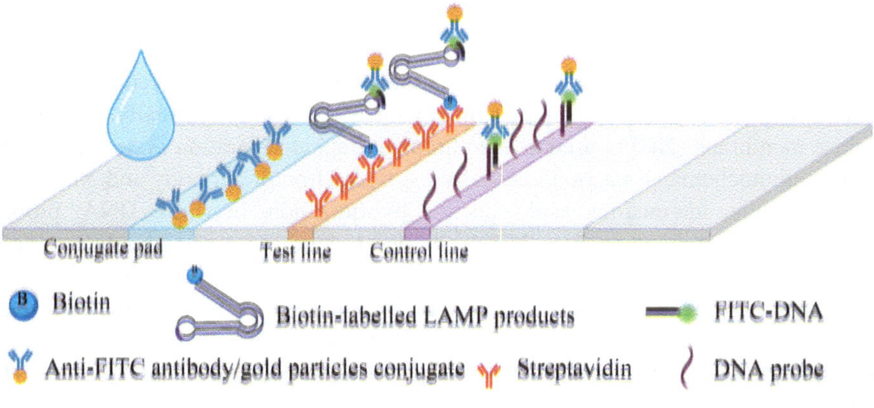

Fig. 3.8 Schematic diagram of LFA for LAMP product detection

traditional analytical methods, LFA is widely used because of its low cost, simple operation, user-friendliness, field response, and intuitive results [73, 75, 79, 80].

LFA is also used for the rapid detection of LAMP amplification products. Generally, biotin-labeled primers and a FITC-labeled DNA capture probe (FITC-DNA) are designed. As displayed in Fig. 3.8, FITC-DNAs can specifically hybridize with LAMP amplification products. On the LFA test strip, a biotin-specific antibody is immobilized on the detection line, which can capture biotin-labeled LAMP amplicon/FITC-DNA hybrids. A DNA probe complementary to FITC-DNA is immobilized on the control line. To generate a readable output signal, a gold-labeled anti-FITC antibody is introduced. A ternary complex of gold-labeled anti-FITC antibody/FITC-DNA/LAMP amplicon is captured on the detection line. Unhybridized FITC-DNA binds to the gold-labeled anti-FITC antibody, forming a binary complex without the LAMP amplicon, and flows through the detection line to move to the control line, forming the control line [84–91]. Compared to embedded dyes, LFA has very high specificity for detecting LAMP amplification products because the FITC-DNA probe specifically targets LAMP amplification products rather than binding non-specifically to dsDNA [91].

3.3 LAMP Applications

Due to its excellent features, including rapid and high efficiency, high sensitivity and specificity, easy operation, low equipment requirements, and a variety of visual detection modes, LAMP has been widely used for the detection of DNA or RNA as well as proteins in various fields.

3.3.1 Pathogenic Microorganism Detection

Pathogenic microorganisms refer to a class of microorganisms that destroy the tissues and cells of the host by competing nutrients or secreting toxins, including bacteria, viruses, and other pathogenic microorganisms such as fungi, parasites, mycoplasma, chlamydia, rickettsia, and spirochetes. At present, the detection of pathogenic microorganisms mainly relies on pathogen isolation and culture methods, immunological strategies, and PCR techniques. Although the pathogen isolation and culture method is the gold standard, it is cumbersome and time-consuming. The immunological strategies usually have low specificity and sensitivity. The PCR-based assay with high sensitivity and specificity is the gold standard method for pathogenic microorganism detection, but it is not suitable for field rapid detection and application due to the expensive instrument, high testing cost, and complicated steps. LAMP technology not only overcomes the shortcomings of PCR technology but also has higher sensitivity and specificity, shorter turnover time, and easy operation, which is an ideal method for rapid clinical pathogenic microbial detection [3, 6, 26]. By designing LAMP primers and combining different detection methods, LAMP or RT-LAMP can rapidly detect various pathogenic microorganisms with high sensitivity and specificity, showing outstanding advantages in the rapid diagnosis of human and animal diseases caused by pathogenic microorganisms.

1. Virus detection: DNA viruses reported to be detected by LAMP technology include human herpes virus (HHV) [92–95], adenovirus [96, 97], hepatitis B virus (HBV) [1, 10, 15, 98, 99], human papillomavirus (HPV) [61, 100], Capripoxviruses (CaPVs) [101], Pseudorabies virus (PRV) [102], etc. The reported RNA viruses detected by RT-LAMP include Respiratory syncytial virus (RSV) [103, 104], Mumps virus (MV) [105, 106], Influenza virus [37, 107–109], West Nile virus (WNV) [110], Zika virus (ZIKV) [111–113], Japanese encephalitis virus (JEV) [114], Hepatitis C virus (HCV) [98], Coxsackievirus (CV) [34, 115], Norwlk virus [116], Measles virus (MV) [116], Dengue [117, 118], Ebola virus (EBOV) [119–121], Rabies virus (RV) [122], Swine vesicular disease virus (SVDV) [123], Newcastle disease virus (NDV) [124], Tomato spotted wilt virus (TSWV) [125], Immunodeficiency Virus (HIV) [78, 98, 100, 113], Severe Acute Respiratory Syndromes (SRAS) [126, 127], Middle East Respiratory Syndrome (MERS) [51], SRAS-CoV-2 [58, 60, 128, 129], etc.
2. Bacterial detection: LAMP technology is a sensitive, rapid, accurate, and economical method for pathogen detection. Bacteria reported to be detected by LAMP include *Escherichia coli* [66, 70, 130], *Staphylococcus aureus* [70, 131, 132] *Mycobacterium tuberculosis* [133], *Streptococcus pneumoniae* [134], *Bordetella pertussis* [135–137], *Klebsiella pneumoniae* [138], *Listeria monocytogenes* [54, 139], *Yersinia enterocolitica* [68], *Vibrio parahaemolyticus* [140, 141], *Salmonella* [72, 142, 143], *Shigella* [143, 144], *Helicobacter pylori* [145, 146], *Pseudomonas aeruginosa* [147], etc.

3. Other pathogenic microorganisms: LAMP technology is not only widely used in the detection of viruses and bacteria, but also successfully applied in the sensitive and rapid detection of other pathogenic microorganisms, such as fungi [148–150], parasites [43, 44, 151–154], mycoplasma [24], chlamydia [155], malaria [156], etc.

3.3.2 SNP Analysis

Single Nucleotide Polymorphism (SNP), as a third-generation genetic marker, plays an important role in the localization of genetic diseases, association analysis, individual disease susceptibility analysis, and drug resistance analysis. Allele-specific loop-mediated isothermal amplification (AS-LAMP) can be used for low-cost, simple, and rapid SNP analysis, which needs to design the SNP nucleotide at each 3′-end of FIP or BIP primer [157]. The primer matches with wild-type DNA, the mutant DNA cannot be amplified by LAMP. On the contrary, the primer matches with mutant DNA, the wild-type DNA cannot be amplified by LAMP. Thus, in this way, the analysis of the SNP site can be identified and detected. To further improve the detection specificity and accuracy of LAMP for SNP analysis. Ren et al. [158] established a modified loop-mediated isothermal amplification (mLAMP) by introducing an artificial mismatch base into the LAMP primer (FIP) to identify the SNP site. By using mLAMP, a DNA target of 100 aM can be accurately determined, and as low as 0.1% mutant dsDNA can be detected in the presence of a large excess of wild-type dsDNA. However, the strict LAMP primer design, including primer length and inter-primer distance, primer T_m value, end stability, GC content, and secondary structure, greatly limits the wide application of AS-LAMP and mLAMP technologies in SNP analysis. Wang et al. ingeniously designed the stem-loop probes and combined the ligase-catalyzed ligation reaction to develop a highly sensitive, specific, and universal strategy, ligation-LAMP, for DNA and RNA analysis assay. The ligation-LAMP has been successfully used for SNP analysis [16, 17]. In addition, the ligation-LAMP was also successfully used for the detection and quantitative analysis of DNA methylation levels in biological samples [17].

3.3.3 MicroRNA Analysis

MicroRNAs (miRNAs) are a class of endogenous, nonprotein-coding small RNA molecules that play an important regulatory role in gene expression by inhibiting protein translation or inducing mRNA degradation. The abnormal expression of miRNAs is closely associated with a variety of diseases. However, the length of mature miRNA is about 19–24 nucleotides (nt), and thus it is impossible to design LAMP amplification primers for miRNA analysis. To address this problem, Li et al. [42] used target miRNA as LAMP amplification primer (F3 primer) to initiate the

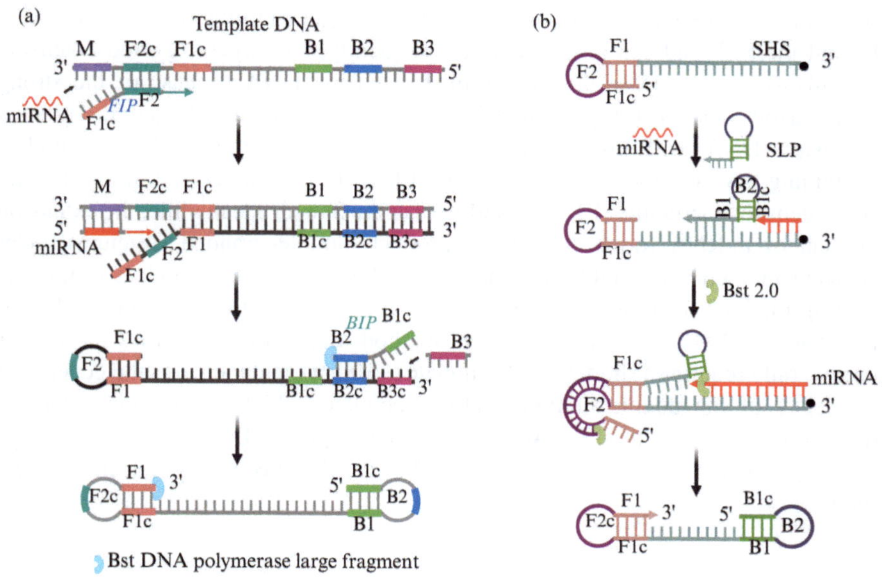

Fig. 3.9 (**a**) Schematic diagram of target miRNA as F3 primer to initiate LAMP amplification reaction [42]. (**b**) Schematic illustration of the TT-LAMP [41]. (Revised and reprinted with permission from Refs. [41, 42])

LAMP amplification reaction, and developed a simple, rapid, specific, and cost-effective miRNA assay (Fig. 3.9a). However, this approach requires a long DNA template (~200 nucleotides) containing six pre-defined regions for primer binding and stem-loop DNA template formation, which makes the primer design extremely complicated. In addition, the overlong DNA template is difficult to chemically synthesize with high purity. Moreover, the low concentration of target miRNA molecules cannot effectively initiate the LAMP amplification reaction, resulting in a relatively lower sensitivity with a detection limit of ~100 fM than classical LAMP in DNA or RNA analysis [1, 110]. To further simplify the probe design and miRNA detection sensitivity, Sun et al. [41].designed target triggered and stem-loop primers-mediated LAMP (TT-LAMP) strategy for one-step miRNA analysis. As shown in Fig. 3.9b, the TT-LAMP system for miRNA assay consisted of SLTD, SLP, FP, and BP DNA probes. The SLTD contains a stem-loop structure (F1c, F2, and F1, F1c complementary to F1) at the 5′-end and a specific hybridization sequence (SHS) at the 3′-end. The 3′-end of the SLTD is modified with a phosphate group to avoid the nonspecific extension. The SLP also consists of a stem-loop structure (B1c, B2, and B1, the B1c complementary to B1) at the 5′-end and an ssDNA sequence complementary to the partial SHS sequence of the SLTD. The other part of the SHS of the SLTD is complementary to the miRNA target. So the ssDNA of the SLP will hybridize with the SHS and then extend along the SLTD to generate a dsDNA (dsDNA) with the catalysis of *Bst* DNA polymerase. In the absence of the miRNA target, the extension will be stopped under isothermal conditions. On the contrary, when the miRNA target is present, the miRNA target can hybridize with the SHS behind the SLP and can also extend. Subsequently, the target-triggered extension will displace and release the SLP-primed extension DNA because of the strong strand displacement activity of *Bst* DNA polymerase. Then, the released SLP-primed extension DNA will immediately form the double dumbbell DNA template containing the stem-loop structure of the SLP at the 5′-end and the newly synthesized stem-loop structure (F1, F2c, and F1c) at the 3′-end. Meanwhile, the extension product of miRNA forms a stable by-product (double-stranded structure) under constant temperature conditions. The double-dumbbell DNA template is just the starting material for rapid and efficient auto-cyclic strand displacement DNA synthesis with LAMP reaction. This novel method remarkably simplifies the primer design, only requires a shorter DNA template (~100 nt), and increases the generation efficiency of the double dumbbell DNA template, which greatly improves the detection sensitivity of TT-LAMP. With one type of DNA polymerase in a one-step reaction, as low as 100 aM miRNA can be specifically detected in real samples.

3.3.4 Analysis of Telomerase Activity

Telomerase, also known as telomere terminal transferase, is a ribonucleoprotein complex composed of telomerase RNA, telomerase catalytic subunit, and telomerase-associated proteins (TEP/TE1 and TP1) with reverse transcriptase activity, which can use its RNA as a template to add repeats to the ends of chromosomes. Telomerase is inactive or has minimal activity in most normal cells, while it is active in about 95% of cancer cells. Telomerase activity is closely associated with cancer development and is used as an important diagnostic biomarker for the disease. Currently, several methods have been established to analyze telomerase activity based on various amplification technologies, but the analysis of telomerase activity in single cell still faces great challenges. Wang et al. [159] innovatively designed a pair of stem-loop DNA probes and constructed a stem-loop probe-mediated LAMP strategy, termed stem-loop primer-mediated exponential amplification (SPEA), which can detect telomerase activity at the single-cell level. The principle of SPEA-based telomerase activity is shown in Fig. 3.10. First, two stem-loop DNA probes (SLTS and SLP) were designed. The 5′ end of SLTS is the stem-loop structure (F1c-F2-F1, F1c complementary to F1), and its 3′ end (arrow represents 3′ direction) is the substrate sequence (TS) of telomerase, which can be recognized by telomerase and extended by adding the repeat sequence (TTAGGG). The 5′ end of the SLP probe is a stem-loop structure (B1c-B2-B1, B1c complementary to), and its 3′ end is a single-stranded DNA with specific recognition function, which can specifically recognize and hybridize with the repetitive sequences in the telomerase extension product. Due to the telomerase extension product containing multiple repetitive sequences, one extended SLTS probe can hybridize with multiple SLP probes. Then *Bst* DNA polymerase will catalyze the self-extension reaction of the

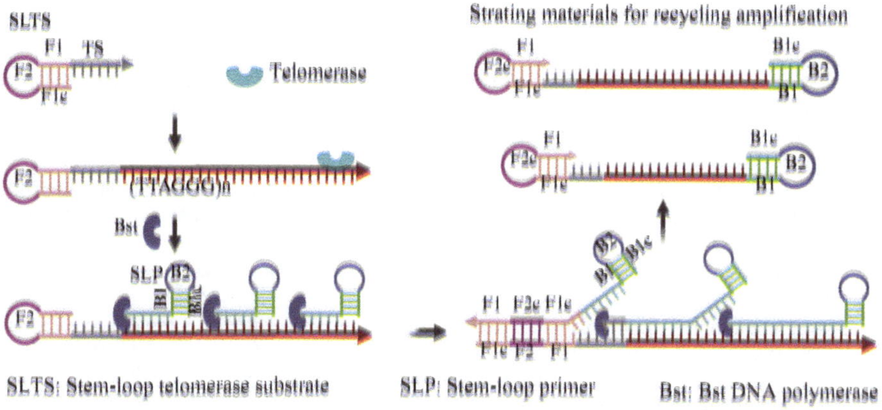

Fig. 3.10 Schematic diagram of the principle of telomerase activity detection based on SPEA reaction [159]. (Revised and reprinted with permission from Ref. [159])

first SLP probe and open the stem-loop structure of the SLTS probe. Meanwhile, the SLP probes on the right side are also extended, and the left-extended SLP probes are released by the strand displacement activity of *Bst* DNA polymerase. The released DNA can form the double-loop DNA template, which can subsequently trigger exponentially isothermal amplification similar to the LAMP amplification. The SPEA can especially and efficiently amplify the telomerase extension product with near-zero nonspecific signal and thus can accurately detect telomerase activity in the crude lysate of a single cell and is suited for detecting the cellular heterogeneity arising from cell-to-cell variations.

3.3.5 Detection of Genetically Modified Foods

Following the introduction of the world's first genetically modified (GM) crop in 1983, the global cultivation of GM varieties has steadily expanded, now encompassing key agricultural products such as soybeans, corn, cotton, canola, and sugar beets. However, the long-term risks and safety of GM crops remain uncertain, causing ongoing consumer concerns regarding their production and use. In response to these issues, the "Management Measures for the Labeling of Agricultural Genetically Modified Organisms" mandates testing and labeling protocols for GM crops.

Transgenic crop and food detection typically involves identifying the introduced foreign promoters, target genes, or terminators. Researchers use these foreign promoters, target genes, and terminators as target sequences to design specific LAMP primers. By integrating various detection techniques, they have developed a series of sensitive, rapid, and user-friendly methods for detecting GM crops and related foods [160–167].

References

1. Notomi T, Okayama H, Masubuchi H, Yonekawa T, Watanabe K, Amino N, Hase T. Loop-mediated isothermal amplification of DNA. Nucleic Acids Res. 2000;28:E63.
2. Zhao Y, Chen F, Li Q, Wang L, Fan C. Isothermal amplification of nucleic acids. Chem Rev. 2015;115:12491–545.
3. Wong YP, Othman S, Lau YL, Radu S, Chee HY. Loop-mediated isothermal amplification (LAMP): a versatile technique for detection of micro-organisms. J Appl Microbiol. 2018;124: 626–43.
4. Peng T. Isothermal amplification of nucleic acids and its applications. China: Science Press; 2009. p. 4–19.
5. Mori Y, Notomi T. Loop-mediated isothermal amplification (LAMP): a rapid, accurate, and cost-effective diagnostic method for infectious. J Infect Chemother. 2009;15:62–9.
6. Li Y, Fan P, Zhou S, Zhang L. Loop-mediated isothermal amplification (LAMP): a novel rapid detection platform for pathogens. Microb Pathog. 2017;107:54–61.
7. Becherer L, Borst N, Bakheit M, Frischmann S, Zengerle R, von Stetten F. Loop-mediated isothermal amplification (LAMP)-review and classification of methods for sequence-specific detection. Anal Methods. 2020;12:717–46.
8. Tomita N, Mori Y, Kanda H, Notomi T. Loop-mediated isothermal amplification (LAMP) of gene sequences and simple visual detection of products. Nat Protoc. 2008;3:877–82.
9. Soliman H, El-Matbouli M. Reverse transcription loop-mediated isothermal amplification (RT-LAMP) for rapid detection of viral hemorrhagic septicaemia virus (VHS). Vet Microbiol. 2006;114:205–13.
10. Nagamine K, Hase T, Notomi T. Accelerated reaction by loop-mediated isothermal amplification using loop primers. Mol Cell Probes. 2002;16:223–9.
11. Yoshida A, Nagashima S, Ansai T, Tachibana M, Kato H, Watari H, Notomi T, Takehara T. Loop-mediated isothermal amplification method for rapid detection of the periodontopathic bacteria Porphyromonas gingivalis, Tannerella forsythia, and treponema denticola. J Clin Microbiol. 2005;43:2418–24.
12. Watts MR, James G, Sultana Y, Ginn AN, Outhred AC. A loop-mediated isothermal amplification (LAMP) assay for Strongyloides stercoralis in stool that uses a visual detection method with SYTO-82 fluorescent dye. Am J Trop Med Hyg. 2014;90:306–11.
13. Primerexplorer v5. Available from: http://primerexplorer.jp/e/. [updated 15 October 2019].
14. A guide to LAMP primer designing (primerexplorer v5). Available from: http://primerexplorer.jp/e/v5_manual/index.html. [updated 15 October 2019].
15. Nagamine K, Watanabe K, Ohtsuka K, Hase T, Notomi T. Loop-mediated isothermal amplification reaction using a nondenatured template. Clin Chem. 2001;47:1742.
16. Wang H, Wang H, Sun Y, Liu X, Liu Y, Wang C, Zhang P, Li Z. A general strategy for highly sensitive analysis of genetic biomarkers at single-base resolution with ligase-based isothermally exponential amplification. Talanta. 2020;212:120754.
17. Wang H, Wang H, Jia Y, Sun R, Hong W, Zhang M, Li Z. Visual detection of fusion genes by ligation-triggered isothermal exponential amplification: a point-of-care testing method for highly specific and sensitive quantitation of fusion genes with a smartphone. Anal Chem. 2019;91:12428–34.
18. Zhang M, Wang H, Wang H, Wang F, Li Z. CRISPR/Cas12a-assisted ligation-initiated loop-mediated isothermal amplification (CAL-LAMP) for highly specific detection of microRNAs. Anal Chem. 2021;93:7942–8.
19. Fu Y, Duan X, Huang J, Huang L, Zhang L, Cheng W, Ding S, Min X. Detection of KRAS mutation via ligation-initiated LAMP reaction. Sci Rep. 2019;9:5955.
20. Du W, Lv M, Li J, Yu R, Jiang J. A ligation-based loop-mediated isothermal amplification (ligation-LAMP) strategy for highly selective. Chem Commun. 2016;52:12721–4.

21. Mori Y, Nagamine K, Tomita N, Notomi T. Detection of loop-mediated isothermal amplification reaction by turbidity derived from magnesium. Biochem Biophys Res Commun. 2001;289:150–4.
22. Zhou D, Guo J, Xu L, Gao S, Lin Q, Wu Q, Wu L, Que Y. Establishment and application of a loop-mediated isothermal amplification (LAMP). Sci Rep. 2014;4:4912.
23. Mori Y, Kitao M, Tomita N, Notomi T. Real-time turbidimetry of lamp reaction for quantifying template DNA. J Biochem Biophys Methods. 2004;59:145–57.
24. Sakai J, Maeda T, Tarumoto N, Misawa K, Tamura S, Imai K, Yamaguchi T, Iwata S, Murakami T, Maesaki S. A novel detection procedure for mutations in the 23s rRNA gene of Mycoplasma pneumoniae with peptide nucleic acid-mediated loop-mediated isothermal amplification assay. J Microbiol Methods. 2017;141:90–6.
25. Lin M, Liu W, Pu W, Tan J, Chen Y. Rapid detection of ermb gene in Clostridium difficile by loop-mediated isothermal amplification. J Med Microbiol. 2015;64:854–61.
26. Mori Y, Notomi T. Loop-mediated isothermal amplification (LAMP): a rapid, accurate, and cost-efficient diagnostic method for infectious. J Infect Chemother. 2009;15:62–9.
27. Parida M, Sannarangaiah S, Dash PK, Rao PV, Morita K. Loop mediated isothermal amplification (LAMP): a new generation of innovative gene amplification technique; perspectives in clinical diagnosis of infectious diseases. Rev Med Virol. 2008;18:407–21.
28. Zhang X, Lowe SB, Gooding JJ. Brief review of monitoring methods for loop-mediated isothermal amplification (LAMP). Biosens Bioelectron. 2014;61:491–9.
29. Be Suschio SA, Mónica LM, Be Natar AF, Severine M, Israel C, Al Be Rt P, Ángeles C, Yutaka K, Wehrendt DP, Paula P. Analytical sensitivity and specificity of a loop-mediated isothermal amplification (LAMP) kit prototype for detection of Trypanosoma cruzi DNA in human blood. PLoS Negl Trop Dis. 2017;11:e0005779.
30. Jens F, Carolin XN, Marcus F, Felix GJ. Shining a light on lamp assays—a comparison of lamp visualization methods including the novel use of berberine. BioTechniques. 2014;58: 189–94.
31. Chou Y, Feng L, Zhang J, Zou B, Zhou G. A systematic review of the research progress and improvement of loop-mediated isothermal amplification. Microbiol China. 2015; 42: 729–35.
32. Goto M, Honda E, Ogura A, Nomoto A, Hanaki K. Colorimetric detection of loop-mediated isothermal amplification reaction by using hydroxy naphthol blue. BioTechniques. 2009;46: 167–72.
33. Srisrattakarn A, Lulitanon DA, Wilailuckana C, Charoensri N, Wonglakorn L, Saenjamla P, Chaimanee P, Daduang J, Chanawong A. Rapid and simple identification of carbapenemase genes, bla ndm, bla oxa-48, bla vim, bla imp-14 and bla kpc groups, in gram-negative bacilli by in-house loop-mediated isothermal amplification with hydroxynaphthol blue dye. World J Microbiol Biotechnol. 2017;33:130.
34. Monazah A, Zeinoddini M, Saeeidinia AR. Evaluation of a rapid detection for coxsackievirus b3 using one-step reverse transcription loop-mediated isothermal amplification (RT-LAMP). J Virol Methods. 2017;246:27–33.
35. Singh R, Singh DP, Savargaonkar D, Singh O, Bhatt RM, Valecha N. Evaluation of SYBR green I based visual loop-mediated isothermal amplification (LAMP) assay for genus and species-specific diagnosis of malaria in P. Vivax and P. Falciparum endemic regions. J Vector Borne Dis. 2017;54:54–60.
36. Chen L, Guo J, Wang Q, Kai G, Yang L. Development of the visual loop-mediated isothermal amplification assays for seven genetically modified maize events and their application in practical samples analysis. J Agric Food Chem. 2011;59:5914–8.
37. Liang C, Cheng S, Chu Y, Wu H, Zou B, Huang H, Xi T, Zhou G. A closed-tube detection of loop-mediated isothermal amplification (LAMP) products using a wax-sealed fluorescent intercalator. J Nanosci Nanotechnol. 2013;13:3999–4005.
38. Tao ZY, Zhou HY, Xia H, Xu S, Zhu HW, Culleton RL, Han ET, Feng L, Qiang F, Gu YP. Adaptation of a visualized loop-mediated isothermal amplification technique for field detection of Plasmodium vivax infection. Parasit Vectors. 2011;4:115.

39. Ma Xueping, Cheng Sijia, Wu Haiping, Song Qinxin, Zou Bingjie, Zhou Guohua. A closed-tube isothermal amplification method for highly sensitive and visualized detection of pine wood nematode (*Bursaphelenchus xylophilus*). J Microbiol. 2014;34:7–14.
40. Jiang Z, Ma X, Yin Z, Lu Y, Cao X, Wu H. A closed-tube isothermal amplification method for highly sensitive and visualized detection of Plasmodium falciparum. Prog Mod Biomed. 2014;14:5014–8.
41. Sun YY, Tian H, Liu CH, Sun YY, Li ZP. One-step detection of microrna with high sensitivity and specificity via target-triggered loop-mediated isothermal amplification (TT-LAMP). Chem Commun. 2017;53:11040–3.
42. Li C, Li Z, Jia H, Yan J. One-step ultrasensitive detection of micrornas with loop-mediated isothermal amplification (LAMP). Chem Commun. 2011;47:2595–7.
43. Sriworarat C, Phumee A, Mungthin M, Leelayoova S, Siriyasatien P. Development of loop-mediated isothermal amplification (LAMP) for simple detection of *leishmania* infection. Parasit Vectors. 2015;8:591.
44. Lucchi NW, Ljolje D, Silva-Flannery L, Udhayakumar V. Use of malachite green-loop mediated isothermal amplification for detection of *Plasmodium* spp. Parasites. PLOS One. 2016;11:e0151437.
45. Nzelu CO, Cáceres AG, Guerrero-Quincho S, Tineo-Villafuerte E, Kato H. A rapid molecular diagnosis of cutaneous leishmaniasis by colorimetric malachite green-loop-mediated isothermal amplification (LAMP) combined with an FTA card as a direct sampling tool. Acta Trop. 2016;153:116–9.
46. Batra K, Kumar A, Kumar V, Nanda T, Maan S. Development and evaluation of loop-mediated isothermal amplification assay for rapid detection of capripoxvirus. Vet World. 2015;8:1286–92.
47. Wassermann M, Mackenstedt U, Romig T. A loop-mediated isothermal amplification (LAMP) method for the identification of species within the *Echinococcus* granulosus complex. Vet Parasitol. 2014;200:97–103.
48. Wozniakowski G, Samorek-Salamonowicz E, Kozdruń W. Comparison of loop-mediated isothermal amplification and PCR for the detection and differentiation of marek's disease virus serotypes 1, 2, and 3. Avian Dis. 2013;57:539–43.
49. Yun L, Jinming T, Hong T, Jie S, Tikang L, Lishan L, Jianli L, Shaoling Z, Chenfu C. Establishment and optimization of an in situ real-time fluorescent reverse transcription loop-mediated isothermal amplification for detection of avian influenza H5 hemagglutinin gene. Chin J Virol. 2013;29:488–94.
50. Mwendwa F, Mbae CK, Kinyua J, Mulinge E, Mburugu GN, Njiru ZK. Stem loop-mediated isothermal amplification test: comparative analysis with classical LAMP and PCR in detection of *Entamoeba histolytica* in Kenya. BMC Res Notes. 2017;10:142.
51. Hee LS, Hee BY, Yang-Hoon K, Young-Ki C, Song MS, Ji-Young A. One-pot reverse transcriptional loop-mediated isothermal amplification (RT-LAMP) for detecting MERS-CoV. Front Microbiol. 2016;7:2166.
52. Oscorbin IP, Belousova EA, Zakabunin AI, Boyarskikh UA, Filipenko ML. Comparison of fluorescent intercalating dyes for quantitative loop-mediated isothermal amplification (qLAMP). BioTechniques. 2016;61:20.
53. Shi W, Li K, Ji Y, Jiang Q, Mi Z. Development and evaluation of reverse transcription-loop-mediated isothermal amplification assay for rapid detection of enterovirus 71. BMC Infect Dis. 2011;11:197.
54. Wu R, Liu X, Guo B, Chen F, Wang X. Development of double loop-mediated isothermal amplification to detect Listeria monocytogenes in food. Curr Microbiol. 2014;69:839–45.
55. Rothberg JM, Hinz W, Rearick TM, Schultz J, Mileski W, Davey M, Leamon JH, Johnson K, Milgrew MJ, Edwards M, Hoon J, Simons JF, Marran D, Myers JW, Davidson JF, Branting A, Nobile JR, Puc BP, Light D, Clark TA, Huber M, Branciforte JT, Stoner IB, Cawley SE, Lyons M, Fu Y, Homer N, Sedova M, Miao X, Reed B, Sabina J, Feierstein E, Schorn M, Alanjary M, Dimalanta E, Dressman D, Kasinskas R, Sokolsky T, Fidanza JA, Namsaraev E, McKernan KJ, Williams A, Roth GT, Bustillo J. An integrated semiconductor device enabling non-optical genome sequencing. Nature. 2011;475:348–52.

56. Toumazou C, Shepherd LM, Reed SC, Chen GI, Patel A, Garner DM, Wang CJ, Ou CP, Amin-Desai K, Athanasiou P, Bai H, Brizido IM, Caldwell B, Coomber-Alford D, Georgiou P, Jordan KS, Joyce JC, La Mura M, Morley D, Sathyavruthan S, Temelso S, Thomas RE, Zhang L. Simultaneous DNA amplification and detection using a pH-sensing semiconductor system. Nat Methods. 2013;10:641–6.
57. Tanner NA, Zhang Y, Evans TC Jr. Visual detection of isothermal nucleic acid amplification using pH-sensitive dyes. BioTechniques. 2015;58:59–68.
58. Lalli MA, Langmade JS, Chen X, Fronick CC, Sawyer CS, Burcea LC, Wilkinson MN, Fulton RS, Heinz M, Buchser WJ, Head RD, Mitra RD, Milbrandt J. Rapid and extraction-free detection of SARS-COV-2 from saliva by colorimetric reverse-transcription loop-mediated isothermal amplification. Clin Chem. 2021;67:415–24.
59. Zhang Y, Odiwuor N, Xiong J, Sun L, Tanner NA. Rapid molecular detection of SARS-CoV-2 (COVID-19) virus RNA using colorimetric LAMP. MedRxiv. 2020.
60. Thi V, Herbst K, Boerner K, Meurer M, Kremer LP, Kirrmaier D, Freistaedter A, Papagiannidis D, Galmozzi C, Stanifer ML. A colorimetric RT-LAMP assay and LAMP-sequencing for detecting SARS-CoV-2 RNA in clinical samples. Sci Transl Med. 2020;12: eabc7075.
61. Daskou M, Tsakogiannis D, Dimitriou TG, Amoutzias GD, Mossialos D, Kottaridi C, Gartzonika C, Markoulatos P. Warmstart colorimetric LAMP for the specific and rapid detection of HPV16 and HPV18 DNA. J Virol Methods. 2019;270:87–94.
62. Daskou M, Dimitriou TG, Alexopoulou DS, Tsakogiannis D, Amoutzias GD, Mossialos D, Kyriakopoulou Z, Markoulatos P. Warmstart colorimetric-LAMP for the rapid, sensitive and specific detection of enteroviruses A-D targeting the 5′UTR region. J Appl Microbiol. 2021;130:292–301.
63. Xie S, Yuan Y, Song Y, Zhuo Y, Li T, Chai Y, Yuan R. Using the ubiquitous pH meter combined with a loop mediated isothermal amplification method for facile and sensitive detection of Nosema bombycis genomic DNA PTP1. Chem Commun. 2014;50:15932–5.
64. Ferguson BS, Buchsbaum SF, Swensen JS, Hsieh K, Lou X, Soh HT. Integrated microfluidic electrochemical DNA sensor. Anal Chem. 2009;81:6503–8.
65. Deféver T, Druet M, Evrard D, Marchal D, Limoges B. Real-time electrochemical PCR with a DNA intercalating redox probe. Anal Chem. 2011;83:1815–21.
66. Safavieh M, Ahmed MU, Tolba M, Zourob M. Microfluidic electrochemical assay for rapid detection and quantification of *Escherichia coli*. Biosens Bioelectron. 2012;31:523–8.
67. Ahmed MU, Hasan Q, Mosharraf Hossain M, Saito M, Tamiya E. Meat species identification based on the loop mediated isothermal and electrochemical DNA sensor. Food Control. 2010;21:599–605.
68. Sun W, Qin P, Gao H, Li G, Jiao K. Electrochemical DNA biosensor based on chitosan/nano-V_2O_5/MWCNTs composite film modified carbon ionic liquid electrode and its application to the LAMP product of yersinia enterocolitica gene sequence. Biosens Bioelectron. 2010;25: 1264–70.
69. Xie S, Chai Y, Yuan Y, Bai L, Yuan R. Development of an electrochemical method for ochratoxin a detection based on aptamer and loop-mediated isothermal amplification. Biosens Bioelectron. 2014;55:324–9.
70. Ahmed MU, Nahar S, Safavieh M, Zourob M. Real-time electrochemical detection of pathogen DNA using electrostatic interaction of a redox probe. Analyst. 2013;138:907–15.
71. Nakamura N, Ito K, Takahashi M, Hashimoto K, Kawamoto M, Yamanaka M, Taniguchi A, Kamatani N, Gemma N. Detection of six single-nucleotide polymorphisms associated with rheumatoid arthritis by a loop-mediated isothermal amplification method and an electrochemical DNA chip. Anal Chem. 2007;79:9484–93.
72. Hsieh K, Patterson AS, Ferguson BS, Plaxco KW, Soh HT. Rapid, sensitive, and quantitative detection of pathogenic DNA at the point of care through microfluidic electrochemical quantitative loop-mediated isothermal amplification. Angew Chem Ent Ed. 2012;51:4896–900.

73. Bahadır EB, Sezgintürk MK. Lateral flow assays: principles, designs and labels. Trends Anal Chem. 2016;82:286–306.
74. Chen YH, Gupta NK, Huang HJ, Lam CH, Huang CL, Tan KT. Affinity-switchable lateral flow assay. Anal Chem. 2021;93:5556–61.
75. Fu Q, Liang J, Lan C, Zhou K, Shi C, Tang Y. Development of a novel dual-functional lateral-flow sensor for on-site detection of small molecule analytes. Sensors Actuators B Chem. 2014;203:683–9.
76. Singh J, Sharma S, Nara S. Evaluation of gold nanoparticle based lateral flow assays for diagnosis of enterobacteriaceae members in food and water. Food Chem. 2015;170:470–83.
77. Meng K, Sun W, Zhao P, Zhang L, Cai D, Cheng Z, Guo H, Liu J, Yang D, Wang S, Chai T. Development of colloidal gold-based immunochromatographic assay for rapid detection of Mycoplasma suis in porcine plasma. Biosens Bioelectron. 2014;55:396–9.
78. Fu X, Cheng Z, Yu J, Choo P, Chen L, Choo J. A sers-based lateral flow assay biosensor for highly sensitive detection of HIV-1 DNA. Biosens Bioelectron. 2016;78:530–7.
79. Yu CY, Ang GY, Chua AL, Tan EH, Lee SY, Falero-Diaz G, Otero O, Rodriguez I, Reyes F, Acosta A, Sarmiento ME, Ghosh S, Ramamurthy T, Yean Yean C, Lalitha P, Ravichandran M. Dry-reagent gold nanoparticle-based lateral flow biosensor for the simultaneous detection of vibrio cholerae serogroups O1 and O139. J Microbiol Methods. 2011;86:277–82.
80. Taranova NA, Berlina AN, Zherdev AV, Dzantiev BB. 'Traffic light' immunochromatographic test based on multicolor quantum dots for the simultaneous detection of several antibiotics in milk. Biosens Bioelectron. 2015;63:255–61.
81. Wang C, Hou F, Ma Y. Simultaneous quantitative detection of multiple tumor markers with a rapid and sensitive multicolor quantum dots based immunochromatographic test strip. Biosens Bioelectron. 2015;68:156–62.
82. Li Z, Liu J, Zhang L, Li J. Quantum dots-based lateral flow test strip for glutathione detection. In: Fontes A, Santos BS, editors. Quantum dots: applications in biology. New York: Springer; 2020. p. 249–57.
83. Hu J, Wang S, Wang L, Li F, Pingguan-Murphy B, Lu TJ, Xu F. Advances in paper-based point-of-care diagnostics. Biosens Bioelectron. 2014;54:585–97.
84. Nimitphak T, Kiatpathomchai W, Flegel TW. Shrimp hepatopancreatic parvovirus detection by combining loop-mediated isothermal amplification with a lateral flow dipstick. J Virol Methods. 2008;154:56–60.
85. Jaroenram W, Kiatpathomchai W, Flegel TW. Rapid and sensitive detection of white spot syndrome virus by loop-mediated isothermal amplification combined with a lateral flow dipstick. Mol Cell Probes. 2009;23:65–70.
86. Khunthong S, Jaroenram W, Arunrut N, Suebsing R, Mungsantisuk I, Kiatpathomchai W. Rapid and sensitive detection of shrimp yellow head virus by loop-mediated isothermal amplification combined with a lateral flow dipstick. J Virol Methods. 2013;188:51–6.
87. Arunrut N, Prombun P, Saksmerprome V, Flegel TW, Kiatpathomchai W. Rapid and sensitive detection of infectious hypodermal and hematopoietic necrosis virus by loop-mediated isothermal amplification combined with a lateral flow dipstick. J Virol Methods. 2011;171:21–5.
88. Lalle M, Possenti A, Dubey JP, Pozio E. Loop-mediated isothermal amplification-lateral-flow dipstick (LAMP-LFD) to detect Toxoplasma gondii oocyst in ready-to-eat salad. Food Microbiol. 2018;70:137–42.
89. Njiru ZK. Rapid and sensitive detection of human African trypanosomiasis by loop-mediated isothermal amplification combined with a lateral-flow dipstick. Diagn Microbiol Infect Dis. 2011;69:205–9.
90. Liu Z, Zhang Q, Yang NN, Xu MG, Xu JF, Jing ML, Wu WX, Lu YD, Shi F, Chen CF. Rapid and sensitive detection of salmonella in chickens using loop-mediated isothermal amplification combined with a lateral flow dipstick. J Microbiol Biotechnol. 2019;29:454–64.
91. Diribe O, North S, Sawyer J, Roberts L, Fitzpatrick N, La Ragione R. Design and application of a loop-mediated isothermal amplification assay for the rapid detection of Staphylococcus pseudintermedius. J Vet Diagn Invest. 2014;26:42–8.

92. Enomoto Y, Yoshikawa T, Ihira M, Akimoto S, Miyake F, Usui C, Suga S, Suzuki K, Kawana T, Nishiyama Y, Asano Y. Rapid diagnosis of herpes simplex virus infection by a loop-mediated isothermal amplification method. J Clin Microbiol. 2005;43:951–5.
93. Ihira M, Yoshikawa T, Enomoto Y, Akimoto S, Ohashi M, Suga S, Nishimura N, Ozaki T, Nishiyama Y, Notomi T, Ohta Y, Asano Y. Rapid diagnosis of human herpesvirus 6 infection by a novel DNA amplification method, loop-mediated isothermal amplification. J Clin Microbiol. 2004;42:140–5.
94. Yoshikawa T, Ihira M, Akimoto S, Usui C, Miyake F, Suga S, Enomoto Y, Suzuki R, Nishiyama Y, Asano Y. Detection of human herpesvirus 7 DNA by loop-mediated isothermal amplification. J Clin Microbiol. 2004;42:1348–52.
95. Ihira M, Akimoto S, Miyake F, Fujita A, Sugata K, Suga S, Ohashi M, Nishimura N, Ozaki T, Asano Y, Yoshikawa T. Direct detection of human herpesvirus 6 DNA in serum by the loop-mediated isothermal amplification method. J Clin Virol. 2007;39:22–6.
96. Wakabayashi T, Yamashita R, Kakita T, Kakita M, Oshika T. Rapid and sensitive diagnosis of adenoviral keratoconjunctivitis by loop-mediated isothermal amplification (LAMP) method. Curr Eye Res. 2004;28:445–50.
97. Ziros PG, Kokkinos PA, Allard A, Vantarakis A. Development and evaluation of a loop-mediated isothermal amplification assay for the rapid detection of porcine cytomegalovirus under field conditions. Food Environ Virol. 2015;7:276–85.
98. Liang C, Chu Y, Cheng S, Wu H, Kajiyama T, Kambara H, Zhou G. Multiplex loop-mediated isothermal amplification detection by sequence-based barcodes coupled with nicking endonuclease-mediated pyrosequencing. Anal Chem. 2012;84:3758–63.
99. Mori Y, Hirano T, Notomi T. Sequence specific visual detection of lamp reactions by addition of cationic polymers. BMC Biotechnol. 2006;6:3.
100. Song J, Liu C, Mauk MG, Peng J, Schoenfeld T, Bau HH. A multifunctional reactor with dry-stored reagents for enzymatic amplification of nucleic acids. Anal Chem. 2018;90: 1209–16.
101. Batra K, Kumar A, Kumar V, Nanda T, Maan NS, Maan S. Development and evaluation of loop-mediated isothermal amplification assay for rapid detection of capripoxvirus. Vet World. 2015;8:1286–92.
102. Fang X, Liu Y, Kong J, Jiang X. Loop-mediated isothermal amplification integrated on microfluidic chips for point-of-care quantitative. Anal Chem. 2010;82:3002–6.
103. Ushio M, Yui I, Yoshida N, Fujino M, Yonekawa T, Ota Y, Notomi T, Nakayama T. Detection of respiratory syncytial virus genome by subgroups-a, b specific reverse transcription loop-mediated isothermal amplification (RT-LAMP). J Med Virol. 2005;77:121–7.
104. Shirato K, Nishimura H, Saijo M, Okamoto M, Noda M, Tashiro M, Taguchi F. Diagnosis of human respiratory syncytial virus infection using reverse transcription loop-mediated isothermal amplification. J Virol Methods. 2007;139:78–84.
105. Okafuji T, Yoshida N, Fujino M, Motegi Y, Ihara T, Ota Y, Notomi T, Nakayama T. Rapid diagnostic method for detection of mumps virus genome by loop-mediated isothermal amplification. J Clin Microbiol. 2005;43:1625–31.
106. Yoshida N, Fujino M, Ota Y, Notomi T, Nakayama T. Simple differentiation method of mumps hoshino vaccine strain from wild strains by reverse transcription loop-mediated isothermal amplification (RT-LAMP). Vaccine. 2007;25:1281–6.
107. Fang X, Chen H, Yu S, Jiang X, Kong J. Predicting viruses accurately by a multiplex microfluidic loop-mediated isothermal amplification chip. Anal Chem. 2011;83:690–5.
108. Poon LL, Leung CS, Chan KH, Lee JH, Yuen KY, Guan Y, Peiris JS. Detection of human influenza A viruses by loop-mediated isothermal amplification. J Clin Microbiol. 2005;43: 427–30.
109. Ito M, Watanabe M, Nakagawa N, Ihara T, Okuno Y. Rapid detection and typing of influenza A and B by loop-mediated isothermal amplification: comparison with immunochromatography and virus isolation. J Microbiol Methods. 2006;135:272–5.

110. Parida M, Posadas G, Inoue S, Hasebe F, Morita K. Real-time reverse transcription loop-mediated isothermal amplification for rapid detection of West Nile Virus. J Clin Microbiol. 2004;42:257–63.
111. Tian B, Qiu Z, Ma J, Gomez Z, de la Torre T, Johansson C, Svedlindh P, Stromberg M. Attomolar Zika virus oligonucleotide detection based on loop-mediated isothermal amplification and ac susceptometry. Biosens Bioelectron. 2016;86:420–5.
112. Song J, Mauk MG, Hackett BA, Cherry S, Bau HH, Liu C. Instrument-free point-of-care molecular detection of Zika virus. Anal Chem. 2016;88:7289–94.
113. Song J, Pandian V, Mauk MG, Bau HH, Cherry S, Tisi LC, Liu C. Smartphone-based mobile detection platform for molecular diagnostics and spatiotemporal disease mapping. Anal Chem. 2018;90:4823–31.
114. Toriniwa H, Komiya T. Rapid detection and quantification of Japanese encephalitis virus by real-time reverse transcription loop-mediated isothermal amplification. Microbiol Immunol. 2013;50:379–87.
115. Shi W, Li K, Ji Y, Jiang Q, Shi M, Mi Z. Development and evaluation of reverse transcription-loop-mediated isothermal amplification assay for rapid detection of enterovirus 71. BMC Infect Dis. 2011;11:197.
116. Fukuda S, Takao S, Kuwayama M, Shimazu Y, Miyazaki K. Rapid detection of norovirus from fecal specimens by real-time reverse transcription-loop-mediated isothermal amplification assay. J Clin Microbiol. 2006;44:1376–81.
117. Neeraja M, Lakshmi V, Lavanya V, Priyanka EN, Parida MM, Dash PK, Sharma S, Rao PV, Reddy G. Rapid detection and differentiation of dengue virus serotypes by NS1 specific reverse transcription loop-mediated isothermal amplification (RT-LAMP) assay in patients presenting to a tertiary care hospital in Hyderabad. J Virol Methods. 2015;211:22–31.
118. Kumar S, Sharma S, Bhardwaj N, Pande V, Savargaonkar D, Anvikar AR. Advanced lyophilised loop mediated isothermal amplification (L-LAMP) based point of care technique for the detection of dengue virus. J Virol Methods. 2021;293:114168.
119. Kurosaki Y, Takada A, Ebihara H, Grolla A, Kamo N, Feldmann H, Kawaoka Y, Yasuda J. Rapid and simple detection of Ebola virus by reverse transcription-loop-mediated isothermal amplification. J Virol Methods. 2007;141:78–83.
120. Kurosaki Y, Magassouba N, Oloniniyi OK, Cherif MS, Sakabe S, Takada A, Hirayama K, Yasuda J. Development and evaluation of reverse transcription-loop-mediated isothermal amplification (RT-LAMP) assay coupled with a portable device for rapid diagnosis of Ebola virus disease in Guinea. PLoS Negl Trop Dis. 2016;10:e0004472.
121. Du Y, Pothukuchy A, Gollihar JD, Nourani A, Li B, Ellington AD. Coupling sensitive nucleic acid amplification with commercial pregnancy test strips. Angew Chem Ent Ed. 2017;56:992–6.
122. Boldbaatar B, Inoue S, Sugiura N, Noguchi A, Yamada A. Rapid detection of rabies virus by reverse transcription loop-mediated isothermal amplification. Jpn J Infect Dis. 2009;62:187–91.
123. Blomstrom AL, Hakhverdyan M, Reid SM, Dukes JP, King DP, Belak S, Berg M. A one-step reverse transcriptase loop-mediated isothermal amplification assay for simple and rapid detection of swine vesicular disease virus. J Virol Methods. 2008;147:188–93.
124. Tian B, Ma J, Gómez Z, de la Torre T, Bálint Á, Donolato M, Hansen MF, Svedlindh P, Strömberg M. Rapid Newcastle disease virus detection based on loop-mediated isothermal amplification and optomagnetic readout. ACS Sensors. 2016;1:1228–34.
125. Fukuta S, Ohishi K, Yoshida K, Mizukami Y, Ishida A, Kanbe M. Development of immunocapture reverse transcription loop-mediated isothermal amplification for the detection of tomato spotted wilt virus from chrysanthemum. J Virol Methods. 2004;121:49–55.
126. Hong TC, Mai QL, Cuong DV, Parida M, Minekawa H, Notomi T, Hasebe F, Morita K. Development and evaluation of a novel loop-mediated isothermal amplification method for rapid detection of severe acute respiratory syndrome coronavirus. J Clin Microbiol. 2004;42:1956–61.

127. Poon LLM, Leung CSW, Tashiro M, Chan KH, Wong BWY, Yuen KY, Guan Y, Peiris JSM. Rapid detection of the severe acute respiratory syndrome (SARS) coronavirus by a loop-mediated isothermal amplification assay. Clin Chem. 2004;50:1050–2.
128. Broughton JP, Deng X, Yu G, Fasching CL, Servellita V, Singh J, Miao X, Streithorst JA, Granados A, Sotomayor-Gonzalez A, Zorn K, Gopez A, Hsu E, Gu W, Miller S, Pan CY, Guevara H, Wadford DA, Chen JS, Chiu CY. CRISPR-Cas12-based detection of SARS-CoV-2. Nat Biotechnol. 2020;38:870–4.
129. Xiong H, Ye X, Li Y, Wang L, Zhang J, Fang X, Kong J. Rapid differential diagnosis of seven human respiratory coronaviruses based on centrifugal microfluidic nucleic acid assay. Anal Chem. 2020;92:14297–302.
130. Maruyama F, Kenzaka T, Yamaguchi N, Tani K, Nasu M. Detection of bacteria carrying the stx2 gene by in situ loop-mediated isothermal amplification. Appl Environ Microbiol. 2003;69:5023–8.
131. Dong J, Xu Q, Li CC, Zhang CY. Single-color multiplexing by the integration of high-resolution melting pattern recognition with loop-mediated isothermal amplification. Chem Commun. 2019;55:2457–60.
132. Zhao X, Li Y, Park M, Wang J, Zhang Y, He X, Forghani F, Wang L, Yu G, Oh DH. Loop-mediated isothermal amplification assay targeting the fema gene for rapid detection of Staphylococcus aureus from clinical and food samples. J Microbiol Biotechnol. 2013;23: 246–50.
133. Iwamoto T, Sonobe T, Hayashi K. Loop-mediated isothermal amplification for direct detection of Mycobacterium tuberculosis complex, m. Avium, and m. Intracellular in sputum samples. J Clin Microbiol. 2003;41:2616–22.
134. Seki M, Yamashita Y, Torigoe H, Tsuda H, Sato S, Maeno M. Loop-mediated isothermal amplification method targeting the lyta gene for detection of Streptococcus pneumoniae. J Clin Microbiol. 2005;43:1581–6.
135. Kamachi K, Moriuchi T, Hiramatsu Y, Otsuka N, Shibayama K. Evaluation of a commercial loop-mediated isothermal amplification assay for diagnosis of Bordetella pertussis infection. J Microbiol Methods. 2017;133:20–2.
136. Brotons P, Paz H, Esteva C, Latorre I, Muoz-Almagro C. Validation of a loop-mediated isothermal amplification assay for rapid diagnosis of pertussis infection in nasopharyngeal samples. Expert Rev Mol Diagn. 2016;16:125–30.
137. Fujino M, Suzuki E, Watanabe M, Nakayama T. Loop-mediated isothermal amplification (LAMP) aids the clinical diagnosis of pertussis. Jpn J Infect Dis. 2015;68:532–3.
138. Nakano R, Nakano A, Ishii Y, Ubagai T, Kikuchi-Ueda T, Kikuchi H, Tansho-Nagakawa S, Kamoshida G, Mu X, Ono Y. Rapid detection of the Klebsiella pneumoniae carbapenemase (KPC) gene by loop-mediated isothermal amplification (LAMP). J Infect Chemother. 2015;21: 202–6.
139. Ye L, Li Y, Zhao J, Zhang Z, Meng H. Development of a real-time loop-mediated isothermal amplification assay for the sensitive and rapid detection of Listeria monocytogenes. Lett Appl Microbiol. 2015;61:85–90.
140. Siddique MP, Jang WJ, Lee JM, Ahn SH, Suraiya S, Kim CH, Kong IS. Groel is a suitable genetic marker for detecting Vibrio parahaemolyticus by loop-mediated isothermal amplification assay. Lett Appl Microbiol. 2017;65:106–13.
141. Zhao X, Wang L, Chu J, Li Y, Li Y, Xu Z, Li L, Shirtliff ME, He X, Liu Y, Wang J, Yang L. Rapid detection of Vibrio parahaemolyticus strains and virulent factors by loop-mediated isothermal amplification assays. Food Sci Biotechnol. 2010;19:1191–7.
142. Mashooq M, Kumar D, Niranjan AK, Agarwal RK, Rathore R. Development and evaluation of probe based real time loop mediated isothermal amplification for Salmonella: a new tool for DNA quantification. J Microbiol Methods. 2016;126:24–9.
143. Shao Y, Zhu S, Jin C, Chen F. Development of multiplex loop-mediated isothermal amplification-rflp (mLAMP-RFLP) to detect *Salmonella* spp. and *Shigella* spp. In milk. Int J Food Microbiol. 2011;148:75–9.

144. Song T, Toma C, Nakasone N, Iwanaga M. Sensitive and rapid detection of Shigella and enteroinvasive Escherichia coli by a loop-mediated isothermal amplification. FEMS Microbiol Lett. 2005;243:259–63.
145. Yari F, Abiri R, Aryan E, Jouybari TA, Alvandi A. Loop-mediated isothermal amplification as a fast noninvasive method of helicobacter pylori diagnosis. J Clin Lab Anal. 2016;30:464–70.
146. Bakhtiari S, Alvandi A, Pajavand H, Navabi J, Najafi F, Abiri R. Development and diagnostic evaluation of loop-mediated isothermal amplification using a new gene target for rapid detection of Helicobacter pylori. Jundishapur J Microbiol. 2016;9:e28831.
147. Kim HJ, Kim HS, Lee JM, Sang SY, Yong D. Rapid detection of Pseudomonas aeruginosa and Acinetobacter baumannii harboring blavim-2, blaimp-1 and blaoxa-23 genes by using loop-mediated isothermal amplification methods. Ann Lab Med. 2016;36:15–22.
148. Endo S, Komori T, Ricci G, Sano A, Yokoyama K, Ohori A, Kamei K, Franco M, Miyaji M, Nishimura K. Detection of gp43 of Paracoccidioides brasiliensis by the loop-mediated isothermal amplification (LAMP) method. FEMS Microbiol Lett. 2004;234:93–7.
149. Nakayama T, Yamazaki T, Yo A, Tone K, Mahdi Alshahni M, Fujisaki R, Makimura K. Detection of fungi from an indoor environment using loop-mediated isothermal amplification (LAMP) method. Biocontrol Sci. 2017;22:97–104.
150. Luo J, Taniwaki MH, Iamanaka BT, Vogel RF, Niessen L. Application of loop-mediated isothermal amplification assays for direct identification of pure cultures of Aspergillus flavus, A. nomius, and A. Caelatus and for their rapid detection in shelled Brazil nuts. Int J Food Microbiol. 2014;172:5–12.
151. Wassermann M, Mackenstedt U, Romig T. A loop-mediated isothermal amplification (LAMP) method for the identification of species within the Echinococcus granulosus. Vet Parasitol. 2014;200:97–103.
152. Besuschio SA, Llano Murcia M, Benatar AF, Monnerat S, Cruz I, Picado A, Curto MLA, Kubota Y, Wehrendt DP, Pavia P, Mori Y, Puerta C, Ndung'u JM, Schijman AG. Analytical sensitivity and specificity of a loop-mediated isothermal amplification (LAMP) kit prototype for detection of Trypanosoma cruzi DNA in human blood samples. PLoS Negl Trop Dis. 2017;11:e0005779.
153. Nzelu CO, Gomez EA, Caceres AG, Sakurai T, Martini-Robles L, Uezato H, Mimori T, Katakura K, Hashiguchi Y, Kato H. Development of a loop-mediated isothermal amplification method for rapid mass-screening of sand flies for leishmania infection. Acta Trop. 2014;132:1–6.
154. Watts MR, James G, Sultana Y, Ginn AN, Outhred AC, Kong F, Verweij JJ, Iredell JR, Chen SCA, Lee R. A loop-mediated isothermal amplification (LAMP) assay for Strongyloides stercoralis in stool that uses a visual detection method with SYTO-82 fluorescent dye. Am J Trop Med Hyg. 2014;90:306–11.
155. Choopara I, Arunrut N, Kiatpathomchai W, Dean D, Somboonna N. Rapid and visual chlamydia trachomatis detection using loop-mediated isothermal amplification and hydroxynaphthol blue. Lett Appl Microbiol. 2017;64:51–6.
156. Pian H, Yang M, Sun X, Zheng Z. Sandwich hybridization-based loop-mediated isothermal amplification (SHB-LAMP) for high-throughput detection of malaria RNA from asymptomatic infections. Sensors Actuators B Chem. 2022;365:131973.
157. Aonuma H, Badolo A, Okado K, Kanuka H. Detection of mutation by allele-specific loop-mediated isothermal amplification (AS-LAMP). In: Kolpashchikov DM, Gerasimova YV, editors. Nucleic acid detection: methods and protocols. Totowa: Humana Press; 2013. p. 121–7.
158. Ren Y, Li Y, Duan X, Wang H, Wang H, Li Z. One-step quantitative single nucleotide polymorphism (SNP) diagnosis by modified loop-mediated isothermal amplification (mLAMP). ChemistrySelect. 2019;4:1423–7.
159. Wang HH, Wang H, Liu CH, Duan XR, Li ZP. Ultrasensitive detection of telomerase activity in a single cell using stem-loop primer-mediated exponential amplification (SPEA) with near zero nonspecific signal. Chem Sci. 2016;7:4945–50.

160. Kaygusuz D, Vural S, Aytekin AÖ, Lucas SJ, Elitas M. DaimonDNA: a portable, low-cost loop-mediated isothermal amplification platform for naked-eye detection of genetically modified organisms in resource-limited settings. Biosens Bioelectron. 2019;141:111409.
161. Li R, Chen J, Zhang X, Cui J, Tao S, Yang L. Mini-disk capillary array coupling with LAMP for visual detection of multiple nucleic acids using genetically modified organism analysis as an example. J Agric Food Chem. 2019;8(3):899–906.
162. Wang C, Li R, Quan S, Shen P, Zhang D, Shi J, et al. GMO detection in food and feed through screening by visual loop-mediated isothermal amplification assays. Anal Bioanal Chem. 2015;407(16):4829–34.
163. Wang X, Fu Z, Chen X, Peng C, Xu X, Wei W, et al. Use of a novel metal indicator to judge loop-mediated isothermal amplification for detecting the 35S promoter. Anal Bioanal Chem. 2016;409(4):881–9.
164. Xing Y, Liang J, Dong F, Wu J, Shi J, Xu J, et al. Rapid visual LAMP method for detection of genetically modified organisms. ACS Omega. 2023;8(32):29608–14.
165. Zhang M, Liu Y, Chen L, Quan S, Jiang S, Zhang D, et al. One simple DNA extraction device and its combination with modified visual loop-mediated isothermal amplification for rapid on-field detection of genetically modified organisms. Anal Chem. 2012;85(1):75–82.
166. Pataer P, Gao K, Zhang P, Li Z. Ultrasensitive and visual detection of genetically modified crops using two primers-induced cascade exponential amplification assay. Talanta. 2024;268: 125282.
167. Pataer P, Gao K, Zhang P, Wang X, Li Z. A simple, portable and multiplex LAMP-based CRISPR/Cas12a assay for visually screening genetically modified crops. Sensors Actuators B Chem. 2024;403:135124.

Chapter 4
Rolling Circle Amplification (RCA)

Fengxia Su and Gaoting Wang

4.1 Introduction

In nature, plasmid DNA and phage virus DNA often exist in the form of circular DNA, and extrachromosomal circular DNA (eccDNA) is also widely present in eukaryotic cells [1]. Unlike the linear DNA replication mechanism, circular DNA replication follows the rolling circle replication mechanism. For example, phage φX174 produces single-stranded progeny DNA through a simplified form of rolling ring replication mechanism. λ phage, on the other hand, uses the replaced leader strand in rolling circle replication as a template to synthesize the delayed strand, thereby forming double-stranded DNA (dsDNA). Based on this mechanism, researchers developed a novel in vitro nucleic acid amplification technique, rolling circle amplification (RCA) [2, 3]. In the process of rolling circle amplification, the primer is continuously extended along the circular DNA template under the catalysis of polymerase to form long single-stranded DNA (ssDNA)-containing tandem repeats. After linear or exponential RCA, the target molecule can be amplified 10^5 times [4] or 10^9 times [2], so the detection of the target molecule can be achieved with high sensitivity using RCA. Compared with the traditional polymerase chain reaction (PCR), RCA can be performed under constant temperature conditions without the need to rely on thermal cycles, thus eliminating the need for precisely controlled thermal cycler and heat-stable DNA polymerase. In addition, RCA can be performed in solutions, solid substrates, or complex biological environments. Since its invention in the mid-1990s, RCA has been widely studied for its unique advantages in detecting a variety of biomolecules, including DNA [5], RNA [6], DNA methylation [7], single-nucleotide polymorphism (SNP) [8, 9], protein [10], and cell [11, 12].

F. Su (✉) · G. Wang
University of Science and Technology Beijing, Beijing, China
e-mail: sufengxia@ustb.edu.cn

Z. Li (ed.), *Nucleic Acid Detection*, https://doi.org/10.1007/978-981-95-7295-3_4

4.2 Fundamentals of RCA

A typical RCA reaction requires four components: (1) a circular DNA template (the template length usually at 50 ~ 200 nucleotides); (2) a short DNA or RNA primer; (3) DNA polymerase with strand displacement activity and suitable reaction buffer; (4) deoxynucleotide triphosphates (dNTPs). Among the many DNA polymerases, Phi29 DNA polymerase is widely used because of its excellent continuous synthesis properties and strand displacement ability. The RCA is carried out under constant temperature conditions. After annealing and hybridizing the DNA or RNA primer with circular DNA, the primer performs extension and strand replacement along the circular DNA under the catalysis of DNA polymerase, resulting in long ssDNA products that complement the DNA template sequence (Fig. 4.1). Because of the flexibility of the design of circular DNA template and primer sequence, the length, sequence, composition, structure, and stiffness of the RCA product can be precisely regulated, making RCA a highly versatile amplification technique [13].

4.3 Classification of RCA

4.3.1 Single-Primer RCA

Single-primer RCA requires only one primer that complements the circular DNA and then extends using the circular DNA as the template under the catalysis of DNA polymerase (usually Phi29 DNA polymerase). DNA polymerase with strand replacement activity can catalyze the primer to replace the synthesized DNA strand along the circular template and continue to extend with the circular DNA as the template, thus producing a large number of repetitive linear long ssDNAs that are complementary to the template sequence. Figure 4.2 shows the principle of this technology. Single-primer RCA is a linear RCA (LRCA), and the length of long DNA product can often reach nearly a thousand times the length of circular DNA template.

Rolling circle transcription (RCT) is an improved method based on LRCA for the synthesis of RNA products. The basic principle is shown in Fig. 4.3. The complementary sequence of T7 promoter is embedded in the circular DNA sequence, and then the T7 promoter hybridizes with the circular DNA and acts as a primer. In the

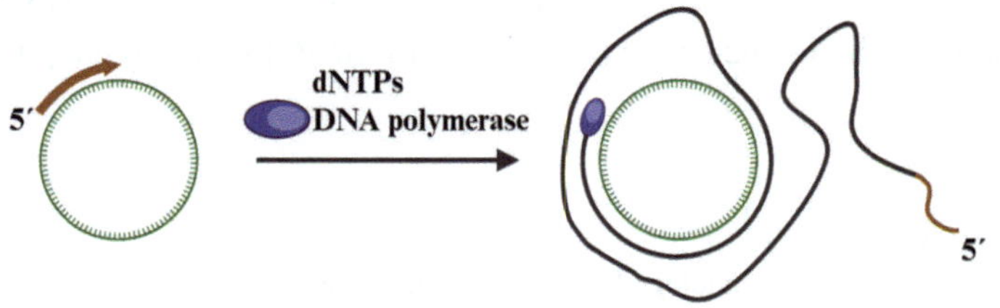

Fig. 4.1 Schematic diagram of the basic principle of RCA

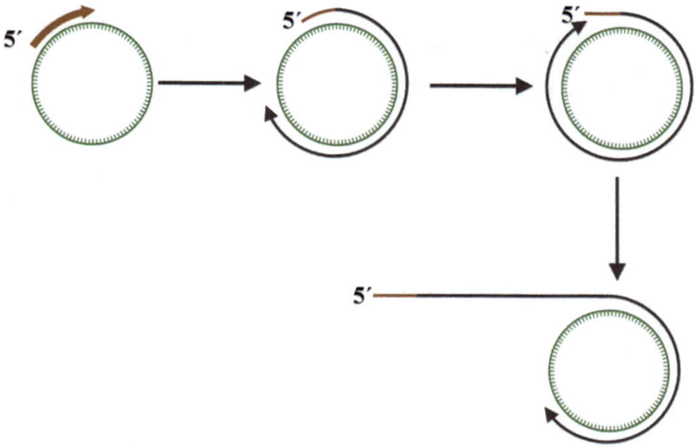

Fig. 4.2 Schematic diagram of LRCA

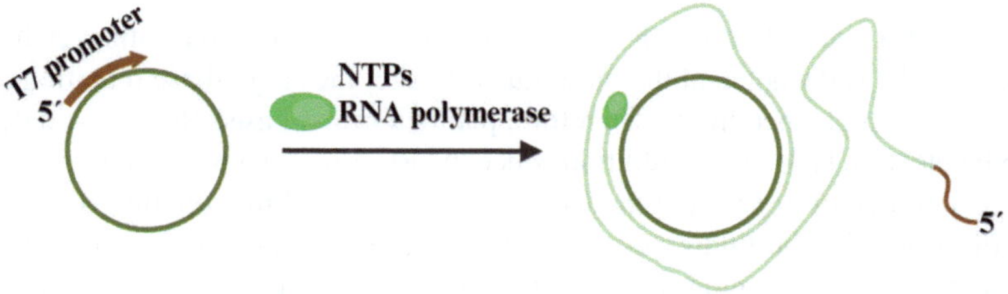

Fig. 4.3 Schematic diagram of RCT

presence of nucleotide triphosphates (NTPs), RCT is performed with the catalysis of T7 RNA polymerase to produce a long-stranded RNA product that complements the template sequence with tandem repeats [14]. RCT technology has been successfully applied to prepare polymeric transcription products, microRNA (miRNA), and small interfering RNA in vitro [15–17].

4.3.2 Padlock Probe-Based RCA

Padlock probe refers to an oligonucleotide fragment whose 5′ end and 3′ end sequence are complementary to the adjacent sequences of the target nucleic acid, respectively, and the 5′ end is phosphorylated. In the presence of the target, the padlock probe pairs complementary with the target through the sequence of its two ends to form a ring structure with a nick, and under the catalysis of DNA ligase, the phosphodiester bond is formed through the dehydration condensation reaction of the 5′-end phosphate group and the 3′-end hydroxyl group to form a single-stranded circular DNA. This circular DNA serves as a template for RCA [18] (Fig. 4.4).

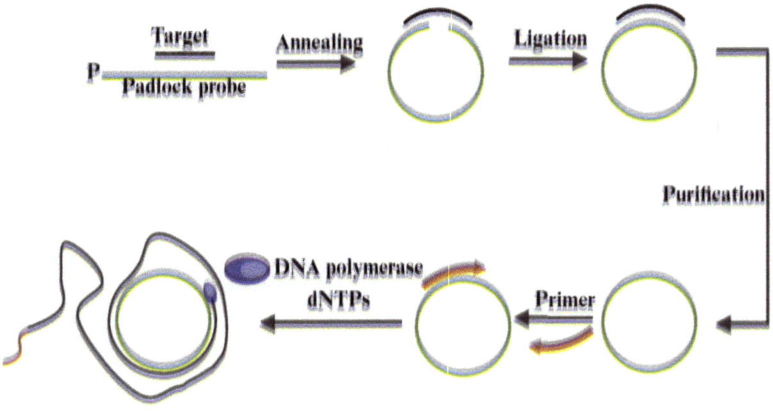

Fig. 4.4 Schematic diagram of padlock probe-based RCA. (Revised and reprinted with permission from Ref. [13])

Because DNA ligase has a high recognition ability for single-base mismatch, if the padlock probe is designed at the base mutation site, the single-base mutation can be detected with high specificity. Therefore, padlock probe-based RCA is widely used in SNP detection [19–21]. In addition, since the RCA product is a ssDNA containing a large number of repeated sequences, it can form submicron fluorophore after hybridization with the fluorescently labeled DNA probe and can obtain high sensitivity detection at the single-molecule level through fluorescence imaging [22].

4.3.3 Hyperbranched RCA

Due to the low amplification efficiency of LRCA with single primer, the detection sensitivity in homogeneous solution is generally insufficient. In order to obtain higher sensitivity, in 1998, Lizardi developed hyperbranched RCA (HRCA) [2], which significantly improved the amplification efficiency of RCA, as shown in Fig. 4.5. In the process of HRCA, two primers P1 and P2 are added to the pre-prepared circular DNA system. Under the strand displacement activity of DNA polymerase, P1 hybridizes with circular DNA and uses it as a template for LRCA. Then, P2 hybridizes with a specific sequence of P1′s LRCA product and extends with that product as a template. Since the tandem repetitive long DNA product generated by P1 contains multiple complementary sites with P2, P2 can hybridize with it and carry out elongation reaction. When upstream P2 is extended to overlap downstream P2, strand displacement reaction occurs, resulting in the displacement of downstream P2 extension products. The resulting single-strand tandem repeat is identical to the original circular DNA sequence and contains multiple P1 binding sites. Then, P1 hybridizes with the replaced DNA and extends with it as a template. Through the continuous strand displacement extension of P1 and P2, the

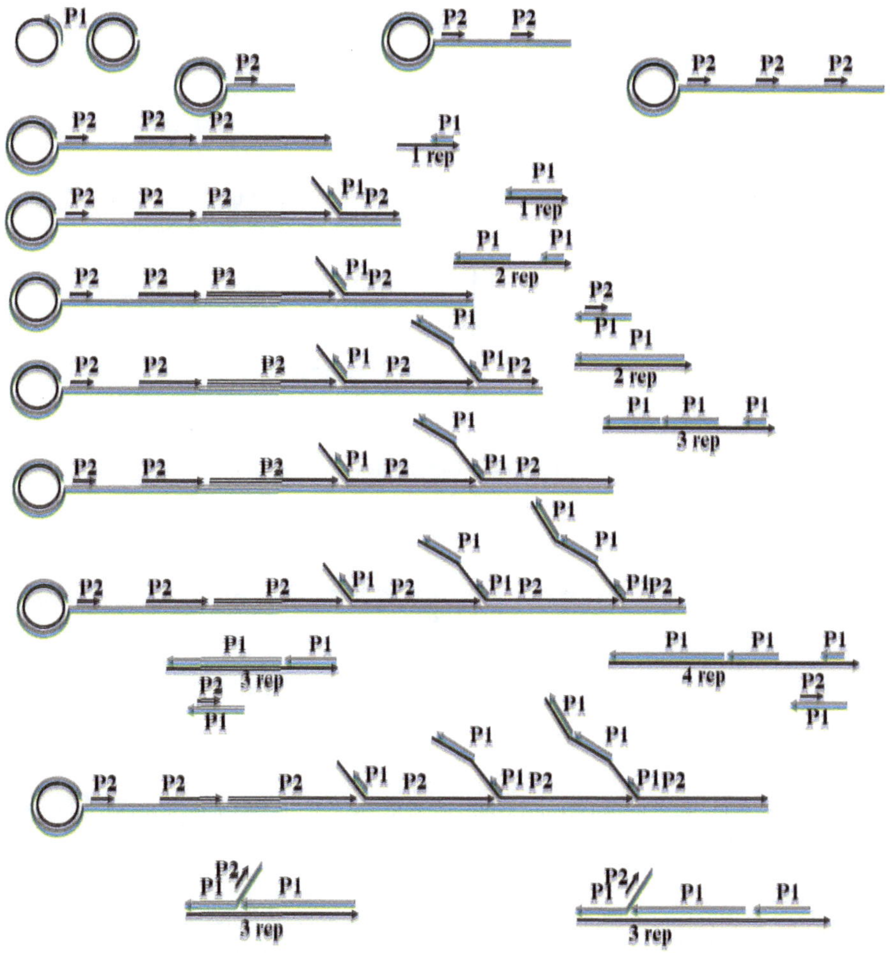

Fig. 4.5 Schematic diagram of HRCA. (Revised and reprinted with permission from Ref. [2])

hyperbranched products are generated, and the exponential amplification is achieved. In view of the exponential amplification pattern shown by HRCA in a short period of time, it can amplify circular DNA 10^9-fold within 90 min, which is suitable for the detection of small amounts of target nucleic acid molecules.

In addition, Cao developed a simple and highly sensitive padlock-based HRCA for analysis of the methylation of specific CpG site [23]. The principle is illustrated in Fig. 4.6. After treated with $NaHSO_3$, unmethylated cytosine (C) is converted to uracil (U) while the methylated cytosine (mC) remains unchanged, resulting in sequence difference between the methylated and unmethylated DNA. The 5′ and 3′ terminal sequences of the padlock probe are designed to complement the methylated DNA after treatment. Under the action of DNA ligase, the padlock probe uses the treated methylated DNA as a template to form circular DNA. Subsequently, the

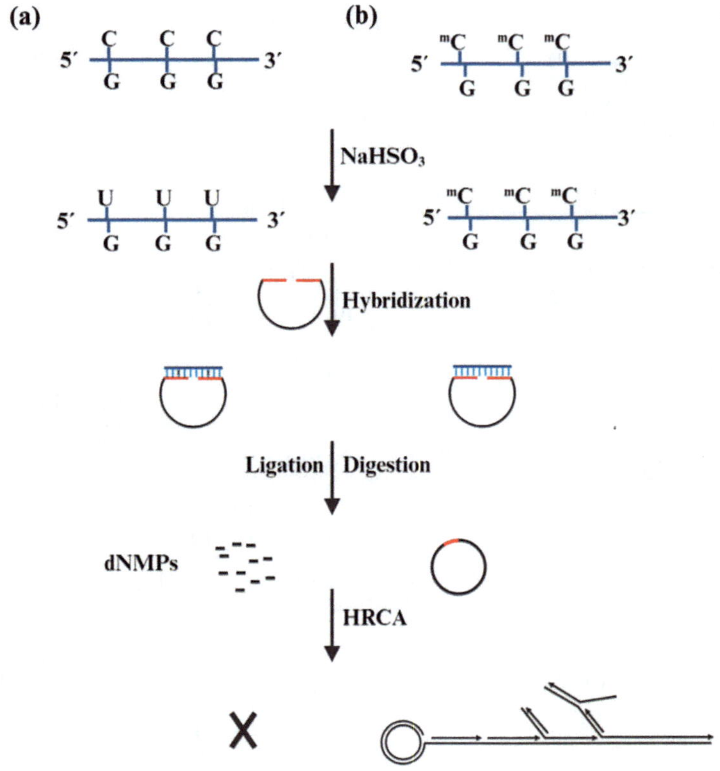

Fig. 4.6 Schematic diagram of padlock probe-based HRCA for DNA methylation detection. (Revised and reprinted with permission from Ref. [23])

uncyclized padlock probes are digested with Exonuclease I and Exonuclease III to purify circular DNA, effectively reducing nonspecific amplification. Under the action of DNA polymerase, dNTPs and two primers, the circular DNA undergoes HRCA and achieves exponential amplification, thus achieving high sensitive detection of methylated DNA. Because the C of unmethylated DNA is converted into U, there is a single mismatched base at the ligation site after hybridization with the padlock probe, and the padlock probe cannot be ligated into a circular DNA and cannot trigger subsequent RCA, so the unmethylated DNA will not be detected. The method is capable of sensitively detecting methylated DNA as low as 0.8 fM and is able to discern methylation levels as low as 0.01%.

4.3.4 Circle-to-Circle Amplification

In RCA, in addition to adding two or more primers to the system to improve the amplification efficiency, it can also improve the amplification efficiency by

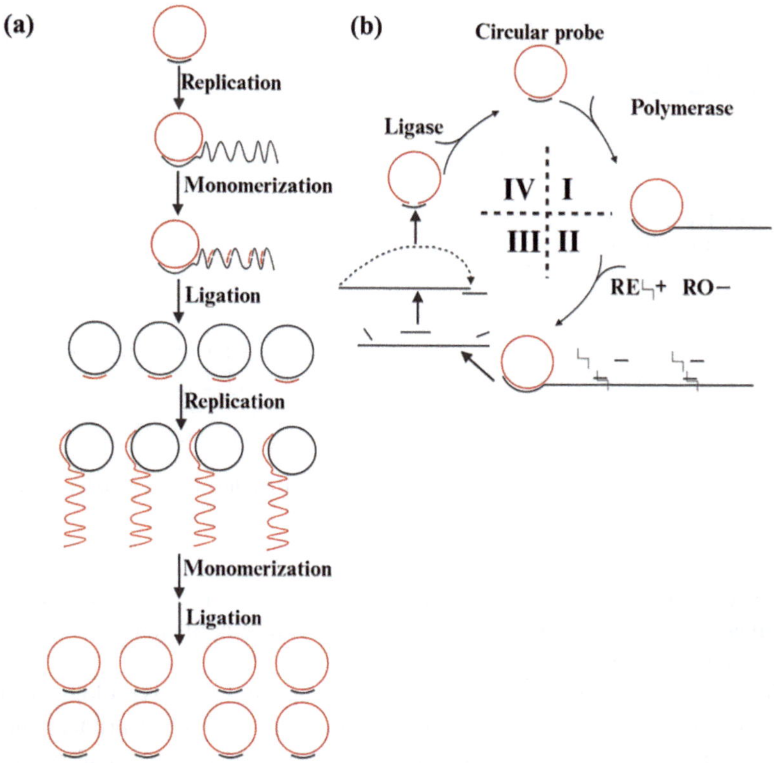

Fig. 4.7 Schematic diagram of C2CA. (Revised and reprinted with permission from Ref. [24])

generating multiple circular DNA templates in the RCA process, so as to achieve exponential amplification mode. In 2004, Dahl developed a circle-to-circle amplification (C2CA) to improve amplification efficiency of RCA [24]. As shown in Fig. 4.7, the padlock probe is firstly hybridized with the target DNA and ligated to form a circular DNA. Subsequently, LRCA is performed using the circular DNA as a template and the target DNA as a primer, resulting in long strand amplification products containing repeat sequences (Fig. 4.7b, I). Next, the DNA polymerase is inactivated by heat and probe RO is added. The probe RO is complementary hybridized with each repeat unit in the long product to generate double-stranded sites specifically recognized by restriction enzyme (RE). Subsequently, the corresponding RE is added, and after a short incubation, the long strand product is cleaved into the monomer fragment product (Fig. 4.7b, II). After RE is deactivated by heating, digested RO fragments are separated from fragmented RCA products. As the temperature decreases, the excess intact RO probes in the solution hybridize with the two ends of the fragmented products (Fig. 4.7b, III). DNA ligase is then added to ligate the fragmented RCA products into circular DNAs (Fig. 4.7b, IV) for a new round of LRCA. Subsequently, the probe RO is hybridized with a new round of RCA products to form the recognition site of RE, and the digestion and ligation steps

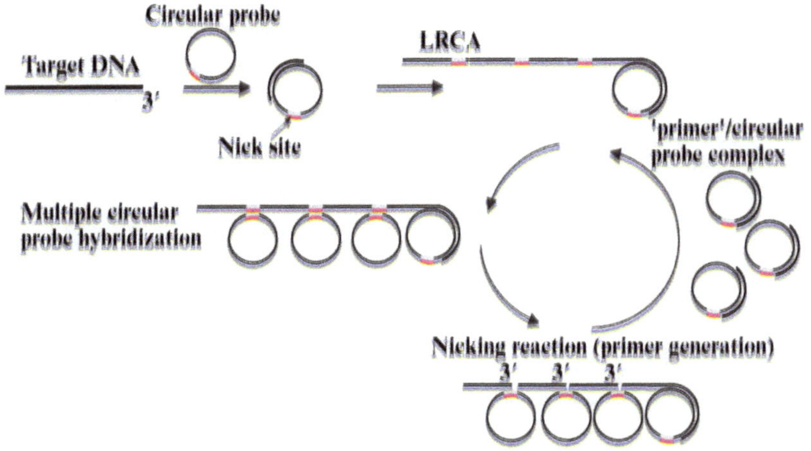

Fig. 4.8 Schematic illustration of PG-RCA. (Revised and reprinted with permission from Ref. [25])

are carried out again to form a cyclic process (Fig. 4.7a) to achieve exponential amplification. The technology can achieve 10^9-fold amplification in 1 h, significantly improving amplification efficiency.

Since C2CA involves multiple steps, the experimental cycle is prolonged, and the operation process is complicated. In 2009, Murakami developed a primer generation RCA (PG-RCA) [25], which significantly simplified the experimental process. By carefully designing circular DNA and introducing nicking endonuclease, the conventional LRCA can be successfully transformed into exponential amplification mode at constant temperature. As shown in Fig. 4.8, the target DNA and the pre-prepared circular probe are annealed and hybridized, and LRCA is performed to produce a long DNA with tandem repeat sequence that is complementary to the circular probe. Subsequently, multiple circular probes are hybridized with multiple repeat sequences of the LRCA product to form double-stranded recognition sites of nicking endonuclease. Unlike restriction endonuclease, nicking endonuclease cuts only one strand of the dsDNA without cutting the complementary strand. After the double-stranded recognition sites are recognized and cut by the nicking endonuclease, multiple primer and circular probe complex are produced. Each complex can initiate a new round of PG-RCA, thus forming an exponential amplification. The amplification method is highly efficient and is able to detect genomic DNA as low as 0.163 pg (about 60 molecules) of *L. monocytogenes*.

PG-RCA is simple to operate and has high sensitivity. Based on this method, Liu developed a padlock probe-based exponential RCA method for the detection of miRNA with high specificity and sensitivity [26]. As shown in Fig. 4.9, the 5′ end and 3′ end sequences of the padlock probe are complementary to the target miRNA sequence. After hybridization between miRNA and padlock probe, T4 RNA ligase 2 catalyzes the padlock probe to form circular DNA. The circular DNA consists of three complementary miRNA sequences (black) and three nicking endonuclease

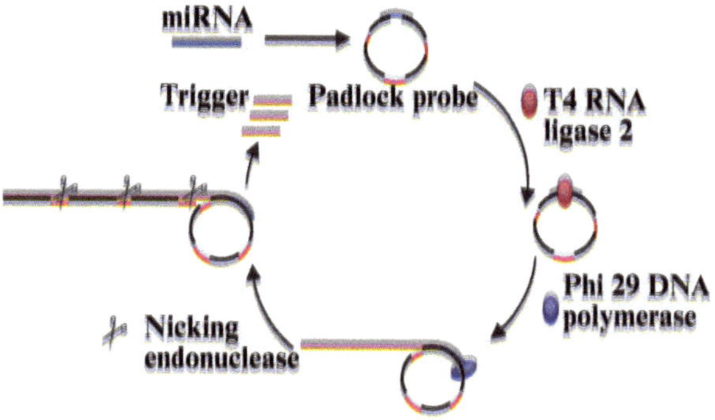

Fig. 4.9 Schematic diagram of exponential RCA based on padlock probe and nicking endonuclease for miRNA detection. (Revised and reprinted with permission from Ref. [26])

recognition sequences (red). Subsequently, LRCA is performed using miRNA as primer, and the resulting long products are hybridized with multiple padlock probes to form multiple double-stranded recognition sites for nicking endonuclease. The nicking endonuclease recognizes and cuts these sites, producing multiple "trigger" that are identical to miRNA sequences. These "triggers" can continue to trigger the padlock probe ligation, LRCA, and cutting reaction, forming a continuous cycles, so as to transform the traditional LRCA into exponential amplification, achieving highly sensitive detection of miRNA. The detection limit of this method is 0.24 zmol, and the target miRNA can be detected in the total RNA extracted from cells.

4.3.5 Netlike RCA

Netlike RCA (NRCA) is developed on the basis of LRCA and HRCA, and a more efficient nucleic acid amplification is formed by introducing nicking endonuclease. As shown in Fig. 4.10, the designed circular DNA contains the nicking endonuclease recognition site, and primers 1 and 2 can trigger LRCA and HRCA. Under the catalysis of DNA polymerase, primer 1 uses circular DNA as the template for LRCA, and then primer 2 uses the extension product of primer 1 as the template for HRCA. After the dsDNA forms through extension reaction, the nicking endonuclease recognition sites are generated, and nicking endonuclease recognizes these sites and cuts to produce nicks. Subsequently, the ssDNA produced by cleavage is displaced under the action of the strand displacement DNA polymerase and acts as a primer for the circular DNA, continuing to trigger the cyclic reaction of LRCA and HRCA. By carefully designing circular DNA containing nicking endonuclease sites, the amplification process achieved further amplification on the basis of the original LRCA and HRCA. Through atomic force microscopy (AFM) characterization, the amplified product showed a relatively unitary network structure, based on which the researchers named it NRCA [27]. Compared with LRCA and HRCA amplification, NRCA further enhances the signal amplification capability without weakening the

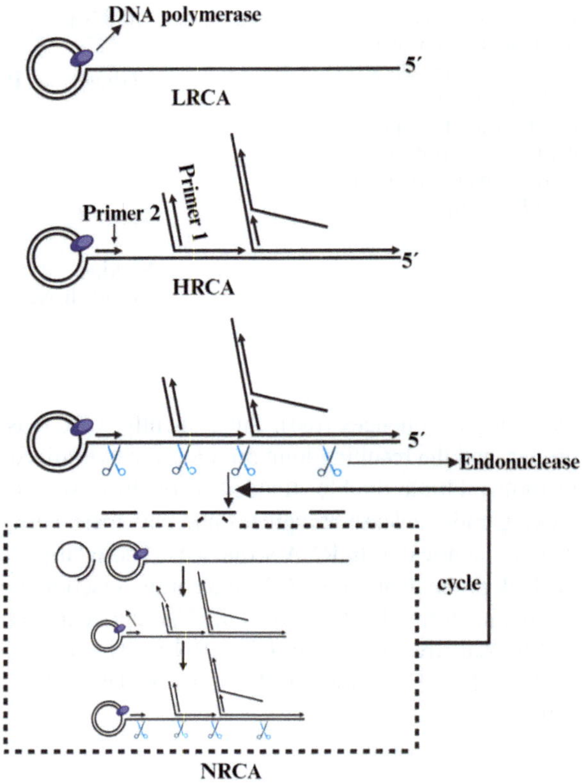

Fig. 4.10 Schematic diagram of NRCA. (Revised and reprinted with permission from Ref. [27])

original advantages such as ease of operation, cost effectiveness, and amplification time. This progress provides a more superior technical support for the analysis and detection of low-abundance nucleic acid samples.

4.4 Detection of RCA Product

4.4.1 Gel Electrophoresis

Gel electrophoresis is a common method for the analysis of RCA products. The ssDNA produced by LRCA is continuously distributed in sequence length, and its electrophoresis image appears as a broad dispersion band (Fig. 4.11a). dsDNA produced by HRCA varies in length, and the electrophoresis image presents a stepped distribution (Fig. 4.11b). The situation is similar with multi-primer RCA. Due to the long length of the RCA product, it is easy to block the sample hole during electrophoresis, so it usually appears that a clear band is observed at the sample hole location. In order to further verify the characteristics of RCA products, specific

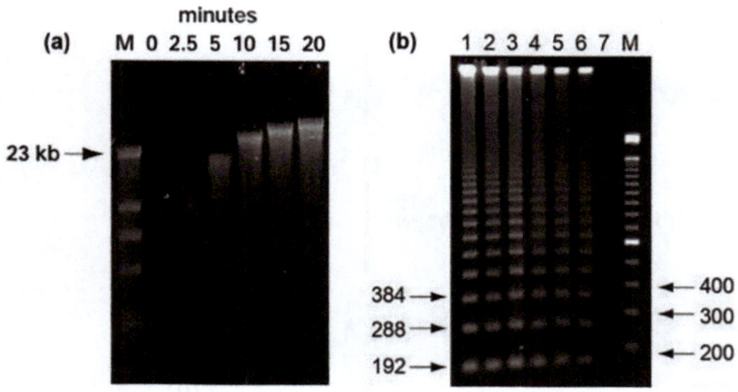

Fig. 4.11 Gel electrophoresis images of (**a**) LRCA product and (**b**) HRCA product

Fig. 4.12 AFM image of RCA product. (Revised and reprinted with permission from Ref. [28])

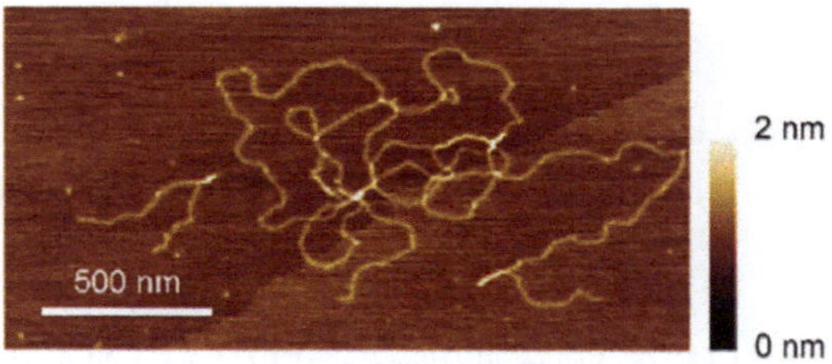

restriction endonuclease can be used to cut the amplified products, and gel electrophoresis can be used to analyze whether the cut products are single bands, so as to achieve semi-quantitative analysis of the amplified products.

4.4.2 Atomic Force Microscopy

Gel electrophoresis can analyze the quantity of RCA product on the macroscopic level, but cannot observe its morphology on the microscopic level. While the shape of RCA product can be directly observed by AFM, and a single RCA product molecule can be detected. As shown in Fig. 4.12, a typical AFM image shows that the product of LRCA is a ssDNA molecule ranging in length from hundreds of nanometers to micrometers [28].

In addition, three-dimensional nanostructures prepared by RCA can also be observed using AFM. Zhao designed RCA reaction on the surface of gold nanoparticles and assembled them into three-dimensional nanostructures [29]. Subsequently, these three-dimensional nanostructures were detected using AFM. As shown in Fig. 4.13a, a DNA primer with the thiol group at 3′ end is fixed to the surface of a 15 nm diameter gold nanoparticles, forming a primer-Au complex. Then, the prefabricated circular DNA template is hybridized with primers on the surface of gold nanoparticles, and RCA is performed by DNA polymerase to obtain

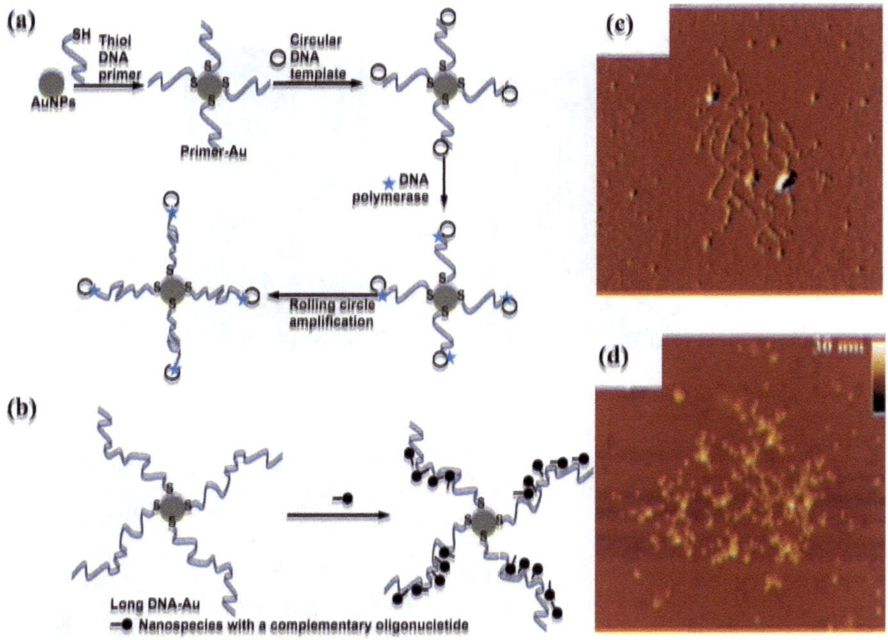

Fig. 4.13 (**a**) Schematic diagram of RCA on gold nanoparticles. (**b**) Schematic diagram of three-dimensional nanostructure assembly. (**c**) AFM image of long DNA-Au complex. (**d**) AFM image of three-dimensional nanostructures. (Revised and reprinted with permission from Ref. [29])

long DNA-Au complex. Finally, a DNA molecule complementary to the RCA product is modified on the surface of 5 nm gold nanoparticles and incubated with the DNA-Au complex overnight to form three-dimensional nanostructures through hybridization between DNA (Fig. 4.13b). The DNA-Au product (Fig. 4.13c) as well as the final assembled three-dimensional nanostructures (Fig. 4.13d) can be clearly observed under AFM.

4.4.3 Fluorescence Method

4.4.3.1 Embedded Fluorescent Dye

DNA-embedded fluorescent dyes are a class of commonly used fluorescent molecules with low intrinsic fluorescence intensity. When these dyes bind to DNA, they emit a significantly enhanced fluorescent signal. According to the type of embedded DNA, these signal molecules can be divided into fluorescent dyes that bind ssDNA and fluorescent dyes that bind dsDNA. Since LRCA products exist in single-strand form, they can be detected using single-strand-specific fluorescent dyes such as SYBR Green II. Takahashi used target messenger RNA (mRNA) as primer and

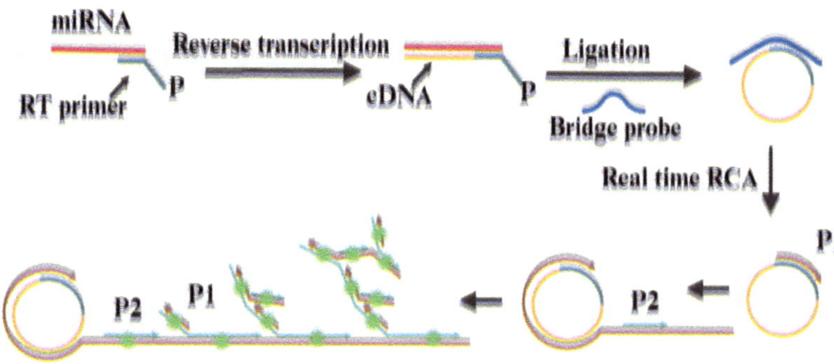

Fig. 4.14 Principle of real-time detection of HRCA products by SYBR Green I. (Revised and reprinted with permission from Ref. [31])

pre-prepared circular DNA as template for LRCA. Subsequently, direct, rapid, and real-time detection of target mRNA is achieved by embedding SYBR Green II into the resulting single-strand product [30]. In addition, double-stranded specific dyes, such as SYBR Green I, can be used to monitor double-stranded products produced during HRCA. Yao used SYBR Green I to conduct real-time monitoring of HRCA product for highly sensitive detection of miRNA [31]. As shown in Fig. 4.14, the reverse transcription primer (RT primer) modified with the phosphate group at the 5′ end is hybridized with the 3′ end of the target miRNA, and the miRNA is used as a template for reverse transcription to generate cDNA. The cDNA hybridizes with the bridge probe as a padlock probe and is ligated to a circular DNA under the action of DNA ligase. Subsequently, primer 1 (P1) and primer 2 (P2) are added for HRCA of the circular DNA, resulting in dsDNA products that could be monitored in real time by SYBR Green I. The technology can detect let-7a as low as 10^3 copies and has a dynamic detection range of up to seven orders of magnitude.

In addition, RCA products can be detected using structurally selective embedded fluorescent dyes, such as thioflavin T (ThT) and its derivatives, which can bind specifically to the G-quadruplex structure to produce a strong fluorescence signal. Based on the specific interaction of ThT with G-quadruplex, a range of biosensing strategies with low background noise and no labeling can be developed. Li used ThT to detect RCA products and established a labeling-free telomerase detection method [32], the principle of which is shown in Fig. 4.15. A telomerase-specific primer (TS) and a padlock probe are specifically designed. In the presence of dNTPs, telomerase can specifically recognize and bind to TS primer and then extends. When only dGTP and dTTP are added to the reaction system, telomerase only extends a specific "GGGTT" five-base sequence at the 3′ end of the TS primer. This sequence is complementary and paired with the sequence at both ends of the padlock probe, and under the catalysis of T4 DNA ligase, the padlock probe is ligated to form a circular structure, which can be used as a template for RCA. Since the two identical regions II of the padlock probe are designed as specific recognition sites for the nicking endonuclease Nb.BbvCI, the LRCA products are then specifically recognized and cut into shorter fragments by Nb.BbvCI. The cut pieces can

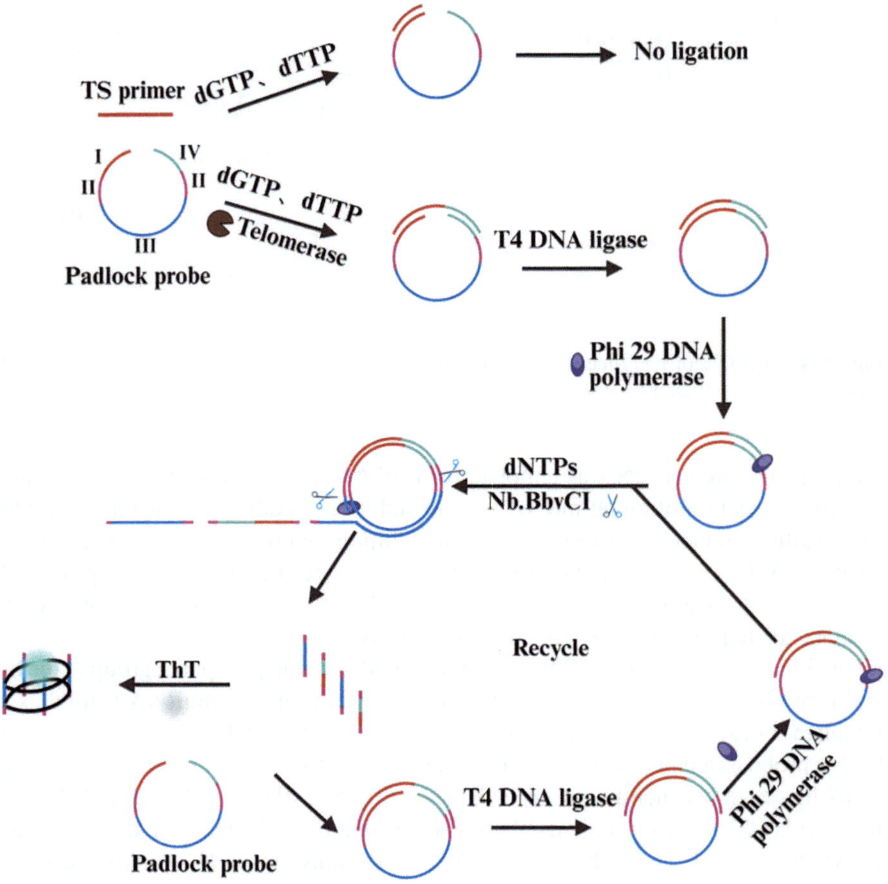

Fig. 4.15 Schematic diagram of detecting RCA products by ThT. (Revised and reprinted with permission from Ref. [32])

continue to act as templates and RCA primers for ligating the padlock probes, prompting the remaining padlock probes to be ligated and trigger new RCA, thereby building an exponential signal amplification pattern. Region III of the padlock probe is designed to be C-rich sequences, so the other cut products contain G-rich sequences that can fold into a G-quadruplex structure. When ThT specifically recognizes and binds to the G-quadruplex, it will release a strong fluorescent signal, which can realize the detection of telomerase. This method can detect telomerase activity in single HeLa cells.

4.4.3.2 Fluorescently Labeled DNA Probes

Fluorescent-labeled DNA probes are widely used in the detection of RCA products, which can realize simultaneous detection of multiple targets by labeling fluorophores of various colors. Fluorescence signals can be generated directly by DNA probes labeled with fluorophores, or by fluorescence resonance energy transfer (FRET) between different probes.

(1) Fluorescently Labeled Probes for In Situ Detection

Fluorescently labeled probes are oligonucleotides that modify fluorescent molecules at one end. When used for the detection of RCA products, the sequence is complementary to the specific sequence of RCA products. Since the LRCA product is a ssDNA containing tandem repeats and can reach hundreds of thousands of bases in length, a single RCA product can be detected by laser scanning confocal microscopy after hybridization with a large number of fluorescently labeled probes. Therefore, the organic combination of RCA technology and fluorescently labeled probes shows high applicability in the field of fluorescence in situ imaging detection and can be used to analyze DNA, RNA, and special epigenetic modifications in cells [33–35].

In 2004, Larsson combined padlock probe-based RCA with fluorescently labeled probes to construct a method for DNA in situ imaging in cells [36]. As shown in Fig. 4.16, since the padlock probe cannot directly hybridize with long dsDNA, the

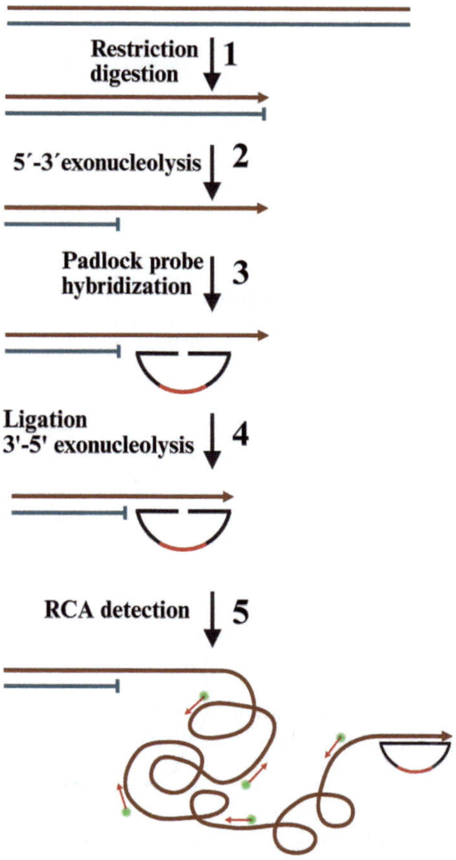

Fig. 4.16 Schematic diagram of the detection of RCA products by fluorescently labeled probes for in DNA in situ imaging. (Revised and reprinted with permission from Ref. [36])

target region of dsDNA needs to be turned into ssDNA. First, dsDNA is cut near the target sequence by restriction endonuclease (λ exonuclease), and then 5′–3′ exonuclease is used to remove the lower strand DNA from the incision, so that the 3′ end sequence of the upper strand is single strand. The fore-tail sequence of the padlock probe is hybridized with the ssDNA and ligated into a circular DNA under the action of DNA ligase. Then the exonuclease activity of Phi29 DNA polymerase hydrolyzes the 3′ end of ssDNA beyond the part of the circular DNA [37], and RCA is performed using the circular probe as a template. Finally, a fluorescently labeled DNA probe is added to hybridize the RCA products to realize the detection of the products. This method uses the specific recognition ability of padlock probe and the amplification ability of RCA to detect the single-base mutation of mitochondrial A3243G site in three kinds of cells.

In 2010, the team also designed a padlock probe and RCA-based method for in situ multiple detection of mRNA in cells [38]. As shown in Fig. 4.17, first, reverse primers modified with locked nucleic acid (LNA) are designed for target mRNA. Among them, LNA, as a nucleic acid analogue, introduces a methylene bridge

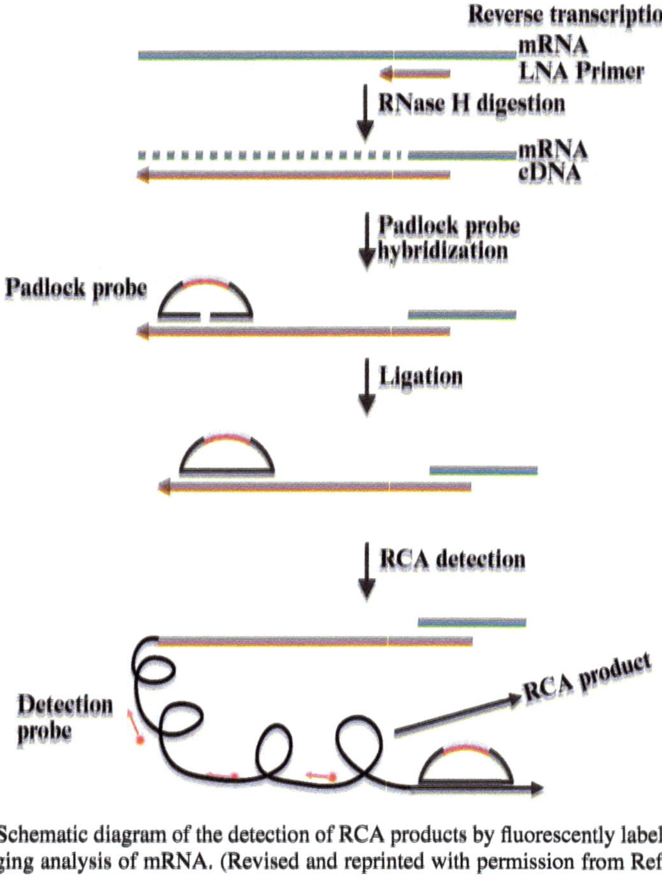

Fig. 4.17 Schematic diagram of the detection of RCA products by fluorescently labeled probes for in situ imaging analysis of mRNA. (Revised and reprinted with permission from Ref. [38])

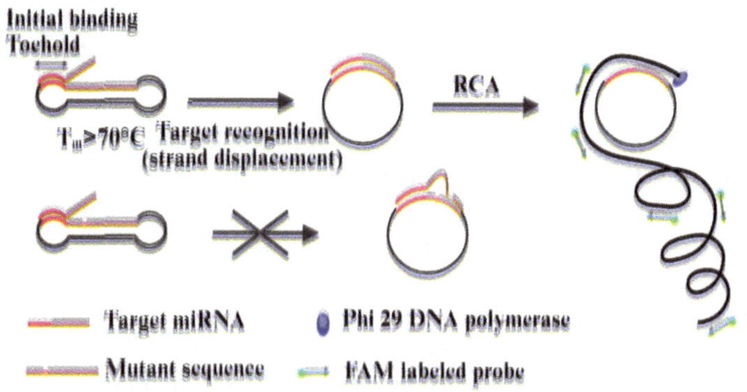

Fig. 4.18 Schematic diagram of RCA products detected by fluorescently labeled probes for miRNA in situ imaging. (Revised and reprinted with permission from Ref. [39])

between the 2′ oxygen atom and the 4′ carbon atom of its five-carbon ring, thus forming a lock-like structure. This structure effectively reduces the flexibility of the ribose structure and significantly enhances the local stability of the phosphate skeleton. Therefore, LNA can improve the T_m value of the primer or probe, thereby enhancing the stability of hybridization with the target nucleic acid. After hybridizing the target mRNA with the LNA-modified reverse transcription primers, cDNA is produced by reverse transcription. RNA in the DNA/RNA complex is degraded by ribonuclease H (RNase H) to obtain single-stranded cDNA. The padlock probe targets cDNA and forms circular DNA by ligase. The circular DNA is used as template and cDNA as primer for LRCA. The fluorescence imaging detection of target mRNA can be realized by hybridizing the fluorescently labeled probes with RCA products. This method successfully realizes single-base mutation detection of mRNA in somatic cells and realizes multiple detection of mRNA in human cells, mouse cells, and tissue samples by designing different padlock probes and fluorescently labeled probes.

In 2014, Deng constructed dumbbell DNA probe-mediated RCA to achieve high sensitivity and in situ imaging detection of miRNA in cells [39]. As shown in Fig. 4.18, the loop region of the dumbbell probe is complementary to the 5′ end of the target miRNA as a toehold. At the same time, the stem region is complementary to the 3′ end of miRNA. At room temperature, the probe can maintain its dumbbell structure stably in solution. When the target miRNA is present, the 5′ end of miRNA hybridizes with the loop region of the dumbbell probe, and subsequently, the 3′ end of miRNA spontaneously replaces the stem sequence of the probe, prompting the complete hybridization of miRNA and dumbbell probe, forming a hybrid complex of miRNA and circular DNA. Subsequently, RCA is triggered with miRNA as primer to generate a series of long-strand DNA products with tandem repeats. The fluorescently labeled probe is added to the system and hybridized with RCA products, and the miRNA could be detected by fluorescence in situ imaging. If the miRNA in the cell does not exactly match the sequence of the dumbbell probe, the structure of the dumbbell probe cannot be opened, and the RCA will not be triggered; thus, the specificity of RCA is improved. This method can not only detect miRNA in single cells, but also has the ability to detect single base mutation of miRNA.

(2) Molecular Beacon (MB)

After hybridizing the fluorescently labeled probe with the RCA product, the fluorescence probes that do not participate in the reaction should be removed through the washing step, so as to detect the fluorescence signal. Since the fluorescently labeled probe cannot distinguish the fluorescence signal intensity before and after hybridization, this method is only suitable for the detection of molecules fixed on the surface of the support. In contrast, in homogeneous solutions, hybridization of fluorescently labeled probes with RCA products requires a change in fluorescence signal to achieve effective detection of RCA products, such as by MB.

MB is an oligonucleotide probe with stem-loop structure, which is modified with fluorophore and quenching group, respectively, at both ends. The stem structure of the probe makes the fluorophore and the quenching group close to each other and then quenches the fluorescence of the fluorophore. When the target nucleic acid is present, the loop of the MB hybridizes with the target, resulting in the separation of the fluorophore from the quenching group, and then the fluorescence is restored. Li combined MB with the cyclic enzyme digestion of nicking endonuclease to detect target DNA [40]. As shown in Fig. 4.19a, the loop sequence of MB is complementary to the target DNA, and the target DNA contains the recognition sequence of nicking endonuclease. After hybridization of target DNA with MB, the recognition site of nicking endonuclease can be formed. Nicking endonuclease cuts MB, generates fluorescence signal, and releases target DNA at the same time. The released target DNA will hybridize again with the intact MB, and the hybridized complex will be cut again, and so on, creating a cycle of hybridization, cutting, and release. After several cycles, one target DNA generates multiple fluorescent signals, achieving linear signal amplification. Compared with the traditional MB method, this technique improves the sensitivity by 3 orders of magnitude. The team combined the padlock-based RCA with the cyclic cleavage reaction of MB to achieve the detection of target DNA (Fig. 4.19b). The head and tail sequence of the padlock probe is complementary to the target DNA, and there is a sequence in the middle of the padlock probe that is the same as MB. When the target DNA is present, the padlock probe is specifically ligated to a circular DNA with the catalysis of ligase. Primers and polymerase are then added for RCA to produce a large number of long-strand DNA products with tandem repeats. The long products contain multiple sequences complementary to MB, which bind to MB to form multiple nicking endonuclease sites. The fluorescence signal is released after the MB is cut by the endonuclease to form a further signal amplification, which increases the sensitivity of the traditional MB method by 5 orders of magnitude.

(3) FRET Probes

Wu designed the FRET probes for the detection of RCA products, realizing the sensitive detection of miRNA [41]. As shown in Fig. 4.20a, two FRET probes are designed, complementary to the two adjacent sequences of the RCA product. Both

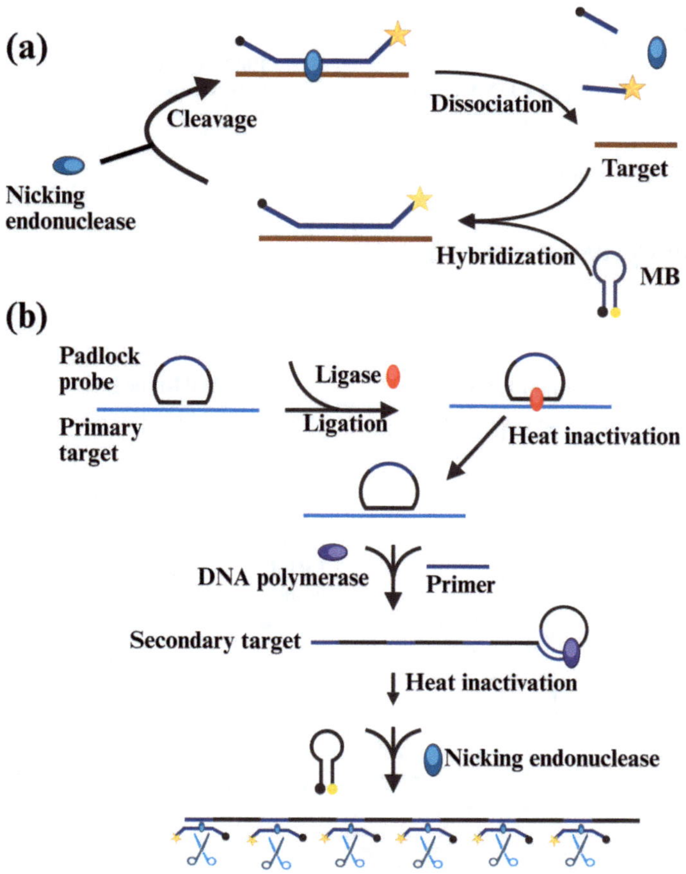

Fig. 4.19 (**a**) Schematic diagram of MB's cyclic enzyme digestion reaction to detect target DNA. (**b**) Schematic diagram of MB's cyclic enzyme digestion reaction to detect RCA products for analysis of target DNA. (Revised and reprinted with permission from Ref. [40])

ends of one probe are modified with Cy3, and both ends of the other probe are modified with Cy5. After the ligation of the padlock probe and RCA are triggered by the target miRNA, the fluorescent probes hybridize with the RCA product and are located in an adjacent position, narrowing the distance between the donor fluorophores Cy3 and the acceptor fluorophores Cy5 (only 1–2 nucleotides), meeting the conditions for FRET. By exciting Cy3 and monitoring the fluorescence signal of Cy5, the concentration of target miRNA can be calculated based on the fluorescence intensity of Cy5. If miRNA is not present, the ligation and RCA cannot be performed, resulting in a long distance between the two probes in the solution, and FRET cannot occur. In this case, even if Cy3 is excited, the fluorescence signal of Cy5 cannot be detected, thus achieving sensitive detection of the target miRNA.

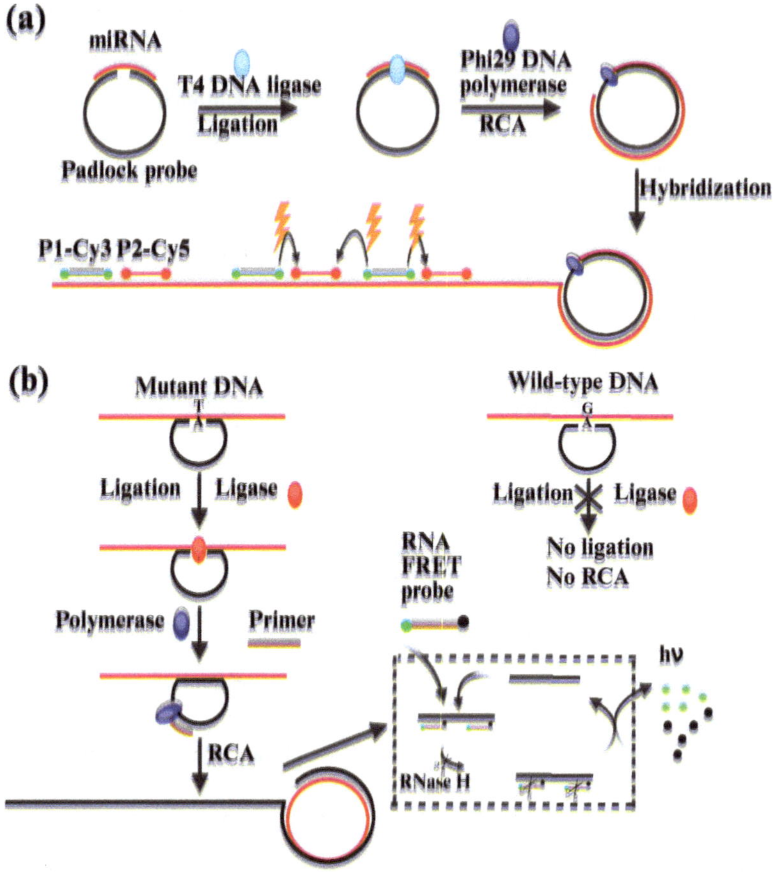

Fig. 4.20 Schematic diagram of FRET probe for detection of RCA products. (Revised and reprinted with permission from Ref. [41, 42])

Zhou developed an RNA FRET probe for the detection of RCA products and realized the analysis of SNP [42], as shown in Fig. 4.20b. First, a padlock probe that complements the mutant DNA is designed. After hybridization with mutant DNA, the probe is ligated to a circular DNA under the action of ligase, and RCA is performed. RNA FRET probe is designed for RCA products, and both ends of the probes are modified with fluorophore and quench groups, respectively. Due to the short length of the probe (14 nucleotides), the fluorescence of the fluorophore is quenched. After the RNA FRET probe forms a DNA/RNA hybrid with the RCA product, RNase H can specifically cut the RNA of the hybrid, resulting in the separation of the fluorophores from the quenching groups, producing a detectable fluorescence signal. After the probe is cut, it is separated from the RCA product. The new RNA FRET probe continues to hybridize with the amplified product and is again degraded by RNase H, cyclically producing a fluorescent signal. Wild-type

DNA cannot initiate the RCA reaction, so the RNA FRET probe in solution is not cut and cannot produce a fluorescence signal. This enables specific detection of mutant DNA, which can detect as low as 100 fM mutant DNA.

4.4.3.3 Fluorescent Labeling

(1) Fluorophores Directly Label dNTPs

The fluorophores directly label dNTPs method is to covalently bind the fluorescent dye to the structure of dNTP and make it participate in RCA as a raw material. This method can label more fluorescent molecules on RCA products, which is more efficient and less prone to fluorescence quenching than the hybridization method based on fluorescently labeled probes. Hao used Cy3-modified dUTP (Cy3-dUTP) to stain DNA hydrogels generated by RCA to achieve the detection of ochratoxin A (OTA) [43]. As shown in Fig. 4.21, a complementary primer is designed for the OTA aptamer sequence. The aptamer targeting of OTA recognizes OTA, and its affinity is higher than the hybridization force between primer and aptamer. Therefore, in the presence of OTA, the aptamer preferentially binds to the OTA, resulting in the aptamer separating from the primer. Subsequently, the free primers guide the ligation of the padlock probe under the action of ligase to form circular DNA. The primer and circular DNA perform RCA in a mixture containing Cy3-dUTP, and Cy3-dUTP is integrated into the amplification product to produce Cy3-labeled RCA products. When these products accumulate to a certain concentration, they are able

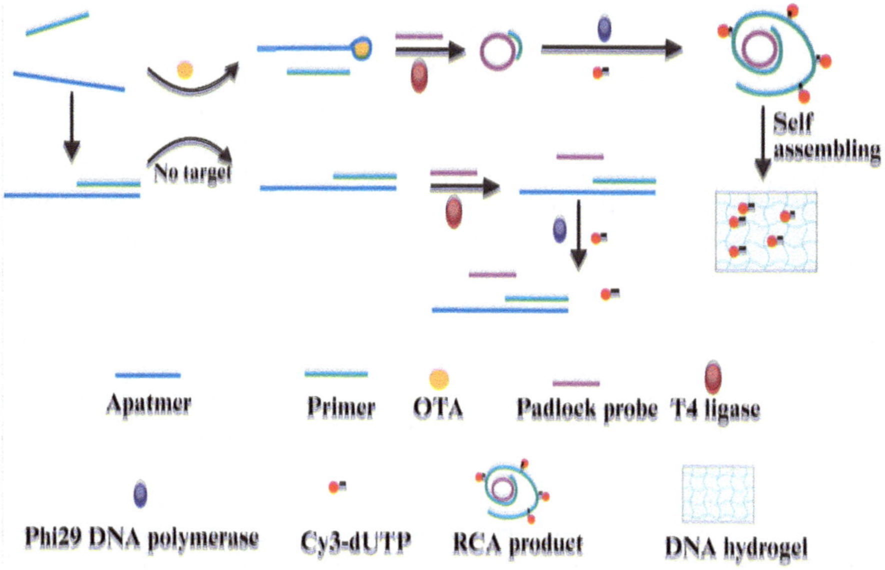

Fig. 4.21 Schematic diagram of detecting RCA products based on Cy3-dUTP. (Revised and reprinted with permission from Ref. [43])

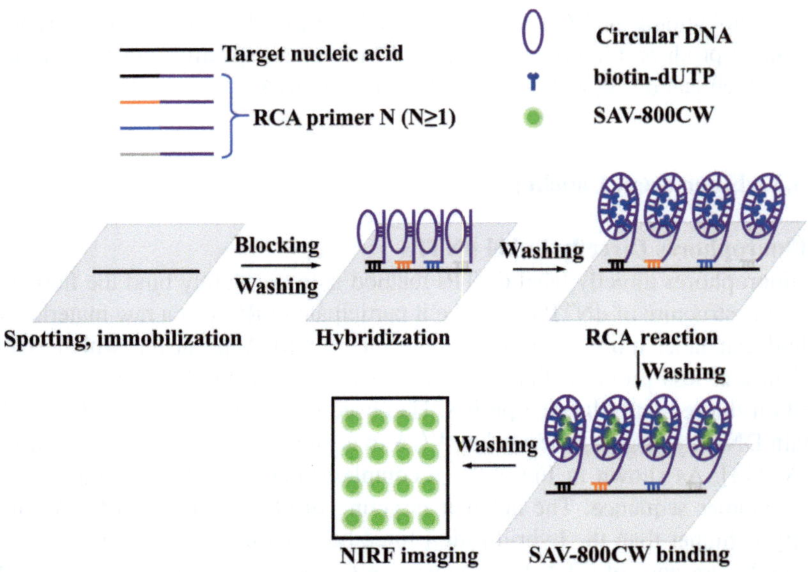

Fig. 4.22 Schematic diagram of detection of RCA products by fluorophores indirect labeling dNTPs. (Revised and reprinted with permission from Ref. [44])

to spontaneously assemble into DNA hydrogels. The quantitative detection of OTA can be achieved by measuring the fluorescence intensity of DNA hydrogels after removing unbound Cy3-dUTP by washing. If OTA is not present, the primer is in a stable hybridization state with the aptamer, unable to trigger ligation and RCA, and does not form fluorescence-labeled DNA hydrogel.

(2) Fluorophores Indirectly Label dNTPs

The fluorophores indirectly label dNTPs method refers to the specific binding between molecules, such as streptavidin (SAV) and biotin, to label fluorophores in dNTPs to achieve the detection of RCA products. As shown in Fig. 4.22, Xu used biotin-dUTP as raw material for RCA reaction and used fluorescently labeled SAV (SAV-800CW) to bind specifically to the biotin of RCA products to achieve staining and detection of RCA products [44]. In this method, the target nucleic acid is fixed on the nylon membrane by UV cross-linking technique. Specific probes containing RCA primer sequences are hybridized with pre-prepared circular DNA to form hybrid complexes. These complexes are then hybridized with target nucleic acid molecules on the nylon membrane. The unhybridized nucleic acid molecules have been removed through washing, and the hybridized complex undergoes RCA and incorporates biotin-dUTP into the RCA product. Following the removal of free dUTP by washing, SAV-800CW is introduced into the RCA product through the binding of SAV to biotin, thereby enriching the fluorescent molecules on the RCA products. After the removal of any unreacted fluorescent molecules, the RCA products can then be detected using a near-infrared fluorescence (NIRF) imager. For the detection of different DNA samples, it is necessary to design one or more primers, meaning the number of RCA primers N should be ≥1.

4.4.4 *Surface-Enhanced Raman Spectroscopy*

Among optical detection methods, surface-enhanced Raman spectroscopy (SERS) has attracted the attention of researchers due to its narrow spectrum band of multiple detection and the ability to greatly reduce photobleaching and background signals. It has been widely used in the detection of various biomolecules [45, 46]. Hu used SERS to detect RCA products for the detection of nucleic acid molecules [47]. In this method, the detection probe is complementary to the RCA product, with 5′ end modifying the Raman signaling molecule roxithromycin (Rox) and 3′ end modifying the thiol group to be fixed to the surface of the gold nanoparticles. The principle is shown in Fig. 4.23. First, the capture probe with 5′ end modified thiol group is fixed on the surface of the gold electrode, and then the hybrid with sandwich structure is formed by adding target DNA and primer to the capture probe. DNA ligase is used to catalyze the ligation between the capture probe and the primer. The primer is hybridized with the pre-prepared circular DNA, and RCA reaction is performed under the action of Phi29 DNA polymerase and dNTPs to form long ssDNA with repeated sequences on the surface of the gold electrode. The gold nanoparticles modified with detection probes are hybridized with the RCA products, and after the unhybridized gold nanoparticles are removed through the washing step, SERS detection is performed. The combination of gold nanoparticles modified with RCA products significantly enhanced SERS signal. Compared with PCR-based SERS

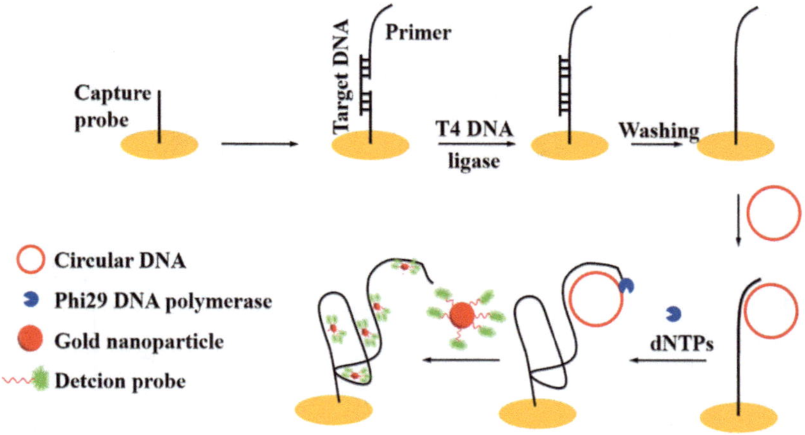

Fig. 4.23 Schematic diagram of RCA product detection based on SERS. (Revised and reprinted with permission from Ref. [47])

technology, the detection sensitivity of this method is improved by three orders of magnitude.

4.4.5 Electrochemical Method

Methylene blue is a widely used organic dye for electrochemical signal output, which can specifically bind to guanine (G) in DNA molecules [48, 49]. Based on this property, researchers have developed a variety of electrochemical methods using methylene blue to detect RCA products. Zhang combined RCA with methylene blue to develop an electrochemical detection method for SNP [50]. The principle is shown in Fig. 4.24. For mutant DNA, a padlock probe that is completely complementary to it is designed. The probe uses mutant DNA as a template to ligate to form a circular DNA, triggering subsequent RCA reaction and producing long ssDNA product (Fig. 4.24a). The capture probe sequence is complementary to the RCA product sequence, and when the mutant DNA triggers the ligation reaction of the padlock probe and the RCA reaction is completed, the capture probe is hybridized with the RCA product to enrich the long ssDNA to the gold electrode surface (Fig. 4.24b). When methylene blue is combined with long ssDNA as an electrochemical indicator, strong electrochemical signal enhancement is generated. The presence of mutant DNA can be detected by changes in the electrochemical signal. In contrast, wild-type DNA has a single-base mismatch with padlock probe, which cannot trigger the ligation and RCA reaction, and cannot produce a detectable electrochemical signal. The assay is able to detect mutant DNA as low as 40 amol,

(a)

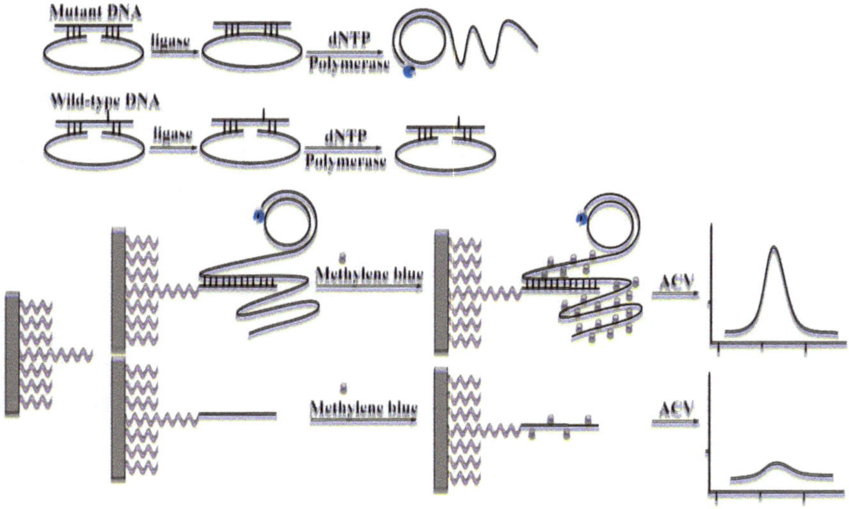

(b)

Fig. 4.24 Schematic diagram of electrochemical method for detecting RCA products. (Revised and reprinted with permission from Ref. [50])

and in the context of wild-type DNA, it is able to detect mutant DNA in a ratio of 1 in 5000.

4.4.6 Magnetization of Magnetic Beads-Based Method

When the target DNA hybridizes with single-stranded oligonucleotide on the surface of the magnetic bead, the hydrodynamic diameter of the magnetic bead increases slightly, resulting in a decrease in the Brownian relaxation frequency of the magnetic bead. The target DNA can be detected by measuring the change of Brownian relaxation frequency of magnetic beads in solution. The magnetization of magnetic beads for the detection of biomolecules was first proposed by Connolly [51]. Strömberg developed a new method based on volume-amplified magnetic nanobeads to detect RCA products [52]. As shown in Fig. 4.25, the target DNA can trigger the ligation of the padlock probe and subsequent RCA, generating ssDNA of random coils with a diameter of about 1 μm in the solution. Nano-magnetic beads (~150 nm) modified with detection probes are added, and the magnetic beads are involved in the RCA products through hybridization of the detection probes and RCA products to form micron-scale magnetic bead aggregates. A significant increase in volume affects the dynamic magnetic response of the magnetic bead in solution, resulting in a decrease in its Brownian relaxation frequency (blue curve in Fig. 4.25). In contrast, in the absence of target DNA, no RCA products can be produced, and the magnetic beads are uniformly dispersed in the solution, with their Brownian relaxation frequency remaining constant (green

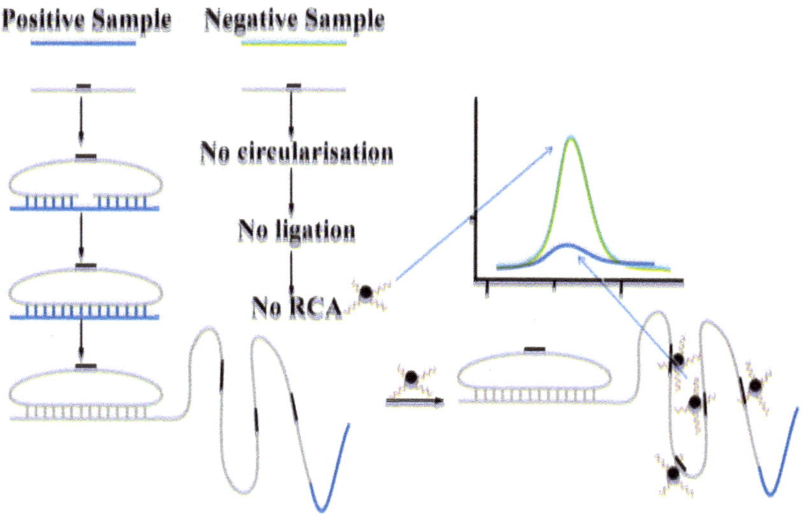

Fig. 4.25 Schematic diagram of detecting RCA products by magnetic bead magnetization. (Revised and reprinted with permission from Ref. [52])

curve). Based on this principle, the method takes the change of the Brown relaxation frequency of the magnetic bead as the signal output to realize the detection of the target DNA.

4.4.7 Colorimetric Method

RCA products can also be detected by colorimetry, such as gold nanoparticle-based colorimetry. In the stable state, the gold nanoparticles remain dispersed in the solution, the distance between the particles is much larger than their average particle size, and the solution appears red. When the gold nanoparticles aggregate and the particle spacing shrinks to less than their average particle size, the color of the solution changes to blue [53, 54]. Based on this property of gold nanoparticles, many colorimetric methods have been developed for the detection of nucleic acid [55–58]. Xu [59] applied the gold nanoparticle colorimetric method to the detection of RCA products and realized the visual analysis of target DNA. As shown in Fig. 4.26, the target DNA triggers the ligation of the padlock probe and subsequent RCA to generate long ssDNA products containing multiple ab repeats. Subsequently, the probe a′b′ is hybridized with the ab region of the RCA product to form a recognition site of nicking endonuclease, and the probe a′b′ is cut and separated from the RCA product. The uncut probe a′b′ continues to hybridize with the product and goes through the cutting process again, resulting in an enzyme cut amplification cycle. Two gold nanoparticles labeled probe a and probe b are added, where probe a and probe b are complementary to probe a′ and b′, respectively. Since the probe a′b′ is cut, the gold nanoparticles cannot be aggregated through hybridization, and the solution color remains red. If no target DNA is present or a base-mismatched probe a′b′ is introduced, the enzyme digestion reaction cannot be initiated, and the

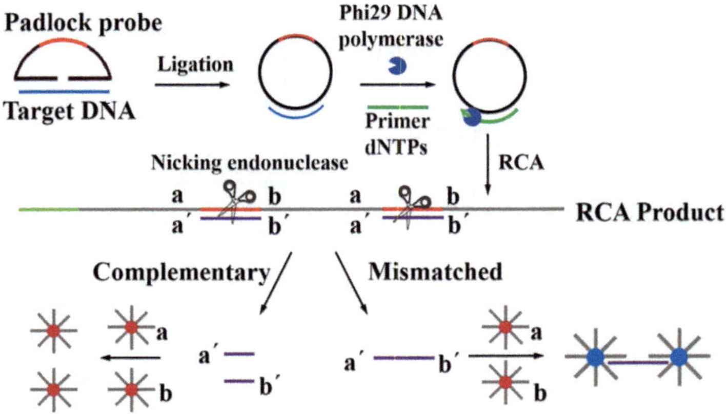

Fig. 4.26 Schematic diagram of detecting RCA products by colorimetric method of gold nanoparticles. (Revised and reprinted with permission from Ref. [59])

intact probe a′b′ hybridizes with the probes on the gold nanoparticles, inducing the gold nanoparticle to aggregate, and the solution turns blue. The target DNA is detected by observing the color change of gold nanoparticles. This method integrates RCA technology on the basis of circular enzyme digestion, and its sensitivity is one order of magnitude higher than that of the previous method used only by this research group [57].

4.5 RCA Technology for Cascade Signal Amplification

In the field of biochemical analysis, it is often difficult for single amplification technique to detect trace target molecules. Therefore, researchers often combine two or more signal amplification strategies into a single analysis process to develop cascade signal amplification techniques. With the cascade signal amplification method, the product of upstream amplification acts as a "bridge", triggering the downstream reaction, thus achieving the cumulative amplification of the signal through the series of different amplification techniques. In view of the simplicity and flexibility of RCA, many researchers choose to combine other amplification methods with RCA for the detection of target molecules, such as enzyme-assisted nucleic acid amplification [60, 61] and enzyme-free nucleic acid amplification [62, 63].

4.5.1 RCA Combining with Enzyme-Assisted Nucleic Acid Amplification

Researchers often integrate enzyme-assisted nucleic acid amplification and RCA into a sensing system to improve the sensitivity of the assay. Cheng developed a new amplification strategy by combining ligase chain reaction (LCR) and RCA [64], and realized labeling-free detection of SNP in homogeneous solution. This method is capable of detecting target DNA molecules as low as 1 fM and allele frequencies as low as 1.0%, see Sect. 2.2.7 for details. In addition, isothermal exponential amplification reaction (EXPAR), strand displacement amplification (SDA) [65, 66], loop-mediated isothermal amplification (LAMP) [67], circulating enzyme amplification [68, 69], CRISPR/Cas system [70, 71], etc. can also be combined with RCA to establish biomolecular analysis methods.

4.5.1.1 EXPAR in Combination with RCA

EXPAR is an isothermal amplification technique invented by Van Ness in 2003 [72]. Due to its simplicity and high amplification efficiency, the amplification

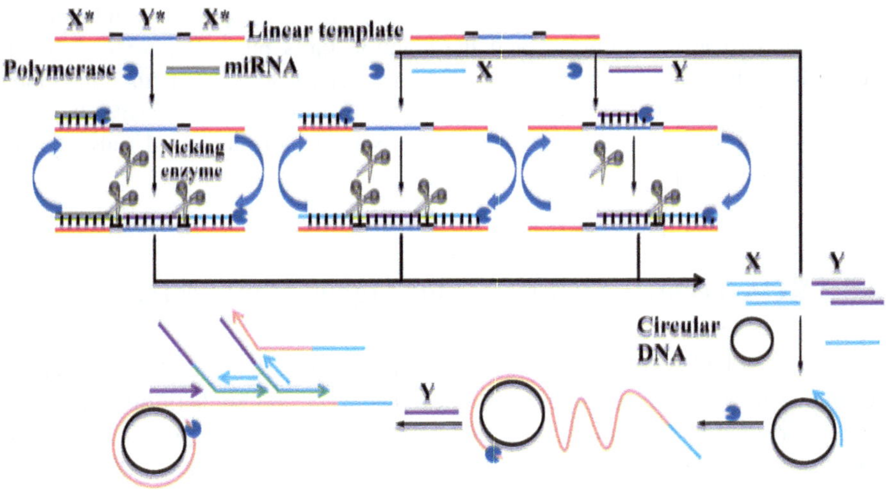

Fig. 4.27 Schematic diagram of EXPAR combined with HRCA for miRNA detection. (Revised and reprinted with permission from Ref. [65])

technique has been widely used in the field of life analysis since its invention, and is often combined with other amplification techniques for nucleic acid detection. Zhang developed a EXPAR-mediated HRCA technology to realize sensitive and label-free detection of miRNA [65]. As shown in Fig. 4.27, the linear template contains three main regions X*-Y*-X*, in which X* is complementary to the target miRNA and X, Y* is complementary to Y, and there are two recognition sites of nicking endonuclease between X* and Y*. The target miRNA hybridizes with the X* region of the linear template and acts as a primer to initiate the elongation under the action of dNTPs and DNA polymerase with strand displacement activity to produce dsDNA. Endonuclease recognizes two specific sites of dsDNA and cuts, while DNA polymerase with strand displacement activity initiates a new extension cycle using the cut DNA as a primer, producing large amounts of X and Y. The displaced X and Y are hybridized specifically with the X* and Y* regions of the linear template, initiating a new strand displacement reaction. Through the extension, cutting, and displacement cycles, exponential amplification is achieved. At the same time, the replaced X and Y act as the first and second primers, respectively, to initiate the HRCA reaction and produce a series of dsDNA of different lengths. The dsDNA is detected by SYBR Green I. In the absence of target miRNA, neither the EXPAR nor HRCA reaction can be initiated, so no fluorescence signal can be detected. This method can detect let-7a as low as 1.8 amol.

4.5.1.2 LAMP in Combination with RCA

LAMP, originally invented by Notomi [73], has been widely used in the detection of a variety of biomolecules due to its low background signal, high sensitivity, and the

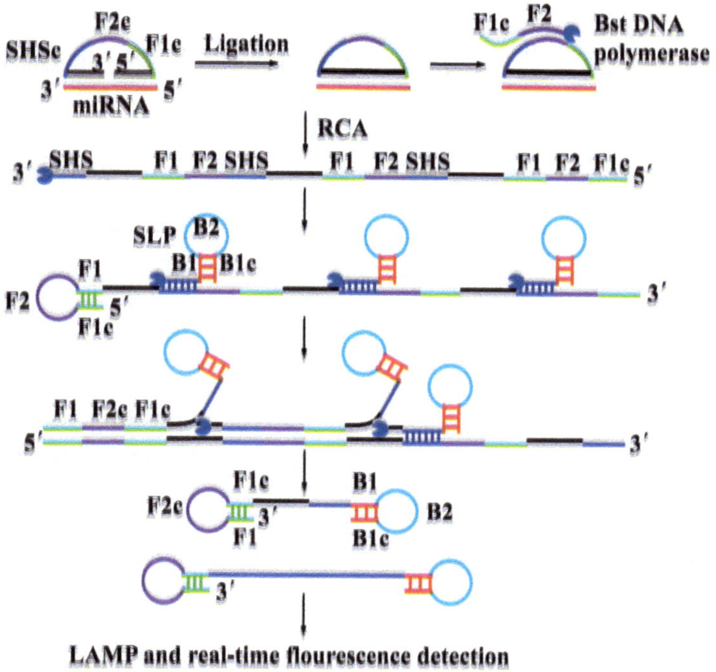

Fig. 4.28 Schematic diagram of miRNA detection by LAMP and RCA. (Revised and reprinted with permission from Ref. [67])

fact that only one DNA polymerase can achieve efficient amplification under constant temperature conditions. Aiming at the relatively low amplification efficiency of LRCA, Tian combined LRCA with LAMP to establish a fast and ultrasensitive miRNA detection method [67]. As shown in Fig. 4.28, the target miRNA, as a template, hybridizes with both ends of the padlock probe and ligates the padlock probe into circular DNA under the catalysis of ligase. The addition of primers to initiate RCA to produce long ssDNA with tandem repeats, and the 5′ end sequence forms stem-loop structure. The SHSc sequence is designed on a padlock probe, and the SHSc complementary sequence SHS is generated in the RCA product. A primer SLP with stem loop structure is designed. The ssDNA part of SLP is hybridized with SHS sequence in RCA products, and the dumbbell DNAs are generated through the process of extension and strand displacement. This structure initiates the following LAMP reaction. Finally, highly sensitive analysis of target molecules can be achieved by real-time fluorescence monitoring of the amplification process. Compared with the traditional RCA-based miRNA detection methods, this method combines the padlock probe-based RCA with LAMP reaction, significantly improving the amplification efficiency and detection sensitivity of target molecules, and the detection limit is 10 aM.

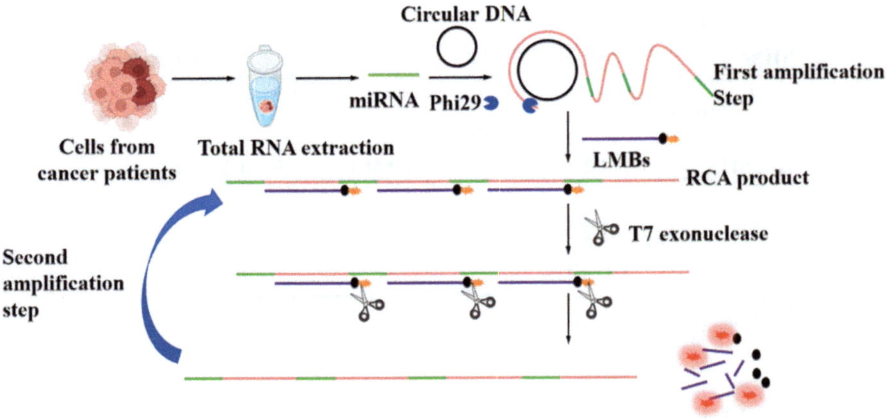

Fig. 4.29 Schematic diagram of miRNA detection by RCA-CEAM. (Revised and reprinted with permission from Ref. [79])

4.5.1.3 Cyclic Enzyme Digestion Amplification in Combination with RCA

In recent years, the cyclic enzyme digestion amplification method has been commonly used for the detection of various biomolecules, such as T7 exonuclease [74] and duplex-specific nuclease (DSN) [75]. Due to the low amplification efficiency of a single method, many researchers combine them with other amplification methods to improve the detection sensitivity of target molecules [76–78]. Cui combined the T7 exonuclease-assisted cyclic enzymatic amplification (CEAM) with RCA to develop an RCA-CEAM double amplification method for the sensitive detection of miRNA [79]. As shown in Fig. 4.29, the target miRNA hybridizes with the circular DNA and acts as a primer to trigger the LRCA reaction, generating repeated long ssDNA. Subsequently, a linear MB probe (LMB) labeled with fluorophore and quenching group at 5′ end and complementary to RCA products is designed to hybridize with RCA products. Since T7 exonuclease can degrade nucleic acids from the 5′ end depression of dsDNA, but has no effect on ssDNA, the added T7 exonuclease degrades the hybrid LMB from the 5′ end, resulting in the separation of the fluorophores from the quenching groups, and the fluorescence signal is emitted. The intact LMB probes hybridize again with the released product, triggering a new round of degradation. This cyclic hybridization and degradation mechanism can release a large number of fluorophores and achieve efficient fluorescence signal amplification. The method can detect let-7a as low as 12 fM and can be applied to analyze the expression level of let-7a in liver cancer cells and normal cells.

In addition, Kim combined magnetic particles, DSN and RCA to design an enzyme-mediated cascade reaction for the detection of miRNA with high sensitivity and specificity [69]. As shown in Fig. 4.30, the detection method consists of three main steps: (1) Using DSN, specific DNA fragments (Tc) are generated through target cycling and magnetic separation techniques; (2) Tc triggers RCA reaction; (3) Using deoxyribonuclease (DNAzyme) substrate to detect RCA products. First,

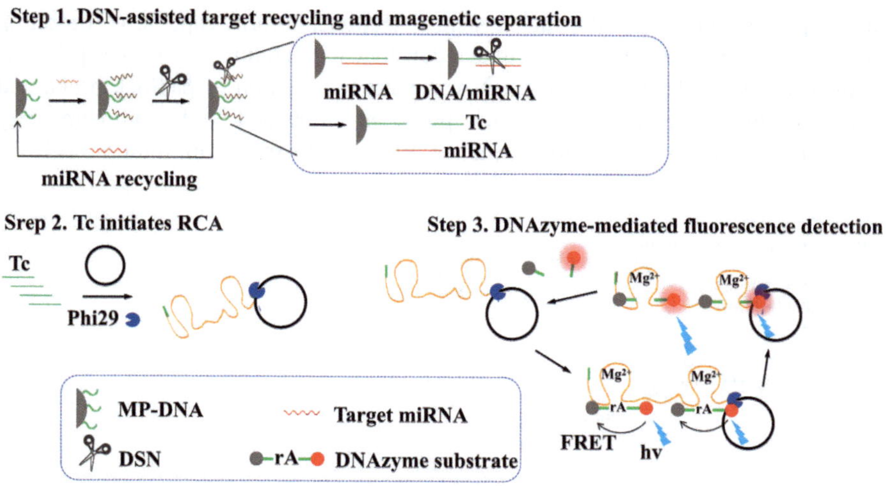

Fig. 4.30 Schematic diagram of DSN-based amplification and RCA for the detection of miRNA. (Revised and reprinted with permission from Ref. [69])

the ssDNA probe with 3′ end modified biotin is fixed on the surface of the SAV-modified magnetic particles. When the miRNA is present, the miRNA hybridizes with the ssDNA probe to form a miRNA/DNA hybrid. The DSN recognizes the hybrid and cuts the DNA in it into segments, but does not cut the miRNA, resulting in the 5′ end of the DNA probe (Tc) being released into the solution while the 3′ end fragment remains fixed to the magnetic particle. The intact miRNA hybridizes with the intact DNA probe on the magnetic particle, triggering new cutting reaction of DSN, forming a cycle process of hybridization and cutting. After the magnetic particles are removed by simple magnetic separation, the pre-prepared circular DNA is added for RCA to produce tandem repeated ssDNA products. Since circular DNA contains a Mg^{2+}-dependent DNAzyme complementary sequence, each repeating unit of the RCA product contains a Mg^{2+}-dependent DNAzyme sequence. Adding a DNAzyme substrate with FRET effect, the RCA product specifically cuts the substrate in the presence of Mg^{2+}, resulting in the dissociation of the fluorophores and the quenching groups, restoring the fluorescence signal. The detection limit of miR-21 in 20% of human serum reaches 1.8 fM, and its sensitivity is nearly two orders of magnitude higher than that of the detection method using only DSN.

4.5.1.4 CRISPR/Cas System in Combination with RCA

Clustered regularly interspaced short palindromic repeats (CRISPR) and its related proteins (CRISPR-associated, Cas) are a common adaptive immune mechanism in bacteria [80]. In recent years, CRISPR/Cas system has attracted wide attention in the field of biosensing due to its excellent performance in nucleic acid detection [81, 82]. By integrating the advantages of CRISPR/Cas system and RCA, Wang developed a highly specific nucleic acid detection platform capable of detecting miRNA in extracellular vesicles under constant temperature conditions [70]. As shown in Fig. 4.31a, the target miRNA triggers the ligation of padlock probe and RCA reaction. Because the padlock probe is embedded with complementary sequence of sgRNA recognition sequence, RCA products contain multiple sgRNA

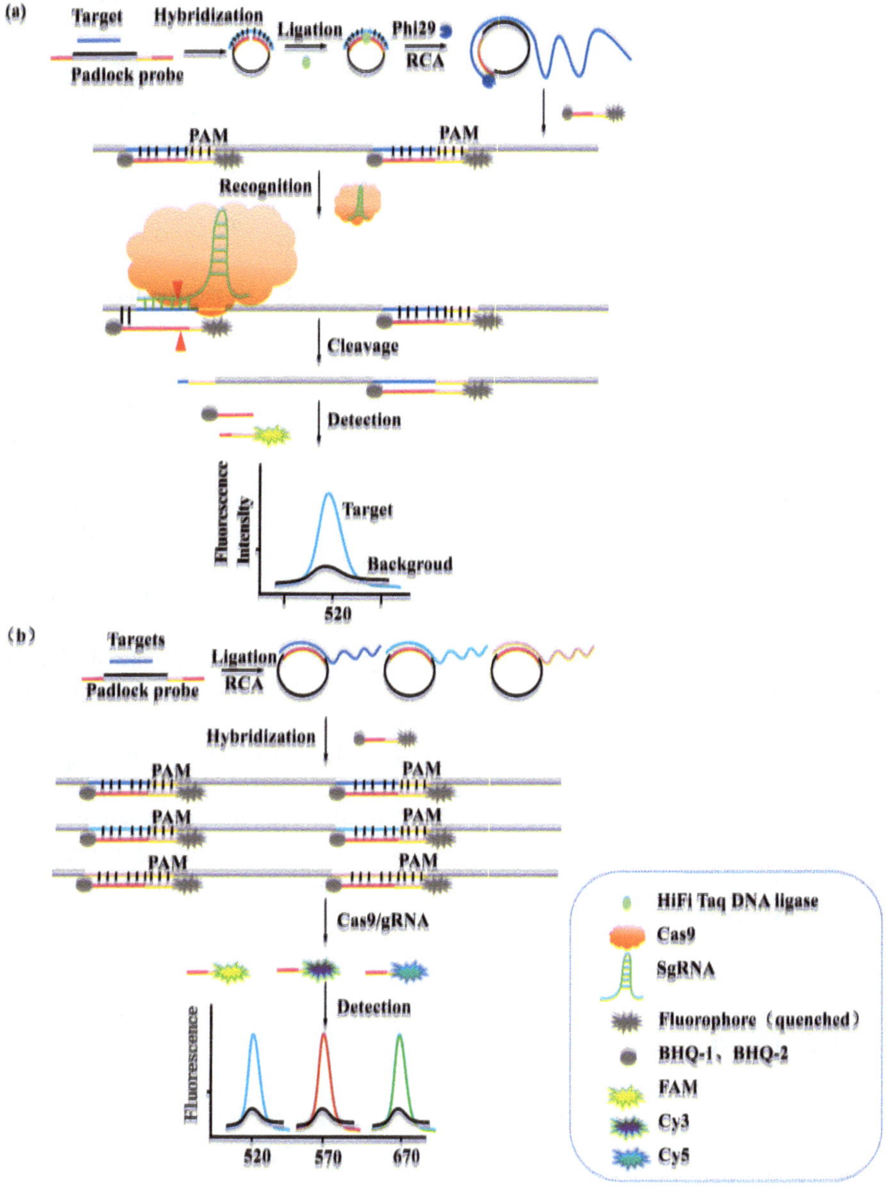

Fig. 4.31 Schematic diagram of CRISPR/Cas9 combined with RCA for multiple miRNA detection. (Revised and reprinted with permission from Ref. [70])

recognition sequences. When the TaqMan probe binds to the recognition sequence on the RCA product to form dsDNA, the Cas9/sgRNA complex recognizes and cuts the dsDNA, cutting the TaqMan probe and restoring the fluorescence signal. To achieve multiple detection of multiple miRNAs, the research team designed specific padlock probes and TaqMan probes for each target miRNA. Only when the corresponding miRNA is present, the corresponding specific TaqMan probe shear reaction will be triggered to generate different fluorescence signals to achieve multiple detection (Fig. 4.31b).

4.5.2 *RCA Combining with Enzyme-Free Nucleic Acid Amplification*

In addition to enzyme-assisted amplification technology, biosensing systems often use the combination of RCA and enzyme-free amplification technology for nucleic acid detection [83–85]. Catalytic hairpin assembly (CHA) and hybridization chain reaction (HCR) are two widely used enzyme-free amplification techniques. CHA uses two carefully designed DNA hairpin probes H1 and H2, which are able to coexist stably in solution. The probe H1 is opened only if the primer is present, and then the single-stranded part of H1 is hybridized with H2 to form a more stable H1/H2 double-stranded structure, which replaces the primer down. The displaced primer continues to trigger the self-assembly of H1 and H2, eventually producing a large number of H1/H2 double-stranded structures. Zhuang combined CHA and RCA to design a novel DNA nanomachine for sensitive detection of miRNA [83]. As shown in Fig. 4.32, the target miRNA triggers the ligation of padlock

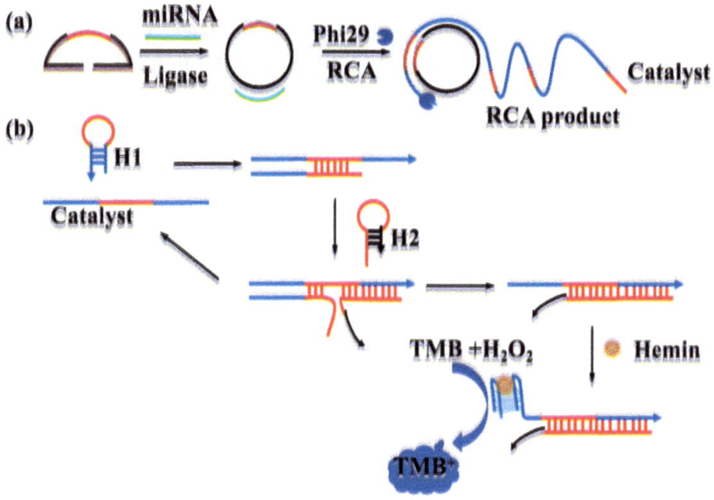

Fig. 4.32 Schematic diagram of miRNA detection by combining CHA and RCA. (Revised and reprinted with permission from Ref. [83])

probe and RCA reaction. By elaborately designing a padlock probe, the primer sequence of CHA activation is embedded into the RCA product. When H1 and H2 are present in the solution, these primers will trigger the cyclic self-assembly of H1 and H2, and then form a large number of H1/H2 double-stranded structures. Since the 5′ end of hairpin H1 has a DNAzyme sequence dependent on Hemin, after adding Hemin, the DNAzyme sequence changes to a G-quadruplex, and the catalytic activity is activated. In the presence of H_2O_2, this enzyme can catalyze 3,3′,5,5′ -tetramethylbenzidine (TMB) solution from colorless to blue to achieve colorimetric detection of miRNA. The detection limit of miRNA by this method is 0.68 fM, which is two orders of magnitude more sensitive than RCA alone.

4.6 Application of RCA in Functional Protein Analysis

The application of RCA in the detection of many functional proteins can be divided into two categories based on solid-phase RCA and homogeneous solution RCA.

4.6.1 Protein Detection Based on Solid-Phase RCA

In 2000, Schweitzer first developed an RCA-based immunoassay for the detection of IgE on a slide [86]. The principle is shown in Fig. 4.33. In the process of immuno RCA, the antibody modifying a primer is specifically bound to the antigen fixed on the slide, and the primer is anchored on the surface of the slide. In the presence of circular DNA, DNA polymerase, and dNTPs, primer triggers the RCA reaction to generate a long strand of DNA containing hundreds of repeats, which greatly improves the sensitivity of immunoassay by detecting RCA products. Therefore, solid-phase RCA technology has been widely used in the field of protein biosensing detection.

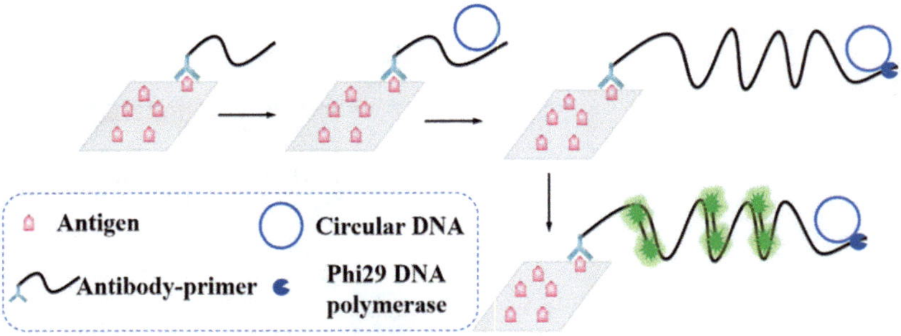

Fig. 4.33 Schematic diagram of immuno RCA. (Revised and reprinted with permission from Ref. [86])

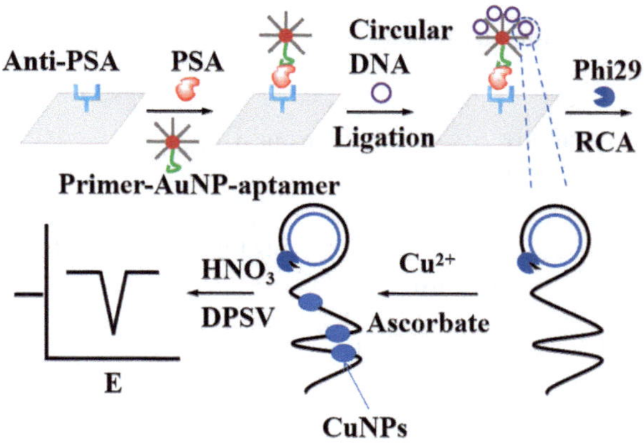

Fig. 4.34 Gold electrode surface-based RCA for the detection of PSA. (Revised and reprinted with permission from Ref. [89])

4.6.1.1 Protein Detection on Electrode Surface

As a solid medium, electrodes provide an ideal platform for RCA. Therefore, RCA strategies based on electrode surface have been widely used in the field of protein detection [87–89]. Zhu developed an electrochemical analysis method combining the RCA with polythymine templated copper nanoparticles (CuNPs) for ultra-sensitive and highly selective detection of prostate-specific antigen (PSA) [89]. As shown in Fig. 4.34, PSA antibodies are first fixed on the surface of the gold electrode, and then PSA aptamers and RCA primers are modified on the surface of the gold nanoparticles. When PSA is present, aptamer, PSA, and PSA antibody on the surface of gold nanoparticles form sandwich structure. Primers on the gold nanoparticles activate RCA using circular DNA that is subsequently added. Because circular DNA contains a large number of contiguous sequences of adenine, RCA products contain hundreds of contiguous thymine fragments. Cu^{2+} and ascorbate are introduced into the system, and due to the strong interaction between Cu^{2+} and thymine, Cu^{2+} forms CuNPs on the template of polythymine. After the unreacted Cu^{2+} is removed by a simple washing step, nitric acid is added to dissolve the CuNPs, releasing large amounts of Cu^{2+}. The resulting electrochemical signals are measured by differential pulse stripping voltammetry (DPSV). In the absence of PSA, RCA cannot be performed, CuNPs cannot be formed, and electrochemical signals cannot be detected. The method uses gold nanoparticles as the carrier to enrich a large number of aptamers and primers, improve the primers/aptamers ratio, and further enhance the signal amplification effect. The method achieves high sensitivity detection of target molecules, and the detection limit of PSA is 0.02 fg/mL.

4.6.1.2 Protein Detection on Microarray Chips

RCA based on microarray chips is often used for protein detection [90]. Lee developed an innovative self-assembled optical diffraction biosensor [91], which

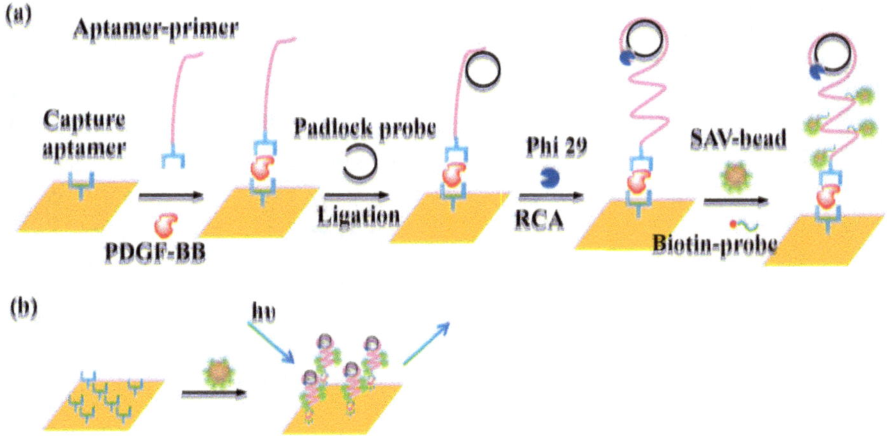

Fig. 4.35 The principle diagram of detecting PDGF-BB by diffraction method based on RCA of chip surface. (Revised and reprinted with permission from Ref. [91])

was based on microarray chip RCA technology and used magnetic microbeads as signal enhancement means to detect platelet-derived growth factor B-chain (PDGF-BB). PDGF-BB is a key protein that plays an important role in cell transformation, tumor growth and progression. For this reason, researchers often use it as a model protein for analysis and detection studies. As shown in Fig. 4.35, the sensor is based on the sandwich structure principle of aptamer and target. First, the SAV is modified periodically at 15 μm intervals on the polydimethylsiloxane (PDMS) chip substrate. Then, the biotinylated capture aptamer is fixed on the chip by affinity. Subsequently, PDGF-BB and the aptamer with the RCA primer sequence bind to the capture aptamer to form a typical sandwich structure. The primers trigger the ligation of the padlock probe and subsequent RCA, producing long strands of DNA. The DNA is then hybridized with biotin-probe, and the SAV-labeled beads are enriched on the RCA products to form regular periodic grating structures. When illuminated by a laser, the structure produces a diffraction pattern signal. Therefore, the quantitative analysis of PDGF-BB can be obtained by diffraction measurement under optical microscope. The detection limit of this method is 10 pM, and the sensitivity of this method is two orders of magnitude higher than that of the method without RCA signal amplification.

4.6.1.3 Protein Detection on Magnetic Microbeads

Magnetic microbeads are widely used in the enrichment, separation, and fixation of DNA or proteins [92–94]. In recent years, researchers have developed many biosensing strategies based on functional magnetic microbeads for the detection of various biomolecules [95–97], especially RCA technology based on magnetic microbeads has been widely used in the field of protein analysis and detection

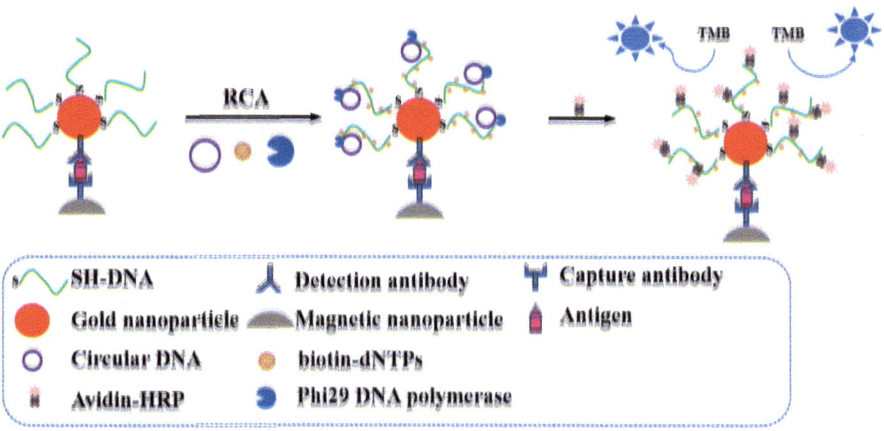

Fig. 4.36 Schematic diagram of CEA detection by colorimetric method based on RCA of magnetic nanoparticles. (Revised and reprinted with permission from Ref. [98])

[98–100]. Yan [98] developed a nano-RCA (nanoRCA) colorimetric method based on magnetic nanoparticles for rapid, convenient, and highly sensitive immunoassay. As shown in Fig. 4.36, the capture antibody is first modified on the magnetic nanoparticle, and the detection antibody and HS-DNA are modified on the surface of the gold nanoparticle. When the target antigen is present, the capture antibody, the antigen, and detection antibody form a sandwich complex. The circular DNA is then hybridized with the DNA on the surface of the gold nanoparticles, and RCA reaction is carried out using this DNA as a primer. In the amplification process, biotin-modified RCA products are produced by bioin-dNTPs. Avidin-modified horseradish peroxidase (Avidin-HRP) is enriched in RCA products by binding with biotin on the product and catalyzes TMB to change from colorless to blue, thus achieving colorimetric detection of target antigen. The method can detect about 30 (5 aM) carcinoembryonic antigen (CEA) molecules in a volume of 10 μL using visual observation, which is at least five orders of magnitude more sensitive than traditional ELISA methods.

In 2009, Konry developed a high-density fiber microsphere array platform capable of detecting both protein and DNA in a single test [99]. The system uses microsphere-based RCA as a signal amplification method for protein and nucleic acid detection, with detection limits of 10 fM for protein and 1 pM for DNA. In addition, in 2020, this team developed a dropcast single molecule assays (dSimoa) that enables counting and analysis of more target molecules by improving sampling efficiency and simplifying the operational process [100]. As shown in Fig. 4.37, the capture antibody on magnetic bead, target protein, and biotin-detection antibody specifically bind to form a single immune sandwich complex. Subsequently, SAV-DNA is bound to biotin and fixed on the surface of the magnetic bead, and prefabricated circular DNA is added for RCA reaction, resulting in long ssDNA fixed by each immune complex. The fluorescently labeled DNA probes are hybridized with the RCA product, followed by magnetic separation. The magnetic beads

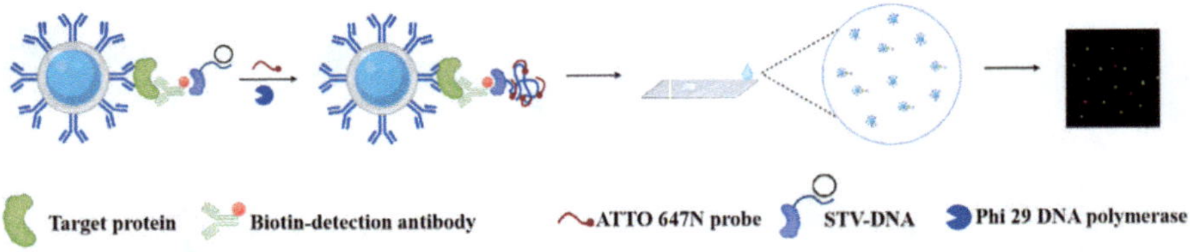

Fig. 4.37 Schematic diagram of single-molecule protein analysis based on RCA of magnetic microspheres. (Revised and reprinted with permission from Ref. [100])

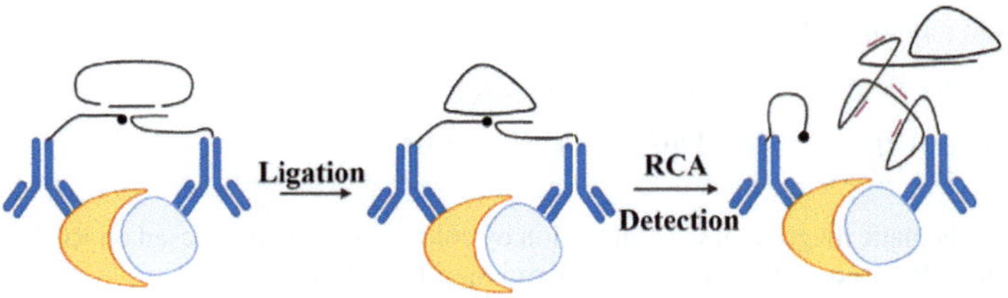

Fig. 4.38 Schematic diagram of protein-protein interaction detection based on proximity ligation and RCA. (Revised and reprinted with permission from Ref. [103])

are coated on a microscope slide, and after drying, a single-layer film is formed. The analysis of individual target protein molecules is realized by counting and analyzing the fluorescent magnetic beads in the film. Each magnetic bead that detects a fluorescence signal represents a protein molecule, thus achieving a single-molecule count analysis of the target protein. The detection limit of the single molecule detection platform of dSimoa is as low as aM, and its sensitivity is 25 times higher than that of the current digital enzyme-linked immunoassay technology [101, 102], and four orders of magnitude higher than that of the traditional immunoassay.

4.6.1.4 Protein Detection Based on RCA and Proximity Ligation

RCA-based proximity ligation is generally applicable to protein detection [103–105]. Söderberg used the method of proximity ligation combined with RCA to detect the interaction between endogenous proteins in cells with single molecular resolution [103], as shown in Fig. 4.38. First, two ssDNAs are modified on two protein-specific antibodies. The two proteins bind to their respective antibodies, and when the two proteins interact, the distance is close, resulting in the close proximity of the two antibodies and ssDNA. Two ssDNAs that are close to each other can be used as a template to ligate the padlock probe to form a circular structure. After the ligation is complete, one of the ssDNA is used as the primer to initiate the RCA, and the resulting long DNA is still attached to the antibody-antigen complex. Subsequently, fluorescence in situ imaging detection is achieved by hybridization of fluorescently labeled probe with RCA products. Liu designed a new scheme for protein detection by proximity ligation, which used three DNAs to trigger RCA to achieve

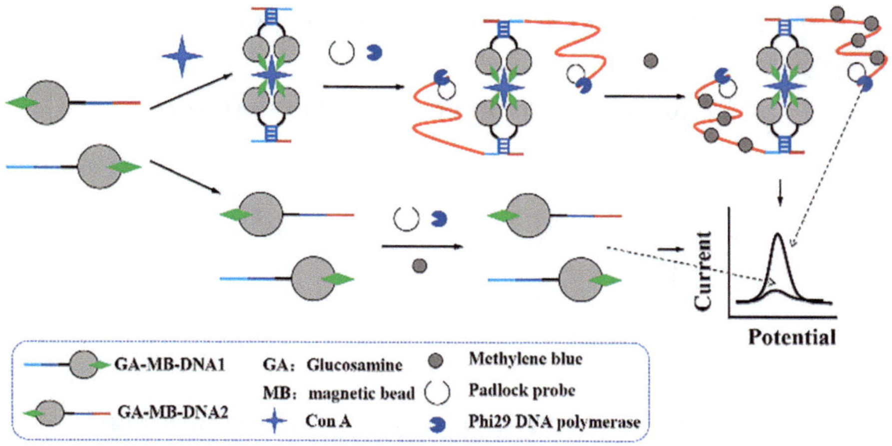

Fig. 4.39 Schematic diagram of Con A detection based on proximity ligation and RCA. (Revised and reprinted with permission from Ref. [104])

ultra-sensitive detection of canavalin A (Con A) [104]. The principle is shown in Fig. 4.39. First, two kinds of magnetic sensing probes are prepared: glucosamine-magnetic bead-DNA1 (GA-MB-DNA1) and glucosamine-magnetic bead-DNA2 (GA-MB-DNA2). Due to the specific interaction of Con A with glucosamine, Con A induces the aggregation of GA-MB-DNA1 and GA-MB-DNA2 when the target Con A is present. The proximity of DNA1 and DNA2 leads to hybridization of complementary sequences, while the padlock probe ends are hybridized with the 3′ end of DNA1 and the 5′ end of DNA2 and are ligated into a circular structure. Subsequently, DNA1 acts as a primer to initiate RCA to produce a long DNA and maintain the RCA product fixed on the magnetic bead. Outside the electrode, a magnet is used to enrich the magnetic beads to the electrode surface. Methylene blue is combined with RCA products on magnetic beads, and then Con A can be detected by monitoring electrochemical signals. The detection method combined with proximity ligation and targeted induction can reduce background signals and improve detection sensitivity to a certain extent, and the detection limit of Con A is 1.5 pM.

4.6.2 *RCA-Based Protein Detection in Homogeneous Solution*

RCA based on homogeneous solution is often combined with nucleic acid aptamer for protein detection [106–111]. Giusto used the proximity of circular DNA probe and linear probe to trigger RCA to achieve real-time detection of thrombin [106], as shown in Fig. 4.40. The 5′ end of the first probe modifies the phosphate group and contains the sequence of the thrombin aptamer. The 5′ end of the second probe modifies the phosphate group and also contains the RCA primer binding site. The

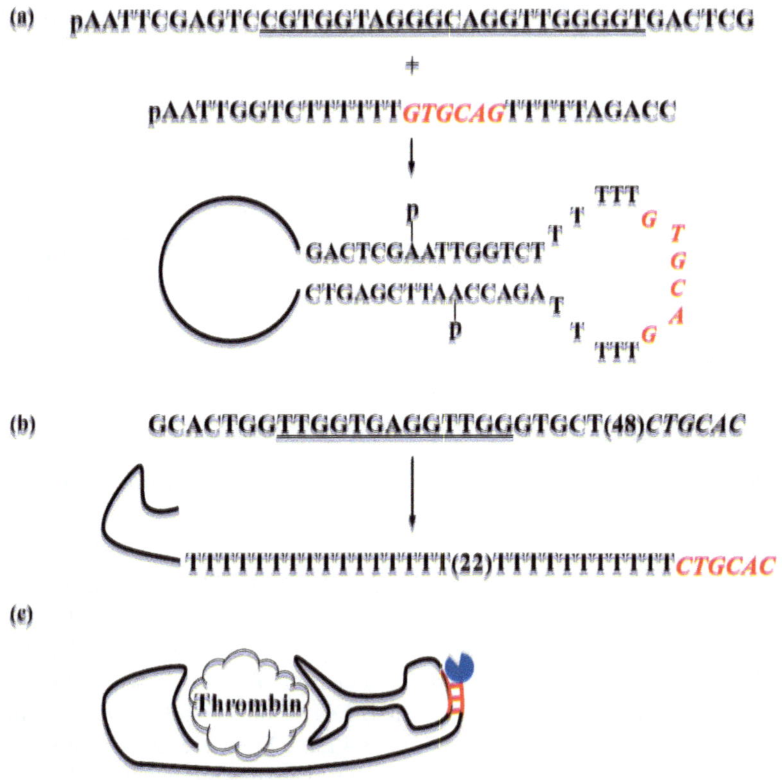

Fig. 4.40 Schematic diagram of thrombin detection by RCA based on proximity aptamer. (Revised and reprinted with permission from Ref. [106])

two probes can hybridize with each other to form a dumbbell-shaped structure with two nicks, which are ligated by ligase to form a circular DNA probe (Fig. 4.40a). As shown in Fig. 4.40b, the 5′ end of the linear probe is the aptamer sequence of thrombin exosome I, the 3′ end is a six-base RCA primer sequence, and the middle part contains 48 spacer sequences of polyT. When thrombin is present, the two aptamer sequences bind specifically to the thrombin, respectively, bringing the two probes close to each other. At this point, the 3′ end sequence of the linear probe hybridizes with the circular DNA probe, triggering the RCA (Fig. 4.40c). Real-time fluorescence monitoring of the amplified product is performed by SYBR Green I to detect thrombin. When thrombin is not present, the free linear probe will not hybridize with the circular DNA probe, cannot initiate the RCA reaction, and thus cannot generate a fluorescence signal.

Wu proposed a protein detection method based on aptamer structure transformation to trigger RCA [108], which was relatively simple to construct and avoided the use of sandwich immunization. The principle is shown in Fig. 4.41. A DNA probe is designed with the target protein aptamer sequence on one end. Part of the aptamer sequence and the remaining sequence together constitute the complementary

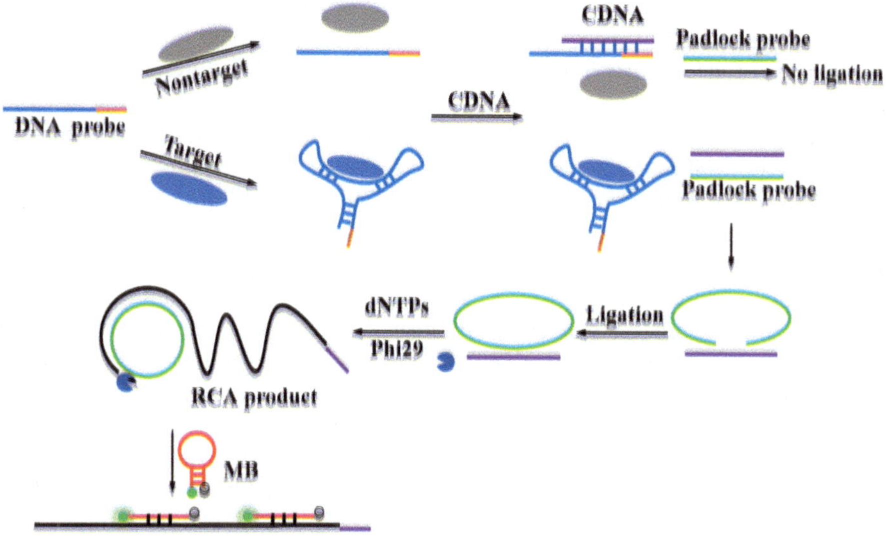

Fig. 4.41 Schematic diagram of protein detection based on aptamer structure switching and RCA. (Revised and reprinted with permission from Ref. [108])

sequence of CDNA. When the target protein is present, the aptamer sequence of the probe binds to it, and the structure changes, resulting in the CDNA being unable to hybridize with the probe and existing in free form. The free CDNA can be used as a template to ligate the padlock probe and as a primer to initiate the RCA. Subsequently, the RCA products are detected using MB to enable the analysis of target proteins. In the absence of the target protein, CDNA hybridizes with the 3′ end of the aptamer probe to form a stable DNA double-stranded structure, which could not hybridize with the padlock probe and thus could not trigger RCA. The probe design of this method is simple, and the detection of target proteins at amol level can be realized.

4.7 Outlook

As a simple and efficient isothermal nucleic acid amplification, RCA has become an important tool in basic research and clinical diagnosis. The well-designed circular DNA template can give the amplified product various functional properties such as aptamer and DNAzyme, so that the combination of this technology with other detection platforms, such as electrochemical method, colorimetric method and SERS, provides a new strategy and idea for the realization of ultra-sensitive detection of various targets. The aptamer sequence encoded in RCA products can recognize multiple types of targets, including nucleic acids, small molecules, proteins, and cells, thus enabling the analysis of many different types of targets.

DNAzyme formed in RCA products with shear or catalytic functions can further enhance signal amplification. At present, RCA technology has shown great application value and potential in genome, SNP analysis, mRNA, miRNA, and protein detection and single cell analysis. In addition, the technology can also be used to detect pathogens and disease-related genes, avoiding the time-consuming and complex procedures of traditional clinical testing methods.

However, up to now, RCA technology still faces several limitations. It is still a difficult task to prepare high purity circular DNA template in large quantities. In particular, when using the enzyme-facilitated ligation method, the ligation efficiency is easily affected by the length of the padlock probe, which is difficult to accurately regulate, and this process may produce linear polymer by-products that affect the detection sensitivity. In addition, given the large molecular weight of RCA products, it may interfere with the ability of aptamer recognition, restriction enzyme recognition, cell targeting, or drug delivery, increasing the risk of nonspecific binding. However, with the continuous progress of RCA technology, this powerful tool will be more widely used in the field of diagnosis and treatment of diseases, and provide technical support for revealing the mechanism of related diseases and deepening the understanding of diseases.

References

1. Mehta D, Cornet L, Hirsch-Hoffmann M, Zaidi S S-e-A, Vanderschuren H. Full-length sequencing of circular DNA viruses and extrachromosomal circular DNA using CIDER-Seq. Nat Protoc. 2020;15:1673–89.
2. Lizardi PM, Huang X, Zhu Z, Bray-Ward P, Thomas DC, Ward DC. Mutation detection and single-molecule counting using isothermal rolling-circle amplification. Nat Genet. 1998;19: 225–32.
3. Fire A, Xu SQ. Rolling replication of short DNA circles. Proc Natl Acad Sci USA. 1995;92: 4641–5.
4. Gusev Y, Sparkowski J, Raghunathan A, Ferguson H, Montano J, Bogdan N, Schweitzer B, Wiltshire S, Kingsmore SF, Maltzman W, Wheeler V. Rolling circle amplification: a new approach to increase sensitivity for immunohistochemistry and flow cytometry. Am J Pathol. 2001;159:63–9.
5. Bi S, Li L, Zhang S. Triggered polycatenated DNA scaffolds for DNA sensors and aptasensors by a combination of rolling circle amplification and DNAzyme amplification. Anal Chem. 2010;82:9447–54.
6. Christian AT, Pattee MS, Attix CM, Reed BE, Sorensen KJ, Tucker JD. Detection of DNA point mutations and mRNA expression levels by rolling circle amplification in individual cells. Proc Natl Acad Sci USA. 2001;98:14238–43.
7. Liu H, Luo J, Fang L, Huang H, Deng J, Huang J, Zhang S, Li Y, Zheng J. An electrochemical strategy with tetrahedron rolling circle amplification for ultrasensitive detection of DNA methylation. Biosens Bioelectron. 2018;121:47–53.
8. Li J, Deng T, Chu X, Yang R, Jiang J, Shen G, Yu R. Rolling circle amplification combined with gold nanoparticle aggregates for highly sensitive identification of single-nucleotide polymorphisms. Anal Chem. 2010;82:2811–6.

9. Xiang Y, Deng K, Xia H, Yao C, Chen Q, Zhang L, Liu Z, Fu W. Isothermal detection of multiple point mutations by a surface plasmon resonance biosensor with Au nanoparticles enhanced surface-anchored rolling circle amplification. Biosens Bioelectron. 2013;49:442–9.
10. Schweitzer B, Roberts S, Grimwade B, Shao W, Wang M, Fu Q, Shu Q, Laroche I, Zhou Z, Tchernev VT, Christiansen J, Velleca M, Kingsmore SF. Multiplexed protein profiling on microarrays by rolling-circle amplification. Nat Biotechnol. 2002;20:359–65.
11. Xu L, Jiang Z, Mu Y, Zhang Y, Zhan Q, Cui J, Cheng W, Ding S. Colorimetric assay of rare disseminated tumor cells in real sample by aptamer-induced rolling circle amplification on cell surface. Sens Actuators B Chem. 2018;259:596–603.
12. Ding C, Liu H, Wang N, Wang Z. Cascade signal amplification strategy for the detection of cancer cells by rolling circle amplification and nanoparticles tagging. Chem Commun. 2012;48:5019–21.
13. Ali MM, Li F, Zhang Z, Zhang K, Kang DK, Ankrum JA, Le XC, Zhao W. Rolling circle amplification: a versatile tool for chemical biology, materials science and medicine. Chem Soc Rev. 2014;43:3324–41.
14. Lee JB, Hong J, Bonner DK, Poon Z, Hammond PT. Self-assembled RNA interference microsponges for efficient siRNA delivery. Nat Mater. 2012;11:316–22.
15. Wang X, Li C, Gao X, Wang J, Liang X. Preparation of small RNAs using rolling circle transcription and site-specific RNA disconnection. Mol Ther Nucl Acids. 2015;4:e215.
16. Ha JS, Lee JS, Jeong J, Kim H, Byun J, Kim SA, Lee HJ, Chung HS, Lee JB, Ahn DR. Poly-sgRNA/siRNA ribonucleoprotein nanoparticles for targeted gene disruption. J Control Release. 2017;250:27–35.
17. Wang Y, Kim E, Lin Y, Kim N, Kit-Anan W, Gopal S, Agarwal S, Howes PD, Stevens MM. Rolling circle transcription-amplified hierarchically structured organic–inorganic hybrid RNA flowers for enzyme immobilization. ACS Appl Mater Interfaces. 2019;11:22932–40.
18. Banér J, Nilsson M, Mendel-Hartvig M, Landegren U. Signal amplification of padlock probes by rolling circle replication. Nucleic Acids Res. 1998;26:5073–8.
19. Zou Z, Qing Z, He X, Wang K, He D, Shi H, Yang X, Qing T, Yang X. Ligation-rolling circle amplification combined with γ-cyclodextrin mediated stemless molecular beacon for sensitive and specific genotyping of single-nucleotide polymorphism. Talanta. 2014;125:306–12.
20. Qi X, Bakht S, Devos KM, Gale MD, Osbourn A. L-RCA (ligation-rolling circle amplification): a general method for genotyping of single nucleotide polymorphisms (SNPs). Nucleic Acids Res. 2001;29:e116.
21. Pickering J, Bamford A, Godbole V, Briggs J, Scozzafava G, Roe P, Wheeler C, Ghouze F, Cuss S. Integration of DNA ligation and rolling circle amplification for the homogeneous, end-point detection of single nucleotide polymorphisms. Nucleic Acids Res. 2002;30:e60.
22. Jarvius J, Melin J, Göransson J, Stenberg J, Fredriksson S, Gonzalez-Rey C, Bertilsson S, Nilsson M. Digital quantification using amplified single-molecule detection. Nat Methods. 2006;3:725–7.
23. Cao A, Zhang CY. Sensitive and label-free DNA methylation detection by ligation-mediated hyperbranched rolling circle amplification. Anal Chem. 2012;84:6199–205.
24. Dahl F, Baner J, Gullberg M, Mendel-Hartvig M, Landegren U, Nilsson M. Circle-to-circle amplification for precise and sensitive DNA analysis. Proc Natl Acad Sci USA. 2004;101:4548–53.
25. Murakami T, Sumaoka J, Komiyama M. Sensitive isothermal detection of nucleic-acid sequence by primer generation-rolling circle amplification. Nucleic Acids Res. 2009;37:e19.
26. Liu H, Li L, Duan L, Wang X, Xie Y, Tong L, Wang Q, Tang B. High specific and ultrasensitive isothermal detection of microRNA by padlock probe-based exponential rolling circle amplification. Anal Chem. 2013;85:7941–7.
27. Zhu X, Feng C, Zhang B, Tong H, Gao T, Li G. A netlike rolling circle nucleic acid amplification technique. Analyst. 2015;140:74–8.
28. Beyer S, Nickels P, Simmel FC. Periodic DNA nanotemplates synthesized by rolling circle amplification. Nano Lett. 2005;5:719–22.

29. Zhao W, Gao Y, Kandadai SA, Brook MA, Li Y. DNA polymerization on gold nanoparticles through rolling circle amplification: towards novel scaffolds for three-dimensional periodic nanoassemblies. Angew Chem Int Ed. 2006;45:2409–13.
30. Takahashi H, Matsumoto A, Sugiyama S, Kobori T. Direct detection of green fluorescent protein messenger RNA expressed in *Escherichia coli* by rolling circle amplification. Anal Biochem. 2010;401:242–9.
31. Yao B, Li J, Huang H, Sun C, Wang Z, Fan Y, Chang Q, Li S, Xi J. Quantitative analysis of zeptomole microRNAs based on isothermal ramification amplification. RNA. 2009;15:1787–94.
32. Li X, Cui Y, Du Y, Tang A, Kong D. Label-free telomerase detection in single cell using a five-base telomerase product-triggered exponential rolling circle amplification strategy. ACS Sens. 2019;4:1090–6.
33. Gyllborg D, Langseth CM, Qian X, Choi E, Salas SM, Hilscher MM, Lein ES, Nilsson M. Hybridization-based *in situ* sequencing (HybISS) for spatially resolved transcriptomics in human and mouse brain tissue. Nucleic Acids Res. 2020;48:e112.
34. Bai M, Cao X, Huang W, Chen F, Zhao Y, Xue J, Li Y, Cheng Y, Zhang L, Zhao Y. Visualizing newly synthesized RNA by bioorthogonal labeling-primed DNA amplification. Anal Chem. 2020;92:8444–9.
35. Chen F, Xue J, Zhang J, Bai M, Yu X, Fan C, Zhao Y. Differentiated visualization of single-cell 5-hydroxymethylpyrimidines with microfluidic hydrogel encoding. J Am Chem Soc. 2020;142:2889–96.
36. Larsson C, Koch J, Nygren A, Janssen G, Raap AK, Landegren U, Nilsson M. *In situ* genotyping individual DNA molecules by target-primed rolling-circle amplification of padlock probes. Nat Methods. 2004;1:227–32.
37. de Vega M, Lázaro JM, Salas M, Blanco L. Mutational analysis of ∅29 DNA polymerase residues acting as ssDNA ligands for 3′-5′ exonucleolysis. J Mol Biol. 1998;279:807–22.
38. Larsson C, Grundberg I, Söderberg O, Nilsson M. *In situ* detection and genotyping of individual mRNA molecules. Nat Methods. 2010;7:395–7.
39. Deng R, Tang L, Tian Q, Wang Y, Lin L, Li J. Toehold-initiated rolling circle amplification for visualizing individual microRNAs in situ in single cells. Angew Chem Int Ed. 2014;53:2389–93.
40. Li JJ, Chu Y, Lee BY, Xie XS. Enzymatic signal amplification of molecular beacons for sensitive DNA detection. Nucleic Acids Res. 2008;36:e36.
41. Wu X, Zhu S, Huang P, Chen Y. Highly specific quantification of microRNA by coupling probe-rolling circle amplification and Förster resonance energy transfer. Anal Biochem. 2016;502:16–23.
42. Zhou H, Wang H, Liu C, Wang H, Duan X, Li Z. Ultrasensitive genotyping with target-specifically generated circular DNA templates and RNA FRET probes. Chem Commun. 2015;51:11556–9.
43. Hao L, Wang W, Shen X, Wang S, Li Q, An F, Wu S. A fluorescent DNA hydrogel aptasensor based on the self-assembly of rolling circle amplification products for sensitive detection of ochratoxin A. J Agric Food Chem. 2020;68:369–75.
44. Xu X, Zhang B, Gan P, Wu J, Dai W, Zhang L, Wang J. On-nylon membrane detection of nucleic acid molecules by rolling circle amplification. Anal Biochem. 2017;533:26–33.
45. Ye S, Guo Y, Xiao J, Zhang S. A sensitive SERS assay of l-histidine *via* a DNAzyme-activated target recycling cascade amplification strategy. Chem Commun. 2013;49:3643–5.
46. Guven B, Dudak FC, Boyaci IH, Tamer U, Ozsoz M. SERS-based direct and sandwich assay methods for mir-21 detection. Analyst. 2014;139:1141–7.
47. Hu J, Zhang C-y. Sensitive detection of nucleic acids with rolling circle amplification and surface-enhanced Raman scattering spectroscopy. Anal Chem. 2010;82:8991–7.
48. Meric B, Kerman K, Ozkan D, Kara P, Erensoy S, Akarca US, Mascini M, Ozsoz M. Electrochemical DNA biosensor for the detection of TT and Hepatitis B virus from PCR amplified real samples by using methylene blue. Talanta. 2002;56:837–46.

49. Zhu N, Chang Z, He P, Fang Y. Electrochemically fabricated polyaniline nanowire-modified electrode for voltammetric detection of DNA hybridization. Electrochim Acta. 2006;51:3758–62.
50. Zhang S, Wu Z, Shen G, Yu R. A label-free strategy for SNP detection with high fidelity and sensitivity based on ligation-rolling circle amplification and intercalating of methylene blue. Biosens Bioelectron. 2009;24:3201–7.
51. Connolly J, St Pierre TG. Proposed biosensors based on time-dependent properties of magnetic fluids. J Magn Magn Mater. 2001;225:156–60.
52. Strömberg M, Göransson J, Gunnarsson K, Nilsson M, Svedlindh P, Strømme M. Sensitive molecular diagnostics using volume-amplified magnetic nanobeads. Nano Lett. 2008;8:816–21.
53. Kreibig U, Genzel L. Optical absorption of small metallic particles. Surf Sci. 1985;156:678–700.
54. Dusemund B, Hoffmann A, Salzmann T, Kreibig U, Schmid G. Cluster matter: the transition of optical elastic scattering to regular reflection. Z Phys D-Atoms, Mol Clus. 1991;20:305–8.
55. Elghanian R, Storhoff JJ, Mucic RC, Letsinger RL, Mirkin CA. Selective colorimetric detection of polynucleotides based on the distance-dependent optical properties of gold nanoparticles. Science. 1997;277:1078.
56. Ma X, Chen X, Tang Y, Yan R, Miao P. Triple-input molecular and logic gates for sensitive detection of multiple miRNAs. ACS Appl Mater Interfaces. 2019;11:41157–64.
57. Xu W, Xue X, Li T, Zeng H, Liu X. Ultrasensitive and selective colorimetric DNA detection by nicking endonuclease assisted nanoparticle amplification. Angew Chem Int Ed. 2009;48: 6849–52.
58. Hu B, Guo J, Xu Y, Wei H, Zhao G, Guan Y. A sensitive colorimetric assay system for nucleic acid detection based on isothermal signal amplification technology. Anal Bioanal Chem. 2017;409:4819–25.
59. Xu W, Xie X, Li D, Yang Z, Li T, Liu X. Ultrasensitive colorimetric DNA detection using a combination of rolling circle amplification and nicking endonuclease-assisted nanoparticle amplification (NEANA). Small. 2012;8:1846–50.
60. Li D, Zhang T, Yang F, Yuan R, Xiang Y. Efficient and exponential rolling circle amplification molecular network leads to ultrasensitive and label-free detection of microRNA. Anal Chem. 2020;92:2074–9.
61. Xu H, Wu D, Zhang Y, Shi H, Ouyang C, Li F, Jia L, Yu S, Wu Z-S. RCA-enhanced multifunctional molecule beacon-based strand-displacement amplification for sensitive microRNA detection. Sens Actuators B Chem. 2018;258:470–7.
62. Bi S, Zhao T, Luo B, Zhu J-J. Hybridization chain reaction-based branched rolling circle amplification for chemiluminescence detection of DNA methylation. Chem Commun. 2013;49:6906–8.
63. Hu X, Zhang H, Zhou Y, Liang W, Yuan R, Chen S. A fluorometric lead(II) assay by using a DNA dendrimer as a carrier for the immobilization of the signal probe. Microchim Acta. 2019;186:582.
64. Cheng Y, Zhao J, Jia H, Yuan Z, Li Z. Ligase chain reaction coupled with rolling circle amplification for high sensitivity detection of single nucleotide polymorphisms. Analyst. 2013;138:2958–63.
65. Zhang L-r, Zhu G, Zhang C-y. Homogeneous and label-free detection of microRNAs using bifunctional strand displacement amplification-mediated hyperbranched rolling circle amplification. Anal Chem. 2014;86:6703–9.
66. Wang W, Peng Y, Wu J, Zhang M, Li Q, Zhao Z, Liu M, Wang J, Cao G, Bai J, Gao Z. Ultrasensitive detection of 17β-estradiol (E2) based on multistep isothermal amplification. Anal Chem. 2021;93:4488–96.
67. Tian W, Li P, He W, Liu C, Li Z. Rolling circle extension-actuated loop-mediated isothermal amplification (RCA-LAMP) for ultrasensitive detection of microRNAs. Biosens Bioelectron. 2019;128:17–22.

68. Zhan X, Yang S, Huang G, Yang L, Zhang Y, Tian H, Xie F, Lamy de la Chapelle M, Yang X, Fu W. Streptavidin-functionalized terahertz metamaterials for attomolar exosomal microRNA assay in pancreatic cancer based on duplex-specific nuclease-triggered rolling circle amplification. Biosens Bioelectron. 2021;188:113314.
69. Kim E, Howes PD, Crowder SW, Stevens MM. Multi-amplified sensing of microRNA by a small DNA fragment-driven enzymatic cascade reaction. ACS Sens. 2017;2:111–8.
70. Wang R, Zhao X, Chen X, Qiu X, Qing G, Zhang H, Zhang L, Hu X, He Z, Zhong D, Wang Y, Luo Y. Rolling circular amplification (RCA)-assisted CRISPR/Cas9 cleavage (RACE) for highly specific detection of multiple extracellular vesicle microRNAs. Anal Chem. 2020;92: 2176–85.
71. Qing M, Chen SL, Sun Z, Fan Y, Luo HQ, Li NB. Universal and programmable rolling circle amplification-CRISPR/Cas12a-mediated immobilization-free electrochemical biosensor. Anal Chem. 2021;93:7499–507.
72. Van Ness J, Van Ness LK, Galas DJ. Isothermal reactions for the amplification of oligonucleotides. Proc Natl Acad Sci USA. 2003;100:4504–9.
73. Notomi T, Okayama H, Masubuchi H, Yonekawa T, Watanabe K, Amino N, Hase T. Loop-mediated isothermal amplification of DNA. Nucleic Acids Res. 2000;28:e63.
74. Liu Y, Zhang J, Tian J, Fan X, Geng H, Cheng Y. Multiplex detection of microRNAs by combining molecular beacon probes with T7 exonuclease-assisted cyclic amplification reaction. Anal Bioanal Chem. 2016;409:107–14.
75. Li Y, Zhang J, Zhao J, Zhao L, Cheng Y, Li Z. A simple molecular beacon with duplex-specific nuclease amplification for detection of microRNA. Analyst. 2016;141:1071–6.
76. Du Y-C, Cui Y-X, Li X-Y, Sun G-Y, Zhang Y-P, Tang A-N, Kim K, Kong D-M. Terminal deoxynucleotidyl transferase and T7 exonuclease-aided amplification strategy for ultrasensitive detection of uracil-DNA glycosylase. Anal Chem. 2018;90:8629–34.
77. Ma F, Liu W-j, Liang L, Tang B, Zhang C-Y. Sensitive detection of alkaline phosphatase by dephosphorylation-initiated transcription reaction-mediated dual signal amplification. Chem Commun. 2018;54:2413–6.
78. Tian B, Qiu Z, Ma J, Donolato M, Hansen MF, Svedlindh P, Strömberg M. On-particle rolling circle amplification-based core–satellite magnetic superstructures for microRNA detection. ACS Appl Mater Interfaces. 2018;10:2957–64.
79. Cui L, Zhu Z, Lin N, Zhang H, Guan Z, Yang CJ. A T7 exonuclease-assisted cyclic enzymatic amplification method coupled with rolling circle amplification: a dual-amplification strategy for sensitive and selective microRNA detection. Chem Commun. 2014;50:1576–8.
80. Hsu PD, Lander ES, Zhang F. Development and applications of CRISPR-Cas9 for genome engineering. Cell. 2014;157:1262–78.
81. Liu W, Yu H, Zhou X, Xing D. *In vitro* evaluation of CRISPR/Cas9 function by an electrochemiluminescent assay. Anal Chem. 2016;88:8369–74.
82. Gootenberg JS, Abudayyeh OO, Kellner MJ, Joung J, Collins JJ, Zhang F. Multiplexed and portable nucleic acid detection platform with Cas13, Cas12a, and Csm6. Science. 2018;360: 439–44.
83. Zhuang J, Lai W, Chen G, Tang D. A rolling circle amplification-based DNA machine for miRNA screening coupling catalytic hairpin assembly with DNAzyme formation. Chem Commun. 2014;50:2935–8.
84. Song W, Zhang Q, Sun W. Ultrasensitive detection of nucleic acids by template enhanced hybridization followed by rolling circle amplification and catalytic hairpin assembly. Chem Commun. 2015;51:2392–5.
85. Xu S, Lou Z. Ultrasensitive detection of nasopharyngeal carcinoma-related miRNA through garland rolling circle amplification integrated catalytic hairpin assembly. ACS Omega. 2021;6: 6460–5.
86. Schweitzer B, Wiltshire S, Lambert J, O'Malley S, Kukanskis K, Zhu Z, Kingsmore SF, Lizardi PM, Ward DC. Immunoassays with rolling circle DNA amplification: a versatile platform for ultrasensitive antigen detection. Proc Natl Acad Sci USA. 2000;97:10113–9.

87. Zhou L, Ou L-J, Chu X, Shen G-L, Yu R-Q. Aptamer-based rolling circle amplification: a platform for electrochemical detection of protein. Anal Chem. 2007;79:7492–500.
88. Huang L, Wu J, Zheng L, Qian H, Xue F, Wu Y, Pan D, Adeloju SB, Chen W. Rolling chain amplification based signal-enhanced electrochemical aptasensor for ultrasensitive detection of ochratoxin A. Anal Chem. 2013;85:10842–9.
89. Zhu Y, Wang H, Wang L, Zhu J, Jiang W. Cascade signal amplification based on copper nanoparticle-reported rolling circle amplification for ultrasensitive electrochemical detection of the prostate cancer biomarker. ACS Appl Mater Interfaces. 2016;8:2573–81.
90. Yan J, Su S, He S, He Y, Zhao B, Wang D, Zhang H, Huang Q, Song S, Fan C. Nano rolling-circle amplification for enhanced SERS hot spots in protein microarray analysis. Anal Chem. 2012;84:9139–45.
91. Lee J, Icoz K, Roberts A, Ellington AD, Savran CA. Diffractometric detection of proteins using microbead-based rolling circle amplification. Anal Chem. 2010;82:197–202.
92. Berensmeier S. Magnetic particles for the separation and purification of nucleic acids. Appl Microbiol Biotechnol. 2006;73:495–504.
93. Lee IS, Lee N, Park J, Kim BH, Yi Y-W, Kim T, Kim TK, Lee IH, Paik SR, Hyeon T. Ni/NiO core/shell nanoparticles for selective binding and magnetic separation of histidine-tagged proteins. J Am Chem Soc. 2006;128:10658–9.
94. Hill HD, Mirkin CA. The bio-barcode assay for the detection of protein and nucleic acid targets using DTT-induced ligand exchange. Nat Protoc. 2006;1:324–36.
95. Zhang GY, Deng SY, Cai WR, Cosnier S, Zhang XJ, Shan D. Magnetic zirconium hexacyanoferrate(II) nanoparticle as tracing tag for electrochemical DNA assay. Anal Chem. 2015;87:9093–100.
96. Wu Z, Hu J, Zeng T, Zhang ZL, Chen J, Wong G, Qiu X, Liu W, Gao GF, Bi Y, Pang DW. Ultrasensitive Ebola virus detection based on electroluminescent nanospheres and immunomagnetic separation. Anal Chem. 2017;89:2039–48.
97. Zhu L, Chen D, Lu X, Qi Y, He P, Liu C, Li Z. An ultrasensitive flow cytometric immunoassay based on bead surface-initiated template-free DNA extension. Chem Sci. 2018;9:6605–13.
98. Yan J, Song S, Li B, Zhang Q, Huang Q, Zhang H, Fan C. An on-nanoparticle rolling-circle amplification platform for ultrasensitive protein detection in biological fluids. Small. 2010;6:2520–5.
99. Konry T, Hayman RB, Walt DR. Microsphere-based rolling circle amplification microarray for the detection of DNA and proteins in a single assay. Anal Chem. 2009;81:5777–82.
100. Wu C, Garden PM, Walt DR. Ultrasensitive detection of attomolar protein concentrations by dropcast single molecule assays. J Am Chem Soc. 2020;142:12314–23.
101. Rissin DM, Kan CW, Campbell TG, Howes SC, Fournier DR, Song L, Piech T, Patel PP, Chang L, Rivnak AJ, Ferrell EP, Randall JD, Provuncher GK, Walt DR, Duffy DC. Single-molecule enzyme-linked immunosorbent assay detects serum proteins at subfemtomolar concentrations. Nat Biotechnol. 2010;28:595–9.
102. Yelleswarapu V, Buser JR, Haber M, Baron J, Inapuri E, Issadore D. Mobile platform for rapid sub–picogram-per-milliliter, multiplexed, digital droplet detection of proteins. Proc Natl Acad Sci USA. 2019;116:4489.
103. Söderberg O, Gullberg M, Jarvius M, Ridderstrale K, Leuchowius KJ, Jarvius J, Wester K, Hydbring P, Bahram F, Larsson LG, Landegren U. Direct observation of individual endogenous protein complexes *in situ* by proximity ligation. Nat Methods. 2006;3:995–1000.
104. Liu B, Zhang B, Chen G, Yang H, Tang D. Proximity ligation assay with three-way junction-induced rolling circle amplification for ultrasensitive electronic monitoring of concanavalin A. Anal Chem. 2014;86:7773–81.
105. Du M, Mao G, Tian S, Liu Y, Zheng J, Ke X, Zheng Z, Wang H, Ji X, He Z. Target-induced cascade amplification for homogeneous virus detection. Anal Chem. 2019;91:15099–106.
106. Di Giusto DA, Wlassoff WA, Gooding JJ, Messerle BA, King GC. Proximity extension of circular DNA aptamers with real-time protein detection. Nucleic Acids Res. 2005;33:e64.

107. Yang L, Fung CW, Cho EJ, Ellington AD. Real-time rolling circle amplification for protein detection. Anal Chem. 2007;79:3320–9.
108. Wu Z-S, Zhang S, Zhou H, Shen G-L, Yu R. Universal aptameric system for highly sensitive detection of protein based on structure-switching-triggered rolling circle amplification. Anal Chem. 2010;82:2221–7.
109. Wu Z-S, Zhou H, Zhang S, Shen G, Yu R. Electrochemical aptameric recognition system for a sensitive protein assay based on specific target binding-induced rolling circle amplification. Anal Chem. 2010;82:2282–9.
110. Zhang S, Zheng L, Hu X, Shen G, Liu X, Shen G, Yu R. Highly sensitive fluorescent aptasensor for thrombin detection based on competition triggered rolling circle amplification. Chin J Anal Chem. 2015;43:1688–94.
111. Li J, Yang F, Jiang B, Zhou W, Xiang Y, Yuan R. The synchronization of multiple signal amplifications for label-free and sensitive aptamer-based sensing of a protein biomarker. Analyst. 2020;145:7858–63.

Chapter 5
Isothermal Exponential Amplification Reaction (EXPAR)

Honghong Wang

5.1 Principle of EXPAR and Technical Points

In 2003, David J. Galas et al. first reported the Isothermal Exponential Amplification Reaction (EXPAR) [1]. EXPAR utilizes DNA templates, deoxyribonucleotide triphosphate (dNTPs), DNA polymerase, and nicking endonuclease to achieve exponential amplification of target sequences. The basic principle of a classical single-template EXPAR is shown in Fig. 5.1 [2]. First, an amplification template (X′-X′) is designed based on the target sequence (Target). The 3′-end and 5′-end of the amplification template are sequences X′ (blue) with the same bases, which are complementary to the target. The intermediate sequence (black) of the amplification template is the nicking recognition sequence of the nicking endonuclease (NEase). When the target is present, it hybridizes with the 3′-end of the amplification template (i). By the catalytic action of DNA polymerase, the target sequence undergoes an elongation reaction along the template, resulting in the formation of double-stranded DNA (dsDNA) (ii). The NEase specifically recognizes the nicking recognition sequence on dsDNA (Nicking sequence) and creates a cut on the upper strand (iii). The DNA polymerase continues the elongation reaction at the incision site, which releases short DNA (iv) through strand displacement of the polymerase and forms new dsDNA. The new dsDNA undergoes continuous enzymatic cleavage, elongation, and strand displacement under the joint action of the NEase and polymerase, continuously releasing new short DNA and forming a linear amplification mechanism.

On the other hand, due to the same base sequence at both ends of the amplification template, the released short DNA is identical to the target sequence. The released short DNA hybridizes with the 3′-end sequence of the new amplification

H. Wang (✉)
University of Science and Technology Beijing, Beijing, China
e-mail: wanghh@ustb.edu.cn

Z. Li (ed.), *Nucleic Acid Detection*, https://doi.org/10.1007/978-981-95-7295-3_5

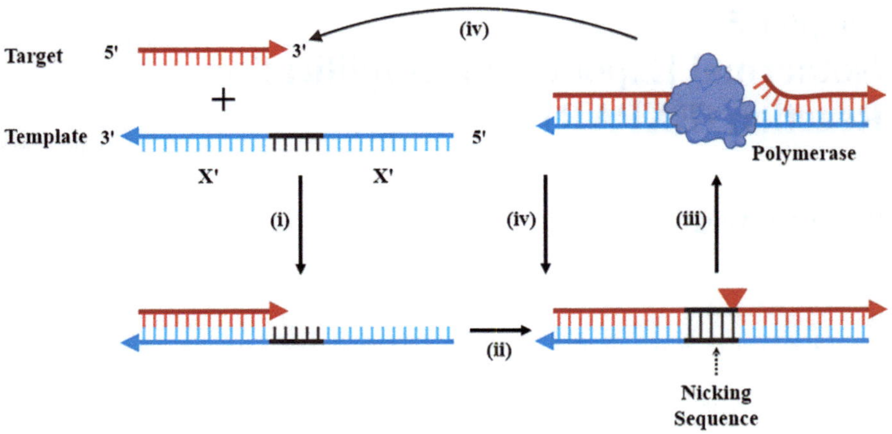

Fig. 5.1 Schematic diagram of classical single-template EXPAR principle [2]. (Revised and reprinted with permission from Ref. [2])

template and then undergoes continuous enzymatic cleavage, elongation, and a strand displacement process to release new short DNA. In this way, a target DNA produces multiple short-stranded DNAs, and each short-stranded DNA hybridizes with the amplification template to continue the extension and cleavage reactions, producing more short-stranded DNA and forming a rapid exponential amplification reaction.

5.1.1 Template Design of EXPAR

A classical single-template EXPAR amplification system only requires designing amplification templates based on the target sequence and the recognition sequence of NEase. Although designing a classical single-template EXPAR amplification system is relatively simple, to achieve efficient target sequence amplification, some key factors need to be considered. A classical single-template EXPAR template (i.e., X′-X′) contains two repeated X′ regions. First, it is necessary to ensure that the bases in the X′ region are complementary to the target sequence. Secondly, the length of the X′ region should be rationally designed to ensure its Tm value is close to the amplification reaction temperature, so that the target sequence can effectively hybridize with the EXPAR amplification template and be easily released from the 5′-end of the template. If the Tm value is lower than the amplification reaction temperature, the target cannot effectively hybridize with the EXPAR amplification template [1], which affects the amplification efficiency of EXPAR. If the NEase produces a cleavage at the nicking site, the released short DNA will have several additional spacer bases added to its 5′-end compared to the target sequence. For example, the recognition sequence of the Nb.BbvCI enzyme is 5′-GCTGAGG-3′, and its cutting site is 5′-GC↓TGAGG-3′. The replacing sequence is five bases longer than the target sequence, with the extra TGAGG bases. These additional bases cause the replacing DNA to preferentially hybridize with the 5′-end of the EXPAR amplification template, preventing the replacement DNA from initiating further amplification, thus reducing amplification efficiency. To prevent this from happening, these extra bases can be added to the 3′-end of the template, allowing the replacing DNA to effectively initiate the subsequent amplification cycles.

To reduce nonspecific amplification, the design of a classical EXPAR template should minimize nonspecific interactions of the template to ensure high amplification efficiency. This includes avoiding secondary structures of the template, minimizing hybridization between templates, and reducing interactions between the template and other nucleic acid molecules in the sample. For target sequences with secondary structures, classical X′-X′ templates exhibit complex secondary structures that prevent both the target DNA and the replacing DNA from effectively hybridizing with the template, leading to poor amplification efficiency. In addition, because the template concentration used in the EXPAR reaction is relatively high, interactions between templates can also occur, resulting in high background amplification. To address these issues in the classical single-template EXPAR system, a dual-template EXPAR system has been developed [3]. As shown in Fig. 5.2, the first amplification template (X′-Y′) is designed based on the target sequence. The 3′-end (X′) of the template is complementary to the target sequence, and the black region represents the NEase cleavage site and spacer bases. When the target is present, the target hybridizes with the 3′-end of the X′-Y′ template. Under the catalysis of DNA polymerase, a double-stranded DNA (i) is formed. The NEase recognizes the cleavage site in the middle of the dsDNA and cuts the upper strand (ii). Through

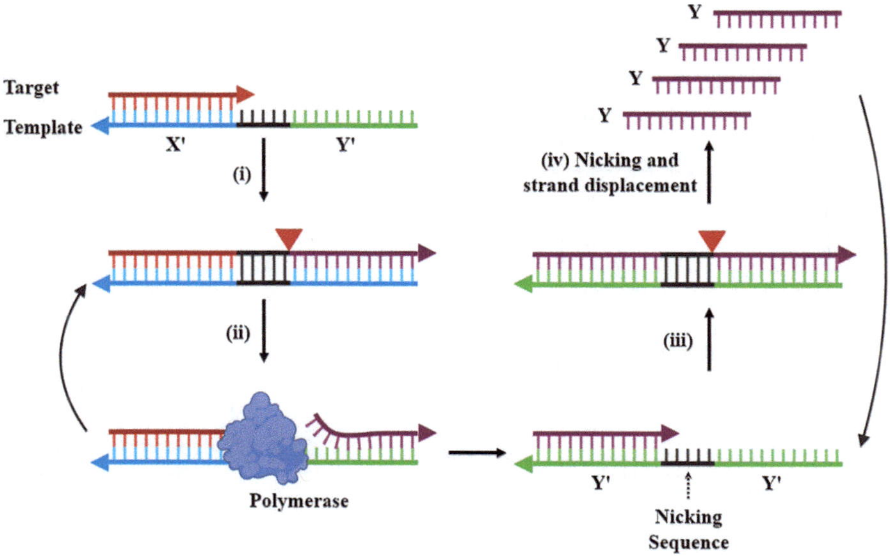

Fig. 5.2 Schematic diagram of the dual-template EXPAR system [2]. (Revised and reprinted with permission from Ref. [2])

the polymerization and replacement activity of DNA polymerase, Y is released while forming new dsDNA. The newly generated dsDNA, under the combined action of the NEase and DNA polymerase, continuously releases new Y, thus establishing a linear amplification mechanism. On the other hand, the released Y can hybridize with the 3′-end of the second amplification template (Y′-Y′), initiating an exponential amplification mechanism. The amplification mechanism is the same as the classical single-template EXPAR principle described above. The dual-template EXPAR system significantly avoids the issues of low amplification efficiency and high nonspecific background signals associated with the X′-X′ template. Meanwhile, the short-chain DNA (Y) generated by the amplification of the second template is unrelated to the target DNA sequence, allowing for the design of functionalized DNA sequences for subsequent detection.

Similar to the dual-template EXPAR system, a second endonuclease recognition sequence and a complementary sequence of Y can be added to the 5′-end of the typical single template EXPAR amplification template X′-X′, indicating that the amplification template has the X′-X′-Y′ form (Fig. 5.3). The X′-X′-Y′ template contains two cleavage sites for endonucleases. The detected target DNA hybridizes with the 3′-end of X′-X′-Y′, following the same principle as single template EXPAR. Using X′-X′ in X′-X′-Y′ as the template, DNA-X can be continuously generated under the action of polymerase and endonuclease. At the same time, the endonuclease recognizes the second cleavage site and can continuously produce DNA-Y, forming an exponential amplification mechanism. Due to the independence of the DNA-Y sequence from the target DNA, a functionalized DNA sequence can be designed to generate subsequent detection signals.

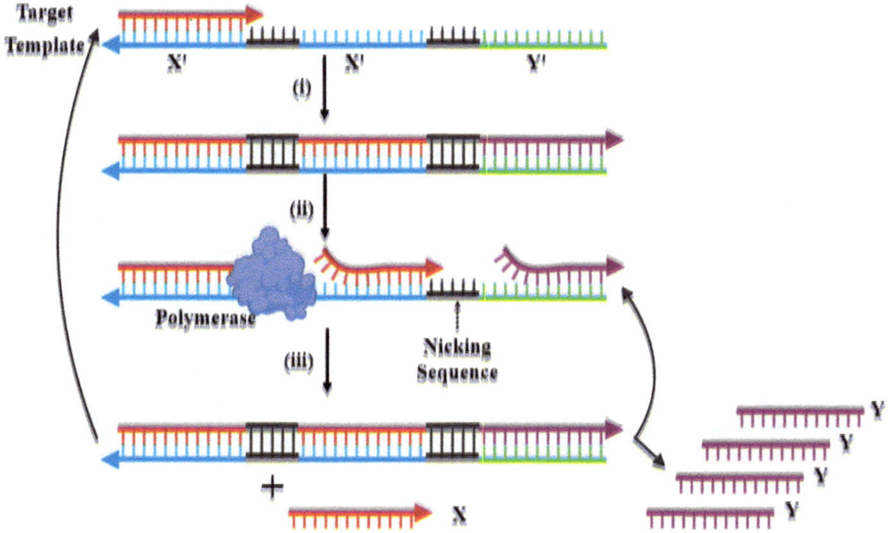

Fig. 5.3 Schematic diagram of the EXPAR principle for the X′-X′-Y′ template [2]. (Revised and reprinted with permission from Ref. [2])

5.1.2 Selection of Polymerase and Nicking Endonuclease

EXPAR amplification technology utilizes two types of enzymatic tools: DNA polymerase and Nicking endonuclease (NEase). DNA polymerase has two functions: one is polymerization, which includes the extension of the primer X and the extension at the 3′-end of the incision; the other is replacement, which replaces previously synthesized chains through strand displacement activity. The NEase recognizes the cleavage site and creates a cut in dsDNA. Since both DNA polymerase and NEase are present in the same reaction mixture, suitable reaction conditions (such as temperature and buffer) are crucial for ensuring the high catalytic activity of both enzymes.

Commercially available DNA polymerases with strand displacement activity include Klenow Fragment (3′-5′ exo^{-}), phi 29, Bst polymerase, and Vent (3′-5′ exo^{-}), where 3′-5′ exo^{-} indicates the absence of 3′-5′ exonuclease activity. The use of 3′ -5′ exo^{-} DNA polymerase can prevent the degradation of DNA after the incision during the DNA elongation process. Most EXPAR reactions are designed to be carried out between 55 and 60 °C. Commonly used enzymes with high catalytic activity include Vent (exo^{-}) and Bst polymerase. Commercially available NEases that can be used at higher temperatures include Nt.BstNBI, Nt.BspQI, Nb.BsmI, and Nb.BsrDI. The recognition sequences, cleavage sites, and optimal reaction temperatures for these enzymes are shown in Table 5.1.

Table 5.1 Recognition sequences, cleavage sites, and recommended temperatures for NEases

Enzyme	Recognition sequences and cleavage sites	Recommended temperature
Nt.BstNBI	5′...GAGTCNNNN▼N...3′ 3′...CTCAGNNNNN...5′	55 °C
Nt.BspQI	5′...GCTCTTCN▼...3′ 3′...CGAGAAGN...5′	50 °C
Nb.BsmI	5′...GAATGCN...3′ 3′...CTTAC▼GN...5′	65 °C
Nb.BsrDI	5′...GCAATG NN...3′ 3′...CGTTAC▼NN...5′	65 °C
Nt.BbvCI	5′...CC▼TCAGC...3′ 3′...GGAGTCG...5′	37 °C
Nb.BbvCI	5′...CCTCAGC...3′ 3′...GGAGT▼CG...5′	37 °C
Nt.BsmAI	5′...GTCTCN▼N...3′ 3′...CAGAGNN...5′	37 °C
Nt.CviPII	5′...▼CCD...3′ 3′...GGH...5′	37 °C
Nt.AlwI	5′...GGATCNNNN▼N...3′ 3′...CCTAGNNNNN...5′	37 °C
Nb.BssSI	5′...CACGAG...3′ 3′...GTGCT▼C...5′	37 °C
Nb.BtsI	5′...GCAGTGNN...3′ 3′...CGTCAC▼NN...5′	37 °C

▼ Represents the enzyme cleavage site

The EXPAR reaction can also be designed to be carried out at 37 °C, with the most commonly used enzyme being Klenow Fragment ($3' \rightarrow 5'$ exo$^-$) polymerase, which has an optimal reaction temperature of 37 °C. The NEases suitable for use at 37 °C include Nt.BbvCI, Nb.BbvCI, Nt.BsmAI, Nt.CviPII, Nt.AlwI, Nb.BssSI, and Nb.BtsI. The recognition sequences, cleavage sites, and recommended reaction temperatures for these enzymes are shown in Table 5.1.

The reaction temperature is a key parameter in EXPAR amplification, which not only affects the activity of polymerase and NEase but also affects amplification efficiency and specificity. Higher temperatures can also effectively reduce nonspecific DNA interactions. Furthermore, EXPAR is best performed with a DNA polymerase that has strong strand displacement activity to achieve high amplification efficiency.

5.1.3 Nonspecific Background Amplification and Sequence Dependence

Nonspecific interactions between the EXPAR amplification template and other nucleic acid molecules in the sample can lead to nonspecific amplification, making it difficult to distinguish nonspecific amplification products from the target-specific products. Most amplification techniques use short primers to amplify long DNA or RNA sequences, while EXPAR technology utilizes specially designed amplification templates to amplify short target sequences. As a result, EXPAR templates are used at high concentrations, which increases the likelihood of nonspecific interactions between template strands. These nonspecific interactions can trigger nonspecific background signals through DNA polymerase-mediated extension reactions of the templates.

Niemz and colleagues attempted to elucidate the source of nonspecific background amplification in EXPAR. Through experiments, they found that preincubating the amplification template with the polymerases accelerated background amplification, while “hot start” reactions reduced background amplification [4]. These findings are consistent with the explanation that background signals mainly arise from nonspecific interactions between templates. At lower temperatures, preincubating the templates with the polymerase increases nonspecific interactions between amplification templates, whereas “hot start” reactions effectively reduce the background signals caused by template-template interactions. Their experiments also revealed that the Y′-X′ template showed no significant nonspecific background signals [4]. Additionally, they tested the performance of 384 EXPAR templates in target DNA amplification reactions and found that background amplification depends on the template sequence, indicating that EXPAR has a significant sequence dependence [5]. Moreover, their experimental results indicated that background amplification is related to specific base sequences. Typically, DNA polymerase has a strong affinity for purine oligonucleotides, so amplification templates rich in GA sequences often exhibit more pronounced nonspecific background signals.

The formation of hairpin structures at the 3′-end of the template can also trigger template extension reactions, which can generate target (primer) sequences and thus generate high background signals. People are attempting to chemically seal the 3′-end of the template to prevent it from being extended by polymerase. Niemz and his colleagues compared the differences between closed and unclosed templates and found that there was no statistically significant difference in background amplification results [4]. These seemingly theoretical violations can be understood by considering the following estimates: if the total concentration of the template is 100 nM (10^{-7} M) and the blocking efficiency of the template is 99%. Then the concentration of the unclosed template is 1 nM (10^{-9} M). Compared to the concentration of the target to be analyzed (e.g., 10^{-15} M), an unclosed template concentration of 10^{-9} M or 10^{-7} M is sufficient to cause background amplification.

5.2 Detection of EXPAR Products

5.2.1 Fluorescence

Fluorescence is the most commonly used method for detecting EXPAR amplification products, and researchers have achieved real-time or endpoint detection of EXPAR amplification products using various fluorescent dyes and probes. In Galas and his colleagues' earliest study, they monitored the EXPAR amplification process using the dsDNA-specific SYBR Green I (SG) fluorescent dye (Fig. 5.4a) [1]. As the reaction progresses, the EXPAR template transforms from single-stranded DNA (ssDNA) to dsDNA, leading to an exponential increase in dsDNA and, consequently, an exponential increase in the fluorescence signal. By using real-time fluorescence detection instruments (such as real-time fluorescence PCR) to monitor the fluorescence curves during the EXPAR reaction, one can obtain the real-time fluorescence intensity curve of EXPAR (Fig. 5.4b). Similar to PCR real-time fluorescence curves, the POI value (Point of Inflection)—which corresponds to the reaction time at the maximum slope point of the real-time fluorescence curve—can be used as a quantitative basis. By plotting the POI value on the y-axis and the logarithm of the target concentration on the x-axis, a wide detection range can often be achieved (Fig. 5.4c).

The method of detecting EXPAR products using SYBR Green I (SG) has the advantages of high sensitivity, simplicity, and low cost. However, since all dsDNA emits a fluorescent signal upon binding with SG, the dye cannot distinguish between specific and nonspecific amplification products. As a result, nonspecific amplification can lead to background signals.

Researchers have also designed various fluorescent probes to detect EXPAR amplification products. Ye et al. developed an EXPAR system capable of generating a reporter oligonucleotide that forms fluorescent silver nanoclusters (AgNCs) in the

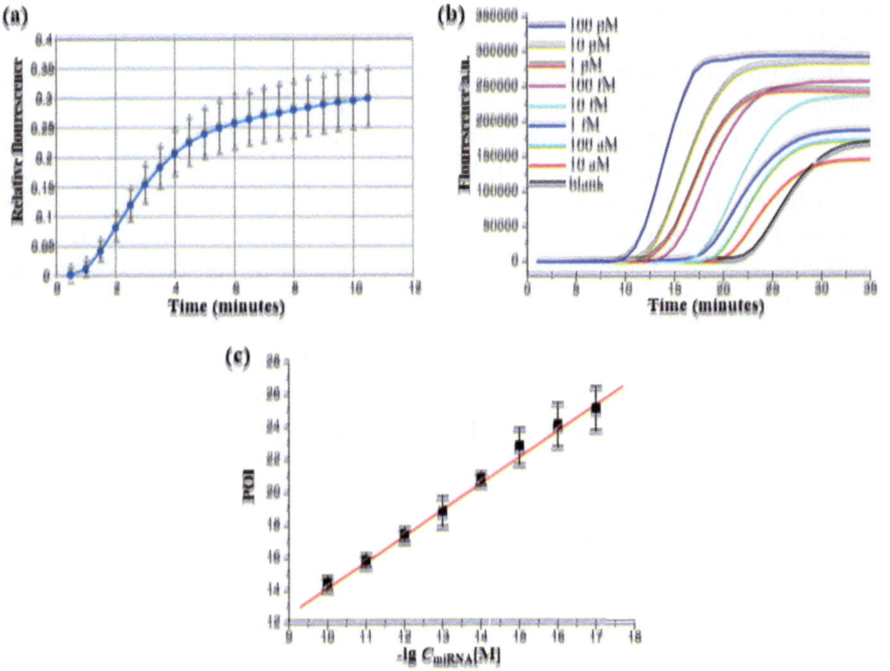

Fig. 5.4 Real-time monitoring of EXPAR amplification process using SYBR Green I [1]. (Revised and reprinted with permission from Ref. [1])

presence of Ag^+ and $NaBH_4$, thereby enabling the detection of EXPAR amplification products [6, 7]. As shown in Fig. 5.5, the amplification template in this EXPAR system consists of five regions (AXAXB). Two repeated sequences (A) are complementary to the target miRNA, and two repeated sequences (X) serve as recognition sequences for NEase. The B sequence is complementary to the reporter oligonucleotide (R), which acts as a scaffold for the synthesis of fluorescent silver nanoclusters. The target miRNA initiates EXPAR, leading to a cyclic amplification reaction involving extension, enzymatic cleavage, and strand displacement, ultimately producing a large number of reporter oligonucleotide sequences (R). Using R as the scaffold, fluorescent silver nanoclusters (DNA-AgNCs) are synthesized in the presence of Ag^+ and reduced by $NaBH_4$. By detecting the fluorescence intensity, the target nucleic acid can be both detected and quantitatively analyzed.

Chunyang Zhang's research team designed a dual-template EXPAR system (Fig. 5.6) [8], which produces a functional DNA (Y). Y can hybridize with a pair of capture probes (modified with biotin) and reporter probes (modified with Cy5 fluorophore). The biotin on the capture probe can specifically bind to streptavidin modified on the surface of quantum dots (QDs). This allows the Cy5-labeled reporter probes to aggregate on the QD surface, forming a complex. When excited at 488 nm, the QD transfers energy to Cy5 via fluorescence resonance energy transfer (FRET), enabling the highly sensitive detection of miRNA.

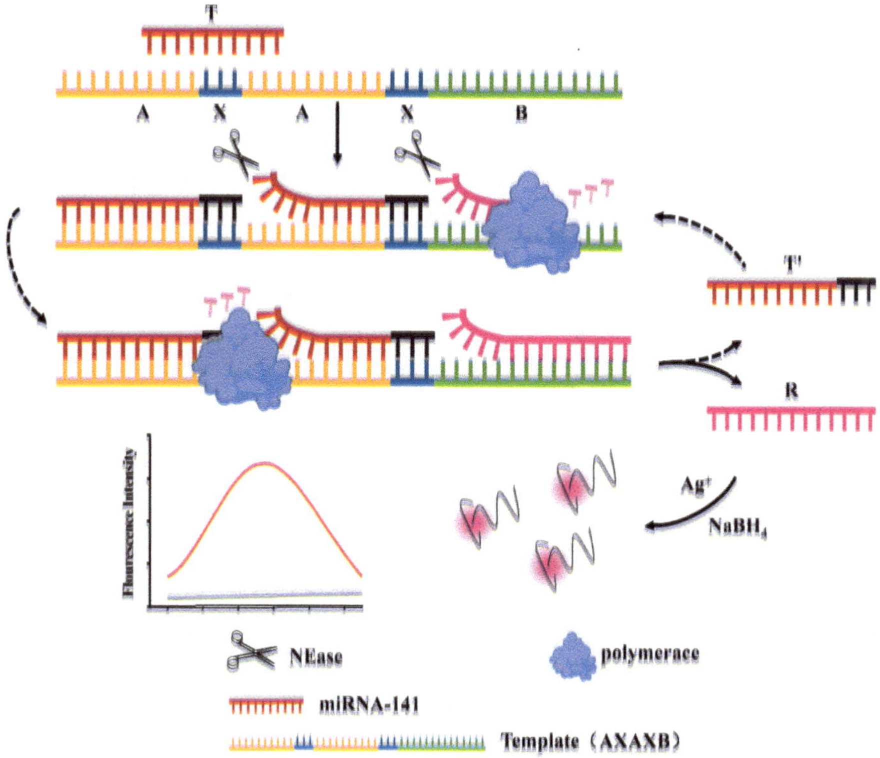

Fig. 5.5 The schematic diagram of miRNA detection based on EXPAR combined with silver nanocluster fluorescence signal amplification [6]. (Revised and reprinted with permission from Ref. [6])

Zinc(II)-protoporphyrin IX (ZnPPIX) is a porphyrin compound with zinc as its central metal, exhibiting inherent fluorescence. When bound to a G-quadruplex, it forms a supramolecular ZnPPIX/G-quadruplex structure, which significantly enhances the fluorescence of ZnPPIX [9]. The ZnPPIX/G-quadruplex structure offers several advantages, including excellent photostability, non-toxicity, biocompatibility, and low background signals. Similarly, an EXPAR template sequence can be designed to produce functional oligonucleotide sequences capable of forming G-quadruplexes. In the presence of ZnPPIX, the compound binds to the generated G-quadruplex, enhancing its fluorescence signal. By detecting the fluorescence signal, the target can be analyzed [10, 11].

In addition, Chun-yang Zhang's team designed another dual-template system utilizing an integrated molecular beacon on the second template [12]. After the target primer initiates the amplification reaction of the first template, a large number of primers are generated. These primers further initiate the EXPAR amplification reaction on the second template. The sequences produced by the second template's EXPAR reaction continue to trigger amplification on the first template, thereby

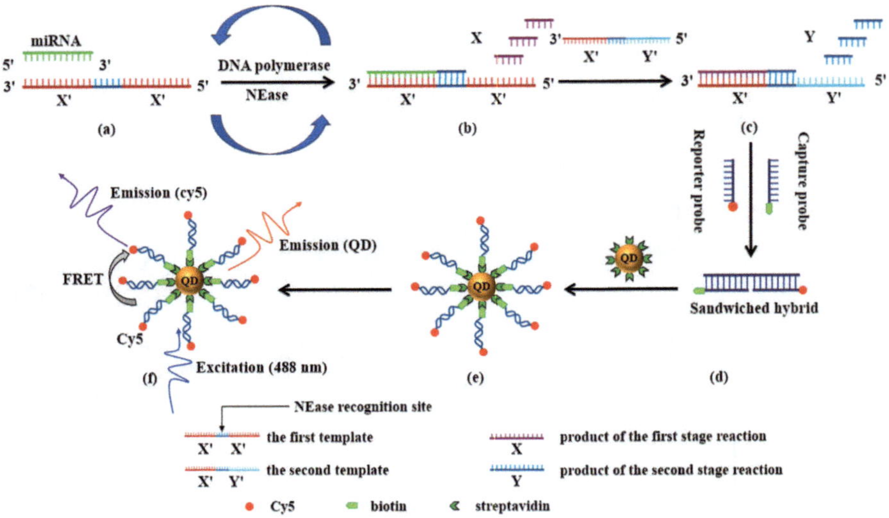

Fig. 5.6 MiRNA detection scheme based on dual-template EXPAR system and quantum dot nanosensor [8]. (Revised and reprinted with permission from Ref. [8])

forming a cyclic process. The molecular beacon structure on the second template is opened during polymerase extension, causing the fluorophore and quencher on the molecular beacon to separate. This separation restores the fluorescence signal of the fluorophore, enabling detection.

Argonaute (Ago) proteins are gene-editing tools that use short DNA sequences with 5′-phosphate as guide strands (gDNA), which can precisely recognize and cleave complementary gDNA sequences with high specificity and efficient target cleavage capability. Due to their unique molecular recognition mechanism, Ago proteins hold great potential in molecular diagnostics, particularly in precise detection and gene manipulation. The research team led by Ji-Lie Kong combined the EXPAR amplification strategy with the *Thermus thermophilus* Ago (TtAgo) protein cleavage mechanism [13], as illustrated in Fig. 5.7. The EXPAR initiated by miRNA as a primer can generate a large number of short nucleic acid sequences with 5′-phosphate. These sequences serve as gDNA to activate the Ago protein cleavage activity. To achieve multiplexed miRNA detection, the strategy involves designing various hairpin probes labeled with fluorescent and quencher groups, which are complementary to the EXPAR products (gDNA). The design of the hairpin probe ensures that the Ago protein cleavage site is located in the loop region of the hairpin structure (between the tenth and 11th nucleotides from the 3′ to 5′ direction). Upon cleavage of the hairpin probe, the fluorescent group and quencher group are separated, generating the fluorescence signal.

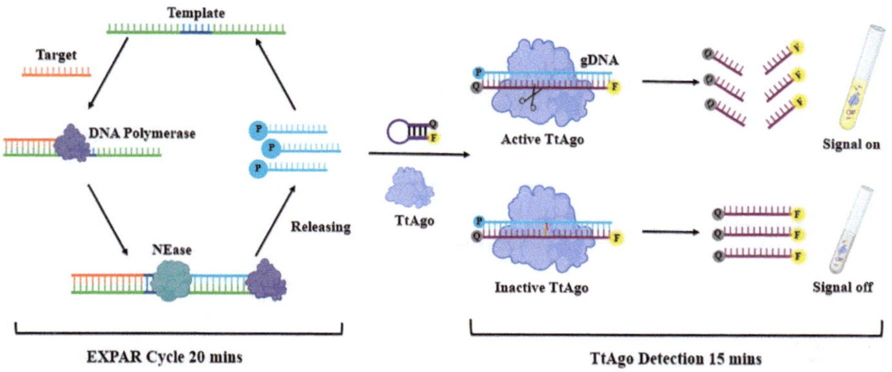

Fig. 5.7 Multiple detection of miRNAs based on Argonaute protein combined with EXPAR [13]. (Revised and reprinted with permission from Ref. [13])

5.2.2 *Colorimetric*

Colorimetric detection methods allow direct detection with the naked eye without requiring instrumentation. This method utilizes target nucleic acid molecules to initiate the EXPAR reaction, generating a functional ssDNA that can aggregate gold nanoparticles (AuNPs) and produce a color change [14–16]. As shown in Fig. 5.8, Niemz and colleagues constructed a dual-template EXPAR system (X′-X′ and Y′-X′). The target nucleic acid X initiates classical EXPAR amplification using the first template (X′-X′), while the second template (Y′-X′) converts the target nucleic acid sequence into functional ssDNA (Y). The ssDNA Y induces the aggregation of DNA-functionalized AuNPs, leading to a color change in the solution from red to purple. The presence of the target nucleic acid can be visually determined based on this color change. In a similar manner, a dual-template system has been applied in lateral flow strip assays (Fig. 5.8d). Functional ssDNA anchors AuNPs to the test line (T), where a red color appears, enabling the visual detection of the target X [17].

Zhang et al. combined a dual-template EXPAR system with a G-quadruplex/hemin DNAzyme to achieve label-free colorimetric DNA analysis [18]. As shown in Fig. 5.9, this method designs the functional ssDNA Y in the dual-template EXPAR system as G-rich DNA. The dual-template EXPAR system, initiated by the target sequence, ultimately generates a large number of G-rich DNA fragments. These DNA fragments can bind with hemin to form a G-quadruplex structure complex. The resulting complex exhibits peroxidase-like catalytic activity, similar to that of horseradish peroxidase (HRP), and is referred to as the G-quadruplex/hemin DNAzyme (G4/hemin DNAzyme), also known as G-quadruplex peroxidase. This enzyme subsequently catalyzes the colorimetric reaction of TMB-H_2O_2 (3,3′,5,5′--tetramethylbenzidine and hydrogen peroxide). In this reaction, TMB, a colorless substrate, is oxidized under the catalytic action of the DNAzyme, with H_2O_2 serving as the oxidizing agent, to produce a visible optical signal. Leveraging the advantages

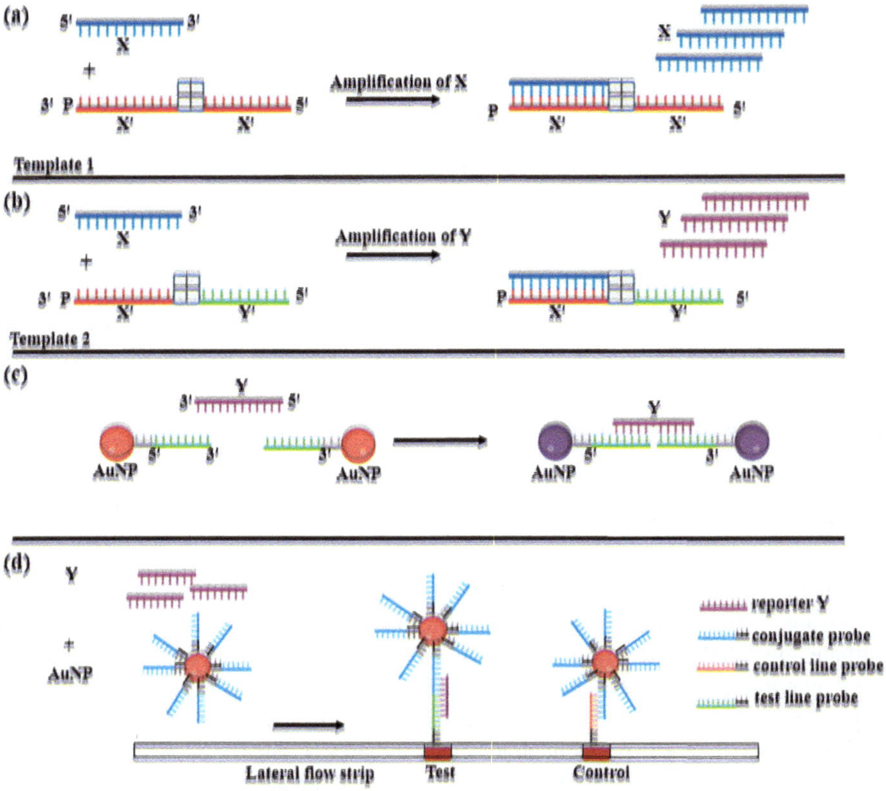

Fig. 5.8 Schematic illustration of visual detection of EXPAR products using AuNPs. (**a**) Exponential amplification reaction of target sequence X. (**b**) Amplification reaction of functional sequence Y. (**c**) AuNP aggregation induced by sequence Y. (**d**) Lateral flow strip detection of EXPAR products using AuNPs [15, 17]. (Revised and reprinted with permission from Refs. [15, 17])

of isothermal, label-free, and homogeneous analysis, and without requiring complex synthesis or additional operations, the entire process can be completed within 1 h.

5.2.3 Chemiluminescence and Raman

Chemiluminescence (CL), a widely used biomarker detection method in clinical diagnostics, is applied to detect EXPAR products due to its advantages of simple instrumentation, rapid reactions, and low cost [19]. The research team led by Huang-Xian Ju combines the dual-template EXPAR system with chemiluminescence (CL) to detect EXPAR products, as shown in Fig. 5.10. The first template is designed in a hairpin structure, including regions I, II, and III [19]. Region I is the stem structure, Region II is the recognition sequence for the NEase, and Region III is the

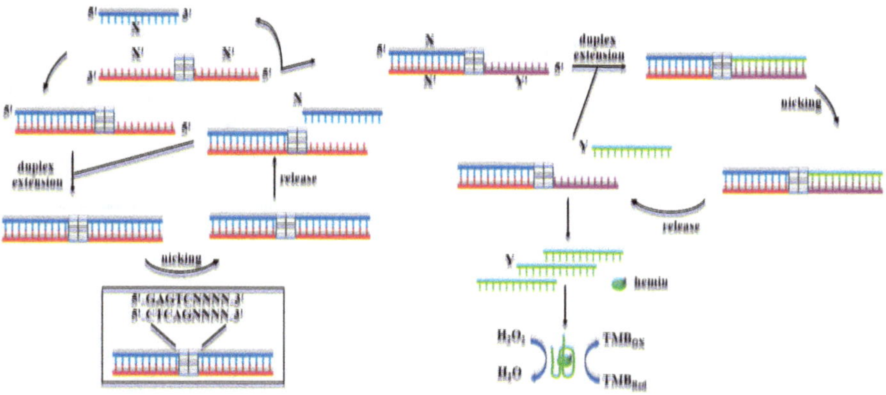

Fig. 5.9 Schematic diagram of specific detection of DNA using EXPAR [18]. (Revised and reprinted with permission from Ref. [18])

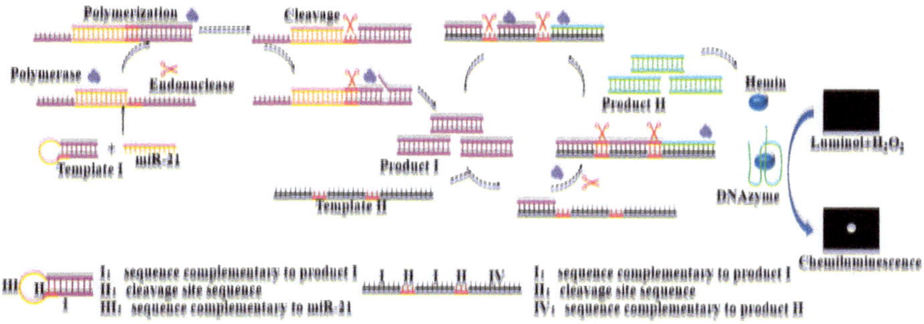

Fig. 5.10 Schematic diagram of chemiluminescence imaging strategy based on cascade EXPAR mechanism for miRNA detection [19]. (Revised and reprinted with permission from Ref. [19])

sequence complementary to the target miRNA. Template II consists of two repeat regions (Region I) and two repeat NEase recognition sequences (Region II), and a complementary sequence to the functional G-quadruplex sequence at the 5′ end of the template. In the presence of miR-21, the hairpin structure in Region III of the template hybridizes with the target miR-21. Under the catalytic activity of the polymerase and NEase, an extension/cleavage mechanism is initiated, producing product I. Product I then serves as a primer to further trigger the EXPAR amplification reaction, generating large amounts of products I and II. Product II is a G-rich sequence that can self-assemble with hemin to form a G-quadruplex structure complex with DNAzyme activity. This DNAzyme exhibits catalytic activity similar to that of peroxidases (e.g., horseradish peroxidase, HRP). In the presence of hydrogen peroxide (H_2O_2), the DNAzyme catalyzes the oxidation of luminol, producing light (usually within the wavelength range around 425 nm), enabling the detection of miR-21.

To enhance the specificity and multiplex detection capability of EXPAR, the research team led by Chun-yang Zhang utilizes surface-enhanced Raman

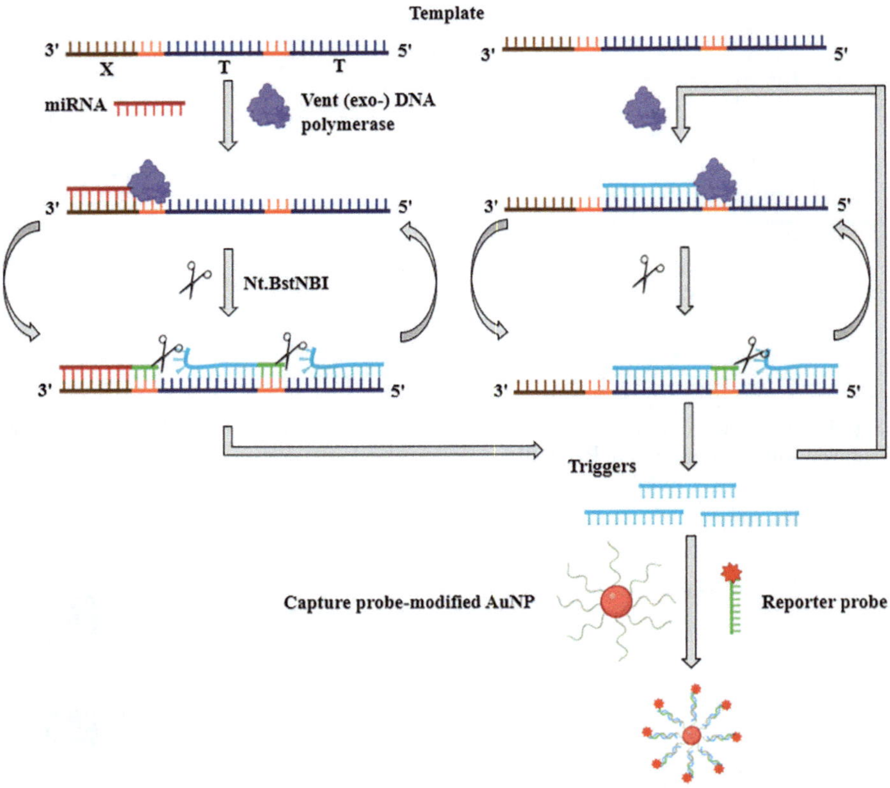

Fig. 5.11 Schematic of miRNA detection based on EXPAR and surface-enhanced Raman spectroscopy [20]. (Revised and reprinted with permission from Ref. [20])

spectroscopy (SERS) to detect EXPAR products, enabling the simultaneous and sensitive detection of multiple miRNAs in non-small cell lung cancer (NSCLC) [20]. As shown in Fig. 5.11, in the presence of the target miRNA, the miRNA hybridizes with the template sequence X, which can trigger linear amplification under the catalytic activity of Vent (exo^{-}) DNA polymerase and the endonuclease Nt.BstNBI. The linear amplification produces functional oligonucleotide Y. The generated Y can further trigger EXPAR amplification, producing large amounts of Y. Y will bind the reporter probes, capable of generating Raman signals, to the DNA-modified AuNP surface, producing a SERS signal [21]. More importantly, due to the strong characteristic nature of Raman spectra, different fluorescent groups can be modified on the reporter probes for different miRNAs, allowing the detection of multiple miRNAs.

It is worth noting that the sensitivity of the EXPAR-based SERS method has been significantly improved. Compared to the unlabeled SERS method (28 nM) [22], the sensitivity of the EXPAR-based SERS method (0.5 fM) has increased by 8 orders of magnitude. Compared to previously reported SERS-based direct detection methods (0.36 nM) [23], the sensitivity has improved by 6 orders of magnitude.

5.2.4 *Electrochemical Methods*

Ding et al. developed an electrochemical method for detecting EXPAR products [24]. As shown in Fig. 5.12, the amplification template with a thiol group at the 3′ end is fixed on the gold electrode surface via gold-sulfur bonds. The template contains three regions (ABA). Two repeated A regions are complementary to the target miRNA, and the B region contains the recognition sequence for the Nt.BstNBI endonuclease. The miRNA hybridizes with the A region of the template, and in the presence of DNA polymerase and dNTPs, the miRNA (as a primer) can extend along the template to form dsDNA. The Nt.BstNBI then recognizes the cleavage site in the B region, generating cuts. Therefore, extension, cleavage, and strand displacement can be repeatedly triggered, leading to exponential amplification of the target miRNA. Finally, a large amount of dsDNA can be generated on the electrode surface. After EXPAR, the remaining single-stranded template on the electrode surface hybridizes with biotin-labeled capture probes, which are then connected to streptavidin-modified alkaline phosphatase (ST-AP). The more remaining single-stranded templates on the electrode surface, the more AP is enriched on the electrode surface. The generation of the electrical signal is mainly completed by the following two reactions:

1. Enzymatic hydrolysis reaction: AP catalyzes the hydrolysis of α-naphthyl phosphate (α-NP), producing α-naphthol and phosphate ions (α-NP + H_2O → α-naphthol + phosphate).
2. Electrochemical oxidation reaction of α-naphthol: on the electrode surface, α-naphthol undergoes oxidation (α-naphthol → α-naphthoquinone + $2H^+$ + $2e^-$), generating a measurable current signal.

Chen et al. immobilized an oligonucleotide with a G-quadruplex structure onto the gold electrode surface, as shown in Fig. 5.13 [25]. They used an EXPAR template that hybridizes with the oligonucleotide to form an arched probe, which

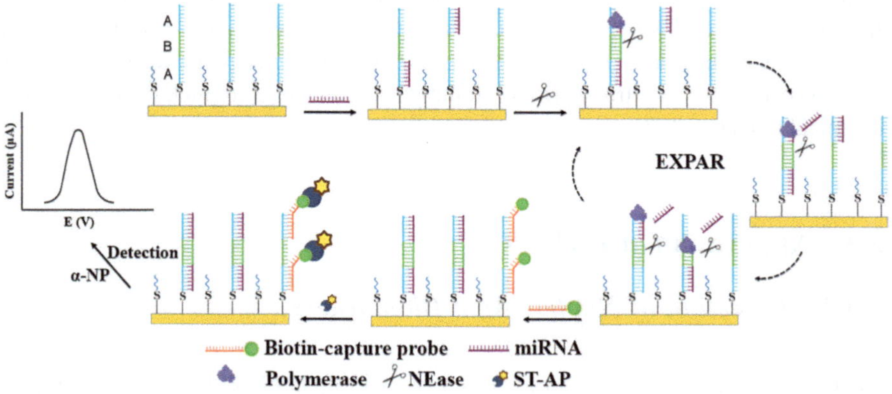

Fig. 5.12 Schematic diagram of electrochemical biosensor based on target-initiated isothermal exponential amplification reaction [24]. (Revised and reprinted with permission from Ref. [24])

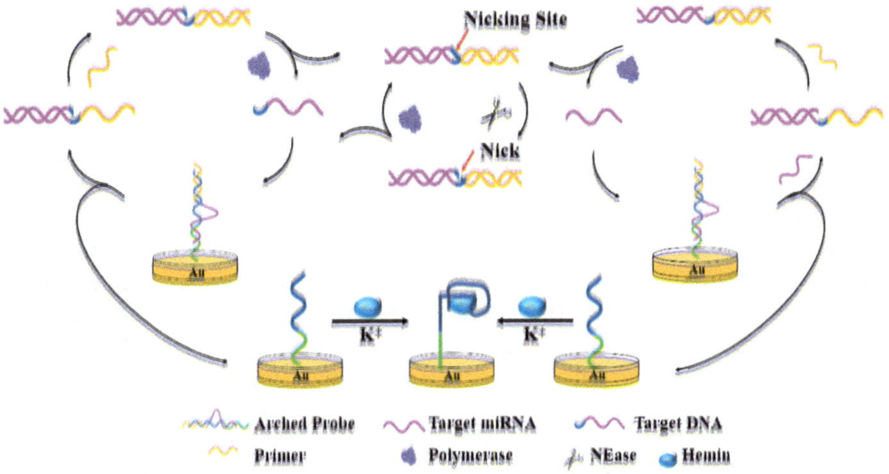

Fig. 5.13 Schematic of miRNA detection using the arched probe-mediated EXPAR strategy [25]. (Revised and reprinted with permission from Ref. [25])

can prevent the G-quadruplex from forming its G4 structure. The target miRNA hybridizes with the template, triggering EXPAR, which causes the template to detach from the electrode surface. The oligonucleotide fixed on the electrode surface forms a G-quadruplex complex with hemin, exhibiting DNAzyme activity similar to horseradish peroxidase. The catalytic reaction of the DNAzyme affects the redox process of $Fe(CN)_6^{3-/4-}$, and the electrochemical signal of $Fe(CN)_6^{3-/4-}$ reflects the catalytic activity of the DNAzyme, enabling the detection of miRNA [26, 27]. This approach is simple and rapid, allowing for ultrasensitive detection of miRNA at femtomolar levels.

Additionally, Zhang et al. developed a dual-template system (Y′-X′ and Y′-Y′), where Y is a G-rich sequence that forms a G-quadruplex complex with hemin, acting as a DNAzyme. The DNAzyme catalyzes the hydroquinone-H_2O_2 reaction (H_2O_2 + hydroquinone → quinone + $2H_2O$), generating an electrochemical signal. In this method, the DNAzyme that catalyzes the redox reaction is not immobilized on the microelectrode, but instead, a miniaturized electrochemical device is employed to achieve ultra-sensitive, homogeneous, label-free nucleic acid electrochemical detection [28].

5.3 Application of EXPAR

5.3.1 Detection of Nucleic Acid Biomarkers

Nucleic acid biomarkers detections play a crucial role in biological function studies and disease diagnosis. EXPAR technology is particularly suitable for the analysis of short-chain nucleic acid biomarkers, especially miRNA detection. Hong-Xia Jia

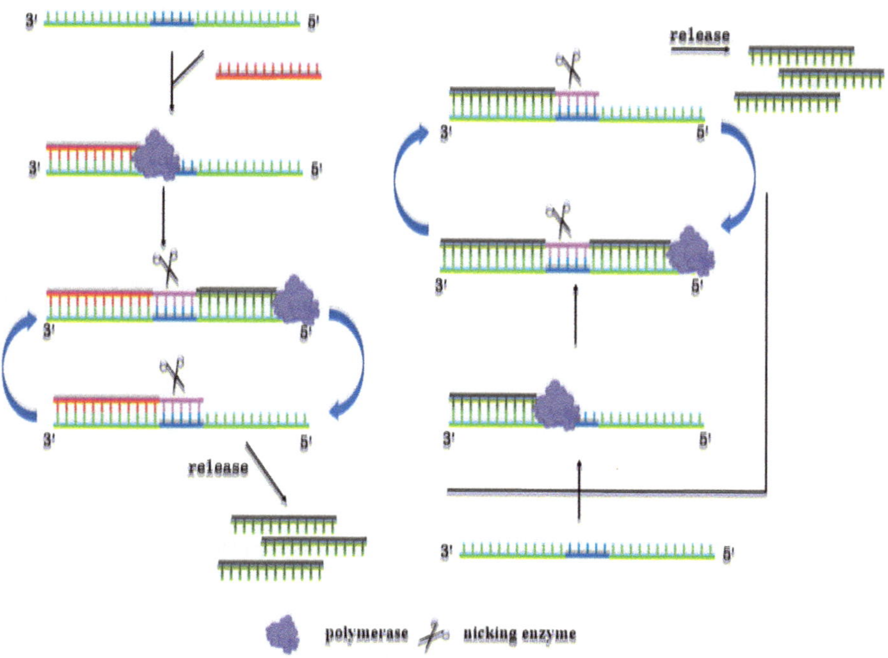

Fig. 5.14 Schematic diagram of EXPAR triggered by let-7a miRNA as primer [3]. (Revised and reprinted with permission from Ref. [3])

et al. first demonstrated that short RNA (miRNA) can act as a primer to initiate EXPAR [3], establishing an ultra-sensitive miRNA analysis method. The principle is shown in Fig. 5.14. First, an amplification template (X′-X′) is designed based on the target miRNA sequence (Let-7a as an example). This template contains identical sequences at both the 3′ and 5′ ends, complementary to the target miRNA sequence. The middle black region is the recognition sequence for Nt.BstNBI endonuclease. To prevent the amplification template from self-elongating, its 3′ end is blocked, such as by phosphorylation modification. When the target miRNA hybridizes with the 3′ end of the amplification template, Vent (exo^-) DNA polymerase catalyzes the miRNA extension, forming a double-stranded template. The NEase specifically recognizes the cleavage site on the dsDNA and generates a nick in the upper strand. The DNA polymerase extends at the nick position and displaces the previously extended DNA, continually releasing DNA fragments, thereby forming a linear amplification mechanism. On the other hand, the released DNA fragments hybridize with the 3′-end of the new amplification template, initiating extension-cleavage-strand displacement. This cycle repeats, forming a rapid exponential amplification reaction. Using a dsDNA-specific fluorescent dye, SG, real-time monitoring of fluorescence intensity enables ultra-sensitive miRNA quantitative analysis.

In addition, the 3′-end of plant miRNAs is 2′-O-methylated, which reduces the efficiency of extension reactions catalyzed by DNA polymerase. Xiang-Dong Wang et al. designed the 3WJ-EXPAR system for detecting plant miRNAs, as shown in

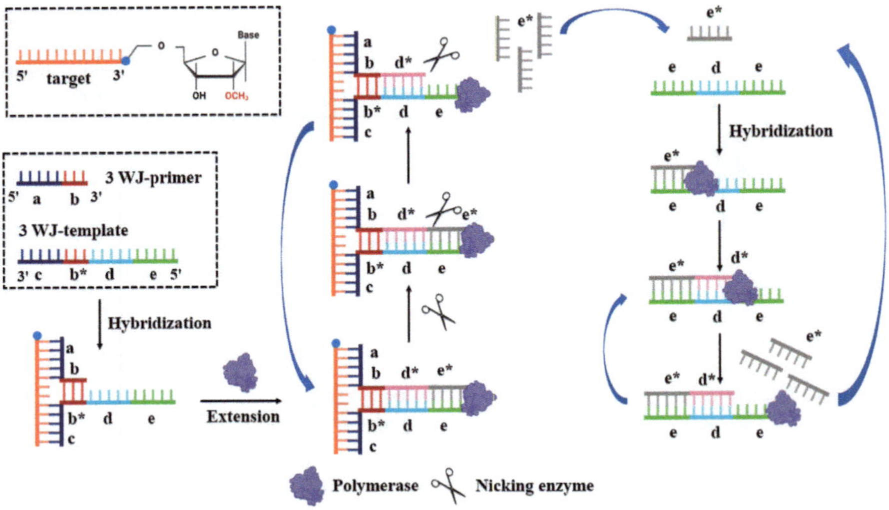

Fig. 5.15 Schematic diagram of the 3WJ-EXPAR system for detecting plant miRNA [29]. (Revised and reprinted with permission from Ref. [29])

Fig. 5.15 [29]. Based on the target plant miRNA sequence, they designed 3WJ primers and 3WJ templates. The sequence "a" in the 3WJ primer and the sequence "c" in the 3WJ template are complementary to half of the miRNA sequence, respectively. Meanwhile, the sequence "b" in the 3WJ primer and the sequence "b*" in the 3WJ template are designed as complementary sequences of 7 base pairs. In the presence of target miRNA, the sequence "a" in the 3WJ primer and the sequence "c" in the 3WJ template will hybridize with the target miRNA, and in this case, the short sequence "b" in the 3WJ primer and the sequence "b*" in the 3WJ template will also form a stable double-stranded structure, generating a 3WJ structure connected to the miRNA. Then, under the catalysis of Bst DNA polymerase, sequence "b" in the 3WJ primer will extend along the sequence "d–e" in the 3WJ template, where sequence "d" is designed as the recognition sequence for the endonuclease Nt.BstNBI. After extension, the endonuclease recognizes the site and cleaves the upstream DNA strand. The cleaved strand containing the recognition sequence will extend again along the 3WJ template, and the short single-stranded "e*" will be displaced by the strand displacement activity of Bst DNA polymerase. As a result, extension, cleavage, and strand displacement can be continuously repeated, producing a large amount of "e*" sequence. Meanwhile, the "e*" sequence acts as a primer to initiate subsequent EXPAR amplification. In contrast, in the absence of target miRNA, the melting temperature (Tm) of the 7 base pairs of the "b/b*" sequences is about 24 °C, which is lower than the reaction temperature of 37 °C. Thus, the 3WJ primer and 3WJ template cannot stably bind to initiate the polymerization reaction, and EXPAR does not occur.

Theoretically, efficient exponential amplification techniques can amplify a single nucleic acid molecule to detectable levels, but background signals resulting from nonspecific amplification limit their sensitivity. To effectively reduce the background of nucleic acid amplification and achieve single-molecule and single-cell analysis, Zheng-Ping Li's research team adopted a single microsphere strategy [30], where the amplification template is fixed on the surface of a single microsphere for the EXPAR reaction. This strategy prevents the hybridization and extension reactions between amplification templates, effectively reducing nonspecific amplification. At the same time, it allows the fluorescence signals generated from the amplification reaction to be concentrated in a very small area, significantly improving the sensitivity of detection.

In addition, various specific EXPAR designs combined with multiple detection methods have been used for miRNA detection, such as fluorescence probe methods [6, 8, 12], multiplex detection methods for Argonaute proteins [13], chemiluminescence [19], surface-enhanced Raman spectroscopy [20], electrochemical detection methods [24, 25], and so on. These methods provide diverse options for the ultrasensitive detection of miRNAs.

For long-chain mRNA, the 3′ end of the target sequence to be detected is located in the middle of the RNA and cannot directly initiate EXPAR. To address this issue, as shown in Fig. 5.16, Hui Wang and colleagues designed a DNA probe that can specifically hybridize with sequences I and II in the target mRNA, forming a single-

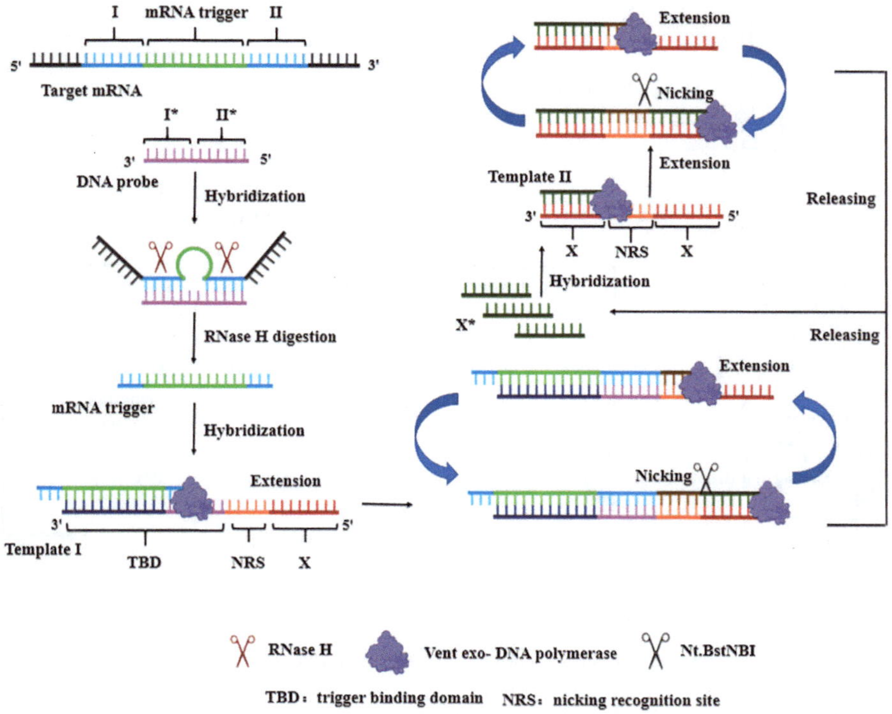

Fig. 5.16 Schematic diagram of mRNA detection of EXPAR based on RNase H assistance [31]. (Revised and reprinted with permission from Ref. [31])

stranded loop structure [31]. RNase H specifically digests the mRNA hybridized with the DNA probe, releasing the single-stranded mRNA trigger. The mRNA trigger can then act as a primer to initiate the dual-template EXPAR (Y′-X′, X′-X′) amplification system. This design overcomes the limitations of EXPAR in long-chain RNA detection, greatly expanding the scope of EXPAR applications.

mRNA plays a crucial role as a bridge between DNA genetic information and functional protein synthesis in gene expression. Mutations at specific sites in mRNA can alter the structure and function of the corresponding expressed protein, which is closely related to the onset and progression of various diseases. The RNase H specifically cleaves RNA sequences in DNA/RNA hybrid duplexes at random sites, meaning that the aforementioned method cannot be used for single-base mutation analysis of RNA. Cheng-Hui Liu and colleagues proposed an isothermal exponential amplification (EXPAR) system based on RNase H site-specific cleavage, which enables high-specificity detection of single-base mutations in mRNA without the need for a reverse transcription step [32]. As shown in Fig. 5.17, this

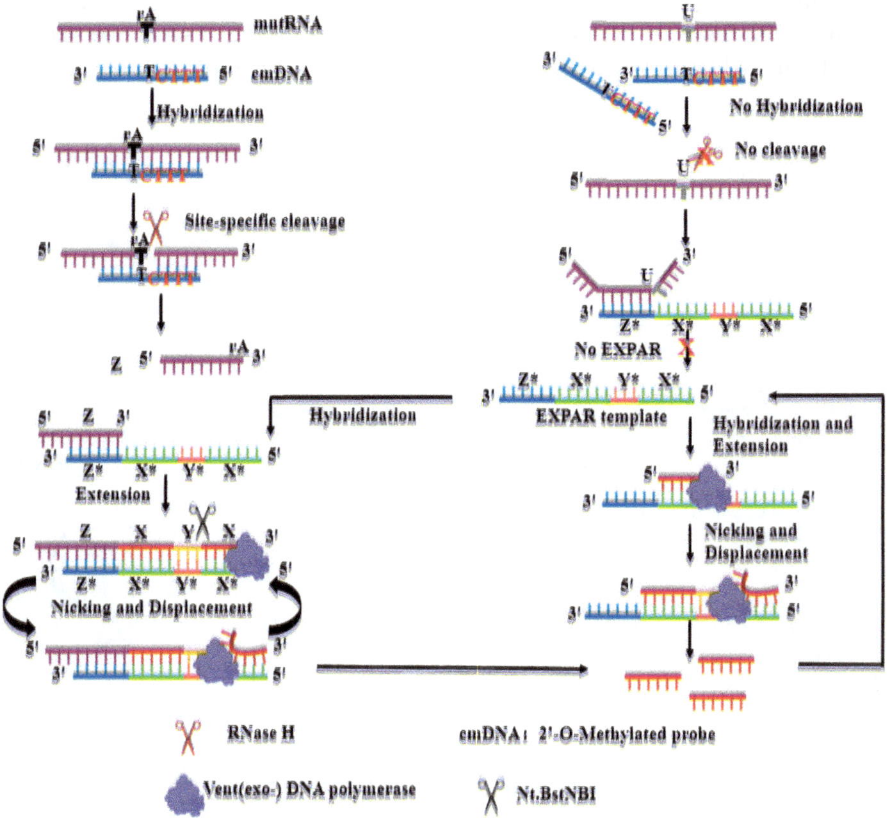

Fig. 5.17 The principle of the EXPAR method based on the site-specific cleavage mechanism for analyzing RNA single-nucleotide mutation. The CTTT highlighted in red in the chimeric DNA probe represents the 2′-O-methylated nucleotides [32]. (Revised and reprinted with permission from Ref. [32])

method designs a chimeric DNA probe (cmDNA) with oxygen-methylated bases to specifically recognize mRNA mutation sites and guide RNase H to perform precise cleavage only at the mutation site. The cmDNA probe can fully pair with mutant mRNA (mutRNA) but has a single-base mismatch with wild-type mRNA (wtRNA). At the appropriate temperature, cmDNA hybridizes exclusively with mutRNA, enabling RNase H to perform site-specific cleavage at the mutation site on mutRNA, exposing the 3′-OH terminus and triggering subsequent efficient EXPAR. Due to the Tm value difference caused by the single-base mismatch, wtRNA cannot effectively hybridize with cmDNA. Even if a wtRNA/cmDNA hybrid structure is occasionally formed, the mismatch prevents RNase H from performing site-specific cleavage, thus preventing the initiation of the EXPAR reaction. This cmDNA-assisted RNase H site-specific cleavage mechanism effectively avoids false-positive interference caused by wtRNA, and in combination with the high sensitivity of EXPAR, successfully achieves accurate detection of specific mutation sites in mRNA. This method also has good versatility, as adjusting the base sequence of the 2′-O-methylated DNA probe allows it to be applied to the detection of different mRNA mutation sites.

On the other hand, studies have found that the nuclease BstNI can selectively cleave DNA fragments in DNA/RNA hybrid duplexes. Timothy R. Dafforn and colleagues utilized the site-specific cleavage activity of BstNI to develop a rapid method for converting RNA into short-chain DNA without the need for reverse transcription, as shown in Fig. 5.18. This method, combined with the EXPAR amplification technique, was applied for ultra-fast detection of COVID-19 RNA [33]. The specific recognition sequence of the BstNI enzyme is CCWGG, where W represents A or T. Upon recognizing this sequence, BstNI cleaves between the two bases, specifically cutting between C↓CWGG. The cleaved DNA fragment then serves as an EXPAR primer, initiating the EXPAR amplification reaction for COVID-19 detection.

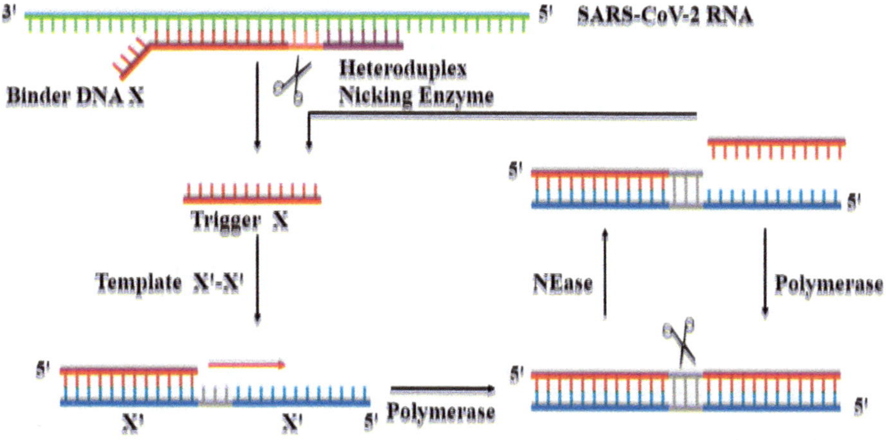

Fig. 5.18 Schematic diagram of rapid detection of COVID-19 RNA based on BstNI enzyme-assisted EXPAR [33]. (Revised and reprinted with permission from Ref. [33])

According to the principle of EXPAR, the EXPAR technology cannot directly detect double-stranded genomic DNA, so additional designs and steps are required to generate ssDNA as primers to initiate the EXPAR amplification. For example, by locating recognition sequences for NEase on the genomic DNA, and using the strand displacement activity of polymerase and the nicking activity of NEase, ssDNA sequences can be generated from the genomic DNA to trigger the EXPAR process, thus enabling the analysis of genomic DNA (Fig. 5.19) [1].

In addition, Xiao-Yong Zou et al. used the $NaIO_4$/LiBr oxidation-hot piperidine method combined with EXPAR for methylation analysis [34]. As shown in Fig. 5.20, due to the slightly lower redox potential of 5-methylcytosine (5-mC) compared to cytosine, the C5–C6 double bond of 5-mC is selectively epoxidized under the action of $NaIO_4$. Subsequently, bromination occurs at the C6 position in the presence of LiBr, making it prone to cleavage. Piperidine, a strongly alkaline chemical, specifically cleaves the oxidized nucleotide residues. Finally, under high-temperature conditions, piperidine interacts with the oxidized DNA, causing the cleavage of the phosphodiester bonds, thereby localizing the methylated regions. The resulting DNA fragments can hybridize with the EXPAR template to trigger the EXPAR amplification, enabling quantitative analysis of DNA methylation levels.

In addition, Yue-Ying Sun et al. utilized methylation-dependent restriction endonucleases, such as GlaI, to specifically cleave the recognition sites on genomic DNA. The cleaved DNA can then initiate EXPAR amplification, enabling methylation analysis of specific sites [35]. As shown in Fig. 5.21, this endonuclease specifically recognizes methylated DNA sites, 5′-GmCGmC-3′/3′-mCGmCG-5′, and cleaves at the middle of the GmCGmC sequence. The cleaved DNA fragments can serve as

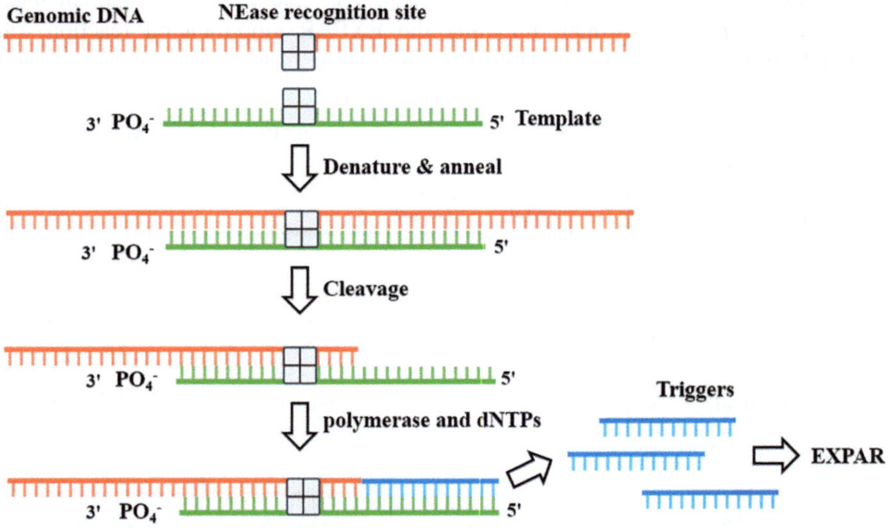

Fig. 5.19 Schematic diagram of genomic DNA analysis based on EXPAR [1]. (Revised and reprinted with permission from Ref. [1])

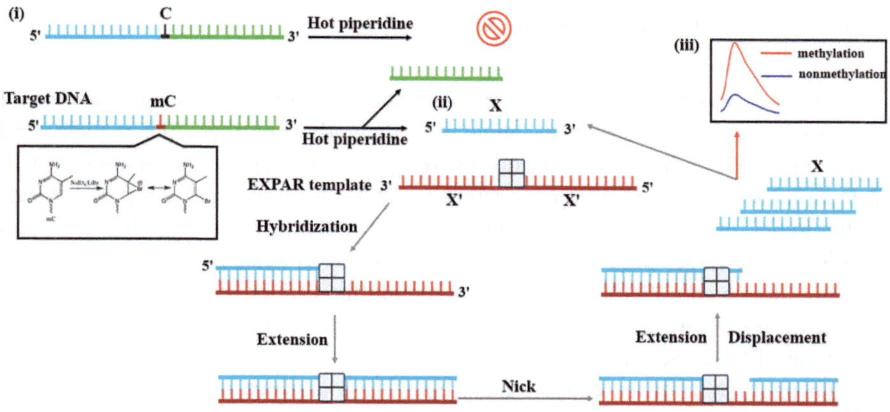

Fig. 5.20 The schematic diagram for methylation analysis based on the $NaIO_4$/LiBr oxidation-hot piperidine method combined with EXPAR [34]. (Revised and reprinted with permission from Ref. [34])

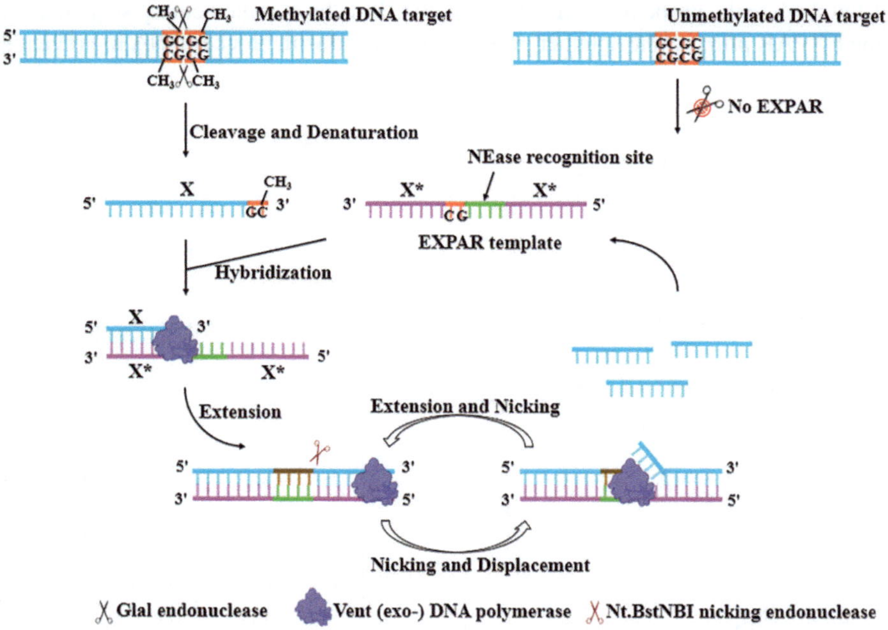

Fig. 5.21 The schematic diagram for DNA methylation analysis based on the restriction endonuclease combined with isothermal exponential amplification reaction (EXPAR) [35]. (Revised and reprinted with permission from Ref. 35

primers to initiate the EXPAR amplification reaction, thus enabling methylation analysis of specific DNA sites. In contrast, unmethylated 5′-GCGC-3′/3′-CGCG-5′ sites are not recognized by GlaI and remain intact, preventing the initiation of the subsequent EXPAR reaction.

5.3.2 Protein Detection

Highly sensitive protein detection is fundamental to biomedical research and clinical diagnostics. However, proteins cannot be efficiently amplified like nucleic acid molecules, which limits the sensitivity of protein detection. To enable protein detection using EXPAR, a target protein recognition mechanism is needed. Some technologies utilize nucleic acid aptamers that undergo structural changes upon binding to the target protein, triggering EXPAR. Other systems use target proteins that bind to specific DNA sequences, protecting the primer sequences from degradation by nucleases, allowing the primers to initiate EXPAR and thus enable the detection of the target protein.

The research team of Chun-Yang Zhang used a structural transition aptamer combined with EXPAR to detect PDGF-BB (Platelet-derived growth factor BB) protein. PDGF-BB is a low molecular weight cytokine that plays a critical role in cell transformation and tumorigenesis. As shown in Fig. 5.22 [36], the aptamer probe consists of three regions: region I is the specific binding region for PDGF-BB. Region II is the core of the probe, which forms a recognition sequence for Nt. BbvCI when the probe forms dsDNA. Region III is the complementary sequence for the EXPAR primer. The aptamer probe is hybridized with blocker DNA to prevent it

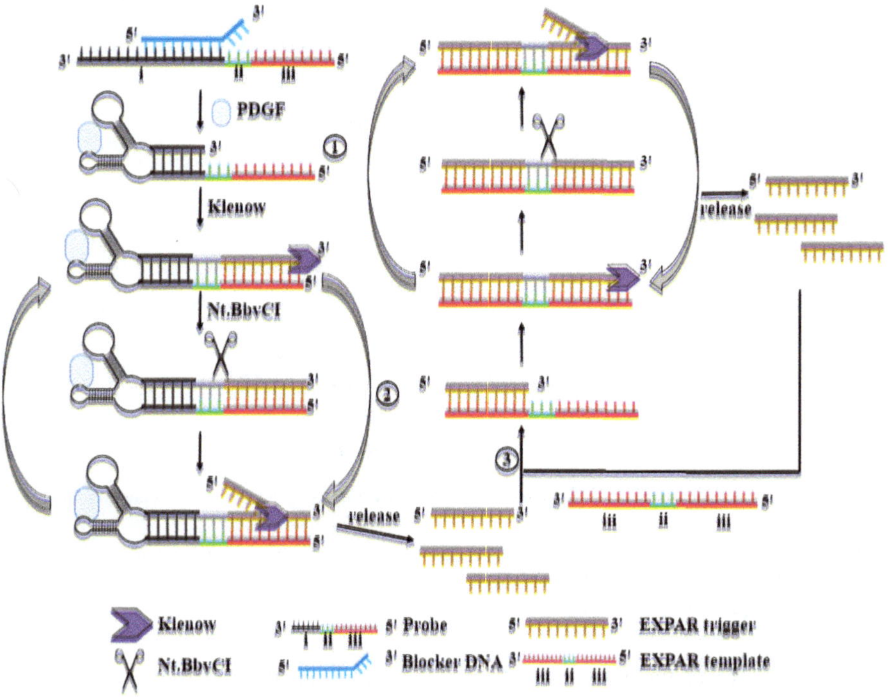

Fig. 5.22 The schematic diagram for the sensitive detection of PDGF-BB based on aptamer-initiated EXPAR [36]. (Revised and reprinted with permission from Ref. [36])

from folding into the PDGF-BB aptamer structure in the absence of the target protein. When PDGF-BB is present, it specifically binds to the aptamer, releasing the blocker DNA from the aptamer probe and allowing the aptamer to fold into a compact structure. At this point, the 3′-end of the aptamer probe can act as a primer, initiating an extension reaction in the presence of Klenow DNA polymerase and dNTPs, resulting in dsDNA with a Nt.BbvCI recognition sequence. The recognition sequence is subsequently cleaved, generating new replication sites for the polymerase. Through cycles of polymerization, cleavage, and substitution, short oligonucleotides are produced. These short oligonucleotides serve as primers for the EXPAR reaction, resulting in exponential amplification and the production of large amounts of dsDNA. In this study, SYBR Green I was used to detect the dsDNA. This detection method has high specificity and sensitivity, with a detection limit of 9.04×10^{-13} M and a detection range greater than five orders of magnitude. The proposed approach has advantages such as being isothermal, simple, rapid, cost-effective, without the need for multiple separations and washes, and requiring no labeled DNA probes. Additionally, the structural transition aptamer can be used to bind to target thrombin, releasing an oligonucleotide primer to initiate EXPAR, enabling thrombin detection [37]. Such methods can also be extended to the sensitive detection of other biomolecules (aptamers capable of undergoing similar conformational changes).

Chun-Yang Zhang's research team also developed two methods for detecting transcription factors (TFs). In each system, TFs bind to a dsDNA probe, protecting the probe from cleavage by exonucleases, and subsequently generating primers from the dsDNA probe to initiate the EXPAR amplification reaction [14, 38]. As shown in Fig. 5.23, the transcription factor (NF-κB p50) binds to the probe, which is a dsDNA containing a specific transcription factor recognition element and a T7 promoter sequence at the 5′ end of the upstream strand, and a 6-base protruding end at the 3′ end to prevent exonuclease III (EXO III) degradation. In the presence of the transcription factor, the binding of the TF to the dsDNA probe prevents it from being degraded by EXO III (which removes single nucleotides from the 3′-hydroxyl end of dsDNA). After heat-inactivating EXO III and the transcription factor, exonuclease I (EXO I) is added to degrade the 6-base protruding end of the dsDNA probe (EXO I degrades ssDNA from the 3′-hydroxyl end). At this point, both ends of the dsDNA probe become blunt ends. The blunt-ended probe can then be efficiently transcribed by T7 RNA polymerase, producing a large amount of RNA complementary to region I of the EXPAR template, which subsequently triggers the EXPAR reaction. In contrast, in the absence of the transcription factor, EXO III can remove a single nucleotide from the 3′-hydroxyl blunt end of the dsDNA probe. After heat-inactivating EXO III, EXO I is added to further degrade the remaining ssDNA. In this case, the dsDNA probe is completely degraded and cannot trigger the subsequent EXPAR reaction.

They also developed a similar method to detect folate receptors (FR) [11]. As shown in Fig. 5.24, this method designs a hairpin structure probe, with the 3′-end

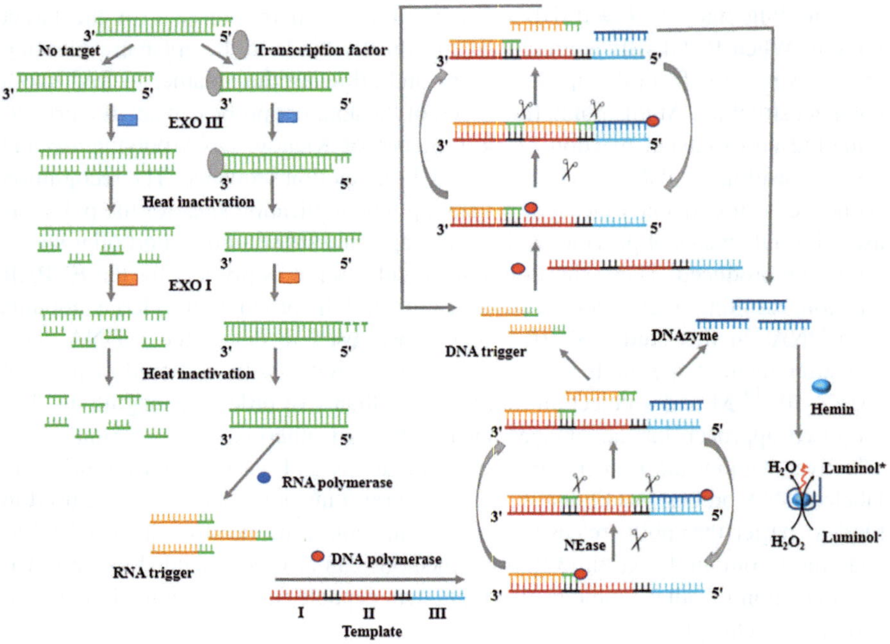

Fig. 5.23 Schematic diagram of transcription factor detection based on EXPAR [14, 38]. (Revised and reprinted with permission from Refs. [14, 38])

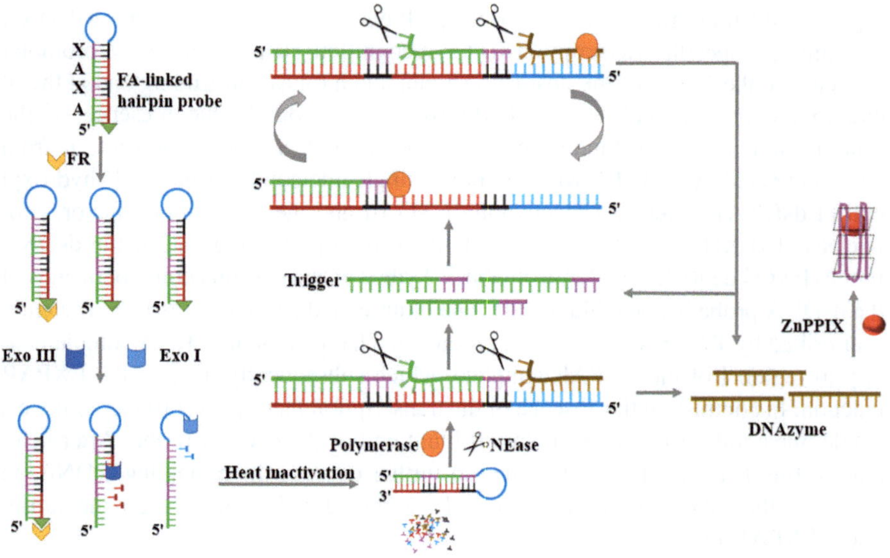

Fig. 5.24 Schematic diagram of folate receptor detection based on EXPAR method [11]. (Revised and reprinted with permission from Ref. [11])

covalently linked to folic acid (FA) molecules. FA specifically binds to FR. The binding protects the hairpin probe from degradation by EXO III and EXO I, maintaining the stem-loop structure of the probe. NEase recognizes the recognition sequence on the stem region of the hairpin probe and induces cleavage. Under the catalysis of the polymerase, an extension and replacement reaction is triggered, which subsequently initiates the EXPAR reaction. The detection limit (LOD) for folate receptors is 0.23 fM, with a recovery rate greater than 95% in the spiked recovery test. Further analysis of cell membrane extracts confirmed the accuracy of the experiment.

5.3.3 Enzyme Activity Assay

Human telomerase is a ribonucleoprotein complex that adds the TTAGGG repeat sequence to the ends of chromosome DNA. Telomerase activity is present in the majority of human cancer cells, while most human somatic cells lack telomerase activity. Therefore, telomerase is considered a biomarker for early cancer diagnosis and a therapeutic target. Chun-Yang Zhang's team developed a simple, rapid, and highly sensitive method for detecting telomerase activity at the single-cell level using telomerase-induced two-step isothermal amplification [39]. The principle of this method is shown in Fig. 5.25. The designed telomerase substrate primer sequence consists of seven parts (AXBX′CX′D).

1. Telomerase Extension Reaction: In the presence of telomerase and dNTPs, telomerase adds the (TTAGGG)n repeat unit to the 3′ end of the telomerase substrate primer (D), generating a telomerase extension product. The amount of extension product is proportional to the telomerase activity.
2. Telomerase-Induced Two-Step Isothermal Amplification: The extended telomerase substrate primer serves as the template for isothermal amplification, hybridizing with the reverse primer. In the presence of Bst 2.0 WarmStart DNA polymerase and dNTPs, an extension reaction is initiated, generating stable dsDNA containing three recognition sequences for NEase. Through cyclic amplification, a large amount of C, R, and A′ is produced.
3. Chemiluminescent Analysis: Both C and R are G-rich sequences, which bind to hemin to form a hemin-G-quartet nanostructure. In the presence of luminol/H_2O_2, chemiluminescent signals are generated.

By leveraging the high amplification efficiency of the two-step isothermal amplification reaction and the high sensitivity of DNAzyme-driven chemiluminescence, this method enables rapid, simple, and sensitive detection of telomerase activity.

Uracil-DNA glycosylase (UDG) is an enzyme responsible for repairing uracil-induced DNA damage to maintain genomic integrity, and its abnormal activity is associated with various cancers. Therefore, accurate detection of UDG activity is

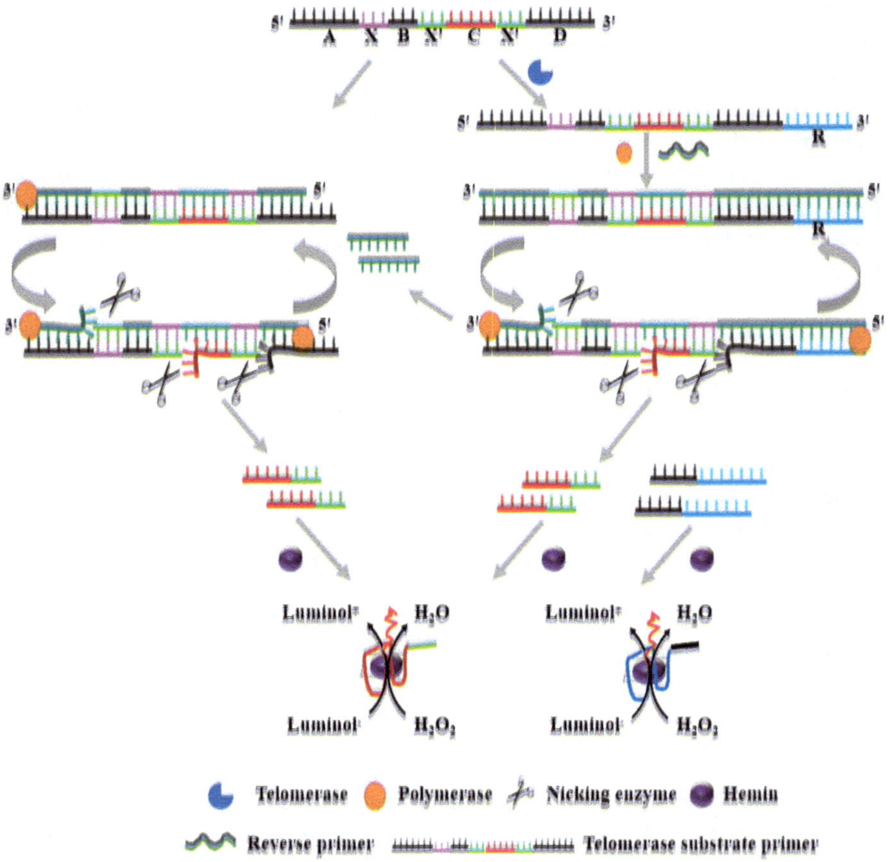

Fig. 5.25 Schematic diagram of telomerase activity detection using telomerase-induced isothermal amplification combined with chemiluminescence [39]. (Revised and reprinted with permission from Ref. [39])

crucial for biomedical research and clinical diagnosis. Chun-Yang Zhang's research team developed a fluorescence-based EXPAR method for ultra-sensitive analysis of UDG activity [40]. The principle of this method is shown in Fig. 5.26, and it involves three steps for UDG activity analysis.

1. UDG-Driven Uracil Excision Repair Reaction: A stem-loop DNA substrate containing uracil (U) is designed. UDG specifically recognizes uracil and removes it from the DNA substrate, releasing uracil and creating an apyrimidinic (AP) site. Endo IV, an AP endonuclease, can recognize the AP sites and cleave at the 5′ end of the AP site, opening the DNA substrate's loop.
2. Dual-Template EXPAR Nucleic Acid Amplification: The opened DNA substrate acts as both a template and a primer to initiate the dual-template EXPAR system in the presence of DNA polymerase and NEase, ultimately generating a large amount of trigger.

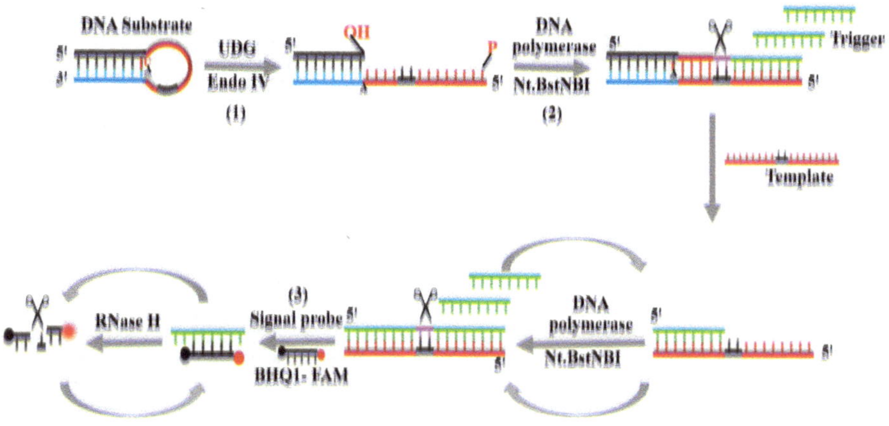

Fig. 5.26 Schematic diagram of UDG detection using dual-template EXPAR. (1) UDG-driven uracil excision repair reaction. (2) Dual-template EXPAR nucleic acid amplification. (3) RNase H-induced signal probe hydrolysis and fluorescence signal generation [40]. (Revised and reprinted with permission from Ref. [40])

3. RNase H-Induced Signal Probe Hydrolysis and Fluorescent Signal Generation: The signal probe hybridizes with the generated trigger. RNase H selectively hydrolyzes the RNA in the RNA-DNA hybrid, causing the separation of the fluorescent and quencher groups, thereby generating a fluorescent signal.

This method has significant advantages, including being isothermal, fast, simple, and highly sensitive.

Using a similar principle, the EXPAR system has also been designed for the analysis and detection of methyltransferase (MTase) activity [10]. As shown in Fig. 5.27, Xue et al. designed a hairpin probe, with the green sequence (5′-G-A-T-C-3′) in the stem region serving as the substrate for DNA Adenine Methyltransferase (Dam MTase) reaction. Dam MTase specifically recognizes and methylates the 5′-G-A-T-C-3′ sequence in dsDNA. It methylates the N6 position of adenine (A), generating N6-methyladenine (m6A). Moreover, Dam MTase symmetrically methylates the GATC sequence on both strands of DNA, creating double-stranded methylation sites on the GATC sequences. In the presence of S-Adenosyl Methionine (SAM), the base A in the hairpin probe substrate sequence is methylated by Dam MTase (5′-G-mA-T-C-3′). DpnI is a methylation-sensitive restriction endonuclease that specifically recognizes dsDNA with the methylated form 5′-G-(m6A)-T-C-3′ and cleaves at the center of the AT sequence, generating sticky ends. The methylated hairpin probe can be cleaved by DpnI into two parts: one is a new short hairpin DNA, and the other is a heteroduplex with a 5′-overhang of 6 bases. However, this double-stranded fragment is unstable and dissociates into two separate ssDNA fragments. 19-base ssDNA fragment is designed as the EXPAR primer, which hybridizes with the EXPAR template and initiates the EXPAR amplification reaction.

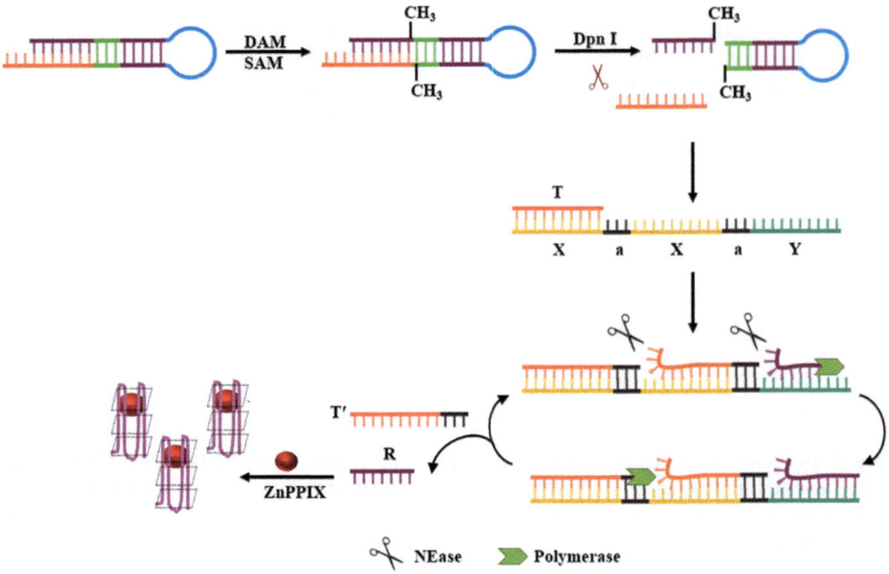

Fig. 5.27 Schematic diagram of DNA methyltransferase activity analysis using methylation-sensitive cleavage-triggered EXPAR [10]. (Revised and reprinted with permission from Ref. [10])

5.3.4 Detection of Bacteria, Cells, and Metal Ions

EXPAR has been applied to detect bacteria, tumor cells, and metal ions. Qiu et al. developed an EXPAR system for detecting pathogenic bacteria using a structure-switching aptamer [41]. The aptamer initially hybridizes with EXPAR primers. Upon binding to the target pathogenic bacteria, the aptamer folds and releases the EXPAR primers, triggering a dual-template EXPAR system. Tang et al. developed a similar technique to detect tumor cells [42]. The experiment used a structure conversion aptamer fixed on magnetic beads to identify target tumor cells. The aptamer first hybridizes with the EXPAR primers, which are released upon the aptamer binding to the target tumor cells and then amplified via a dual-template EXPAR system. This technique achieved a limit of detection (LOD) of 10 tumor cells. Jia et al. introduced a poly(dT)-based Hg^{2+} recognition mechanism into a dual-template EXPAR system (Y′-X′ and Y′-Y′) for Hg^{2+} detection [43]. The T-Hg^{2+}-T base interaction allows a $(T)_{10}$ primer to hybridize with the first template and trigger EXPAR. The method achieved an LOD for Hg^{2+} of 100 pM.

5.4 Conclusions

EXPAR employs a simple template and two enzymes to achieve isothermal amplification, resulting in amplification of 10^6–10^8 times in less than 30 min. EXPAR is particularly suitable for the amplification and detection of short nucleotides, such as miRNA. Through specific design, long-chain DNA/RNA can produce fragments that hybridize with the EXPAR template, allowing EXPAR to be used for the analysis and detection of various DNA/RNA types. In addition, through the interaction of nucleic acid aptamers with target proteins, bacteria, cells, etc., the aptamers undergo structural changes, which can initiate the EXPAR amplification reaction, enabling the analysis of various target molecules. However, EXPAR suffers from high background amplification, and the amplification efficiency is closely related to both the background signal and sequence dependence. The sensitivity of the EXPAR system varies greatly for different nucleic acid sequences, limiting its widespread application. To improve EXPAR, it is crucial to gain a deeper understanding of the mechanisms driving background signals and sequence dependence, and to develop effective strategies to reduce or eliminate background amplification.

References

1. Ness JV, Ness LV, Galas DJ. Isothermal reactions for the amplification of oligonucleotides. Proc Natl Acad Sci USA. 2003;100:4504–9.
2. Reid MS, Le XC, Zhang H. Exponential isothermal amplification of nucleic acids and assays for proteins, cells, small molecules, and enzyme activities: an EXPAR example. Angew Chem Int Edit. 2018;57:11856–66.
3. Jia H, Li Z, Liu C, Cheng Y. Ultrasensitive detection of microRNAs by exponential isothermal amplification. Angew Chem Int Edit. 2010;49:5498–501.
4. Tan E, Erwin B, Dames S, Ferguson T, Buechel M, Irvine B, Voelkerding K, Niemz A. Specific versus nonspecific isothermal DNA amplification through thermophilic polymerase and nicking enzyme activities. Biochemistry. 2008;47:9987–99.
5. Qian J, Ferguson TM, Shinde DN, Ramirez-Borrero AJ, Hintze A, Adami C, Niemz A. Sequence dependence of isothermal DNA amplification via EXPAR. Nucleic Acids Res. 2012;40:e87.
6. Liu YQ, Zhang M, Yin BC, Ye BC. Attomolar ultrasensitive microRNA detection by DNA-scaffolded silver-nanocluster probe based on isothermal amplification. Anal Chem. 2012;84:5165–9.
7. Petty JT, Jie Z, Hud NV, Dickson RM. DNA-templated Ag nanocluster formation. J Am Chem Soc. 2004;126:5207–12.
8. Zhang Y, Zhang CY. Sensitive detection of microRNA with isothermal amplification and a single-quantum-dot-based nanosensor. Anal Chem. 2012;84:224–31.
9. Zhang Z, Sharon E, Freeman R, Liu X, Willner I. Fluorescence detection of DNA, adenosine-5′-triphosphate (ATP), and telomerase activity by zinc(II)-protoporphyrin IX/G-quadruplex labels. Anal Chem. 2012;84:4789–97.
10. Xue Q, Lv Y, Xu S, Zhang Y, Wang L, Li R, Yue Q, Li H, Gu X, Zhang S, Liu J. Highly sensitive fluorescence assay of DNA methyltransferase activity by methylation-sensitive cleavage-based primer generation exponential isothermal amplification-induced G-quadruplex formation. Biosens Bioelectron. 2015;66:547–53.

11. Wang LJ, Zhang Y, Zhang CY. A target-triggered exponential amplification-based DNAzyme biosensor for ultrasensitive detection of folate receptors. Chem Commun. 2014;50:15393–6.
12. Wang GL, Zhang CY. Sensitive detection of microRNAs with hairpin probe-based circular exponential amplification assay. Anal Chem. 2012;84:7037–42.
13. Lin Q, Han G, Fang X, Chen H, Weng W, Kong J. Programmable analysis of microRNAs by *Thermus thermophilus* argonaute-assisted exponential isothermal amplification for multiplex detection (TEAM). Anal Chem. 2022;94:11290–7.
14. Zhang Y, Hu J, Zhang CY. Sensitive detection of transcription factors by isothermal exponential amplification-based colorimetric assay. Anal Chem. 2012;84:9544–9.
15. Tan E, Wong J, Nguyen D, Zhang Y, Erwin B, Ness L, Baker S, Galas D, Niemz A. Isothermal DNA amplification coupled with DNA nanosphere-based colorimetric detection. Anal Chem. 2005;77:7984–92.
16. Lode P, Syrjala A, Hagren V, Kojola H, Soukka T, Lovgren T, Nurmi J. Isothermal DNA amplification with gold nanosphere-based visual colorimetric readout for herpes simplex virus detection. Clin Chem. 2007;53:2017–20.
17. Roskos K, Hickerson AI, Lu HW, Ferguson TM, Shinde DN, Klaue Y, Niemz A. Simple system for isothermal DNA amplification coupled to lateral flow detection. PLoS One. 2013;8: e69355.
18. Nie J, Zhang DW, Tie C, Zhou YL, Zhang XX. G-quadruplex based two-stage isothermal exponential amplification reaction for label-free DNA colorimetric detection. Biosens Bioelectron. 2014;56:237–42.
19. Xu Y, Li D, Cheng W, Hu R, Sang Y, Yin Y, Ding S, Ju H. Chemiluminescence imaging for microRNA detection based on cascade exponential isothermal amplification machinery. Anal Chim Acta. 2016;936:229–35.
20. Ye LP, Hu J, Liang L, Zhang CY. Surface-enhanced Raman spectroscopy for simultaneous sensitive detection of multiple microRNAs in lung cancer cells. Chem Commun. 2014;50: 11883–6.
21. Sonntag MD, Klingsporn JM, Zrimsek AB, Sharma B, Ruvuna LK, Duyne RPV. Molecular plasmonics for nanoscale spectroscopy. Chem Soc Rev. 2014;43:1230–47.
22. Driskell JD, Tripp RA. Label-free SERS detection of microRNA based on affinity for an unmodified silver nanorod array substrate. Chem Commun. 2010;46:3298–300.
23. Guven B, Dudak FC, Boyaci IH, Tamer U, Ozsoz MJA. SERS-based direct and sandwich assay methods for mir-21 detection. Analyst. 2014;139:1141–7.
24. Yan Y, Zhao D, Yuan T, Hu J, Zhang D, Cheng W, Zhang W, Ding S. A simple and highly sensitive electrochemical biosensor for microRNA detection using target-assisted isothermal exponential amplification reaction. Electroanalysis. 2013;25:2354–9.
25. Yu Y, Chen Z, Shi L, Yang F, Pan J, Zhang B, Sun D. Ultrasensitive electrochemical detection of microRNA based on an arched probe mediated isothermal exponential amplification. Anal Chem. 2014;86:8200–5.
26. Liu S, Wang C, Zhang C, Wang Y, Tang B. Label-free and ultrasensitive electrochemical detection of nucleic acids based on autocatalytic and exonuclease III-assisted target recycling strategy. Anal Chem. 2013;85:2282–8.
27. Yang N, Cao Y, Han P, Zhu X, Sun L, Li G. Tools for investigation of the RNA endonuclease activity of mammalian Argonaute2 protein. Anal Chem. 2012;84:2492–7.
28. Nie J, Zhang DW, Zhang FT, Yuan F, Zhou YL, Zhang XX. Reporter-triggered isothermal exponential amplification strategy in ultrasensitive homogeneous label-free electrochemical nucleic acid biosensing. Chem Commun. 2014;50:6211–3.
29. Wang X, Wang H, Liu C, Wang H, Li Z. A three-way junction structure-based isothermal exponential amplification strategy for sensitive detection of 3′-terminal 2′-O-methylated plant microRNA. Chem Commun. 2017;53:1124–7.
30. Zhang X, Liu C, Sun L, Duan X, Li Z. Lab on a single microbead: an ultrasensitive detection strategy enabling microRNA analysis at the single-molecule level. Chem Sci. 2015;6:6213–8.

31. Wang H, Wang H, Duan X, Wang X, Sun Y, Li Z. Sensitive detection of mRNA by using specific cleavage-mediated isothermal exponential amplification reaction. Sensor Actuat B-Chem. 2017;252:215–21.
32. Chang F, Sun Y, Yang D, Yang W, Sun Y, Liu C, Li Z. Specific detection of RNA mutation at single-base resolution by coupling the isothermal exponential amplification reaction (EXPAR) with chimeric DNA probe-aided precise RNA disconnection at the mutation site. Chem Commun. 2019;55:6934.
33. Carter JG, Iturbe LO, Duprey J-LHA, Carter IR, Southern CD, Rana M, Whalley CM, Bosworth A, Beggs AD, Hicks MR, Tucker JHR, Dafforn TR. Ultrarapid detection of SARS-CoV-2 RNA using a reverse transcription–free exponential amplification reaction, RTF-EXPAR. Proc Natl Acad Sci USA. 2021;118:1–6.
34. Xu Y, Niu C, Xiao X, Zhu W, Dai Z, Zou X. Chemical-oxidation cleavage triggered isothermal exponential amplification reaction for attomole gene-specific methylation analysis. Anal Chem. 2015;87:2945–51.
35. Sun Y, Sun Y, Tian W, Liu C, Gao K, Li Z. A novel restriction endonuclease GlaI for rapid and highly sensitive detection of DNA methylation coupled with isothermal exponential amplification reaction. Chem Sci. 2018;9:1344–51.
36. Zhang ZZ, Zhang CY. Highly sensitive detection of protein with aptamer-based target-triggering two-stage amplification. Anal Chem. 2012;84:1623–9.
37. Xie S, Chai Y, Yuan Y, Bai L, Yuan R. A novel electrochemical aptasensor for highly sensitive detection of thrombin based on the autonomous assembly of hemin/G-quadruplex horseradish peroxidase-mimicking DNAzyme nanowires. Anal Chim Acta. 2014;832:51–7.
38. Ma F, Yang Y, Zhang CY. Ultrasensitive detection of transcription factors using transcription-mediated isothermally exponential amplification-induced chemiluminescence. Anal Chem. 2014;86:6006–11.
39. Wang LJ, Zhang Y, Zhang CY. Ultrasensitive detection of telomerase activity at the single-cell level. Anal Chem. 2013;85:11509–17.
40. Wang LJ, Ren M, Zhang Q, Tang B, Zhang CY. Excision repair-initiated enzyme-assisted bicyclic cascade signal amplification for ultrasensitive detection of uracil-DNA glycosylase. Anal Chem. 2017;89:4488–94.
41. Qiu T, Wang Y, Yu J, Liu S, Wang H, Guo Y, Huang J. Label-free, homogeneous, and ultrasensitive detection of pathogenic bacteria based on target-triggered isothermally exponential amplification. RSC Adv. 2016;6:62031–7.
42. Tang W, Zhang T, Li Q, Wang H, Wang H, Li Z. Sensitive detection of tumor cells based on aptamer recognition and isothermal exponential amplification. RSC Adv. 2016;6:89888–94.
43. Jia H, Wang Z, Wang C, Chang L, Li Z. Real-time fluorescence detection of Hg^{2+} ions with high sensitivity by exponentially isothermal oligonucleotide amplification. RSC Adv. 2014;4:9439–44.

Chapter 6
Helicase- and Recombinase-Dependent DNA Isothermal Amplification (HDA and RPA)

Hui Wang and Weixiang Hong

6.1 Helicase-Dependent DNA Isothermal Amplification (HDA)

Helicase-dependent Isothermal DNA Amplification (HDA) is an isothermal amplification technology developed by Vincent and colleagues at New England Biolabs (NEB), utilizing *Escherichia Coli* UvrD helicase, DNA polymerase, and single-stranded DNA-binding protein (SSB) [1]. This technique mimics the natural process of DNA replication in vivo, enabling DNA amplification in vitro under constant temperature conditions. The amplification mechanism involves the following steps. First, helicase unwinds the double-stranded DNA (dsDNA) under isothermal conditions. Then, SSB binds to the unwound single-stranded DNA (ssDNA), stabilizing it and providing a template for primer binding. Subsequently, DNA polymerase catalyzes the synthesis of double-stranded DNA. The newly synthesized dsDNA is unwound by helicase to generate single strands, which serve as templates for further cycles of DNA synthesis, resulting in exponential amplification of the target nucleic acid sequence. Compared to traditional PCR, HDA utilizes helicase and SSB to generate DNA templates under isothermal conditions, overcoming the need for thermal cycling to denature dsDNA into ssDNA. In terms of primer design and product detection, HDA is like conventional PCR. The simplicity of its principle, ease of operation, and low equipment requirements make it particularly suitable for on-site diagnostic testing, offering promising applications in this field.

H. Wang (✉) · W. Hong
University of Science and Technology Beijing, Beijing, China
e-mail: winscavin@ustb.edu.cn

Z. Li (ed.), *Nucleic Acid Detection*, https://doi.org/10.1007/978-981-95-7295-3_6

6.1.1 Principle of HDA

6.1.1.1 Amplification Mechanism of HDA

The core of the HDA amplification technology lies in utilizing the strand-unwinding activity of helicase and the stabilizing effect of single-stranded DNA-binding protein (SSB) on single-stranded DNA (ssDNA) to cyclically generate ssDNA templates. These templates allow primer binding and the exponential amplification of the target sequence. The HDA amplification process, illustrated in Fig. 6.1, consists of five stages: helicase unwinding of double-stranded DNA (dsDNA), binding and stabilization of ssDNA by SSB, primer hybridization to ssDNA, DNA polymerase-catalyzed synthesis of the target sequence, and helicase-mediated strand unwinding for template recycling. At the start of the reaction, with the assistance of the MutL cofactor and ATP [1], UvrD helicase binds to dsDNA and unwinds it into ssDNA. The resulting ssDNA is immediately bound by SSB to stabilize its structure and prevent re-annealing into dsDNA. Once the single-stranded DNA amplification template is formed, primers randomly bind to the target DNA sequence. DNA polymerase then extends and synthesizes the target sequence. The newly synthesized dsDNA target sequence enters a new cycle of HDA amplification, involving helicase unwinding, SSB binding and stabilization, primer annealing, and DNA polymerase extension. This cycle enables the exponential amplification of the target sequence. Like conventional PCR products, the final amplification products are dsDNAs.

Additionally, by incorporating reverse transcriptase, the HDA technique can also be applied to the amplification of RNA templates [2]. Compared to conventional HDA, the RT-HDA reaction system requires the addition of thermostable reverse transcriptase, as well as thermostable DNA polymerase and single-stranded DNA-binding protein (SSB). As shown in Fig. 6.2, the RT-HDA amplification

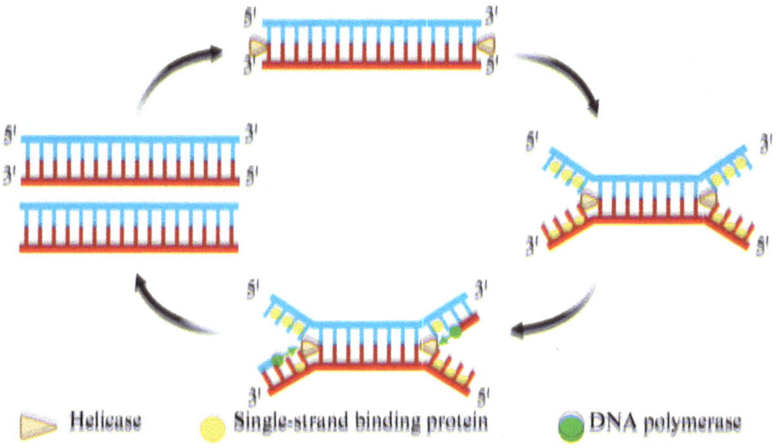

Fig. 6.1 The amplification principle of HDA [1]. (Revised and reprinted with permission from Ref. [1])

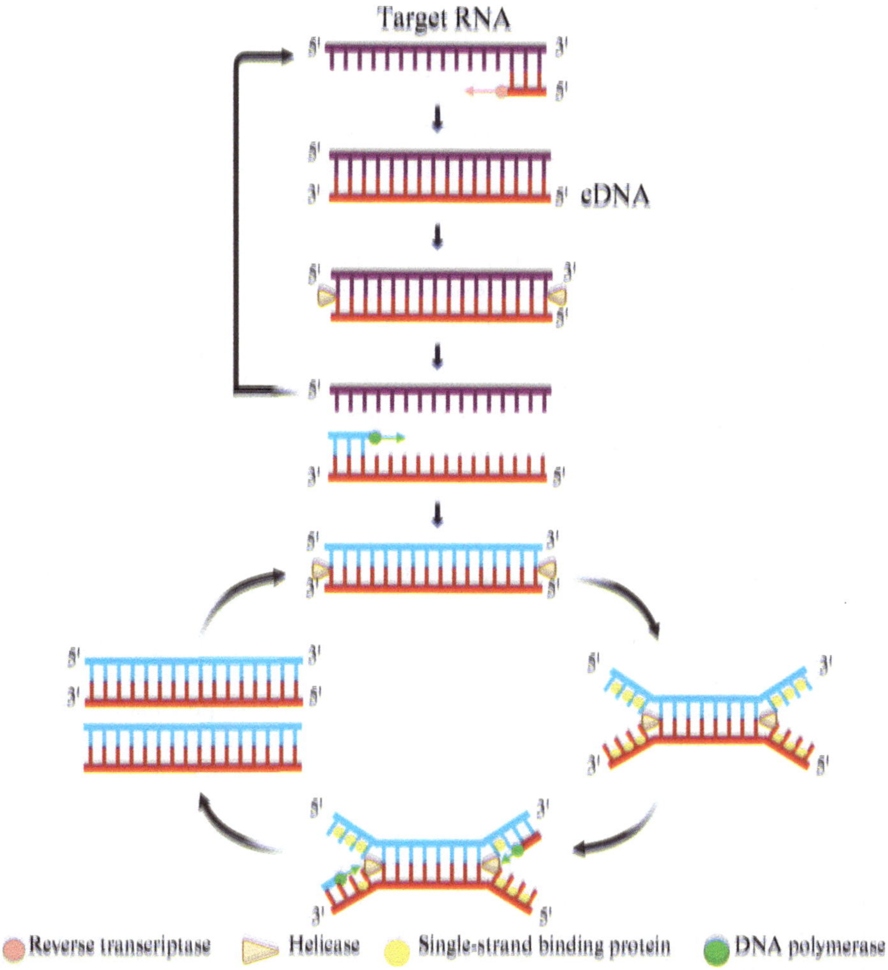

Fig. 6.2 The principle of RT-HDA [2]. (Revised and reprinted with permission from Ref. [2])

process begins with a reverse transcription primer binding to the single-stranded RNA template. Under the action of reverse transcriptase, complementary DNA (cDNA) is synthesized, forming a hybrid double-stranded structure. This hybrid double-strand is unwound by the DNA helicase into an RNA single strand and a DNA single strand. The RNA single strand continues to serve as a template for the reverse transcription cycle, while the DNA single strand is used by DNA polymerase to synthesize double-stranded DNA. The newly synthesized double-stranded DNA serves as a template for subsequent rounds of HDA amplification cycles, including helicase unwinding, SSB binding and stabilization, primer binding, and DNA polymerase extension. This process enables rapid exponential amplification of the RNA template.

6.1.1.2 Design of HDA Amplification Systems

1. Target sequence selection and primer design.

The selection of the HDA amplification sequence and primer design follows principles similar to those of PCR, targeting conserved regions of specific genes. Due to limitations imposed by the helicase's binding to the target sequence and its unwinding rate, the length of the target sequence for HDA amplification should not be too long, typically ranging from 70 to 120 bp. Amplification efficiency significantly decreases when the target sequence exceeds 400 bp, and for RT-HDA, effective amplification is usually not achieved if the target sequence exceeds 120 bp. Generally, the target sequence Tm value is within 68–75 °C, and primer Tm values range from 62 to 74 °C. Primer lengths are typically 24–30 bp. For HDA performed at higher temperatures, primers with lower Tm values hinder primer binding. The GC content of primers should typically range from 30% to 60%.

2. Selection of Helicase and DNA Polymerase.

Helicase plays a critical role in HDA amplification, as its unwinding reaction is the rate-limiting step in the HDA system. The unwinding efficiency of helicase directly affects the amplification efficiency and the achievable length of the target sequence. The unwinding capability of a DNA helicase depends on factors such as its unwinding speed and processivity. The first helicase used in the HDA system was *Escherichia Coli* UvrD helicase [3], which requires auxiliary proteins (MutL protein) and ATP to complete the unwinding process [1, 4]. In addition, single-stranded DNA-binding proteins (SSB) are essential in the HDA system to stabilize the unwound single-stranded DNA, facilitating primer binding. Commonly used SSBs include the gp32 protein from the T4 phage or the gp32 protein from the RB49 phage [2]. UvrD helicase exhibits relatively low unwinding capability, with an unwinding speed of 20 bp/s and a processivity of approximately 100 bp [5], which limits its ability to amplify long target sequences effectively. Leveraging the replication mechanism of T7 phage, researchers developed a faster and more efficient HDA system. This system employs the gp4B protein from T7 phage as the helicase, with an unwinding speed of up to 300 bp/s and a processivity reaching 75 kb. Its high processivity enables the in vitro amplification of long target sequences [6–8]. Additionally, Kong et al. [9] fused the UvrD helicase gene from thermophilic anaerobes with the large fragment gene of Geobacillus stearothermophilus DNA polymerase I, expressing a bifunctional fusion protein with both helicase and polymerase activities. When applied directly to the HDA system, this fusion protein exhibited higher DNA amplification efficiency than systems using helicase and polymerase separately, successfully amplifying target sequences up to 2.3 kb in length.

The DNA polymerases commonly used in HDA include *Klenow* DNA polymerase lacking 3′-5′ exonuclease activity, T7 DNA polymerase, and thermally stable *Bst* DNA polymerase. These enzymes operate at different reaction temperatures: *Klenow* DNA polymerase at 37 °C, T7 DNA polymerase at 25 °C, and *Bst* DNA polymerase at 60–65 °C.

6.1.2 Detection Methods of HDA Products

HDA amplification products are similar to those of conventional PCR, consisting of double-stranded DNA typically ranging from 70 to 400 bp in length. Therefore, the detection methods for HDA products are essentially the same as those for conventional PCR products, including endpoint detection and real-time detection. For endpoint detection, a 2% agarose gel is used to separate the amplification products by electrophoresis, and the results are analyzed using a gel fluorescence imaging system. This method provides a straightforward assessment of whether amplification was successful and confirms the product size. Real-time detection typically employs double-stranded DNA-binding dyes (such as SYBR Green I) or dual-labeled probes (such as TaqMan probes) to monitor the amplification process in real time, enabling quantitative analysis of the amplification products. The principle is the same as that of quantitative PCR (qPCR). In addition, classical methods such as enzyme-linked immunosorbent assay (ELISA) [10], electrochemical techniques [11, 12], and lateral flow assays (LFA) [13, 14] have also been applied for the detection of HDA amplification products.

6.1.3 Applications of HDA

The HDA technique mimics the natural DNA replication process by utilizing helicase to continuously unwind double-stranded DNA, providing single-stranded templates for DNA polymerase to synthesize new DNA. HDA requires no complex primer design and has a simple mechanism, significantly simplifying the reaction system. Similar to other isothermal nucleic acid amplification methods, the entire amplification process in HDA is completed at a constant temperature, eliminating the need for the thermal cycling required in PCR and LCR. This greatly reduces the complexity and cost of instrumentation. Combined with various detection technologies, HDA has been widely applied in the analysis of pathogens [13–15], genetically modified crops [12], and other biomarkers [16–19].

6.2 Recombinase Polymerase Amplification (RPA)

Recombinase Polymerase Amplification (RPA) technology was developed in 2006 and is an efficient isothermal nucleic acid amplification strategy [20]. RPA utilizes recombinase to form complexes with specific primers, enabling homologous sequence recognition and localization within the target DNA. Subsequently, with the synergistic action of single-stranded DNA-binding protein (SSB) and strand-displacing DNA polymerase, a strand exchange reaction is completed, initiating DNA synthesis and achieving exponential amplification of the target region. RPA operates under constant temperatures ranging from 25 to 42 °C, eliminating the need for complex thermal cycling equipment. The reaction is rapid, reaching detectable levels within 10–20 min. This technology is characterized by high sensitivity and simplicity of operation [21, 22]. The amplified products can be detected through various methods, including gel electrophoresis, real-time fluorescence quantification, and lateral flow analysis [23]. Currently, RPA has demonstrated broad application potential in clinical diagnostics, food safety, environmental monitoring, and agriculture and livestock industries [24, 25].

6.2.1 *Principle of RPA*

6.2.1.1 Primers and Amplification Mechanism of RPA

Recombinase Polymerase Amplification (RPA) is based on the nucleic acid replication mechanism of the T4 bacteriophage and relies on three key proteins: recombinase (UvsX) [26, 27], single-stranded DNA-binding protein (Gp32), and strand-displacement DNA polymerase (Bsu). The amplification reaction requires a pair of primers, consisting of an upstream primer and a downstream primer. The mechanism of the RPA reaction, illustrated in Fig. 6.3 [20], involves five steps:

(a) Formation of the recombinase-primer complex: With the assistance of adenosine triphosphate (ATP) and the accessory factor (UvsY), recombinase binds to the upstream and downstream primers to form a recombinase-primer complex. This

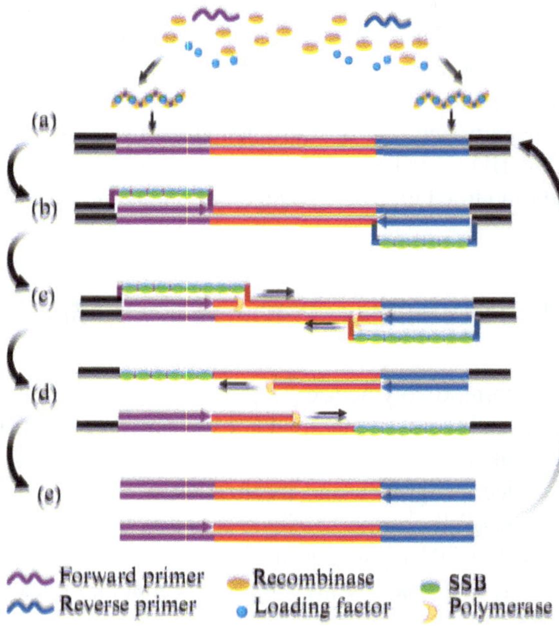

Fig. 6.3 The mechanism of the RPA reaction [20]. (Revised and reprinted with permission from Ref. [20])

complex scans the double-stranded DNA (dsDNA) template to identify target sequences homologous to the primers.

(b) Initiation of strand exchange reaction: Once the recombinase-primer complex locates the homologous target sequence on the DNA template, the recombinase unwinds the dsDNA structure, facilitating the primers to engage in a strand exchange reaction and insert into the dsDNA template, forming a D-loop structure. Meanwhile, single-stranded DNA-binding protein (Gp32) rapidly binds to the displaced single-stranded DNA (ssDNA), preventing its re-annealing and ensuring the primers stably bind to the template.

(c) Recombinase dissociation and polymerase binding: As ATP in the recombinase-primer complex is hydrolyzed, the complex undergoes a conformational change, leading to the dissociation of the recombinase and exposing the 3′-end of the primer. Meanwhile, DNA polymerase recognizes and binds to the 3′-end of the primer, initiating strand-displacement DNA polymerization.

(d) Strand displacement polymerization reaction: During the polymerization reaction, DNA polymerase continues to unwind the dsDNA template, with the displaced ssDNA strands bound by Gp32 to prevent re-annealing. This process allows DNA polymerase to continuously synthesize new complementary strands along the template, simultaneously driving the strand-displacement reaction.

(e) Amplification Product Formation and Exponential Amplification: The upstream and downstream primers simultaneously participate in the above reactions, ultimately generating complete amplification products. The newly synthesized products can serve as templates for subsequent reactions, enabling exponential amplification of the DNA template.

6.2.1.2 Reverse Transcription RPA (RT-RPA)

Reverse Transcription RPA (RT-RPA) is achieved by adding reverse transcriptase to the classic RPA reaction system, enabling RNA detection [28, 29]. The target RNA serves as a template for synthesizing complementary DNA (cDNA) via reverse transcription. The cDNA then acts as an amplification template for RPA. One-step RT-RPA can detect target RNA under isothermal conditions (40–42 °C) [30].

6.2.1.3 Target Sequence Selection and Primer Design

The selection of target sequences and primer design is crucial for RPA amplification efficiency and specificity. The target sequence should have a relatively uniform nucleotide composition, with a GC content ideally between 40% and 60%. Excessively high or low GC content may reduce amplification efficiency, compromising the sensitivity and specificity of RPA. Additionally, regions containing long stretches of purines (e.g., poly-A or poly-G) or pyrimidines (e.g., poly-T or poly-C) should be avoided, as these sequences may result in unstable primer binding or secondary structure formation, thereby interfering with the amplification process. Forward repeats, reverse repeats, and palindromic sequences should also be excluded as much as possible. These complex sequence structures can form hairpin loops during amplification or cause nonspecific amplification, reducing the efficiency and accuracy of the reaction. Specifically, palindromic sequences may form stable hairpin structures in double-stranded DNA, obstructing effective primer binding and extension. The ideal target sequence length ranges from 100 to 200 bp. Although RPA technology can amplify longer fragments, it is recommended to limit the target sequence length to 500 bp to ensure optimal sensitivity and reaction time. Longer fragments may lower amplification efficiency and extend reaction times, affecting the reliability and reproducibility of the experimental results [31].

Primer design for RPA amplification is similar to that for PCR but includes notable differences. Key parameters for RPA primer design include primer length, primer sequence, and secondary structure [31, 32].

1. Primer length: RPA primers are generally longer than PCR primers because RPA relies on recombinase-primer complexes to recognize the target sequence. RPA primers are typically 30–35 bases in length. Longer primers enhance recombinase activity and specificity, but excessively long primers (>45 bases) may increase the risk of secondary structure formation, reducing amplification efficiency. Therefore, overly long primers are not recommended.
2. Primer sequence: The GC content of primers should be controlled between 30% and 70%. Extreme GC content can lead to unstable primer binding or reduced amplification efficiency. An ideal GC range ensures effective primer-template binding while minimizing nonspecific amplification. The 5′ end of primers should avoid multiple consecutive guanines (G), as these may interfere with the formation of the recombinase-primer complex. Conversely, cytosine (C) at the 5′ end is beneficial, promoting recombination and enhancing primer binding. At the 3′ end, guanine (G) and cytosine (C) are preferred as they facilitate stable binding by polymerase, improving primer amplification performance. The stability of the 3′ end is critical for initiating efficient DNA synthesis. Primer sequences should avoid excessive forward repeats, reverse repeats, or palindromic structures, as these complex sequences may form hairpin loops or cause nonspecific amplification during the reaction, reducing efficiency and specificity. Unlike PCR primers, RPA primers do not require strict melting temperature (Tm) considerations, as primer annealing and extension in RPA are enzyme-mediated rather than temperature-dependent. Therefore, RPA primer design should focus on sequence specificity and enzyme binding efficiency rather than Tm.
3. Primer secondary structure: Like PCR primers, RPA primers should avoid forming secondary structures or complementary sequences to prevent self-hairpin structures or primer-dimer formation. These structures can interfere with primer-template binding, reducing amplification efficiency or producing nonspecific

products. Bioinformatics tools (e.g., Oligo Analyzer) should be used to evaluate the secondary structure of primers, ensuring their effectiveness in the reaction.

6.2.1.4 Design of Probes for PRA Detection

Most PCR detection probes, such as TaqMan probes, are unsuitable for use in RPA reactions primarily due to their short length, which results in low binding efficiency with recombinases. Additionally, TaqMan probes rely on the 5′-3′ nuclease activity of polymerases for cleavage, a feature absent in the RPA system, rendering them incompatible. To ensure the effectiveness and specificity of RPA reactions, probe design must carefully consider their length and structure. Detailed guidelines for RPA probe design are available on the official websites of RPA reagent providers, such as TwistAmp. Common RPA probe types, as shown in Fig. 6.4, include Exo probes, Fpg probes, and Nfo probes [23, 31]. These probes typically range from 46 to 52 nucleotides in length and are designed with blocking groups at the 3′-end to prevent DNA extension. They also include specific cleavage sites within the probe that, when bound to the template DNA, are cleaved by specific endonucleases, resulting in fluorescence generation.

1. Exo probes (Fig. 6.4a) are typically 46–52 nucleotides long. A blocking group (e.g., a phosphate group or dideoxynucleotide) is attached to the 3′ end of the probe to prevent it from acting as a primer in the amplification reaction. A tetrahydrofuran (THF) residue replaces a base between the fluorophore and quencher groups, located at least 30 nucleotides from the 5′ end and 15 nucleotides from the 3′ end of the probe. When the probe binds to template DNA, Exonuclease III (Exo) recognizes the THF site and removes nucleotides one by one starting from the 3′-end, ultimately separating the fluorophore from the

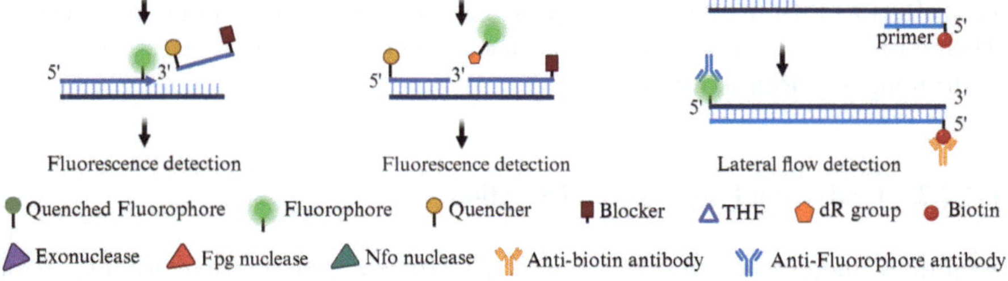

Fig. 6.4 The detection probes of RPA

quencher, and generating a fluorescent signal. The cleaved probe creates a free 3′ end, which can act as a forward primer for polymerase extension.

2. Fpg Probes (Fig. 6.4b) are 46–52 nucleotides long. The 3′ end of the probe is blocked to prevent extension. A fluorophore-labeled deoxyribose (dR), connected by a C-O-C bond, replaces a base in the probe, positioned at least 15 nucleotides from the 3′-end. The quencher group is located 1–3 nucleotides downstream of the fluorophore. When the probe binds to template DNA, formamidopyrimidine-DNA glycosylase (Fpg, also known as 8-oxoguanine DNA glycosylase) recognizes and cleaves the C-O-C bond, releasing the fluorophore and generating fluorescence. Notably, the 3′ end generated after Fpg probe cleavage is non-extendable, ensuring it does not interfere with subsequent amplification reactions.
3. Nfo Probes (Fig. 6.4c) are generally 46–52 nucleotides long. A blocking group is attached to the 3′ end, and a fluorophore is attached to the 5′ end, with a THF residue positioned at least 15 nucleotides from the 3′ end and more than 30 nucleotides from the 5′ end. When the probe binds to template DNA, Endonuclease IV (Nfo) recognizes the THF site and cleaves it. The cleaved probe generates a free 3′ end that can be extended and further participate in the amplification reaction. Nfo probes, in combination with biotin-labeled RPA primers, are commonly used for lateral flow detection of RPA amplification products.

6.2.2 Methods for Detecting RPA Amplification Products

The detection methods for RPA products include gel electrophoresis, real-time fluorescence quantification, lateral flow assay (LFA), flocculation-based detection, and CRISPR-based detection.

6.2.2.1 Gel Electrophoresis

Gel electrophoresis is one of the conventional methods for specific analysis of RPA amplification products [33, 34]. The final RPA products, similar to those of PCR, are double-stranded DNA of specific lengths. However, single-stranded DNA and recombinases present in the RPA reaction can interact with the amplification products, affecting their migration in the gel and causing tailing in electrophoresis results. Therefore, simple purification or heat inactivation of proteins is necessary before performing gel electrophoresis.

6.2.2.2 Real-Time Fluorescence Detection

Real-time fluorescence quantification is an efficient and sensitive technique for monitoring the RPA amplification process. By incorporating specially designed fluorescent probes into the reaction system, the amplification process can be monitored in real time. Since traditional PCR probes are incompatible with RPA due to differences in length and fluorescence signaling mechanisms, RPA-specific fluorescent probes, such as Exo and Fpg probes, are commonly used [32].

The exo probe method involves adding Exonuclease III (Exo) and Exo probes (Fig. 6.4a) to the RPA reaction. When the Exo probe binds to the target DNA to form a double-stranded DNA, the Exo enzyme recognizes and cleaves the THF site, removing nucleotides one by one from the 3′ end. This separates the fluorophore from the quencher, generating a cumulative fluorescent signal. The Fpg probe method involves adding formamidopyrimidine-DNA glycosylase (Fpg) and Fpg probes (Fig. 6.4b) to the RPA reaction. When the Fpg probe binds to the target DNA sequence, the Fpg enzyme cleaves the C-O-C bond in the dR fluorophore group, releasing the fluorophore and generating fluorescence. The fluorescence accumulation is synchronized with the quantity of amplification products, ensuring high sensitivity and specificity.

Fluorescent signals are collected and analyzed using instruments with fluorescence detection capabilities, such as real-time PCR systems or portable fluorescence detectors. These instruments monitor fluorescence changes during the reaction, record amplification curves, and track the progress of the reaction. RPA real-time fluorescence detection is typically completed within 10–20 min and offers excellent sensitivity and specificity [35, 36].

6.2.2.3 Lateral Flow Assay (LFA)

Lateral flow assays (LFA) are widely used in molecular diagnostics due to their simplicity, rapidity, and visual results [37]. To detect RPA amplification products using LFA, the reaction system incorporates Endonuclease IV (Nfo) and Nfo probes. Downstream primers are labeled with biotin or digoxigenin [38]. The Nfo probe structure (Fig. 6.4c) includes a fluorophore (e.g., FAM or FITC) at the 5′ end and a blocking group at the 3′ end. Nfo recognizes and cleaves the THF residue in the probe, allowing DNA extension at the 3′ cleavage site. This process generates amplification products labeled with both a fluorophore and biotin (FITC-dsDNA-Biotin). In LFA, biotin-specific antibodies (streptavidin) are immobilized on the test line to capture biotin-labeled RPA products. A control line with secondary antibodies captures gold nanoparticle-labeled anti-FITC antibodies. When RPA products are present, a complex of anti-FITC antibodies and FITC-dsDNA-Biotin is captured on the test line, forming a visible band. Unreacted probes flow to the control line and are captured, forming a second visible band (Fig. 6.5b). LFA can detect results within 30 min, requiring no complex instruments, making it ideal for rapid on-site testing [39, 40].

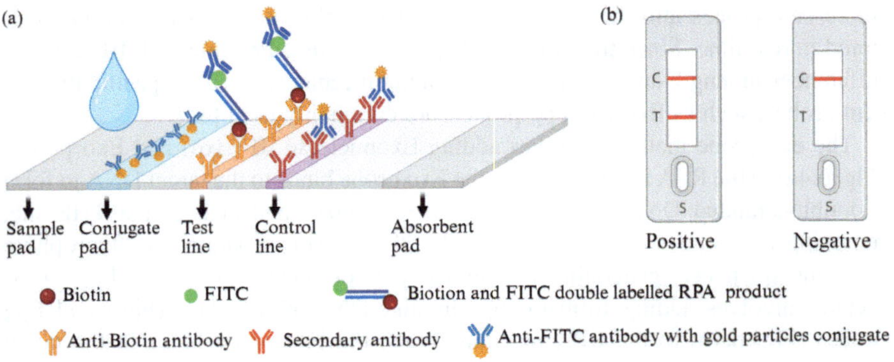

Fig. 6.5 Schematic diagram of LFA for RPA amplification products

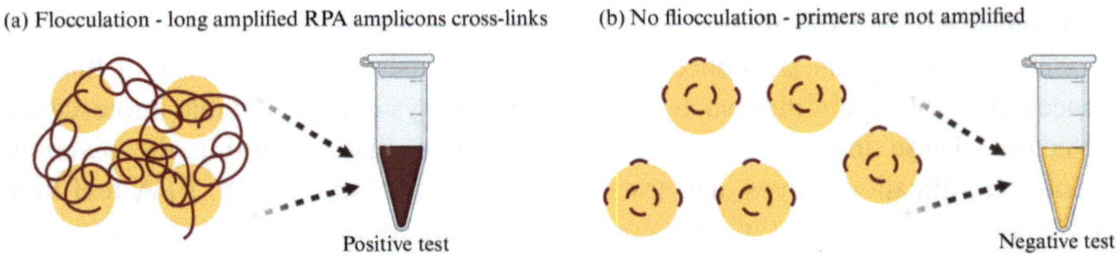

Fig. 6.6 Schematic diagram of detecting RPA amplification products using the flocculation method [43]. (Revised and reprinted with permission from Ref. [43])

6.2.2.4 Flocculation-Based Detection

Flocculation-based detection relies on the bridge flocculation phenomenon in colloid chemistry. Soluble long-chain polymers are introduced into a sol system to crosslink multiple colloidal particles, inducing rapid sedimentation, and forming loose, cotton-like precipitates under specific buffer conditions (e.g., pH and salt concentration) [41, 42]. Wee et al. first utilized flocculation for visual detection of RPA products [43, 44]. As shown in Fig. 6.6, long-chain DNA (>100 bp) from RPA reactions can bridge colloidal particles, triggering flocculation, while shorter RPA primers cannot. This enables direct visual detection. The reaction takes less than 10 min, and small amounts of RPA products induce significant flocculation without requiring DNA labeling or modification. Flocculation depends on salt concentration, pH, and DNA length, necessitating optimization of reaction conditions.

6.2.2.5 Enzyme-Linked Immunosorbent Assay (ELISA)

Enzyme-linked immunosorbent assay (ELISA) strategies can be used for detecting RPA products via colorimetric [45], electrochemical [46], or chemiluminescent [47] methods. The basic principle of ELISA-mediated RPA detection is illustrated in Fig. 6.7a. The upstream RPA primer is immobilized on a solid-phase carrier (e.g.,

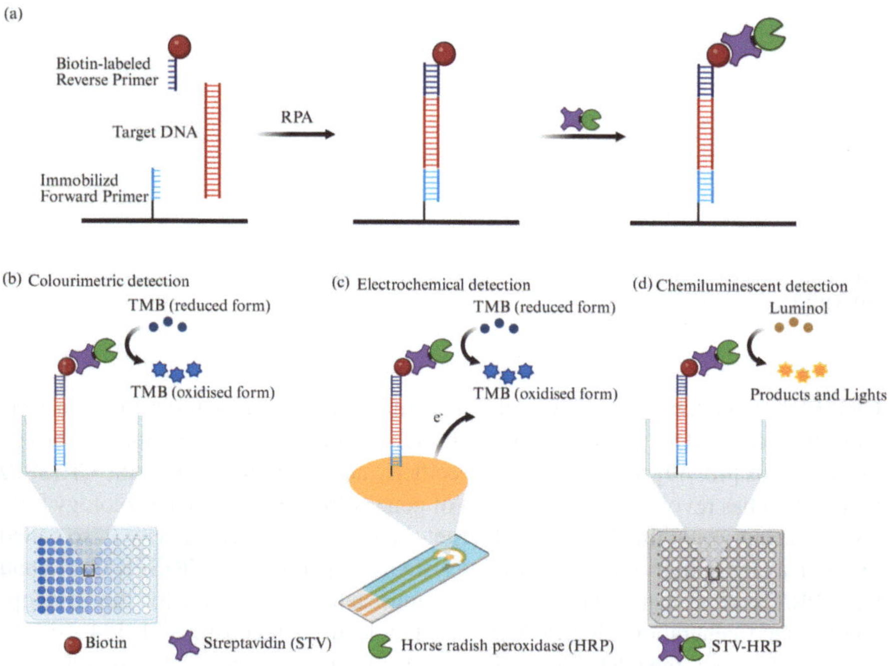

Fig. 6.7 Schematic diagram of detecting RPA amplification products using Enzyme-Linked Immunosorbent Assay (ELISA)-based methods [45–47]. (Revised and reprinted with permission from Refs. [45–47])

microplate or working electrode surface). When the target DNA is present, the downstream primer labeled with biotin at the 5′ end facilitates RPA amplification, producing products anchored to the carrier and labeled with biotin. Streptavidin-conjugated horseradish peroxidase (HRP) binds to the biotin-labeled RPA products.

For colorimetric detection, HRP catalyzes the oxidation of TMB substrate in the presence of hydrogen peroxide, causing a color change measurable by absorbance (Fig. 6.7b). In electrochemical detection, the HRP-catalyzed redox reaction of TMB generates an electrochemical signal detectable by portable devices (Fig. 6.7c). And chemiluminescent detection, HRP catalyzes the luminol-hydrogen peroxide system to produce luminescence for detecting RPA products (Fig. 6.7d). Electrochemical methods offer higher sensitivity, faster detection, and lower cost and are well-suited for on-site applications using small, affordable components (e.g., screen-printed carbon electrodes) with portable signal measurement devices.

6.2.2.6 Detection of RPA Amplification Products Based on the CRISPR/Cas12 System

The Clustered Regularly Interspaced Short Palindromic Repeats (CRISPR) and its associated proteins (CRISPR-associated proteins, Cas), collectively referred to as the

Fig. 6.8 Schematic diagram of the *cis*- and *trans*-cleavage activities of CRISPR/Cas12 and CRISPR/Cas13 systems

CRISPR/Cas system, represent an adaptive immune system found in most bacteria and all archaea. This system enables the recognition and degradation of foreign nucleic acids [48, 49]. Due to its high specificity and programmability, the CRISPR/Cas system has revolutionized the fields of gene editing and molecular biology [50–53]. In recent years, the discovery of target-dependent nuclease activity (also called *trans*-cleavage activity) of CRISPR/Cas systems, in particularly CRISPR/Cas12 and CRISPR/Cas13 systems, has opened up significant opportunities for the development of next-generation CRISPR/Cas-based molecular diagnostic tools [54–56]. As shown in Fig. 6.8, the CRISPR/Cas12 system specifically recognizes target double-stranded DNA (dsDNA) or single-stranded DNA (ssDNA), activating the *trans*-cleavage activity of Cas12, which non-specifically cleaves any ssDNA in the surrounding environment. Similarly, the CRISPR/Cas13 system recognizes target RNA and activates the ribonuclease activity of Cas13, resulting in continuous cleavage of any ssRNA. By leveraging the target-dependent *trans*-cleavage activities of the CRISPR/Cas12 and CRISPR/Cas13 systems, along with various reporter probe designs, the CRISPR/Cas system has been widely applied for the specific detection of RPA amplification products.

As shown in Fig. 6.9a, target nucleic acids (ssRNA or dsDNA) are amplified by RPA to generate specific dsDNA amplicons. Subsequently, the CRISPR/Cas12 system specifically recognizes and binds to the RPA amplification products, triggering the *trans*-cleavage activity of the system. When using a fluorophore and quencher dual-labeled single-stranded DNA (F-ssDNA-Q) as a fluorescent reporter probe, the RPA-generated amplification products activate the *trans*-cleavage activity of CRISPR/Cas12 system to continuously cleave the F-ssDNA-Q probe, separating the fluorophore from the quencher and generating a fluorescence signal. Multiple fluorescence detection methods can be used to detect the RPA amplification products. As illustrated in Fig. 6.9b, real-time fluorescence quantitative systems can monitor the fluorescence signal generated by the CRISPR/Cas system's *trans*-cleavage reaction in real-time, enabling dynamic analysis of the RPA amplification products over time [57, 58]. Additionally, portable fluorescence detection devices can be used for endpoint fluorescence signal detection [59], or color changes can be observed with the naked eye under portable light sources [60, 61], enabling visual detection of RPA amplification products.

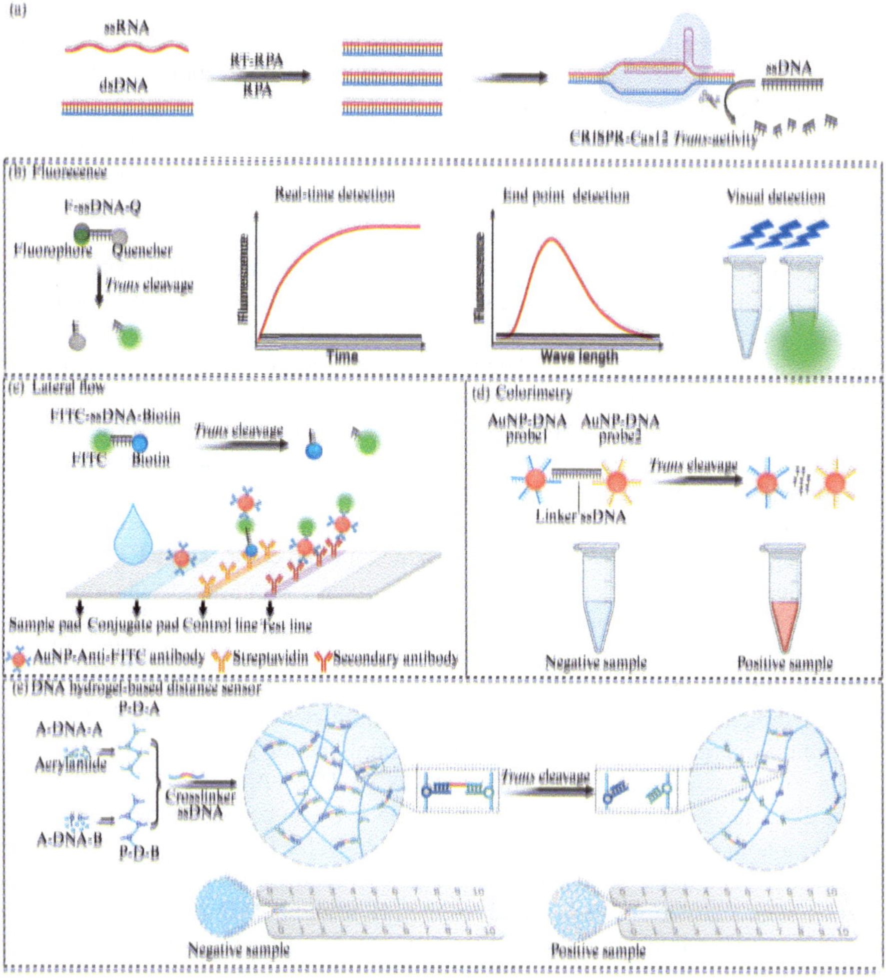

Fig. 6.9 Detection methods of RPA amplification products based on the CRISPR/Cas12 system

Alternatively, using fluorescein (e.g., FITC) and biotin dual-labeled single-stranded DNA (FITC-ssDNA-Biotin) as a reporter probe, RPA amplification products can be detected using lateral flow test strips [62]. The principle of lateral flow analysis for detecting RPA amplification products is shown in Fig. 6.9c. The conjugate pad of the lateral flow test strip contains gold nanoparticles (AuNPs) labeled with anti-FITC antibodies (AuNP-anti-FITC antibodies), while the control line is fixed with streptavidin, and the test line contains a secondary antibody. In the absence of specific RPA amplification products, the AuNP-anti-FITC antibodies interact immunologically with the FITC-ssDNA-Biotin reporter probe, forming AuNP-anti-FITC/FITC-ssDNA-Biotin complexes. Subsequently, the interaction between streptavidin and biotin fixes the complexes at the control line, forming the control line. In the presence of specific RPA amplification products, the CRISPR/Cas12 system specifically recognizes and binds to the RPA amplicons, activating *trans*-cleavage activity that cleaves the FITC-ssDNA-Biotin reporter probe, separating the FITC and biotin. The AuNP-anti-FITC/FITC complexes are captured by the secondary antibody at the test line, forming a visible test line, while uncut probes are captured by streptavidin, forming the control line. Thus, positive samples will form both the control and test lines, while negative samples will only form the control line.

Colorimetric detection based on the aggregation state of gold nanoparticles can also be used to detect RPA amplification products [63]. The detection principle of this method is shown in Fig. 6.9d. Gold nanoparticles-DNA probe 1 (AuNP-DNA1), gold nanoparticle-DNA probe 2 (AuNP-DNA2), and a linker ssDNA probe (Linker ssDNA) are designed. The gold nanoparticles form an aggregated state via hybridization with AuNP-DNA1, AuNP-DNA2, and the Linker ssDNA, resulting in a colorless or light blue appearance. When specific RPA amplification products are present, the CRISPR/Cas12 system activates *trans*-cleavage activity, cleaving the ssDNA sequence in the Linker ssDNA, causing the gold nanoparticles to transition from an aggregated to a dispersed state, producing a red color. The RPA amplification products can be detected by measuring absorbance with a spectrophotometer or visually observing the color change.

Coupling the target-dependent *trans*-cleavage activity of CRISPR/Cas12 and capillary action, our team developed a capillary-driven CRISPR/Cas12a-based DNA hydrogel distance sensor, which successfully enables high-sensitivity, high-specificity, and rapid visual detection of RPA amplification products [64]. As shown in Fig. 6.9e, two acrylamide-modified DNAs (A-DNA-A and A-DNA-B) and a linker DNA are designed. A-DNA-A and A-DNA-B can co-polymerize with acrylamide monomers to form linear polyacrylamide-DNA (P-D-A and P-D-B). The linker ssDNA contains three functional regions: two regions at the 3′ and 5′ ends are complementary to A-DNA-A and A-DNA-B, bridging P-D-A and P-D-B, while the "TTTATT" sequence in the middle serves as the substrate for Cas12a *trans*-cleavage. The ends of the linker DNA hybridize with A-DNA-A and A-DNA-B, forming a three-dimensional porous structure for the CRISPR/Cas12a-responsive DNA hydrogel. Utilizing the thermo-reversibility of the DNA hydrogel and capillary action [65, 66], we loaded the CRISPR/Cas12a-responsive DNA hydrogel film (approximately 0.2 mm thick) at the end of the capillary tube, creating a DNA hydrogel distance sensor. When specific RPA amplification products are present (i.e., positive samples), the *trans*-cleavage activity of the CRISPR/Cas system is activated, and the activated Cas12a enzyme enters the DNA hydrogel film through capillary action, continuously cleaving the single-stranded sequences in the linker DNA, enhancing the permeability of the DNA hydrogel film. Positive sample solutions show greater migration distances in the capillary tube, while negative samples exhibit shorter migration distances. Therefore, by reading the migration distance of the solution in the capillary tube, RPA amplification products can be detected visually.

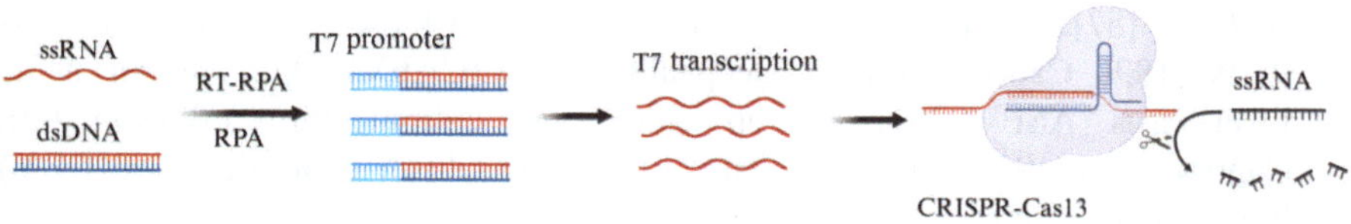

Fig. 6.10 Schematic diagram of RPA amplification product detection based on the CRISPR/Cas13 system

6.2.2.7 Detection of RPA Amplification Products Based on the CRISPR/Cas13 System

The principle of RPA amplification product detection using the CRISPR/Cas13 system is illustrated in Fig. 6.10. Initially, RPA amplification primers containing the T7 promoter sequence are used to amplify the target nucleic acid (either dsDNA or ssRNA) through RPA or RT-RPA, resulting in amplification products with terminal T7 promoter sequences. Subsequently, a T7 transcription reaction is employed to generate target-specific RNA (i.e., T7-transcribed products). These RNA transcripts are then specifically recognized and bound by the CRISPR/Cas13 system, activating the *trans*-cleavage activity of Cas13, which cleaves the single-stranded RNA reporter probe. Similar to the CRISPR/Cas12 system, by replacing the single-stranded DNA reporter probe with a single-stranded RNA reporter probe, the RPA amplification products can be detected using various methods, such as fluorescence [59], lateral flow strips [67], colorimetry [63], and distance sensors [65].

6.2.3 Applications of RPA

Benefiting from the advantages of RPA technology, including isothermal conditions, sensitivity, simplicity, and rapid results, combined with various detection methods, RPA has found widespread applications in nucleic acid detection across diverse fields such as clinical diagnostics, food safety, environmental monitoring, and agriculture and livestock industries.

1. Viral Nucleic Acid Detection: Various viruses have been detected using RPA technology, including human papillomavirus (HPV) [64], Human Immunodeficiency virus (HIV) [68–70], Severe acute respiratory syndrome coronavirus 2 (SARS-CoV-2) [71, 72], Ebola virus (EBOV) [73], Infectious subcutaneous and Hematopoietic necrosis virus (IHHNV) [74], Smallpox virus [75], Rift Valley fever virus (RVFV) [76], H7N9 avian influenza virus (H7N9) [77], Zika virus (ZIKV) [78], Dengue virus (DENV) [79], Yellow fever virus (YFV) [80], Nodular skin disease virus (LSDV) [81, 82], Foot-and-mouth disease virus (FMDV) [83], Carp herpesvirus (CYHV-2) [84], Feline herpesvirus (FHV-1) [85], Feline panleukopenia virus (FPV) [86], Bovine respiratory syncytial virus

(BRSV) [87], Avian enteric coronavirus (AECoV) [88], Canine parvovirus (CPV) [89], Caprine arthritis-encephalitis virus (CAEV) [90], African swine fever virus (ASFV) [91], Shrimp white spot syndrome virus (WSSV) [92], Capripoxvirus [93], Birnavirus (BBV) [34], and Clover dwarfing virus (CDV) [94], among others.

2. Bacterial Detection Using RPA: Bacteria that have been detected using RPA technology include *Brucella* [95], *Rickettsia typhi* (epidemic typhus) [96], *Group B Streptococcus* [97, 98], *Borrelia* [99], *Shiga toxin-producing Escherichia coli* [100], *Neisseria gonorrhoeae* [101], *Methicillin-resistant Staphylococcus aureus* (MRSA) [101], *Listeria monocytogenes* [102], *Streptococcus pneumoniae* [103], *Vibrio parahaemolyticus* [104], *Leptospira* [105, 106], *Vibrio cholerae* [107], *Mycobacterium tuberculosis* [36, 108, 109], *Campylobacter jejuni* [110], *Salmonella* [111–114], *Salmonella Typhimurium* [115], *Yersinia pestis* (plague) [75], *Bacillus anthracis* (anthrax) [75], *Tularaemia* [116], *Orientia tsutsugamushi* (scrub typhus) [96], *Yersinia enterocolitica* (enterocolitis) [117], *Nocardia asteroides* [118], *Staphylococcus aureus* [119], *Mycobacterium avium subspecies paratuberculosis* (Johne's disease) [120, 121], *and Rhodococcus equi* [122], *among others.*

3. Other Pathogens: RPA technology has also been applied to detect other pathogens such as *Cryptosporidium* [123], *Giardia lamblia* [124], *Entamoeba histolytica* [124], *Plasmodium falciparum* (malignant malaria) [125], *Schistosoma mansoni* (Egyptian schistosomiasis) [126, 127], *Schistosoma japonicum* (Japanese schistosomiasis) [128], *Leishmania* [129], *Chlamydia trachomatis* (trachoma) [130], *Toxoplasma gondii* (toxoplasmosis) [131], *Mycoplasma ovipneumonia* (ovine pneumonia) [132], *Mycoplasma capricolum* (caprine contagious pleuropneumonia) [133], and *Babesia gibsoni* (dog babesiosis) [134], among others.

RPA's ability to rapidly and accurately detect a wide range of pathogens, including viruses, bacteria, and other microorganisms, demonstrates its versatility and potential for widespread use in various diagnostic and surveillance applications.

References

1. Vincent M, Xu Y, Kong HM. Helicase-dependent isothermal DNA amplification. EMBO Rep. 2004;5:795–800.
2. Goldmeyer J, Kong H, Tang W. Development of a novel one-tube isothermal reverse transcription thermophilic helicase-dependent amplification platform for rapid RNA detection. J Mol Diagn. 2007;9:639–44.
3. Runyon GT, Lohman TM. Escherichia coli helicase II (uvrD) protein can completely unwind fully duplex linear and nicked circular DNA. J Biol Chem. 1989;264:17502–12.
4. An LX, Tang W, Ranalli TA, Kim HJ, Wytiaz J, Kong HM. Characterization of a thermostable UvrD helicase and its participation in helicase-dependent amplification. J Biol Chem. 2005;280:28952–8.

5. Ali JA, Maluf NK, Lohman TM. An oligomeric form of E. coli UvrD is required for optimal helicase activity. J Mol Biol. 1999;293:815–34.
6. Xu Y. Simultaneous amplification and screening of whole plasmids using the T7 bacteriophage replisome. Nucleic Acids Res. 2006;34:e98.
7. Nakai H, Richardson CC. Leading and lagging strand synthesis at the replication fork of bacteriophage T7. Distinct properties of T7 gene 4 protein as a helicase and primase. J Biol Chem. 1988;263:9818–30.
8. Kim YT, Tabor S, Churchich JE, Richardson CC. Interactions of gene 2.5 protein and DNA polymerase of bacteriophage T7. J Biol Chem. 1992;267:15032–40.
9. Motre A, Li Y, Kong H. Enhancing helicase-dependent amplification by fusing the helicase with the DNA polymerase. Gene. 2008;420:17–22.
10. Gill P, Amini M, Ghaemi A, Shokouhizadeh L, Abdul-Tehrani H, Karami A, Gilak A. Detection of *Helicobacter pylori* by enzyme-linked immunosorbent assay of thermophilic helicase-dependent isothermal DNA amplification. Diagn Microbiol Infect Dis. 2007;59:243–9.
11. Kivlehan F, Mavré F, Talini L, Limoges B, Marchal D. Real-time electrochemical monitoring of isothermal helicase-dependent amplification of nucleic acids. Analyst. 2011;136:3635.
12. Moura-Melo S, Miranda-Castro R, de-los-Santos-Alvarez N, Miranda-Ordieres AJ, Ribeiro Dos Santos Junior J, da Silva Fonseca RA, Jesus Lobo-Castanon M. Targeting helicase-dependent amplification products with an electrochemical genosensor for reliable and sensitive screening of genetically modified organisms. Anal Chem. 2015;87:8547–54.
13. Chow WHA, McCloskey C, Tong Y, Hu L, You Q, Kelly CP, Kong H, Tang Y-W, Tang W. Application of isothermal helicase-dependent amplification with a disposable detection device in a simple sensitive stool test for toxigenic Clostridium difficile. J Mol Diagn. 2008;10:452–8.
14. Du X-J, Zhou T-J, Li P, Wang S. A rapid *Salmonella* detection method involving thermophilic helicase-dependent amplification and a lateral flow assay. Mol Cell Probes. 2017;34:37–44.
15. Andresen D, von Nickisch-Rosenegk M, Bier FF. Helicase dependent OnChip-amplification and its use in multiplex pathogen detection. Clin Chim Acta. 2009;403:244–8.
16. Cao A, Zhang C-Y. Real-time detection of transcription factors using target-converted helicase-dependent amplification assay with zero-background signal. Anal Chem. 2013;85:2543–7.
17. Chen F, Zhang D, Zhang Q, Zuo X, Fan C, Zhao Y. Zero-background helicase-dependent amplification and its application to reliable assay of telomerase activity in cancer cell by eliminating primer–dimer artifacts. Chembiochem. 2016;17:1171–6.
18. Ma F, Liu M, Tang B, Zhang C-Y. Sensitive quantification of microRNAs by isothermal helicase-dependent amplification. Anal Chem. 2017;89:6183–8.
19. Liu M, Li C-C, Luo X, Ma F, Zhang C-Y. 5-Hydroxymethylcytosine glucosylation-triggered helicase-dependent amplification-based fluorescent biosensor for sensitive detection of β-glucosyltransferase with zero background signal. Anal Chem. 2020;92:16307–13.
20. Piepenburg O, Williams CH, Stemple DL, Armes NA. DNA detection using recombination proteins. PLoS Biol. 2006;4:e204.
21. Zhao Y, Chen F, Li Q, Wang L, Fan C. Isothermal amplification of nucleic acids. Chem Rev. 2015;115:12491–545.
22. Munawar MA. Critical insight into recombinase polymerase amplification technology. Expert Rev Mol Diagn. 2022;22:725–37.
23. Li J, Macdonald J, Von Stetten F. Review: a comprehensive summary of a decade development of the recombinase polymerase amplification. Analyst. 2019;144:31–67.
24. Tan M, Liao C, Liang L, Yi X, Zhou Z, Wei G. Recent advances in recombinase polymerase amplification: principle, advantages, disadvantages and applications. Front Cell Infect Microbiol. 2022;12:1019071.

25. Mota DS, Guimaraes JM, Gandarilla AMD, Filho J, Brito WR, Mariúba LAM. Recombinase polymerase amplification in the molecular diagnosis of microbiological targets and its applications. Can J Microbiol. 2022;68:383–402.
26. Yonesaki T, Ryo Y, Minagawa T, Takahashi H. Purification and some of the functions of the products of bacteriophage T4 recombination genes, uvsX and uvsY. Eur J Biochem. 1985;148: 127–34.
27. Formosa T, Alberts BM. Purification and characterization of the T4 bacteriophage uvsX protein. J Biol Chem. 1986;261:6107–18.
28. Nguyen VD, Nguyen HQ, Bui HK, Kang Y-J, Seo TS. A smartphone-controllable molecular diagnostic platform for SARS-CoV-2 detection by reverse-transcription recombinase polymerase amplification. Sensors Actuat B-Chem. 2024;398:134728.
29. Kumar R, Kaundal P, Tiwari RK, Lal MK, Kumari H, Kumar R, Naga KC, Kumar A, Singh B, Sagar V, Sharma S. Development of reverse transcription recombinase polymerase amplification (RT-RPA): a methodology for quick diagnosis of potato leafroll viral disease in potato. Int J Mol Sci. 2023;24:2511.
30. Ma Y, Wu H, Chen S, Xie C, Hu J, Qi X, Ma X, Chu Y, Shan J, Lu Y, Cui L, Zou B, Zhou G. FEN1-aided recombinase polymerase amplification (FARPA) for one-pot and multiplex detection of nucleic acids with an ultra-high specificity and sensitivity. Biosens Bioelectron. 2023;237:115456.
31. Twistamp assay design manual. https://www.twistdx.co.uk/wp-content/uploads/2023/12/twistamp-assay-design-manual_rev1_v3.pdf. Accessed 27 Oct 2023.
32. Daher RK, Stewart G, Boissinot M, Bergeron MG. Recombinase polymerase amplification for diagnostic applications. Clin Chem. 2016;62:947–58.
33. Magrina Lobato I, O'Sullivan CK. Recombinase polymerase amplification: basics, applications and recent advances. Trac-Trends Analyt Chem. 2018;98:19–35.
34. Londoño MA, Harmon CL, Polston JE. Evaluation of recombinase polymerase amplification for detection of begomoviruses by plant diagnostic clinics. Virol J. 2016;13:48.
35. Ying J, Mao L, Tang Y, Fassatoui M, Song W, Xu X, Tang X, Li J, Liu H, Jian F, Du Q, Wong G, Feng W, Berthet N. Development and validation of real-time recombinase polymerase amplification-based assays for detecting HPV16 and HPV18 DNA. Microbiol Spectrum. 2023;11:e0120723.
36. Xu Y, Wu P, Zhang H, Li J. Rapid detection of Mycobacterium tuberculosis based on antigen 85B via real-time recombinase polymerase amplification. Lett Appl Microbiol. 2021;72:106–12.
37. Boehringer HR, O'Farrell BJ. Lateral flow assays in infectious disease diagnosis. Clin Chem. 2022;68:52–8.
38. James A, Macdonald J. Recombinase polymerase amplification: emergence as a critical molecular technology for rapid, low-resource diagnostics. Expert Rev Mol Diagn. 2015;15: 1475–89.
39. Liu D, Shen H, Zhang Y, Shen D, Zhu M, Song Y, Zhu Z, Yang C. A microfluidic-integrated lateral flow recombinase polymerase amplification (MI-IF-RPA) assay for rapid COVID-19 detection. Lab Chip. 2021;21:2019–26.
40. Wang L, Wang Y, Wang F, Zhao M, Gao X, Chen H, Li N, Zhu Q, Liu L, Zhu W, Liu X, Chen Y, Zhou P, Lu Y, Wang K, Zhao W, Liang W. Development and application of rapid clinical visualization molecular diagnostic technology for Cryptococcus neoformans/C. gattii based on recombinase polymerase amplification combined with a lateral flow strip. Front Cell Infect Microbiol. 2022;11:802798.
41. Ruehrwein RA, Ward DW. Mechanism of clay aggregation by polyelectrolytes. Soil Sci. 1952;73:485–92.
42. Healy TW, La Mer VK. The energetics of flocculation and redispersion by polymers. J Colloid Sci. 1964;19:323–32.
43. Wee EJH, Ha Ngo T, Trau M. A simple bridging flocculation assay for rapid, sensitive and stringent detection of gene specific DNA methylation. Sci Rep. 2015;5:15028.

44. Wee EJH, Lau HY, Botella JR, Trau M. Re-purposing bridging flocculation for on-site, rapid, qualitative DNA detection in resource-poor settings. Chem Commun. 2015;51:5828–31.
45. Koo KM, Wee EJH, Trau M. Colorimetric *TMPRSS2-ERG* gene fusion detection in prostate cancer urinary samples via recombinase polymerase amplification. Theranostics. 2016;6: 1415–24.
46. del Río JS, Yehia Adly N, Acero-Sánchez JL, Henry OYF, O'Sullivan CK. Electrochemical detection of Francisella tularensis genomic DNA using solid-phase recombinase polymerase amplification. Biosens Bioelectron. 2014;54:674–8.
47. Kunze A, Dilcher M, Abd El Wahed A, Hufert F, Niessner R, Seidel M. On-chip isothermal nucleic acid amplification on flow-based chemiluminescence microarray analysis platform for the detection of viruses and bacteria. Anal Chem. 2016;88:898–905.
48. Mojica FJM, Díez-Villaseñor C, García-Martínez J, Soria E. Intervening sequences of regularly spaced prokaryotic repeats derive from foreign genetic elements. J Mol Evol. 2005;60: 174–82.
49. Makarova KS, Haft DH, Barrangou R, Brouns SJJ, Charpentier E, Horvath P, Moineau S, Mojica FJM, Wolf YI, Yakunin AF, van der Oost J, Koonin EV. Evolution and classification of the CRISPR-Cas systems. Nat Rev Microbiol. 2011;9:467–77.
50. Doudna JA, Charpentier E. The new frontier of genome engineering with CRISPR-Cas9. Science. 2014;346:1258096.
51. Knott GJ, Doudna JA. CRISPR-Cas guides the future of genetic engineering. Science. 2018;361:866–9.
52. Pickar-Oliver A, Gersbach CA. The next generation of CRISPR-Cas technologies and applications. Nat Rev Mol Cell Biol. 2019;20:490–507.
53. Tang Y, Gao L, Feng W, Guo C, Yang Q, Li F, Le XC. The CRISPR–Cas toolbox for analytical and diagnostic assay development. Chem Soc Rev. 2021;50:11844–69.
54. Kaminski MM, Abudayyeh OO, Gootenberg JS, Zhang F, Collins JJ. CRISPR-based diagnostics. Nature Biomed Eng. 2021;5:643–56.
55. Feng W, Newbigging AM, Tao J, Cao Y, Peng H, Le C, Wu J, Pang B, Li J, Tyrrell DL, Zhang H, Le XC. CRISPR technology incorporating amplification strategies: molecular assays for nucleic acids, proteins, and small molecules. Chem Sci. 2021;12:4683–98.
56. Weng Z, You Z, Yang J, Mohammad N, Lin M, Wei Q, Gao X, Zhang Y. CRISPR-Cas biochemistry and CRISPR-based molecular diagnostics. Angew Chem Int Ed. 2023;62: e202214987.
57. Gootenberg JS, Abudayyeh OO, Lee JW, Essletzbichler P, Dy AJ, Joung J, Verdine V, Donghia N, Daringer NM, Freije CA, Myhrvold C, Bhattacharyya RP, Livny J, Regev A, Koonin EV, Hung DT, Sabeti PC, Collins JJ, Zhang F. Nucleic acid detection with CRISPR-Cas13a/C2c2. Science. 2017;356:438–42.
58. Chen JS, Ma E, Harrington LB, Da Costa M, Tian X, Palefsky JM, Doudna JA. CRISPR-Cas12a target binding unleashes indiscriminate single-stranded DNase activity. Science. 2018;360:436–9.
59. Tian T, Qiu Z, Jiang Y, Zhu D, Zhou X. Exploiting the orthogonal CRISPR-Cas12a/Cas13a trans-cleavage for dual-gene virus detection using a handheld device. Biosens Bioelectron. 2021;196:113701.
60. Feng W, Peng H, Xu J, Liu Y, Pabbaraju K, Tipples G, Joyce MA, Saffran HA, Tyrrell DL, Babiuk S, Zhang H, Le XC. Integrating reverse transcription recombinase polymerase amplification with CRISPR technology for the one-tube assay of RNA. Anal Chem. 2021;93: 12808–16.
61. Ding X, Yin K, Li Z, Lalla RV, Ballesteros E, Sfeir MM, Liu C. Ultrasensitive and visual detection of SARS-CoV-2 using all-in-one dual CRISPR-Cas12a assay. Nat Commun. 2020;11:4711.
62. Lee RA, De Puig H, Nguyen PQ, Angenent-Mari NM, Donghia NM, McGee JP, Dvorin JD, Klapperich CM, Pollock NR, Collins JJ. Ultrasensitive CRISPR-based diagnostic for

field-applicable detection of Plasmodium species in symptomatic and asymptomatic malaria. Proc Natl Acad Sci USA. 2020;117:25722–31.
63. Yuan C, Tian T, Sun J, Hu M, Wang X, Xiong E, Cheng M, Bao Y, Lin W, Jiang J, Yang C, Chen Q, Zhang H, Wang H, Wang X, Deng X, Liao X, Liu Y, Wang Z, Zhang G, Zhou X. Universal and naked-eye gene detection platform based on the clustered regularly interspaced short palindromic repeats/Cas12a/13a system. Anal Chem. 2020;92:4029–37.
64. Pian H, Wang H, Wang HH, Tang F, Li ZP. Capillarity-powered and CRISPR/Cas12a-responsive DNA hydrogel distance sensor for highly sensitive visual detection of HPV DNA. Biosens Bioelectron. 2024;264:116657.
65. Wang H, Wang H, Pian H, Su F, Tang F, Chen D, Chen J, Wen Y, Le XC, Li Z. CRISPR/Cas13a-responsive and RNA-bridged DNA hydrogel capillary sensor for point-of-care detection of RNA. Anal Chem. 2024;96:12022–9.
66. Wang H, Wang H, Li Y, Jiang C, Chen D, Wen Y, Li Z. Capillarity self-driven DNA hydrogel sensor for visual quantification of microRNA. Sensors Actuators B Chem. 2020;313:128036.
67. Multiplexed and portable nucleic acid detection platform with Cas13, Cas12a, and Csm6. Science. 2018;360:439–44.
68. Crannell ZA, Rohrman B, Richards-Kortum R. Quantification of HIV-1 DNA using real-time recombinase polymerase amplification. Anal Chem. 2014;86:5615–9.
69. Rohrman BA, Richards-Kortum RR. A paper and plastic device for performing recombinase polymerase amplification of HIV DNA. Lab Chip. 2012;12:3082–8.
70. Crannell ZA, Rohrman B, Richards-Kortum R. Equipment-free incubation of recombinase polymerase amplification reactions using body heat. PLoS One. 2014;9:e112146.
71. Abd El Wahed A, Patel P, Maier M, Pietsch C, Rüster D, Böhlken-Fascher S, Kissenkötter J, Behrmann O, Frimpong M, Diagne MM, Faye M, Dia N, Shalaby MA, Amer H, Elgamal M, Zaki A, Ismail G, Kaiser M, Corman VM, Niedrig M, Landt O, Faye O, Sall AA, Hufert FT, Truyen U, Liebert UG, Weidmann M. Suitcase lab for rapid detection of SARS-CoV-2 based on recombinase polymerase amplification assay. Anal Chem. 2021;93:2627–34.
72. Patchsung M, Jantarug K, Pattama A, Aphicho K, Suraritdechachai S, Meesawat P, Sappakhaw K, Leelahakorn N, Ruenkam T, Wongsatit T, Athipanyasilp N, Eiamthong B, Lakkanasirorat B, Phoodokmai T, Niljianskul N, Pakotiprapha D, Chanarat S, Homchan A, Tinikul R, Kamutira P, Phiwkaow K, Soithongcharoen S, Kantiwiriyawanitch C, Pongsupasa V, Trisrivirat D, Jaroensuk J, Wongnate T, Maenpuen S, Chaiyen P, Kamnerdnakta S, Swangsri J, Chuthapisith S, Sirivatanauksorn Y, Chaimayo C, Sutthent R, Kantakamalakul W, Joung J, Ladha A, Jin X, Gootenberg JS, Abudayyeh OO, Zhang F, Horthongkham N, Uttamapinant C. Clinical validation of a Cas13-based assay for the detection of SARS-CoV-2 RNA. Nature Biomed Eng. 2020;4:1140–9.
73. Faye O, Faye O, Soropogui B, Patel P, El Wahed AA, Loucoubar C, Fall G, Kiory D, Magassouba N, Keita S, Kondé MK, Diallo AA, Koivogui L, Karlberg H, Mirazimi A, Nentwich O, Piepenburg O, Niedrig M, Weidmann M, Sall AA. Development and deployment of a rapid recombinase polymerase amplification Ebola virus detection assay in Guinea in 2015. Eurosurveillance. 2015;20:10–8.
74. Xia XM, Yu YX, Hu LH, Weidmann M, Pan YJ, Yan SL, Wang YJ. Rapid detection of infectious hypodermal and hematopoietic necrosis virus (IHHNV) by real-time, isothermal recombinase polymerase amplification assay. Arch Virol. 2015;160:987–94.
75. Euler M, Wang YJ, Heidenreich D, Patel P, Strohmeier O, Hakenberg S, Niedrig M, Hufert FT, Weidmann M. Development of a panel of recombinase polymerase amplification assays for detection of biothreat agents. J Clin Microbiol. 2013;51:1110–7.
76. Euler M, Wang YJ, Nentwich O, Piepenburg O, Hufert FT, Weidmann M. Recombinase polymerase amplification assay for rapid detection of Rift Valley fever virus. J Clin Virol. 2012;54:308–12.
77. Abd El Wahed A, Weidmann M, Hufert FT. Diagnostics-in-a-Suitcase: development of a portable and rapid assay for the detection of the emerging avian influenza A (H7N9) virus. J Clin Virol. 2015;69:16–21.

78. Wand NIV, Bonney LC, Watson RJ, Graham V, Hewson R. Point-of-care diagnostic assay for the detection of Zika virus using the recombinase polymerase amplification method. J Gen Virol. 2018;99:1012–26.
79. Xi Y, Xu CZ, Xie ZZ, Zhu DL, Dong JM. Rapid and visual detection of dengue virus using recombinase polymerase amplification method combined with lateral flow dipstick. Mol Cell Probes. 2019;46:101413.
80. Escadafal C, Faye O, Sall AA, Faye O, Weidmann M, Strohmeier O, von Stetten F, Drexler J, Eberhard M, Niedrig M, Patel P. Rapid molecular assays for the detection of yellow fever virus in low-resource settings. PLoS Negl Trop Dis. 2014;8:e2730.
81. Shalaby MA, El-Deeb A, El-Tholoth M, Hoffmann D, Czerny CP, Hufert FT, Weidmann M, Abd El Wahed A. Recombinase polymerase amplification assay for rapid detection of lumpy skin disease virus. BMC Vet Res. 2016;12:244.
82. Jiang CW, Tao DG, Geng YC, Yang H, Xu BR, Chen YY, Hu CM, Chen HC, Xie SS, Guo AZ. Sensitive and specific detection of lumpy skin disease virus in cattle by CRISPR-Cas12a fluorescent assay coupled with recombinase polymerase amplification. Genes. 2022;13:734.
83. Yang Y, Qin XD, Wang GX, Jin JX, Shang YJ, Zhang ZD. Development of an isothermal amplification-based assay for rapid visual detection of an Orf virus. Virol J. 2016;13:46.
84. Prescott MA, Reed AN, Jin L, Pastey MK. Rapid detection of cyprinid herpesvirus 3 in latently infected koi by recombinase polymerase amplification. J Aquat Anim Health. 2016;28:173–80.
85. Liu MZ, Han XH, Yao LQ, Zhang WK, Liu BS, Chen ZL. Development and application of a simple recombinase polymerase amplification assay for rapid point-of-care detection of feline herpesvirus type 1. Arch Virol. 2019;164:195–200.
86. Wang ZH, Wang XJ, Hou SH. Development of a recombinase polymerase amplification assay with lateral flow dipstick for rapid detection of feline parvovirus. J Virol Methods. 2019;271:113679.
87. Hou PL, Wang HM, Zhao GM, He CQ, He HB. Rapid detection of infectious bovine Rhinotracheitis virus using recombinase polymerase amplification assays. BMC Vet Res. 2017;13:386.
88. Ma L, Shi HF, Zhang ML, Song YW, Zhang KP, Cong F. Establishment of a real-time recombinase polymerase amplification assay for the detection of avian reovirus. Front Vet Sci. 2020;7:551350.
89. Wang JC, Liu LB, Li RW, Wang JF, Fu Q, Yuan WZ. Rapid and sensitive detection of canine parvovirus type 2 by recombinase polymerase amplification. Arch Virol. 2016;161:1015–8.
90. Tu PA, Shiu JS, Lee SH, Pang VF, Wang DC, Wang PH. Development of a recombinase polymerase amplification lateral flow dipstick (RPA-LFD) for the field diagnosis of caprine arthritis-encephalitis virus (CAEV) infection. J Virol Methods. 2017;243:98–104.
91. Mao G, Luo X, Ye S, Wang X, He J, Kong J, Dai J, Yin W, Ma Y. Fluorescence and colorimetric analysis of African swine fever virus based on the RPA-assisted CRISPR/Cas12a strategy. Anal Chem. 2023;95:8063–9.
92. Xia XM, Yu YX, Weidmann M, Pan YJ, Yan SL, Wang YJ. Rapid detection of shrimp white spot syndrome virus by real time, isothermal recombinase polymerase amplification assay. PLoS One. 2014;9:e104667.
93. Liu LB, Wang JF, Nie FP, Li RW, Gao YX, Sun XX, Yuan WZ, Wang JC. Development of the isothermal recombinase polymerase amplification assays for rapid detection of the genus Capripoxvirus. J Virol Methods. 2023;320:114788.
94. Cao YH, Yan DK, Wu XY, Chen ZQ, Lai YC, Lv LQ, Yan F, Chen JP, Zheng HY, Song XM. Rapid and visual detection of milk vetch dwarf virus using recombinase polymerase amplification combined with lateral flow strips. Virol J. 2020;17:102.
95. Ren H, Yang MJ, Zhang GX, Liu SW, Wang XH, Ke YH, Du XY, Wang ZJ, Huang LY, Liu C, Chen ZL. Development of a rapid recombinase polymerase amplification assay for detection of Brucella in blood samples. Mol Cell Probes. 2016;30:122–4.

96. Chao CC, Belinskaya T, Zhang ZW, Ching WM. Development of recombinase polymerase amplification assays for detection of Orientia tsutsugamushi or Rickettsia typhi. PLoS Negl Trop Dis. 2015;9:e0003884.
97. Clarke C, O'Connor L, Carré-Skinner H, Piepenburg O, Smith TJ. Development and performance evaluation of a recombinase polymerase amplification assay for the rapid detection of group B streptococcus. BMC Microbiol. 2016;16:221.
98. Daher RK, Stewart G, Boissinot M, Bergeron MG. Isothermal recombinase polymerase amplification assay applied to the detection of group B streptococci in vaginal/anal samples. Clin Chem. 2014;60:660–6.
99. Liu W, Liu HX, Zhang L, Hou XX, Wan KL, Hao Q. A novel isothermal assay of Borrelia burgdorferi by recombinase polymerase amplification with lateral flow detection. Int J Mol Sci. 2016;17:1250.
100. Murinda SE, Ibekwe AM, Zulkaffly S, Cruz A, Park S, Razak N, Paudzai FM, Ab Samad L, Baquir K, Muthaiyah K, Santiago B, Rusli A, Balkcom S. Real-time isothermal detection of Shiga toxin–producing Escherichia coli using recombinase polymerase amplification. Foodborne Pathog Dis. 2014;11:529–36.
101. Kersting S, Rausch V, Bier FF, von Nickisch-Rosenegk M. Multiplex isothermal solid-phase recombinase polymerase amplification for the specific and fast DNA-based detection of three bacterial pathogens. Microchim Acta. 2014;181:1715–23.
102. Gao WF, Huang HL, Zhang Y, Zhu P, Yan XJ, Fan JZ, Chen XF. Recombinase polymerase amplification-based assay for rapid detection of Listeria monocytogenes in food samples. Food Anal Methods. 2017;10:1972–81.
103. Clancy E, Higgins O, Forrest MS, Boo TW, Cormican M, Barry T, Piepenburg O, Smith TJ. Development of a rapid recombinase polymerase amplification assay for the detection of Streptococcus pneumoniae in whole blood. BMC Infect Dis. 2015;15:481.
104. Geng YY, Tan K, Liu LB, Sun XX, Zhao BH, Wang JC. Development and evaluation of a rapid and sensitive RPA assay for specific detection of Vibrio parahaemolyticus in seafood. BMC Microbiol. 2019;19:186.
105. Ahmed A, van der Linden H, Hartskeerl RA. Development of a recombinase polymerase amplification assay for the detection of pathogenic Leptospira. Int J Environ Res Public Health. 2014;11:4953–64.
106. Jirawannaporn S, Limothai U, Tachaboon S, Dinhuzen J, Kiatamornrak P, Chaisuriyong W, Bhumitrakul J, Mayuramart O, Payungporn S, Srisawat N. Rapid and sensitive point-of-care detection of Leptospira by RPA-CRISPR/Cas12a targeting lipL32. PLoS Negl Trop Dis. 2022;16:e0010112.
107. Tang YY, Cao YQ, Yu YX, Yan SQ, Wang YJ, Pan YJ, Zhang WJ. Real-time recombinase polymerase amplification assay for the detection of Vibrio cholerae in seafood. Food Anal Methods. 2017;10:2657–66.
108. Shan C, Wu P, Su R, Gao C, Yu Y, Zhang H, Xu Y. Detection and analysis of Mycobacterium tuberculosis by recombinase polymerase amplification. Chin J Microecol. 2019;31:164.
109. Ng BYC, Wee EJH, West NP, Trau M. Rapid DNA detection of Mycobacterium tuberculosis-towards single cell sensitivity in point-of-care diagnosis. Sci Rep. 2015;5:15027.
110. Geng YY, Liu GH, Liu LB, Deng QE, Zhao LW, Sun XX, Wang JF, Zhao BH, Wang JC. Real-time recombinase polymerase amplification assay for the rapid and sensitive detection of Campylobacter jejuni in food samples. J Microbiol Methods. 2019;157:31–6.
111. Li JL, Ma B, Fang JH, Zhi AT, Chen EJ, Xu Y, Yu XP, Sun CX, Zhang MZ. Recombinase polymerase amplification (RPA) combined with lateral flow immunoassay for rapid detection of Salmonella in food. Foods. 2020;9:27.
112. Kim JY, Lee JL. Rapid detection of Salmonella enterica serovar Enteritidis from eggs and chicken meat by real-time recombinase polymerase amplification in comparison with the two-step real-time PCR. J Food Saf. 2016;36:402–11.
113. Liu HB, Zang YX, Du XJ, Li P, Wang S. Development of an isothermal amplification-based assay for the rapid visual detection of Salmonella bacteria. J Dairy Sci. 2017;100:7016–25.

114. Zhao LW, Wang JC, Sun XX, Wang JF, Chen ZM, Xu XD, Dong MY, Guo YN, Wang YY, Chen PP, Gao WJ, Geng YY. Development and evaluation of the rapid and sensitive RPA assays for specific detection of Salmonella spp. in food samples. Front Cell Infect Microbiol. 2021;11:631921.
115. Hu JQ, Huang RN, Sun YT, Wei XK, Wang Y, Jiang CP, Geng Y, Sun XC, Jing JZ, Gao H, Wang ZC, Dong CW. Sensitive and rapid visual detection of Salmonella typhimurium in milk based on recombinase polymerase amplification with lateral flow dipsticks. J Microbiol Methods. 2019;158:25–32.
116. Euler M, Wang YJ, Otto P, Tomaso H, Escudero R, Anda P, Hufert FT, Weidmann M. Recombinase polymerase amplification assay for rapid detection of Francisella tularensis. J Clin Microbiol. 2012;50:2234–8.
117. Zhang HJ, Zhao M, Hu SY, Ma KR, Li JX, Zhao J, Wei X, Tong LN, Li SQ. Establishment of a real-time recombinase polymerase amplification for rapid detection of pathogenic Yersinia enterocolitica. Pathogens. 2023;12:226.
118. Ahmed SA, van de Sande WWJ, Desnos-Ollivier M, Fahal AH, Mhmoud NA, de Hoog GS. Application of isothermal amplification techniques for identification of Madurella mycetomatis, the prevalent agent of human mycetoma. J Clin Microbiol. 2015;53:3280–5.
119. Geng YY, Liu SY, Wang JF, Nan HZ, Liu LB, Sun XX, Li DY, Liu M, Wang JC, Tan K. Rapid detection of Staphylococcus aureus in food using a recombinase polymerase amplification-based assay. Food Anal Methods. 2018;11:2847–56.
120. Zhao GM, Wang HM, Hou PL, He CQ, He HB. Rapid visual detection of Mycobacterium avium subsp paratuberculosis by recombinase polymerase amplification combined with a lateral flow dipstick. J Vet Sci. 2018;19:242–50.
121. Hansen S, Schäfer J, Fechner K, Czerny CP, Abd El Wahed A. Development of a recombinase polymerase amplification assay for rapid detection of the Mycobacterium avium subsp paratuberculosis. PLoS One. 2016;11:e0168733.
122. Ghosh DK, Kokane SB, Kokane AD, Warghane AJ, Motghare MR, Bhose S, Sharma AK, Reddy MK. Development of a recombinase polymerase based isothermal amplification combined with lateral flow assay (HLB-RPA-LFA) for rapid detection of "Candidatus Liberibacter asiaticus". PLoS One. 2018;13:e0208530.
123. Crannell ZA, Castellanos-Gonzalez A, Irani A, Rohrman B, White AC, Richards-Kortum R. Nucleic acid test to diagnose cryptosporidiosis: lab assessment in animal and patient specimens. Anal Chem. 2014;86:2565–71.
124. Crannell Z, Castellanos-Gonzalez A, Nair G, Mejia R, White AC, Richards-Kortum R. Multiplexed recombinase polymerase amplification assay to detect intestinal protozoa. Anal Chem. 2016;88:1610–6.
125. Kersting S, Rausch V, Bier FF, von Nickisch-Rosenegk M. Rapid detection of Plasmodium falciparum with isothermal recombinase polymerase amplification and lateral flow analysis. Malar J. 2014;13:99.
126. Rosser A, Rollinson D, Forrest M, Webster BL. Isothermal recombinase polymerase amplification (RPA) of Schistosoma haematobium DNA and oligochromatographic lateral flow detection. Parasit Vectors. 2015;8:446.
127. Rostron P, Pennance T, Bakar F, Rollinson D, Knopp S, Allan F, Kabole F, Ali SM, Ame SM, Webster BL. Development of a recombinase polymerase amplification (RPA) fluorescence assay for the detection of Schistosoma haematobium. Parasit Vectors. 2019;12:514.
128. Sun K, Xing WW, Yu XL, Fu WL, Wang YY, Zou MJ, Luo ZH, Xu DG. Recombinase polymerase amplification combined with a lateral flow dipstick for rapid and visual detection of Schistosoma japonicum. Parasit Vectors. 2016;9:476.
129. Saldarriaga OA, Castellanos-Gonzalez A, Porrozzi R, Baldeviano GC, Lescano AG, de Los Santos MB, Fernandez OL, Saravia NG, Costa E, Melby PC, Travi BL. An innovative field-applicable molecular test to diagnose cutaneous Leishmania Viannia spp. infections. PLoS Negl Trop Dis. 2016;10:e0004638.

130. Krolov K, Frolova J, Tudoran O, Suhorutsenko J, Lehto T, Sibul H, Mager I, Laanpere M, Tulp I, Langell O. Sensitive and rapid detection of Chlamydia trachomatis by recombinase polymerase amplification directly from urine samples. J Mol Diagn. 2014;16:127–35.
131. Wu YD, Xu MJ, Wang QQ, Zhou CX, Wang M, Zhu XQ, Zhou DH. Recombinase polymerase amplification (RPA) combined with lateral flow (LF) strip for detection of Toxoplasma gondii in the environment. Vet Parasitol. 2017;243:199–203.
132. Wang JF, Li RW, Sun XX, Liu LB, Hao XP, Wang JC, Yuan WZ. Development and validation of the isothermal recombinase polymerase amplification assays for rapid detection of Mycoplasma ovipneumoniae in sheep. BMC Vet Res. 2020;16:172.
133. Liljander A, Yu MY, O'Brien E, Heller M, Nepper JF, Weibel DB, Gluecks I, Younan M, Frey J, Falquet L, Jores J. Field-applicable recombinase polymerase amplification assay for rapid detection of Mycoplasma capricolum subsp. capripneumoniae. J Clin Microbiol. 2015;53:2810–5.
134. Cui J, Zhao YN, Sun YL, Yu L, Liu Q, Zhan XY, Li MX, He L, Zhao JL. Detection of Babesia gibsoni in dogs by combining recombinase polymerase amplification (RPA) with lateral flow (LF) dipstick. Parasitol Res. 2018;117:3945–51.

Chapter 7
CRISPR-Cas-Based Nucleic Acid Detection Technologies

Zhian Hu, Song Wang, Mai Zhang, and Yanfei Che

7.1 Overview of the CRISPR-Cas Gene Editing System

Clustered Regularly Interspaced Short Palindromic Repeats (CRISPR) and CRISPR-associated proteins (Cas) form an adaptive immune system evolved by prokaryotes. This system allows efficient recognition and cleavage of foreign genetic material, silencing foreign genes and defending against bacteriophage invasion [1, 2]. Due to its precise nucleic acid targeting ability, the CRISPR-Cas system has evolved into a programmable and highly specific gene-editing tool. It represents a new generation of "genome-targeted editing technology," following the first-generation Zinc-Finger Nuclease (ZFN) and second-generation Transcription Activator-Like Effector Nuclease (TALEN) technologies. In recent years, CRISPR-Cas9 and other novel gene-editing technologies have advanced rapidly, leading to a surge in research interest. In October 2020, Doudna and Charpentier were awarded the Nobel Prize in Chemistry for their pioneering work in developing CRISPR-Cas as a site-specific nucleic acid cleavage tool. As a powerful gene-editing technology, CRISPR-Cas has been successfully applied in gene screening, gene knockout, gene activation, disease model construction, and early diagnosis [3–5].

In nature, the CRISPR-Cas system acts as an RNA-guided nuclease, where Cas proteins and RNA form a complex (RNP) that recognizes and cleaves target genes through base-pair complementarity. To defend against constantly evolving viral nucleic acid sequences, prokaryotes have developed various Cas proteins [6, 7]. Type I Cas proteins consist of multiple effector subunits that form a complete Cas protein complex upon detecting invading foreign nucleic acids, leading to DNA

Z. Hu (✉) · S. Wang · M. Zhang · Y. Che
The School of Chemistry and Biological Engineering, University of Science and Technology Beijing, Beijing, China
e-mail: zhianhu@ustb.edu.cn; m202410934@xs.ustb.edu.cn; 15712581556@163.com; m202421096@xs.ustb.edu.cn

Z. Li (ed.), *Nucleic Acid Detection*, https://doi.org/10.1007/978-981-95-7295-3_7

degradation. In contrast, Type II systems rely on a single effector protein for nucleic acid degradation. As a result, Type II CRISPR-Cas nucleases, such as CRISPR-Cas9, CRISPR-Cas12, and CRISPR-Cas13, have become the primary tools for site-specific nucleic acid cleavage [2, 7–9]. Another component, CRISPR RNA (crRNA), consists of a genomic targeting sequence, called the spacer sequence, and a conserved repeat sequence known as the direct repeat [6, 7]. The spacer sequence of crRNA is typically about 20 bases long. Altering these bases allows for broad nucleic acid recognition. This RNA hybridization-based targeting strategy simplifies nucleic acid recognition compared to earlier methods like ZFNs or TALENs, which require multiple proteins [1, 10]. The CRISPR-Cas system also relies on a specific sequence called the Protospacer Adjacent Motif (PAM), typically 2–5 bases long, located just upstream of the recognition region for each Cas protein. The system cleaves target DNA only when both the PAM sequence and the complementary sequence in the guide RNA are present, ensuring precise recognition [11, 12].

The CRISPR-Cas system's sequence recognition capability and programmability have enabled its application across various life science fields, including gene therapy, drug development, biosensing, nucleic acid detection, and disease diagnosis. The discovery of its nonspecific trans-cleavage activity has led to the development of innovative nucleic acid detection platforms, such as SHERLOCK and DETECTR [13–15], which show great promise in molecular diagnostics and biosensing. The following sections provide a detailed overview of CRISPR-Cas-based nucleic acid detection technologies, categorized by Cas protein types.

7.2 CRISPR-Cas9-Based Nucleic Acid Detection Technology

The CRISPR-Cas9 protein is an RNA-guided double-stranded DNA nuclease and the key protein in the Type II CRISPR system [2]. The adaptive immune process mediated by CRISPR-Cas9 is illustrated in Fig. 7.1. When a virus infects a host cell, foreign DNA, such as phage DNA, is recognized by the Cas1 and Cas2 protein complex and incorporated into the CRISPR loci of the host genome as new spacer sequences. This process allows the host to store fragments of viral DNA, thereby establishing an "immune memory." Upon reinvasion by bacteriophages, prokaryotic organisms express the Cas9 protein, transcribe the CRISPR sequence into pre-crRNA, and produce the trans-activating CRISPR RNA (tracrRNA). The pre-crRNA and tracrRNA then hybridize through complementary base pairing and are processed by ribonuclease III (RNase III) to form a mature tracrRNA-crRNA complex. This complex associates with the Cas9 protein to form the Cas9-tracrRNA-crRNA ternary complex, which has double-stranded DNA (dsDNA) cleavage activity. The complex then identifies and cleaves exogenous target sequences

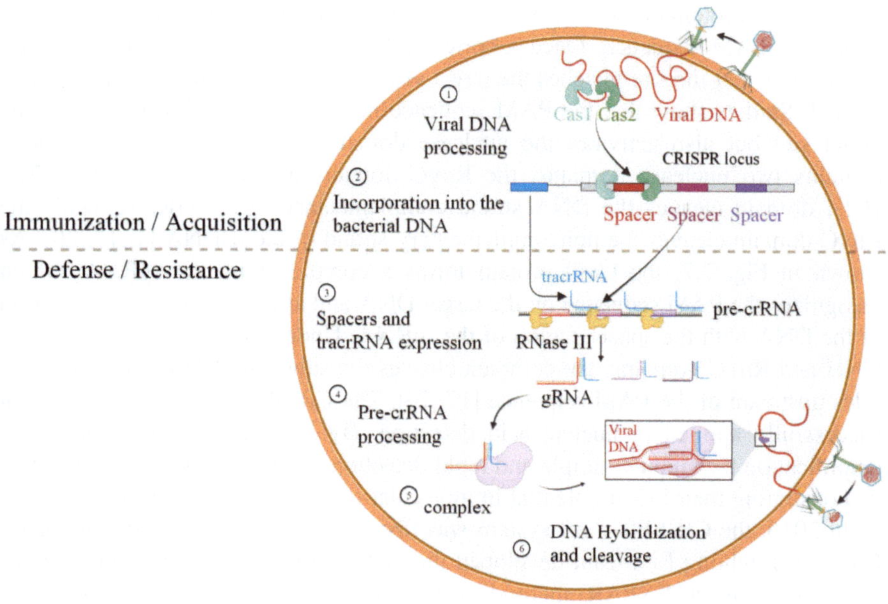

Fig. 7.1 The bacterial immune process mediated by the CRISPR-Cas9 system [17]

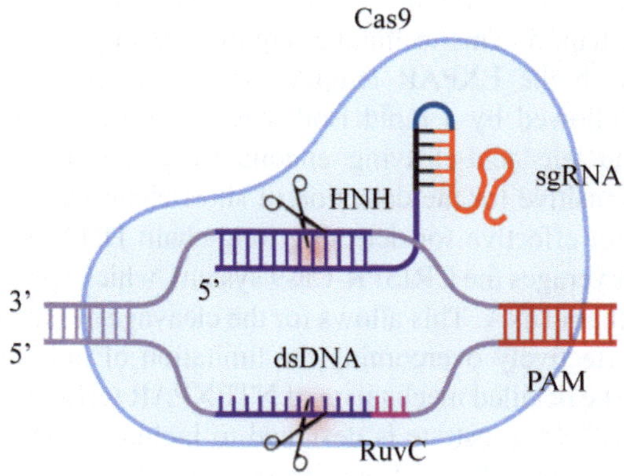

Fig. 7.2 The cleavage mechanism of the CRISPR-Cas9 system

complementary to the crRNA, defending against the invasion of foreign nucleic acids [6, 16–18].

Through artificial modification, tracrRNA and crRNA can be fused into a single-stranded guide RNA (sgRNA) [17]. The sgRNA directs the Cas9 protein to recognize and cleave the target DNA (Fig. 7.2). The protospacer adjacent motif (PAM) sequence is a G-rich DNA sequence that must be precisely recognized by the Cas9 protein for its activation. The PAM sequence, represented as "5′-NGG-3′" or "5′-NAG-3′" (where N denotes any nucleotide), is located on the noncomplementary strand of the target double-stranded DNA (dsDNA).

The Cas9 protein initiates activity by using its PAM-binding domain to search for a suitable PAM sequence. Once a correct PAM is identified, the sgRNA region verifies whether the site matches the target sequence, enabling subsequent cleavage [7, 16]. Studies show that the PAM sequence not only guides Cas proteins to the target site but also activates the nuclease domains [16, 19]. The Cas9 protein contains two nuclease domains, the RuvC domain and the HNH domain. The HNH domain cleaves the DNA strand complementary to the crRNA, while the RuvC domain cleaves the noncomplementary strand of the dsDNA [7, 17, 20]. As shown in Fig. 7.2, the Cas9 protein forms a complex with the sgRNA, which recognizes the PAM sequence on the target DNA and aligns the protospacer region of the DNA with the spacer region of the sgRNA. Under the catalytic action of the HNH and RuvC domains, the complex cleaves the substrate dsDNA three nucleotides upstream of the PAM sequence [17–20]. The CRISPR-Cas9 system has been successfully applied in nucleic acid detection. By integrating with nucleic acid amplification techniques, simple and rapid detection methods have been developed, demonstrating their broad potential in molecular diagnostics and biosensing.

In 2016, the CRISPR-Cas9 system was first applied to nucleic acid diagnostics [21, 22]. Jinghong Li's team developed the nicking-triggered exponential amplification reaction (NTEXPAR) method, which combines the CRISPR-Cas9 system's specific cleavage of double-stranded DNA (dsDNA) with exponential amplification reaction (EXPAR). This method was successfully used to detect long-chain dsDNA [22]. The principle underlying the EXPAR amplification reaction is outlined in Chap. 5. This technique requires the target nucleic acid (DNA/RNA) to hybridize with the EXPAR template DNA and undergo a 3′ extension reaction. This is followed by a rapid isothermal exponential amplification process catalyzed by a nucleic acid-cleaving endonuclease. As a result, EXPAR technology is highly sensitive for the detection of short-chain DNA or RNA (e.g., microRNA), but it is not effective for detecting long-chain DNA or RNA [23]. NTEXPAR technology leverages the CRISPR-Cas9 system, which specifically recognizes and cleaves long-chain DNA. This allows for the cleavage of the target long-chain DNA at the 3′ end, effectively overcoming the limitation of initiating EXPAR with long-chain DNA. The detailed mechanism of NTEXPAR technology is illustrated in Fig. 7.3. First, the dsDNA substrate is designed to include a nicking site, target sequence, and PAM sequence. The nicking site generates the single-stranded DNA (ssDNA) required for the subsequent amplification. After cleavage by Cas9, the dsDNA is nicked by a nicking endonuclease, leading to the continuous production of short ssDNA by DNA polymerase. This short ssDNA serves as a primer for EXPAR amplification. In contrast, dsDNA substrates not cleaved by Cas9 produce only longer ssDNA, which cannot initiate the EXPAR process. Additionally, the EXPAR amplification template contains two repetitive sequences (complementary to ssDNA), separated by cleavage site sequences. The resulting short ssDNA can hybridize with the amplification template and undergo extension reactions along the template under the action of DNA polymerase. The newly formed dsDNA is recognized by the nicking endonuclease, initiating a new cycle of nicking, strand displacement, and extension. This process triggers exponential amplification, producing large quantities of dsDNA.

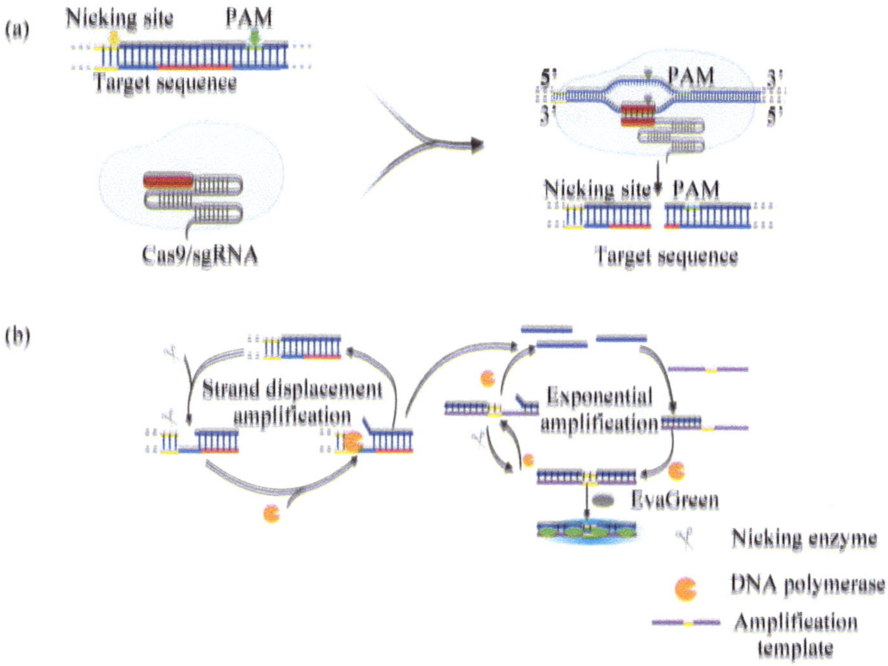

Fig. 7.3 Schematic diagram of the NTEXPAR technique workflow. (Revised and reprinted with permission from Ref. [22])

The dsDNA binds to the fluorescent dye EvaGreen, generating a detectable fluorescence signal in real time. Under optimal conditions, fluorescence monitoring enables the quantification of dsDNA in a linear range of 10 fmol/L to 100 pmol/L within 30 min.

Under normal conditions, the CRISPR-Cas9 protein targets double-stranded DNA (dsDNA) but cannot act on single-stranded DNA (ssDNA) or RNA. However, based on the principle of PAM sequence recognition, Doudna's team designed PAM-presenting oligonucleotides (PAMmers) with a PAM sequence (5′-NGG-3′) at the 5′ end. When complementary ssDNA or RNA is present, the Cas9/sgRNA complex can cleave the ssDNA or RNA [24]. Therefore, by using specially designed PAMmers, Cas9 can perform site-specific cleavage of ssDNA or RNA.

Building on this capability, Huang et al. [25] combined PAMmer-mediated site-specific cleavage with the rapid amplification of EXPAR to develop the CRISPR/Cas9-triggered exponential amplification method (CAS-EXPAR). The CAS-EXPAR technique can detect and differentiate single-base mismatches at the cleavage site, as mismatches prevent the extension of EXPAR primers and inhibit signal generation [25]. This method demonstrates high sensitivity in detecting DNA methylation and Listeria monocytogenes RNA.

As shown in Fig. 7.4, the Cas9/sgRNA complex first cleaves the target ssDNA at specific sites, resulting in the generation of cleavage fragment X. This fragment X

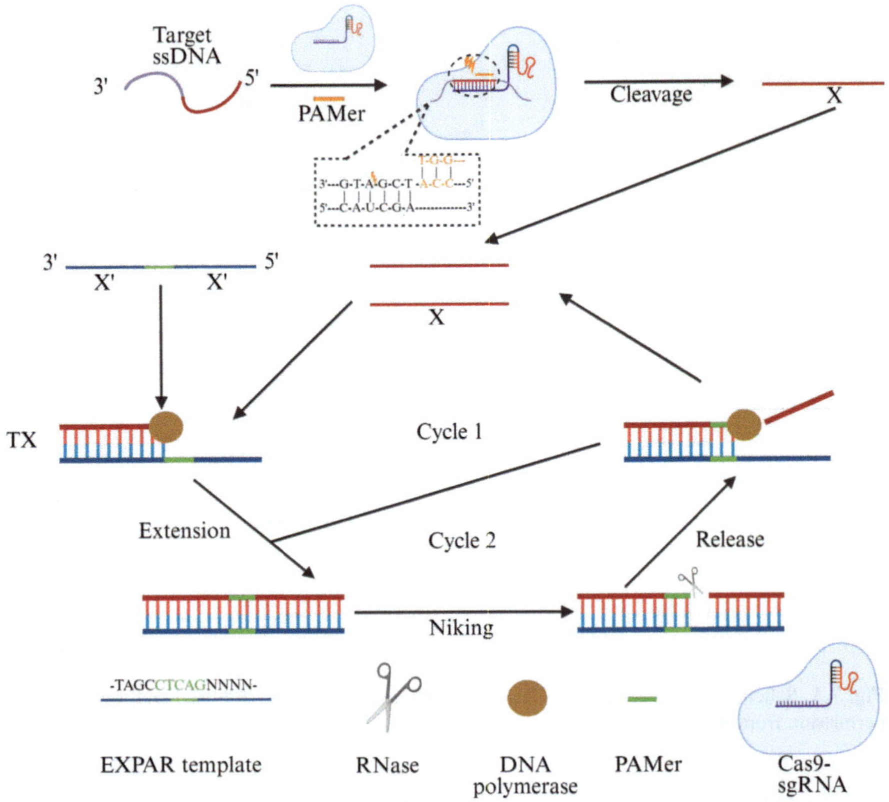

Fig. 7.4 Schematic diagram of the CAS-EXPAR mechanism. (Revised and reprinted with permission from Ref. [25])

serves as a primer for subsequent EXPAR reactions. The EXPAR amplification template contains a nicking endonuclease (NEase) recognition site in the center, flanked by two identical sequences (X′, complementary to fragment X).

Fragment X hybridizes with the 3′ flanking sequence (X′) on the template. DNA polymerase then extends fragment X along the template, synthesizing double-stranded DNA containing the NEase recognition site. This double-stranded DNA is nicked by NEase, releasing fragment X, which hybridizes with another EXPAR template to trigger a new amplification cycle. This process leads to exponential amplification.

By coupling with real-time fluorescence analysis, CAS-EXPAR enables highly specific ssDNA detection within 1 h, with a detection limit as low as 0.82 amol of target DNA. This strategy, based on CRISPR-Cas9's precise recognition and cleavage of target ssDNA, significantly reduces false-positive signals from nonspecific amplification. Furthermore, the integration of CAS-EXPAR with bisulfite conversion reactions allows for site-specific DNA methylation detection. Thus,

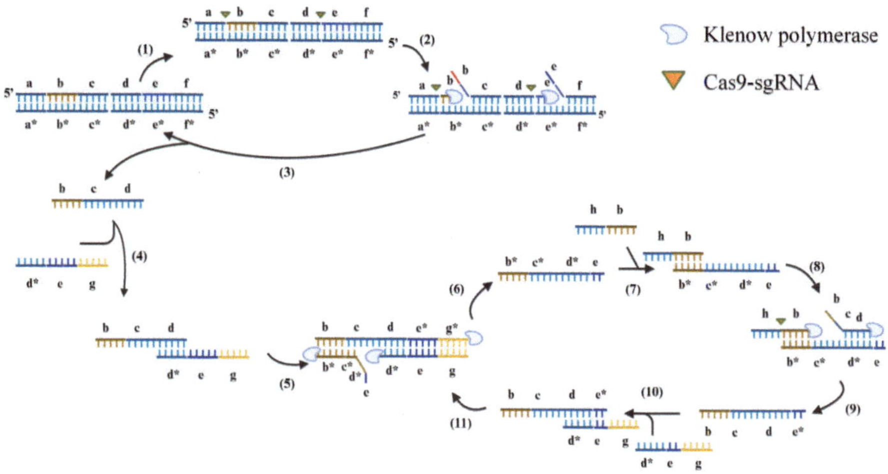

Fig. 7.5 Schematic diagram of the Cas9nAR isothermal amplification technique. (Revised and reprinted with permission from Ref. [26])

CAS-EXPAR shows promise as a molecular diagnostic tool and a versatile nucleic acid detection method.

Cas9n, or Cas9 nickase, is a mutant form of the Cas9 nuclease, created by introducing a D10A point mutation in the RuvC domain or an H840A mutation in the HNH domain. This results in Cas9n cleaving only one strand of double-stranded DNA (dsDNA) or leaving the DNA completely uncleaved [17]. Wang et al. [26] developed an isothermal amplification method based on Cas9 nickase, called the Cas9 Nickase-Based Method for Universal Isothermal DNA Amplification (Cas9nAR). This method combines the single-strand nicking activity of Cas9n with the strand displacement capability of DNA polymerase, as shown in Fig. 7.5.

The process consists of two cycles. In the first cycle (Steps 1–3), the goal is to extract the desired ssDNA sequence from genomic DNA. Step 1 involves designing a pair of sgRNAs that are complementary to regions b and e (protospacer regions, highlighted in red). The Cas9n complex introduces two nicks in the target genomic DNA, releasing the ssDNA fragment (region bcd). Simultaneously, exo@Klenow polymerase extends from the nick sites to synthesize a complete dsDNA molecule, transitioning from Step 2 to Step 3, thus initiating the next cycle.

In the second cycle (Steps 4–11), the ssDNA (region bcd) released in the first cycle is amplified using primers 1 and 2. Primer 1 consists of regions g, e, and d, where region d is complementary to the d region of the ssDNA released in the first cycle. Region g contains the sequence 5′-CCN-3′ (complementary to the PAM sequence), and region e is complementary to the e region of the sgRNA. In Step 4, the d region of Primer 1 hybridizes with the d region at the 3′ end of the released ssDNA, forming a double-stranded DNA with a 5′ overhang. Exo@Klenow polymerase extends both 3′ ends of the dsDNA, producing a complementary strand of ssDNA (region bcd) and a PAM sequence required to activate Cas9n (Step 5).

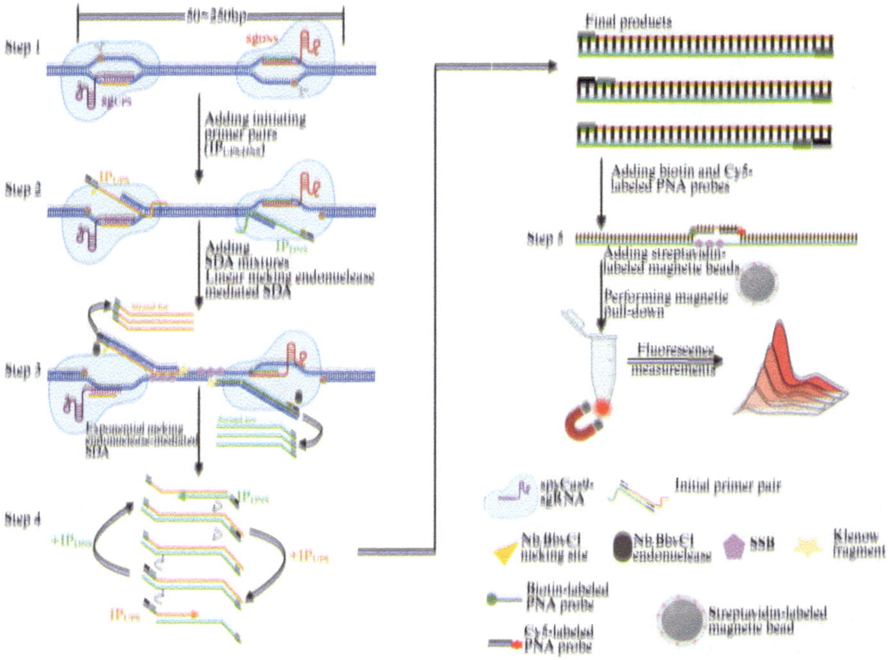

Fig. 7.6 Schematic diagram of the CRISDA technique. (Revised and reprinted with permission from Ref. [27])

Next, the Cas9n/sgRNA complex introduces a nick at region e. Exo@Klenow polymerase extends from the nick site, releasing fragment edcb (Step 6), which serves as the template for Primer 2. Primer 2 has the complete b sequence at its 3′ end and region h at its 5′ end, containing the 5′-CCN-3′ sequence (complementary to the PAM sequence). In Step 7, Primer 2 hybridizes with the displaced fragment edcb, initiating a new extension, nicking, and strand displacement cycle (Steps 8–11).

Under optimal conditions, the sensitivity of Cas9nAR was evaluated using Salmonella Typhimurium genomic DNA. The method detected DNA concentrations as low as 166 zmol/L within 60 min and demonstrated excellent single-base discrimination capabilities.

Zhou et al. developed a strand displacement amplification technique based on CRISPR-Cas9 nickase, called CRISDA (CRISPR-Cas9 Nickase-Triggered Strand Displacement Amplification), as shown in Fig. 7.6 [27]. Typically, Strand Displacement Amplification (SDA) requires heating to 95 °C to denature double-stranded DNA (dsDNA) into single-stranded DNA (ssDNA) for primer binding.

In CRISDA, a pair of Cas9/sgRNA complexes specifically recognizes the boundary regions of the target DNA, introducing nicks in the nontarget strand of the dsDNA. In Step 2, primers hybridize with the exposed nontarget DNA strands. Each primer contains three parts: a 5′ Nb.BbvCI nicking endonuclease recognition site, a middle hybridization region complementary to the exposed nontarget DNA strand,

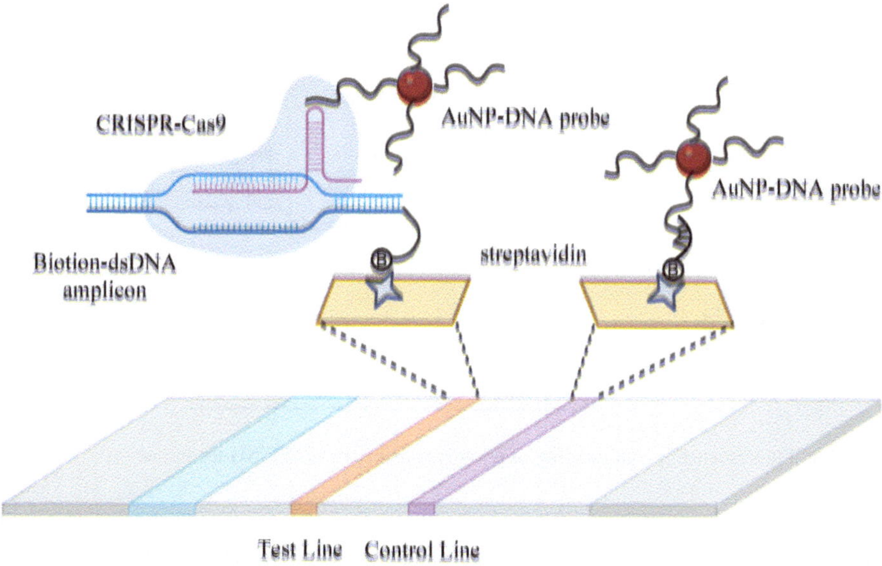

Fig. 7.7 Schematic diagram of the CASLFA technique. (Revised and reprinted with permission from Ref. [28])

and a 3′ overhang that is complementary to the double-stranded region of the nontarget strand.

In Step 3, after adding the SDA reaction mixture, DNA polymerase, assisted by single-stranded DNA binding protein (SSB), extends the primers along the nontarget DNA strand. In Step 4, the newly synthesized strand is nicked by Nb.BbvCI endonuclease, released, and hybridized with primers to initiate another cycle of SDA amplification. Finally, in Step 5, a fluorescently labeled PNA probe, complementary to the middle region of the SDA product, is introduced. The target DNA is quantified by measuring the fluorescence intensity.

Using CRISDA, DNA fragments from human chromosome 9 were analyzed, achieving detection sensitivity as low as the attomolar (aM) level. Moreover, CRISDA is an isothermal nucleic acid detection method that does not require thermal cycling, making it highly suitable for point-of-care diagnostics and field-based analyses.

Lateral flow nucleic acid assay (LFA) is a convenient point-of-care testing technology that has gained widespread use in on-site analysis across various settings. Wang et al. [28] successfully integrated the CRISPR-Cas9 gene-editing system into LFA technology, developing the CRISPR/Cas9-mediated lateral flow nucleic acid assay platform (CASLFA). This platform has been applied to identify pathogenic microorganisms, detect genetically modified organisms, and diagnose African swine fever virus (ASFV) [28].

As shown in Fig. 7.7, the device consists of a sample application zone, binding zone, test line, control line, and sample recovery zone. First, PCR or RPA

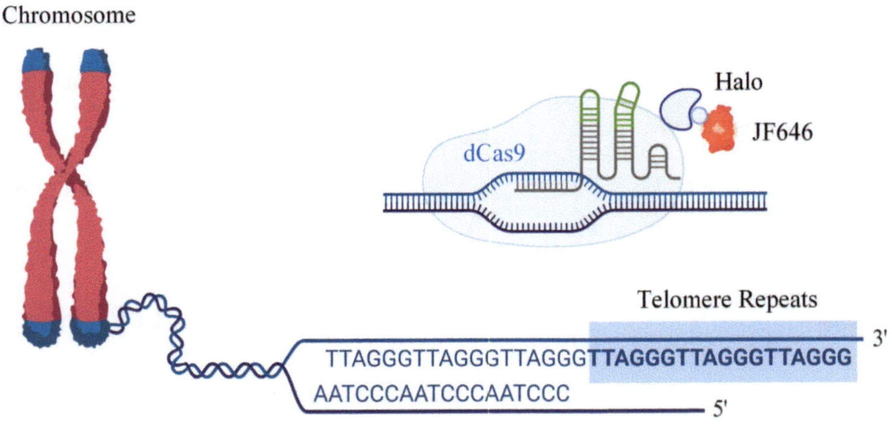

Fig. 7.8 Fluorescence in situ imaging of telomeres using the CASFISH technique

amplification is performed using biotinylated primers to generate biotin-labeled amplicons. After incubating the biotinylated amplicons with Cas9/sgRNA, the resulting Cas9/sgRNA-biotinylated amplicon complex is applied to the sample application zone. Driven by capillary force, the complex moves to the binding zone containing AuNP-DNA probes. Through nucleic acid hybridization, a Cas9/sgRNA-biotinylated dsDNA-AuNP-DNA probe complex forms.

This complex continues to flow through the test line, where it is captured by pre-coated streptavidin. The accumulation of gold nanoparticles generates a visible color band. Excess AuNP-DNA probes flow to the control line, which is coated with streptavidin-biotinylated DNA. These probes are captured via nucleic acid hybridization at the control line, ensuring a valid test.

The entire detection process is completed within 3 min. Due to its simplicity, speed, and lack of need for sophisticated equipment, this technology holds significant potential for nucleic acid analysis in resource-limited settings or non-laboratory environments.

As a versatile DNA-binding tool, dCas9 can be fused with various effectors, such as fluorescent proteins or enzymes, expanding the applications of the CRISPR-Cas system. This innovation marks a key development in applying CRISPR technology to live-cell imaging and nucleic acid sensing. In 2013, Chen et al. [29] introduced a dynamic imaging technique based on the CRISPR-Cas system for genomic loci in live human cells. By fusing enhanced green fluorescent protein (EGFP) to the dead Cas9 (dCas9) protein, which lacks endonuclease activity, they achieved in situ imaging of telomeres and genomic repetitive sequences. Similarly, Deng et al. [30] developed the Cas9-mediated fluorescence in situ hybridization (CASFISH) technique, using fluorescently labeled dCas9 and sgRNA complexes to visualize repetitive genomic loci (Fig. 7.8). However, the CasFISH method requires aggregating up to 73 Cas9-EGFP complexes to produce a strong fluorescence signal for imaging non-repetitive genomic regions. Building on this, Wang et al. [5] reported the CRISPR LiveFISH technology, a real-time DNA imaging technique for live cells.

This method is suitable for prenatal detection of chromosomal abnormalities in amniotic fluid cells and for tracking genetic disease loci in human T cells. The technology involves assembling dCas9 and fluorescently labeled sgRNA, then transfecting them into amniotic fluid cells from Patau syndrome patients. By exploiting the differences in degradation rates between free RNPs and those bound to target DNA, the technique selects an optimal observation window. The free RNPs-GFP protein is rapidly degraded, while the fluorescence signal from the targeted DNA remains, enabling high signal-to-noise ratio imaging of the repetitive region of chromosome 13 (Chr13q34). This approach holds great potential for detecting chromosomal diseases.

However, the CRISPR-Cas9-based cell imaging methods described earlier are limited by highly repetitive genomic loci, such as telomeres and the MUC4 gene. In situ imaging of non-repetitive genomic loci, like single-nucleotide variations (SNVs), remains challenging, as a single Cas9 protein does not produce a strong enough signal to distinguish the target from the background. To address this, Jinghong Li's team developed a CRISPR-Cas9-mediated rolling circle amplification imaging technique (CRISPR-Cas9-mediated proximity ligation assay, CasPLA) [31], which enables single-molecule resolution for SNV imaging.

As illustrated in Fig. 7.9, this method uses two Cas9/sgRNA probes and a set of proximity ligation probes. The Cas9/sgRNA probes target specific mitochondrial DNA (mtDNA) sequences, while the proximity probes hybridize with the loop regions of the sgRNA in the two Cas9/sgRNA probes. Once the Cas9/sgRNA probes bind to the target mtDNA sequence, the proximity probes initiate ligation, forming a circular DNA structure. This circular DNA then serves as a template for rolling circle amplification (RCA), catalyzed by phi29 DNA polymerase, enabling visualization through hybridization with fluorescently labeled oligonucleotide probes. Because the recognition of the target sequence by Cas9/sgRNA depends on base-pair complementarity between the sgRNA and the target sequence, and the Cas9-sgRNA complex is highly sensitive to mismatches, the Cas9/sgRNA probes can accurately distinguish single-nucleotide variations (SNVs) in mtDNA. As a result, the CasPLA technique can precisely map the spatial localization of wild-type and mutant mtDNA within individual cells, achieving single-molecule resolution for in situ SNV imaging.

In summary, integrating the CRISPR-Cas9 system with nucleic acid amplification technology has opened new possibilities in nucleic acid detection, making it a powerful tool for both clinical diagnostics and fundamental research.

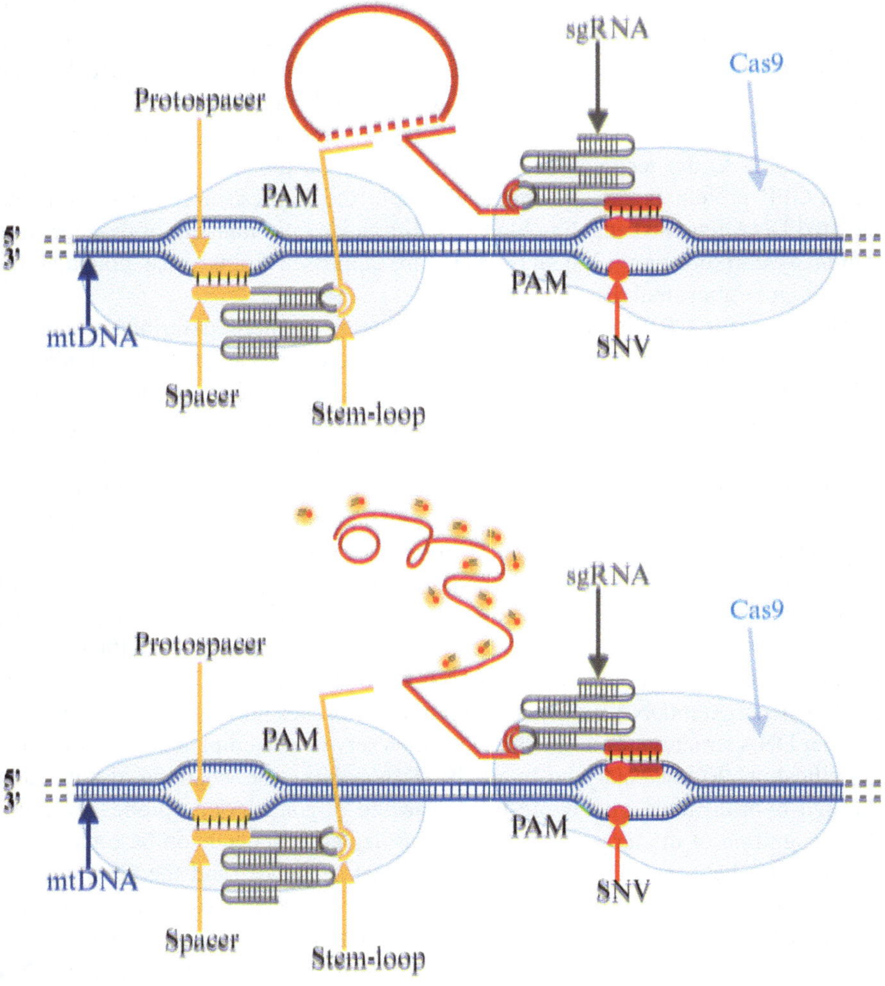

Fig. 7.9 Schematic diagram of the CasPLA technique. (Revised and reprinted with permission from Ref. [31])

7.3 CRISPR-Cas12-Based Nucleic Acid Detection Technology

7.3.1 Nucleic Acid Detection Technology Based on Cas12a Protein

The CRISPR-Cas12a effector protein is a DNA endonuclease guided by a single-stranded guide RNA (sgRNA) [2]. Also known as Cpf1, Cas12a was first introduced by the Feng Zhang team in 2015 [32]. As a key component of the CRISPR-Cas gene-editing system, Cas12a binds with mature crRNA to form a Cas12a-crRNA

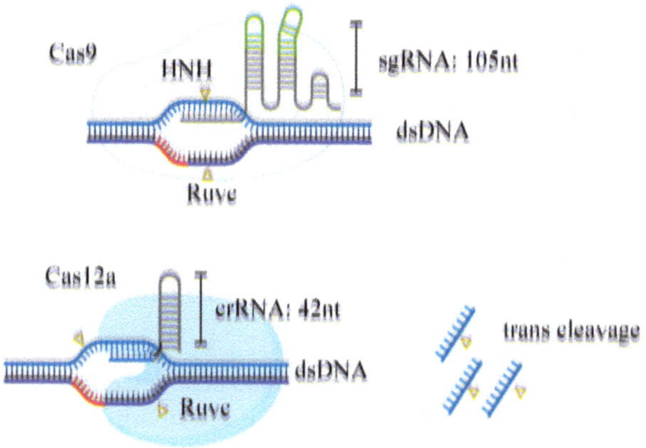

Fig. 7.10 Comparison of Cas12a and Cas9

complex, which precisely recognizes and cleaves exogenous target sequences, providing defense against foreign genetic material [2, 33, 34]. CRISPR-Cas12a and CRISPR-Cas9 differ significantly in terms of substrate specificity, guide RNA design, protein structure, and cleavage mechanisms, each offering distinct advantages (Fig. 7.10).

1. CRISPR-Cas12a is a versatile deoxyribonuclease that cleaves both double-stranded DNA (dsDNA) and single-stranded DNA (ssDNA), with higher cleavage efficiency for dsDNA. In contrast, Cas9 exclusively targets dsDNA [14].
2. The PAM region of Cas12a consists of a 3- to 4-nucleotide sequence rich in thymine (T). The most commonly used Cas12a variants—FnCas12a from *Francisella novicida*, AsCas12a from *Acidaminococcus* sp., and LbCas12a from *Lachnospiraceae* bacterium—recognize PAM sequences of "5′-TTTN-3′" or "5′-TTN-3′" (where N denotes any nucleotide) [35–37]. Protein engineering modifications of Cas12a can expand its PAM recognition region, enabling broader DNA targeting and cleavage, as seen in AsCas12a RVR ("5′-TATN-3′") and Lb2-KY ("5′-CTTN-3′") [38, 39].
3. The crRNA of CRISPR-Cas12a consists of 42 nucleotides, 19 of which are highly conserved and form a stem-loop structure that stabilizes the Cas protein. The remaining nucleotides are involved in target DNA recognition. In contrast, the sgRNA of Cas9 is over 100 nucleotides long, with more than 70 nucleotides supporting the protein complex. Cas12a also has a smaller molecular weight of approximately 130 kDa, compared to the 160 kDa of Cas9, making it a more compact nucleic acid editing tool [18, 40].
4. Cas12a exhibits dual enzymatic activities: RNase III and DNase activities. Unlike Cas9, which requires additional RNases to process pre-crRNA, Cas12a can independently process pre-crRNA into mature, functional crRNA for nucleic acid cleavage [34, 41].
5. The Cas12a protein contains a single RuvC DNA cleavage domain, producing 5′ overhanging sticky ends [1]. In contrast, Cas9 generates blunt-ended cleavage products. This cleavage mechanism makes Cas12a products more compatible

with the HDR (homology-directed repair) pathway in mammalian cells, reducing the undesirable gene mutations and misalignments associated with the NHEJ (non-homologous end joining) pathway [32, 42].

6. When the Cas12a-crRNA complex binds to target DNA, its RuvC domain remains highly active, enabling cleavage of nearby short ssDNA in a process known as trans-cleavage. Cas12a exhibits strong trans-cleavage activity, which can be detected by introducing fluorescent DNA molecular beacons. The target molecule activates this activity, cleaving the beacons and releasing fluorophores for highly sensitive detection of the target [14, 43, 44].

Recently, the CRISPR-Cas12a system has emerged as a promising tool in biosensing, particularly for nucleic acid detection. The Doudna team first combined the nonspecific ssDNase cleavage activity of Cas12a with recombinase polymerase amplification (RPA) to develop DETECTR (DNA Endonuclease Targeted CRISPR Trans Reporter), enabling highly sensitive DNA detection [14]. In this method, the sample is first treated with proteinase K to remove protein interference, followed by RPA amplification of the target DNA. Unlike PCR, which involves thermal cycling for denaturation, annealing, and extension, RPA amplifies nucleic acids rapidly at a constant temperature of 25–42 °C. The Cas12a/crRNA complex then recognizes and cleaves the amplified dsDNA target sequence, activating its trans-cleavage ssDNase activity. This leads to the cleavage of fluorescently and quencher-labeled DNA reporter probes, generating a fluorescent signal.

DETECTR allows for rapid and specific detection of human papillomavirus (HPV) in clinical samples, providing a simple and efficient platform for molecular diagnostics and clinical applications.

Around the same time, Li et al. [45] developed the HOLMES (One-Hour Low-cost Multipurpose Highly Efficient System) technology, as illustrated in Fig. 7.11a. The process begins with PCR amplification of viral DNA or RT-PCR amplification of viral RNA, incorporating a PAM sequence into the amplified product via primers. Subsequently, the Cas12a/crRNA complex recognizes and cleaves the amplified dsDNA product, simultaneously activating the lateral cleavage activity. This leads to the cleavage of the ssDNA reporter probes, which are labeled with fluorescent and quencher groups, resulting in the release of fluorescence. This technology, based on the Cas12a system and PCR amplification, can differentiate viral strains within 1 h from cell lines or clinical samples. It is also applicable for the rapid and straightforward detection of human single-nucleotide polymorphisms (SNPs). Due to the high amplification efficiency, specificity, and sensitivity of loop-mediated isothermal amplification (LAMP), several methods combining LAMP with CRISPR-Cas12a have been developed. For example, James P. Broughton et al. [15] integrated LAMP with CRISPR-Cas12a. In this method, LAMP amplifies target dsDNA, followed by the addition of Cas12a and ssDNA reporter probes. Cas12a recognizes the amplified products, activating its trans-cleavage activity and extensively cleaving the ssDNA reporter probes. Fluorescence detection enables single-molecule dsDNA analysis.

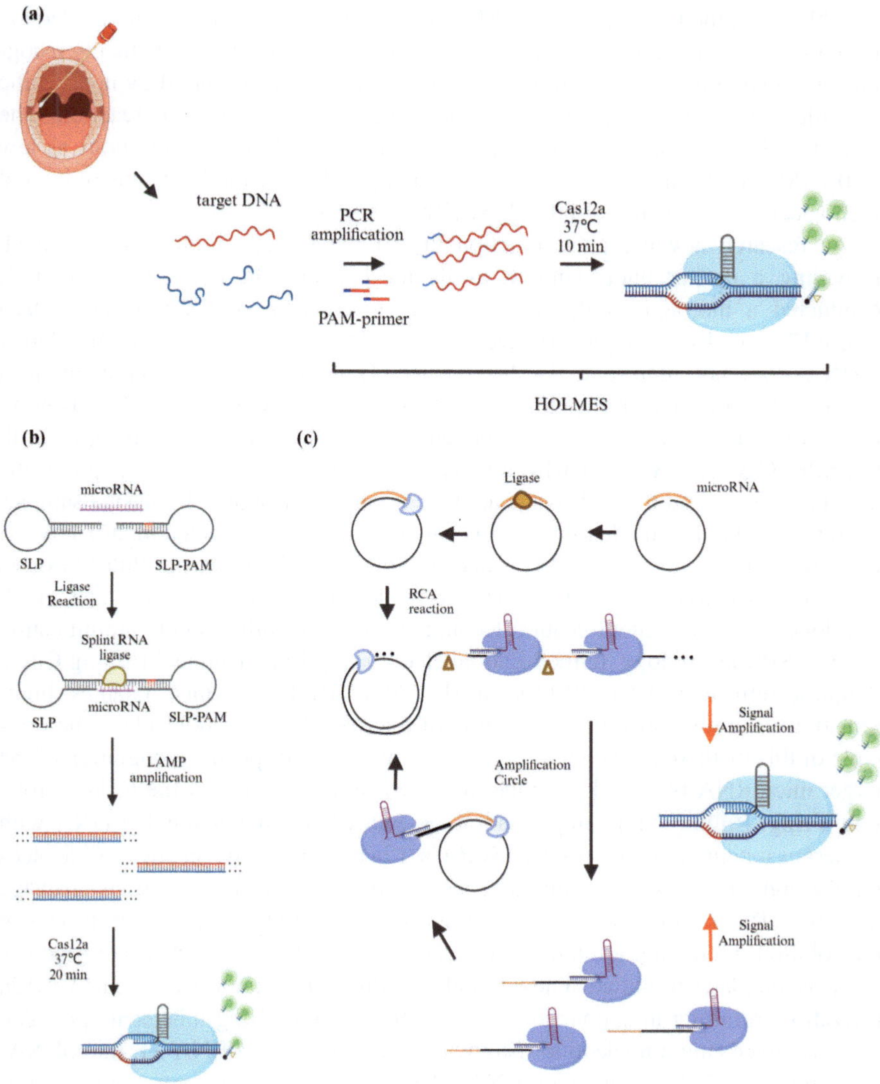

Fig. 7.11 (**a**) Principle of the HOLMES technology [45]; (**b**) Principle of the CAL-LAMP technology [46]; (**c**) Principle of the EXTRA-CRISPR technology [47]. (Revised and reprinted with permission from Refs. [45–47])

In addition, the Li team proposed a CRISPR/Cas12a-assisted ligation-initiated loop-mediated isothermal amplification (CAL-LAMP) technique, which enables single-base resolution detection of microRNA targets [46]. As shown in Fig. 7.11b, this technique involves the design of a pair of target-specific DNA stem-loop probes. These DNA probes use the microRNA target molecules as templates to form double stem-loop DNA, thereby initiating the LAMP reaction.

The high specificity of the Splint RNA ligase reaction enables single-nucleotide resolution. CRISPR/Cas12a facilitates signal amplification while specifically recognizing LAMP amplification products, reducing blank signals caused by nonspecific amplification, and further enhancing the sensitivity of the LAMP reaction. The CAL-LAMP technique can sensitively detect microRNA at concentrations as low as 0.1 fM. Furthermore, this method is also applicable for the high-sensitivity and high-specificity detection of other DNA/RNA targets.

As research advances, the functionality of the Cas12a protein has evolved, providing a deeper understanding of its mechanisms. In 2018, Li et al. [43] conducted a thorough study on the action mechanism of Cas12a. They used FnCas12a to cleave single-stranded DNA (ssDNA) and found that the target ssDNA does not require a PAM sequence. Building on this, Yan et al. [47] developed a cascade amplification technique combining RCA and CRISPR technology, termed Endonucleolytically Exponentiated Rolling Circle Amplification with CRISPR–Cas12a (EXTRA-CRISPR), as illustrated in Fig. 7.11c. RCA amplifies the target sequence, and trans-cleavage activity amplifies the signal. In plasma samples from pancreatic ductal adenocarcinoma patients, this assay detected microRNA-21, microRNA-196a, microRNA-451a, and microRNA-1246 in extracellular vesicles with single-nucleotide specificity at sub-femtomolar concentrations. Yan et al. developed a cascade amplification technique based on rolling circle amplification and CRISPR technology, termed Endonucleolytically Exponentiated Rolling Circle Amplification with CRISPR-Cas12a (EXTRA-CRISPR), which enables high-sensitivity and high-specificity detection of microRNA target molecules. The principle of this method, as shown in Fig. 7.11c, involves the specific recognition of the target microRNA by a locked probe, followed by the ligation of the locked probe into a ring. Rolling circle amplification generates long single-stranded DNA with numerous repetitive sequences. CRISPR-Cas12a recognizes the repetitive sequences on the long DNA strand, simultaneously activating its cis- and trans-cleavage activities. The trans-cleavage activity directly cleaves ssDNA fluorescent probes in the solution, producing a fluorescent signal. Meanwhile, the cis-cleavage products serve as templates to initiate a new round of rolling circle amplification and Cas12a activation, resulting in exponential signal amplification. Using this technique, Yan et al. achieved single-nucleotide specificity in detecting microRNA-21, microRNA-196a, microRNA-451a, and microRNA-1246 at femtomolar concentrations in extracellular vesicles from the plasma of pancreatic cancer patients.

The first-generation CRISPR-Cas12a-based nucleic acid analysis technologies require separate amplification and detection steps. This workflow involves opening reaction tubes, which can lead to amplicon aerosol contamination and false-positive results. To address this, an "all-in-one detection" strategy has been developed to reduce contamination risks and improve detection efficiency. However, directly combining CRISPR-Cas12a with nucleic acid amplification poses challenges, as Cas12a's cleavage activity can degrade dsDNA amplification templates. A key challenge in advancing "all-in-one detection" technology is balancing Cas12a's cleavage activity with nucleic acid amplification.

To solve this, Liu et al. [48] developed a photoactivated CRISPR-Cas12a system. In this method, crRNA is modified with a photoresponsive RNA caging group, which suppresses Cas12a's cleavage activity during amplification. After a set period, ultraviolet (UV) light exposure releases the crRNA, allowing the CRISPR-Cas12a system to recognize the amplicons and cleave fluorescent probes. This "all-in-one" strategy significantly shortens detection time while achieving highly sensitive and rapid nucleic acid biomarker analysis.

In another approach, the Le team [49] introduced a spatial separation strategy, placing RT-LAMP reagents in the sample vial and CRISPR-Cas12a reagents in the vial cap. This setup prevents the CRISPR-Cas12a system from interfering with the early stages of nucleic acid amplification. After 30 min of RT-LAMP amplification, the vial is inverted and gently tapped to mix the detection reagents with the amplicons. They tested SARS-CoV-2 N and E genes in 100 human respiratory swab samples, achieving 100% clinical specificity with no false positives.

Chen et al. [50] developed the Cas12aVDet technique based on the DETECTR platform. In this improved methodology, Cas12a proteins are immobilized on the tube wall, followed by a 15-min recombinase polymerase amplification (RPA) reaction. Subsequent direct mixing of the Cas12a complexes with RPA amplicons via centrifugation enabled significant time reduction, achieving detection within 30 min compared to the original 1-h protocol required by conventional DETECTR technology. Furthermore, Lu et al. [51] investigated the PAM sequence and crRNA binding sites of CRISPR-Cas12a. They found that designing crRNA targets based on suboptimal PAM sequences effectively suppresses Cas12a's dsDNA cleavage activity without inhibiting its trans-cleavage activity, thus preventing interference with amplification. Based on this, they developed the sPAMC (suboptimal PAM of Cas12a-based test) strategy for "all-in-one detection." The sPAMC method directly mixes amplification and CRISPR reagents without requiring centrifugation, light activation, or other extra steps. In a study of 204 nasopharyngeal samples, this method achieved 100% specificity for SARS-CoV-2 detection, with no need for RNA extraction, and the single-tube process was completed in under 20 min (Fig. 7.12).

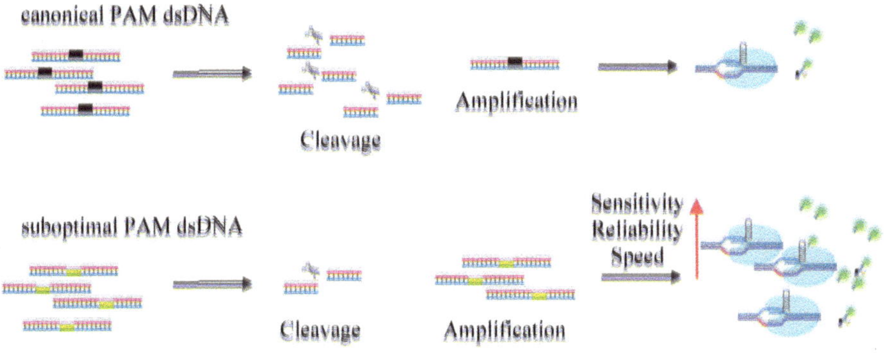

Fig. 7.12 Suboptimal PAM sequences effectively suppress the dsDNA cleavage activity of Cas12a. (Revised and reprinted with permission from Ref. [51])

Overall, the CRISPR-Cas12a system offers significant advantages over the Cas9 system in nucleic acid analysis. Its superiority arises from the dual functions of the Cas12a protein: in addition to its high specificity, Cas12a possesses trans-cleavage activity that enables signal amplification. This unique capability allows Cas12a to act as both an amplifier and a detector in nucleic acid amplification techniques. As a result, CRISPR-Cas12a-based detection technologies provide faster molecular diagnostics, higher sensitivity, and greater specificity. These qualities underscore the immense potential of CRISPR-Cas12a-based methods in biomedical research and clinical diagnostics.

7.3.2 Nucleic Acid Detection Technology Based on Cas12b Protein

Cas12b, like Cas12a, recognizes the 5′ TTN protospacer adjacent motif (PAM) sequence, generates sticky ends during DNA cleavage, and exhibits trans-cleavage activity. However, Cas12b offers several advantages over Cas12a, including broader temperature and pH tolerance, a lower off-target rate, and a smaller molecular size, making it more suitable for clinical applications. For example, Cas12b derived from *Alicyclobacillus acidiphilus* (AacCas12b, also known as C2c1) maintains high nuclease activity between 31 and 59 °C. It efficiently cleaves double-stranded DNA (dsDNA) and retains robust trans-cleavage activity for single-stranded DNA (ssDNA) [52, 53]. Li et al. combined loop-mediated isothermal amplification (LAMP) with Cas12b protein to develop the HOLMESv2 system, which simplifies the workflow and reduces the risk of cross-contamination (Fig. 7.13) [54]. The

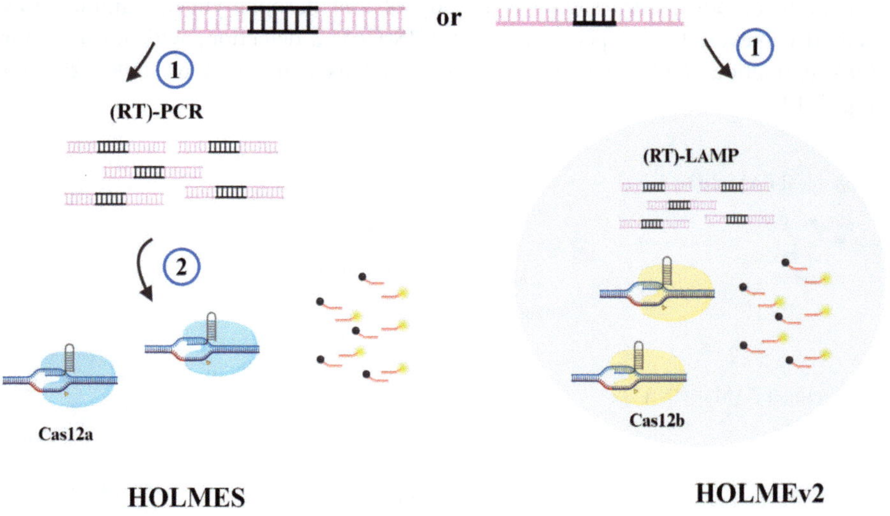

Fig. 7.13 Schematic illustration of HOLMES and HOLMESv2 detection methods. (Revised and reprinted with permission from Ref. [54])

thermal stability of C2c1 allows compatibility with reverse transcription-mediated isothermal amplification (RT-LAMP), enabling target RNA amplification and detection at a constant temperature of 55 °C in a single reaction tube. This streamlined process is simple and effective for precise quantification of DNA targets and their methylation modifications in various human cell lines. Teng et al. [55] developed another Cas12b-based DNA detection strategy, called CDetection. By integrating optimized guide RNA (tgRNA), CDetection can distinguish single-nucleotide variations and perform SNP genotyping in tumor cells. Building on the genomic sequence of SARS-CoV-2, the researchers optimized the CDetection workflow to create CASdetec, a viral detection platform that enables RNA detection from samples in a single reaction tube, offering a straightforward and efficient approach for viral diagnostics.

7.3.3 Nucleic Acid Detection Technology Based on Cas12c Protein

Since its discovery in 2015, Cas12c (also known as C2c3) [56] has attracted attention for its unique antiviral mechanism. Unlike typical CRISPR-Cas12 systems, Cas12c does not cleave the invader's DNA. Instead, it inhibits viral transcription by binding tightly to the invader's DNA [57]. Thus, Cas12c functions as an RNA-guided DNA-binding enzyme rather than a DNA-cleaving enzyme.

Three Cas12c proteins have been identified: Cas12c1, Cas12c2, and OspCas12c. Cas12c1 and OspCas12c recognize the PAM sequence 5′-TG-3′, while Cas12c2 recognizes 5′-TN-3′ [58]. These PAM sequences have a broader recognition range compared to other Cas12 systems. Additionally, studies show that Cas12c1 exhibits trans-cleavage activity for single-stranded DNA (ssDNA), similar to the Cas12a system. Building on this property, Wang et al. [59] developed the Cas12c-DETECTR technology, replacing Cas12a with Cas12c1 in the DECTER system. Cas12c1 has shown higher sensitivity and specificity in nucleic acid detection than Cas12a. This technology has been successfully applied to detect SNP mutations in the COVID-19 spike protein gene and to identify plant pathogens.

7.3.4 Nucleic Acid Detection Technology Based on Cas12f Protein

The large size of the CRISPR-Cas system has been a major challenge limiting its widespread application. To address this, scientists have been exploring smaller, more efficient alternatives for DNA cleavage. Significant progress has been made, including the discovery of nucleases such as CasX (Cas12e) [60], CasΦ (Cas12j) [61], and Cas14 (Cas12f) [62], all of which are smaller than 1000 amino acids.

Among these, Cas12f stands out due to its compact size (400–700 amino acids) and high cleavage efficiency.

Studies show that Cas12f (Cas14) not only enables PAM-dependent double-stranded DNA cleavage in vitro [63] but also demonstrates strong gene-editing potential in both bacterial and mammalian cells [64]. By engineering single-guide RNAs and the Cas12f protein, the editing efficiency of the CRISPR-Cas12f system at specific genomic loci has been made comparable to that of SpCas9 [65]. Furthermore, novel Cas12f variants identified through gene mining have expanded the range of recognizable PAM sequences, enhancing the system's application potential [66]. Building on these advantages, the Doudna team [67] replaced Cas12a with Cas12f in the DETECTR technology, creating a new diagnostic platform, Cas12f-DETECTR. This system successfully detected two previously undetectable HERC2 gene variants that Cas12a-DETECTR could not identify. The CRISPR-Cas12f system thus advances both the CRISPR-based detection of critical single-stranded DNA pathogens and the precise identification of single-nucleotide polymorphisms (SNPs).

7.4 Nucleic Acid Detection Technology Based on Cas13a Protein

CRISPR-Cas13, an RNA-guided bacterial immune system, exhibits single-stranded ribonuclease activity and is a highly efficient RNA-targeting gene-editing platform within the CRISPR-Cas family. Many of the world's most prevalent or lethal human pathogens, including the Ebola virus, novel coronavirus, HIV, and influenza, are RNA viruses. Thus, Cas13, which specifically targets RNA instead of DNA, holds significant potential for molecular diagnostics [68].

In June 2016, Feng Zhang's team first reported the type VI-A CRISPR-Cas effector protein, Cas13a (formerly C2c2), demonstrating its ability to cleave specific RNA sequences in *Escherichia coli*. They later introduced mutations in the protein's arginine residues to generate a catalytically inactive variant, termed dead Cas13a (dCas13a), which can bind to target RNA without cleaving it [69]. Since then, the RNA-targeting and analytical capabilities of Cas13a have attracted global attention and spurred rapid advancements in the field.

The bacterial immune process mediated by CRISPR-Cas13a is similar to that of the CRISPR-Cas9 and Cas12a systems. When a phage invades, prokaryotes transcribe the CRISPR array into pre-crRNA and translate Cas13a protein. Cas13a uses its RNase activity to process pre-crRNA into mature crRNA. The mature crRNA then associates with Cas13a to form a complex that identifies and cleaves the invading RNA, defending against phage infection. As shown in Fig. 7.14, crRNA consists of a hairpin structure and a 28–30 nt spacer for target recognition [69, 70]. When the Cas13a/crRNA complex binds to target RNA, the conformation of Cas13a changes, exposing its RNA cleavage domains and activating its cleavage

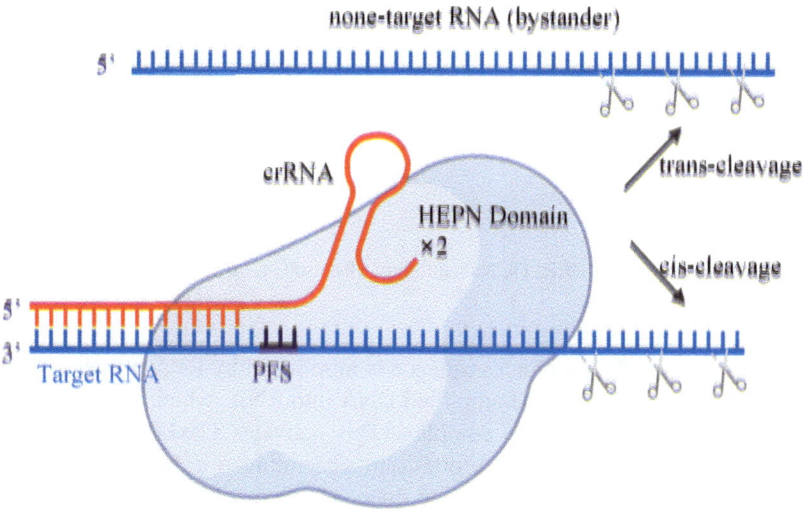

Fig. 7.14 Cleavage mechanism of the CRISPR/Cas13a system

activity. This activated complex can perform specific cleavage of the target RNA (cis-cleavage) and nonspecific cleavage of surrounding RNA (trans-cleavage). Cas13a's RNase activity is facilitated by its two Higher Eukaryotes and Prokaryotes Nucleotide-binding (HEPN) domains. The Cas13 family consists of four subtypes (VI-A, B, C, and D), with Cas13a being the earliest developed and most commonly used RNA gene-editing tool. Cas13a variants, such as LbaCas13a, LwaCas13a, LbuCas13a, and LshCas13a, are categorized based on their bacterial origin. Unlike the type II CRISPR/Cas systems (CRISPR/Cas9), Cas13a does not rely on a protospacer adjacent motif (PAM) for sequence recognition. Instead, it recognizes a protospacer-flanking site (PFS) [69, 71]. Studies have shown that if the PFS contains a guanine (G), the nuclease activity of certain Cas13a HEPN domains is significantly inhibited [71]·

CRISPR-Cas13 RNA cleavage technology has significantly advanced RNA tracking and detection, leading to a new generation of nucleic acid analysis methods [3, 4, 72, 73]. In 2017, Feng Zhang's team developed SHERLOCK (Specific High-sensitivity Enzymatic Reporter Unlocking) [13], a nucleic acid detection platform based on Cas13a protein and recombinase polymerase amplification (RPA). This pioneering work integrated isothermal amplification, in vitro transcription, and highly specific LwCas13a detection into a single workflow, advancing nucleic acid analysis technologies. SHERLOCK achieves single-molecule detection sensitivity and can identify single-nucleotide mismatches, enabling rapid detection of pathogens such as Zika virus (ZIKV) and dengue virus (DENV) in blood and saliva samples.

As shown in Fig. 7.15, the SHERLOCK mechanism starts with extracting target dsDNA or RNA from the sample. The target DNA undergoes RPA pre-amplification, or target RNA is reverse transcribed before RPA. During

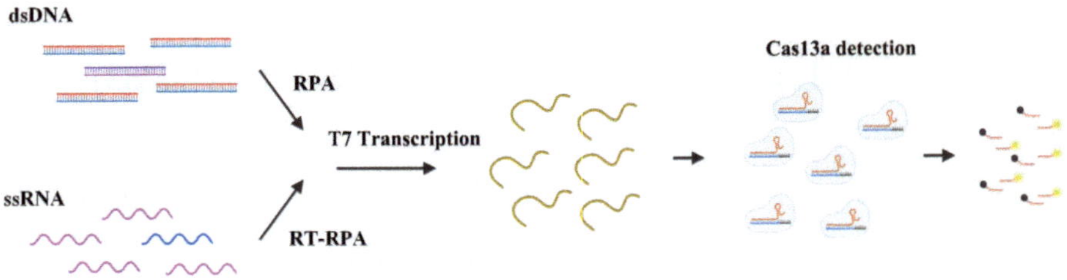

Fig. 7.15 Principle of the SHERLOCK technology. (Revised and reprinted with permission from Ref. [13])

pre-amplification, a T7 promoter sequence is introduced by RPA primers. T7 RNA polymerase then transcribes the amplified DNA into RNA, which is recognized and cleaved by the Cas13a/crRNA complex. This activates Cas13a's trans-cleavage activity, cleaving RNA reporter probes and generating a detectable fluorescence signal by separating the fluorescent and quencher groups. Cas13a's trans-cleavage activity amplifies the signal from a single RNA target by up to 10^5-fold. The SHERLOCK platform can detect DNA or RNA at concentrations as low as aM levels, enabling rapid detection of ZIKV and DENV, pathogen differentiation, and identification of DNA mutations associated with human cancers in a short time [74].

In 2018, Feng Zhang's team optimized the first-generation SHERLOCK technology to develop multiplex SHERLOCK (SHERLOCKv2) [75]. Unlike the original SHERLOCK, which detected only one target per reaction, SHERLOCKv2 uses Cas13 and Cas12a proteins from different bacterial species to simultaneously detect up to four distinct targets. The researchers also integrated SHERLOCKv2 with a portable lateral flow strip for visual detection. By using a FAM-biotin reporter molecule and lateral flow chromatography, results can be visually observed within a short time, enabling rapid and portable nucleic acid analysis. In the same year, Zhang's team introduced HUDSON (Heating Unextracted Diagnostic Samples to Obliterate Nucleases) [76], a simplified method that uses heat to inactivate nucleases in unprocessed clinical samples. This technique combines thermal treatment with the SHERLOCK platform. By heating patient body fluid samples, nucleases are deactivated, and RNA is released for detection by SHERLOCK. Results are displayed on a lateral flow strip, allowing visual interpretation within 2 h. HUDSON is efficient, sensitive, and straightforward, making it ideal for point-of-care diagnostics and rapid pathogen detection.

The SHERLOCK system developed by Zhang et al. laid the foundation for integrating nucleic acid amplification with Cas13-based detection. Building on this, Huang et al. [77] developed HyperCas, a hyper-branching rolling circle amplification CRISPR-Cas13a biosensor for ultra-sensitive and specific microRNA detection (Fig. 7.16). In HyperCas, a padlock probe hybridizes with the target microRNA and is ligated into a circular structure by SplintR ligase. Primer 1, containing a T7 promoter sequence, binds to the circular probe and is extended by Bst DNA polymerase, generating long single-stranded DNA (ssDNA) with repeated sequences. Primer 2 then binds to the repeats in the ssDNA and initiates strand-

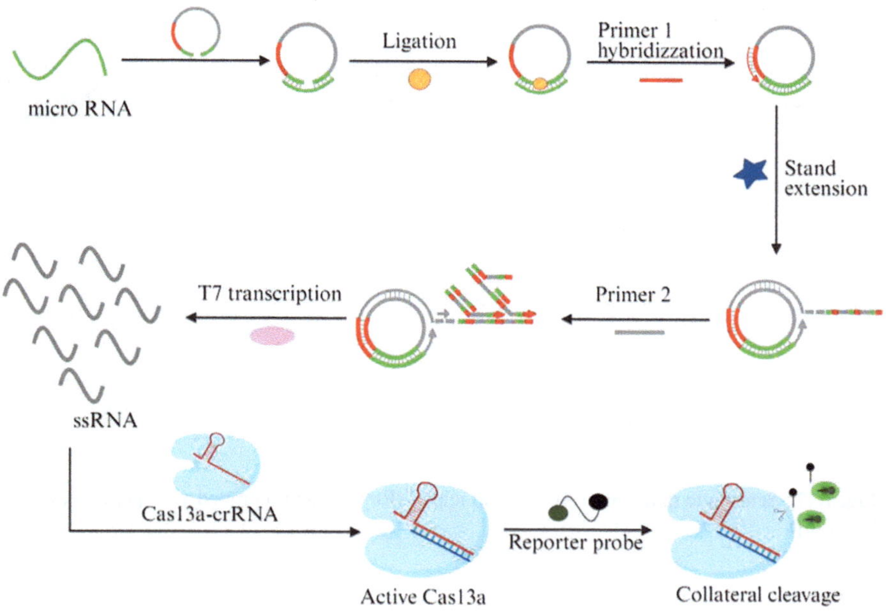

Fig. 7.16 Schematic diagram of the HyperCas technology principle. (Revised and reprinted with permission from Ref. [77])

displacement reactions, producing dendritic double-stranded DNA (dsDNA).T7 RNA polymerase transcribes the dsDNA into single-stranded RNA (ssRNA). When the Cas13a/crRNA complex recognizes and cleaves the target ssRNA, Cas13a's trans-cleavage activity is activated, cleaving fluorophore-quencher-labeled RNA reporter probes and generating a fluorescence signal. HyperCas combines the high amplification efficiency of hyper-branching rolling circle amplification with Cas13a's robust trans-cleavage activity, ensuring high sensitivity and specificity in microRNA analysis. Experimental results show that HyperCas can distinguish highly homologous microRNA-17 family members with a detection limit of 200 aM. It accurately detects microRNA-17 from cancer cells and demonstrates strong anti-interference capability and adaptability in human serum samples.

Jayeon Song et al. introduced SCOPE (Self-amplified and CRISPR-aided Operation to Profile EVs), a nucleic acid detection method that combines CRISPR-Cas13a lateral cleavage signal amplification with RNA amplification [78]. This method allows for the detection of mRNA in exosomes (EVs). As shown in Fig. 7.17, SCOPE works by using the Cas13a/crRNA complex to recognize the target mRNA sequence, activating Cas13a's lateral cleavage activity, and cleaving RNA fragments on the signal template. This releases a fluorescence signal and exposes a T7 RNA polymerase binding site, which then replicates the target RNA. This reactivates the CRISPR-Cas13a system, triggering multiple amplification cycles and resulting in ultrasensitive RNA detection. SCOPE has important applications in cancer liquid biopsy, including early detection, mutation monitoring,

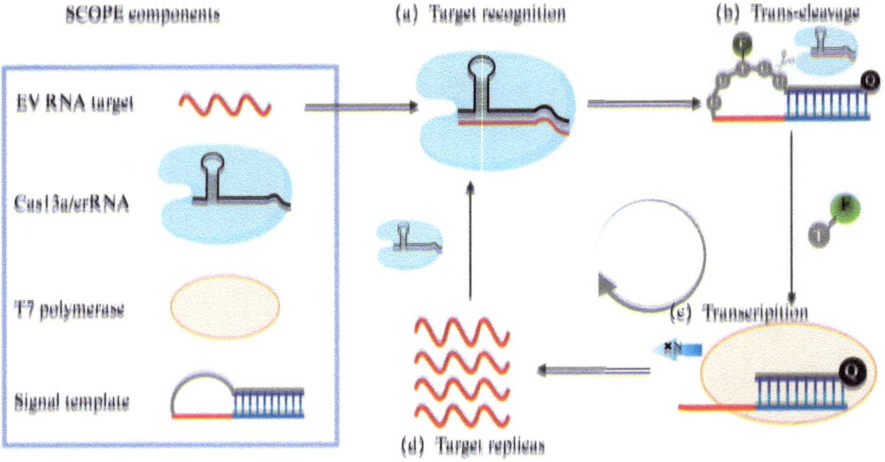

Fig. 7.17 Schematic diagram of SCOPE principle. (Revised and reprinted with permission from Ref. [78])

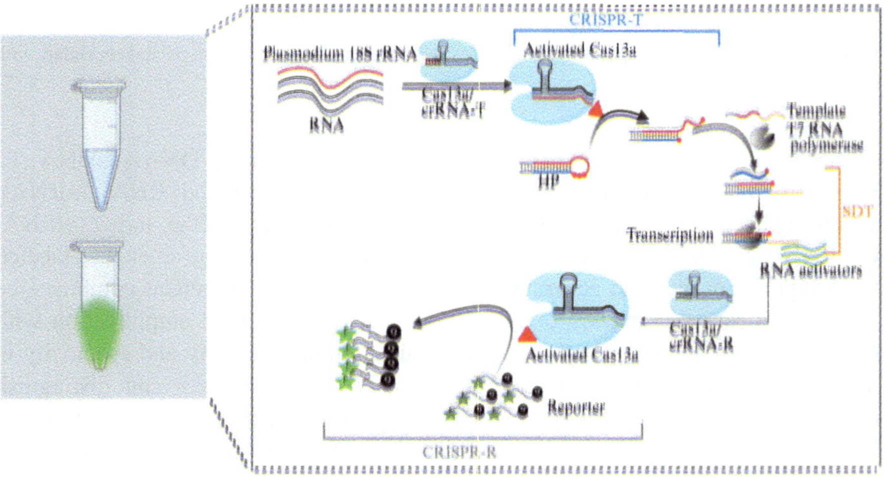

Fig. 7.18 Schematic diagram of Cas13 SDT technology principle. (Revised and reprinted with permission from Ref. [79])

personalized treatment assessment, and tumor recurrence tracking. It has been successfully applied to detect exosomal mRNA in patients with lung cancer, colorectal cancer, and glioblastoma (GBM). By analyzing mutations in genes such as KRAS, BRAF, EGFR, and IDH1, SCOPE provides precise molecular insights that support clinical decision-making.

The Li team developed a sequence recognition and signal amplification method called Cas13-SDT, based on dual Cas13a cleavage reactions and strand displacement-triggered transcription (SDT) cascades (Fig. 7.18) [79]. This method shows high specificity for RNA detection and meets the World Health Organization (WHO) sensitivity standards for effective malaria detection. Notably, it has been successfully used in nucleic acid screening for malaria in asymptomatic carriers using clinical samples. The Cas13-SDT strategy involves: CRISPR-Cas13a cleavage of a hairpin probe (CRISPR-T) activated by target RNA, strand displacement-triggered transcription (SDT), and CRISPR-Cas13a cleavage of a fluorescence-labeled single-stranded RNA reporter molecule (CRISPR-R) activated by transcription product RNA. Initially, a crRNA-T is designed to target the conserved sequence of Plasmodium 18S rRNA. The hairpin probe (HP) consists of a T7 promoter sequence (red), three uridine nucleotides (U), and a blocking sequence at the stem end (black). When Plasmodium 18S rRNA is present, the crRNA-T-guided Cas13a specifically binds to the target RNA, activating its trans cleavage activity. Cas13a then cleaves the uridine nucleotides in the HP, producing multiple double-stranded structures with a 3′ end guide region. The cleaved HP hybridizes with the template via the guide region, forming a double-stranded T7 promoter that initiates the SDT reaction, generating a large amount of transcription product RNA. This RNA subsequently reactivates Cas13a's lateral cleavage activity, which continuously cleaves the fluorescence self-quenched single-stranded RNA reporter molecule, resulting in a strong fluorescence signal.

Zhou et al. [80] developed a portable electrochemiluminescence chip system powered by the CRISPR-Cas13a system (PECL-CRISPR). The platform operates as shown in Figs. 7.18 and 7.19. It features a primary trigger (Pretrigger, PT) containing two uridine nucleotides (U), with 16 complementary bases at the 5′ end that match the EXpar amplification template, and a 5-base random sequence at the other end. The Cas13a/crRNA complex first recognizes and cleaves the target microRNA, activating its trans lateral cleavage activity. This cleaves the uridine nucleotide sequence in PT, generating two fragments. The 5′ fragment acts as a mature trigger (Mature-trigger, MT), which hybridizes with the EXpar template and initiates the EXpar amplification reaction in the presence of DNA polymerase and nicking enzymes, producing abundant dsDNA.Next, $[Ru(phen)_2dppz]^{2+}$ molecules are added to the Cas13a-mediated EXpar (CAS-EXpar) system along with tripropylamine (TPA) solution. Finally, electrochemiluminescence detection is used to analyze the results.

Unlike traditional methods, which amplify the target nucleic acid before using the CRISPR/Cas system for signal amplification, the PECL-CRISPR technique directly employs the Cas13a/crRNA complex to recognize and cleave the target microRNA. This activation triggers exponential amplification, fully utilizing the RNA-specific recognition ability of the CRISPR/Cas13a system. This approach reduces nonspecific amplification, significantly improving detection specificity and sensitivity. Experimental results showed that the technique achieved a detection limit of 1×10^{-15} mol/L under optimal conditions and demonstrated the ability to distinguish highly homologous microRNA families, allowing for single-base resolution microRNA detection.

Existing nucleic acid analysis methods typically rely on nucleic acid amplification. RNA detection often involves a reverse transcription step, which introduces

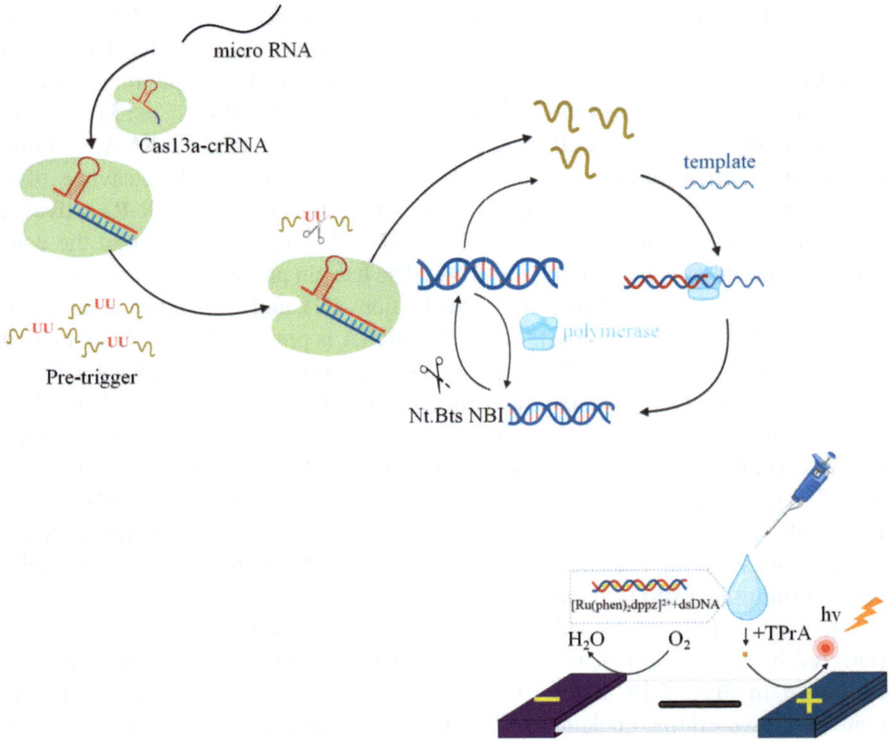

Fig. 7.19 Schematic diagram of the PECL-CRISPR technology principle. (Revised and reprinted with permission from Ref. [80])

amplification biases, cross-contamination, and nonspecific amplification, reducing the sensitivity and specificity of detection. Since Cas13a's trans-cleavage activity is 10^5 times more efficient than its cis-cleavage activity, it acts as an exceptionally effective signal amplifier. Therefore, developing CRISPR-Cas13a-based detection methods without amplification holds significant potential.

To address these limitations, Tian et al. [81] developed a fluorescence imaging-based microdroplet CRISPR-Cas13a technique for amplification-free, single-molecule RNA diagnostics, enabling absolute digital quantification of RNA. In this system, a mixture of target RNA, the Cas13a/crRNA complex, and fluorescent RNA reporter probes is processed using a microfluidic droplet chip to generate stable, uniform microdroplets. These microdroplets self-assemble into a high-density array. Inside the droplets, Cas13a/crRNA specifically recognizes and cleaves the target RNA, activating its trans-cleavage activity. This cleaves the RNA reporter probes, separating the fluorophore from the quencher and generating fluorescence. The fluorescence is detected, and droplets are counted using fluorescence microscopy. This method allows absolute digital quantification of single RNA molecules without the need for reverse transcription or amplification. It has been successfully

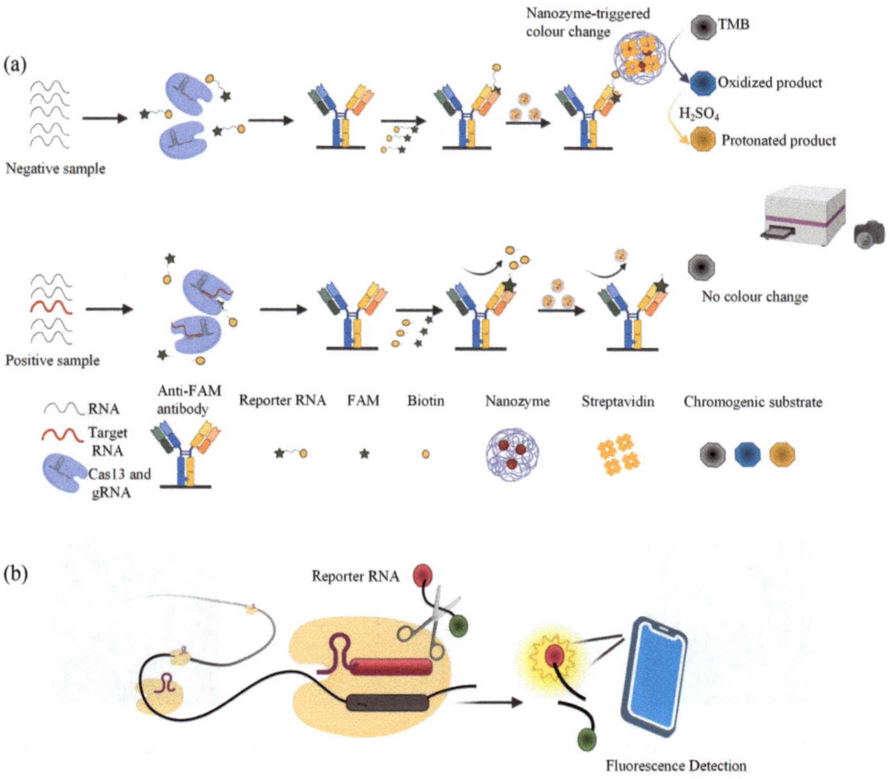

Fig. 7.20 (**a**) Schematic diagram of CRISPRZyme technology [82]; (**b**) CRISPR-Cas13a multi-target RNA detection technology [83]. (Revised and reprinted with permission from Refs. [82, 83])

applied to various RNA samples, such as microRNA, 16sRNA, and SARS-CoV-2 RNA, proving its effectiveness for single-molecule RNA diagnostics.

Marta et al. [82] developed CRISPRZyme, a CRISPR-based diagnostic method that incorporates a nanozyme. This approach effectively addresses challenges associated with traditional target pre-amplification techniques. CRISPRZyme combines Cas-based reactions with nanozyme-linked immunosorbent assays (NLISA), using a bimetallic nanozyme of platinum and gold (Pt@Au) to catalyze the oxidation of 3,3′,5,5′-tetramethylbenzidine (TMB), generating a color change as the detection signal (Fig. 7.20a).

Additionally, the Doudna team [83] proposed another amplification-free CRISPR-Cas13a detection method that improves sensitivity by combining multiple crRNAs. Upon activating Cas13a's trans-cleavage activity, fluorescence signals from cleaved ssRNA reporter probes are detected using a smartphone camera. This method enables rapid SARS-CoV-2 RNA detection from nasal swabs within 5 min and highly sensitive analysis at 100 copies/μL within 30 min. By eliminating RNA-to-DNA conversion and amplification, this approach combines the advantages of Cas13a detection and smartphone-based home testing. It provides an efficient,

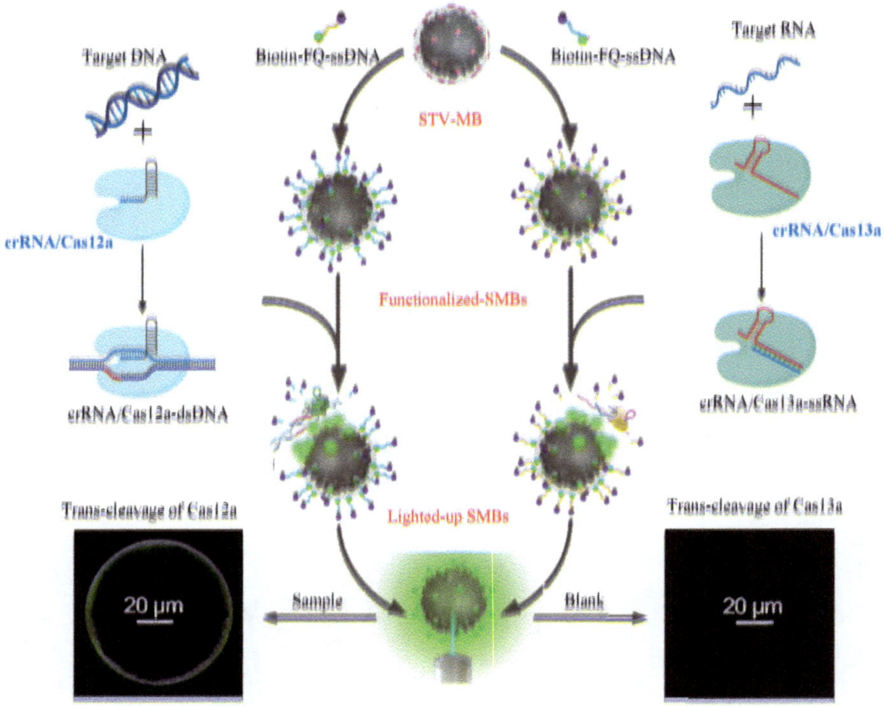

Fig. 7.21 CRISPR-Cas-driven single micromotor technology. (Revised and reprinted with permission from Ref. [86])

highly sensitive RNA virus detection platform with potential for rapid, low-cost, on-site SARS-CoV-2 diagnostics and screening for future pandemic pathogens (Fig. 7.20b).

Recently, the Li team developed a series of amplification-free nucleic acid analysis techniques based on the CRISPR-Cas13a system [84–87]. In 2023, they introduced the CRISPR-Cas-driven single micromotor technology (Cas-DSM), which eliminates the need for target nucleic acid pre-amplification. This method enables direct, high-specificity detection of nucleic acid targets at the single-molecule level [86]. As shown in Fig. 7.21, Cas-DSM is created by immobilizing biotin-labeled, fluorescently self-quenched single-stranded RNA reporter molecules (Biotin-FQ-ssRNA) onto streptavidin-coated magnetic beads (STV-MB). These microbeads serve as microreactors for RNA-activated CRISPR-Cas systems with trans-cleavage activity and act as signal platforms that concentrate fluorescence onto a small surface area. In the presence of target RNA, the crRNA guides the CRISPR-Cas complex to recognize the RNA via the spacer region, activating the trans-cleavage activity. This cleaves the reporter molecules on the microbead surface, separating the fluorophore (FAM) from the quencher (BHQ1) and leaving FAM on the microbead. This causes the microreactor to emit light under laser excitation. The fluorescence is highly concentrated in the small region, significantly enhancing detection sensitivity and enabling detection of RNA targets at the aM level. This microreactor can also be adapted to the CRISPR-Cas12a system for detecting dsDNA. The Cas-DSM platform has shown strong performance in detecting various nucleic acids, including SARS-CoV-2 RNA, fusion gene transcripts, HPV16 DNA, and Staphylococcus aureus DNA, in clinical samples such as swabs, RNA from cells, serum, and heat-treated cell lysates.

The growing demand for portable, cost-effective, and sensitive point-of-care nucleic acid detection has driven the development of innovative diagnostic methods. To address this need, Li et al. introduced a CRISPR/Cas13a-responsive DNA hydrogel capillary sensor for direct, visual detection of specific RNA targets [84]. The sensor features an RNA-crosslinked DNA hydrogel film at the end of a capillary tube. Upon recognizing the target RNA, Cas13a cleaves the RNA within the hydrogel, increasing its permeability. Varying concentrations of target RNA activate different amounts of Cas13a, resulting in the cleavage of different RNA quantities within the hydrogel, thus altering its permeability. When a capillary is inserted into the sample solution, samples with different RNA concentrations will travel different distances due to capillary action. This variation in distance, easily observable, enables quantitative RNA detection without the need for amplification or additional equipment. The method has been successfully applied to detect SARS-CoV-2 RNA in clinical nasopharyngeal swab and saliva samples. The measurable distance, which can be quantified with a ruler, eliminates the need for optical or electrochemical devices, making the method suitable for point-of-care and field applications (Fig. 7.22).

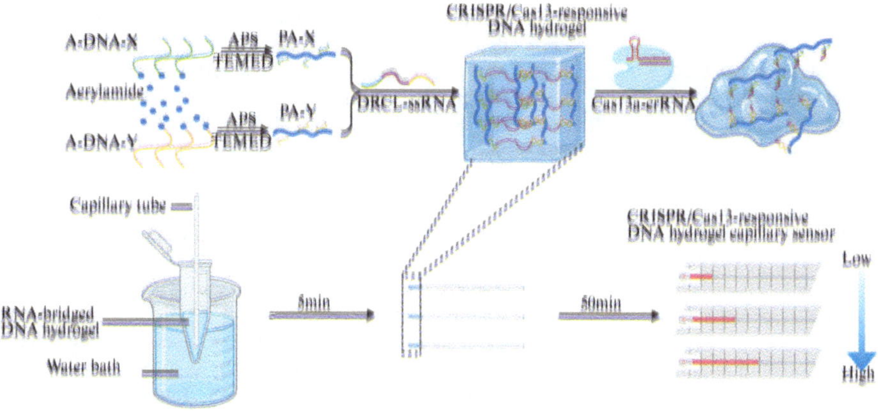

Fig. 7.22 DNA hydrogel capillary sensor based on CRISPR/Cas13a response. (Revised and reprinted with permission from Ref. [84])

In conclusion, contemporary nucleic acid detection technologies are essential in fields such as medicine, agriculture, environmental science, and life science research. The demand for cost-effective, rapid, and accurate methods continues to drive innovation. CRISPR-Cas-based diagnostic methods effectively address these needs. Effector proteins like Cas9, Cas12a, and Cas13a leverage efficient cleavage activity and precise nucleic acid targeting. Combined with amplification techniques, they enable highly sensitive and specific diagnostic platforms. As a result, the CRISPR-Cas system has evolved from a gene-editing tool into a versatile nucleic acid detection technology. It is widely used in disease diagnosis and medical research, offering significant potential for future advancements in biotechnology and biomedicine.

References

1. Knott GJ, Doudna JA. CRISPR-Cas guides the future of genetic engineering. Science. 2018;361:866–9.
2. Makarova KS, Zhang F, Koonin EV. SnapShot: class 2 CRISPR-Cas systems. Cell. 2017;168: 328.
3. Cox DBT, Gootenberg JS, Abudayyeh OO, Franklin B, Kellner MJ, Joung J, Zhang F. RNA editing with CRISPR-Cas13. Science. 2017;358:1019–27.
4. Abudayyeh OO, Gootenberg JS, Franklin B, Koob J, Kellner MJ, Ladha A, Joung J, Kirchgatterer P, Cox DBT, Zhang F. A cytosine deaminase for programmable single-base RNA editing. Science. 2019;365:382–6.
5. Wang H, Nakamura M, Abbott TR, Zhao D, Luo K, Yu C, Nguyen CM, Lo A, Daley TP, La Russa M, Liu Y, Qi LS. CRISPR-mediated live imaging of genome editing and transcription. Science. 2019;365:1301–5.
6. Wiedenheft B, Sternberg SH, Doudna JA. RNA-guided genetic silencing systems in bacteria and archaea. Nature. 2012;482:331–8.
7. Jinek M, Chylinski K, Fonfara I, Hauer M, Doudna JA, Charpentier E. A programmable dual-RNA-guided DNA endonuclease in adaptive bacterial immunity. Science. 2012;337:816–21.
8. Makarova KS, Wolf YI, Alkhnbashi OS, Costa F, Shah SA, Saunders SJ, Barrangou R, Brouns SJ, Charpentier E, Haft DH, Horvath P, Moineau S, Mojica FJ, Terns RM, Terns MP, White MF, Yakunin AF, Garrett RA, van der Oost J, Backofen R, Koonin EV. An updated evolutionary classification of CRISPR-Cas systems. Nat Rev Microbiol. 2015;13:722–36.
9. Pickar-Oliver A, Gersbach CA. The next generation of CRISPR-Cas technologies and applications. Nat Rev Mol Cell Biol. 2019;20:490–507.
10. Cox DB, Platt RJ, Zhang F. Therapeutic genome editing: prospects and challenges. Nat Med. 2015;21:121–31.
11. Gong S, Zhang S, Lu F, Pan W, Li N, Tang B. CRISPR/Cas-based in vitro diagnostic platforms for cancer biomarker detection. Anal Chem. 2021;93:11899–909.
12. Dai Y, Wu Y, Liu G, Gooding JJ. CRISPR mediated biosensing toward understanding cellular biology and point-of-care diagnosis. Angew Chem Int Ed Engl. 2020;59:20754–66.
13. Gootenberg JS, Abudayyeh OO, Lee JW, Essletzbichler P, Dy AJ, Joung J, Verdine V, Donghia N, Daringer NM, Freije CA, Myhrvold C, Bhattacharyya RP, Livny J, Regev A, Koonin EV, Hung DT, Sabeti PC, Collins JJ, Zhang F. Nucleic acid detection with CRISPR-Cas13a/C2c2. Science. 2017;356:438–42.

14. Chen JS, Ma E, Harrington LB, Da Costa M, Tian X, Palefsky JM, Doudna JA. CRISPR-Cas12a target binding unleashes indiscriminate single-stranded DNase activity. Science. 2018;360:436–9.
15. Broughton JP, Deng X, Yu G, Fasching CL, Servellita V, Singh J, Miao X, Streithorst JA, Granados A, Sotomayor-Gonzalez A, Zorn K, Gopez A, Hsu E, Gu W, Miller S, Pan C-Y, Guevara H, Wadford DA, Chen JS, Chiu CY. CRISPR–Cas12-based detection of SARS-CoV-2. Nat Biotechnol. 2020;38:870–4.
16. Sternberg SH, Redding S, Jinek M, Greene EC, Doudna JA. DNA interrogation by the CRISPR RNA-guided endonuclease Cas9. Nature. 2014;507:62–7.
17. Jiang F, Doudna JA. CRISPR-Cas9 structures and mechanisms. Annu Rev Biophys. 2017;46: 505–29.
18. Swarts DC, Jinek M. Cas9 versus Cas12a/Cpf1: structure-function comparisons and implications for genome editing. Wiley Interdiscip Rev RNA. 2018;9:1481.
19. Hsu PD, Lander ES, Zhang F. Development and applications of CRISPR-Cas9 for genome engineering. Cell. 2014;157:1262–78.
20. Shmakov S, Smargon A, Scott D, Cox D, Pyzocha N, Yan W, Abudayyeh OO, Gootenberg JS, Makarova KS, Wolf YI, Severinov K, Zhang F, Koonin EV. Diversity and evolution of class 2 CRISPR-Cas systems. Nat Rev Microbiol. 2017;15:169–82.
21. Pardee K, Green AA, Takahashi MK, Braff D, Lambert G, Lee JW, Ferrante T, Ma D, Donghia N, Fan M, Daringer NM, Bosch I, Dudley DM, O'Connor DH, Gehrke L, Collins JJ. Rapid, low-cost detection of Zika virus using programmable biomolecular components. Cell. 2016;165:1255–66.
22. Zhang K, Deng R, Li Y, Zhang L, Li J. Cas9 cleavage assay for pre-screening of sgRNAs using nicking triggered isothermal amplification. Chem Sci. 2016;7:4951–7.
23. Jia H, Li Z, Liu C, Cheng Y. Ultrasensitive detection of microRNAs by exponential isothermal amplification. Angew Chem Int Ed Engl. 2010;49:5498–501.
24. O'Connell MR, Oakes BL, Sternberg SH, East-Seletsky A, Kaplan M, Doudna JA. Programmable RNA recognition and cleavage by CRISPR/Cas9. Nature. 2014;516:263–6.
25. Huang M, Zhou X, Wang H, Xing D. Clustered regularly interspaced short palindromic repeats/Cas9 triggered isothermal amplification for site-specific nucleic acid detection. Anal Chem. 2018;90:2193–200.
26. Wang T, Liu Y, Sun HH, Yin BC, Ye BC. An RNA-guided Cas9 nickase-based method for universal isothermal DNA amplification. Angew Chem Int Ed Engl. 2019;58:5382–6.
27. Zhou W, Hu L, Ying L, Zhao Z, Chu PK, Yu XF. A CRISPR-Cas9-triggered strand displacement amplification method for ultrasensitive DNA detection. Nat Commun. 2018;9:5012.
28. Wang X, Xiong E, Tian T, Cheng M, Lin W, Wang H, Zhang G, Sun J, Zhou X. Clustered regularly interspaced short palindromic repeats/Cas9-mediated lateral flow nucleic acid assay. ACS Nano. 2020;14:2497–508.
29. Chen B, Gilbert LA, Cimini BA, Schnitzbauer J, Zhang W, Li GW, Park J, Blackburn EH, Weissman JS, Qi LS, Huang B. Dynamic imaging of genomic loci in living human cells by an optimized CRISPR/Cas system. Cell. 2013;155:1479–91.
30. Deng W, Shi X, Tjian R, Lionnet T, Singer RH. CASFISH: CRISPR/Cas9-mediated in situ labeling of genomic loci in fixed cells. Proc Natl Acad Sci USA. 2015;112:11870–5.
31. Zhang K, Deng R, Teng X, Li Y, Sun Y, Ren X, Li J. Direct visualization of single-nucleotide variation in mtDNA using a CRISPR/Cas9-mediated proximity ligation assay. J Am Chem Soc. 2018;140:11293–301.
32. Zetsche B, Gootenberg JS, Abudayyeh OO, Slaymaker IM, Makarova KS, Essletzbichler P, Volz SE, Joung J, van der Oost J, Regev A, Koonin EV, Zhang F. Cpf1 is a single RNA-guided endonuclease of a class 2 CRISPR-Cas system. Cell. 2015;163:759–71.
33. East-Seletsky A, O'Connell MR, Knight SC, Burstein D, Cate JH, Tjian R, Doudna JA. Two distinct RNase activities of CRISPR-C2c2 enable guide-RNA processing and RNA detection. Nature. 2016;538:270–3.

34. Fonfara I, Richter H, Bratovič M, Le Rhun A, Charpentier E. The CRISPR-associated DNA-cleaving enzyme Cpf1 also processes precursor CRISPR RNA. Nature. 2016;532:517–21.
35. Swiat MA, Dashko S, den Ridder M, Wijsman M, van der Oost J, Daran JM, Daran-Lapujade P. FnCpf1: a novel and efficient genome editing tool for Saccharomyces cerevisiae. Nucleic Acids Res. 2017;45:12585–98.
36. Yamano T, Nishimasu H, Zetsche B, Hirano H, Slaymaker IM, Li Y, Fedorova I, Nakane T, Makarova KS, Koonin EV, Ishitani R, Zhang F, Nureki O. Crystal structure of Cpf1 in complex with guide RNA and target DNA. Cell. 2016;165:949–62.
37. Yamano T, Zetsche B, Ishitani R, Zhang F, Nishimasu H, Nureki O. Structural basis for the canonical and non-canonical PAM recognition by CRISPR-Cpf1. Mol Cell. 2017;67:633–45.
38. Gao L, Cox DBT, Yan WX, Manteiga JC, Schneider MW, Yamano T, Nishimasu H, Nureki O, Crosetto N, Zhang F. Engineered Cpf1 variants with altered PAM specificities. Nat Biotechnol. 2017;35:789–92.
39. Tran MH, Park H, Nobles CL, Karunadharma P, Pan L, Zhong G, Wang H, He W, Ou T, Crynen G, Sheptack K, Stiskin I, Mou H, Farzan M. A more efficient CRISPR-Cas12a variant derived from Lachnospiraceae bacterium MA2020. Mol Ther Nucleic Acids. 2021;24:40–53.
40. Paul B, Montoya G. CRISPR-Cas12a: functional overview and applications. Biom J. 2020;43:8–17.
41. Deltcheva E, Chylinski K, Sharma CM, Gonzales K, Chao Y, Pirzada ZA, Eckert MR, Vogel J, Charpentier E. CRISPR RNA maturation by trans-encoded small RNA and host factor RNase III. Nature. 2011;471:602–7.
42. Stella S, Alcón P, Montoya G. Structure of the Cpf1 endonuclease R-loop complex after target DNA cleavage. Nature. 2017;546:559–63.
43. Li SY, Cheng QX, Liu JK, Nie XQ, Zhao GP, Wang J. CRISPR-Cas12a has both cis- and trans-cleavage activities on single-stranded DNA. Cell Res. 2018;28:491–3.
44. Xiong Y, Zhang J, Yang Z, Mou Q, Ma Y, Xiong Y, Lu Y. Functional DNA regulated CRISPR-Cas12a sensors for point-of-care diagnostics of non-nucleic-acid targets. J Am Chem Soc. 2020;142:207–13.
45. Li S-Y, Cheng Q-X, Wang J-M, Li X-Y, Zhang Z-L, Gao S, Cao R-B, Zhao G-P, Wang J. CRISPR-Cas12a-assisted nucleic acid detection. Cell Discov. 2018;4:20.
46. Zhang M, Wang H, Wang H, Wang F, Li Z. CRISPR/Cas12a-assisted ligation-initiated loop-mediated isothermal amplification (CAL-LAMP) for highly specific detection of microRNAs. Anal Chem. 2021;93:7942–8.
47. Yan H, Wen Y, Tian Z, Hart N, Han S, Hughes SJ, Zeng Y. A one-pot isothermal Cas12-based assay for the sensitive detection of microRNAs. Nat Biomed Eng. 2023;7:1583–601.
48. Meng H, Zhi Q, Zi B, Xiao Z. Photocontrolled crRNA activation enables robust CRISPR-Cas12a diagnostics. PNAS. 2022;119:e2202034119.
49. Pang B, Xu J, Liu Y, Peng H, Feng W, Cao Y, Wu J, Xiao H, Pabbaraju K, Tipples G, Joyce MA, Saffran HA, Tyrrell DL, Zhang H, Le XC. Isothermal amplification and ambient visualization in a single tube for the detection of SARS-CoV-2 using loop-mediated amplification and CRISPR technology. Anal Chem. 2020;92:16204–12.
50. Wang B, Wang R, Wang D, Wu J, Li J, Wang J, Liu H, Wang Y. Cas12aVDet: a CRISPR/Cas12a-based platform for rapid and visual nucleic acid detection. Anal Chem. 2019;91:12156–61.
51. Lu S, Tong X, Han Y, Zhang K, Zhang Y, Chen Q, Duan J, Lei X, Huang M, Qiu Y, Zhang DY, Zhou X, Zhang Y, Yin H. Fast and sensitive detection of SARS-CoV-2 RNA using suboptimal protospacer adjacent motifs for Cas12a. Nat Biomed Eng. 2022;6:286–97.
52. Teng F, Cui T, Feng G, Guo L, Xu K, Gao Q, Li T, Li J, Zhou Q, Li W. Repurposing CRISPR-Cas12b for mammalian genome engineering. Cell Discov. 2018;4:63–73.

53. Teng F, Li J, Cui T, Xu K, Guo L, Gao Q, Feng G, Chen C, Han D, Zhou Q, Li W. Enhanced mammalian genome editing by new Cas12a orthologs with optimized crRNA scaffolds. Genome Biol. 2019;20:15–27.
54. Li L, Li S, Wu N, Wu J, Wang G, Zhao G, Wang J. HOLMESv2: a CRISPR-Cas12b-assisted platform for nucleic acid detection and DNA methylation quantitation. ACS Synth Biol. 2019;8: 2228–37.
55. Teng F, Guo L, Cui T, Wang XG, Xu K, Gao Q, Zhou Q, Li W. CDetection: CRISPR-Cas12b-based DNA detection with sub-attomolar sensitivity and single-base specificity. Genome Biol. 2019;20:132.
56. Shmakov S, Abudayyeh OO, Makarova KS, Wolf YI, Gootenberg JS, Semenova E, Minakhin L, Joung J, Konermann S, Severinov K, Zhang F, Koonin EV. Discovery and functional characterization of diverse class 2 CRISPR-Cas systems. Mol Cell. 2015;60:385–97.
57. Huang CJ, Adler BA, Doudna JA. A naturally DNase-free CRISPR-Cas12c enzyme silences gene expression. Mol Cell. 2022;82:2148–60.
58. Yan WX, Hunnewell P, Alfonse LE, Carte JM, Keston-Smith E, Sothiselvam S, Garrity AJ, Chong S, Makarova KS, Koonin EV, Cheng DR, Scott DA. Functionally diverse type V CRISPR-Cas systems. Science. 2019;363:88–91.
59. Wang Z, Zhong C. Cas12c-DETECTOR: a specific and sensitive Cas12c-based DNA detection platform. Int J Biol Macromol. 2021;193:441–9.
60. Liu JJ, Orlova N, Oakes BL, Ma E, Spinner HB, Baney KLM, Chuck J, Tan D, Knott GJ, Harrington LB, Al-Shayeb B, Wagner A, Brötzmann J, Staahl BT, Taylor KL, Desmarais J, Nogales E, Doudna JA. CasX enzymes comprise a distinct family of RNA-guided genome editors. Nature. 2019;566:218–23.
61. Pausch P, Al-Shayeb B, Bisom-Rapp E, Tsuchida CA, Li Z, Cress BF, Knott GJ, Jacobsen SE, Banfield JF, Doudna JA. CRISPR-CasΦ from huge phages is a hypercompact genome editor. Science. 2020;369:333–7.
62. Makarova KS, Wolf YI, Iranzo J, Shmakov SA, Alkhnbashi OS, Brouns SJJ, Charpentier E, Cheng D, Haft DH, Horvath P, Moineau S, Mojica FJM, Scott D, Shah SA, Siksnys V, Terns MP, Venclovas Č, White MF, Yakunin AF, Yan W, Zhang F, Garrett RA, Backofen R, van der Oost J, Barrangou R, Koonin EV. Evolutionary classification of CRISPR-Cas systems: a burst of class 2 and derived variants. Nat Rev Microbiol. 2020;18:67–83.
63. Karvelis T, Bigelyte G, Young JK, Hou Z, Zedaveinyte R, Budre K, Paulraj S, Djukanovic V, Gasior S, Silanskas A, Venclovas Č, Siksnys V. PAM recognition by miniature CRISPR-Cas12f nucleases triggers programmable double-stranded DNA target cleavage. Nucleic Acids Res. 2020;48:5016–23.
64. Wu Z, Zhang Y, Yu H, Pan D, Wang Y, Wang Y, Li F, Liu C, Nan H, Chen W, Ji Q. Programmed genome editing by a miniature CRISPR-Cas12f nuclease. Nat Chem Biol. 2021;17:1132–8.
65. Kim DY, Lee JM, Moon SB, Chin HJ, Park S, Lim Y, Kim D, Koo T, Ko JH, Kim YS. Efficient CRISPR editing with a hypercompact Cas12f1 and engineered guide RNAs delivered by adeno-associated virus. Nat Biotechnol. 2022;40:94–102.
66. Kong X, Zhang H, Li G, Wang Z, Kong X, Wang L, Xue M, Zhang W, Wang Y, Lin J, Zhou J, Shen X, Wei Y, Zhong N, Bai W, Yuan Y, Shi L, Zhou Y, Yang H. Engineered CRISPR-OsCas12f1 and RhCas12f1 with robust activities and expanded target range for genome editing. Nat Commun. 2023;14:2046.
67. Harrington LB, Burstein D, Chen JS, Paez-Espino D, Ma E, Witte IP, Cofsky JC, Kyrpides NC, Banfield JF, Doudna JA. Programmed DNA destruction by miniature CRISPR-Cas14 enzymes. Science. 2018;362:839–42.
68. Tang T, Han Y, Wang Y, Huang H, Qian P. Programmable system of Cas13-mediated RNA modification and its biological and biomedical applications. Front Cell Dev Biol. 2021;9: 677587.

69. Abudayyeh OO, Gootenberg JS, Konermann S, Joung J, Slaymaker IM, Cox DB, Shmakov S, Makarova KS, Semenova E, Minakhin L, Severinov K, Regev A, Lander ES, Koonin EV, Zhang F. C2c2 is a single-component programmable RNA-guided RNA-targeting CRISPR effector. Science. 2016;353:5573–92.
70. East-Seletsky A, O'Connell MR, Burstein D, Knott GJ, Doudna JA. RNA targeting by functionally orthogonal type VI-A CRISPR-Cas enzymes. Mol Cell. 2017;66:373–83.
71. Tambe A, East-Seletsky A, Knott GJ, Doudna JA, O'Connell MR. RNA binding and HEPN-nuclease activation are decoupled in CRISPR-Cas13a. Cell Rep. 2018;24:1025–36.
72. Jing X, Xie B, Chen L, Zhang N, Jiang Y, Qin H, Wang H, Hao P, Yang S, Li X. Implementation of the CRISPR-Cas13a system in fission yeast and its repurposing for precise RNA editing. Nucleic Acids Res. 2018;46:90–105.
73. Konermann S, Lotfy P, Brideau NJ, Oki J, Shokhirev MN, Hsu PD. Transcriptome engineering with RNA-targeting type VI-D CRISPR effectors. Cell. 2018;173:665–76.e614.
74. Kellner MJ, Koob JG, Gootenberg JS, Abudayyeh OO, Zhang F. SHERLOCK: nucleic acid detection with CRISPR nucleases. Nat Protoc. 2019;14:2986–3012.
75. Gootenberg JS, Abudayyeh OO, Kellner MJ, Joung J, Collins JJ, Zhang F. Multiplexed and portable nucleic acid detection platform with Cas13, Cas12a, and Csm6. Science. 2018;360: 439–44.
76. Myhrvold C, Freije CA, Gootenberg JS, Abudayyeh OO, Metsky HC, Durbin AF, Kellner MJ, Tan AL, Paul LM, Parham LA, Garcia KF, Barnes KG, Chak B, Mondini A, Nogueira ML, Isern S, Michael SF, Lorenzana I, Yozwiak NL, MacInnis BL, Bosch I, Gehrke L, Zhang F, Sabeti PC. Field-deployable viral diagnostics using CRISPR-Cas13. Science. 2018;360:444–8.
77. Huang M, Huang R, Yue H, Shan Y, Xing D. Ultrasensitive and high-specific microRNA detection using hyper-branching rolling circle amplified CRISPR/Cas13a biosensor. Sensors Actuators B Chem. 2020;325:128799–810.
78. Song J, Cho MH, Cho H, Song Y, Lee SW, Nam HC, Yoon TH, Shin JC, Hong J-S, Kim Y, Ekanayake E, Jeon J, You DG, Im SG, Choi G-S, Park JS, Carter BC, Balaj L, Seo AN, Miller MA, Park SY, Kang T, Castro CM, Lee H. Amplifying mutational profiling of extracellular vesicle mRNA with SCOPE. Nat Biotechnol. 2024;2:1–11.
79. Pian H, Wang H, Wang H, Li Z. Dual CRISPR/Cas13a cascade strand displacement-triggered transcription for point-of-care detection of plasmodium in asymptomatic malaria. Anal Chem. 2024;96:7524–31.
80. Zhou T, Huang R, Huang M, Shen J, Shan Y, Xing D. CRISPR/Cas13a powered portable electrochemiluminescence chip for ultrasensitive and specific MiRNA detection. Adv Sci (Weinh). 2020;7:1903661–82.
81. Tian T, Shu B, Jiang Y, Ye M, Liu L, Guo Z, Han Z, Wang Z, Zhou X. An ultralocalized Cas13a assay enables universal and nucleic acid amplification-free single-molecule RNA diagnostics. ACS Nano. 2021;15:1167–78.
82. Broto M, Kaminski MM, Adrianus C, Kim N, Greensmith R, Dissanayake-Perera S, Schubert AJ, Tan X, Kim H, Dighe AS, Collins JJ, Stevens MM. Nanozyme-catalysed CRISPR assay for preamplification-free detection of non-coding RNAs. Nat Nanotechnol. 2022;17:1120–6.
83. Fozouni P, Son S, Díaz de León Derby M, Knott GJ, Gray CN, D'Ambrosio MV, Zhao C, Switz NA, Kumar GR, Stephens SI, Boehm D, Tsou CL, Shu J, Bhuiya A, Armstrong M, Harris AR, Chen PY, Osterloh JM, Meyer-Franke A, Joehnk B, Walcott K, Sil A, Langelier C, Pollard KS, Crawford ED, Puschnik AS, Phelps M, Kistler A, DeRisi JL, Doudna JA, Fletcher DA, Ott M. Amplification-free detection of SARS-CoV-2 with CRISPR-Cas13a and mobile phone microscopy. Cell. 2021;184:323–33.
84. Wang H, Wang H, Pian H, Su F, Tang F, Chen D, Chen J, Wen Y, Le XC, Li Z. CRISPR/Cas13a-responsive and RNA-bridged DNA hydrogel capillary sensor for point-of-care detection of RNA. Anal Chem. 2024;96:12022–9.

85. Pian H, Wang H, Wang H, Tang F, Li Z. Capillarity-powered and CRISPR/Cas12a-responsive DNA hydrogel distance sensor for highly sensitive visual detection of HPV DNA. Biosens Bioelectron. 2024;264:116657–63.
86. Chen D, Liang Y, Wang H, Wang H, Su F, Zhang P, Wang S, Liu W, Li Z. CRISPR-Cas-driven single micromotor (Cas-DSM) enables direct detection of nucleic acid biomarkers at the single-molecule level. Anal Chem. 2023;95:5729–37.
87. Wang H, Wang S, Wang H, Tang F, Chen D, Liang Y, Li Z. Amplification-free detection of telomerase activity at the single-cell level via Cas12a-lighting-up single microbeads (Cas12a-LSMBs). Lab Chip. 2023;23:4674–9.

Part II
Application Section

Chapter 8
DNA Single Nucleotide Polymorphism (SNP) Analysis

Pengbo Zhang

8.1 Overview

Single nucleotide polymorphism (SNP) refers to the variation of DNA sequences caused by a single base change at certain specific sites on the genomic DNA. SNPs are the most prevalent type of all heritable variations in humans, and more than 3 million SNP loci have been identified. On average, one SNP locus exists for every 600–1000 bases in the human genome [1]. Numerous diseases, such as sickle cell anemia [2] and β-thalassemia [3] are caused by SNP. Additionally, certain genetic mutations can significantly increase disease risks, including those occurring at specific loci of the breast cancer genes *BRCA1* and *BRCA2* that increase the risk of breast cancer [4].

Depending on base changes, SNP can be divided into two forms: transitions and transversions. If a purine is replaced by another purine or a pyrimidine is replaced by another pyrimidine, this is termed a transition. In contrast, a transversion involves the substitution of a purine for a pyrimidine or vice versa. Since DNA contains four bases, there may be four kinds of transitions (C $\longleftrightarrow$ T, and on its complementary strand G $\longleftrightarrow$ A) and eight kinds of transversions (C $\longleftrightarrow$ A, G $\longleftrightarrow$ T, C $\longleftrightarrow$ G, A $\longleftrightarrow$ T). The incidence of transition is significantly higher than that of other variants, with about 2/3 of SNPs having transition-type variants and similar odds of occurrence for several other variants. The higher transition rate may stem from cytosine residues at CpG dinucleotides that represent the most mutagenic regions in the human genome, most of which are methylated and can spontaneously deaminate to form thymine [5]. Either transition or transversion may alter the amino acid composition or order in the polypeptide chain, thereby affecting protein or enzyme biological functions and causing abnormal phenotypes. SNPs are classified

P. Zhang (✉)
University of Science and Technology Beijing, Beijing, China
e-mail: zhangpengbo@ustb.edu.cn

Z. Li (ed.), *Nucleic Acid Detection*, https://doi.org/10.1007/978-981-95-7295-3_8

according to the effect of individual base mutations on the amino acid sequence in the polypeptide chain as follows.

Synonymous mutation: After base substitution, although the codon changes, the amino acids remain unchanged due to the degeneracy of the codon. As a result, the mutation does not lead to any change in the protein product. For example, if the third G in the DNA codon GCG is replaced by A and becomes GCA, the corresponding mRNA codon changes from CGC to CGU. Since both CGC and CGU encode arginine, the resulting protein remains identical after the mutation. These mutations account for approximately 25% of all base substitution mutations.

Missense mutation: A mutation that changes a codon in mRNA to encode another amino acid is called a missense mutation. Missense mutation can lead to structural or functional abnormalities in protein or enzyme, potentially leading to diseases. A classic example is the mutation underlying sickle cell anemia: the sixth position of the β chain of normal human hemoglobin is glutamic acid, whose codon is GAA or GAG. If the second base A is replaced by U, it becomes GUA or GUG, where glutamic acid is substituted by valine, forming abnormal hemoglobin that causes sickle cell anemia, producing a mutation effect. However, if the amino acids are conserved (e.g., leucine becomes isoleucine), they may be harmless.

Nonsense mutation: When a codon encoding an amino acid is mutated to a termination codon, the synthesis of a polypeptide chain is terminated prematurely, resulting in a biologically inactive polypeptide fragment, which is called a nonsense mutation. For example, when the G in the ATG is replaced by T, the corresponding mRNA codon changes from UAC to UAA, thus halting translation and shortening the polypeptide chain. Such mutations affect the function of proteins or enzymes in most cases.

Terminator codon mutation: A mutation in a gene in which a termination codon is mutated to one encoding an amino acid is called a termination codon mutation. It is also called a lengthening mutation because the synthesis of the peptide chain proceeds until the downstream terminator codon appears, producing an abnormally extended polypeptide. A classic example is the hemoglobin variant Hb Constant Spring, in which the α-globin chain is synthesized with an additional 31 amino acids beyond the normal C-terminus.

SNPs are widely used as a class of genetic markers in genetic analysis, mainly due to several key characteristics: (I) High density. SNPs exhibit high density in the human genome, with an average of one SNP locus every 600–1000 bases and a distribution of 3×10^6 across the genome. This density surpasses that of microsatellite markers and can provide a range of markers within or near any genes. (II) High representativeness. Certain SNPs located within genes may directly affect protein structure or expression levels, potentially representing functional factors in disease inheritance mechanisms. (III) Genetic stability. Compared with repetitive sequence polymorphic markers like microsatellites, SNPs demonstrate higher genetic stability. (IV) Easy to automate. SNP markers present only two allele types in population, requiring detection through simple "+/−" or "yes/no" methods. Unlike restriction fragment length polymorphism or microsatellite analysis, which require fragment length measurements, this simplicity facilitates automated SNP-based detection and analysis.

8.2 Triggers of SNP

The occurrence of SNPs is a combination of genetic and environmental factors. Errors during DNA replication and spontaneous DNA damage are important sources of SNPs. Cytosine, adenine, and guanine are susceptible to spontaneous deamidation, such as adenine deamidation to hypoxanthine and cytosine deamidation to guanine, both of which can introduce errors in subsequent DNA replication. DNA can also undergo depurination through spontaneous hydrolysis of the N-glycosidic bond, creating a deacylation site. Environmental factors also contribute to SNPs. Prolonged exposure to radiation and chemical mutagens such as formaldehyde, acetaldehyde, nitrosamines, and radiation, as well as various unhealthy lifestyles, may also induce SNPs. SNPs play an important role in molecular evolution, with deleterious mutations being eliminated by natural selection while beneficial ones accumulate. Over time, accumulated SNPs across different populations within species ultimately form distinct species. This explains why comparative genomic analyses consistently reveal representative SNPs that characterize specific species.

8.3 Relationship Between SNP and Diseases

8.3.1 *SNP and Sickle Cell Anemia*

Sickle cell anemia is an autosomal recessive hemoglobinopathy and the most typical example of a disease caused by a single-base mutation. As shown in Fig. 8.1, normal human red blood cells maintain their characteristic double concave disc shape. When blood oxygen levels drop, hemoglobin molecules polymerize into helical fibers, distorting red blood cells to twist into sickle cells (the sickle transformation phenomenon) [6]. These rigid, sickle-shaped cells are unable to pass through narrow capillaries, leading to blockage and rupture of blood vessels and localized ischemia

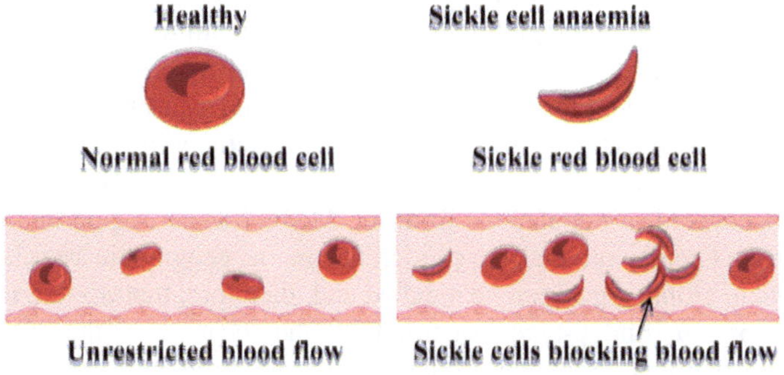

Fig. 8.1 Normal red blood cell (left) and sickle red blood cell (right) and their states in capillaries

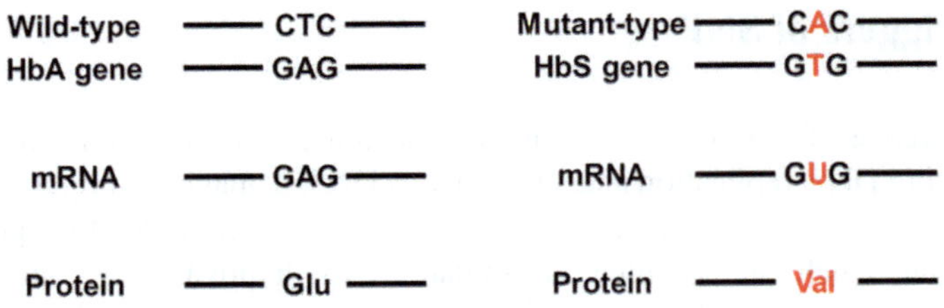

Fig. 8.2 Mutation sites and protein alterations in sickle cells

in the body, which in turn result in internal bleeding and pain [7]. The frequent rupture of sickle cells leads to anemia, and without timely medical intervention, patients with this condition face life-threatening risks.

To investigate the pathogenesis of sickle cell anemia, Vernon Ingram in 1957 performed a comparative analysis of partial amino acid sequences in normal and sickle cells using the protein sequencing method invented by Frederick Sanger [8]. A single amino acid difference was found in the peptide at the terminal amino acid of normal hemoglobin (HbA) and hemoglobin of sickle cells (HbS), i.e., glutamate at the sixth position of HbA was replaced by valine in HbS. The two codons encoding glutamic acid (Glu) are GAA and GAG, while the four codons encoding valine (Val) contain GUA and GUG. As shown in Fig. 8.2, if the DNA sequence of HbA is GAG and becomes GTG after a single-base mutation, the transcribed mRNA sequence becomes GUG, thereby converting the amino acid at the insertion site of HbS from glutamic acid to valine. Similarly, if the initial DNA sequence is GAA and mutated to GTA, its mRNA becomes GUA, which also results in the amino acid at the insertion site of HbS being converted from glutamic acid to valine. Thus, sickle cell anemia is caused by a genetic missense mutation.

8.3.2 SNP and β-Thalassemia

Thalassemia, like sickle cell anemia, is a hemoglobin disorder caused by abnormal hemoglobin. Thalassemia arises from defective synthesis of the globin chain due to deletions or mutations in the globin gene, resulting in an imbalanced production of α- or β-chains. In affected individuals, free, unbound α- or β-hemoglobin chains accumulate on the red blood cell membrane, which causes oxidative damage, leading to abnormal erythrocyte metabolism and extremely low intracellular hemoglobin levels that prevent oxygen-carrying capacity (Fig. 8.3). This genetic disorder was first discovered in populations from the Mediterranean region and derives its name from the Greek word thalassa (sea), reflecting its geographic association. Both α- and β-chain globin genes can be mutated and, therefore, thalassemia is classified into two types, α- and β-type, according to the location of the mutated gene [9, 10]. The majority of β-chain thalassemia stem from point mutations in the β-chain globin gene. These mutations cause nonfunctional or unstable mRNA that

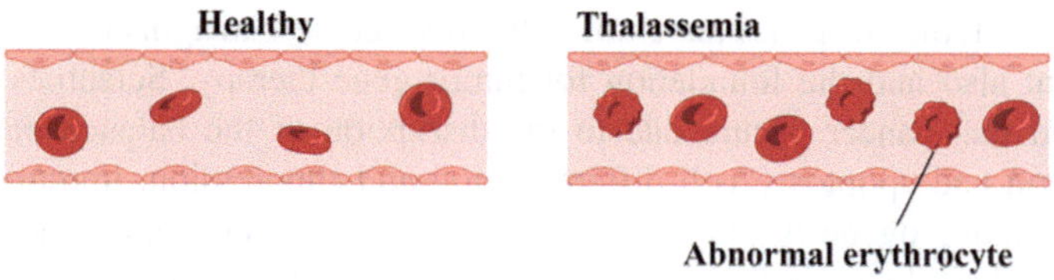

Fig. 8.3 Normocyte (left) and red blood cell of a patient with β-thalassemia (right)

cannot produce normal β-chain globin, such as the ATG → AGG codon mutation and the A-T mutation at codon 17 [11]. Additionally, these mutations may prevent proper processing of mRNA precursors, leading to abnormal mRNAs or producing structurally defective β-chains during translation. All these abnormalities disrupt β-chain globin synthesis, resulting in β-thalassemia.

8.3.3 SNP and Cancer

Cancer is essentially a "genetic disease" in which a gene goes wrong and a "genetic mutation" occurs. However, a single mutation is insufficient to cause cancer, and it requires multiple independent rare mutations occurring within the same cell lineage. Cancer develops through the combined effects of both congenital and acquired risk factors. Acquired risk factors include aging, various unhealthy lifestyles, like smoking, drinking, and even direct exposure to "carcinogens," such as nicotine, aflatoxin, acetaldehyde, nitrosamines, formaldehyde, radiation, etc. These cumulative exposures over time may induce gene mutation leading to cancer. These mutations occur later in life, which are not inherited by offspring, and are termed as "acquired mutations". Numerous factors known to induce cancer through genetic mutations have been identified, including chemical carcinogens that alter nucleotide sequences, ionizing radiation like X-rays causing DNA breaks, and viruses that introduce carcinogenic exogenous DNA into cells. In most cases, humans are not simply exposed to a single carcinogen but rather to complex mixtures of carcinogens through various lifestyle factors.

Another category of cancer stems from inborn genetic defects. These genes normally protect us from cancer by correcting errors or damage to DNA replication that occur naturally during cell division. Gene defects mean they do not effectively repair damaged DNA in cells, increasing the chances of the cells becoming cancerous. Such cancer-prone genetic defects are either inherited directly from parents or developed during DNA replication in the ovum or sperm, and are becoming prevalent in all body cells, known as "inherited cancer genes", or "cancer susceptibility genes". While inheriting defective gene does not guarantee cancer, carriers do face higher risks of specific cancer and tend to develop them at younger ages. Some people exposed to specific carcinogens develop cancer, others remain unaffected. This phenomenon demonstrates the genetic susceptibility to cancer.

Intensive studies of genes have not only provided new insights into the nature of cancer, but also laid the foundation for tumor gene therapy. Scientists have now recognized that cancer occurs due to the disruption of the balance between cell proliferation and apoptosis, cell differentiation and suppression, immune response and evasion, angiogenesis and inhibition, and metastasis and suppression of metastasis, which boil down to the balance of oncogenes and oncogenes.

8.3.3.1 Proto-oncogene

Proto-oncogenes are normal genes present in the genome and whose encoded products act as positive regulatory signals to promote cell proliferation and growth. Under normal conditions, proto-oncogenes in the genome remain low-expressed or non-expressed, performing essential physiological functions. However, under certain conditions, such as viral infection, chemical carcinogens, or radiation exposure, proto-oncogenes can be activated abnormally through point mutations, gene rearrangements, chromosomal translocations, amplifications, etc., transforming into oncogenes to induce cell carcinogenesis, as shown in Fig. 8.4a. The relationship between the activation modes of common proto-oncogenes in human tumors and the associated cancers is shown in Table 8.1 [12]. Proto-oncogenes exhibit the following characteristics: highly conserved gene sequences, widespread distribution across organisms, precise regulation of normal cell growth, reproduction, development and differentiation by expression products, and abnormal proliferation and differentiation caused by abnormal gene structure or uncontrolled expression. The expression products of proto-oncogenes are classified according to their localization and function in cells: *ras* family, *myc* family, *src* family, *sis* family, *myb* family.

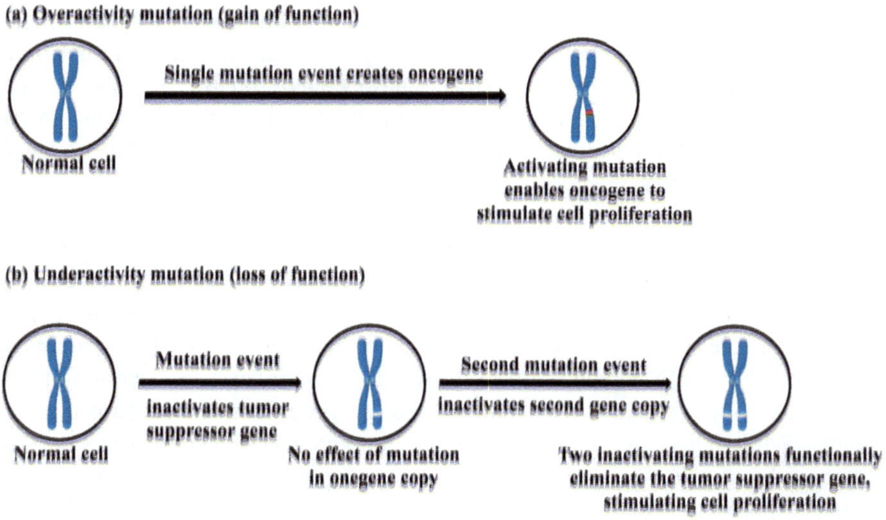

Fig. 8.4 Mechanism of proto-oncogene and oncogene

Table 8.1 Human tumor-associated proto-oncogenes

Proto-oncogene	Activation method	Related tumors
K-ras	Point mutation	Lung cancer, colon cancer
H-ras	Point mutation	Bladder Cancer
N-ras	Point mutation	Leukemia
Myc	Amplification, chromosome translocation	Lymphoma
N-myc	Amplification	Nerve cell carcinoma
L-myc	Amplification	Small cell lung cancer
Raf	Chromosome rearrangement	Gastric cancer
Erbb2	Amplification	Breast cancer, ovarian cancer, gastric cancer
Erbb	Amplification	Breast cancer, glioma
Bcl2	Chromosomal translocations	Lymphoma
Gsp	Point mutation	Pituitary tumor
Hst	Chromosome rearrangement	Gastric cancer
Ret	Chromosome rearrangement	Thyroid Cancer

1. *Ras* family

The *Ras* (rat sarcoma) family, first identified in human bladder cancer cells in 1982, represents the first human oncogene discovered to play a crucial role in regulating normal cell proliferation and differentiation, making it the most conserved oncogene family known to date. It serves as a central player in cellular signaling, controlling transcription and translation at the molecular level, while modulating cell proliferation, differentiation, aging, and apoptosis at the cellular level [13]. With mutations in *Ras* accounting for approximately 30% of human cancers, it has long been regarded as a key therapeutic target. The *Ras* gene family contains three tumor-associated genes: *H-ras*, *K-ras*, and *N-ras*, localized on chromosomes 11, 12, and 1, respectively. The first two are transformation genes derived from rat sarcoma viruses, while the latter is isolated from human neuroblastoma.

Ras genes are activated by various mechanisms including point mutation, gene overexpression, gene insertion and transposition, with point mutations being the most common form, accounting for 10–15% of solid tumors. When activated, the proto-oncogene *Ras* transforms into a cancer-causing gene. Its product, the Ras protein, remains perpetually activated and continuously transmits growth and differentiation signals, leading to uncontrolled cell proliferation and tumor formation. Ras protein exhibits GTP hydrolase activity and function as a molecular switch: it becomes "activated" when binding GTP and "deactivated" upon GTP hydrolysis [14]. Studies have shown that *Ras* gene mutations are highly prevalent in tumors, with the most recognized *Ras* mutation sites occurring at amino acid codons 12, 13, or 61 of the three *Ras* genes (Table 8.2). These mutations enable prolonged GTP binding by the Ras protein, continuously activating downstream effector signals that drive tumor development [15].

Table 8.2 Point mutation activation of the *Ras* gene locus

Genes	Codon	Point mutation	Amino acid changes	Related tumors
H-ras	12	GGC → GTC	Gly → Val	Bladder cancer, gastric cancer, nasopharyngeal cancer, cervical cancer
		GGC → GAC	Gly → Asp	Breast cancer, bowel cancer
	61	CAG → CTG	Gln → Leu	Bladder cancer, lung cancer, melanoma
		CAG → CGG	Gln → Arg	Kidney cancer
K-ras	12	GGT → CGT	Gly → Arg	Bladder cancer
		GGT → AGT	Gly → Ser	Gastric adenocarcinoma, gallbladder cancer
		GGT → TGT	Gly → Cys	Lung cancer, bowel cancer, chronic myeloid leukemia
		GGT → GAT	Gly → Asp	Pancreatic cancer
	13	GGG → GAC	Gly → Asp	Breast cancer
	61	GAA → CTA	Gln → His	Lung cancer
		CAA → CTA	Gln → Leu	Lung cancer
N-ras	12	GGT → AGT	Gly → Ser	Myelodysplastic syndrome
		GGT → GTT	Gly → Val	Acute myelogenous leukemia
		GGT → GAT	Gly → Asp	Acute leukemia, teratoma
	13	GGT → CGT	Gly → Arg	Lung adenocarcinoma
		GGT → GTT	Gly → Val	Acute myelogenous leukemia
		GGT → GAT	Gly → Asp	Acute myelogenous leukemia
	61	CAA → AAA	Gln → Lys	Neuroblastoma, fibrosarcoma, melanoma
		CAA → CAT	Gln → His	Transverse myxosarcoma
		CAA → CTA	Gln → Leu	Promyelocytosis

Numerous studies have shown that the *Ras* genes play a pivotal role in the development and progression of a variety of malignancies. Abnormalities in the *Ras* oncogene have been detected in bladder, breast, colon, kidney, liver, lung, pancreatic, gastric, and hematopoietic tumors [16]. The mutation rate of *Ras* oncogene varies significantly across tumor types, with 90% for pancreatic cancer, 53% for thyroid cancer, and 47% for colon cancer. Once the *Ras* gene is mutated, its expression product, Ras protein, undergoes conformational change and remains perpetually activated. This enables binding with downstream effector proteins, leading to uncontrolled cell proliferation, reduced apoptosis, and malignant transformation. The type of mutated *Ras* oncogene is closely related to specific disease patterns, i.e., there is a dominant activation phenomenon. *K-ras* mutations are predominant in pancreatic, colon, and lung cancers, while mutations in *N-ras* are mostly found in hematopoietic tumors and *H-ras* mutations are predominant in urologic tumors. Conversely, the expression of *H-ras* and *K-ras* is closely related not only to bladder, renal pelvis, lung, and colon cancers, but also significantly associated with gallbladder cancer, pancreatic cancer, renal blastoma, chronic lymphocytic leukemia, and melanoma formation. Although elevated *N-ras* expression primarily occurs in hematopoietic malignancies, it also shows increased levels in neuroblastoma, fibrosarcoma, and rhabdomyosarcoma. Therefore, detection of *Ras* mutations is of great significance for understanding tumor development and monitoring the therapeutic efficacy, providing crucial guidance for clinical practice.

2. *Myc* family

Myc family is the most extensively studied class of nucleoprotein proto-oncogenes, mainly including *C-myc*, *N-myc*, and *L-myc*, which are located on chromosome 8, 2, and 1, respectively. They are closely related to the occurrence and development of different human tumors [17]. *Myc* genes were first identified in Burkitt's lymphoma [18], and can be activated by chromosomal translocations. *Myc* is most commonly activated through translocation between chromosome 8 and 14, which fuses the *myc* gene or its adjacent region on chromosome 8 with the immunoglobulin heavy chain gene on chromosome 14. As a key transcriptional regulator, *Myc* controls a variety of cellular processes to maintain normal cellular life activities [19]. Once cellular regulation fails, *myc* is overexpressed, which may induce tumorigenesis. Various tumors, such as breast cancer, lung cancer, stomach cancer, breast cancer, colon cancer, cervical cancer, and certain other neuroblastoma, granulocytic leukemia, retinoblastoma, osteogenic sarcoma, chondrosarcoma, chordoma, liposarcoma, rhabdomyosarcoma, Hodgkin's disease, and head tumors all exhibit *myc* gene amplification or hyperexpression [20]. Studies have shown that the *myc* expression in most tumor tissues significantly exceeds that in their paraneoplastic tissues, suggesting that overexpression of *myc* contributes to cancer progression. Given the important role of *myc* in cancer pathogenesis, the development of inhibitors targeting *myc* or its pathway to regulate *myc* expression and activity has become a major research focus.

C-myc gene is one of the important members of *myc* gene family, which is primarily distributed in the nucleus and regulates about 15–20% of genomic genes, and regulates gene expression by promoting and repressing transcriptional functions, respectively. Thus, it affects biological functions such as cell proliferation, death, metabolism, differentiation, and cell multifunctionality [21]. *C-myc* gene can rapidly increase the energy supply and anabolism of tumor cells to meet the demand of rapid proliferation and growth of tumor cells by regulating multiple critical genes of substance and energy metabolism within cancer cells.

3. *Src* family

Src is a class of non-receptor protein tyrosine kinases activated by many extracellular signaling molecules and belongs to the proto-oncogene. *Src* has shown remarkable evolutionary conservation with high homology across species. Src proteins primarily exist in two forms: the viral oncogene expression protein v-Src and the cellular oncogene expression protein c-Src. The abnormal expression of c-Src protein contributes to the occurrence and development of tumors, while v-Src protein itself induces cellular transformation and tumorigenesis. Both proteins play crucial roles in intracellular signal transduction pathways.

The *Src* family of non-receptor tyrosine kinases is presented in almost all multicellular organisms and plays a crucial role in regulating various cellular functions including cell proliferation, differentiation, apoptosis, angiogenesis, and metastasis. Studies have shown that *Src* gene expression in breast cancer patients is closely associated with breast cancer. The expression and activity of Src protein are significantly increased in breast cancer tissues compared to normal breast tissue, which is associated with Src protein expression and regulation rather than genetic variation [22]. Activation of the proto-oncogene receptor downstream signaling pathway plays a pivotal role in the growth and progression of breast cancer, and the *Src* gene activates the proto-oncogene receptor downstream thereby activating the mitogen-activated protein kinase pathway. Clinical studies have shown a correlation between aberrant *c-Src* expression and cancer progression, with *c-Src* showing high expression in human malignancies, such as intestinal, pancreatic, prostate, colon, lung cancers, and other malignant tumors [23]. Elevated levels of c-Src protein or kinase have been reported in 50–80% of lung cancer patients. Most patients with colon cancer show a five- to eightfold increase in *Src* activity. As disease progresses, *Src* activity gradually increases, with primary colon cancer demonstrating higher *Src* activity than that observed in polyps and metastatic liver cancers. The effect of *Src* on colon cancer can be inferred by detecting *Src* levels at different stages of colon cancer development. Lutz et al. found significantly higher Src protein levels and kinase activity in pancreatic ductal carcinoma and pancreatic cell lineage carcinoma [24].

4. *Sis* family

Sis is one of the common oncogene families, with only one member of the *sis* gene. The proto-oncogene *sis* encodes the platelet-derived growth factor B chain. Studies have shown that *sis* promotes cell proliferation and inhibits apoptosis, thereby promoting tissue repair. Insertion of exogenous DNA, point mutations in the gene coding region, and certain compounds can activate the *sis* proto-oncogene, leading to cell transformation. *Sis* proto-oncogene is located on the long arm of human chromosome 22, which translocates from chromosome 22 to chromosome 9 in 90% of chronic myeloid leukemias [25]. In Ewing's sarcoma-derived cell lines and neuroepithelioma cell lines, the gene translocates from chromosome 22 to 11 [26]. However, no rearrangements or amplifications of sis proto-oncogene are observed in these translocated cell lines, nor are its transcripts detected. The *sis* proto-oncogene is not expressed in normal fibroblasts and glial cells, whereas the expression product is found in fibrosarcoma or glioblastoma, and multiple *sis* transcripts are also detected in U-2OS human osteosarcoma, suggesting that activation of the proto-oncogene is associated with tumor formation of connective tissue origin.

5. *Myb* family

The *myb* family comprises three members: *A-myb*, *B-myb*, and *C-myb*. *A-myb* is primarily expressed in highly proliferative tissues, such as B-lymphocytes and developing central nervous system. *B-myb* is not limited to specific tissue types and co-expresses with other myb proteins, while *C-myb* predominantly exists in early hematopoietic cells. Studies on hematopoiesis-related cells and various tissues revealed that the mRNA expression level of *C-myb* was highest in immature blood cells, and decreases during cell differentiation. The proto-oncogene *C-myb* transcription factor shares similarities with the extensively studied *C-myc* transcription factor in regulating cellular growth and differentiation. *C-myb*, as an intranuclear oncogene, is closely related to the development of malignant tumors. Studies indicates that artificial upregulation of *C-myb* expression effectively inhibits tumor cell differentiation and promotes cell proliferation, whereas reducing the expression levels of intracellular *C-myb* enables cancer cells to differentiate into benign cells or undergo apoptosis [27].

The current gene activation mechanisms of *C-myb* include: overexpression of *myb* genes, gene and protein structural alterations and chromosomal rearrangements. *C-myb* overexpression is associated with the development of many cancers, such as leukemia, breast cancer, colorectal cancer, etc. [28, 29]. MLL-ENL and BCR-ABL are leukemia-associated fusion genes associated with the development and metastasis of leukemia. In the pathogenesis of leukemia induced by MLL-ENL, *c-myb* serves as the target sequence. Silencing of *c-myb* gene inhibits the formation of leukemia produced by MLL-ENL [30]. Studies on breast cancer reveals that estrogen has a stimulatory effect on *c-myb* expression with significantly elevated levels observed in breast cancer cells containing estrogen receptors. High expression of *c-myb* impairs differentiation and apoptosis of breast cancer cells [31]. Colorectal cancer cells exhibit sequence mutations at the transcriptional regulatory sequence of *c-myb*, allowing cellular overexpression that disrupts differentiation. Studies show that over 80% of colorectal cancer cases demonstrate *C-myb* overexpression, which is generally consistent with the incidence of APC mutations in colorectal cancer patients [32].

8.3.3.2 Tumor Suppressor Genes

Tumor suppressor genes, also known as tumor inhibitory genes, are negative regulatory genes present in normal cells that inhibit cell proliferation, as shown in Fig. 8.4b. These genes play a crucial negative regulatory role in regulating cell growth, proliferation, and differentiation, working in concert with proto-oncogenes. When these genes become inactivated by mutations, deletions, etc., it leads to the loss of growth inhibition of cells, leading to malignant transformation and tumor formation. Some of the common suppressor genes associated with human cancers are listed in Table 8.3 [12].

1. *TP53* gene

The tumor suppressor gene *TP53*, also known as *p53* gene, is a crucial tumor suppressor gene closely related to the occurrence and development of human tumors. The *p53* gene encodes a protein with a molecular weight of 43.7 kDa, which is slowed down by electrophoresis due to the high proline content, causing the protein band to appear at 53 kDa, hence the name p53. The primary function of the *p53* gene involves regulating cell cycle arrest, apoptosis, aging, DNA repair, and metabolic changes by regulating the expression of target genes. In the cell cycle, when DNA

Table 8.3 Tumor suppressor genes

Tumor suppressor genes	Chromosome location	Function	Related tumors
P53	17p	Encoding p53 protein (transcription factor)	Leukemia, breast cancer, stomach cancer, lung cancer, osteogenic sarcoma, glioblastoma, astrocytoma, and
APC	5q21	Wnt signal transduction components	Colon cancer
DCC	18q21	Encodes surface glycoproteins (Cell adhesion factor)	Colon cancer, stomach cancer
Rb	13q14	Encodes p105Rb1 protein (Transcription factor)	Small cell lung cancer, retinoblastoma, osteosarcoma
WT1	11p13	Encoding zinc finger protein	Renal cell carcinoma
NF1	17q12.2	GTPase activator	Neurofibroma
NF2	22q	Connecting the membrane to the cytoskeleton	Nerve sheath tumor, meningioma
BRCA1	17q21	DNA repair factor	Breast cancer, Ovarian cancer
BRCA2	13q12-13	DNA repair factor	Breast cancer, pancreatic cancer

damage or defects occur in cells, p53 protein halts cell division in G1 and G2 phases and initiates corresponding repair mechanisms to repair DNA defects. If repair fails, p53 protein initiates apoptosis mechanisms to eliminate damaged cells, thus regulating cell metabolism, controlling normal gene expression and preventing cell carcinogenesis. Deficiencies in *p53* function caused by gene deletion or mutation lead to DNA expression damage and uncontrolled cell proliferation, ultimately inducing cancer development.

In recent years, the *p53* gene has been the focus in scientific research. When *p53* gene undergoes mutations, its spatial conformation changes lead to impaired regulation of cell growth, apoptosis, and DNA repair, making *p53* gene changes from a tumor suppressor gene to oncogene [33]. *p53* gene mutation can be classified as either destructive mutation or nondestructive mutation. Destructive mutations completely disable the p53 protein's function, while nondestructive mutations allow partial normal activity. Approximately 70–80% of *p53* mutations are missense mutations that occur in the DNA-binding region, as shown in Table 8.4, and over 90% of *p53* point mutations occur at the highly conserved 175, 248, 273, and 282 loci [34]. Studies indicate that over 50% of human tumors harbor p53 mutations, including those in lung, liver, breast, bladder, stomach, colon, prostate, soft tissue sarcoma, ovarian, brain, lymphoma, esophageal, and osteogenic sarcomas [35]. *p53* gene mutations are detected in 35–60% of patients with non-small cell lung cancer, and particularly in patients with squamous lung cancer and a smoking history. In comparison with nonsmoking patients, the transversion of G-T caused by tobacco

Table 8.4 High frequency mutation sites of *p53* gene

Ciphers	Amino acids	Mutation	Protein structure effects
175	Arginine	Histidine	Break the hydrogen bond between L2 and L3 rings
248	Arginine	Glutamine	Disruption of DNA junctions in the minor groove region
273	Arginine	Histidine	Disrupting the connection of the DNA major grove region
248	Arginine	Tryptophan	Disruption of DNA junctions in the minor groove region
273	Arginine	Cysteine	Disrupting the connection of the DNA major grove region
282	Arginine	Tryptophan	Making the binding of H2 helix and DNA sulci region unstable and breaking the connection of β hairpin structure

carcinogens was found in smoking lung cancer patients [36]. Studies indicate that advanced non-small cell lung cancer patients carrying *EGFR/TP53* co-mutations are less effective to TKIs-targeted therapy compared to those with wild-type *TP53*, suggesting that *TP53* may mediate EGFR-TKIs resistance [37]. The majority of *p53* gene mutations in colon cancer happens at the CpG dinucleotide sites with over 50% of conversion mutations occurring at the CpG regions in structural domains 3–5 (located at codons 175, 248, and 273). In breast cancer, only 13% of conversions are found at the CpG site. In addition, G-T transversion accounts for 1/4 of the breast cancer cases, but is extremely rare in colon cancer. Lymphoma and leukemia exhibit similar p53 mutation patterns to colorectal cancer: most mutations are transversions at the CpG site with lower G → T transversion and higher A:T → G:C mutation at the A:T site. The *p53* mutation profile of Burkitt's lymphoma is similar to that of other B-cell lymphomas and T-lymphocyte malignancies, but Burkitt's lymphoma has a higher transversion mutation rate. In esophageal cancer there is a high transversion rate, with comparable mutation frequencies at the G:C and A:T loci.

2. *Rb* gene

The tumor suppressor gene *Rb* was initially discovered in retinoblastoma in children, hence the name *Rb* gene. The anticancer properties of *Rb* gene have two aspects: firstly, Rb protein inhibits cell growth in normal cells. Secondly, *Rb* gene suppresses both tumor cell growth and tumorigenicity in tumor cells. The introduction of normal *Rb* genes into cancer cell lines such as retinoblastoma, bladder cancer, and lung cancer by retroviral infection revealed that *Rb* gene could inhibit the oncogenicity of these cancer cell lines in nude mice to varying degrees, providing direct evidence for the anticancer properties of *Rb* gene.

Rb gene, a tumor suppressor gene, causes excessive cell proliferation and tumor formation when mutated or deleted. Studies show that while the Rb gene maintains normal structure and expression in healthy human tissues, it becomes absent or mutated in cancerous tissues, resulting in defective Rb protein function. In nearly all human tumors, the *Rb* gene either undergoes direct mutations, deletions, or hypermethylation, resulting in their functional inactivation, or abnormal expression or altered activity of Rb proteins by altering their upstream transcriptional regulators. Inactivation of *Rb* genes in oocytes disrupts follicular growth regulation, driving the formation of ovarian teratomas [38]. The E7 proto-oncoprotein expressed by high-risk HPV16 regulates the interaction between the transcriptional activator CBP and the Rb protein, leading to abnormal cell cycle control and carcinogenesis. Mutation and inactivation of the *Rb* gene also play critical roles in the progression of osteosarcoma, with approximately 70% of adult osteosarcoma samples showing altered *Rb* gene expression levels [39].

3. *BRCA* gene

The *BRCA1/2* genes are currently the most well-established susceptibility gene for breast and ovarian cancers, with genetic mutations significantly increasing breast cancer risk. First identified in 1994 by Miki et al., susceptibility gene 1 (*BRCA1*) was cloned and isolated as the first gene associated with familial breast cancer and ovarian cancer [40]. Subsequently, Wooste et al. identified a second gene associated with familial breast and ovarian cancer, breast cancer susceptibility gene 2 (*BRCA2*) [41]. Both *BRCA1* and *BRCA2* genes are tumor suppressor genes that encode proteins involved in double-stranded DNA damage repair through a homologous recombination pathway [42]. Functionally active BRCA proteins regulate cell growth and prevent the division of abnormal cells that may lead to tumorigenesis.

Mutations in *BRCA1* and *BRCA2* genes primarily result in protein truncation, impairing of homologous recombination function and reducing tissue repair capacity. These genetic alterations lead to abnormal gene expression, thus inducing cancer [43]. *BRCA1* and *BRCA2* gene mutation carriers increase the risk of breast and ovarian cancer. The risk of breast cancer in *BRCA1* mutation carriers is 57–60% by age 70, while *BRCA2* mutation carriers show a 49–55% risk (Chinese Expert Consensus on BRCA1/2 Testing and Clinical Application in Breast Cancer). In addition, carrying *BRCA1* and *BRCA2* gene mutations significantly increases the risk of developing not only breast and ovarian cancer in women, but also ovarian cancer, contralateral breast cancer, prostate cancer, pancreatic cancer, and male breast cancer. Studies indicate that the mutation rate of *BRCA* is 16.9% in familial breast cancer patients, 5.2% in early-onset breast cancer patients, and 2.0% in sporadic breast cancer patients [44]. In addition, a correlation analysis between clinical case characteristics of breast cancer and *BRCA* mutations revealed that patients with *BRCA1* and *BRCA2* mutations had a significantly younger age of onset than non-mutated individuals, with higher incidence of bilateral breast cancer and more advanced histological grade. Li et al. conducted a large-scale *BRCA* gene testing study involving 1331 Chinese ovarian cancer patients and 1763 healthy controls [45]. The results showed that the *BRCA* gene mutation rate was 0.45% in healthy individuals and about 21.64% in ovarian cancer patients. Therefore, genetic testing for *BRCA1* and *BRCA2* genes enables early detection of these mutations, which helps assess the risk of developing *BRCA*-related cancers and facilitates the development of targeted preventive measures.

4. *APC* gene

The adenomatous polyposis coli (*APC*) gene, identified by Herrear in 1986 in patients with Gardner's syndrome (GS), is a novel oncogene located on chromosome 5q21-22. The *APC* gene is a large housekeeping gene [46], and its mRNA transcripts are expressed in many cells and tissues. APC proteins are an important component of the Wnt signaling pathway. Studies demonstrate that all APC proteins function within this network to regulate intercellular signaling, influence cell adhesion, cause abnormal cellular activity, and participate in a range of cellular processes, including cell cycle regulation, apoptosis, and cell adhesion [47].

Through in-depth research on *APC* gene mutation and protein in tumor tissues, it has been found that *APC* gene exhibits a certain mutation rate in malignant tumors such as colorectal cancer, gastric cancer, breast cancer, esophageal cancer, and other malignant tumors. The occurrence of *APC* gene mutation in colorectal cancer precedes mutations in *K-ras* and *p53* genes. Therefore, *APC* gene is called the gatekeeper gene of colorectal carcinogenesis [48]. Wang et al. performed *APC* gene testing on 46 patients with esophageal cancer, identifying 29% with *APC* heterozygous loss [49]. Lu et al. compared the expression of *APC* gene in gastric cancer, precancerous lesions, and gastric adenomas, revealing differences among the three types [50]. It was hypothesized that the expression of *APC* gene was related to the degree of gastric cancer development. Studies on non-small cell lung cancer showed that APC protein expression in non-small cell lung cancer tissues differed from normal lung tissues. Although *APC* gene mutations are unrelated to tumor type, pathological grade, or degree of differentiation, APC protein expression may indeed be a significant contributing factor of non-small cell lung carcinogenesis [51]. Saito et al. performed *APC* gene testing on synovial sarcoma and found that the *APC* gene mutation rate was 91.8%, indicating a correlation between *APC* gene mutations and synovial sarcoma occurrence [52].

8.4 Detection of Unknown SNPs

With the rapid advancement of molecular biology, the existence of whole genome sequence information in various bioinformatics databases has provided great convenience for the detection of gene mutations. However, identifying specific mutations in normal tissues remains a daunting challenge akin to searching for a needle in a haystack. Current detection methods primarily fall into two categories: identifying unknown mutations and detecting known ones. Through diverse sequencing technologies and amplification techniques, various approaches have emerged with distinct advantages in sensitivity, specificity, throughput efficiency, and cost-effectiveness.

8.4.1 Single-Strand Conformation Polymorphism Analysis

Single-strand conformation polymorphism (SSCP) can be used to rapidly screen for unknown mutations by combining PCR technology with single-stranded DNA (ssDNA) or RNA conformation polymorphism, which is widely used in the

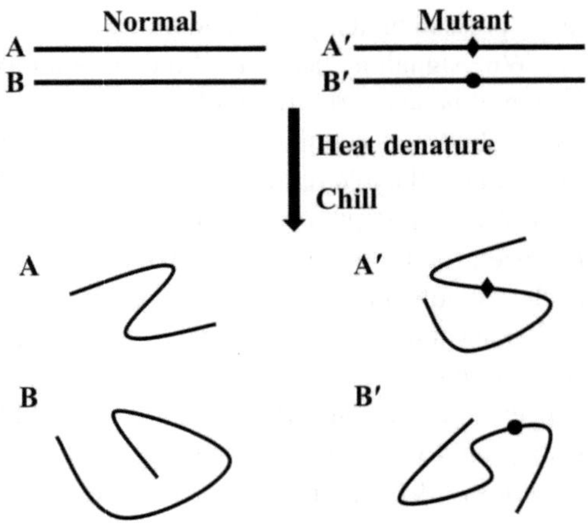

Fig. 8.5 Schematic representation of SNP detection using SSCP. (Revised and reprinted with permission from Ref. [53])

detection of SNPs and diagnosis of diseases [53]. The principle of SSCP is shown in Fig. 8.5. Under non-denaturing conditions, a ssDNA or RNA can self-fold into specific spatial conformations determined by their base sequences, stabilized by intramolecular interactions (primarily hydrogen bonds). Even single-base variations within identical-length strands can result in distinct conformations. Given the high resolution of polyacrylamide gel electrophoresis (PAGE), denature-free neutral PAGE is used for differentiation during SNP detection, where ssDNA migration rate primarily depends on its conformation stability. When PCR products undergo denaturation, changes caused by single-base substitutions, insertions, or deletions lead to altered single-strand conformations and the migration rate in electrophoresis changes, thus distinguishing mutant DNA from normal DNA. SSCP technique for gene mutation detection is usually combined with PCR technique, so it is called PCR-SSCP technique. This technique is extensively applied in the detection of oncogenes and tumor suppressor gene mutations, analysis of pathogenic genes for genetic diseases, and genetic diagnosis.

Jacquemier et al. detected *p53* gene mutation using PCR-SSCP technique [54]. First, PCR was used to amplify the exon 4–9 region of the *p53* gene. After SSCP analysis, the top band in the gel electrophoresis image represented the undenatured double-stranded DNA (dsDNA), with normal ss DNA migrating slower than mutated DNA. By mixing different proportions of wild-type and mutant *p53*, both DNA fragments could be detected simultaneously. Lin et al. established a noninvasive colorectal cancer screening method using PCR-SSCP technique to achieve early cancer diagnosis by detecting *APC* and *MCC* gene mutations in stool samples [55]. This approach has played a crucial role in improving the cure rate of colon and rectal cancer, extending the survival life and enhancing the quality of patients' survival.

8.4.2 Heteroduplex Analysis

There is only one or a few bases differences between wild-type and mutant DNA. Therefore, when wild-type and mutant DNA complement each other, they form heterozygous dsDNA with mismatched bases, creating a localized loop of unmatched region. When analyzed using non-denaturing gel electrophoresis, the perfectly matched dsDNA and heterozygous dsDNA produce bands with different mobility, thus enabling differentiation of mutant genes. This technique, known as heteroduplex analysis (HA), is illustrated in Fig. 8.6. The resulting heteroduplexes can be classified into two types: [56] bubble-type heteroduplexes induced by one or a few base mutations, and bulge-type heteroduplexes caused by gene insertion or deletion. In PAGE electrophoresis, bulge-type heteroduplexes show higher resolution than bubble-type heteroduplexes [57]. Scientists studied the factors affecting the heteroduplex analysis of SNPs. The results showed that target length, GC content, and mismatch position did not affect the detection, but the factors affecting the separation quality were the hybridization efficiency of the mismatching bases, in which G:G/C:C > A:C/T:G = A:G/T:C > A:A/T:T [53]. This method is similar to SSCP and serves as an effective tool for gene mutations screening. Unlike SSCP, which isolated and detected ssDNA, HA targeted dsDNA and was only suitable for the analysis of small fragments, which could be applied to those mutations that cannot be detected by SSCP.

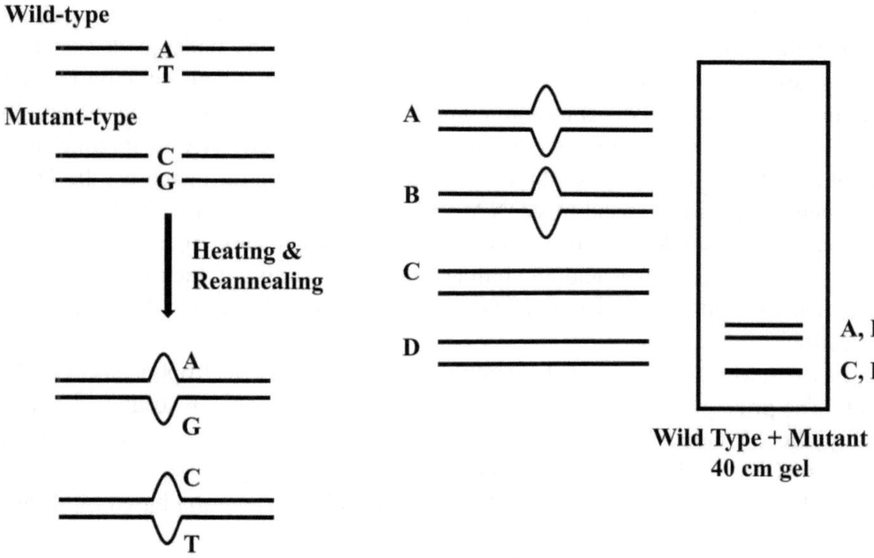

Fig. 8.6 Schematic representation of SNP detection by heteroduplex analysis. (Revised and reprinted with permission from Ref. [53])

Pignatti et al. used both SSCP and heteroduplex analysis to detect four genes with known mutations [58]. Although neither method achieved 100% accuracy, the heteroduplex analysis detected 14/15 mutations compared to 10 out of 15 with SSCP, indicating the superior sensitivity of heteroduplex analysis. Coupling the heteroduplex analysis with SSCP can increase the mutation detection rate to nearly 100%.

8.4.3 Denaturing Gradient Gel Electrophoresis Analysis

Denaturing gradient gel electrophoresis (DGGE) was originally invented by Fischer and Lerman et al. in 1983 by adding denatured gradient gel agents (urea and formamide) during electrophoresis. DGGE is mainly used to detect point mutations in DNA fragments [59].

The size and charge of dsDNA molecules determine their migration behavior in PAGE. DNA fragments of different lengths can be distinguished, those of identical length exhibit indistinguishable migration patterns within the gel. The DGGE technique enhances conventional PAGE by incorporating a denaturant gradient, enabling differentiation between DNA fragments of identical length but different sequences. Each DNA fragment's unique sequence determines its unwind behavior. DNA fragments of several hundred base pairs typically contain multiple unwind regions, each consisting of one consecutive sequence of base pair. As the concentration of denaturant gradually increases, dissociation begins at the lowest concentration threshold for the initial region. Subsequent regions undergo dissociation progressively as denaturant levels rise. When the denaturant concentration reaches the highest unwound concentration, complete double-strand unwinding occurs. In DGGE, initial low denaturant concentrations allow intact DNA to migrate as it would in standard PAGE. As the concentration of denaturant increases, the DNA fragments gradually unwind to ssDNA, which continue to migrate through electrophoresis. Any single-base mutation within these fragments triggers synergistic effects, altering T_m values and unwinding behavior, resulting in the different migration rate of wild-type and mutant DNA fragments in DGGE for the purpose of detection.

In 1993, Muyzer et al. invented temperature gradient gel electrophoresis (TGGE) [60]. The principle is similar to that of DGGE except that DGGE forms a denaturant gradient, whereas TGGE forms a temperature gradient on the gel. The T_m values between DNA and mutated DNA are different, and a single-base difference can cause distinct unwinding behavior of dsDNA, enabling separation based on differential migration rates during electrophoresis.

Lu et al. examined the mitochondrial DNA D-loop region in leukemia patients before treatment and during remission using DNA sequence analysis and PCR-DGGE method [61], respectively, and compared the ability of these two methods to detect mutations in the mitochondrial DNA D-loop region of leukocytes before treatment. The results showed that the PCR-DGGE method had 100%

specificity and 97.1% sensitivity for detecting base mutations, with the PCR-DGGE method capable of identifying single-base variations. Mitani et al. first amplified two *FUT2* gene fragments using PCR with lengths of 259 and 426 nt [62], and subsequently analyzed the genotype of *FUT2* by TGGE. The results revealed that TGGE could be used not only to detect SNPs but also to determine whether genetic variants exist in homozygous form or heterozygous form.

8.4.4 Enzymatic Mutation Detection

Both T4 endonuclease VII and T7 endonuclease I originate from bacteriophages, and typically exist as homodimers, which can recognize abnormal DNA structures, such as base mismatches, base deletions and insertions, DNA loops, and multiple DNA strand connections. Then the enzymes cleave DNA from the 3′ end within 6 bases of the abnormal base to form two short strands of DNA [63, 64]. T4 endonuclease VII and T7 endonuclease I have excellent recognition and cleavage capabilities for all eight possible mismatches (C/C, A/A, G/G, T/T, G/T, G/A, T/C, A/C) [65]. Therefore, when the enzymatic mutation detection (EMD) method is applied to SNP detection, the target sequence is first amplified by PCR, and then T4 endonuclease VII and T7 endonuclease I are used to identify and cleave the sequence near the mismatched bases. The enzymatic cut site can be determined according to the length of the cut DNA strand, which helps deduce the mutation location.

Benjamin et al. used EMD to detect *p53* gene mutation, followed by capillary electrophoresis for detection of digested DNA [66]. The experimental principle is shown in Fig. 8.7, and the whole process was divided into four steps. In the first step, identical primers were used to amplify both wild-type and mutant targets simultaneously. In the second step, the target was denatured at high temperature and then renatured, and the presence of a single mutated base in the target DNA causes mismatched hybridization during reannealing. In the third step, T4 endonuclease VII was introduced to locate mismatched bases along the dsDNA, cleaving the DNA from the 3′ end within 6 bases of the mismatched bases to form two short fragments. The fourth step was separation and detection using electrophoresis analysis. The wild-type DNA of *β-globin* gene was 560 bp in length. The DNA was cut into 101 and 459 bp fragments in the first sample, while 60 and 500 bp fragments were obtained in the second sample. The length of the wild-type DNA of *p53* gene was 480 bp, and the first sample DNA was cut into 236 and 244 bp fragments, while the second sample DNA was cut into 200 and 280 bp fragments. From these results, distinct mutation sites of the *β-globin* gene and the *p53* gene could be inferred. Applying this method to clinical DNA mutation detection, 100% sensitivity and 94% specificity were achieved, enabling specific identification of mutant DNA from over 20-fold excess wild-type DNA. Further studies revealed that T4 endonuclease VII can recognize substrates exceeding 1000 bp and cleave at multiple sites [67].

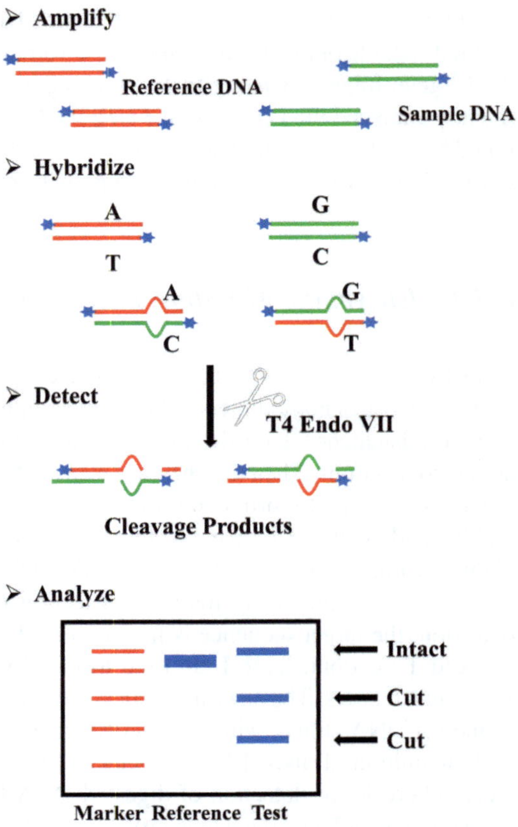

Fig. 8.7 Schematic representation of SNP detection by enzymatic mutation detection. (Revised and reprinted with permission from Ref. [66])

8.4.5 *High-Throughput Sequencing Method*

Although the above detection methods can detect unknown SNP, they generally suffer from low detection sensitivity and limited throughput. To uncover more novel and significant mutations, sequencing methods have gained widespread attention. The completion of the first draft of the human genome in 2000 marked the dawn of high-throughput sequencing. As a new generation sequencing technology, next-generation sequencing (NGS) can detect multiple mutation types of multiple genes simultaneously and can be quantitatively analyzed.

8.5 Detection of Known SNPs

8.5.1 *Allele-Specific Oligonucleotide Hybridization*

Allele-specific oligonucleotide hybridization (ASOH) is a hybridization-based technique for detecting known SNPs. In this method, a 20 bp-long oligonucleotide fragment was designed according to the target sequence, namely, allele-specific oligonucleotide (ASO). To ensure specificity, the mutation site must be positioned in the middle of ASO. After PCR amplification, the PCR products are immobilized onto the substrate. ASO serves as the probe to hybridize with the amplified product for detection, and is designed separately according to different mutation types. A positive hybridization band indicates the presence of corresponding mutation in the sample. The detection of single-base mutations can be achieved by strictly controlling the hybridization conditions.

Howell et al. improved the ASOH method by controlling the temperature of the hybridization system and detecting the temperature change simultaneously to achieve SNP detection [68]. The principle is shown in Fig. 8.8a. Firstly, primers modified with biotin at the 5′ end were used for amplification during PCR. Therefore, the 5′ end of the PCR amplification products were labeled with biotin, which could bind specifically to avidin-labeled microplate. Unlabeled DNA strands were removed during washing. ASO designed for wild/mutant bases was added separately. ASO bound to the corresponding target to form a double-stranded structure. After combining with the embedded fluorescent dye, the produced fluorescence was proportional to the amount of dsDNA. The fluorescence signal-temperature curve was obtained by gradually increasing the temperature and monitoring the fluorescence change of the system, as shown in Fig. 8.8b. After the derivation, the T_m value

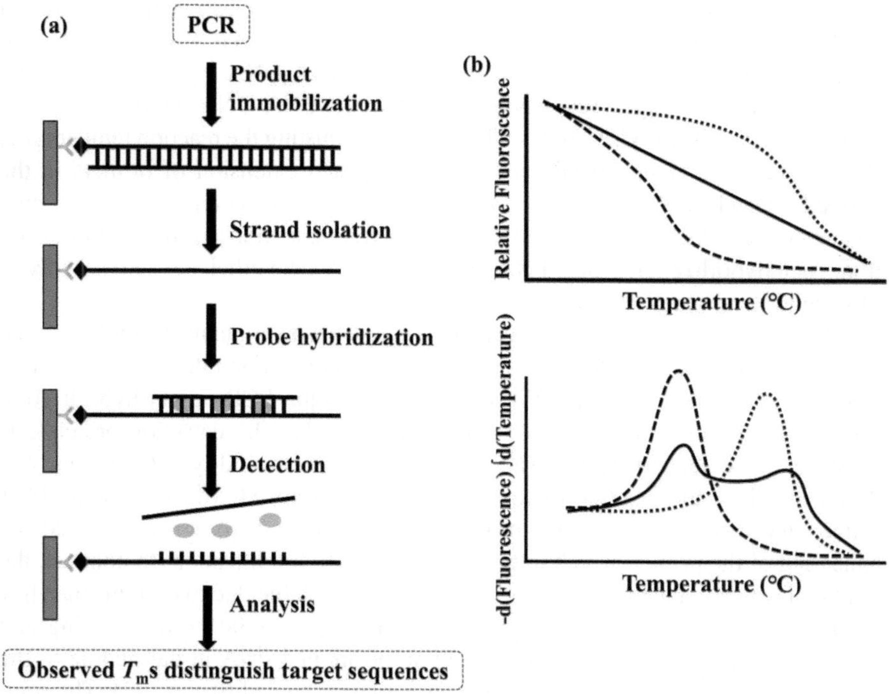

Fig. 8.8 (**a**) Schematic representation of SNP detection by ASOH method, (**b**) fluorescence-temperature curve of the system, (**c**) fluorescence-temperature curve obtained by deriving the curves in (**b**). (Revised and reprinted with permission from Ref. [68])

of dsDNA in the system was obtained (Fig. 8.8c). When the system was all wild-type targets, the T_m value of dsDNA was about 72 °C. When the system was all mutant targets, the T_m value of dsDNA was about 64 °C. For hybrid targets, the T_m values of both DNA appeared simultaneously, indicating heterozygotes. By ASOH analysis, not only SNPs could be detected, but also the genotyping of the target could be analyzed.

8.5.2 Clamp-PCR-Based Assays

In 1991, Danish scientist Nielsen developed a DNA analogue called peptide nucleic acid (PNA) by replacing the sugar phosphate backbone of DNA with a polypeptide skeleton [69]. As an artificially synthesized nucleic acid analogue, PNA is uncharged and interacts with DNA through base complementarity without electrostatic repulsion, ensuring high stability and specificity when binding to DNA. In addition, PNA resists degradation by DNA enzymes and proteases, maintaining exceptional stability in both in vivo and in vitro environments. Given these advantages, PNA has been successfully employed as a clamp for PCR-based gene mutation detection. When used as a clamp during PCR, PNA exhibits two critical characteristics: (1) PNA cannot serve as PCR primer, and remains resistant to exonuclease activity from Taq DNA polymerase; (2) The T_m values of perfectly matched PNA-DNA hybrid duplexes are significantly higher than those of DNA-DNA hybrid strands with identical length. Even a single-base mismatch in the PNA-DNA hybrid can reduce its T_m value by 10–18 °C. Therefore, by precisely adjusting the reaction temperature, PNA serves as a clamp to inhibit the annealing and extension of primers in the perfectly matched PNA-DNA strand, enabling SNP detection. Due to its own characteristics, PNA can regulate DNA replication, gene transcription, and translation. As a hybridization probe, PNA greatly improves the efficiency and sensitivity of genetic testing and medical diagnosis.

Clamp-PCR method for SNP detection uses PNA as a clamp to inhibit signal generation by wild-type DNA, which leads to enrichment and detection of mutant DNA [70]. Thiede et al. used PNA as a clamp to inhibit PCR primer hybridization for the detection of mouse proto-oncogene *K-ras* [71]. The detection principle is shown in Fig. 8.9a, where a 15 bp PNA probe targeting the wild-type *K-ras* gene was designed. Since the position of the PNA probe overlapped with that of the PCR primer, the binding of the PNA probe to the wild-type *K-ras* gene inhibited the annealing of the primer, which in turn inhibited PCR amplification. In contrast, the binding between mutant *K-ras* gene and PNA was unstable due to base mismatch at the middle position of the PNA probe, allowing preferential primer binding and specific detection of mutations. Taback et al. positioned PNA in the middle of the PCR primer to detect *K-ras* mutation in formalin-embedded tissues [72]. The principle was shown in Fig. 8.9b, where the PNA probe specifically hybridized to wild-type *K-ras*. When the PCR primers extended to the PNA binding site, the stronger PNA-DNA binding toward DNA-DNA prevented primer extension and

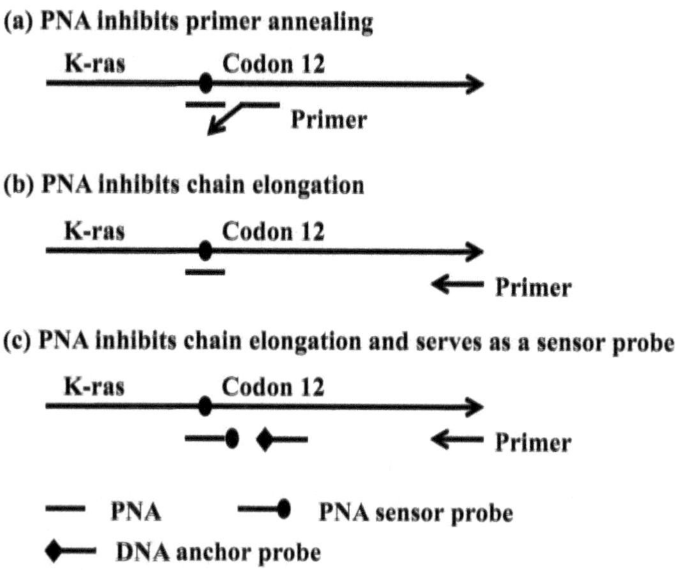

Fig. 8.9 Schematic representation of the Clamp-PCR method for SNP detection. (Revised and reprinted with permission from Ref. [70])

amplification of wild-type *K-ras*, making it unable to generate amplification signal. In contrast, the PNA probe did not exactly match the mutant *K-ras*, and the binding force was weakened so that it could not block the primer amplification, thus achieving specific detection of mutant *K-ras*. It was shown that when the PCR primers were located on both sides of the PNA probe, the distance between the reverse primer and the PNA probe was greater than 50 bp.

To enhance the detection specificity, the PNA probe was further used as a signal detection probe in addition to serving as PCR clamp. SNP detection was achieved by fluorescence resonance energy transfer (FRET) between PNA probe and adjacent response probe [73]. A fluorescent group was modified at one end of the PNA (PNA sensor probe), and a response probe carrying another fluorescent group (DNA anchor probe) was introduced to the system. When the two probes were close to each other, the fluorescent group on the PNA sensor probe was excited and transferred energy to the fluorophore modified on the anchor probe. When the two probes were far apart, the fluorescence signal remained unexcited, resulting in no fluorescence signal (Fig. 8.9c). The target could be detected by the FRET signal between the two probes. For wild-type targets, both the PNA sensor probe and the anchor probe could bind, and the fluorescence emitted by the anchor probe could be detected. Whereas for mutant targets, the PNA sensor probe was unstable due to base mismatches between PNA and target, and the anchor probe could not be excited. By adjusting the system's temperature, the PNA sensor probe was detached, and thus the anchor probe becomes unexcited and undetectable. In addition, the fluorescence-temperature curve could be obtained by programmed heating and fluorescence monitoring, which enabled the detection of mutant targets. At lower initial temperatures, both the PNA sensor probe and the anchor probe bind to the target, enabling anchor probe fluorescence detection. As the temperature increased, the PNA sensor probe preferentially detached from the mutant target. When the temperature raised to the T_m at which PNA bound DNA, the PNA sensor probe detached from the wild-type target, resulting in reduced fluorescence signal. The difference in response to temperature change between wild-type and mutant targets was used to differentiate the detection. Chen et al. used this method to simultaneously detect multiple *BRAF V600* mutations and applied it to melanoma patient samples [74]. This method enabled the detection of SNPs by identifying the peak of the melting curve without additional procedures, demonstrating significantly improved detection sensitivity and specificity compared with conventional PCR.

8.5.3 Nucleic Acid Invasion Reaction-Based Assays

The nucleic acid invasion reaction is catalyzed by a specialized endonuclease that specifically recognizes the unique structure formed by the 3′ end of the nucleic acid invading the downstream double strand and cleaves the 5′ end flap of the invaded downstream nucleic acid. This reaction plays a crucial role in DNA replication, Okazaki fragments excision, and DNA repair. The principle of the nucleic acid invasion reaction is shown in Fig. 8.10. Two probes are designed according to the target: an upstream invader probe (UP) and a downstream flap probe (DP). The two probes hybridize to the target after recognition to form a three-base overlap structure. The 5′ end of the DP is not complementary to the target, creating a flapped structure after hybridization. A single base at the 3′ end of the UP invades into the double-strand formed by DP and the target. Endonuclease recognizes this structure and specifically cleaves the flap part of DP. The cleaved DP hybridizes with the target and is replaced by a new DP probe for a new round of recognition cleavage reaction, resulting in a system with multiple flapped fragments. The endonucleases used in the reaction are members of the flap endonuclease 1 (FEN1) family isolated from archaea, including AfuFEN 1, MjaFEN1, PfuFEN 1, MthFEN 1. These enzymes demonstrate excellent structural recognition specificity, specifically targeting and cleaving at the invasion site where the 3′ end of upstream oligonucleotide fragments invade downstream regions by one base [75]. Leveraging the high-precision structural recognition of nucleic acid invasion reactions and probe hybridization specificity, this method enables highly sensitive detection of SNPs.

The advantage of the nucleic acid invasion reaction is that it converts diverse target sequences into identical flap fragments, which enables versatile integration with various nucleic acid amplification techniques, simplifying experimental design while maintaining detection specificity. To improve SNP detection sensitivity, Hessner et al. developed a cascade nucleic acid invasion technique to detect the G1691A mutation in factor V associated with deep venous embolism [76]. The principle is shown in Fig. 8.11, where flap probes and universal invader probes were

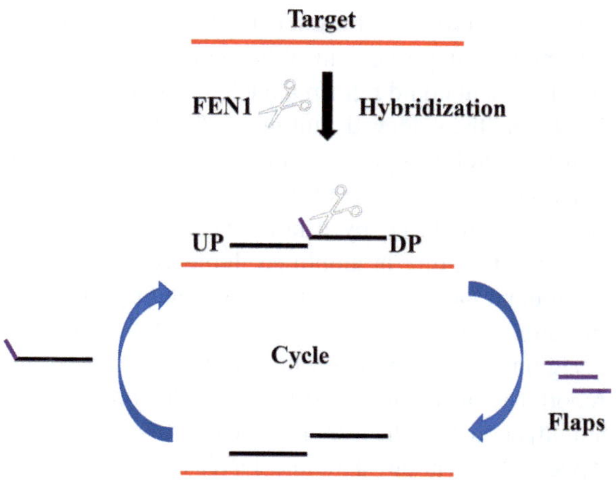

Fig. 8.10 Schematic representation of nucleic acid invasion reaction

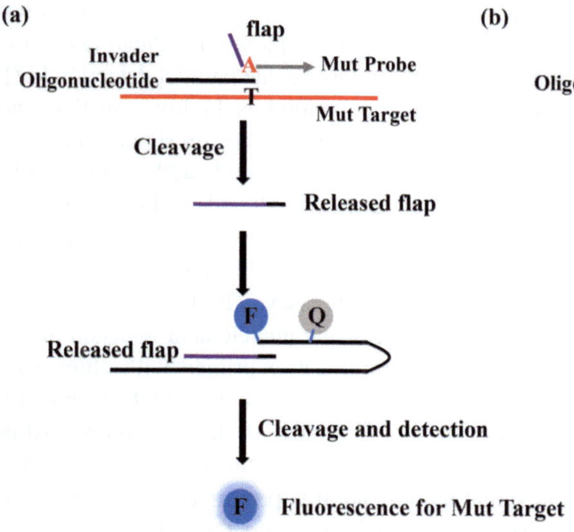

Fig. 8.11 Schematic representation of SNP detection by cascade nucleic acid invasion reaction. (Revised and reprinted with permission from Ref. [76])

designed for wild-type and mutant targets, respectively, and the mutation site was designed at the invaded base in the nucleic acid invasion structure. When both the mutant target and the mutant flap probe coexisted, a single base at the 3′ end of the invader probe invaded into the double-stranded structure formed by the mutant flap probe and the mutant target. FEN1 recognized this structure and cleaved the flap fragment from the probe along with the invaded base. Subsequently, the cleaved invader probe was replaced by a new one through a cyclic cleavage reaction that eventually yielded abundant flap fragments. The system also contained another hairpin fluorescent reporter probe with the enzymatic cleavage site labeled with a fluorescent group and a quenching group at both sides. The 3′ end of this probe matched the flap fragment. When combined with the fluorescent reporter probe, the flap fragment could reform a nucleic acid invasion structure that was recognized by FEN1 and then cleaved, cutting off the fluorescent group at the 5′ end of the hairpin structure probe to generate fluorescence signals. The temperature of the cleavage reaction used was close to the T_m value of the flap fragment, so the cleaved hairpin probe was replaced by a new one that could perform a new round of cleavage reaction, resulting in amplified fluorescence signals. The wild-type flap probe and the mutant target failed to be recognized by FEN1 due to the base mismatch at the invasion site, preventing flap fragment formation and subsequent reactions. By designing hairpin probes with different flap fragments corresponding to fluorescent reporter groups emitting distinct wavelengths, multiple SNP sites could be detected simultaneously. The cascade nucleic acid invasion method was able to directly detect SNP without genomic DNA amplification, yielding results comparable to allele-specific PCR. Olivier et al. applied the cascade nucleic acid invasion reaction for high-throughput SNP detection, which achieved 99.7% accuracy across 1500 samples with 20 SNP loci through automated reagent preparation, indicating that the nucleic acid invasion reaction was well suited for large-scale parallel detection [77].

Although the cascade nucleic acid invasion method is highly sensitive and enables SNP detection without template amplification, it relies on signal amplification with a signal amplification of only 10^7, requiring to couple additional amplification methods to detect low-concentration templates. Invader Plus technology integrates PCR with nucleic acid invasion assay; the highly sensitive and specific detection of SNP can be achieved by using the efficient amplification of PCR and the specific recognition ability of nucleic acid invasion assay. The Invader Plus assay first amplifies the target by PCR, followed by cascade nucleic acid invasion detection. To simplify the experimental procedure, amplification primers with a higher T_m value than the nucleic acid invasion flap probe are designed. PCR is initially performed at a temperature higher than the T_m value of the flap probe. After amplification, the reaction temperature is lowered to near the T_m value of the flap probe, and nucleic acid invasion signal amplification is performed to achieve SNP detection by monitoring the fluorescence signal of the system. Or in the cycle of PCR amplification, the reaction temperature is lowered to near the T_m value of the flap probe, and nucleic acid invasion signal amplification is performed while amplifying. Real-time quantitative detection of amplification products can be achieved by monitoring fluorescence signals at each PCR cycle. Kani et al. analyzed the SNPs of *IL28B* gene (rs8099917 and rs12979860) in chronic hepatitis C patients using Invader Plus technology and could detect genomic DNA as low as 1 ng [78]. By comparing detection results with sequencing methods, this method achieved 99% (rs8099917) and 100% (rs12979860) accuracy, respectively, and achieved the genotyping of *IL28B*. Allawi et al. improved the Invader Plus technique by integrating PCR upstream primers with invader probes [79], simplifying the experimental design to just two primers and one flap probe. The PCR amplification was first performed at a higher temperature, then lowers the reaction temperature to initiate a cascade of nucleic acid invasion reaction using the upstream primers and flap probe, releasing the fluorescence signal. This method detects herpes virus type I and II in cerebrospinal fluid with a minimum detectable target molecule of ten copies. Compared with the PCR colorimetric methods, it demonstrates 100% sensitivity and 98.6% specificity, respectively. The combination of cascade nucleic acid invasion reaction with PCR enables high sensitivity and specificity for SNP detection.

The nucleic acid invasion reaction imposes no sequence requirements on the target. The upstream invader probe is designed according to the target sequence, which can hybridize with the downstream flap probe to complete invasion. Then the FEN1 enzyme can cut the flap probe, and the sequence of the cleaved flap fragment is independent of the target sequence. Therefore, it can be applied to endonuclease signal amplification reaction to enhance signal amplification efficiency, while addressing the limitation of requiring templates with endonuclease recognition sites for such amplification. By combining the nucleic acid invasion reaction, the endonuclease-based nucleic acid signal amplification reaction and the aggregation-induced color change of gold nanoparticles (AuNPs), Zhou's group used the nucleic acid invasion reaction to convert the target into a flap fragment, and through the linker reaction to convert the flap fragment into a substrate for cleavage endonuclease, continuously cleaving the linker molecule to achieve the cascade signal amplification [80]. Finally, the visual detection of SNP was realized by the aggregation-induced colorimetric reaction of AuNPs. As shown in Fig. 8.12, the Dp and Up were designed for the mutant target sequence. The 3′ end of Up invaded one base from Dp to form an invasion structure. FEN recognized this structure and cleaved to produce the flap fragment. The intact Dp replaced the cleaved Dp in a new round of cleavage reaction, producing a large number of flap fragments. A linker molecule that could hybridize with two probes modified on AuNPs and a P-oligo with 5′ end modified phosphate group were added to the system. The flap fragment hybridized with the P-oligo molecule to form a double-stranded structure with a nick on the linker molecule, which was ligated to a dsDNA under the catalysis of ligase. The linker contained an endonuclease recognition site, so the endonuclease recognized the dsDNA and cleaved the linker molecule. Subsequently, the new linker bound to the ligated flap, initiating a new round of cleavage reaction. As the linker molecules are continuously cut off, the two AuNPs failed to aggregate through linker-mediated aggregation and remained red. When the target was wild-type, Up and Dp could not form an invasive structure and could not be recognized for cleavage by FEN due to the single-base difference from the wild-type target. Therefore, no flap fragment was generated and could not trigger subsequent ligation and cleavage reactions. The linker molecule in the system remained intact and hybridized with the probe labeled on the AuNPs, resulting in the aggregation of AuNPs and purple discoloration. The presence of the mutant target could be determined by visual observation of the color of the system. This method was applied to detect the *EGFR* mutation site 2573T > G. By detecting the characteristic absorption peak of the AuNPs, i.e., the absorbance at 530 nm, it can detect mutant targets down to 1% and achieve the detection of mutations in real samples. However, for the detection of individual SNPs, this method required four reaction steps: nucleic acid invasion reaction, ligation reaction, endonuclease cleavage reaction,

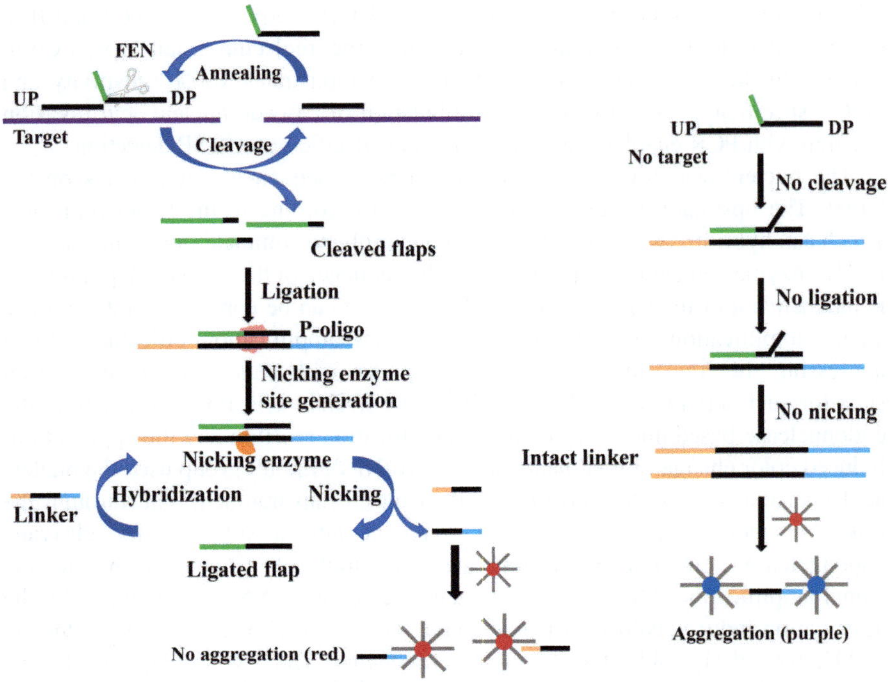

Fig. 8.12 Schematic representation of SNP detection by nucleic acid invasion reaction in combination with endonuclease and AuNPs-induced colorimetric reaction. (Revised and reprinted with permission from Ref. [80])

and AuNPs aggregation, which was a complicated experimental operation and not conducive to wide application.

8.5.4 CRISPR-Based Assays

CRISPR is a powerful gene editing technology, and the principle has been described in detail in Chapter 7. Given its programmability, ease of operation, simplicity of operation, and strong resistance to interference, CRISPR technology has gained widespread application in fields such as in vitro nucleic acid molecules and disease diagnosis. Up to now, CRISPR technology is usually combined with nucleic acid amplification technology, and the amplification products are used to activate the CRISPR/Cas system, so as to detect the target. These methods have very high sensitivity and specificity.

Courtney et al. utilized the feature of PAM-dependent Cas9 to cleavage dsDNA, and designed the PAM sequence at the SNP site to detect *KRT12* mutation [81]. As shown in Fig. 8.13, the *KRT12* mutation was a T-C single-base mutation, which was manifested as a CTT-CCT change in the DNA sequence, and its reverse

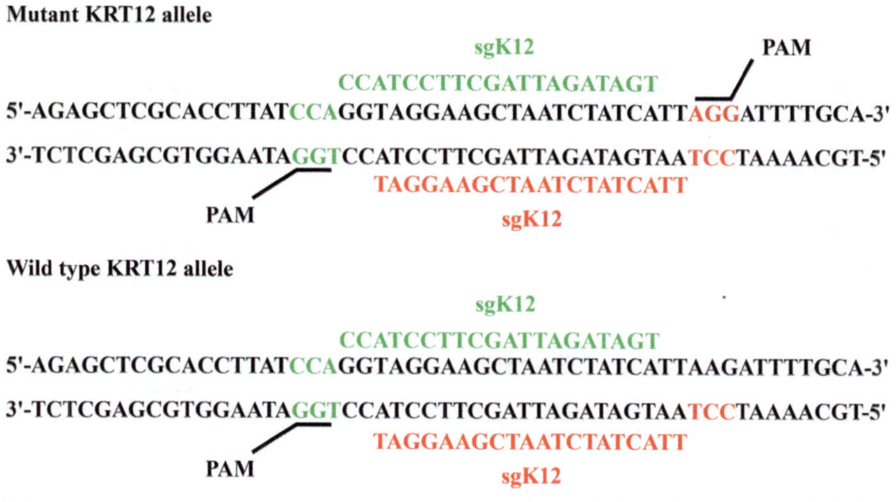

Fig. 8.13 SNP detection by CRISPR/Cas9 system based on PAM sequence recognition. (Revised and reprinted with permission from Ref. [81])

complementary sequence was AAG-AGG. Since Cas9 required an NGG sequence for PAM recognition, the mutated *KRT12* perfectly serves as a PAM sequence to guide Cas9 toward target DNA. After the PAM sequence was selected, the sgRNA (sgK12LP) was designed to target the complementary sequence adjacent to the PAM sequence to facilitate the targeted cleavage of mutant DNA by locating Cas9 near the PAM sequence via base pairing principles. For region outside the SNP sites, another NGG sequence (TGG) was selected as the universal PAM sequence to recognize both wild-type and mutant DNA. The sgRNA (sgK12) designed for this PAM sequence allowed Cas9 to simultaneously recognize both wild-type and mutant targets through the binding of sgK12, which was used as a positive control for analysis and detection.

Chen et al. used PCR primers with PAM sequences to amplify the target genes, positioning the PAM sequence upstream of the mutation base in the amplification products [82]. Subsequently, the amplification products then activated the Cas12a system, which cleaved the ssDNA probe to generate fluorescence signal for specific detection of SNP and analysis of genotyping. As shown in Fig. 8.14, PCR primers with PAM sequence at the 5′ end were used for simultaneous PCR amplification of wild-type and mutant target genes. The PCR products of both wild-type and mutant targets contained PAM sequences. Take *CYP1A1*2* mutation as an example, *CYP1A1*2* mutation involved an A-to-G transition at position 4889. crRNA was designed for wild-type and mutant *CYP1A1*2*, respectively. The crRNA of wild-type *CYP1A1*2* was uA-crRNA and for mutant *CYP1A1*2* was uG-crRNA. After PCR amplification, the products were equally divided into two groups and added to Cas12a/crRNA system, respectively. If wild-type *CYP1A1*2* presented in the system, uA-crRNA could target the wild-type amplification products, activate Cas12a to cleave the signal probe and generate fluorescence signal. If mutant *CYP1A1*2*

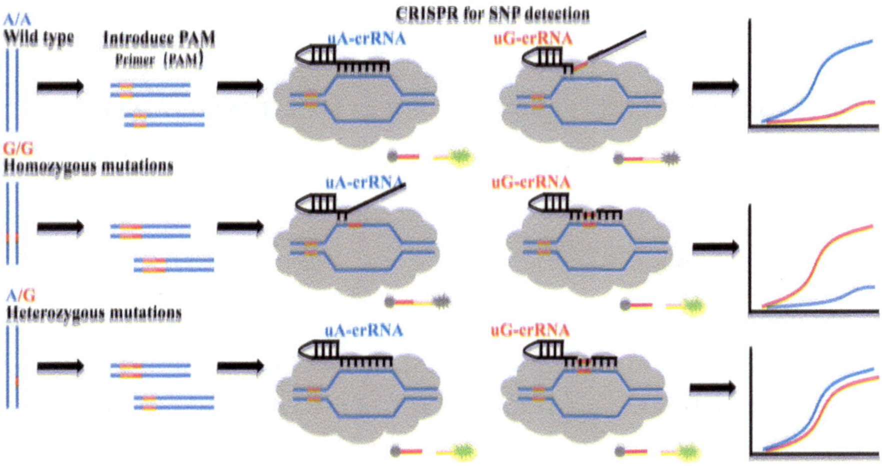

Fig. 8.14 Schematic representation of SNP detection based on Cas12a system. (Revised and reprinted with permission from Ref. [82])

existed, the amplification products could activate Cas12a after binding with uG-crRNA to generate fluorescence signal, thus enabling SNP detection. For homozygous targets, only the corresponding Cas12a/crRNA system could generate fluorescence signal. For heterozygous targets, both Cas12a/crRNA systems could produce fluorescence signal. This method enabled specific detection of *CYP1A1*2*, *CYP2C19*2*, *CYP2C9*3*, and *CYP2C19*3* mutations in clinical blood samples.

Ke et al. improved the cas13a digestion system by designing crRNA with hairpin structures to specifically detect the B genotype of hepatitis B virus [83]. This system tripled detection specificity without compromising sensitivity. As shown in Fig. 8.15, the B and C genotypes of hepatitis B virus were first amplified simultaneously using recombinase polymerase amplification (RPA), followed by transcription of the amplification products using T7 RNA polymerase. Only when the crRNA recognized the RNA transcribed from the B genotype could Cas13a be activated to cleave any single-stranded RNA, thereby cutting the fluorescent probe in the system and releasing the signal. In contrast, for the C genotype, the crRNA could not fully match the transcribed RNA, failing to activate Cas13a activity and producing no fluorescence. When using cas13a to detect SNPs, artificial mismatch of 1–2 bases was usually designed in the spacer region of crRNA to improve the ability of cas13a to detect single-base differences [84, 85].

In the above methods, the target sequence needs to be amplified by PCR or other amplification methods, followed by CRISPR-based endpoint detection of the amplification products to achieve high sensitivity detection of SNPs. To improve the detection sensitivity, Balderston et al. first pioneered CRISPR microarray technology in 2021 to achieve the detection of SNPs without amplification [86]. As shown in Fig. 8.16, the authors used graphene transistors as electrical signal carriers and combined with CRISPR technology to achieve a seamless convergence of biology and electronics. Taking advantage of the specific recognition ability of the CRISPR

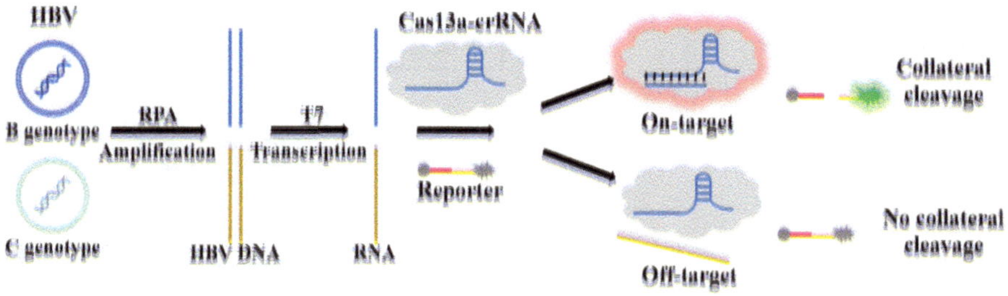

Fig. 8.15 Schematic representation of SNP detection by Cas13a system. (Revised and reprinted with permission from Ref. [83])

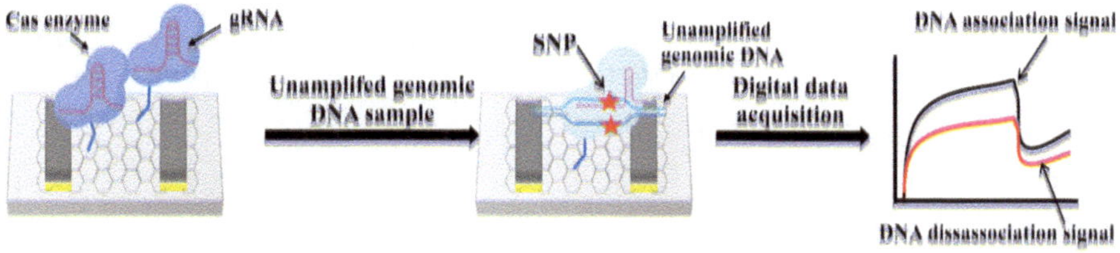

Fig. 8.16 Schematic representation of SNP detection by CRISPR microarray technology. (Revised and reprinted with permission from Ref. [86])

system, SNP was detected in samples from patients with sickle cell disease and amyotrophic lateral sclerosis (ALS). CRISPR microarray technology can directly, rapidly, and accurately analyze SNPs, opening new avenues for disease diagnosis.

8.5.5 Allele-Specific PCR Assays

In 1989, Newton et al. firstly pioneered allele-specific PCR and applied it to the detection of human insulin resistance gene defects [87]. Compared with SSCP, HA, and other detection methods, this method is more sensitive, user-friendly and efficient for large-scale sample processing. Subsequently, allele-specific PCR was widely used for the detection of SNP.

Allele-specific PCR leverages the characteristic of Taq DNA polymerase lacking 3′–5′ exonuclease activity to design suitable primers for different known mutations. When the 3′ terminal of the primer does not match with the target, PCR amplification is inhibited, resulting in specific distinction between wild-type and mutant targets. As shown in Fig. 8.17, allele-specific PCR generally requires the design of three primers. Primers P1 and P2 share identical hybridization sites on the target with their 3′ ends designed at the mutation site, differing only by the last base at the 3′ end. Primer P1 is complementary to the wild-type target with terminal base A, and primer P2 is complementary to the mutant target with a G-terminal base. Primer P3 serves as the normal counter-primer. The target is divided into two parts, and the PCR

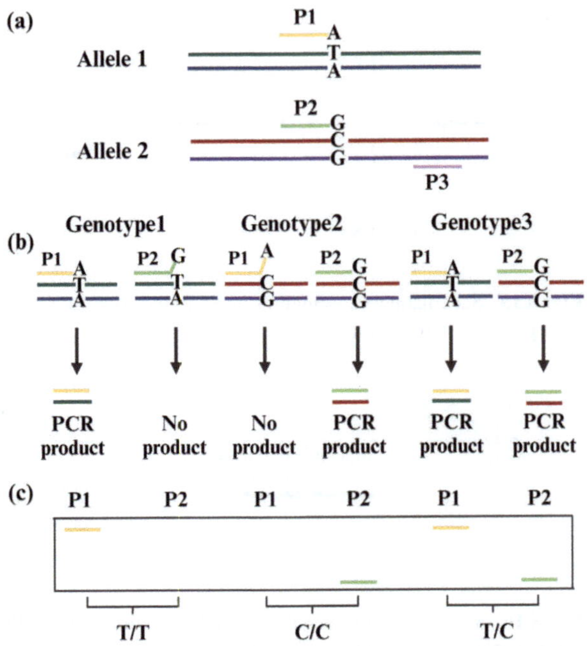

Fig. 8.17 Principal representation of allele-specific PCR for SNP detection

amplification is performed by adding primers P1 and P3, and primers P2 and P3, respectively. Primer P1 and P3 only amplify wild-type target, whereas primer P2 and P3 only amplify mutant target. The amplification products are analyzed by electrophoresis to verify the presence of mutant target. At the same time, this method enables genotype analysis. If only one primer pair generates amplification products, the target is homozygosity; and if both primer pairs produce amplification products, the target is heterozygous.

When designing primers for allele-specific PCR, special attention should be paid to ensure the final base at the 3′ end of the primer is located at the mutation site, otherwise it will affect the discrimination between wild-type and mutant targets. If a single-base mismatch at the 3′ end of the primer fails to achieve the desired resolution, artificial mismatches can be introduced at the second or third base at the 3′ end of the primer to enhance resolution [88]. In addition, Koizumi et al. performed allele-specific PCR by modifying the last three bases at the 3′ end of the primer with 2′-O, 4′-C-ethylene nucleic acid (ENA) [89]. The results showed that modification at the second base of 3′ end with EVA achieved optimal differentiation between wild-type and mutant targets. The presence of mismatched bases significantly hindered Taq DNA polymerase from extending at the end of the ENA primers, thereby enabling highly specific identification of wild-type and mutant targets.

To simplify the experimental operation, Litos et al. pioneered the visual detection of SNP using allele-specific PCR combined with dry test strips [90]. This method could distinguish SNP types by visible bands within 10 min without requiring

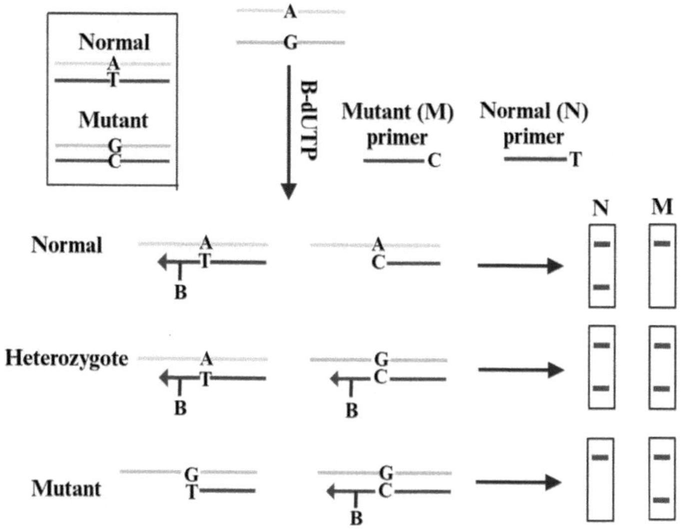

Fig. 8.18 Schematic representation of allele-specific PCR-based visualization for SNP detection. (Revised and reprinted with permission from Ref. [90])

specialized equipment, offering rapid and cost-effective results consistent with DNA sequencing. As shown in Fig. 8.18, the primers were designed for the wild-type and mutant targets with poly(dA) tails at their 5′ end and mutation site at their 3′ end. The only difference lied in the final base at the 3' end, T for wild-type and C for mutant primers. The target was divided into two portions, and both primers were added during PCR amplification using biotin-labeled dUTP, so the amplification products all carried biotin labels. The amplification products were added dropwise to the test strips separately with the bottom of the strip immersed in the buffer. Driven by capillary action, the amplification products migrated to the detection zone of the strip pre-immobilized with streptavidin. Since the amplification products were biotin labeled, they could bind to the streptavidin and thus be immobilized to the detection zone. The poly(A) at the 5′ end of the products bound to the poly(T) on gold nanoparticles, which immobilized the gold nanoparticles to the detection zone, resulting in a red band in the detection zone. The strip also had a positive control zone, which was pre-fixed with oligo(A). When the gold nanoparticles modified with poly(T) migrate to the control zone, they bound to the fixed oligo(A), producing a red band. This method allowed the detection of six mutations in the *MBL2* gene using the patient's genomic DNA with results matching those obtained through DNA sequencing.

Duan et al. used the fluorescence resonance energy transfer (FRET) between fluorescent conjugated polymers (CPs) and fluorescent groups to detect allele-specific PCR amplification products, which could detect as low as 1 fM targets [91]. This method had also been applied to the detection of real samples. As shown in Fig. 8.19, G primers and C primers were designed for PCR amplification based on

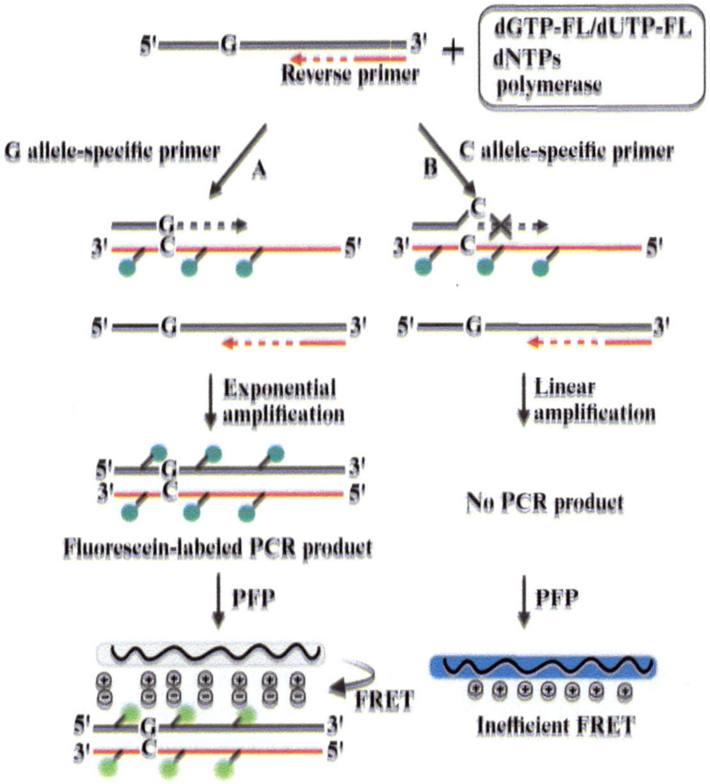

Fig. 8.19 Schematic representation of SNP detection by FRET based on CPs. (Revised and reprinted with permission from Ref. [91])

different targets. Then dUTP and dGTP with fluorescein modification (dUTP-Fl and dGTP-Fl) were added during amplification, resulting in fluorescein-labeled products. The maximum excitation wavelength of PFP was 380 nm and the maximum emission wavelength was about 424 nm, while fluorescein exhibited an excitation wavelength of about 490 nm and an emission wavelength of 525 nm. The emission spectrum of PFP partially overlapped with the excitation spectrum of fluorescein, which satisfied the condition of FRET. Therefore, when PFP and fluorescein labeled on amplification products were close enough, PFP transfers energy to fluorescein after excitation, thereby enhancing the fluorescence of fluorescein. The fluorescence of the system was proportional to the amount of dsDNA, allowing determination of target initial concentration through fluorescence measurement.

To further improve the specificity, Myakishev et al. combined allelic-specific PCR with real-time fluorescence PCR to establish a high-throughput SNP detection method using fluorescence detection, eliminating the need for a real-time quantitative PCR instrument [92]. The principle was shown in Fig. 8.20. When designing the primers, a stem-loop structure similar to a molecular beacon was added at the 5′ end of the primer, featuring fluorescent group and quenching group at both ends. Primer

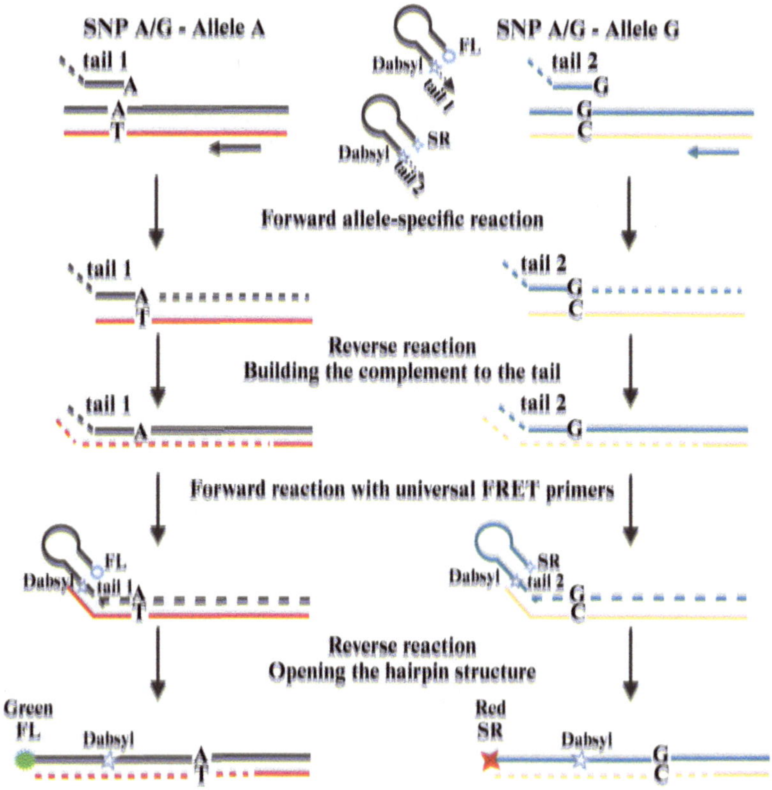

Fig. 8.20 Schematic representation of fluorometric allele-specific PCR for SNP detection. (Revised and reprinted with permission from Ref. [92])

A was designed for the wild-type target modified with a green fluorescent group at tail 1, while primer G was designed for the mutant target modified with a red fluorescent group at tail 2. In the free state, the fluorophores of both primers showed no fluorescence. After PCR amplification, different primers bound to different targets and open the stem-loop structure, producing fluorescence signals of different colors. Under fluorescence microscopy, the genotypes of the targets could be identified according to the fluorescence color. The entire process was performed in 96-well plates, enabling high-throughput detection.

Allele-specific PCR can also be combined with pyrophosphate sequencing to monitor the base extension by detecting the pyrophosphate accumulation during the DNA polymerization, thus enabling SNP detection. When the primer binds to a specific target, one molecule of pyrophosphate (PPi) is released for each binding of dNTP as the extension proceeds. PPi is converted to adenosine triphosphate (ATP) through enzymatic reactions mediated by ATP sulfurylase. With ATP as the energy source, firefly luciferase catalyzes the oxidation of firefly luciferin, generating chemiluminescent signal that determine SNP genotypes. Zhou et al. designed

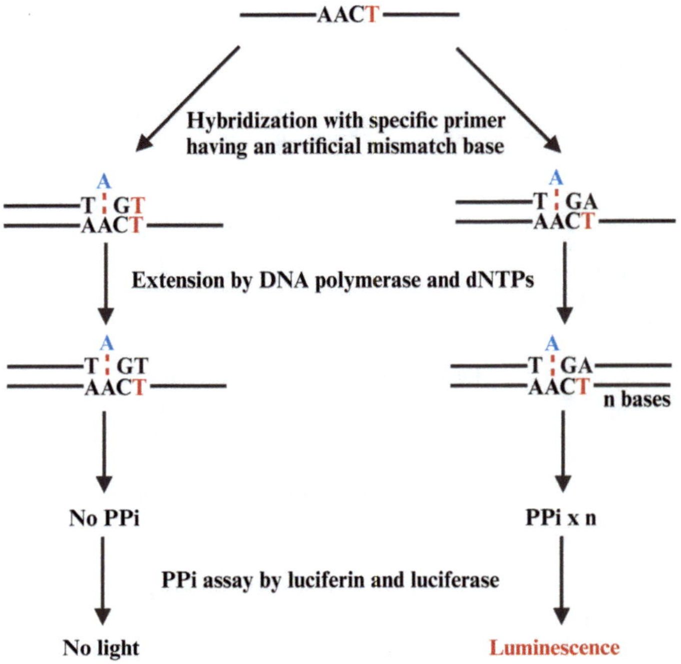

Fig. 8.21 Schematic representation of allele-specific PCR combined with chemiluminescence for SNP detection. (Revised and reprinted with permission from Ref. [93])

primers targeting the exon 8 fragment of wild-type and mutant *p53* genes, respectively [93]. The 3′ end of the primers was strategically positioned at the mutation site with artificial mismatches at the third base of the 3′ end to improve the detection resolution (Fig. 8.21). In the system, the wild-type primer bound exclusively to the wild-type target and released PPi during polymerization, while the mutant primer bound only to the mutant target and similarly released PPi. The released PPi undergoes a series of enzymatic reactions for chemiluminescence detection, achieving the detection of as low as 14 amol target and 0.02% of mutation frequency. Although this homogeneous method offers operational simplicity, its complex experimental conditions and reliance on multiple enzymes result in higher costs. Since deoxyadenosine triphosphate (dATP) also serves as the substrate for ATP sulfatase, it is not possible to use dATP or ddATP directly during allele-specific primer extension. Ronaghi et al. discovered that analogues of dATPαS or ddATPαS exhibited minimal activity for luciferase, making them suitable as alternatives to dATP to reduce background signals [94, 95].

Allele-specific PCR provides a novel method for detecting SNP by using different primers to achieve separate detection of wild-type and mutant targets. Compared with other detection methods, this method is simple, rapid, and suitable for large-scale sample analysis. However, allele-specific PCR relies on the lack of 3′–5′ exonuclease activity of Taq DNA polymerase, which sets the distinction of SNP at the 3′ terminus base difference of primers. Conventional allele-specific PCR requires multiple trials to determine the optimal reaction conditions. Therefore, controlling false-positive results remains a critical challenge.

8.5.6 Single Nucleotide Primer Extension Assays

Single nucleotide primer extension (SNuPE), also known as single-base extension (SBE), involves two key steps: First, a segment of DNA containing SNP site is amplified by PCR. Then a specific primer for single-base extension is introduced, and the SNP type can be determined by analyzing the extended bases. Based on the position of the 3′ terminus base of the specific primer, they can be classified into the following two types.

In the first method, the 3′ terminus of the specific primer is designed to be one base ahead of the mutation site, enabling simultaneous binding to both wild-type and mutant targets' PCR products. The primers are then subjected to a single-base extension reaction by adding different fluorescence-labeled ddNTP. Finally, the fluorescence of the extension products is examined to determine the target type. As shown in Fig. 8.22a, the wild-type target has a G-base and the primer binds to ddCTP with fluorescein labeling (FL-ddCTP), while the mutant target has an A-base and the primer binds to ddTTP with Cy5 labeling (Cy5-ddTTP). During capillary electrophoresis, the genotyping of the target can be analyzed by detecting the fluorescence of FL and Cy5, and the target can also be quantified by fluorescence intensity. This method allows quantitative analysis of both wild-type and mutant targets in a single experiment, featuring simple experimental design. However, the high cost of the fluorescent-labeled ddNTPs and the necessity of two distinct fluorescent-labeled ddNTPs make the experiment more expensive.

The second method involves designing the 3′ terminus of the specific primer at the mutation site, and the primers are designed for the wild-type and mutant targets, namely, wild-type primer and mutant primer. The appropriate ddNTP is selected

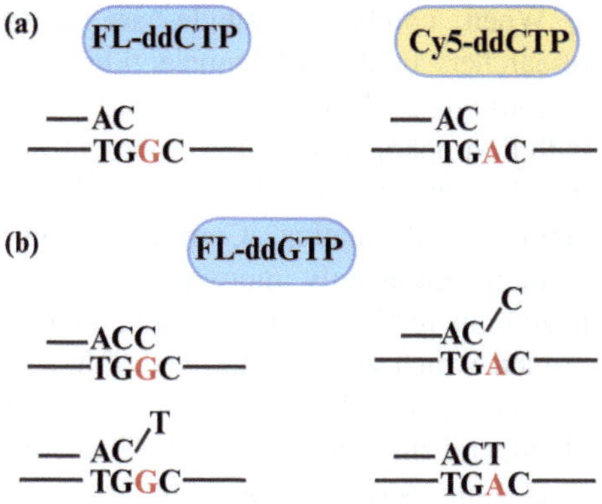

Fig. 8.22 Schematic representation of single-base extension method for SNP detection

according to the nucleotide type following the mutation site. As shown in Fig. 8.22b, when the SNP site is followed by a C base, the fluorescent-labeled ddGTP (FL-ddGTP) was selected. The PCR amplification products are both divided into two groups. One group is added to the wild-type primer and FL-ddGTP, at which time the wild-type primer can only bind to the wild-type target and extend one G base. However, the mutant target does not match the terminal base of the wild-type primer, preventing reaction and generating fluorescence signals exclusively from wild-type targets. Similarly, after adding the mutant primer and FL-ddGTP to another group of the target, only the mutant target generates fluorescence signal. This allows the simultaneous detection of wild-type and mutant targets through parallel experiments. Although this method requires two separate operations, only one type of fluorescent-labeled ddNTP is used, greatly reducing the experimental cost.

The single-base extension method uses the fundamental DNA polymerization reaction in living organisms and the specific recognition of different single nucleotides by DNA polymerase to distinguish target genotypes. The introduction of DNA polymerase ensures the specificity of the method while reducing requirements for primer design, which makes it a highly specific and reliable method for SNP identification that has gained widespread application. It is now integrated with various detection technologies such as DNA microarray chip, chip electrophoresis, fluorescence polarization, pyrophosphate sequencing (bioluminescence), FRET method, flow cytometry, mass spectrometry, and so on. When combined with DNA microarray technology for SNP detection, the detection mainly relies on DNA probe hybridization [96]. Since the specificity of DNA probe hybridization depends on the sequence composition and the conditions of hybridization reaction, achieving high-density DNA array chips with consistent probe specificity under identical conditions poses significant challenges for probe design. In contrast, primer extension method reduces the dependence on temperature conditions, ensuring method reliability. Subsequent multiplex PCR and reverse transcriptase-based extension reaction enable rapid, accurate, and high-throughput detection [41, 47]. While DNA microarray technology can achieve high-throughput SNP detection and also automation, the initial investment cost is relatively high, making it suitable only for large-scale research projects primarily focused on single-base polymorphism.

When combining single-base extension with microarray electrophoresis, researchers modify oligonucleotides of different lengths as labels on primers for various alleles. During the extension reaction, the corresponding primers are labeled with fluorophores. Then the probes are separated by microarray electrophoresis, and the fluorophores are detected by laser-induced fluorescence detector to obtain genotype information. Compared with the traditional gel electrophoresis, this method significantly improves resolution, sensitivity, and specificity. Li et al. optimized this method by introducing a mismatch at the third base of 3′ terminus of the primer, which greatly improved SNP detection efficiency to detect as low as 0.1% mutant targets [97]. As shown in Fig. 8.23, genomic DNA was firstly extracted from blood, and amplified through PCR to obtain DNA fragments from exon 8 of the *p53* gene. Primers were designed for both wild-type and mutant targets with their 3′ end

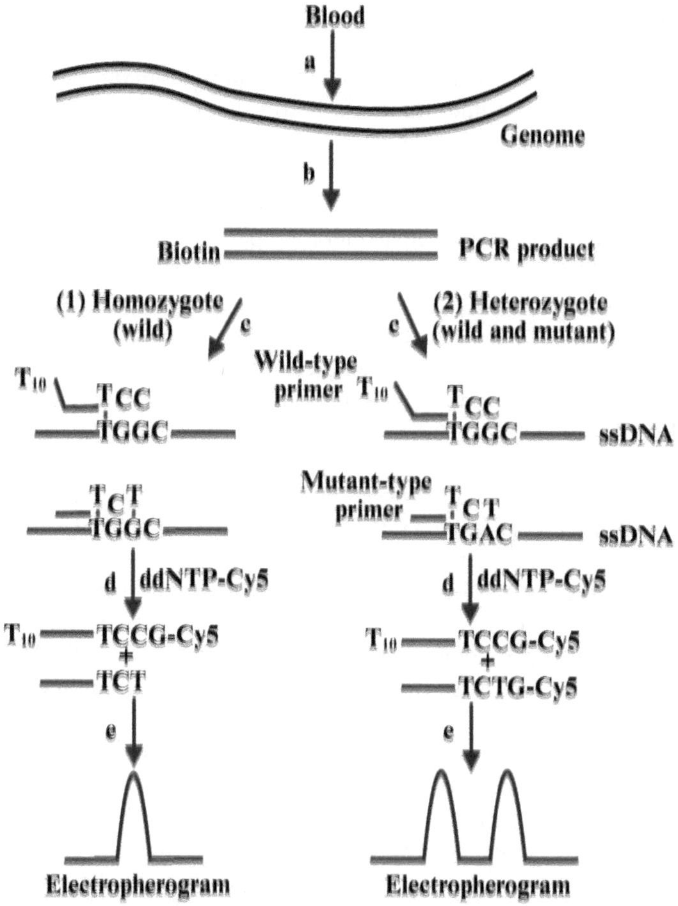

Fig. 8.23 Schematic representation of SNP detection by single-base extension based on mismatched bases. (Revised and reprinted with permission from Ref. [97])

positioned at the mutation site, resulting in C-terminal base for wild-type primer and T-terminal base for mutant primer. To improve the specificity, a mismatch (T-T mismatch) was introduced at the third base of the 3′ end of the primer, and 10 poly (dT) was modified at the 5′ end of the wild-type primer to ensure separation of amplification products during electrophoresis. The wild-type primer, mutant primer, and Cy5-modified ddGTP (ddGTP-Cy5) were added to the system simultaneously. The wild-type primer matched only with the wild-type target and extended G base after binding, while the mutant primer matched only with the mutant target and extended G bases after binding, which were separated and detected using capillary electrophoresis. Since the two primers had different lengths and migration rates in capillary electrophoresis, SNP-specific detection and SNP genotype analysis could be realized based on the peak positions and peak area of the primers.

When the single-base extension reaction is combined with the fluorescence polarization technique, the single-base polymorphism can be analyzed by utilizing the change in fluorescence polarization signal of fluorophore on a single nucleotide before and after binding primer. In 1926, Perrin first described his observation of fluorescence polarization phenomenon in his paper. If the excited fluorescent substances remain stationary, they retain their original excitation light polarization characteristics [98]. If the excited fluorescent substances were in motion, the emitted polarized light exhibited distinct polarization properties from the original excitation light, known as fluorescence depolarization phenomenon. Small molecules rotated rapidly and produced fluorescence depolarization phenomenon with nearly zero polarization, while large molecules rotated slowly and produced varying degrees of fluorescence polarization phenomenon according to their rotation speed, which could be quantitatively detected by measuring fluorescence polarization intensity. As shown in Fig. 8.24, before single-base extension reaction, the fluorescent group was initially bound to ddNTP. Due to the small molecular weight of individual nucleotide and the high rotation speed in solution, the fluorescence polarization was very low. When the fluorescent-labeled ddNTP was bound to the primer, the rotation of the fluorophore became restricted and slowed down, leading to significantly increased fluorescence polarization. Chen et al. used this method to detect the C282Y mutation in *HFE* gene, employing four different fluorescently labeled ddNTPs to simultaneously identify four distinct mutant bases [99]. However, the intensity of polarized light was lower than that of common fluorescence detection method, resulting in limited sensitivity.

When the single-base extension reaction is combined with the FRET method, the excited state energy transfer from the fluorescence donor to the receptor is used to achieve the detection within a specific distance (10 nm). FRET occurs when the donor's fluorescence is quenched while the acceptor's fluorescence appears, conversely no change in donor fluorescence is observed. Through allele-specific primer extension, two different fluorophore-labeled single nucleotides that may undergo energy transfer bind to the primer. Because the distance satisfies the FRET requirement, the FRET occurs. If the extension fails, the two different fluorophore-labeled mononucleotides remain distant in solution, so no FRET occurs. Based on the signal amplification effect of the CPs, Shu Wang's group used the CPs as fluorescence donor and realized the highly sensitive and specific detection of SNP by using the FRET between the CPs and the fluorescent group [100]. As shown in Fig. 8.25, the target DNA was a fragment in exon 8 of the human *p53* gene, which contained a SNP site. At this site, the wild-type sequence contained a G base, while the mutant variant had an A base. The probe with a T at its 3′ terminus was designed to perfectly match the mutant target DNA, while mismatching with the wild-type target DNA at this site. For mutant target DNA, dGTP-FL bound to the probe in the presence of DNA polymerase. By adding the water-soluble conjugated polymer P1, the strong electrostatic interactions between the positively charged polymer and DNA reduced the distance between fluorescein and the polymer's fluorescent backbone, triggering FRET from the polymer to the fluorescein. After excitation of P1, the FRET-induced emission intensity of the fluorescein was approximately ten times higher than that

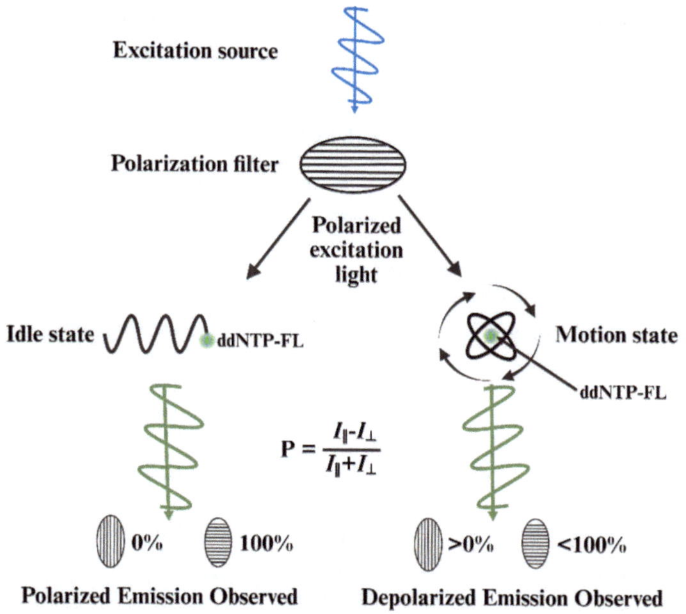

Fig. 8.24 Schematic representation of fluorescence polarization method for detecting different states of fluorescent groups. (Revised and reprinted with permission from Ref. [99])

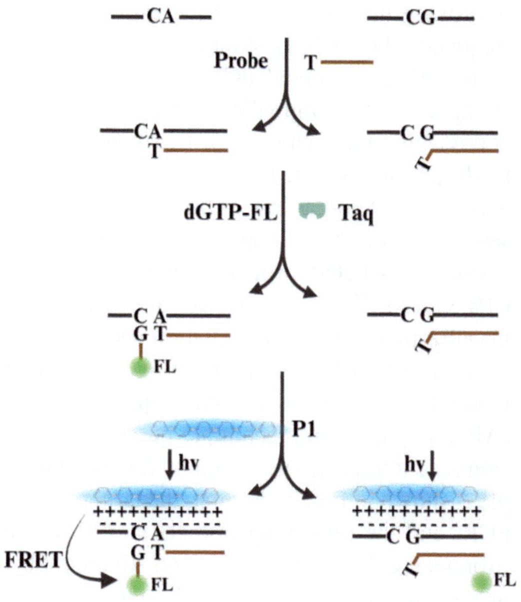

Fig. 8.25 Schematic representation of single-base extension method combined with CPs-based FRET for SNP detection. (Revised and reprinted with permission from Ref. [100])

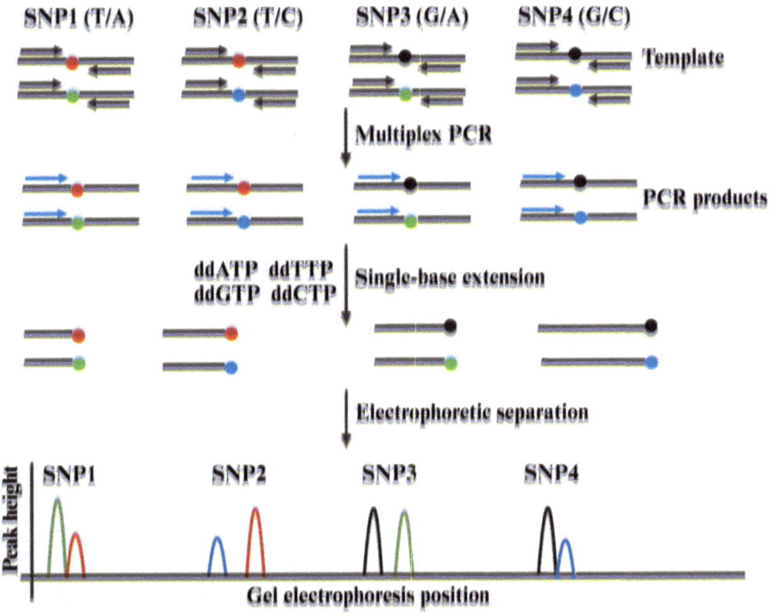

Fig. 8.26 Schematic representation of SNaPshot technology for SNP detection

obtained by direct excitation, showing the signal amplification effect of the CPs and improving the sensitivity of the assay, which could detect as low as 2% mutation frequency. For wild-type target DNA, the 3′ end of the probe mismatches with the target DNA prevented dGTP-FL from binding. After addition of the positively charged P1, the minimal negative charges on dGTP-FL prevented fluorescein from approaching the backbone of P1, thereby inhibiting effective FRET. Applying this method to the detection of real biological samples, SNP genotyping of 76 samples was achieved and as low as 50 ng DNA samples could be detected [101]. The high sensitivity of this homogeneous detection approach eliminated the need for fluorescent-labeled primers, significantly reducing costs. Further application of CPs-based FRET to multicolor fluorescence coding detection enabled simultaneous visualization of multiple mutation sites in a single tube [102, 103].

To further advance high-throughput SNP detection, the SNaPshot genotyping method, also known as mini-sequencing, is developed by Applied Biosystems (ABI), a genotyping technique based on the fluorescent-labeled single-base extension reaction, primarily designed for medium-throughput SNP analysis. The principle is shown in Fig. 8.26. First, the target SNP fragment is amplified using multiplex PCR, and the remaining primers and dNTPs are digested by adding nucleic acid exonuclease I (Exo I) and Shrimp Alkaline Phosphatase (SAP). The purified amplification products are used as the template for single-based extension using DNA polymerase, four fluorescent-labeled ddNTP, and primers positioned adjacent to the SNP site at its 3′ end. For different SNP sites, specific primers with varying lengths of poly(dT) modified at the 5′ end can be used. After detection by ABI sequencing, peak positions identify the corresponding SNP site, while peak colors indicate the incorporated base type to determine genotype. With the multiplex PCR reaction system, multiple SNP sites can be detected simultaneously, and typically 10–35 SNP sites can be analyzed simultaneously.

8.5.7 Ligation-Based Methods

Due to DNA polymerase's limited specificity for single-base mismatches, primers with mismatched bases at the ends may induce nonspecific extension reactions. Once formed, these products can be amplified through subsequent PCR exponential amplification, leading to false-positive results that significantly compromise SNP diagnosis [104, 105]. Chen et al. [106] systematically investigated how various 3′ end mismatched bases affect DNA polymerase-mediated extension reactions. The findings revealed that mismatched base extension efficiency depends on both mismatch types and polymerase species. All twelve 3′ end single-base mismatched primers triggered extension reactions, though only a few mismatches (primer-template: C-C, G-A, A-G, and A-A) exhibited varying degrees of inhibitory effects. To enhance detection specificity, researchers typically employ methods like single-base extension assays, which first amplify the DNA fragment containing the target SNP through PCR before detecting single-base differences. Considering that ligase-mediated ligation reactions demonstrate strong recognition of single-base mismatches, researchers aim to further improve SNP detection specificity leveraging this reaction mechanism.

The principle of the ligation reaction is based on the strong distinguishing ability of ligase for mismatched bases, which can only ligate two strictly complementary bases. In ligation reaction, two probes are typically designed: a common probe and a complementary probe to the mutation site, whose 3′ terminus locates at the mutation site. Only when the probe is completely complementary to the target DNA can DNA ligase ligate the two probes. Even if there is a single-base mismatch between the probe and target DNA, the ligation efficiency of the ligase reaction will be greatly reduced. Based on this principle, the occurrence of ligation reactions can determine whether a gene has undergone mutations. The ligation reaction can be combined with various nucleic acid amplification reactions to achieve highly sensitive detection of SNPs, such as PCR, RPA, strand displacement amplification reaction (SDA), ligase chain reaction (LCR), loop-mediated isothermal amplification reaction (LAMP), and rolling amplification reaction (RCA).

8.5.7.1 Ligation-Based Assays

This method involves designing two probes near the mutation site, with one probe's 5′ end anchored to the vector and 3′ end located at the mutation site. The other probe has the 5′ end exactly connected to the mutation site with a fluorescent label attached to its 3′ end. Therefore, when the mutant target is present, the two perfectly matched probes can hybridize with the target and form a nicked double-stranded structure, which is ligated into complete dsDNA under the catalysis of DNA ligase. The unligated probes are washed away, allowing specific detection of mutated sequences through fluorescence signals retained on the substrate after ligation. Murray et al. combined multiplex PCR, probe hybridization, ligation reaction, and microarray technology to detect stool samples from colorectal cancer patients, and for the first time achieved the simultaneous detection of 28 relevant SNPs, including *APC*, *TP53*, *K-ras*, and *BRAF* in the same sample [107]. However, this method exhibited lower sensitivity due to the absence of nucleic acid amplification method.

8.5.7.2 Allele-Specific Ligation

Allele-specific ligation (ASL), similar to multiplex ligation-dependent probe amplification (MLPA, as shown in Chapter 1), involves designing two DNA probes for ligation reaction followed by amplification using the ligation products. The key difference lies in the SNP site at the allele-specific ligation method. As shown in Fig. 8.27, one probe has a universal primer sequence at its 5′ end and a target-specific sequence at its 3′ end, with the terminal base located at the mutation site. Therefore, primers with distinct terminus bases are designed for wild-type and mutant targets, namely, ASL probe A and ASL probe C. The other probe features a universal primer sequence at its 3′ end and a target-specific sequence at its 5′ end, with the terminal base adjacent to the mutation site. This probe is identical for both wild-type and mutant targets. Each probe binds to its corresponding target to form a nicked double-stranded structure, which is ligated into long-stranded DNA by DNA ligase. The ligated product is used as the template for amplification, enabling SNP detection and genotyping through quantitative analysis of amplification products. To simultaneously detect wild-type and mutant targets, length or fluorescent coding is incorporated to one of the probes (a fluorescent coding is added between the universal primer sequence of ASL probe A and the target-specific sequence, such as FAM and Cy5, as shown in Fig. 8.27). After amplification, the amplification products corresponding to different types of SNPs exhibited varying lengths or fluorescence signals, which were separated and detected using electrophoresis or fluorescence detection. This method typically completed the entire detection within 3 h and has a throughput of 40–100 loci, which is higher than the SNaPshot method and can detect SNP mutations in as low as 1% of tumor cells. Lázaro et al. combined allele-specific ligation with RPA and DNA microarray technology to capture allele-specific amplification products by probe hybridization reaction to different regions of the microarray to achieve SNP genotyping [108]. Park et al. used a mass spectrometry-based assay integrating allele-specific ligation with SDA to detect multi-gene SNP genotyping [109]. In the allele-specific ligation step, a probe with a nicking enzyme recognition site at its end was used. After ligation, extensive short-stranded DNA was formed by SDA, which could be used as a mass spectrometry marker for analysis. By adjusting the length of the mass spectrometry marker in the probe,

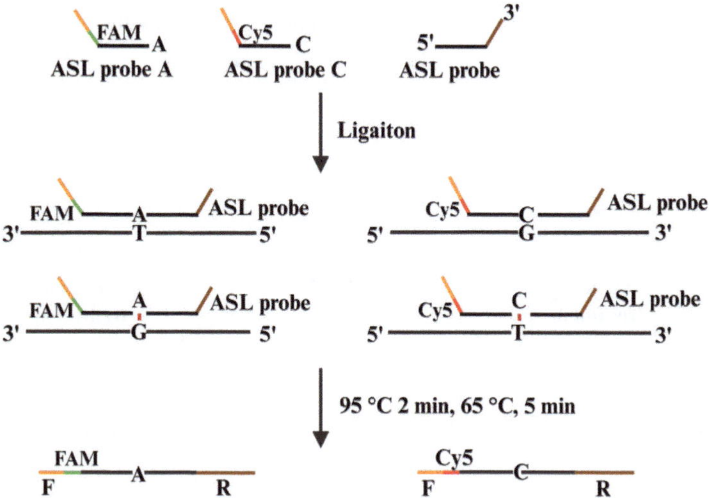

Fig. 8.27 Schematic representation of allele-specific ligation method for SNP detection

multiplex SNP genotyping was achieved, and multiple *BRCA* gene mutations were identified by mass spectrometry analysis.

In addition, ligase chain reaction (LCR) achieves exponential amplification based on the thermal cycle of ligase reaction. As previously mentioned, the ligase reaction can identify single-base differences with high specificity, making LCR particularly suitable for SNP detection (see Chap. 2 for details) [110–113]. Padlock probe-based rolling circle amplification (RCA) requires the formation of circular structures by ligating linear padlock probes to target DNA. By strategically positioning the ligation site at the SNP site, this method also enables highly specific SNP detection (see Chapter 4 for details) [114, 115].

8.5.8 *Enzymatic Digestion-Based Methods*

8.5.8.1 Restriction Fragment Length Polymorphism Analysis

Restriction fragment length polymorphism (RFLP) is a reaction proposed by Botstein et al. in 1980 that uses a restriction endonuclease to cleave dsDNA at a specific recognition site [116]. If the SNP site locates at the recognition site of the enzyme, the base differences between mutation and wild-type target are no longer recognized and cleaved by the endonuclease. This allows detection of such variations through electrophoretic separation. RFLP offers high accuracy and operational simplicity, but when applying RFLP for SNP detection, the SNP site of the DNA molecule is required to be located exactly at the restriction endonuclease recognition site, limiting the widespread application of this method [117].

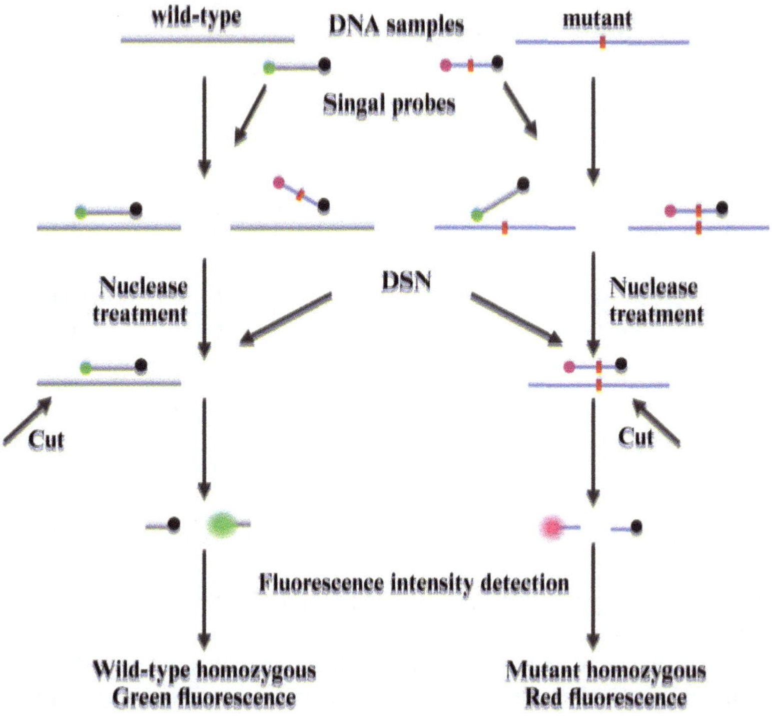

Fig. 8.28 Schematic representation of SNP detection based on DSN digestion. (Revised and reprinted with permission from Ref. [118])

8.5.8.2 Detection of SNPs Using Duplex-Specific Nuclease

Duplex-specific nuclease (DSN) is a specific nuclease from the Kamchatka crab that specifically cleaves dsDNA or the DNA strand in DNA-RNA hybrid, but remains inactive on ssDNA or RNA. DSN demonstrates significantly higher degradation efficiency for short, perfectly matched dsDNA than those of equivalent length, non-perfectly matched dsDNA, making it ideal for distinguishing single-base variations in DNA. Shagin et al. used the excellent single-base recognition ability of DSN and combined with FRET of the signal probe to achieve *p53* gene mutation detection, as shown in Fig. 8.28 [118]. First, the target was amplified by PCR, and signal probes were designed for wild-type and mutant targets, respectively. The mutation site was positioned in the middle of the signal probe to maximize discrimination. The probe had a fluorescent group at one end and a quenching group at the other end. In the free state, both probes showed no fluorescence. Upon enzymatic reaction, DSN recognized and cleaved the perfectly matched dsDNA. Therefore, the wild-type target could be cleaved by DSN after hybridization with the wild-type signal probe, and release green fluorescence, while the mutant signal probe failed to be recognized by DSN due to the single-base mismatch and could not produce green fluorescence. Conversely, the mutant signal probe could only be cleaved by DSN after hybridization with the mutant target, releasing red fluorescence. Since the wild-type and mutant signal probes carry different fluorescent groups, simultaneous detection of wild-type and mutant targets could be achieved.

In summary, different analytical methods have their own advantages and disadvantages. In practical application of nucleic acid analysis, it is crucial to select the appropriate analytical method, which requires the comprehensive consideration of sample properties and characteristics. For general SNP analysis, the selection of the appropriate analytical method and detection mode is usually based on key factors such as experimental objectives, requirements, and laboratory conditions to achieve maximum detection efficiency and accuracy. Alternatively, combining different analytical methods can leverage their respective strengths to obtain satisfactory results. In view of the rapid development of computerized analysis, many optimal combinations of detection techniques and new methodologies continue to emerge. Future SNP analysis methods will undoubtedly become simpler, faster, and more effective.

8.6 SNP and Precision Medicine

In 2015, former US President Barack Obama summarized the achievements of the Human Genome Project and proposed the "Precision Medicine Initiative," sparking a precision medicine revolution. In the Human Genome Project, about 30,000 DNAs of the human genome were identified, and the sequences of about 2.5 billion nucleotide pairs that made up the human genome DNA were determined. Concurrently, various omics, such as proteomics, pharmacogenomics, cytomics, pharmacogenetics, and chemical genomics, have emerged. Pharmacogenomics focuses on how DNA affects drug response, while pharmacogenetics examines how DNA variants affect drug response. Both groups aim to study genetic susceptibility to drug response, enable personalized medication, optimize drug selection and dosing based on individual genetic profiles, and drive innovative approaches in conventional therapies. Individual differences in treatment efficacy and toxicity are influenced by factors such as height/weight, gender, age, complications, genetic polymorphisms, and environmental exposures, with genetic factors playing a decisive role.

Precision medicine, evolving from the concept of personalized medicine, focused more on individualized care by developing tailored treatment plans based on the characteristics of each patient. This innovative medical paradigm has emerged through rapid advancements in genomic sequencing technology and the integration of bioinformatics with big data science. Its essence is to analyze, identify, validate, and apply biomarkers for large populations and specific disease types through omics technologies, such as genomics and proteomics and cutting-edge medical innovations, so as to accurately identify disease origins and therapeutic targets, and precisely classify disease states and progression stages, and finally realize personalized precision treatment for both diseases and specific patients, thereby enhancing diagnostic efficiency and preventive outcomes. Unlike personalized medicine, precision medicine places greater emphasis on the comprehensive characterization of diseases and the high precision of medicine. It represents an advanced medical technology developed through deep understanding of human biology, disease mechanisms, and pharmaceutical applications.

The concept of precision medicine is not new; in fact, blood type analysis has been used to guide blood transfusions over a century ago. With advancements in sequencing technology and big data analysis and processing capability, precision medicine has been gradually implemented. Meanwhile, domestic policies continue to evolve. In the Technical Guidelines for Tumor Individualized Therapy (Trial), the competent authorities reaffirm the practice of predicting drug efficacy and evaluating prognosis by detecting gene mutations, SNP genotyping, gene and its protein expression status in tumor patients' biological samples to guide clinical individualized therapy. It recommends pretreatment genetic testing for cancer patients to assist doctors in formulating medication plans. Traditional drug use follows an empirical approach: starting with drug A, then switching to drug B if ineffective, then switching to C drug if ineffective again. This "one-size-fits-all" model not only increases drug resistance and complicates subsequent treatments but also wastes resources. Precision medicine adopts personalized medication, i.e., different drugs are used for group A, B, and C to achieve equivalent therapeutic effect. Precision medicine adopts personalized medication, tailoring drugs for different groups (A, B, C) to achieve equivalent therapeutic effects. In practice, it combines individual genomic information with various omics-related internal environmental data to design optimal treatment plans for patients. By addressing individual differences in response to therapeutic effects or side effects, it aims to maximize therapeutic benefits while minimizing adverse reactions.

Modern medical research has proven that the root cause of most diseases can be traced to genetic factors. Genes regulate protein synthesis, determine how efficiently the body processes nutrients, eliminates toxins, and fights infections. Just as blood types vary among people, human genes also exist in different variants known as genetic polymorphisms, each with varying sensitivity to environmental triggers. Studies reveal that cancers exhibit distinct genetic signatures, tumor markers, and variants. This new understanding of carcinogenesis mechanisms aids in the evaluation of targeted therapies, such as cancer risk, classification and diagnosis, and treatment strategies. The so-called targeted therapy uses antitumor drugs to target specific tumor sites, destroying cancer cells while minimizing damage to healthy tissues. The complexity of SNP determines the polymorphism of drug response, and drug efficacy and optimal dose depend largely on drug-related genetic variants. Therefore, personalized drug use and personalized medicine have become a cornerstone of clinical treatment for many diseases, such as diabetes, hypertension, and various cancers.

Colorectal cancer, a common malignant tumor of the digestive system, is closely associated with *K-ras* gene mutations in its development. Studies indicate that the mutation status of *K-ras* gene significantly influences colorectal cancer prognosis, and serves as a crucial prognostic indicator [119]. The effect of cetuximab, a molecularly targeted growth factor receptor inhibitor, in treating patients with advanced colorectal cancer is strongly linked to *K-ras* mutation status. Wild-type *K-ras* patients are sensitive to cetuximab, while mutated *K-ras* patients show resistance [120, 121]. Therefore, the mutation status of the *K-ras* gene plays a vital role in the formulation of clinical treatment for colon cancer. The Ministry of Health first issued the "Colorectal Cancer Diagnosis and Treatment Specification" in 2010, which explicitly stipulates that the relevant gene status should be tested for recurrent or metastatic colorectal cancer patients to develop personalized treatment strategies. The status of *K-ras* gene in tumor tissues must be assessed when receiving treatment with epirubicin and panitumumab (anti-*EGFR* monoclonal antibodies). Tumor cells possess signaling pathways that respond to different antibodies. Anti-*EGFR* monoclonal antibodies inhibit tumor cell survival and proliferation by blocking *EGFR* dimer formation and downstream intracellular signaling. If *K-ras* mutations activate cellular signaling pathways, they may render anti-*EGFR* antibodies ineffective. Thus, *K-ras* mutation detection enables targeted drug administration.

Non-small cell lung cancer accounts for 80–85% of lung cancers, and 1/3 of these patients are diagnosed at locally advanced stages, missing the optimal surgical window that typically extends survival to 4–5 months. Targeted therapies using TKIs—specifically tyrosine kinase inhibitors (TKIs)—such as gefitinib (Iressa), erlotinib (Tarceva), and nilotinib, can significantly improve survival rates. Importantly, TKI efficacy is closely linked to the status of EGFR or *K-ras* genes [122], with mutations in *EGFR* gene exon 20 T790 and *K-ras* gene being key indicators of TKI resistance. Therefore, pretreatment testing for *EGFR/K-ras* genetic profiles becomes crucial for guiding targeted therapy decisions.

For the treatment of coronary artery disease, the antiplatelet drug such as clopidogrel is usually used in combination with aspirin. Clopidogrel is a precursor drug that is metabolized to active metabolites by the hepatic cytochrome P450 enzyme system after oral administration. The *CYP2C19* functional deficiency allele in the clopidogrel metabolic pathway is associated with reduced clopidogrel responsiveness and poor clinical prognosis, causing interindividual and interethnic differences in drug metabolism capacity [123, 124]. Currently, clinical testing primarily focuses on two *CYP2C19* loci (636GG and 681GG), of which wild-type *CYP2C19* individuals belong to the fast metabolizing type and clopidogrel demonstrates better therapeutic efficacy. If a heterozygous mutation occurs at either sites, *CYP2C19* becomes a medium metabolism type, resulting in suboptimal clinical response at standard doses requiring increased dosage. Other cases belong to the slow-metabolizing type with no efficacy of clopidogrel and significantly increased risks of resistance, which is recommended to switch to other antiplatelet drugs such as Ticagrelorl. Studies indicate that patients with myocardial infarction after stent implantation obtained comparable results to those with strong antithrombotic standard therapy when treated with *CYP2C19* genotype-guided individualized antiplatelet drugs, with stronger efficacy than Ticagrelor or Prasugrel strong standard therapy in preventing thrombosis within 1 year, and lower bleeding rates [125]. Results from a large-scale randomized controlled trial using *CYP2C19* gene-guided clopidogrel therapy also demonstrate that individualized antiplatelet drug therapy guided by *CYP2C19* genotype is more effective than standard clopidogrel therapy for 3-month ischemia time with no difference in bleeding risk [126]. The above findings further validate the significance of *CYP2C19* gene testing and provide robust evidence for pharmacogenomic guidance of personalized medication.

With the growing availability of clinical evidence, molecular diagnosis by tumor biomarkers to guide personalized precision therapy has gradually become a consensus in the field of oncology clinical treatment. Tumor biomarkers are unique products of tumor cells, which are biochemical substances that indicate the tumor presence and reflect their biological characteristics. From a clinical perspective, they mainly refer to substances detectable in blood, body fluids, and tissues that are associated with tumors. American scientists have developed a method called CancerSEEK, which can simultaneously assess the levels of cancer proteins and the presence of cancer mutations in circulating DNA in the blood. It is a blood test that can be used to screen for eight common cancers and aid tumor localization [127]. Circulating tumor DNA mutations serve as highly specific cancer markers, and the combination of DNA mutations and altered protein levels facilitates the early detection of cancer [128]. Newly developed small input liquid volume extracellular RNA sequencing (SILVER-seq) requires less than a single drop of blood to distinguish breast cancer patients from healthy individuals [129]. As the progressive understanding of tumor occurrence and development deepens, the role of tumor biomarkers in these processes becomes increasingly evident. Given the unique growth patterns of tumor cells within each patient, a comprehensive understanding and testing of tumor biomarkers allow every patient to receive precise treatment plans at the earliest stage.

8.7 Development and Application of SNP

8.7.1 Application of SNP in Disease Diagnosis

DNA serves as the genetic foundation of living organisms. After transcription, DNA is converted into mRNA, which undergoes splicing and various modifications before being translated into functional proteins. Therefore, DNA sequence directly determines protein conformation and functionality. SNPs create subtle variations in both population and individual genetic profiles. When SNPs occur in protein-coding genes, they can alter protein structure and function, making individuals more susceptible to specific diseases or modifying their response to certain drugs. This discovery has paved the way for SNP-based genetic diagnosis and therapy, opening new avenues for advancing human disease research at the genetic level.

Genetic differences among individuals primarily reside in SNP. DNA sequence analysis enables disease diagnosis, including association studies between SNP and related diseases, as well as disease prediction and risk assessment. Multiple significant genetic diseases have been identified through SNP research, such as sickle cell anemia, heart disease, diabetes, hypertension, and cancer. Liu et al. selected

400 patients with esophageal squamous carcinoma and 400 controls from a non-related Han population in Chongqing [130]. Combined with data from UCSC and HapMap database, six SNPs from their tumor suppressor *p53* that showed potential disease relevance were screened. Then the six SNPs were analyzed and examined using the multiple SNaPshot method, and three SNPs were found to be significantly associated with the incidence of esophageal squamous carcinoma. These three SNPs hold promise for early prediction and diagnosis of esophageal squamous carcinoma, though further research is needed due to limited sample size. Chen et al. developed a disease-associated macrogenomic SNP analysis model, and screened and identified high-quality SNP loci by sequencing the human intestinal bacterial macrogenome [131]. Comparative experiments revealed species with significant interspecies SNP expression differences, analyzed SNP pattern variations across genes, and annotated significantly different SNP patterns in target genes. This approach uncovered distinct microbial SNP patterns between healthy individuals and patients with type II diabetes and liver cirrhosis. Kruglyak et al. used computer simulations to estimate genomic ranges and predicted that a genome-wide map containing 500,000 SNPs would be useful for identifying pathogenic genes of complex diseases [132]. Two prostate cancer-related genes have now been identified through SNP association, and steroid hormones have been linked to the development of breast and prostate cancers.

8.7.2 Application of SNP in Animal Husbandry

SNPs not only offer advantages such as stable inheritance but also effectively facilitate molecular marker-assisted breeding. In animal breeding, SNP primarily serves to determine genotype. Zhong et al. utilized PCR-SSCP to detect polymorphisms in the 8th exon of the *BMRB-IB* gene in yellow cattle, identifying its polymorphism and genotype to analyze the relationship between this gene and body measurements in different breeds [133]. Ye et al. examined *IGF2* and *DRD2* genes related to follicular development in Muscovy ducks, pinpointing loci closely related to egg-laying traits [134]. Cai et al. analyzed SNP polymorphisms combined with beef cattle traits, revealing correlations between SNP polymorphisms and multiple economic traits [135]. Through SNP detection, they identified genes strongly associated with specific cattle traits, which became candidate genes influencing these characteristics. Li et al. detected *PmCatL*, *PmTry*, and *PmAmy* genes in Penaeus monodon, from which 56 SNP loci were found [136]. Comparing the growth traits controlled by different genotypes of these SNP loci, significant differences were found. These results suggest that SNP polymorphism within a gene can serve as a basis for genotype identification, while SNP technology also aids in detecting individual genotypes to enhance superior breed development.

8.7.3 Application of SNP in Plant Breeding

With the development of technology, conventional plant breeding technologies have become increasingly inadequate to meet modern demands. The application of molecular biology techniques enables targeted modification, screening, and identification for crop improvement. The implementation of SNP technology has driven advancements in molecular marker technology, enabling rapid, efficient, and large-scale breeding. Liu et al. conducted genetic analysis on blueberry varieties and two endemic wild lingonberries in China, identifying genes related to cold resistance and cold hardiness [137]. Through database comparison, SNP marker loci were identified and developed, successfully screening superior hybrid plants via SNP marker detection. Shen et al. performed high-throughput sequencing of characteristic SNP markers of Perilla, utilizing TaqMan SNP genotyping technology to screen and breed new pure varieties with excellent quality, which provided a novel model of rapid and effective breeding for medicinal plant breeding built on SNP technology [138].

At present, SNP detection technology has been applied in various fields, with ongoing research; SNP has emerged as a key biomarker in molecular biology, attracting widespread attention across medicine, biology, and pharmacology. Given its stable correlation with diseases, further exploration of SNP will certainly advance research in human genetics, molecular genetics, pharmacogenetics, forensic medicine, as well as disease diagnosis and treatment.

References

1. Kruglyak L, Nickerson DA. Variation is the spice of life. Nat Genet. 2001;27:234–6.
2. Ingram VM. A specific chemical difference between the globins of normal human and sickle-cell anaemia haemoglobin. Nature. 1956;178:792–4.
3. Chang JC, Kan YW. Beta 0 thalassemia, a nonsense mutation in man. Proc Natl Acad Sci USA. 1979;76:2886–9.
4. Hall MJ, Reid JE, Burbidge LA, Pruss D, Deffenbaugh AM, Frye C, Wenstrup RJ, Ward BE, Scholl TA, Noll WW. BRCA1 and BRCA2 mutations in women of different ethnicities undergoing testing for hereditary breast-ovarian cancer. Cancer. 2009;115:2222–33.
5. Nei M. Molecular evolutionary genetics. Columbia University Press; 1987. p. 28–9.
6. Barabino GA, Platt MO, Kaul DK. Sickle cell biomechanics. Annu Rev Biomed Eng. 2010;12:345–67.
7. Usami S, Chien S, Scholtz PM, Bertles JF. Effect of deoxygenation on blood rheology in sickle cell disease. Microvasc Res. 1975;9:324–34.
8. Ingram VM. Gene mutations in human haemoglobin: the chemical difference between normal and sickle cell haemoglobin. Nature. 1957;180:326–8.
9. Galanello R, Cao A. Alpha-thalassemia. Genet Med. 2011;13:83–8.
10. Galanello R, Origa R. Beta-thalassemia. Orphanet J Rare Dis. 2010;5:1–15.
11. Rahbar S, Nozari G. A novel initiation codon mutation (ATG→ ATT) in a β-thalassemia patient. Hemoglobin. 1993;17:557–62.
12. Weinberg RA. Oncogenes and tumor suppressor genes. CA Cancer J Clin. 1994;44:160–70.

13. Prior IA, Lewis PD, Mattos C. A comprehensive survey of Ras mutations in cancer. Cancer Res. 2012;72:2457–67.
14. Scheffzek K, Ahmadian MR, Kabsch W, Wiesmüller L, Lautwein A, Schmitz F, Wittinghofer A. The Ras-RasGAP complex: structural basis for GTPase activation and its loss in oncogenic Ras mutants. Science. 1997;277:333–9.
15. Barbacid M. Ras genes. Annu Rev Biochem. 1987;56:779–827.
16. Bos JL. Ras oncogenes in human cancer: a review. Cancer Res. 1989;49:4682–9.
17. Levens DL. Reconstructing MYC. Genes Dev. 2003;17:1071–7.
18. Magrath I. The pathogenesis of Burkitt's lymphoma. Adv Cancer Res. 1990;55:133–270.
19. Dang CV, Le A, Gao P. MYC-induced cancer cell energy metabolism and therapeutic opportunities. Clin Cancer Res. 2009;15:6479–83.
20. Trop-Steinberg S, Azar Y. Is Myc an important biomarker? Myc expression in immune disorders and cancer. Am J Med Sci. 2018;355:67–75.
21. Adhikary S, Eilers M. Transcriptional regulation and transformation by Myc proteins. Nat Rev Mol Cell Biol. 2005;6:635–45.
22. Biscardi JS, Ishizawar RC, Silva CM, Parsons SJ. Tyrosine kinase signaling in breast cancer: epidermal growth factor receptor and c-Src interactions in breast cancer. Breast Cancer Res. 2000;2:203–10.
23. Irby RB, Yeatma TJ. Role of Src expression and activation in human cancer. Oncogene. 2000;19:5636–42.
24. Lutz MP, Esser IB, Flossmann-Kast BB, Vogelmann R, Lührs H, Friess H, Büchler MW, Adler G. Overexpression and activation of the tyrosine kinase Src in human pancreatic carcinoma. Biochem Biophys Res Commun. 1998;243:503–8.
25. Groffen J, Heisterkamp N, Stephenson JR, Van Kessel AG, De Klein A, Grosveld G, Bootsma D. c-sis is translocated from chromosome 22 to chromosome 9 in chronic myelocytic leukemia. J Exp Med. 1983;158:9–15.
26. Van Kessel AG, Turc-Carel C, De Klein A, Grosveld G, Lenoir G, Bootsma D. Translocation of oncogene c-sis from chromosome 22 to chromosome 11 in a Ewing sarcoma-derived cell line. Mol Cell Biol. 1985;5:427–9.
27. Sarvaiya PJ, Schwartz JR, Hernandez CP, Rodriguez PC, Vedeckis WV. Role of c-Myb in the survival of pre B-cell acute lymphoblastic leukemia and leukemogenesis. Am J Hematol. 2012;87:969–76.
28. Nakano K, Uchimaru K, Utsunomiya A, Yamaguchi K, Watanabe T. Dysregulation of c-Myb pathway by aberrant expression of proto-oncogene MYB provides the basis for malignancy in adult T-cell leukemia/lymphoma cells. Clin Cancer Res. 2016;22:5915–28.
29. Drabsch Y, Hugo H, Zhang R, Dowhan DH, Miao YR, Gewirtz AM, Barry SC, Ramsay RG, Gonda TJ. Mechanism of and requirement for estrogen-regulated MYB expression in estrogen-receptor-positive breast cancer cells. Proc Natl Acad Sci USA. 2007;104:13762–7.
30. Hess JL, Bittner CB, Zeisig DT, Bach C, Fuchs U, Borkhardt A, Frampton J, Slany RK. c-Myb is an essential downstream target for homeobox-mediated transformation of hematopoietic cells. Blood. 2006;108:297–304.
31. Fan C, Wang LH, Jiao BH. Relation between c-myb transcription factor and cell proliferation and differentiation. Chem Life. 2011;31:697–701.
32. Hugo H, Cures A, Suraweera N, Drabsch Y, Purcell D, Mantamadiotis T, Phillips W, Dobrovic A, Zupi G, Gonda TJ, Lacopetta B, Ramsay RG. Mutations in the MYB intron I regulatory sequence increase transcription in colon cancers. Genes Chromosom Cancer. 2006;45:1143–54.
33. Surget S, Khoury MP, Bourdon JC. Uncovering the role of p53 splice variants in human malignancy: a clinical perspective. Onco Targets Ther. 2014;7:57–68.
34. Kato S, Han SY, Liu W, Otsuka K, Shibata H, Kanamaru R, Ishioka C. Understanding the function-structure and function -mutation relationships of p53 tumor suppressor protein by high-resolution missense mutation analysis. Proc Natl Acad Sci USA. 2003;100:8424–9.
35. Hainaut P, Hollstein M. p53 and human cancer: the first ten thousand mutations. Adv Cancer Res. 2000;77:81–137.

36. Halvorsen AR, Silwal-Pandit L, Meza-Zepeda LA, Vodak D, Vu P, Sagerup C, Hovig E, Myklebost O, Børresen-Dale A, Brustugun OT, Helland Å. TP53 mutation spectrum in smokers and never smoking lung cancer patients. Front Genet. 2016;7:85.
37. Canale M, Petracci E, Delmonte A, Chiadini E, Dazzi C, Papi M, Capelli L, Casanova C, Luigi ND, Mariotti M, Gamboni A, Chiari R, Bennati C, Calistri D, Ludovini V, Crinò L, Amadori D, Ulivi P. Impact of TP53 mutations on outcome in EGFR-mutated patients treated with first-line tyrosine kinase inhibitors. Clin Cancer Res. 2017;23:2195–202.
38. Li Y, Mair DC, Schuller RM, Li L, Wu J. Genetic mechanism of human neutrophil antigen 2 deficiency and expression variations. PLoS Genet. 2015;11:e1005255.
39. Sosa-García B, Gunduz V, Vázquez-Rivera V, Cress WD, Wright G, Bian H, Hinds PH, Santiago-Cardona PG. A role for the retinoblastoma protein as a regulator of mouse osteoblast cell adhesion: implications for osteogenesis and osteosarcoma formation. PLoS One. 2010;5: e13954.
40. Futreal PA, Liu Q, Shattuck-Eidens D, Cochran C, Harshman K, Tavtigian S, Bennett LM, Haugen-Strano A, Swensen J, Miki Y. BRCA1 mutations in primary breast and ovarian carcinomas. Science. 1994;266:120–2.
41. Wooster R, Bignell G, Lancaster J, Swift S, Seal S, Mangion J, Collins N, Gregory S, Gumbs C, Micklem G, Barfoot R, Hamoudi R, Patel S, Rices C, Biggs P, Hashim Y, Smith A, Connor F, Arason A, Gudmundsson J, Ficenec D, Kelsell D, Ford D, Bishop DT, Spurr NK, Ponder BAJ, Eeles R, Peto J, Devilee P, Cornelisse C, Lynch H, Narod S, Lenoir G, Egilsson V, Barkadottir RB, Easton DF, Bentley DR, Futreal PA, Ashworth A, Stratton MR. Identification of the breast cancer susceptibility gene BRCA2. Nature. 1995;378:789–92.
42. Gudmundsdottir K, Ashworth A. The roles of BRCA1 and BRCA2 and associated proteins in the maintenance of genomic stability. Oncogene. 2006;25:5864–74.
43. Yoshida K, Miki Y. Role of BRCA1 and BRCA2 as regulators of DNA repair, transcription, and cell cycle in response to DNA damage. Cancer Sci. 2004;95:866–71.
44. Zhang J, Sun J, Chen J, Yao L, Ouyang T, Li J, Wang T, Fan Z, Fan T, Lin B, Xie Y. Comprehensive analysis of BRCA1 and BRCA2 germline mutations in a large cohort of 5931 Chinese women with breast cancer. Breast Cancer Res Treat. 2016;158:455–62.
45. Li A, Xie R, Zhi Q, Deng Y, Wu Y, Li W, Yang L, Jiao Z, Luo J, Zi Y, Sun G, Zhang J, Shi Y, Liu J. BRCA germline mutations in an unselected nationwide cohort of Chinese patients with ovarian cancer and healthy controls. Gynecol Oncol. 2018;151:145–52.
46. Herrera L, Kakati S, Gibas L, Pietrzak E, Sandberg AA, Reynolds JF. Gardner syndrome in a man with an interstitial deletion of 5q. Am J Med Genet. 1986;25:473–6.
47. Rubinfeld B, Albert I, Porfiri E, Munemitsu S, Polakis P. Loss of β-catenin regulation by the APC tumor suppressor protein correlates with loss of structure due to common somatic mutations of the gene. Cancer Res. 1997;57:4624–30.
48. Chop AM, Abraham CL, Adrian TE, Murphy R, Boman BM. Immunodetection of the presence or absence of full-length APC gene product in human. Anticancer Res. 1995;15: 991–7.
49. Wang M, Wang H. Loss of heterozygous at APC and MCC genetic loci in esophageal cancer. Acta Acad Med Militaris Tertiae. 2000;22:653–5.
50. Lu Y, Xue H, Lu Y, Chi B. Significance of expression of APC gene in gastric cancer, precancerous lesion and gastric adenoma. J Harbin Med Univ. 2007;41:160–2.
51. Yao H, Pan S, Wei Y, Zhang Z, Suo Y, Huang P, Tong M. Expression of APC gene protein in non-small cell lung cancer. Jiangsu Med J. 2006;32:1123–4.
52. Saito T, Oda Y, Sakamoto A, Kawaguchi KI, Tanaka K, Matsuda S, Tamiya S, Iwamoto Y, Tsuneyoshi M. APC mutations in synovial sarcoma. J Pathol. 2002;196:445–9.
53. Nataraj AJ, Olivos-Glander I, Kusukawa N, Highsmith WE Jr. Single-strand conformation polymorphism and heteroduplex analysis for gel-based mutation detection. Electrophoresis. 1999;20:1177–85.

54. Jacquemier J, Moles JP, Penault-Llorca F, Adelaide J, Torrente M, Viens P, Birnbaum D, Theillet C. p53 immunohistochemical analysis in breast cancer with four monoclonal antibodies: comparison of staining and PCR-SSCP results. Br J Cancer. 1994;69:846–52.
55. Lin J. Application of silver-stained PCR-SSCP for detection of MCC mutations in colorectal cancer tissues. Med J Chin PLAV. 1998;23:198–9.
56. Bhattacharyya A, Lilley DM. The contrasting structures of mismatched DNA sequences containing looped-out bases (bulges) and multiple mismatches (bubbles). Nucleic Acids Res. 1989;17:6821–40.
57. Bhattacharyya A, Lilley DM. Single base mismatches in DNA long- and short-range structure probed by analysis of axis trajectory and local chemical reactivity. J Mol Biol. 1989;209:583–97.
58. Rossetti S, Corrà S, Biasi MO, Turco AE, Pignatti PF. Comparison of heteroduplex and single-strand conformation analyses, followed by ethidium fluorescence visualization, for the detection of mutations in four human genes. Mol Cell Probes. 1995;9:195–200.
59. Fischer SG, Lerman LS. DNA fragments differing by single base-pair substitutions are separated in denaturing gradient gels: correspondence with melting theory. Proc Natl Acad Sci USA. 1983;80:1579–83.
60. Muyzer G, De Waal EC, Uitterlinden AG. Profiling of complex microbial populations by denaturing gradient gel electrophoresis analysis of polymerase chain reaction-amplified genes coding for 16S rRNA. Appl Environ Microbiol. 1993;59:695–700.
61. Lu X, Jia Y, Fan H, Zhang L, Jia J. Methodological evaluation on PCR-DGGE technique in detecting DNA mutation and single nucleotide polymorphism. J Sichuan Univ (Med Sci Edi). 2007;38:882–4.
62. Mitani T, Akane A. Detection of single nucleotide polymorphism in the FUT2 gene by temperature gradient gel electrophoresis. Legal Med. 2005;7:217–21.
63. Youil R, Kemper BW, Cotton RG. Screening for mutations by enzyme mismatch cleavage with T4 endonuclease VII. Proc Natl Acad Sci USA. 1995;92:87–91.
64. Mashal RD, Koontz J, Sklar J. Detection of mutations by cleavage of DNA heteroduplexes with bacteriophage resolvases. Nat Genet. 1995;9:177–83.
65. Babon JJ, McKenzie M, Cotton RGH. The use of resolvases T4 endonuclease VII and T7 endonuclease I in mutation detection. Mol Biotechnol. 2003;23:73–81.
66. Del Tito Jr BJ, Poff HE III, Novotny MA, Cartledge DM, Walker RI, Earl CD, Bailey AL. Automated fluorescent analysis procedure for enzymatic mutation detection. Clin Chem. 1998;44:731–9.
67. Babon JJ, McKenzie M, Cotton RG. Mutation detection using fluorescent enzyme mismatch cleavage with T4 endonuclease VII. Electrophoresis. 1999;20:1162–70.
68. Howell WM, Jobs M, Gyllensten U, Brookes AJ. Dynamic allele-specific hybridization. Nat Biotechnol. 1999;17:87–8.
69. Nielsen PE, Egholm M, Berg RH, Buchardt O. Sequence-selective recognition of DNA by strand displacement with a thymine-substituted polyamide. Science. 1991;254:1497–500.
70. Luo JD, Chan EC, Shih CL, Chen TL, Liang Y, Hwang TL, Chiou CC. Detection of rare mutant K-ras DNA in a single-tube reaction using peptide nucleic acid as both PCR clamp and sensor probe. Nucleic Acids Res. 2006;34:e12.
71. Thiede C, Bayerdorffer E, Blasczyk R, Wittig B, Neubauer A. Simple and sensitive detection of mutations in the ras proto-oncogenes using PNA-meidated PCR clamping. Nucleic Acids Res. 1996;24:983–4.
72. Taback B, Bilchik AJ, Saha S, Nakayama T, Wiese DA, Turner RR, Kuo CT, Hoon DS. Peptide nucleic acid clamp PCR: a novel K-ras mutation detection assay for colorectal cancer micrometastases in lymph nodes. IntJ Cancer. 2004;111:409–14.
73. Chiou CC, Luo JD, Chen TL. Single-tube reaction using peptide nucleic acid as both PCR clamp and sensor probe for the detection of rare mutations. Nat Protoc. 2006;1:2604–12.
74. Chen TL, Chang JWC, Hsieh JJ, Cheng HY, Chiou CC. A sensitive peptide nucleic acid probe assay for detection of BRAF V600 mutations in melanoma. Cancer Genomics Proteomics. 2016;13:381–6.

75. Lyamichev VI, Kaiser MW, Lyamicheva NE, Vologodskii AV, Hall JG, Ma WP, Allawi HT, Neri BP. Experimental and theoretical analysis of the invasive signal amplification reaction. Biochemistry. 2000;39:9523–32.
76. Hessner MJ, Budish MA, Friedman KD. Genotyping of factor V G1691A (Leiden) without the use of PCR by invasive cleavage of oligonucleotide probes. Clin Chem. 2000;46:1051–6.
77. Olivier M, Chuang LM, Chang MS, Chen YT, Pei D, Ranade K, Witte AD, Allen J, Tran N, Curb D, Pratt R, Neefs H, Indig MDA, Law S, Neri B, Wang L, Cox DR. High-throughput genotyping of single nucleotide polymorphisms using new biplex invader technology. Nucleic Acids Res. 2002;30:e53.
78. Kani S, Tanaka Y, Matsuura K, Watanabe T, Yatsuhashi H, Orito E, Inose K, Motojuku N, Wakimoto Y, Mizokami M. Development of new IL28B genotyping method using Invader Plus assay. Microbiol Immunol. 2012;56:318–23.
79. Allawi HT, Li H, Sander T, Aslanukov A, Lyamichev VI, Blackman A, Elagin S, Tang YW. Invader plus method detects herpes simplex virus in cerebrospinal fluid and simultaneously differentiates types 1 and 2. J Clin Microbiol. 2006;44:3443–7.
80. Zou B, Cao X, Wu H, Song Q, Wang J, Kajiyama T, Kambara H, Zhou G. Sensitive and specific colorimetric DNA detection by invasive reaction coupled with nicking endonuclease-assisted nanoparticles amplification. Biosens Bioelectron. 2015;66:50–4.
81. Courtney DG, Moore JE, Atkinson SD, Maurizi E, Allen EHA, Pedrioli DML, Mclean WHI, Nesbit MA, Moore CBT. CRISPR/Cas9 DNA cleavage at SNP-derived PAM enables both in vitro and in vivo KRT12 mutation-specific targeting. Gene Ther. 2016;23:108–12.
82. Chen Y, Mei Y, Jiang X. Universal and high-fidelity DNA single nucleotide polymorphism detection based on a CRISPR/Cas12a biochip. Chem Sci. 2021;12:4455–62.
83. Ke Y, Huang S, Ghalandari B, Li S, Warden AR, Dang J, Kang L, Zhang Y, Wang Y, Sun Y, Wang J, Cui D, Zhi X, Ding X. Hairpin-spacer crRNA-enhanced CRISPR/Cas13a system promotes the specificity of single nucleotide polymorphism (SNP) identification. Adv Sci. 2021;8:2003611.
84. Myhrvold C, Freije CA, Gootenberg JS, Abudayyeh OO, Metsky HC, Durbin AF, Kellner MJ, Tan AL, Paul LM, Parham LA, Garcia KF, Barnes KG, Chak B, Mondini A, Nogueira ML, Iserm S, Michael SF, Lorenzana I, Yozwiak NL, Maclnnis BL, Bosch I, Gehrke L, Zhang F, Sabeti PC. Field-deployable viral diagnostics using CRISPR-Cas13. Science. 2018;360:444–8.
85. Gootenberg JS, Abudayyeh OO, Lee JW, Essletzbichler P, Dy AJ, Joung J, Verdine V, Donghia N, Daringer NM, Freije CA, Myhrvold C, Bhattacharyya RP, Livny J, Regev A, Koonin EV, Hung DT, Sabeti PC, Collins JJ, Zhang F. Nucleic acid detection with CRISPR-Cas13a/C2c2. Science. 2017;356:438–42.
86. Balderston S, Taulbee JJ, Celaya E, Fung K, Jiao A, Smith K, Hajian R, Gasiunas G, Kutanovas S, KiM D, Parkinson J, Dickerson K, Ripoll J, Peytavi R, Lu H, Barron F, Goldsmith BR, Collins PG, Conboy IM, Siksnys V, Aran K. Discrimination of single-point mutations in unamplified genomic DNA via Cas9 immobilized on a graphene field-effect transistor. Nat Biomed Eng. 2021;5:713–25.
87. Newton CR, Graham A, Heptinstall LE, Powell SJ, Summers C, Kalsheker N, Smith JC, Markham AF. Analysis of any point mutation in DNA. The amplification refractory mutation system (ARMS). Nucleic Acids Res. 1989;17:2503–16.
88. Qiang YZ, Qin T, Fu W, Cheng WP, Li YS, Yi G. Use of a rapid mismatch PCR method to detect gyrA and parC mutations in ciprofloxacin-resistant clinical isolates of Escherichia coli. J Antimicrob Chemother. 2002;49:549–52.
89. Koizumi M, Morita K, Takagi M, Yasumo H, Kasuya A. SNP genotyping by allele-specific PCR using ENA® primers. Nucleic Acids Symp Ser Oxf Univ Press. 2005;49:47–8.
90. Litos IK, Ioannou PC, Christopoulos TK, Traeger-Synodinos J, Kanavakis E. Genotyping of single-nucleotide polymorphisms by primer extension reaction in a dry-reagent dipstick format. Anal Chem. 2007;79:395–402.

91. Duan X, Liu L, Wang S. Homogeneous and one-step fluorescent allele-specific PCR for SNP genotyping assays using conjugated polyelectrolytes. Biosens Bioelectron. 2009;24:2095–9.
92. Myakishev MV, Khripin Y, Hu S, Hamer DH. High-throughput SNP genotyping by allele-specific PCR with universal energy-transfer-labeled primers. Genome Res. 2001;11:163–9.
93. Zhou GH, Kamahori M, Okano K, Chuan G, Harada K, Kambara H. Quantitative detection of single nucleotide polymorphisms for a pooled sample by a bioluminometric assay coupled with modified primer extension reactions (BAMPER). Nucleic Acids Res. 2001;29:e93.
94. Ronaghi M, Elahi E. Discovery of single nucleotide polymorphisms and mutations by pyrosequencing. Comp Funct Genomics. 2002;3:51–6.
95. Ronaghi M. Pyrosequencing sheds light on DNA sequencing. Genome Res. 2001;11:3–11.
96. LaFramboise T. Single nucleotide polymorphism arrays: a decade of biological, computational and technological advances. Nucleic Acids Res. 2009;37:4181–93.
97. Li Z, Tsunoda H, Okano K, Nagai K, Kambara H. Microchip electrophoresis of tagged probes incorporated with one-colored ddNTP for analyzing single-nucleotide polymorphisms. Anal Chem. 2003;75:3345–51.
98. Perrin F. Polarisation de la lumière de fluorescence. vie moyenne des molécules dans l' etat excité. J Phys Radium. 1926;7:390–401.
99. Chen X, Levine L, Kwok PY. Fluorescence polarization in homogeneous nucleic acid analysis. Genome Res. 1999;9:492–8.
100. Duan X, Li Z, He F, Wang S. A sensitive and homogeneous SNP detection using cationic conjugated polymers. J Am Chem Soc. 2007;129:4154–1455.
101. Duan X, Yue W, Liu L, Li Z, Li Y, He F, Zhu D, Zhou G, Wang S. Single-nucleotide polymorphism (SNP) genotyping using cationic conjugated polymers in homogeneous solution. Nat Protoc. 2009;4:984–91.
102. Duan X, Wang S, Li Z. Conjugated polyelectrolyte-DNA complexes for multi-color and one-tube SNP genotyping assays. Chem Commun. 2008;11:1302–4.
103. Song J, Yang Q, Lv F, Liu L, Wang S. Visual detection of DNA mutation using multicolor fluorescent coding. ACS Appl Mater Interfaces. 2012;4:2885–90.
104. Huang MM, Arnheim N, Goodman MF. Extension of base mispairs by Taq DNA polymerase: implications for single nucleotide discrimination in PCR. Nucleic Acids Res. 1992;20:4567–73.
105. Germer S, Holland MJ, Higuchi R. High-throughput SNP allele-frequency determination in pooled DNA samples by kinetic PCR. Genome Res. 2000;10:258–66.
106. Chen F, Zhao Y, Fan C, Zhao Y. Mismatch extension of DNA polymerases and high-accuracy single nucleotide polymorphism diagnostics by gold nanoparticle-improved isothermal amplification. Anal Chem. 2015;87:8718–23.
107. Murray HA, Latten MJ, Cartwright A, Mcaleer D, Fitzgerald SP. Simultaneous detection of colorectal cancer mutations in stool samples with biochip arrays. J Med Biochem. 2009;28: 285–92.
108. Lázaro A, Yamanaka ES, Maquieira A, Tortajada-Genaro LA. Allele-specific ligation and recombinase polymerase amplification for the detection of single nucleotide polymorphisms. Sensors Actuators B Chem. 2019;298:126877.
109. Park JH, Jang H, Jung YK, Jung YL, Shin I, Cho DY, Park HG. A mass spectrometry-based multiplex SNP genotyping by utilizing allele- specific ligation and strand displacement amplification. Biosens Bioelectron. 2017;91:122–7.
110. Shen W, Deng H, Teo AKL, Gao Z. Colorimetric detection of single-nucleotide polymorphisms with a real-time PCR-like sensitivity. Chem Commun. 2012;48:10225–7.
111. Cheng Y, Du Q, Wang L, Jia H, Li Z. Fluorescently cationic conjugated polymer as an indicator of ligase chain reaction for sensitive and homogeneous detection of single nucleotide polymorphism. Anal Chem. 2012;84:3739–44.
112. Sun Y, Lu X, Su F, Wang L, Liu C, Duan X, Li Z. Real-time fluorescence ligase chain reaction for sensitive detection of single nucleotide polymorphism based on fluorescence resonance energy transfer. Biosens Bioelectron. 2015;74:705–10.

113. Zhou L, Du F, Zhao Y, Yameen A, Chen H, Tang Z. DNAzyme based gap-LCR detection of single-nucleotide polymorphism. Biosens Bioelectron. 2013;45:141–7.
114. Cheng Y, Li Z, Du B, Zhang X. Homogeneous and label-free bioluminescence detection of single nucleotide polymorphism with rolling circle amplification. Analyst. 2008;133:750–2.
115. Zhou H, Wang H, Liu C, Wang H, Duan X, Li Z. Ultrasensitive genotyping with target-specifically generated circular DNA templates and RNA FRET probes. Chem Commun. 2015;51:11556–9.
116. Botstein D, White RL, Skolnick M, Davis RW. Construction of a genetic linkage map in man using restriction fragment length polymorphisms. Am J Hum Genet. 1980;32:314.
117. Soler S, Rittore C, Touitou I, Philibert L. A comparison of restriction fragment length polymorphism, tetra primer amplification refractory mutation system PCR and unlabeled probe melting analysis for LTA+252 C>T SNP genotyping. Clin Chim Acta. 2011;412: 430–4.
118. Shagin DA, Rebrikov DV, Kozhemyako VB, Altshuler IM, Shcheglov AS, Zhulidov PA, Bogdanova EA, Staroverov DB, Rasskazov VA, Lukyanov S. A novel method for SNP detection using a new duplex-specific nuclease from crab hepatopancreas. Genome Res. 2002;12:1935–42.
119. Finocchiaro G, Cappuzzo F, Janne PA, Bencardino K, Carnaghi C, Franklin WA, Crinò L, Santoro A, Varella-Garcia M. EGFR, HER2 and Kras as predictive factors for cetuximab sensitivity in colorectal cancer. J Clin Oncol. 2007;25:4021.
120. Lievre A, Bachet JB, Le Corre D, Boige V, Landi B, Emile JF, Côté JF, Tomasic G, Penna C, Ducreux M, Rougier P, Penault-Llorca F, Laurent-Puig P. KRAS mutation status is predictive of response to cetuximab therapy in colorectal cancer. Cancer Res. 2006;66:3992–5.
121. Karapetis CS, Khambata-Ford S, Jonker DJ, O'Callaghan CJ, Tu D, Tebbutt NC, Simes RJ, Chalchal H, Shapiro JD, Robitaille S, Price TJ, Shepherd L, Zalcberg JR. K-ras mutations and benefit from cetuximab in advanced colorectal cancer. N Engl J Med. 2008;359:1757–65.
122. Metro G, Chiari R, Duranti S, Siggillino A, Fischer MJ, Giannarelli D, Ludovini V, Bennati C, Marcomigni L, Baldi A, Giansanti M, Minotti V, Crinò L. Impact of specific mutant KRAS on clinical outcome of EGFR-TKI-treated advanced non-small cell lung cancer patients with an EGFR wild type genotype. Lung Cancer. 2012;78:81–6.
123. Sibbing D, Stegherr J, Latz W, Koch W, Mehilli J, Dörrler K, Morath T, Schömig A, Kastrati A, Beckerath NV. Cytochrome P450 2C19 loss -of-function polymorphism and stent thrombosis following percutaneous coronary intervention. Eur Heart J. 2009;30:916–22.
124. Simon T, Verstuyft C, Mary-Krause M, Quteineh L, Drouet E, Méneveau N, Steg PG, Ferrières J, Danchin N, Becquemont L. Genetic determinants of response to clopidogrel and cardiovascular events. N Engl J Med. 2009;360:363–75.
125. Claassens DM, Vos GJ, Bergmeijer TO, Hermanides RS, Van't Hof AW, Van Der Harst P, Barbato E, Morisco C, Tjon Joe Gin RM, Asselbergs FW, Mosterd A, Herrman JR, Dewilde WJM, Janssen PWA, Kelder JC, Postma MJ, Boer AD, Boersma C, Deneer VHM, Ten Berg JM. A genotype-guided strategy for oral P2Y12 inhibitors in primary PCI. N Engl J Med. 2019;381:1621–31.
126. Pereira NL, Farkouh ME, So D, Lennon R, Geller N, Mathew V, Bell M, Bae JH, Jeong MH, Chavez I, Gordon P, Abbott JD, Cagin C, Baudhuin L, Fu YP, Goodman SG, Hasan A, Iturriaga E, Lerman A, Sidhu M, Tanguay JF, Wang L, Weinshilboum R, Welsh R, Rosenberg Y, Bailey K, Rihal C. Effect of genotype-guided oral P2Y12 inhibitor selection vs conventional clopidogrel therapy on ischemic outcomes after percutaneous coronary intervention: the TAILOR-PCI randomized clinical trial. JAMA. 2020;324:761–71.
127. Cohen JD, Li L, Wang Y, Thoburn C, Afsari B, Danilova L, Douville C, Javed AA, Wong F, Mattox A, Hruban RH, Wolfgang CL, Goggins MG, Molin MD, Richard Roden TW, Klein AP, Ptak J, Dobbyn L, Schaefer J, Silliman N, Popoli M, Vogelstein JT, Browne JD, Schoen RE, Brand RE, Tie J, Wong PG, Mansfield AS, Jen J, Hanash SM, Falconi M, Allen PJ, Zhou S, Bettegowda C, Diaz LA, Tomasetti C, Kinzler KW, Bogelstein B, Lennon AM,

Papadopoulos N. Detection and localization of surgically resectable cancers with a multi-analyte blood test. Science. 2018;359:926–30.

128. Cohen JD, Javed AA, Thoburn C, Wong F, Tie J, Gibbs P, Schimidt CM, Yip-Schneider MT, Allen PJ, Schattner M, Brand RE, Singhi AD, Petersen GM, Hong SM, Kim SC, Falconi M, Doglioni C, Weiss MJ, Ahuja N, He J, Makary MA, Maitra A, Hanash SM, Molin MD, Wang Y, Li L, Ptak J, Dobbyn L, Schaefer J, Silliman N, Popoli M, Goggins MG, Hruban RH, Wolfgang CL, Klein AP, Tomasetti C, Papadopoulos N, Kinzler KW, Vogelstein B, Lennon AM. Combined circulating tumor DNA and protein biomarker-based liquid biopsy for the earlier detection of pancreatic cancers. Proc Natl Acad Sci USA. 2017;114:10202–7.
129. Zhou Z, Wu Q, Yan Z, Zheng H, Chen CJ, Liu Y, Qi Z, Calandrelli R, Chen Z, Chien S, Su HI, Zhong S. Extracellular RNA in a single droplet of human serum reflects physiologic and disease states. Proc Natl Acad Sci USA. 2019;116:19200–8.
130. Liu B. Association between SNPs in P53 binding sites and risk of esophageal squamous cell carcinoma. Third Military Medical University; 2012.
131. Chen Y. Research on the association between human gut metagenomic SNP pattern and diseases. Chinese People's Liberation Army Academy of Military Medical Sciences; 2017.
132. Kruglyak L. Prospects for whole-genome linkage disequilibrium mapping of common disease genes. Nat Genet. 1999;22:139–44.
133. Zhong X, Zan L, Wang H, Lin Q, Hu Y, Li C, Huang L, Xin Y. Polymorphism of exon 8 of BMPR-IB gene and its relationship with body measurement traits in cattle. Acta Agriculturae Boreali-occidentalis Sinica. 2010;19:12–6.
134. Ye Q. Association of genetic variation of the IGF2, IGFBP2 and DRD2 gene with muscovy duck egg production traits. South China Agricultural University; 2016.
135. Cai M, Yang K, Wang L, Zuo F. SNP and its application in beef cattle breeding. Chin Agric Sci Bull. 2012;28:7–10.
136. Li Y. Molecular cloning of growth trait-related genes in penaeus monodon and screening and verification of their SNPs makers. Shanghai Ocean University; 2016.
137. Liu X. Study on molecular marker-assisted breeding with single nucleotide polymorphisms linked to cold hardiness and chilling requirement in blueberry. Beijing Forestry University; 2013.
138. Shen Q, Zhang D, Sun W, Zhang Y, Shang Z, Chen S. Medicinal plant DNA marker assisted breeding (II) the assistant identification of SNPs assisted identification and breeding research of high yield Perilla frutescens new variety. China J Chin Mater Med. 2017;42:1668–72.

Chapter 9
Biological Functions and Analytical Methods of DNA Methylation

Fengxia Su and Weiliang Liu

Genetics refers to the study of hereditary changes in gene phenotype or function due to changes in DNA sequence. Such alterations include point mutation, deletion, insertion, and translocation. In contrast, epigenetics is the study of heritable changes in gene activity or function with no change in DNA sequence [1]. In the early twentieth century, many scientists believed that the differentiation of progenitor cells into different somatic cells was due to an increase or decrease in their genomic DNA. However, Conrad Hal Waddington argued that the entire set of genomic DNA of a progenitor cell remains unchanged during this process and that all the changes is the differential regulation and expression of cells between different types. These changes are a response to the environment in which the organism finds itself. To explain the relationship between the intrinsic genes of a species and its environment, Conrad Hal Waddington was the first to introduce the concept of epigenetics [2–5]. After years of in-depth research, the understanding of gene regulation in eukaryotic cells continues to deepen, and the connotation of epigenetics continues to be updated. Epigenetics now focuses on the study of heritable changes in gene function during mitosis and meiosis that cannot be explained by changes in DNA sequence [1]. Epigenetic mechanisms mediate diverse gene expression profiles in a variety of cells and tissues in multicellular organisms and can regulate a wide range of complex phenotypes. Current epigenetic research is focused on three main directions: DNA methylation, histone modification, and chromatin remodeling [6–9].

F. Su (✉) · W. Liu
University of Science and Technology Beijing, Beijing, China
e-mail: sufengxia@ustb.edu.cn; lwliang@tzc.edu.cn

Z. Li (ed.), *Nucleic Acid Detection*, https://doi.org/10.1007/978-981-95-7295-3_9

9.1 DNA Methylation Process

Currently, DNA methylation is one of the most widely studied epigenetic modifications. In the mammalian genome, DNA methylation is the process of methyl transfer from S-adenosyl methionine (SAM) to the C5 position of cytosine to form 5-methylcytosine (5mC) catalyzed by a series of DNA methyltransferases (DNMT), occurring primarily on the cytosine of CpG dinucleotides [10] (Fig. 9.1), while SAM is converted to S-adenosylhomocysteine (SAH).

Seventy percent of mammalian adult tissue cell DNA is methylated [11], and 5mC is non-randomly distributed throughout the genome [12]. When analyzing CpG dinucleotide pairs in genomic DNA, some regions were found to have a higher proportion of CG sequences than expected [13]. This CG-rich region, known as the CpG island, spans approximately 1000 base pairs and is typically located in the promoter or 5′-terminal region of human genome. It is estimated that about 60% of human gene promoters contain CpG islands [6, 14]. The methylation state of CpG island located in the gene promoter plays an important role in regulating gene expression and establishing gene imprinting [15–17]. In germ cells and early embryos, CpG islands are unmethylated [18], and methylated CpG dinucleotides mainly occur outside the CpG islands.

During development, DNA methylation patterns in the genome change as a result of dynamic processes of DNA methylation and demethylation [7, 10]. Based on the process of DNA methylation establishment, removal and recognition, DNA methylation enzymes can be classified into three categories: writers, erasers, and readers. The writers of methylation are DNMT including DNMT1, DNMT3a, and DNMT3b. These enzymes are structurally similar, but they have their own unique expression patterns and functions [7, 19]. DNMT1, also known as a maintenance DNMT, comprises a C-terminal catalytic domain, and an N-terminal domain responsible for localizing the protein to the nucleus and replication sites. It preferentially

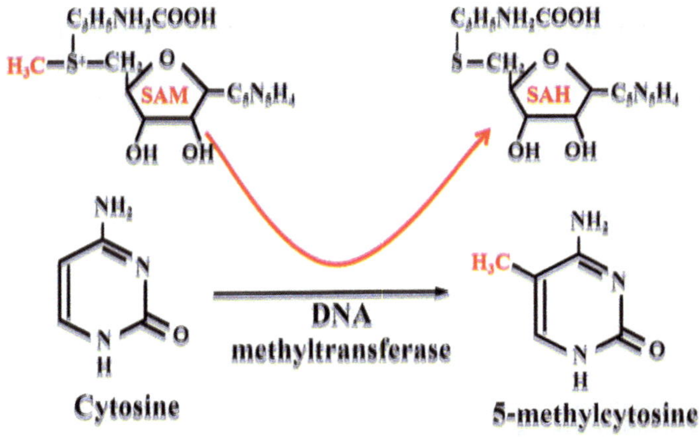

Fig. 9.1 Methylation of cytosine in DNA. (Revised and reprinted with permission from Ref. [10])

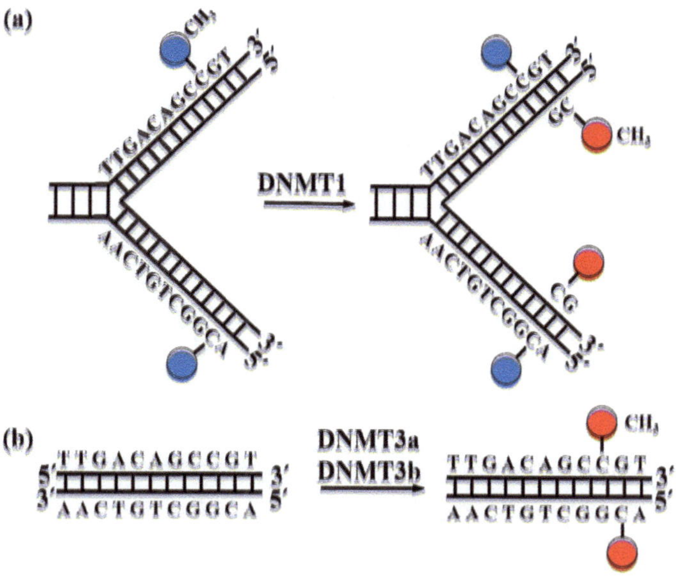

Fig. 9.2 Pathways of DNA methylation. (Revised and reprinted with permission from Ref. [7])

methylates hemimethylated DNA, and during DNA replication, maintains the established methylation patterns by copying them from the parental strand to the newly synthesized daughter strand (Fig. 9.2a), ensuring replication of the methylation pattern during each round of cell division. Thus, DNMT1 plays a key role in cell differentiation and division [7, 20]. DNMT3a and DNMT3b can establish new methylation patterns on unmodified DNA (Fig. 9.2b) and are therefore called de novo DNMT. DNMT3a or DNMT3b are highly active in embryonic stem cells and early embryonic cells [21]. Mice lacking both enzymes exhibit distinct developmental defects or die at various stages depending on the specific deficiency. DNMT3b methylates a wider range of target sequences during early development and is able to specifically methylate centromeric satellite repeat sequence de novo. Its expression is low in most differentiated tissues, with the exception of the thyroid, testis, and bone marrow [22]. In contrast, DNMT3a is responsible for methylating gene sequences involved in later stages of development or postnatal expression [21].

Methylation erasers modify and remove methyl [7]. The process of demethylation can be divided into active demethylation and passive demethylation. Passive demethylation tends to occur in dividing cells, where failure to maintain methylation due to functional inhibition of DNMT1 results in reduced levels of methylated cytosines in the genome. Active demethylation can occur in dividing and undividing cells. The active demethylation process requires an enzymatic reaction to process 5mC to obtain an intermediate product that is ultimately recognized and replaced with unmodified cytosine by the base excision repair (BER) mechanism [23–26]. The BER mechanism has been identified as the final step in the reaction to generate unmodified cytosine, but the enzymes and chemical intermediates in the

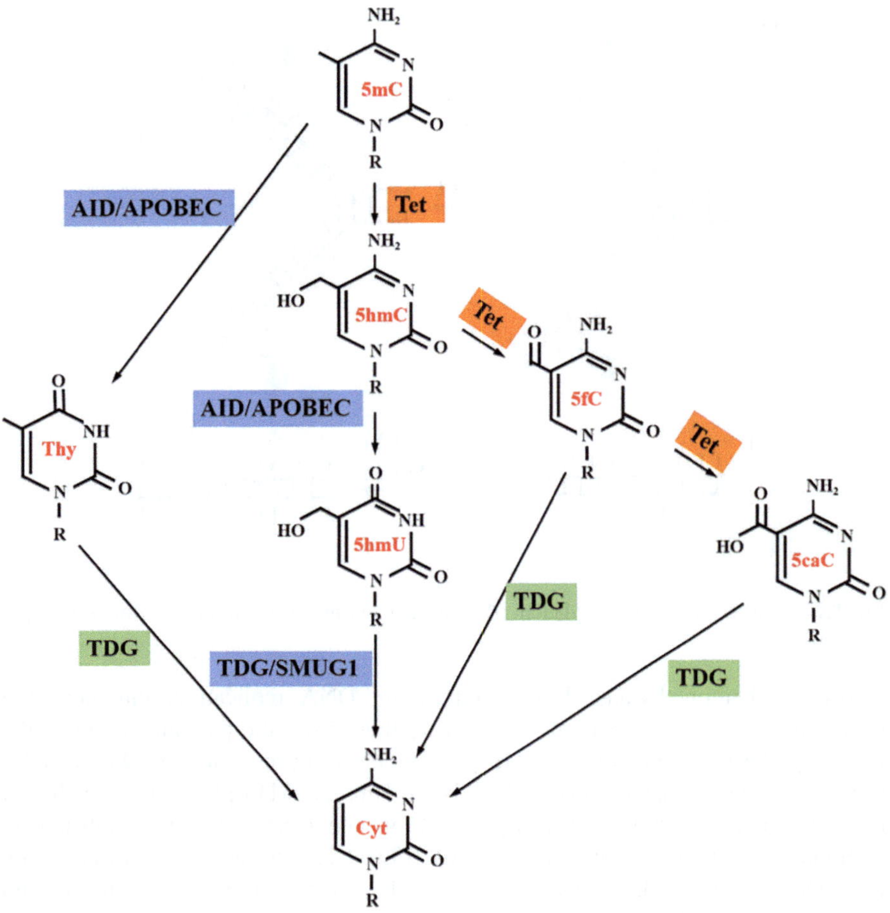

Fig. 9.3 Several possible demethylation pathways. (Revised and reprinted with permission from Ref. [7])

intermediate series of reactions remain controversial [27]. Several possible mechanisms of active demethylation exist [7] as shown in Fig. 9.3. The first pathway is that the amine group of 5mC is removed by activation-induced cytidine deaminase/apolipoprotein B mRNA-editing enzyme complex (AID/APOBEC), converting 5mC to thymine (Thy). Thy is recognized by thymine DNA glycosylase (TDG) and replaced with unmodified cytosine via the BER pathway. Other active demethylation mechanisms are mediated by the ten-eleven translocation (Tet) enzymes. The second of these pathways involves the addition of hydroxyl group to the methyl group of 5mC mediated by Tet enzymes to produce 5-hydroxymethylcytosine (5hmC) [28, 29]. 5hmC undergoes AID/APOBEC deamination to produce 5-hydroxymethyl-uracil (5hmU) [30], which in turn is recognized by TDG or single-strand-selective monofunctional uracil-DNA glycosylase1 (SMUG1), and replaced with unmodified cytosine. In the third pathway, 5mC is further oxidized by Tet enzymes to form 5-formylcytosine (5fC) after Tet enzyme-mediated generation of 5hmC, which is then recognized and replaced by TDG with unmodified cytosine. In the fourth pathway, 5mC is mediated by Tet enzyme to produce 5hmC, which is oxidized by Tet enzyme to form 5fC and then form 5-carboxycytosine (5caC) [31], which is eventually recognized and replaced by TDG as unmodified cytosine.

It has been reported that the 5mC to 5hmC transition inhibits the binding of the methyl-CpG-binding protein 2 (MeCP2) [32]. Whether 5hmC acts only as an intermediate product of DNA demethylation is unclear; however, it is well established that 5hmC is contained in mammalian tissues and plays an important role in regulating DNA demethylation and gene expression. In the DNA demethylation mechanism mentioned above, TDG can recognize and repair abnormal T/G base mismatches through the action of BER, and carry out base-cutting repair on abnormal bases produced by deamination such as Thy and 5hmU [33]. 5caC is a further oxidation product of 5fC, and TDG excises 5fC and 5caC and restores unmodified G:C pairing through different mechanisms [34]. Thus, TDG is essential for maintaining genomic stability during DNA demethylation and is required for normal development. Knockdown or inactivation of TDG results in the death of mouse embryos. Furthermore, these abnormal embryos exhibit hypermethylation, particularly in imprinted genes such as *Igf2* and *H19*, suggesting that an TDG-dependent active demethylation process can inhibit imprinted gene methylation [35]. SMUG1, a BER enzyme of the same family of uracil DNA glycosylases as TDG, has also been shown to be involved in the DNA demethylation process [30, 33–36].

Methyl-binding proteins (MBP) have the ability to recognize and specifically bind methyl-CpG (mCpG) and include three separate families: mCpG-binding domain (MBD) proteins, ubiquitin-like, containing PHD and RING finger domain (UHRF) proteins and zinc finger proteins. The first to be identified was the MBD protein family, which contains a conserved MBD that has a higher affinity for a single mCpG [37]. This family includes the first identified methyl-binding proteins MeCP1 and MeCP2, as well as MBD1, MBD2, MBD3, and MBD4 [38, 39]. MeCP1 is a large multi-protein complex, while MeCP2 is a polypeptide with affinity for a single mCpG. MeCP2 has an important role in maintaining methylation in addition to acting as a transcriptional repressor. MeCP2 binds to DNMT1 through its transcriptional repression domain (TRD) and can recruit DNMT1 to hemimethylated DNA for maintenance of methylation [40]. In the following, researchers successfully identified the smaller MBD [41, 42] and its TRD. MBD1 contains a TRD that binds to DNA containing mCpG and represses gene transcription away from its binding site. Ng reported that MBD2 is actually a component of the previously identified MeCP1 complex [43], which exhibits histone deacetylase activity and can specifically bind methylated DNA. Due to mutations in MBD3's mCpG-binding domain, it cannot bind directly to DNA [44, 45]; however, as a complete subunit of the mammalian histone deacetylation and nucleosome remodeling (Mi-2/NuRD) complex, MBD3 can be recruited to the methylated DNA region by interacting with MBD2. MBD4 does not inhibit gene transcription, and after binding to DNA, it has G/T mismatch glycosylase activity and 5mC DNA glycosylase activity [46].

UHRF proteins, including UHRF1 and UHRF2, are multi-structural domain proteins that bind to 5mC via a SET and RING-associated DNA-binding domain [47]. The UHRF protein family first binds to DNMT1 and then targets it to hemimethylated DNA in order to maintain DNA methylation, a process that occurs primarily during DNA replication [48, 49]. Similar to the results of knockdown of DNMT1, knockdown of UHRF1 results in embryonic death and genome-wide hypomethylation [50, 51].

The zinc finger family, consisting of Kaiso, ZBTB4, and ZBTB38, binds to methylated DNA through the zinc finger structural domain. Kaiso proteins use three zinc finger structures to bind at least two consecutive mCpG- or Kaiso-binding site TCCTGCNA [52]. Unlike Kaiso, ZBTB4 and ZBTB38 can bind a single mCpG [51]. ZBTB4 and ZBTB38 have significant tissue expression specificity and are highly expressed in the brain. Similar to the MBD protein family, zinc finger proteins repress transcription in a DNA methylation-dependent manner [53].

9.2 Biological Functions of DNA Methylation

9.2.1 Repressing Transcription of Genomic DNA

The role of DNA methylation in gene regulation was first proposed in the 1970s [54, 55], and as epigenetic studies have continued over the years, supporting evidence has accumulated. Low methylation levels of CpG dinucleotides within the promoter region allow transcription factor binding and normal transcription to occur (Fig. 9.4a). DNA methylation represses gene transcription in three general ways. First, mCpG dinucleotides may directly interfere with the binding of transcription factors to their recognition sequences by protruding into the DNA major

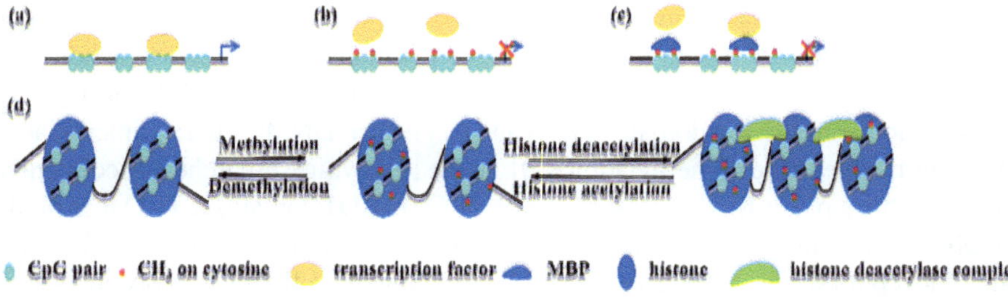

Fig. 9.4 Mechanisms by which DNA methylation represses gene transcription. (**a**) Hypomethylation of CpG dinucleotides within the promoter region allows transcription factor binding and transcription to occur. (**b**) mCpG directly inhibits transcription factor binding, thereby repressing gene transcription. (**c**) Within the promoter region, MBP binding prevents transcription factors from entering their binding sites. (**d**) MBP, which binds to mCpG in the promoter region, forms complexes with histone deacetylases and corepressors, resulting in histone deacetylation, chromatin coagulation, and non-transcriptional chromatin structure. (Revised and reprinted with permission from Ref. [10])

groove (Fig. 9.4b). In the case of direct inhibition of transcription, the methylation site must be located in or near the promoter region or other *cis*-acting regulatory element. AP-2 is the transcription factor that recognizes and binds to the promoter sequence of the proencephalin gene. In vitro studies have shown that methylation of the CCGG site in this promoter sequence inhibits the binding of AP-2, indicating that methylation of this site can directly inhibit gene expression [56]. Another study indicated that although the binding of transcription factor Sp1 to recognition sequences was also inhibited by DNA methylation [57], Sp1 was able to bind to the synthetic recognition sequences when they contained mCpG [58], suggesting that in addition to the direct interference of mCpG with transcription factor binding DNA, MBP may also prevent the binding of transcription factors to their promoter sequences [57] (Fig. 9.4c).

MBP, histone deacetylase complexes can indirectly inhibit transcription factor binding [59]. Methylated DNA regions close to gene promoter sites may recruit MBP, which in turn can bind to large protein complexes containing corepressors and histone deacetylases (Fig. 9.4d). The binding of these complexes to DNA results in a shift in chromatin structure from the active form to the inactive form. The most typical is the methylcytosine-binding protein MeCP2 concentrated in juxtacentromeric heterochromatin, which has a TRD between amino acids 221 and 249 [60, 61] and contains an mCpG-binding region within an 85 amino acid domain near the N terminus [37] that can bind to DNA containing at least one mCpG [39]. The Sin3A histone deacetylase complex contains seven proteins, including the transcriptional repressor Sin3A, the histone deacetylases HDAC1 and HDAC2 [62]. Mouse MeCP2 binds to this complex, thereby linking DNA methylation to histone deacetylation. Histone deacetylation represses gene expression by removing acetyl groups from histones, which in turn leads to more compact, transcriptionally inactive chromatin structures [63]. A similar linkage exists between homologous proteins in *Xenopus laevis* oocytes [64]. However, transcriptional repression of MeCP2 may also occur independently of histone deacetylases [59], suggesting that MeCP2 may compete directly with transcription factors to block gene expression (Fig. 9.4c). Due to the large number of mCpG dinucleotides in the genome, MeCP2 may function as a genome-wide transcriptional repressor [60].

Two additional histone deacetylase complexes were identified. The first complex includes the MBD2-containing MeCP1 [38, 43], the histone deacetylases HDAC1 and HDAC2, and the retinoblastoma protein-binding proteins RbAp46 and RbAp48, which exist in the Sin3A histone deacetylase complex. The second histone deacetylase complex is the Mi-2 (NuRD) complex, which contains at least seven proteins, including Mi-2, which is a protein thought to have chromatin remodeling activity, HDAC1, HDAC2, RbAp46, RbAp48, and MBD3 [65]. It has been found that MBD2 is associated with Sin3A [66] and the Mi-2/NuRD complex [65] and may play a role in Sin3A and Mi2/NuRD histone deacetylase complex-mediated gene repression.

DNMT1 was identified in a complex containing MBD2 and MBD3, capable of binding to hemimethylated DNA and located at the replication site during late DNA synthesis [67]. MBD3 knockout mice die early in embryonic development, whereas

MBD2 knockout mice appear to be essentially normal except for a defect in maternal behavior [68]. In addition to MBD2 and MBD3, the MBD family includes two other members, MBD1 and MBD4. MBD1 contains a TRD that binds to DNA containing mCpG and represses gene transcription away from its binding site [69]. And its ability to repress transcription can be inhibited by the histone deacetylase inhibitor trichostatin A, suggesting that deacetylases are involved in MBD1-mediated gene repression. However, MBD1 was not associated with HDAC1 as MeCP2, MBD2, and MBD3, suggesting that the mode of action of MBD1 may be different [70]. Unlike other MBD family members, MBD4 does not repress gene transcription, but has G/T mismatch glycosylase activity as well as 5-methylcytosine DNA glycosylase activity [46].

9.2.2 DNA Methylation in Embryonic Development

Genomic DNA exhibits varying levels of methylation at different stages of embryonic development. Two dramatic changes in methylation levels of genomic DNA occur during primordial germ cell formation and during pre-implantation embryonic development. Extensive demethylation occurs during the pre-implantation period and this change in methylation levels is of key importance for the generation of embryonic pluripotency. At fertilization, highly methylated sperm undergo a rapid, almost complete demethylation process, and this demethylation is caused by active demethylation mechanisms [71]. Genomic CpG methylation levels in sperm are 80–90%, and after a pulse of paternally targeted demethylation, genome-wide methylation is further cleared after subsequent divisions and reaches a minimum before embryonic differentiation in blastocysts [72, 73]. In contrast, the maternal genome exhibits lower levels of methylation, and thus both show typical epigenetic asymmetries during early embryonic development [24, 70].

The paternal methylome is oxidized by Tet3 to become 5hmC and preferentially demethylated to give unmodified cytosine [74–76]. It was found that Tet3 is specifically localized to the male pronucleus. If Tet3 is absent, 5hmC is not detected and may also cause delayed activation of genes associated with embryonic development on the paternal chromosome and result in delayed establishment of pluripotency [76]. The maternal genome appears to be generally static during reprogramming of the paternal genome. This quiescent process occurs because the maternal pronucleus genome in fertilized zygotes exhibits strong *H3K9* methylation. The Stella protein recognizes and binds to methylated *H3K9* [77], thereby altering the chromatin conformation and protecting the maternal genome from Tet3 oxidation. If the Stella protein is absent, the maternal genome also undergoes events such as Tet3 oxidation [73, 75–79]. During early embryonic development, the specific methylation pattern of the maternal genome is largely dependent on the loss of a passive demethylation process. This is attributed to nuclear rejection by DNMT1 and also indicates an antagonistic effect of Stella protein binding to *H3K9* methylation on Tet3 oxidation. An oocyte-specific DNMT1 isoform, called DNMT1O, is not expressed until the four-cell stage and is responsible for maintaining its methylation levels and plays an important role in maintaining methylation of imprinted genes [80–83].

9.2.3 DNA Methylation and X-Chromosome Inactivation

Human DNA is composed of 23 pairs of chromosomes, including 22 pairs of autosomes and one pair of sex chromosomes. Of each pair of chromosomes, one originates from the father and the other from the mother. Males and females have the same autosomes, with sexual differences arising from disparities in their sex chromosomes. Females have two X chromosomes, while males have one X and one Y chromosome. In males, the X chromosome always originates from the mother and the Y chromosome from the father; in females, the two X chromosomes originate from the father and mother, respectively. This chromosomal difference between the sexes is common in mammals and other organisms. Because the two sexes carry different numbers of X-linked genes, twice as many in females as in males, this can lead to an imbalance in the number of RNAs and proteins. To avoid this phenomenon, dosage compensation mechanisms are used to balance the X gene products between the sexes.

In 1961, Lyon proposed dosage compensation in mammals by silencing one of the two X chromosomes in female cells to achieve transcriptional homeostasis with XY in males [84]. There are two modes of X-chromosome inactivation, imprinted and random inactivation. The imprinted mode is the earlier discovered mechanism in marsupial mammals, such as kangaroos, that undergo non-random X-chromosome inactivation and preferentially shut down the X chromosome of the paternal line [85, 86]. The random X-inactivation takes place in eutherians (placental mammals), so that paternally or maternally inherited X can be silenced. In some cases, traces of imprinted inactivation can be observed in extraembryonic tissues [87, 88], while random inactivation occurs in embryonic cell lines. In recent years, significant progress has been made in understanding both forms of X-chromosome inactivation [89].

In the mid-1990s, transgenic analysis in embryonic stem cells confirmed the concept of the "X-inactivation center" (XIC/*Xic*), which controls the initiation of X-chromosome inactivation [90–94]. Researchers have presented different estimates of how many sequences are present in *Xic*; however, almost everyone agrees that Hunt Wilard's group first proposed that *Xic* contains the noncoding RNA known as X inactivation-specific transcript (*Xist*) [95–97]. It was shown that *Xist* is characterized by *cis* binding and accumulates along the entire chromosome starting from the transcription site [95, 98, 99]. Chromosomes covered with *Xist* RNA provide preparation for the initiation of X chromosome silencing [90, 100]. If the 5′ end of the *Xist* gene is in an unmethylated state, the gene can be transcribed normally to form RNA. This noncoding RNA gradually spreads along the chromosome to wrap around the chromosome, resulting in gene silencing and inactivation of that chromosome. The inactivated X chromosome maintains a low transcriptional level and lags replication, which is similar to structural heterochromatin. If the *Xist* gene is fully methylated at the 5′ end, repressing transcription fails to form RNA and cannot create wrapping and silencing effects on the chromosome, therefore the chromosome remains active [101, 102].

9.2.4 DNA Methylation in Genomic Imprinting

Mammalian cells have two pairs of chromosomes that are inherited from each parent, so there are two copies of each gene. The maternal and paternal versions of each gene have the same potential to be activated in all cells. Genomic imprinting, an epigenetic mechanism independent of DNA sequence, alters this potential by restricting the expression of a gene on only one of the two chromosomes. Imprinted marks are established in gametes during primordial germ cell development in the same manner as the original parent. In mammals, DNA methylation and histone modifications play a key role in establishing imprinted marks [103–105]. Imprinting locus contains imprinting control regions (ICRs) that control the expression of imprinted genes by a unique mechanism depending on their location. ICRs are generally located in intergenic regions or promoter regions of specific gene imprinting sites (promoter ICRs) [106, 107]. Many promoter ICRs control the expression of noncoding RNAs that regulate the imprinting of other genes at imprinted sites. For example, ICR2 is located in the promoter region of *KCNQ1OT1* and regulates imprinting at the *CDKN1C/KCNQ1* locus. KCNQ1OT1 is a long-stranded noncoding RNA that silences genes through *cis* control and functions similarly to *Xist* RNA [108–110]. Normally, the maternal line of ICR2 is methylated while the paternal line is unmethylated. Thus, *KCNQ1OT1* RNA is silenced in the maternal genome and expressed in the paternal genome, resulting in maternal expression of imprinted genes at the *CDKN1C/KCNQ1* locus and paternal silencing [111].

The intergenic ICR regulates gene imprinting in different ways by acting as a transcriptional insulator in the imprinting cluster. For example, ICR1 is located between the *H19* and *Igf2* genes and controls the *H19/Igf2* cluster. The hypomethylated state of ICR1 promotes the binding of a transcriptional repressor protein to this region, which in turn isolates *Igf2* from downstream enhancers. Without transcriptional repressor protein binding, downstream enhancers can contact the *Igf2* promoter, allowing *Igf2* expression [112, 113]. Typically, the paternal ICR1 of the *H19/*Igf2 cluster is hypermethylated, while the maternal ICR1 is hypomethylated. Thus, *Igf2* is expressed monoallelized in the paternal allele and silenced in the maternal allele. It was shown that most promoter ICRs are methylated in the maternal genome, while some intergenic ICRs are methylated in the paternal genome [114, 115].

Any defect in the imprinting process may lead to loss of imprinting at key imprinting loci. In most cases, loss of imprinting is detrimental to fetal development and may lead to neurological, developmental, and metabolic disorders [105]. Loss of imprinting is due to four specific causes, deletion, mutation, uniparental disomy, and epimutations. Mutations or deletions are common causes of imprinting disorders and may affect genes within imprinted loci, promoters, intergenic regions, and ICRs at imprinted loci. Uniparental disomy is the abnormal acceptance of two paternal chromosomes or two maternal chromosomes [116]. Epimutations may be caused by aberrant methylation of the ICR or disruption of histone modifications within the imprinted cluster. These causes can affect imprinted loci and lead to abnormal silencing of the active allele or expression of the inactive allele.

9.2.5 DNA Methylation and Tumorigenesis

Cancer has become a class of diseases that seriously threaten human life and health due to its high incidence and lethality. Attempts have been made to find the relationship between gene regulation and tumorigenesis development at the molecular level. The critical role of mutation in the generation of human cancers has been well established in the last decades. Recently, the potential of epigenetic processes to complement genetic changes has been demonstrated. In addition to harboring multiple mutations, almost all human cancers contain a large number of epigenetic abnormalities that in concert with genetic lesions to drive the cancer phenotypes. Among these, alterations in DNA methylation levels and patterns are a hallmark, including the hypermethylation of cancer suppressor gene promoters and the hypomethylation of oncogenes.

In normal cells, the CpG islands within tumor-suppressor gene promoters are hypomethylated or unmethylated. This unmethylated state is associated with active transcription, allowing for continuous gene expression that helps suppress tumorigenesis. In contrast, in tumor cells, hypermethylation of CpG islands within tumor-suppressor gene promoter leads to transcriptional silencing. This epigenetic alteration can trigger chromatin conformational changes and a cascade of oncogenic events, including entry into the cell cycle, loss of apoptosis, defective DNA repair, angiogenesis, and loss of cell adhesion [117], ultimately contributing to tumorigenesis (Fig. 9.5, left). The literature [118] lists the correlation between genes with highly methylated promoter silencing and cancers.

Erroneous gene activation is also a hallmark of cancer, as shown on the right of Fig. 9.5. Some genes and repetitive sequences in normal cells are in a highly methylated state and may play a role in protecting chromosomal integrity from chromosomal instability, translocation, and destruction. If their methylation levels are reduced, expression of these genes will occur and repetitive sequences will be activated, leading to loss of gene imprinting, excessive cell growth, increased genomic fragility, and ultimately, tumorigenesis as well [119–121]. In addition, hypomethylation of oncogenes may lead to activation of oncogenes and thus be associated with tumorigenesis. Hypomethylation of *raf, c-myc, c-fos, c-H-ras*, and *c-K-ras* genes was observed in chemical carcinogenesis-induced rat liver cancer nodules, demonstrating their association with tumor formation [122]. Among them,

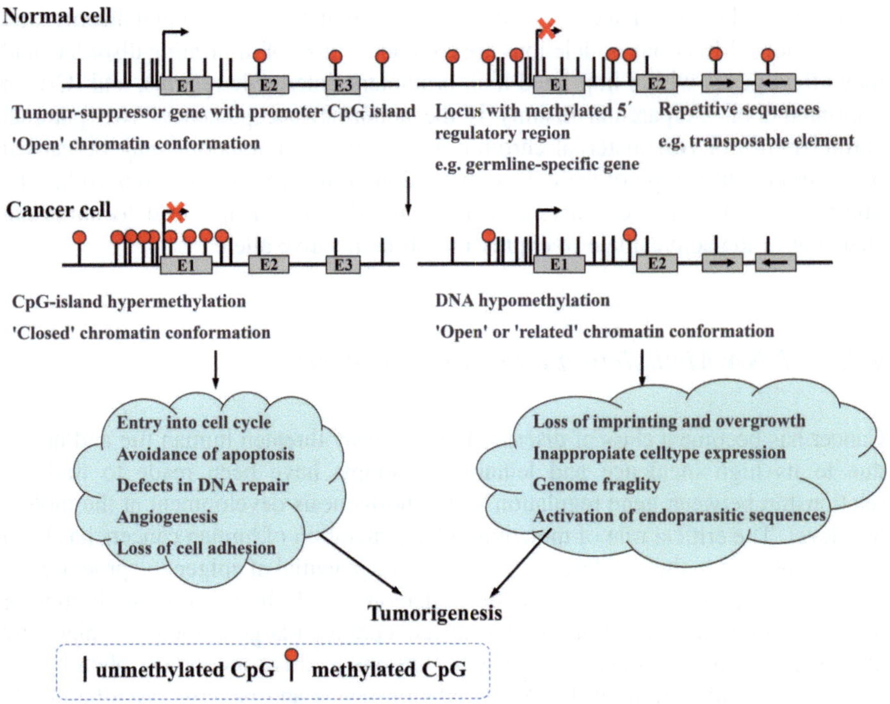

Fig. 9.5 DNA methylation patterns during tumorigenesis. (Revised and reprinted with permission from Ref. [118])

the third exon of *c-myc* gene is hypomethylated in a variety of human tumor cells, while it is fully methylated in normal cells [123].

9.3 Analytical Methods of DNA Methylation

DNA methylation does not involve changes in DNA sequence and cannot be distinguished by conventional nucleic acid analysis methods, so it is necessary to use special mCpG identification methods for identification and analysis. A variety of methods have been developed to detect DNA methylation, which can be divided into three categories according to the difference between methylated and unmethylated DNA means: sodium bisulfite ($NaHSO_3$) treatment, restriction enzyme method, and immunoprecipitation method. In general, DNA methylation is first discovered through large-scale, genome-wide methylation sequencing and/or microarray-based analytical strategies. Then, the methylation sites are analyzed and verified. Given the close link between DNA methylation and tumorigenesis, research into DNA methylation-based biomarkers is advancing rapidly and holds promise for a wide range of clinical applications. In recent years, the rapid development of high-throughput sequencing technology has made it possible to map genome-wide DNA methylomics at single-base resolution, an application that continues to undergo rapid refinement [124]. This momentum, coupled with decreasing cost, has facilitated a growing number of disease-related methylation studies. Against this backdrop, a key priority now is to establish effective detection methods for specific DNA methylation targets.

9.3.1 $NaHSO_3$ Treatment-Based Methods

$NaHSO_3$ treatment of DNA is the most common way to distinguish methylated from unmethylated cytosine. After $NaHSO_3$ treatment, the unmethylated cytosine deamidates into uracil (Fig. 9.6), while the methylated cytosine remains unchanged [125]. The gold standard for detecting 5mC involves the treatment of target DNA with $NaHSO_3$, followed by polymerase chain reaction (PCR) amplification, and then genome sequencing. This method can directly indicate the methylation status of each CpG site. However, due to the large number of site-specific amplification and sequencing involved, high-throughput methylation analysis is not currently the preferred method. The changes from site-specific epigenetic modification to nucleotide sequence can be amplified by PCR, ligase chain reaction (LCR), loop-mediated isothermal amplification (LAMP), or rolling circle amplification

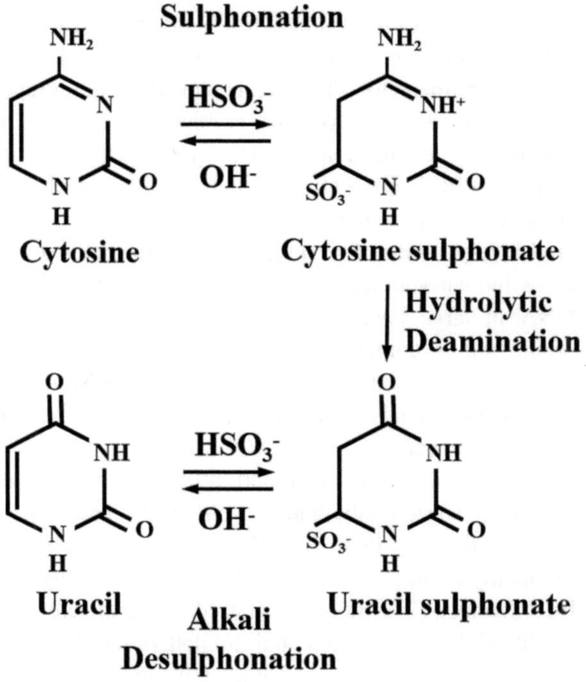

Fig. 9.6 Reaction process of conversion of cytosine by $NaHSO_3$. (Revised and reprinted with permission from Ref. [125])

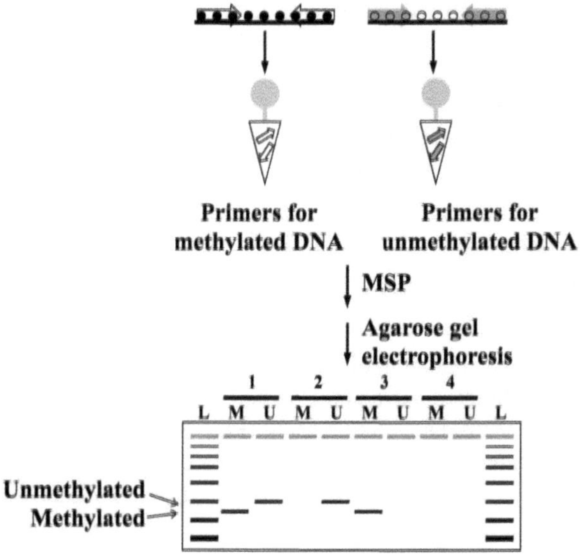

Fig. 9.7 Schematic diagram of MSP. (Revised and reprinted with permission from Ref. [126])

(RCA) to reflect the initial methylation pattern. During nucleic acid amplification, unmethylated cytosine is amplified to thymine and methylated cytosine to cytosine. The resulting amplification products are then analyzed by different means to distinguish between methylated and unmethylated DNA.

Methylation-specific PCR (MSP) [126] is a sensitive method to detect specific methylation sites. Genomic DNA is first treated with $NaHSO_3$, where unmethylated cytosines are converted to uracils and methylated cytosines remain unchanged. Two pairs of primers for PCR are designed to be complementary with the $NaHSO_3$-treated sequences, i.e., one pair recognizes the treated methylated DNA and the other pair recognizes the treated unmethylated DNA. Following bisulfite treatment and PCR, the PCR products are analyzed by gel electrophoresis. If amplification is achieved using primers specific for methylated DNA, the target site is considered methylated; conversely, amplification with primers specific for unmethylated DNA indicates an unmethylated status (Fig. 9.7). MSP can detect 0.1% methylated DNA in large background of unmethylated sequences, requires minimal DNA input, and is compatible with suboptimal specimens, including DNA extracted from paraffin-embedded tissues. However, MSP relies on post-PCR gel electrophoresis, which is labor-intensive and limits the technique to qualitative rather than quantitative analysis.

Worm reported the methylation sensitive-melting curve analysis (MS-MCA) for analyzing DNA methylation levels [127]. The DNA is treated with $NaHSO_3$ and then amplified by PCR. SYBR Green I is added into the reaction system to analyze the methylation level of the initial target DNA by measuring the melting curve of PCR products. The analysis is based on the difference in T_m values of PCR products produced by methylated and unmethylated DNA due to different GC content (Fig. 9.8). If all alleles are completely free of 5mC, all cytosines will be converted

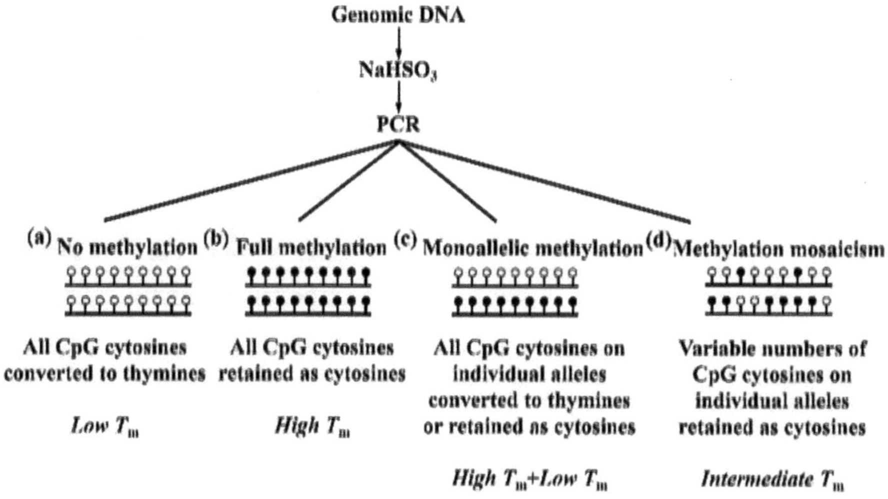

Fig. 9.8 Schematic diagram of MS-MCA. (Revised and reprinted with permission from Ref. [127])

to thymine, resulting in PCR product with a relatively low T_m (Fig. 9.8a). If all alleles contain 5mC on all CpG sites, the T_m of the PCR product will be significantly higher (Fig. 9.8b). If the DNA sample contains a mixture of unmethylated or fully methylated alleles, amplification will result in the presence of both high and low T_m products (Fig. 9.8c). If the target sequence exhibits methylation mosaicism, that is, different amounts of 5mC for different alleles in the same sample, molecules with different T_m together form the PCR product, resulting in a homogeneous T_m of the product at an intermediate temperature (Fig. 9.8d). This method has been successfully applied to identify methylation patterns in samples from Prader–Willi syndrome patients. The subsequently reported melting curve methylation-specific PCR (McMSP) [128], methylation-sensitive high resolution melting (MS-HRM) [129], and sensitive melting analysis after real-time methylation-specific PCR (SMART-MSP) [130] are based on the characteristics of different T_m values of PCR products after DNA treatment with $NaHSO_3$ to quantitatively analyze methylation levels. Among them, the MS-HRM method achieves higher resolution and can detect subtle differences in PCR products. SMART-MSP realizes real-time fluorescence detection based on MS-HRM. [131]

In order to simplify the operation process of MSP and realize quantitative analysis of methylation, MethyLight method employed real-time PCR based on TaqMan probe to detect methylation sites [132]. The method eliminates the need for additional operations after PCR and can detect methylated DNA as low as 0.01% in the presence of unmethylated DNA. The genomic DNA is first treated with $NaHSO_3$ to produce methylation-dependent sequence differences on CpG dinucleotides, followed by TaqMan probe-dependent PCR. PCR primers and TaqMan probes can be designed for differential sequence recognition, respectively. As shown in Fig. 9.9a, the target sequence without CpG sites is used as a blank, and neither PCR primers nor TaqMan probe have the ability to recognize differential sequences.

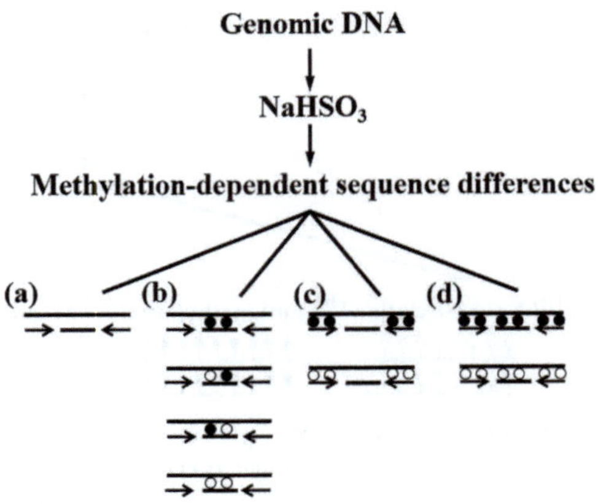

Fig. 9.9 Schematic diagram of MethyLight. (Revised and reprinted with permission from Ref. [132])

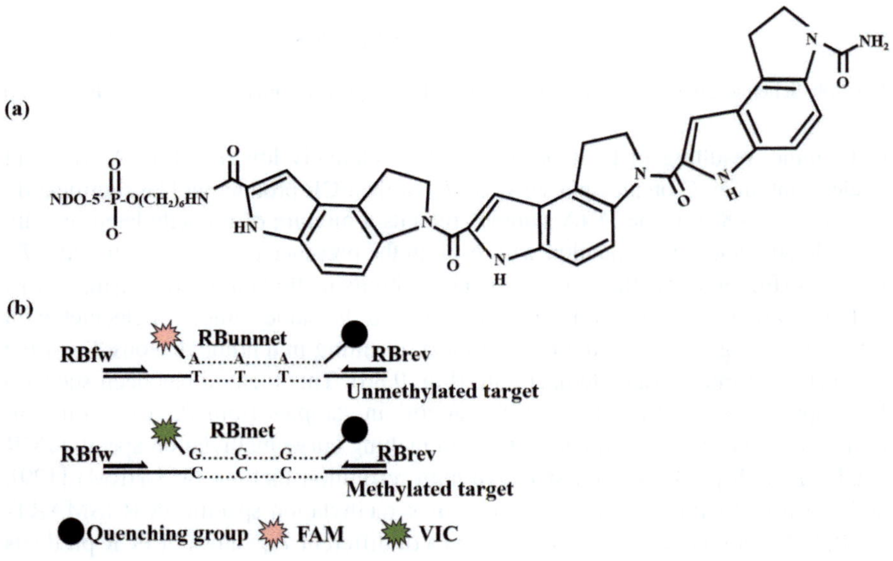

Fig. 9.10 (**a**) Structural formula of $CDPI_3$. (**b**) Schematic diagram of real-time quantitative PCR based on MGB TaqMan probe to detect methylation. (Revised and reprinted with permission from Refs. [133, 134])

In Fig. 9.9b, PCR primers do not have the recognition capability, while TanMan probes do. The situation in Fig. 9.9c is exactly the opposite of that in Fig. 9.9b. In Fig. 9.9d, both PCR primers and TanMan probes can recognize the differential sequences. Thus, differential sequence recognition can occur either at the PCR amplification process level, at the TaqMan probe hybridization level, or at both. Since both PCR primers and TaqMan probe hybridizations can be designed for methylation sites, there are multiple possibilities to detect methylation sites in the genome.

$CDPI_3$ (Fig. 9.10a) is a DNA minor groove binder (MGB). When MGB is coupled to a short oligodeoxynucleotide, this short oligodeoxynucleotide forms an unusually stable double strand with the complementary DNA, where the $CDPI_3$ group is located in a minor groove. Based on this, Afonina developed a TaqMan probe based on MGB technology for real-time monitoring of the PCR [133]. Compared with the traditional TaqMan probe, the sequence specificity of the probe is significantly improved, and even single-base pair changes can be detected. Zeschnigk used MGB TaqMan PCR for quantitative analysis of methylated alleles (QAMA) in retinoblastoma gene promoters in real time [134] (Fig. 9.10b). The target DNA is treated with $NaHSO_3$ and amplified by PCR using the same primer pair (RBfw, RBrev). Two MGB TaqMan probes, RBmet and RBunmet, are designed. One end of RBmet labels VIC and the other end labels MGB, which are used to specifically identify PCR products from methylated DNA. RBunmet labels FAM on one end and MGB on the other for specific identification of PCR products from unmethylated DNA. The hybridization region between the two probes and the target contains four CpG sites, and when the two probes bind to their respective targets, they are cleaved by the 5′-3′ end exonuclease activity of *Taq* DNA polymerase. According to the fluorescence intensity of VIC and FAM released during the PCR process, the content of methylated or unmethylated genes is analyzed. The technique uses MGB TaqMan probes labeled with two different dyes to identify sequences derived from methylated and unmethylated DNA, enabling simultaneous amplification and quantitative analysis of both sequences in a single tube. These sensitive detection techniques have become the most common methods for candidate gene methylation, which can analyze small amounts of DNA extracted from tissues. However, since only fully methylated or completely unmethylated DNA amplification is available in these techniques, the exact pattern of methylation is not reflected.

After treatment with $NaHSO_3$, the differences in methylation status of a single site are transformed into differences in a single base. Therefore, methods for detecting single nucleotide polymorphism (SNP), such as the single-base extension method, can be combined with the $NaHSO_3$ conversion to detect the methylation status of specific sites. For example, in the methylation-sensitive single-nucleotide primer extension (Ms-SNuPE) [135, 136] (Fig. 9.11), genomic DNA is first treated with $NaHSO_3$ and then amplified by PCR to generate a strand-specific DNA template suitable for Ms-SNuPE analysis. PCR products purified by gel electrophoresis are subjected to Ms-SNuPE primers, ^{32}P-dCTP (or ^{32}P-dTTP) and *Taq* DNA polymerase single-base elongation reaction. The ^{32}P-labeled extension products are separated by electrophoresis on 15% denatured polyacrylamide gel and are qualitatively and quantitatively analyzed by phosphorescence images. In addition, the length of Ms-SNuPE primers can be encoded to realize simultaneous analysis of multiple CpG sites.

To overcome the disadvantages of radioactive processing in Ms-SNuPE analysis, SNuPE analysis can be combined with other techniques, such as ion pair reverse-phase high-performance liquid chromatography [137] and matrix-assisted laser desorption/ionization time-of-flight mass spectrometry [138], for multiple and quantitative analysis of methylation at specific sites. Feng [139] combined SNuPE with

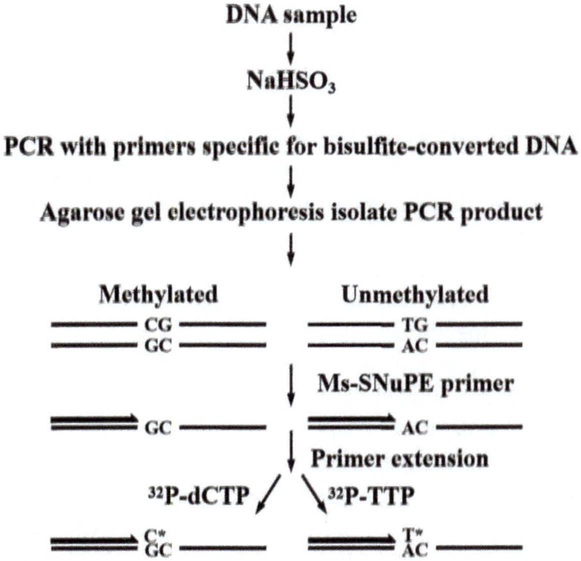

Fig. 9.11 Schematic diagram of Ms-SNuPE for methylation site determination. (Revised and reprinted with permission from Ref. [135])

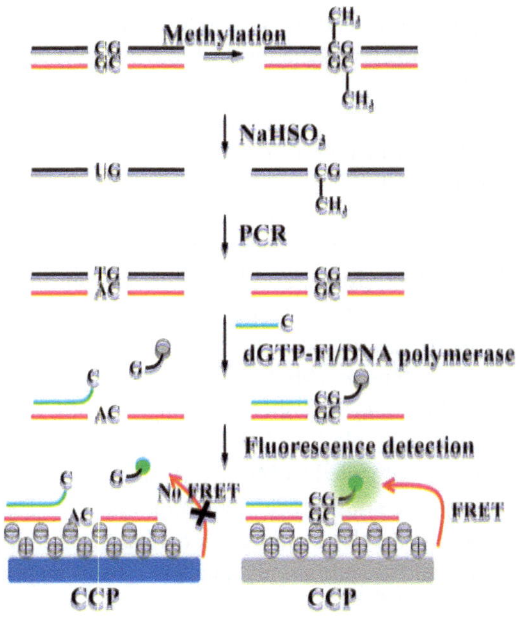

Fig. 9.12 Schematic diagram of CpG site methylation detection based on SNuPE and CCP. (Revised and reprinted with permission from Ref. [139])

cationic conjugated polymer (CCP) to detect methylation levels at three CpG sites within the promoter of *p16* gene by fluorescence resonance energy transfer (FRET) (Fig. 9.12). CCP can bind to negatively charged DNA molecules, and can also be used as a FRET donor in fluorescence analysis, so that the fluorescence signal of the receptor is significantly amplified, which is helpful to improve the detection sensitivity. In this method, genomic DNA is treated with $NaHSO_3$ and amplified by PCR.

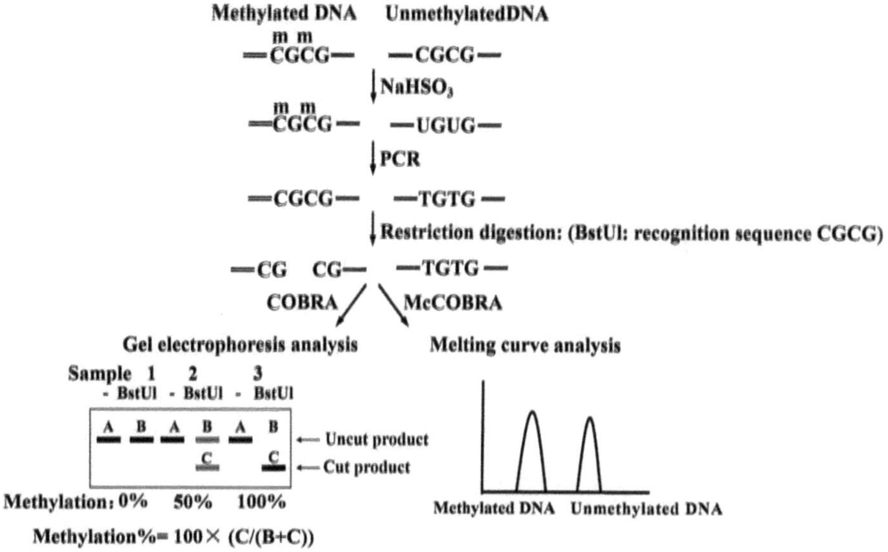

Fig. 9.13 Schematic diagram of COBRA and McCOBRA. (Revised and reprinted with permission from Refs. [128, 140])

SNuPE primers are designed to complement the PCR products. One base at the primer's 3′ end is complementary to the PCR product from methylated DNA, and not complementary to the unmethylated DNA product. In the presence of DNA polymerase and fluorescently labeled dGTP, PCR products from methylated DNA can produce fluorescently labeled SNuPE products. Finally, the fluorescently labeled SNuPE products bind to CCP and FRET occurs. While unmethylated DNA PCR products lack the single-base extension products, failing to binding with CCP and no FRET occurs.

The procedure of the combined bisulfite restriction analysis (COBRA) [140] is shown in Fig. 9.13. After treatment of genomic DNA using $NaHSO_3$, the CpG-containing sites generated or retained in the sequence reflect the original methylation status. PCR amplification is performed using primers that do not distinguish between methylated and unmethylated sites. The PCR product is subsequently digested using an endonuclease that recognizes sequences with cytosine only at the CpG. For example, the restriction enzyme BstUI specifically recognizes CGCG. If the cytosine within this site is methylated, the sequence remains CGCG following $NaHSO_3$ treatment and PCR, allowing for enzyme recognition and cleavage. If unmethylated, the cytosine is converted to thymine, resulting in a TGTG sequence, which disrupts the recognition site and prevents cleavage. Digestion products are analyzed using electrophoresis or hybridization experiments, enabling the calculation of the methylation ratio (Fig. 9.13 left). Similar to McMSP, the melting curve approach is capable of analyzing COBRA products (McCOBRA, Fig. 9.13 right). Compared to COBRA, this method is a simpler, and quantitative and high-throughput method that can be used to accurately estimate the percentage of

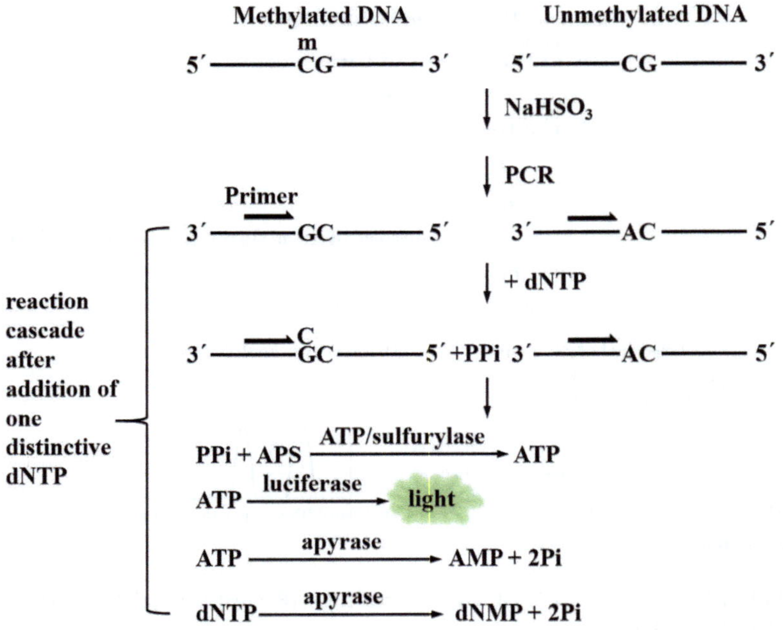

Fig. 9.14 Schematic diagram of PyroMeth. (Revised and reprinted with permission from Ref. [141])

methylated DNA at specific sites [128]. The COBRA-related methods are all based on restriction endonuclease recognition with a low probability of false positives; however, these methods cannot detect all CpG sites due to enzymatic site limitation.

Pyrosequencing methylation (PyroMeth) [141, 142] is a method that combines $NaHSO_3$ and pyrosequencing to detect methylation at specific CpG sites. As shown in Fig. 9.14, the target DNA is amplified by PCR after $NaHSO_3$ treatment, and the methylated site is transformed into CG and the unmethylated site into CA. When the primers are hybridized with PCR products, four dNTPs are added in a predetermined order. Only when the provided dNTP matches the products of $NaHSO_3$ and PCR, will the dNTP bind to the 3′ end of the primer under the action of polymerase and release a molecule of inorganic pyrophosphate (PPi), which triggers the subsequent cascade reaction. ATP sulfurylase catalyzes PPi to bind to 5′-phosphoacylsulfuric acid (APS) to form ATP, and luciferase combines ATP with luciferin to form oxyluciferin, which emits a visible light. Unincorporated dNTP and ATP are degraded by apyrase, which quench the light signal and regenerate the reaction system. By cycling this process, combined with the type of dNTP added and the intensity of the resulting fluorescence signal, information on the methylation status of the initial DNA can be obtained in real time. PyroMeth is able to detect methylation levels at multiple adjacent CpG sites in the gene, unrestricted by recognition site of restriction endonuclease.

The chromogenic reaction can be used for qualitative and quantitative analysis of DNA by visual observation or measurement of the absorbance, which has the

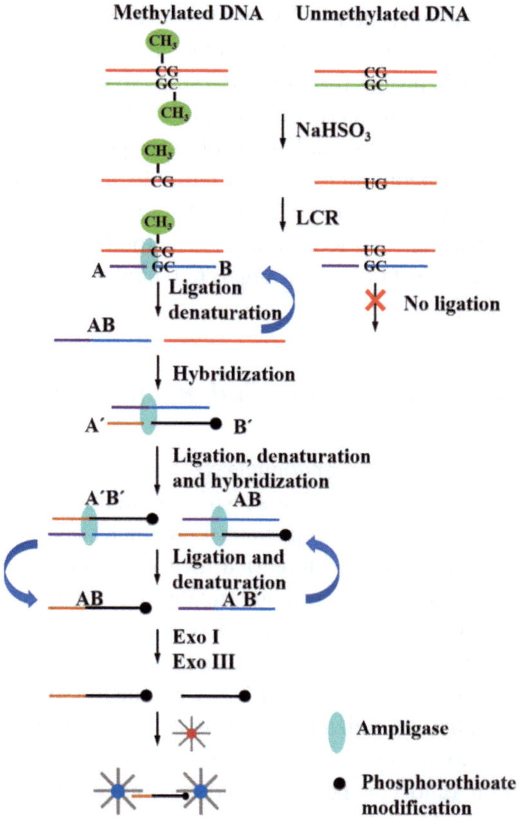

Fig. 9.15 Schematic diagram of DNA methylation detection based on chromogenic reaction of gold nanoparticles and LCR. (Revised and reprinted with permission from Ref. [143])

advantages of simple operation and cost saving. Ligase has a high ability to recognize single-base changes, and the ligation-based reaction can effectively distinguish the polymorphic sites generated by methylated and unmethylated DNA after $NaHSO_3$ treatment. Su combined LCR with color reaction of gold nanoparticles to detect methylation level in the promoter region of *TIMP-3* gene [143] (Fig. 9.15). After treating the target DNA with $NaHSO_3$, probes A and B recognize the methylated CpG site at the junction site and undergo the ligation reaction and subsequent LCR to generate products AB and A′B′. Because the 3′ end of probe B′ is modified by phosphorothioate, probe B′ and product A′B′ can effectively resist the hydrolysis of Exonuclease I (Exo I) and Exonuclease III (Exo III). After the LCR product is treated with Exo I and Exo III, only A′B′ and probe B′ are left. Complementary DNA probes of A′ and B′ are modified on the surface of the two gold nanoparticles, respectively. The product A′B′ is hybridized with two complementary DNA probes, which narrowed the distance between the two gold nanoparticles, and the color of the solution changes from red to blue. The unmethylated DNA could not undergo ligation and LCR, the gold nanoparticles are uniformly dispersed, and the color of the solution remains red. The method enables visual detection of methylated target as low as 10 amol/L and is able to detect 0.1% methylated DNA in an excessive amount of unmethylated DNA. However, chromogenic detection can only detect methylation levels at a single CpG site in a single tube. Subsequently, Su encoded the length of LCR probes for three CpG sites, so that the LCR products at the three methylation sites have different lengths [144]. Three LCR products are separated by capillary electrophoresis, and three product peaks appear at different locations, achieving simultaneous detection of multiple methylation sites in complex genomic DNA samples (See Fig. 2.13, Chapter 2). Cao proposed a padlock probe-based hyperbranched RCA (HRCA) method to analyze the methylation status of specific CpG sites [145]. (See Fig. 4.6, Chapter 4).

9.3.2 Restriction Endonuclease-Based Methods

Two types of endonucleases may be used in restriction endonuclease-based methods: methylation-sensitive endonucleases that cut only unmethylated DNA and methylation-dependent endonucleases that cut only methylated DNA. These enzymes are used to enrich methylated or unmethylated DNA in combination with other means to provide qualitative or quantitative results. DNA digested by methylation-sensitive restriction enzymes can be directly detected by Southern blotting. The DNA is digested first, and then are separated by agarose gel electrophoresis. The separated DNA is transferred to the nylon membrane for imprinting. Finally, hybridization reaction is carried out with radioactive probes. Because of the difference in bands produced by methylated and unmethylated DNA, the methylation ratio can be quantified by image analysis. However, the sensitivity of this method is relatively low and often requires DNA extraction from a large number of cells.

In order to improve the detection sensitivity, the methylation-sensitive HpaII (recognition sequence CCGG) combined with PCR can be used to analyze DNA methylation in a small number of cells (100–300) [146]. Feng used CCP-based FRET to detect the products of HpaII treatment and PCR [147], enabling the analysis of the methylation level at specific CpG sites. As shown in Fig. 9.16, in genomic DNA treated with HpaII, the unmethylated DNA is cut and the methylated DNA is enriched. The enriched methylated DNA tiggers PCR in the presence of fluorescently labeled dNTP, generating fluorescently labeled PCR products. The remaining dNTPs and fluorescently labeled dNTP in the reaction system are degraded by shrimp alkaline phosphatase (SAP). Subsequently, the product is electrostatic combined with CCP, FRET occurs. Unmethylated DNA is digested by HpaII, failing to be amplified by PCR. Only fluorescently labeled dNMP remaining in the SAP digestion system exists in free form and no FRET occurs.

The design of universal primers in multiplex ligation-dependent probe amplification (MLPA, Fig. 1.16, Chapter 1) enables multiple detection of targets. Nygren established a methylation-sensitive MLPA (MS-MLPA) method to achieve multiple detection of methylation levels at CpG sites [148]. As shown in Fig. 9.17, the synthetic oligonucleotide probe and the M13-derived oligonucleotide probe are

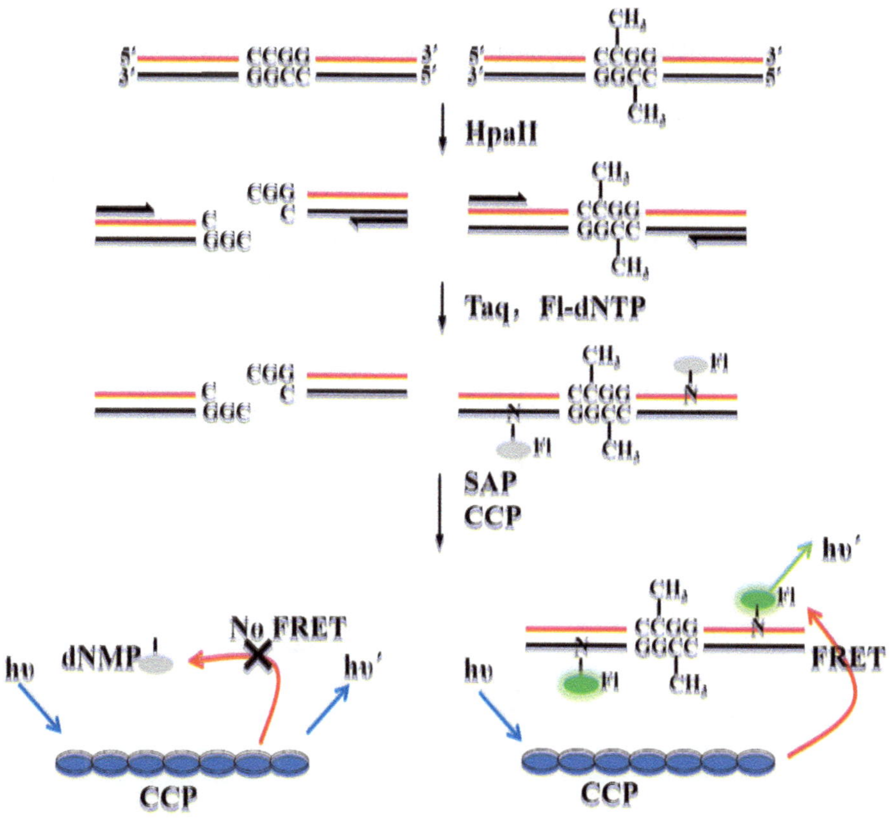

Fig. 9.16 Schematic diagram for detection of methylation level at CpG site based on HpaII and PCR. (Revised and reprinted with permission from Ref. [147])

ligation probes. The synthetic oligonucleotide probe consists of target-specific sequence 1A and primer X-specific sequence. The M13-derived oligonucleotide probe consists of target-specific sequence 1B, stuffer sequence, and primer Y-specific sequence. The ligation reaction and methylation-sensitive restriction enzyme digestion reaction are performed simultaneously, followed by PCR. If the target CpG site is methylated, the two probes hybridize and ligate, and the ligation product cannot be recognized by the methylation-sensitive restriction enzyme, leading to subsequent PCR amplification. If the CpG site is unmethylated, the ligation product is cut off by the enzyme and cannot trigger subsequent PCR. For another CpG site, the target recognition sequence in the probe can be designed as 2A and 2B, respectively, and the ligation probe can be encoded by changing the length of the stuffer sequence, so that the PCR products at different sites have different lengths. Simultaneous analysis of multiple CpG sites is performed by capillary electrophoresis. This method enables simultaneous analysis of methylation levels for up to 40 sequences in a single reaction.

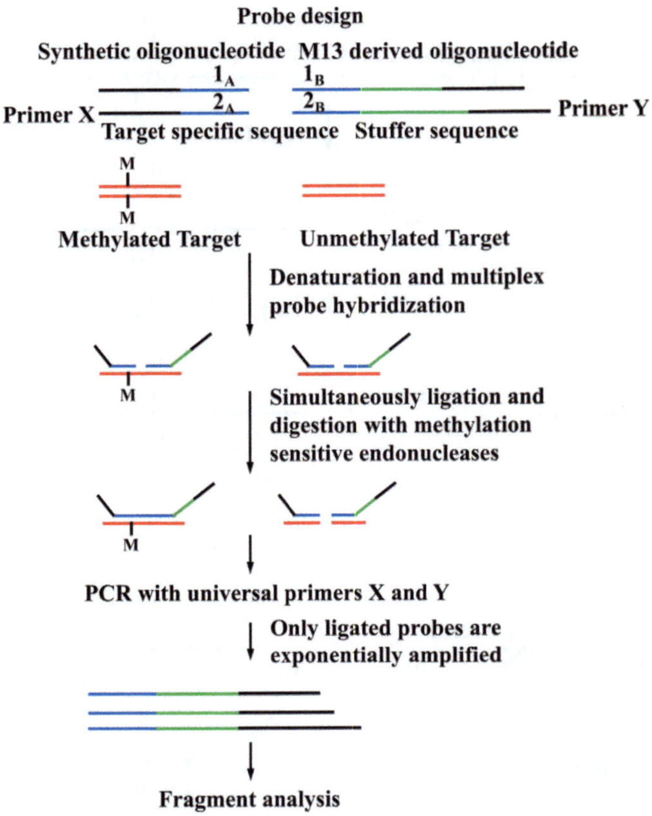

Fig. 9.17 Schematic diagram of MS-MLPA. (Revised and reprinted with permission from Ref. [148])

In addition to the PCR method, the restriction enzyme method can also be combined with other nucleic acid amplification methods to detect site-specific methylation. Wen combined methylation-sensitive restriction endonuclides with LAMP to detect methylation levels at specific sites [149]. As shown in Fig. 9.18, genomic DNA is first treated with HpaII, leading to the digestion of unmethylated DNA, while the methylated DNA unchanged. Intact methylated DNA can trigger LAMP reaction, while cut unmethylated DNA cannot. LAMP products are monitored in real time by SYBR Green I. The method can detect at least 10 aM methylated DNA, and the use of HpaII improves the selectivity of methylated DNA. Recently, Pataer combined methylation-sensitive restriction enzyme with double exponential amplification and used CRISPR/Cas12a system to identify the double exponential amplification products [150], successfully achieving the detection of 1 amol/L methylated DNA and the differentiation of 0.1% methylated DNA. Sun combined a methylation-dependent restriction enzyme GlaI with an exponential

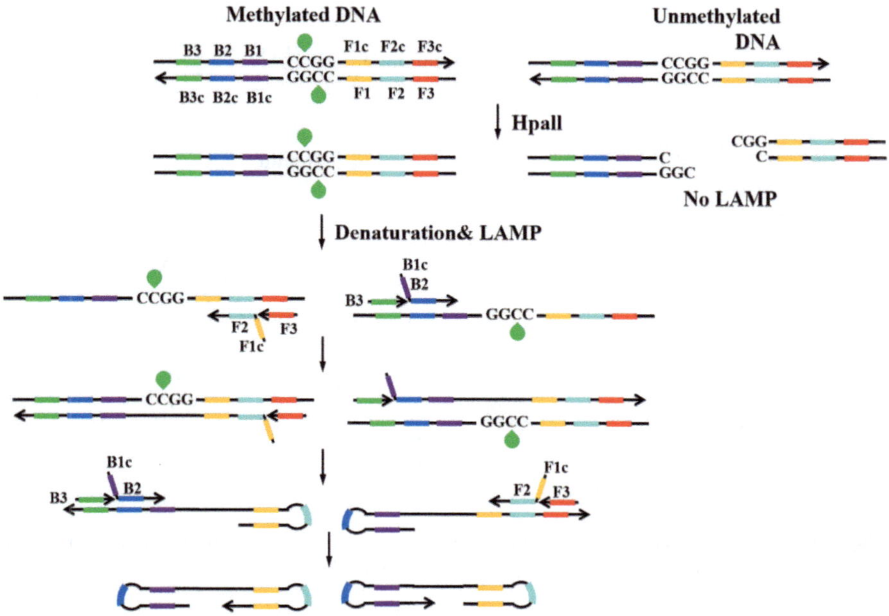

Fig. 9.18 Schematic diagram of the detection of DNA methylation based on HpaII and LAMP. (Revised and reprinted with permission from Ref. [149])

amplification reaction (EXPAR) to detect DNA methylation at specific sites in a highly specific and sensitive manner [151] (see Fig. 5.21, Chapter 5).

Zeng combined restriction endonuclease and RCA to detect the activity of methyltransferase [152] (Fig. 9.19). DpnI is a methylation-dependent restriction endonuclease that can specifically cut methylated DNA. In the hairpin DNA, the cytosine in DpnI's recognition sequence (GATC) is methylated by methyltransferase, allowing DpnI to recognize and cut the sequence. The cut DNA fragment can be used as a primer of RCA on the circular template. The RCA product is then digested by a nicking enzyme to produce numerous short DNA fragments, that harbor a DNAzyme sequence. This DNAzyme, possessing horseradish peroxidase-like activity, catalyzes a chemiluminescent reaction with luminol reagent and Hemin. These short DNAs can also hybridize to the circular template to serve as primers, triggering a new round of RCA and thereby establishing a cyclic amplification process. The amount of methyltransferase determines the methylation level of the hairpin and final luminescence intensity. The detection limit of methyltransferase by this method is 1.29×10^{-4} U/mL.

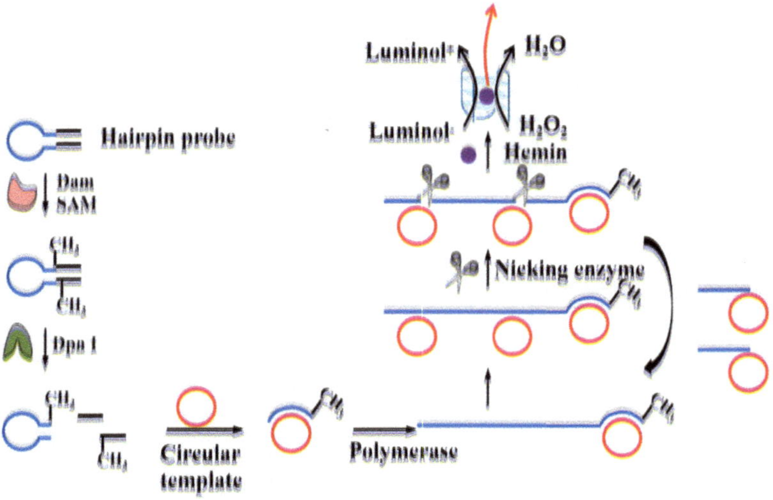

Fig. 9.19 Schematic diagram of detection of methyltransferase activity by DpnI combined with RCA. (Revised and reprinted with permission from Ref. [152])

9.3.3 *Methylated DNA Immunoprecipitation Method*

Methylated DNA immunoprecipitation (MeDIP) is an effective method for enrichment of methylated DNA. By binding 5mC-specific antibodies to methylated DNA fragments, the methylated genome fragments undergo immunoprecipitation, thus achieving the enrichment of methylated DNA. Subsequently, these enriched methylated DNA can be subjected to large-scale DNA methylation analysis by sequencing (MeDIP-Seq) or microarray platforms [153, 154]. MeDIP has been in combination with microarray technology to map the methylation of whole genome DNA of Arabidopsis, human normal and transformed cells [153, 155, 156]. In addition, methylated DNA-binding proteins, such as MBD2 and MBD3, can also be employed to enrich methylated DNA. In the methylated-CpG island recovery assay (MIRA) [157], DNA is first fragmented by ultrasound and then purified. These fragments incubate and bind with MBD2 and MBD3. The unbound DNA is removed through washing steps, and finally methylated DNA is obtained by purification again. These methylated DNA fragments can be amplified by PCR for the analysis. The method can detect 1 ng of DNA or 3000 cells of methylated DNA. These methods, based on affinity, make it possible to assess the average methylation levels of large DNA regions quickly and specifically. However, they could not reveal the methylation status of individual CpG sites. Due to the low sensitivity of these methods, a large amount of DNA are needed in the experiment. In addition, the ability of 5mC antibodies and MBD to bind to methylcytosine may be affected by the density of methylated cytosine as well as the surrounding sequence.

9.4 Analysis and Application of DNA Methylation

1. Early Diagnosis of Tumors

The methylation level of CpG islands within gene promoter regions serves as a key mechanism for regulating gene expression. However, in tumor cells, abnormal methylation of tumor-suppressor genes occurs early in the tumor, thereby silencing the gene expression. Kuester reported that abnormal methylation of *MGMT* gene was a frequent event in the early stage of esophageal adenocarcinoma [158]. Kelavkar found that *ALDX15* gene was abnormally methylated in prostatic intraepithelial neoplasia and prostate cancer, while it was unmethylated in normal prostate cell lines [159], indicating that abnormal methylation of this gene was closely related to the occurrence of prostate cancer. Therefore, abnormal DNA methylation is an early and frequent event in tumor progress and has great potential as a biomarker for the early diagnosis of tumors [160].

2. Prognostic Analysis of Tumors

DNA methylation can also be used for disease classification, prognosis, and therapeutic response [161]. The methylation state of the *RASSF1A* gene in many tumors, including lung, breast, and prostate cancers, has been shown to be associated with poor prognosis [162]. Neuroblastoma with CpG island methylation phenotype is highly associated with poor prognosis [163]. Metastasis potential can be predicted based on methylation levels of E-calmodulin promoter in breast and oral cancers. In terms of response to treatment, the most striking example to date is the high methylation of the *MGMT* gene promoter, which increases glioblastoma sensitivity to alkylating agents [164]. In addition, DNA methylation can be employed as an independent predictor of tumor recurrence, such as oral squamous cell carcinoma, where *RECK* gene methylation has a significantly higher recurrence rate [165].

3. Demethylation therapy for tumors

Based on the close correlation between epigenetic alterations, especially altered DNA methylation status, and tumors, attempts have been made over the years to develop drugs to restore normal gene expression patterns in cancer cells for therapy. Most of the clinically tested DNA methylation inhibitors (DNMTi) are nucleoside analogues. These drugs are converted into deoxynucleotides in the cell and are incorporated into replicated DNA to replace cytosine. Their main mechanism of action may be through the capture of methyltransferase at nucleoside incorporation sites [164, 166], thus depleting the enzyme activity of the cell and producing heritable demethylated DNA. Because nucleoside analogue incorporation occurs during DNA synthesis, only replicating cells are demethylated by DNMTi [167], and thus the reaction may be more preferential to highly proliferative cancer cells, with hypomethylation occurring during subsequent cell division reactivating a variety of silenced tumor-suppressor genes. However, after drug treatment, demethylated DNA sequences exhibit a tendency to remethylate, leading to a progressive blunting of the cellular response [168]. Addressing this issue requires an understanding of what triggers remethylation and the development of a new generation of therapies that can be reused to prevent the remethylation process.

The prototypes of DNMTi are 5-aza-cytidine (azacitidine) and 5-aza-2-′-deoxycytidine (decitabine). Originally called cytotoxic drugs, they were synthesized in Czechoslovakia in the 1960s and were later found to promote DNA demethylation and differentiation, and to reactivate silenced genes at doses lower than cytotoxic [167, 169, 170]. Low doses of DNMTi are now used primarily in hematologic malignancies and can produce better responses and lower toxicity. Both drugs are approved by the US Food and Drug Administration (FDA) for the treatment of myelodysplastic syndromes [171]. Zebraline is a another nucleoside analogue with demethylating properties, which can be administered orally, is less toxic than the previous two drugs, acts preferentially on tumor cells, and inhibits polyp formation in female APC/min-deficient mice [172]. However, the need for high concentrations of zebraline and its limited bioavailability in primates has slowed its progress into clinical trials [173].

Because epigenetic silencing is closely linked to histone deacetylation, various compounds that inhibit histone deacetylases show antitumor capacity, inhibition of tumor growth, pro-apoptotic capacity, and pro-differentiation properties [174]. Two histone deacetylase inhibitors, sialic acid hydroxamic acid (SAHA) and depsipeptide, have been approved by the FDA for the treatment of cutaneous T-cell lymphoma [175]. As interactions between epigenetic pathways are elucidated, there is considerable interest in the development of combination therapies, particularly combinations between DNA demethylation and histone deacetylase inhibitors. Histone deacetylase inhibitors and DNMTi show synergistic effects on transcriptional activation [176], and preliminary clinical trials using both combinations are promising [177], however, rational combinations of these drugs still require further clinical trials.

9.5 Analytical Methods of 5-Hydroxymethylcytosine

5-hydroxymethylcytosine (5hmC) is the hydroxylated form of 5mC, known as the "sixth base," which is widely distributed in mammalian tissues and cells. 5hmC was first discovered in phage DNA in 1952, but was not given enough attention at the time. Until 2009, Kriaucionis found that 5hmC was highly expressed in mouse and human brain cells, embryonic stem cells, and Purkinji neuron cells through thin layer chromatography (TLC) and high performance liquid chromatograin-mass spectrometry (HPLC-MS) [178], indicating that 5hmC played an important role in epigenetic regulation of neuron function. The abnormal expression of 5hmC may also be closely related to the occurrence and development of cancer. In the same year, Tahiliani found that the Tet family of β-D-glycated hydroxymethyluracil-binding proteins homologs (including Tet1, Tet2, and Tet3) can convert 5mC to 5hmC [28]. With the deepening of research, it was found that Tet1 could further oxidize

5hmC to 5-formylcytosine (5fC), and finally oxidize 5fC to 5-carboxycytosine (5caC) [31, 179–181].

Shortly after Tet's discovery, Maiti discovered that thymine DNA glycosylase (TDG) can rapidly remove 5fC and 5caC in vitro or in mammalian cells, which led to the study of 5hmC's involvement in DNA active demethylation [179, 182]. A Tet-mediated DNA active demethylation process was proposed. 5mC is first oxidized to 5hmC by Tet protein, and 5hmC is further oxidized to 5fC, 5caC, and 5fC and 5caC can be resected by TDG, and finally converted to unmodified cytosine by BER pathway [183]. The discovery of Tet protein family not only reveals the production process of 5hmC, but also indicates its role in Tet-mediated active DNA demethylation. This foundational work paved the way for subsequent studies on 5hmC function, ultimately establishing 5hmC as a major research focus. In order to further explore the mechanism of 5hmC and understand its regulatory function in DNA, it is necessary to know the exact location and modification level of 5hmC in DNA. Currently, growing research efforts in this area employ methodologies that can be broadly classified into three categories: single nucleotide analysis of 5hmC, direct detection of 5hmC in DNA, and determination of 5hmC through high-throughput sequencing.

9.5.1 Single Nucleotide Analysis of 5hmC

At first, 5hmC in DNA was analyzed by hydrolyzing DNA fragments into free single nucleotides. The main methods include TLC, high performance liquid chromatography (HPLC), and liquid chromatography-mass spectrometry (LC-MS). TLC was the first technique used to analyze 5hmC [29, 178]. In this method, DNA was first cut at the C↓CGG site using restriction endonuclease MspI, and radioisotopically labeled by T4 DNA kinase. Subsequently, DNase I and phosphodiesterase are used to fully digest the DNA fragments into single nucleotides. The difference in the migration rate among bases on the TLC thin layer stationary phase (such as silica gel) enables the separation. At the same time, samples containing quantitative standards undergo the same separation process. By comparing the position and density of each spot, the 5hmC in the sample can be qualitatively and quantitatively analyzed. Although this method enables bulk analysis of 5hmC, its sensitivity is limited, making accurate quantification challenging. Furthermore, the technique relies on radioisotope labeling, which poses potential health and environmental risks.

HPLC is another important method for 5hmC analysis [179]. In this method, the DNA sample to be analyzed is decomposed into single nucleotides, which are then separated by HPLC and quantitatively analyzed according to the peak area.

LC-MS combines the separation capability of liquid chromatography with the detection ability of mass spectrometry [184]. This technique distinguishes sample components based on their mass charge ratio (m/z), thereby achieving the analysis of 5hmC. LC-MS is a widely used method for the quantitative detection of 5hmC,

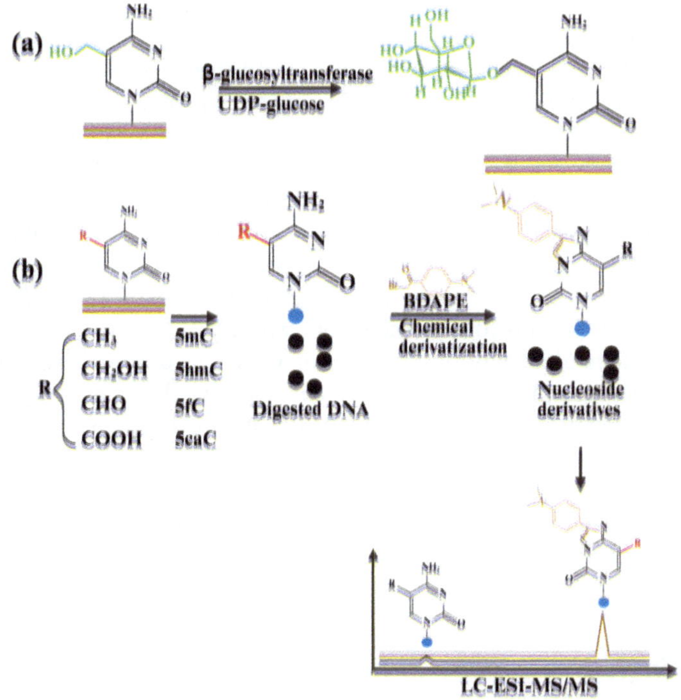

Fig. 9.20 (**a**) Schematic diagram of 5hmC modified by glucose. (**b**) Schematic diagram of reaction of cytosines with derivatizing reagent. (Revised and reprinted with permission from Refs. [185, 186])

capable of not only separating and analyzing 5hmC but also providing detailed information about other DNA components. In order to improve the sensitivity, selective enrichment strategy and chemical derivation methods are generally used to enhance the MS signal of 5hmC bases. For example, a DNA sample is treated with β-glucosyltransferase (Fig. 9.20a), which binds a molecule of glucose to the hydroxymethyl group of 5hmC, converting it to the more hydrophilic β-glucosyl-5-hydroxymethyl-2′-deoxycytidin (5gmC). Subsequently, DNA is hydrolyzed to single nucleotides, and 5gmC is selectively enriched by hydrophilic interaction of amino silica gel, which reduced the interference of other nucleotides. Finally, the enriched 5gmC is analyzed by LC-MS, significantly improving the detection sensitivity [185]. Derivatizing reagent method (Fig. 9.20b) firstly hydrolyzes DNA to mononucleotides and introduces 2-bromo-1-(4-dimethylamino-phenyl)-ethanone (BDAPE) to various cytosines by derivatizing reagent. This method not only significantly improves the separation effect of 5mC and 5hmC, but also improves the sensitivity of the analysis [186]. The above methods are based on the separation and analysis of DNA hydrolyzed to single nucleotides to detect 5hmC. While improve the detection sensitivity through modification, these techniques are limited to quantitative analysis of the overall content of 5hmC in the genome and cannot reveal its specific distribution in DNA sequence.

9.5.2 Direct Detection of 5hmC in DNA

The direct detection method of 5hmC in DNA, that is, the method of qualitative and quantitative analysis of 5hmC in DNA without hydrolysis of the DNA into single nucleotides, mainly includes two categories: targeted detection and non-targeted detection. Non-targeted analysis can quantitatively detect the content of 5hmC in DNA, but cannot determine the location of 5hmC, mainly including gel electrophoresis, FRET, electrochemical methods. Targeted analysis can simultaneously detect the content and site of 5hmC in DNA. These methods are mainly assisted by oxidation reaction or restriction endonucleases such as MSPI and HpaII, combined with nucleic acid amplification to achieve the detection of 5hmC in specific DNA.

In the electrophoretic method, 5hmC is first modified through a reaction catalyzed by 5-cytosine methyltransferase, which replace the hydroxyl group with aliphatic thiols or selenols to generate new cytosine derivatives (Fig. 9.21). These derivatives are subsequently conjugated with a fluorescent reporter group or biotin. Finally, the labeled 5hmC is separated by gel electrophoresis and quantitatively analyzed based on fluorescence intensity [187].

Hong proposed a FRET method based on CCP as an energy donor to detect 5hmC [188], as shown in Fig. 9.22. First 5hmC in genomic DNA is oxidized to 5fC using $KRuO_4$, and then further modified with hydroxylamino-bodipy (4,4-difluoro-4-bora-3a,4a-diaza-s-indacene) on 5fC. Upon the addition of CCP, the fluorescently labeled DNA forms a complex with CCP through electrostatic interaction. This close proximity triggers efficient FRET between CCP and hydroxylamine-BODIPY, thus achieving the detection of 5hmC.

Based on the glucosylation and enzymatic amplification, Zhou established an electrochemical biosensor to detect 5hmC [189]. As shown in Fig. 9.23, the AuNPs/GCE surface is modified with a capture probe to capture the target DNA through hybridization. The 5hmC in the target DNA modifies glucose under the catalysis of

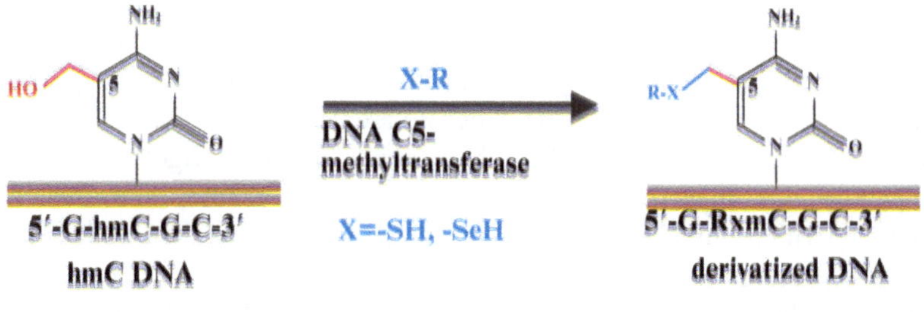

Fig. 9.21 The process of generating cytosine derivatives from 5hmC. (Revised and reprinted with permission from Ref. [187])

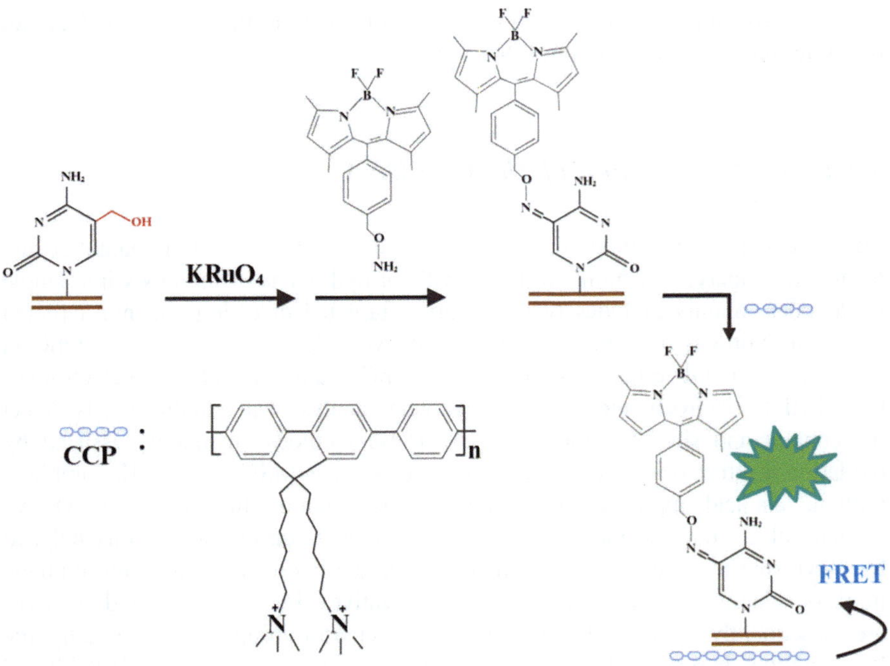

Fig. 9.22 Schematic diagram of 5hmC detection based on CCP. (Revised and reprinted with permission from Ref. [188])

β-glucosyltransferase and then covalently binds to 1,4-phenyldiboric acid to form esters. The product can trap alkaline phosphatase (ALP) to the electrode surface, thereby catalyzing the hydrolysis of p-nitrophenyl phosphate disodium salt to produce the electrochemically active p-nitrophenol. The relationship between the oxidation peak current of p-nitrophenol and the concentration of 5hmC enables the detection of 5hmC, and the detection limit is 1.43 pM. The above methods are non-targeted analysis, which can achieve the determination of 5hmC content in DNA samples or fragments, but cannot determine its specific location in DNA.

HpaII is a methylation-sensitive restriction enzyme that can specifically recognize CCGG sequence. It cleaves between the cytosines (C↓CGG) when the sequence is unmethylated, but is blocked by modification of the cytosine, such as 5mC or 5hmC. Based on this, Zhang established an HpaII-assisted ligation PCR method to detect 5hmC at specific site [190]. As shown in Fig. 9.24, the three dsDNA sequences containing CCGG, C5hmCGG, and C5mCGG are defined as target C, target 5hmC, and target 5mC, respectively. The DNA targets are first digested with HpaII. Subsequently, the digestive product is oxidized by $KRuO_4$ and treated with $NaHSO_3$, in which 5hmC is selectively oxidized to 5fC, and then both C and 5fC are converted to uracil, but 5mC is not changed. Thus, differences in epigenetic modifications between 5mC and 5hmC transforms into sequence differences between U and 5mC. In order to distinguish between 5hmC transformed U and

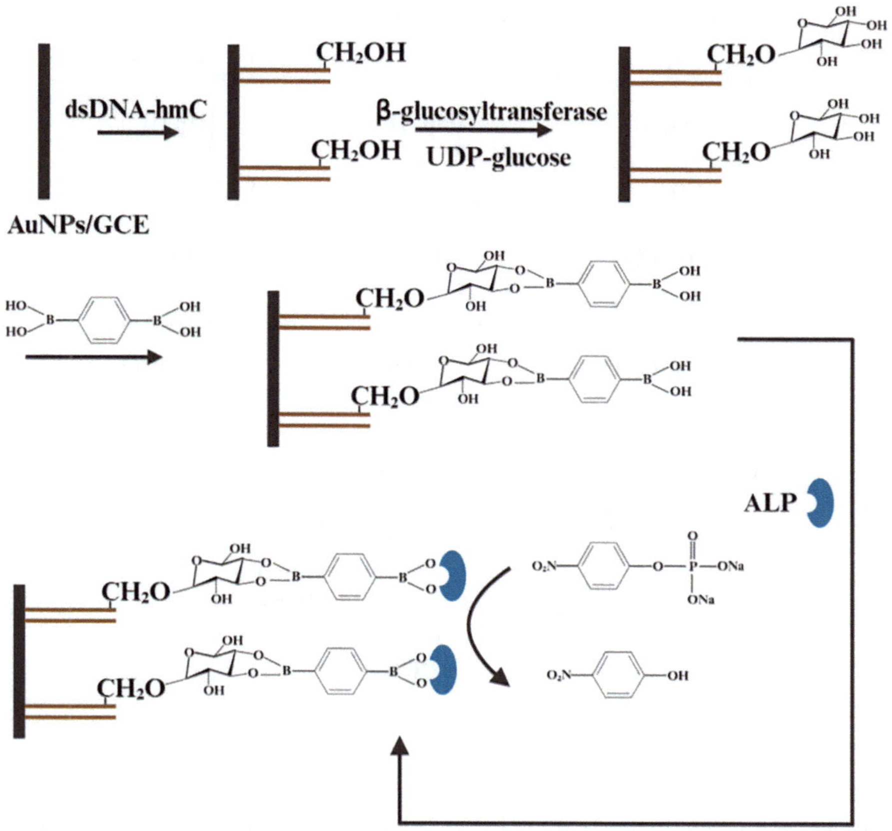

Fig. 9.23 Schematic diagram of electrochemical method to detect 5hmC. (Revised and reprinted with permission from Ref. [189])

5mC, two DNA-binding probes A and B are specially designed to perform the ligation reaction using the treated target 5hmC as a template. Probes A and B, respectively, contain a target-specific sequence, a universal primer specific sequence for subsequent PCR, and a filling sequence in between. The base of the 5′ terminal of probe B is adenine, which is complementary with the 5hmC transformed U. The 5′ end of probe A modifies the phosphate group and the 3′ end modifies the phosphorothioate. Thus, only in the presence of treated target 5hmC, can probes A and B hybridize with the target and ligate to form AB. Exo I and Exo III are then added for digestion. The 3′ -end phosphorothioate modification of probe A and the ligation product AB prevents them from being hydrolyzed, while the unattached probe B is digested. Then PCR amplification of the ligation products is performed and SYBR Green I is used as fluorescent dye to detect the products, thus achieving the quantitative detection of 5hmC at single-base resolution. Since target C has been cut by HpaII and target 5mC has a base 5mC at its 3′ end that does not match

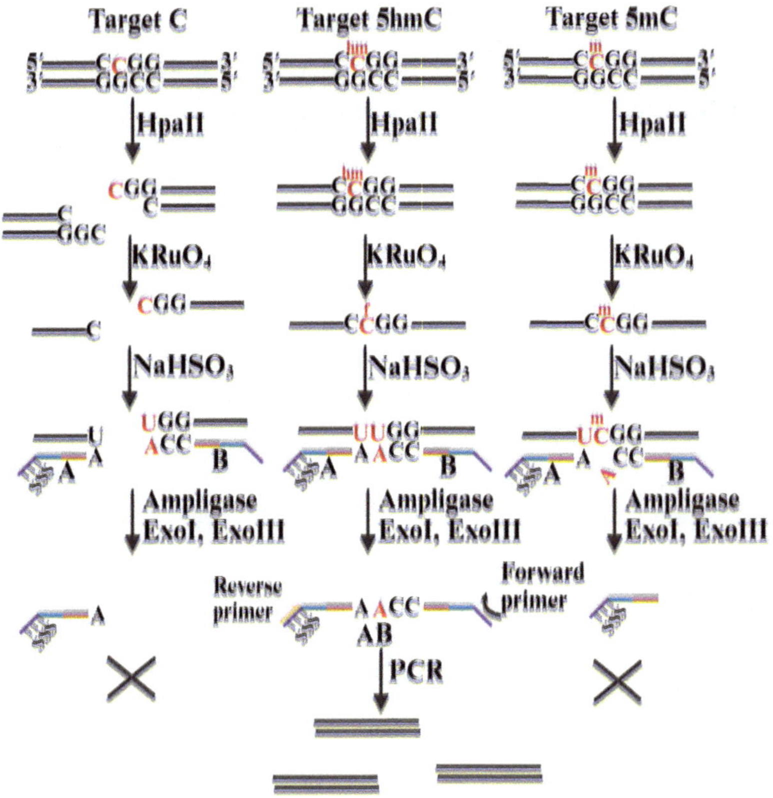

Fig. 9.24 Schematic diagram of 5hmC detection by HpaII-assisted ligation PCR. (Revised and reprinted with permission from Ref. [190])

probe B, these two targets are unable to initiate a ligation reaction and subsequent PCR.

Hayashi found that potassium peroxotungstate can specifically oxidize 5hmC to trihydroxythymine (thT) without reacting with C and 5mC [191]. Based on this, Zhang established a peroxotungstate oxidation-assisted two-phase amplification system (POM-TPAS) to detect 5hmC at specific sites [192]. As shown in Fig. 9.25, the target DNA is first treated with potassium peroxotungstate, 5hmC is specifically oxidized to thT, and C and 5mC remain unchanged. thT is matched with adenine in the hybridization, and the base difference between thT and C/5mC can be distinguished by the ligation reaction. Two probes A and B are designed, where probe A contains two parts: a target-specific sequence and a PCR reverse primer (RP)-specific sequence. Probe B consists of three parts: a target-specific sequence with a 3′ end of adenine, a PCR forward primer (FP) sequence, and a biotin-labeled oligonucleotide T (T20). Probes A and B can be ligated to AB after hybridization with thT-containing sequences catalyzed by ligase. AB and probe B can be fixed on the surface of streptavidin (SAV)-coated single microbeads (SAV-SMB). The unreacted probe A and target DNA are washed away by magnetic separation.

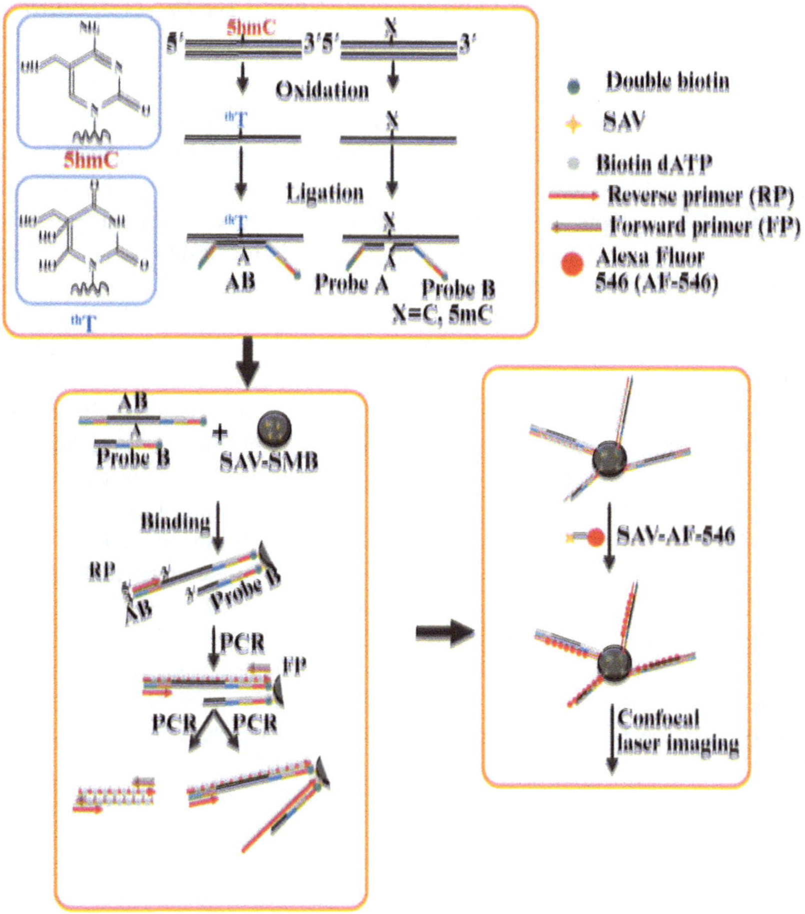

Fig. 9.25 Schematic diagram of POM-TPAS detecting 5hmC at specific site. (Revised and reprinted with permission from Ref. [192])

Then, PCR is performed on SAV-SMB with RP and FP using AB as a template. The amplified products are dispersed in the homogeneous solution for further PCR. Probe B can capture the RP extension product, which acts a template for the probe B's extension reaction. The dNTPs and biotin-modified dATP in the PCR mixture are incorporated into the amplified products. Finally, a SAV-modified fluorescent dye (SAV-AF-546) is used to bind biotin labeled products on the microbeads, enabling the fluorescence record by laser scanning confocal microscopy to quantify 5hmC.

9.5.3 High-Throughput Sequencing Method for the Detection of 5hmC

High-throughput sequencing technology is used to detect 5hmC, which can be divided into general sequencing methods and precise sequencing methods according to the accuracy of the sequencing results. General sequencing methods can only detect the DNA segment where 5hmC is located, but precise sequencing methods can determine the exact site of 5hmC. Early methods for 5hmC analysis generally employed affinity-based enrichment, such as immunoprecipitation or chemical labeling, followed by sequencing. In contrast, more precise approaches now include bisulfite sequencing, Tet-assisted bisulfite sequencing, and restriction endonuclease-based techniques.

In the high-throughput sequencing method based on immunoprecipitation, genomic DNA is first segmented. Fragments containing 5hmC are then enriched using a 5hmC-specific antibody. Finally, sequencing adaptors are ligated to these enriched fragments to construct a DNA sequencing library for high-throughput analysis [193]. Compared with immunoprecipitation sequencing methods, chemical labeling–based sequencing methods are only different in the enrichment of sample DNA. The principle of chemical labeling–based sequencing method is shown in Fig. 9.26. First, N3-5gmC is formed by connecting azido-labeled uridine diphosphate glucose (UDP-6-N3-Glu) with β-glucosyltransferase at the hydroxyl position of 5hmC. Then biotin is modified at the azide position of N3-5gmC by click chemistry, and the DNA fragments with 5hmC are separated by magnetic beads with SAV. After deamination catalyzed by APOBEC enzyme, C and 5mC are converted to uracil, while 5gmC remains unchanged. Finally, the obtained product is tagged with the sequencing adaptor to construct a DNA sequencing library for sequencing [194].

Song applied the single-molecule real-time (SMRT) sequencing method to the detection of 5hmC [195] (Fig. 9.27). In the process of SMRT sequencing, for a specific base in DNA, DNA polymerase will introduce fluorescently labeled dNTP

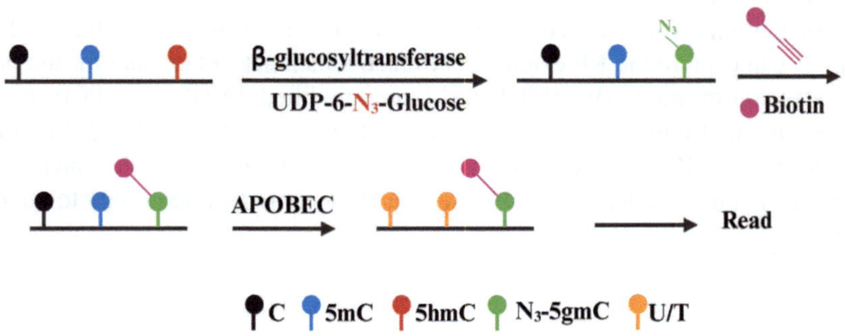

Fig. 9.26 Analysis of 5hmC by high-throughput sequencing based on chemical labeling. (Revised and reprinted with permission from Ref. [194])

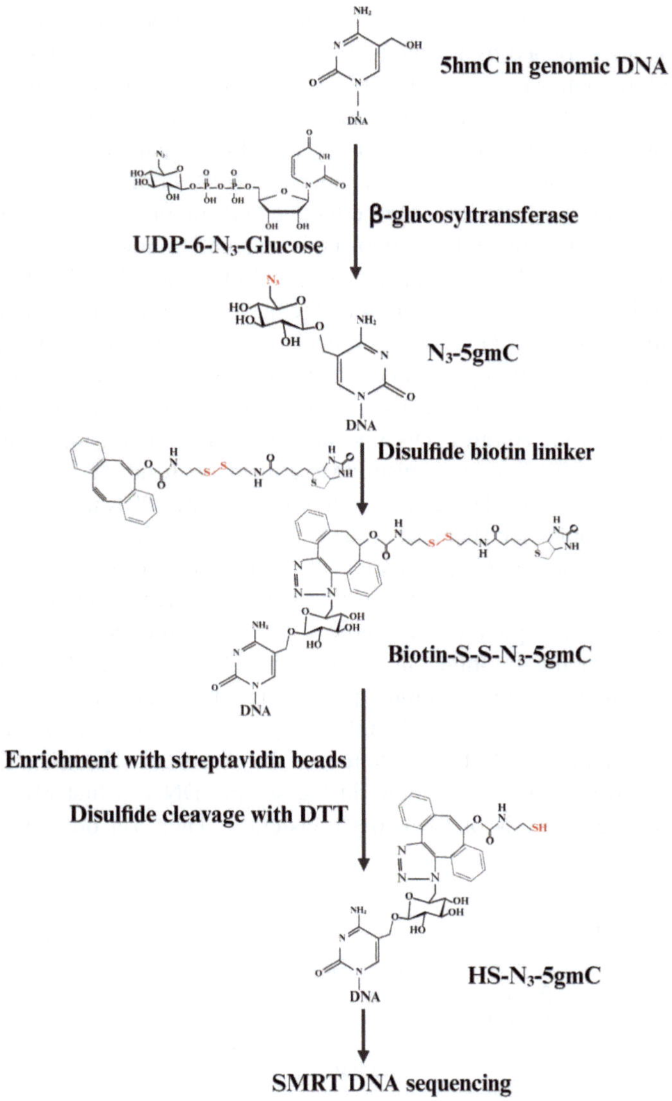

Fig. 9.27 Schematic diagram of SMRT sequencing for detection of 5hmC. (Revised and reprinted with permission from Ref. [195])

paired with it and generate specific fluorescence pulse signals. The interval between two fluorescence pulse signals is called pulse interval time IPD. Compared with unmodified bases, modified bases exhibit an increased IPD due to steric hindrance effect, a feature that enables their detection. In this SMRT sequencing approach, 5hmC is first glycosylated to form N_3-5gmC under the catalysis of β-glucosyltransferase. A disulfide biotin is then attached to the N_3 via Click chemistry, yielding biotin-S-S-N_3-5gmC. The DNA fragments containing biotin-S-S-N_3-5gmC are enriched by magnetic beads with SAV, and then eluted by hydrolyzing disulfide bonds (-S-S-) with DTT. The resulting fragments, which now bear HS-N_3-5gmC, exhibit a greater steric effect, leading to a more pronounced IPD in SMRT sequencing. This amplification of the IPD signal effectively minimizes interference from 5mC, enabling highly specific detection of 5hmC.

Normal $NaHSO_3$ sequencing is not effective in distinguishing between 5mC and 5hmC because only unmethylated cytosine is changed to uracil during the process, while both 5mC and 5hmC retain their original cytosine. To solve this problem, Booth established oxidative bisulfite sequencing (oxBS-Seq) for quantitative mapping of 5hmC in genomic DNA at single-nucleotide resolution [196]. First, the DNA sample is divided into two parts, one part is directly processed with $NaHSO_3$ and sequenced. The original C in the DNA sample becomes T, and the original 5mC or 5hmC appears as C. Another DNA sample is first treated with $KRuO_4$, which oxidizes 5hmC to 5fC. Following bisulfite treatment, the resulting 5fC is deaminated to uracil (U), enabling its detection as a thymine (T) after PCR amplification. After the processed samples are sequenced, the original C and 5hmC become T, and the original 5mC appears as C. Comparing the sequencing results of the two DNA samples, T in both the two sequencing is C in the original DNA sample; C in both the results is 5mC in the original DNA sample; C in the bisulfite sequencing but T in the oxBS-Seq sequencing is 5hmC in the original DNA sample. This method can analyze cytosine and its methylation types at single-base resolution.

Miao established Tet-assisted bisulfite sequencing (TAB-seq) by introducing β-glucosyltransferase and Tet protein on the basis of $NaHSO_3$ sequencing [197]. As shown in Fig. 9.28, 5hmC in the sample DNA is first glycosylated to 5gmC under the catalysis of β-glucosyltransferase. Then Tet protein is added to

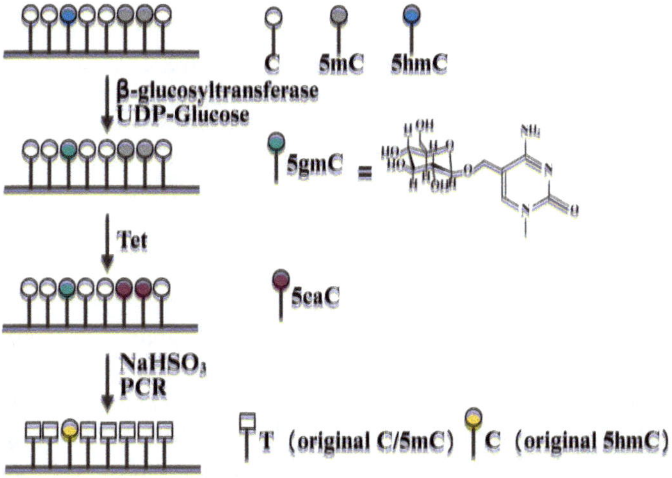

Fig. 9.28 Schematic diagram of TAB-seq. (Revised and reprinted with permission from Ref. [197])

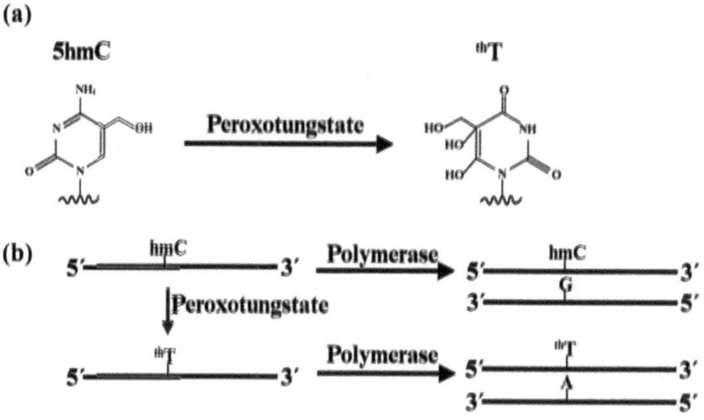

Fig. 9.29 Schematic diagram of peroxotungstate sequencing. (Revised and reprinted with permission from Ref. [191])

initiate oxidation, whereby 5mC is converted to 5caC, while 5gmC remains unoxidized. In the process of $NaHSO_3$ sequencing, both 5caC and C appear as T, while 5gmC still appears as C. Therefore, the 5hmC can be detected at single-base resolution. Compared with oxBS-seq, TAB-seq offers a more streamlined workflow, as it requires only a single sequencing run to identify 5hmC sites.

Since 5gmC resists cleavage of the restriction enzyme Mspl, Sun used this to develop reduced representation hydroxy-methylation profiling (RRHP) for single-base resolution analysis of 5hmC [198]. MspI recognizes and cuts C↓CGG (C includes C, 5mC, and 5hmC). In this method, DNA is first fragmented by MspI digestion at CCGG sites, exposing the target cytosine at the sticky end. Then sequencing adaptors are tagged to both ends of the fragmented DNA, reintroducing the cleavage site adjacent to the target C. The sample is subsequently treated with β-glucosyltransferase to convert 5hmC to 5gmC. A second MspI digestion is performed, during which 5gmC-modified sites remain uncleaved, while unmodified are cut. PCR amplification of uncleaved fragments enables construction of a sequencing library, allowing mapping of 5hmC sites. Although RRHP accurately identifies 5hmC, its application is limited to 5hmC in CCGG and cannot comprehensively detect all 5hmC sites. Hayashi demonstrated an alternative approach using peroxotungstate to specifically oxidize 5hmC to thT. Following PCR amplification, this chemical conversion results in a 5hmC:G to thT:A base-pairing change [191]. Therefore, peroxotungstate is used to treat DNA samples before sequencing, and C-T signal changes can be generated during sequencing, thus determining the 5hmC sites (Fig. 9.29). This method can accurately identify 5hmC sites. Compared with $NaHSO_3$ treatment, peroxotungstate conditions are mild, which can avoid DNA damage during oxidation.

References

1. Eichten SR, Schmitz RJ, Springer NM. Epigenetics: beyond chromatin modifications and complex genetic regulation. Plant Physiol. 2014;165:933–47.
2. Qu Y, Loh KM. Reshaping Waddington's developmental landscape. Nat Rev Genet. 2024;25: 749.
3. Esteller M. Epigenetics in evolution and disease. Lancet. 2008;372:S90–6.
4. Tronick E, Hunter RG. Waddington, dynamic systems, and epigenetics. Front Behav Neurosci. 2016;10:107.
5. Nicoglou A. Waddington′s epigenetics or the pictorial meetings of development and genetics. Hist Phil Life Sci. 2018;40:61.
6. Bird A. DNA methylation patterns and epigenetic memory. Genes Dev. 2002;16:6–21.
7. Moore LD, Le T, Fan G. DNA methylation and its basic function. Neuropsychopharmacology. 2013;38:23–38.
8. Zhao Z, Shilatifard A. Epigenetic modifications of histones in cancer. Genome Biol. 2019;20: 245.
9. Kim S, Kaang BK. Epigenetic regulation and chromatin remodeling in learning and memory. Exp Mol Med. 2017;49:e281.
10. Attwood JT, Yung RL, Richardson BC. DNA methylation and the regulation of gene transcription. Cell Mol Life Sci. 2002;59:241–57.
11. Ehrlich M, Gama-Sosa MA, Huang LH, Midgett RM, Kuo KC, McCune RA, Gehrke C. Amount and distribution of 5-methylcytosine in human DNA from different types of tissues of cells. Nucleic Acids Res. 1982;10:2709–21.
12. Razin A, Cedar H. Distribution of 5-methylcytosine in chromatin. Proc Natl Acad Sci USA. 1977;74:2725–8.
13. McClelland M, Ivarie R. Asymmetrical distribution of CpG in an average mammalian gene. Nucleic Acids Res. 1982;10:7865–77.
14. Saxonov S, Berg P, Brutlag DL. A genome-wide analysis of CpG dinucleotides in the human genome distinguishes two distinct classes of promoters. Proc Natl Acad Sci USA. 2006;103: 1412–7.
15. Wutz A, Smrzka OW, Schweifer N, Schellander K, Wagner EF, Barlow DP. Imprinted expression of the Igf2r gene depends on an intronic CpG Island. Nature. 1997;389:745–9.
16. Caspary T, Cleary MA, Baker CC, Guan XJ, Tilghman SM. Multiple mechanisms regulate imprinting of the mouse distal chromosome 7 gene cluster. Mol Cell Biol. 1998;18:3466–74.
17. Zwart R, Sleutels F, Wutz A, Schinkel AH, Barlow DP. Bidirectional action of the Igf2r imprint control element on upstream and downstream imprinted genes. Genes Dev. 2001;15: 2361–6.
18. Bird A, Taggart M, Frommer M, Miller OJ, Macleod D. A fraction of the mouse genome that is derived from islands of nonmethylated, CpG-rich DNA. Cell. 1985;40:91–9.
19. Yen RW, Vertino PM, Nelkin BD, Yu JJ, El-Deiry W, Cumaraswamy A, Lennon GG, Trask BJ, Celano P, Baylin SB. Isolation and characterization of the cDNA encoding human DNA methyltransferase. Nucleic Acids Res. 1992;20:2287–91.
20. Araujo FD, Croteau S, Slack AD, Milutinovic S, Bigey P, Price GB, Zannis-Hajopoulos M, Szyf M. The DNMT1 target recognition domain resides in the N terminus. J Biol Chem. 2001;276:6930–6.
21. Okano M, Bell DW, Haber DA, Li E. DNA methyltransferases Dnmt3a and Dnmt3b are essential for de novo methylation and mammalian development. Cell. 1999;99:247–57.
22. Xie SP, Wang ZJ, Okano M, Nogami M, Li Y, He WW, Okumura K, Li E. Cloning, expression and chromosome locations of the human DNMT3 gene family. Gene. 1999;236: 87–95.
23. Mayer W, Niveleau A, Walter J, Fundele R, Haaf T. Embryogenesis-demethylation of the zygotic paternal genome. Nature. 2000;403:501–2.

24. Oswald J, Engemann S, Lane N, Mayer W, Olek A, Fundele R, Dean W, Reik W, Walter J. Active demethylation of the paternal genome in the mouse zygote. Curr Biol. 2000;10:475–8.
25. Paroush Z, Keshet I, Yisraeli J, Cedar H. Dynamics of demethylation and activation of the alpha-actin gene in myoblasts. Cell. 1990;63:1229–37.
26. Zhang F, Pomerantz JH, Sen G, Palermo AT, Blau HM. Active tissue-specific DNA demethylation conferred by somatic cell nuclei in stable heterokaryons. Proc Natl Acad Sci USA. 2007;104:4395–400.
27. Bhutani N, Burns DM, Blau HM. DNA demethylation dynamics. Cell. 2011;146:866–72.
28. Tahiliani M, Koh KP, Shen Y, Pastor WA, Bandukwala H, Brudno Y, Agarwal S, Iyer LM, Liu DR, Aravind L, Rao A. Conversion of 5-methylcytosine to 5-hydroxymethylcytosine in mammalian DNA by MLL partner TET1. Science. 2009;324:930–5.
29. Ito S, D'Alessio AC, Taranova OV, Hong K, Sowers LC, Zhang Y. Role of Tet proteins in 5mC to 5hmC conversion, ES-cell self-renewal and inner cell mass specification. Nature. 2010;466:1129–33.
30. Guo JU, Su Y, Zhong C, Ming G-L, Song H. Hydroxylation of 5-methylcytosine TET1 promotes active DNA demethylation in the adult brain. Cell. 2011;145:423–34.
31. Ito S, Shen L, Dai Q, Wu SC, Collins LB, Swenberg JA, He C, Zhang Y. Tet proteins can convert 5-methylcytosine to 5-formylcytosine and 5-carboxylcytosine. Science. 2011;333: 1300–3.
32. Valinluck V, Tsai HH, Rogstad DK, Burdzy A, Bird A, Sowers LC. Oxidative damage to methyl-CpG sequences inhibits the binding of the methyl-CpG binding domain (MBD) of methyl-CpG binding protein 2 (MeCP2). Nucleic Acids Res. 2004;32:4100–8.
33. Cortellino S, Xu J, Sannai M, Moore R, Caretti E, Cigliano A, Le Coz M, Devarajan K, Wessels A, Soprano D, Abramowitz LK, Bartolomei MS, Rambow F, Bassi MR, Bruno T, Fanciulli M, Renner C, Klein-Szanto AJ, Matsumoto Y, Kobi D, Davidson I, Alberti C, Larue L, Bellacosa A. Thymine DNA glycosylase is essential for active DNA demethylation by linked deamination-base excision repair. Cell. 2011;146:67–79.
34. Hashimoto H, Zhang X, Cheng X. Selective excision of 5-carboxylcytosine by a thymine DNA glycosylase mutant. J Mol Biol. 2013;425:971–6.
35. Muller U, Bauer C, Siegl M, Rottach A, Leonhardt H. TET-mediated oxidation of methylcytosine causes TDG or NEIL glycosylase dependent gene reactivation. Nucleic Acids Res. 2014;42:8592–604.
36. Guo JU, Ma DK, Mo H, Ball MP, Jang M-H, Bonaguidi MA, Balazer JA, Eaves HL, Xie B, Ford E, Zhang K, Ming G-L, Gao Y, Song H. Neuronal activity modifies the DNA methylation landscape in the adult brain. Nat Neurosci. 2011;14:1345–51.
37. Nan XS, Meehan RR, Bird A. Dissection of the methyl-CpG binding domain from the chromosomal protein MeCP2. Nucleic Acids Res. 1993;21:4886–92.
38. Meehan RR, Lewis JD, McKay S, Kleiner EL, Bird AP. Identification of a mammalian protein that binds specifically to DNA containing methylated CpGs. Cell. 1989;58:499–507.
39. Lewis JD, Meehan RR, Henzel WJ, Maurerfogy I, Jeppesen P, Klein F, Bird A. Purification, sequence, and cellular-localization of a novel chromosomal protein that binds to methylated DNA. Cell. 1992;69:905–14.
40. Kimura H, Shiota K. Methyl-CpG-binding protein, MeCP2, is a target molecule for maintenance DNA methyltransferase, Dnmt1. J Biol Chem. 2003;278:4806–12.
41. Esteller M. Aberrant DNA methylation as a cancer-inducing mechanism. Annu Rev Pharmacol. 2005;45:629–56.
42. Meehan RR, Lewis JD, Bird AP. Characterization of MeCP2, a vertebrate DNA-binding protein with affinity for methylated DNA. Nucleic Acids Res. 1992;20:5085–92.
43. Ng HH, Zhang Y, Hendrich B, Johnson CA, Turner BM, Erdjument-Bromage H, Tempst P, Reinberg D, Bird A. MBD2 is a transcriptional repressor belonging to the MeCP1 histone deacetylase complex. Nat Genet. 1999;23:58–61.

44. Wade PA, Gegonne A, Jones PL, Ballestar E, Aubry F, Wolffe AP. Mi-2 complex couples DNA methylation to chromatin remodelling and histone deacetylation. Nat Genet. 1999;23: 62–6.
45. Hendrich B, Bird A. Identification and characterization of a family of mammalian methyl-CpG binding proteins. Mol Cell Biol. 1998;18:6538–47.
46. Zhu B, Zheng Y, Angliker H, Schwarz S, Thiry S, Siegmann M, Jost JP. 5-methylcytosine DNA glycosylase activity is also present in the human MBD4 (G/T mismatch glycosylase) and in a related avian sequence. Nucleic Acids Res. 2000;28:4157–65.
47. Smith ZD, Meissner A. DNA methylation: roles in mammalian development. Nat Rev Genet. 2013;14:204–20.
48. Arita K, Ariyoshi M, Tochio H, Nakamura Y, Shirakawa M. Recognition of hemi-methylated DNA by the SRA protein UHRF1 by a base-flipping mechanism. Nature. 2008;455:818–21.
49. Avvakumov GV, Walker JR, Xue S, Li Y, Duan S, Bronner C, Arrowsmith CH, Dhe-Paganon S. Structural basis for recognition of hemi-methylated DNA by the SRA domain of human UHRF1. Nature. 2008;455:822–5.
50. Sharif J, Muto M, Takebayashi S-I, Suetake I, Iwamatsu A, Endo TA, Shinga J, Mizutani-Koseki Y, Toyoda T, Okamura K, Tajima S, Mitsuya K, Okano M, Koseki H. The SRA protein Np95 mediates epigenetic inheritance by recruiting Dnmt1 to methylated DNA. Nature. 2007;450:908–12.
51. Bostick M, Kim JK, Esteve P-O, Clark A, Pradhan S, Jacobsen SE. UHRF1 plays a role in maintaining DNA methylation in mammalian cells. Science. 2007;317:1760–4.
52. Prokhortchouk A, Hendrich B, Jorgensen H, Ruzov A, Wilm M, Georgiev G, Bird A, Prokhortchouk E. The p120 catenin partner Kaiso is a DNA methylation-dependent transcriptional repressor. Genes Dev. 2001;15:1613–8.
53. Filion GJP, Zhenilo S, Salozhin S, Yamada D, Prokhortchouk E, Defossez PA. A family of human zinc finger proteins that bind methylated DNA and repress transcription. Mol Cell Biol. 2006;26:169–81.
54. Holliday R, Pugh JE. DNA modification mechanisms and gene activity during development. Science. 1975;187:226–32.
55. Riggs AD. X-inactivation, differentiation, and DNA methylation. Cytogenet Cell Genet. 1975;14:9–25.
56. Comb M, Goodman HM. CpG methylation inhibits proenkephalin gene-expression and binding of the transcription factor AP-2. Nucleic Acids Res. 1990;18:3975–82.
57. Clark SJ, Harrison J, Molloy PL. Sp1 binding is inhibited by ^{m}Cp ^{m}CpG methylation. Gene. 1997;195:67–71.
58. Harrington MA, Jones PA, Imagawa M, Karin M. Cytosine methylation does not affect binding of transcription factor Sp1. Proc Natl Acad Sci USA. 1988;85:2066–70.
59. Yu F, Thiesen J, Stratling WH. Histone deacetylase-independent transcriptional repression by methyl-CpG-binding protein 2. Nucleic Acids Res. 2000;28:2201–6.
60. Nan XS, Campoy FJ, Bird A. MeCP2 is a transcriptional repressor with abundant binding sites in genomic chromatin. Cell. 1997;88:471–81.
61. Nan XS, Tate P, Li E, Bird A. DNA methylation specifies chromosomal localization of MeCP2. Mol Cell Biol. 1996;16:414–21.
62. Nan XS, Ng HH, Johnson CA, Laherty CD, Turner BM, Eisenman RN, Bird A. Transcriptional repression by the methyl-CpG-binding protein MeCP2 involves a histone deacetylase complex. Nature. 1998;393:386–9.
63. Ng HH, Bird A. Histone deacetylases: silencers for hire. Trends Biochem Sci. 2000;25:121–6.
64. Jones PL, Veenstra GJC, Wade PA, Vermaak D, Kass SU, Landsberger N, Strouboulis J, Wolffe AP. Methylated DNA and MeCP2 recruit histone deacetylase to repress transcription. Nat Genet. 1998;19:187–91.
65. Zhang Y, Ng HH, Erdjument-Bromage H, Tempst P, Bird A, Reinberg D. Analysis of the NuRD subunits reveals a histone deacetylase core complex and a connection with DNA methylation. Genes Dev. 1999;13:1924–35.

66. Boeke J, Ammerpohl O, Kegel S, Moehren U, Renkawitz R. The minimal repression domain of MBD2b overlaps with the methyl-CpG-binding domain and binds directly to Sin3A. J Biol Chem. 2000;275:34963–7.
67. Tatematsu K, Yamazaki T, Ishikawa F. MBD2-MBD3 complex binds to hemi-methylated DNA and forms a complex containing DNMT1 at the replication foci in late S phase. Genes Cells. 2000;5:677–88.
68. Hendrich B, Guy J, Ramsahoye B, Wilson VA, Bird A. Closely related proteins MBD2 and MBD3 play distinctive but interacting roles in mouse development. Genes Dev. 2001;15:710–23.
69. Ng HH, Jeppesen P, Bird A. Active repression of methylated genes by the chromosomal protein MBD1. Mol Cell Biol. 2000;20:1394–406.
70. Zhu J. Epigenetics and precision medicine. 1st ed. Shanghai Jiao Tong University Press; 2017. p. 80–92.
71. Santos F, Hendrich B, Reik W, Dean W. Dynamic reprogramming of DNA methylation in the early mouse embryo. Dev Biol. 2002;241:172–82.
72. Smith ZD, Chan MM, Mikkelsen TS, Gu H, Gnirke A, Regev A, Meissner A. A unique regulatory phase of DNA methylation in the early mammalian embryo. Nature. 2012;484:339–44.
73. Mark W, Julia A, Vittorio S, Konstantin L, Michele B, Richard R, Hans S, Jörn W. Dynamic link of DNA demethylation, DNA strand breaks and repair in mouse zygotes. EMBO J. 2010;29:1877–88.
74. Iqbal K, Jin SG, Pfeifer GP, Szabo PE. Reprogramming of the paternal genome upon fertilization involves genome-wide oxidation of 5-methylcytosine. Proc Natl Acad Sci USA. 2011;108:3642–7.
75. Wossidlo M, Nakamura T, Lepikhov K, Marques CJ, Zakhartchenko V, Boiani M, Arand J, Nakano T, Reik W, Walter J. 5-hydroxymethylcytosine in the mammalian zygote is linked with epigenetic reprogramming. Nat Commun. 2011;2:1089–93.
76. Gu T-P, Guo F, Yang H, Wu H-P, Xu G-F, Liu W, Xie Z-G, Shi L, He X, Jin S, Iqbal K, Shi GY, Deng Z, Szabó PE, Pfeifer GP, Li J, Xu G-L. The role of Tet3 DNA dioxygenase in epigenetic reprogramming by oocytes. Nature. 2011;477:606–10.
77. Toshinobu N, Yu-Jung L, Hiroyuki N, Hiroki U, Kimiko I, Shogo M, Makoto T, Atsuo O, Yoichi S, Toru N. PGC7 binds histone H3K9me2 to protect against conversion of 5mC to 5hmC in early embryos. Nature. 2012;486:415–9.
78. Hajkova P, Jeffries SJ, Lee C, Miller N, Jackson SP, Surani MA. Genome-wide reprogramming in the mouse germ line entails the base excision repair pathway. Science. 2010;329:78–82.
79. Nakamura T, Arai Y, Umehara H, Masuhara M, Kimura T, Taniguchi H, Sekimoto T, Ikawa M, Yoneda Y, Okabe M, Tanaka S, Shiota K, Nakano T. PGC7/stella protects against DNA demethylation in early embryogenesis. Nat Cell Biol. 2007;9:64–71.
80. Smallwood SA, Tomizawa S-I, Krueger F, Ruf N, Carli N, Segonds-Pichon A, Sato S, Hata K, Andrews SR, Kelsey G. Dynamic CpG island methylation landscape in oocytes and preimplantation embryos. Nat Genet. 2011;43:811–4.
81. Hirasawa R, Chiba H, Kaneda M, Tajima S, Li E, Jaenisch R, Sasaki H. Maternal and zygotic Dnmt1 are necessary and sufficient for the maintenance of DNA methylation imprints during preimplantationdevelopment. Genes Dev. 2008;22:1607–16.
82. Ratnam S, Mertineit C, Ding F, Howell CY, Clarke HJ, Bestor TH, Chaillet JR, Trasler JM. Dynamics of Dnmt1 methyltransferase expression and intracellular localization during oogenesis and preimplantation development. Dev Biol. 2002;245:304–14.
83. Howell CY, Bestor TH, Ding F, Latham KE, Mertineit C, Trasler JM, Chaillet JR. Genomic imprinting disrupted by a maternal effect mutation in the Dnmt1 gene. Cell. 2001;104:829–38.
84. Lyon MF. Gene action in the X-chromosome of the mouse (Mus musculus L). Nature. 1961;190:372–3.

85. Cooper DW, Vandeber JL, Sharman GB, Poole WE. Phosphoglycerate kinase polymorphism in kangaroos provides further evidence for paternal X inactivation, vol. 230. Nat.-New Biol; 1971. p. 155–7.
86. Sharman GB. Late DNA replication in paternally derived X chromosome of female kangaroos. Nature. 1971;230:231–2.
87. West JD, Frels WI, Chapman VM, Papaioannou VE. Preferential expression of maternally derived X-chromosome in mouse yolk sac. Cell. 1977;12:873–82.
88. Takagi N, Sasaki M. Preferential inactivation of paternally derived X-chromosome in extra-embryonic membranes of mouse. Nature. 1975;256:640–2.
89. Boumil RM, Lee JT. Forty years of decoding the silence in X-chromosome inactivation. Hum Mol Genet. 2001;10:2225–32.
90. Lee JT, Strauss WM, Dausman JA, Jaenisch R. A 450 kb transgene displays properties of the mammalian X-inactivation center. Cell. 1996;86:83–94.
91. Herzing LBK, Romer JT, Horn JM, Ashworth A. Xist has properties of the X-chromosome inactivation centre. Nature. 1997;386:272–5.
92. Lee JT, Jaenisch R. Long-range cis effects of ectopic X-inactivation centres on a mouse autosome. Nature. 1997;386:275–9.
93. Heard E, Mongelard F, Arnaud D, Avner P. Xist yeast artificial chromosome transgenes function as X-inactivation centers only in multicopy arrays and not as single copies. Mol Cell Biol. 1999;19:3156–66.
94. Lee JT, Lu NF, Han Y. Genetic analysis of the mouse X inactivation center defines an 80-kb multifunction domain. Proc Natl Acad Sci USA. 1999;96:3836–41.
95. Brown CJ, Ballabio A, Rupert JL, Lafreniere RG, Grompe M, Tonlorenzi R, Willard HF. A gene from the region of the human X-inactivation center is expressed exclusively from the inactive X-chromosome. Nature. 1991;349:38–44.
96. Borsani G, Tonlorenzi R, Simmler MC, Dandolo L, Arnaud D, Capra V, Grompe M, Pizzuti A, Muzny D, Lawrence C, Willard HF, Avner P, Ballabio A. Characterization of a murine gene expressed from the inactive X-chromosome. Nature. 1991;351:325–9.
97. Brockdorff N, Ashworth A, Kay GF, Cooper P, Smith S, McCabe VM, Norris DP, Penny GD, Patel D, Rastan S. Conservation of position and exclusive expression of mouse xist from the inactive X-chromosome. Nature. 1991;351:329–31.
98. Brown CJ, Hendrich BD, Rupert JL, Lafreniere RG, Xing Y, Lawrence J, Willard HF. The human XIST gene: analysis of a 17 kb inactive X-specific RNA that contains conserved repeats and is highly localized within the nucleus. Cell. 1992;71:527–42.
99. Brockdorff N, Ashworth A, Kay GF, McCabe VM, Norris DP, Cooper PJ, Swift S, Rastan S. The product of the mouse xist gene is a 15 kb inactive X-specific transcript containing no conserved orf and located in the nucleus. Cell. 1992;71:515–26.
100. Penny GD, Kay GF, Sheardown SA, Rastan S, Brockdorff N. Requirement for Xist in X chromosome inactivation. Nature. 1996;379:131–7.
101. Turner JMA, Mahadevaiah SK, Eillott DJ, Garchon HJ, Pehrson JR, Jaenisch R, Burgoyne PS. Meiotic sex chromosome inactivation in male mice with targeted disruptions of Xist. J Cell Sci. 2002;115:4097–105.
102. Oda M, Yamagiwa A, Yamamoto S, Nakayama T, Tsumura A, Sasaki H, Nakao K, Li E, Okano M. DNA methylation regulates long-range gene silencing of an X-linked homeobox gene cluster in a lineage-specific manner. Genes Dev. 2006;20:3382–94.
103. Ferguson-Smith AC. Genomic imprinting: the emergence of an epigenetic paradigm. Nat Rev Genet. 2011;12:565–75.
104. Sha K. A mechanistic view of genomic imprinting. Annu Rev Genomics Hum Genet. 2008;9: 197–216.
105. Elhamamsy AR. Role of DNA methylation in imprinting disorders: an updated review. J Assist Reprod Genet. 2017;34:549–62.
106. Sleutels F, Zwart R, Barlow DP. The non-coding air RNA is required for silencing autosomal imprinted genes. Nature. 2002;415:810–3.

107. Royo H, Cavaille J. Non-coding RNAs in imprinted gene clusters. Biol Cell. 2008;100:149–66.
108. Pandey RR, Mondal T, Mohammad F, Enroth S, Redrup L, Komorowski J, Nagano T, Mancini-DiNardo D, Kanduri C. Kcnq1ot1 antisense noncoding RNA mediates lineage-specific transcriptional silencing through chromatin-level regulation. Mol Cell. 2008;32: 232–46.
109. Chiesa N, De Crescenzo A, Mishra K, Perone L, Carella M, Palumbo O, Mussa A, Sparago A, Cerrato F, Russo S, Lapi E, Cubellis MV, Kanduri C, Silengo MC, Riccio A, Ferrero GB. The KCNQ1OT1 imprinting control region and non-coding RNA: new properties derived from the study of Beckwith-Wiedemann syndrome and Silver-Russell syndrome cases. Hum Mol Genet. 2012;21:10–25.
110. Brockdorff N. Noncoding RNA and polycomb recruitment. RNA. 2013;19:429–42.
111. Higashimoto K, Soejima H, Saito T, Okumura K, Mukai T. Imprinting disruption of the CDKN1C/KCNQ1OT1 domain: the molecular mechanisms causing beckwith-wiedemann syndrome and cancer. Cytogenet Genome Res. 2006;113:306–12.
112. Ulaner GA, Yang YW, Hu JF, Li T, Vu TH, Hoffman AR. CTCF binding at the insulin-like growth factor-II (IGF2)/H19 imprinting control region is insufficient to regulate IGF2/H19 expression in human tissues. Endocrinology. 2003;144:4420–6.
113. Beygo J, Citro V, Sparago A, De Crescenzo A, Cerrato F, Heitmann M, Rademacher K, Guala A, Enklaar T, Anichini C, Silengo MC, Graf N, Prawitt D, Cubellis MV, Horsthemke B, Buiting K, Riccio A. The molecular function and clinical phenotype of partial deletions of the IGF2/H19 imprinting control region depends on the spatial arrangement of the remaining CTCF-binding sites. Hum Mol Genet. 2013;22:544–57.
114. Demars J, Gicquel C. Epigenetic and genetic disturbance of the imprinted 11p15 region in Beckwith-Wiedemann and Silver-Russell syndromes. Clin Genet. 2012;81:350–61.
115. Jacob KJ, Robinson WP, Lefebvre L. Beckwith-wiedemann and silver-Russell syndromes: opposite developmental imbalances in imprinted regulators of placental function and embryonic growth. Clin Genet. 2013;84:326–34.
116. Yamazawa K, Ogata T, Ferguson-Smith AC. Uniparental disomy and human disease: an overview. Am J Med Genet. 2010;154C:329–34.
117. Hanahan D, Weinberg RA. The hallmarks of cancer. Cell. 2000;100:57–70.
118. Esteller M. Cancer epigenomics: DNA methylomes and histone-modification maps. Nat Rev Genet. 2007;8:286–98.
119. Walsh CP, Chaillet JR, Bestor TH. Transcription of IAP endogenous retroviruses is constrained by cytosine methylation. Nat Genet. 1998;20:116–7.
120. Gaudet F, Hodgson JG, Eden A, Jackson-Grusby L, Dausman J, Gray JW, Leonhardt H, Jaenisch R. Induction of tumors in mice by genomic hypomethylation. Science. 2003;300: 489–92.
121. Esteller M, Almouzni G. How epigenetics integrates nuclear functions. Workshop on epigenetics and chromatin: transcriptional regulation and beyond. EMBO Rep. 2005;6:624–8.
122. Rao PM, Antony A, Rajalakshmi S, Sarma DS. Studies on hypomethylation of liver DNA during early stages of chemical carcinogenesis in rat liver. Carcinogenesis. 1989;10:933–7.
123. Sharrard RM, Royds JA, Rogers S, Shorthouse AJ. Patterns of methylation of the c-myc gene in human colorectal cancer progression. Br J Cancer. 1992;65:667–72.
124. Plongthongkum N, Diep DH, Zhang K. Advances in the profiling of DNA modifications: cytosine methylation and beyond. Nat Rev Genet. 2014;15:647–61.
125. Clark SJ, Harrison J, Paul CL, Frommer M. High sensitivity mapping of methylated cytosines. Nucleic Acids Res. 1994;22:2990–7.
126. Herman JG, Graff JR, Myohanen S, Nelkin BD, Baylin SB. Methylation-specific PCR: a novel PCR assay for methylation status of CpG islands. Proc Natl Acad Sci USA. 1996;93:9821–6.
127. Worm J, Aggerholm A, Guldberg P. In-tube DNA methylation profiling by fluorescence melting curve analysis. Clin Chem. 2001;47:1183–9.

128. Akey DT, Akey JM, Zhang K, Jin L. Assaying DNA methylation based on high-throughput melting curve approaches. Genomics. 2002;80:376–84.
129. Wojdacz TK, Dobrovic A, Hansen LL. Methylation-sensitive high-resolution melting. Nat Protoc. 2008;3:1903–8.
130. Kristensen LS, Mikeska T, Krypuy M, Dobrovic A. Sensitive melting analysis after real time-methylation specific PCR (SMART-MSP): high-throughput and probe-free quantitative DNA methylation detection. Nucleic Acids Res. 2008;36:e42.
131. Olkhov-Mitsel E, Bapat B. Strategies for discovery and validation of methylated and hydroxymethylated DNA biomarkers. Cancer Med. 2012;1:237–60.
132. Eads CA, Danenberg KD, Kawakami K, Saltz LB, Blake C, Shibata D, Danenberg PV, Laird PW. MethyLight: a high-throughput assay to measure DNA methylation. Nucleic Acids Res. 2000;28:e32.
133. Afonina I, Zivarts M, Kutyavin I, Lukhtanov E, Gamper H, Meyer RB. Efficient priming of PCR with short oligonucleotides conjugated to a minor groove binder. Nucleic Acids Res. 1997;25:2657–60.
134. Zeschnigk M, Bohringer S, Price EA, Onadim Z, Masshofer L, Lohmann DR. A novel real-time PCR assay for quantitative analysis of methylated alleles (QAMA): analysis of the retinoblastoma locus. Nucleic Acids Res. 2004;32:e125.
135. Gonzalgo ML, Jones PA. Rapid quantitation of methylation differences at specific sites using methylation-sensitive single nucleotide primer extension (Ms-SNuPE). Nucleic Acids Res. 1997;25:2529–31.
136. Gonzalgo ML, Jones PA. Quantitative methylation analysis using methylation-sensitive single-nucleotide primer extension (Ms-SNuPE). Methods. 2002;27:128–33.
137. El-Maarri O, Herbiniaux U, Walter J, Oldenburg J. A rapid, quantitative, non-radioactive bisulfite-SNuPE-IP-RP HPLC assay for methylation analysis at specific CpG sites. Nucleic Acids Res. 2002;30:e25.
138. Tost J, Schatz P, Schuster M, Berlin K, Gut IG. Analysis and accurate quantification of CpG methylation by MALDI mass spectrometry. Nucleic Acids Res. 2003;31:e50.
139. Feng F, Wang H, Han L, Wang S. Fluorescent conjugated polyelectrolyte as an indicator for convenient detection of DNA methylation. J Am Chem Soc. 2008;130:11338–43.
140. Xiong Z, Laird PW. COBRA: a sensitive and quantitative DNA methylation assay. Nucleic Acids Res. 1997;25:2532–4.
141. Uhlmann K, Brinckmann A, Toliat MR, Ritter H, Nurnberg P. Evaluation of a potential epigenetic biomarker by quantitative methyl-single nucleotide polymorphism analysis. Electrophoresis. 2002;23:4072–9.
142. Colella S, Shen L, Baggerly KA, Issa JPJ, Krahe R. Sensitive and quantitative universal pyrosequencing methylation analysis of CpG sites. BioTechniques. 2003;35:146–50.
143. Su F, Wang L, Sun Y, Liu C, Duan X, Li Z. Highly sensitive detection of CpG methylation in genomic DNA by AuNP-based colorimetric assay with ligase chain reaction. Chem Commun. 2015;51:3371–4.
144. Su F, Wang L, Sun Y, Liu C, Duan X, Li Z. Highly sensitive and multiplexed analysis of CpG methylation at single-base resolution with ligation-based exponential amplification. Chem Sci. 2015;6:1866–72.
145. Cao A, Zhang C-Y. Sensitive and label-free DNA methylation detection by ligation-mediated hyperbranched rolling circle amplification. Anal Chem. 2012;84:6199–205.
146. Singer-Sam J, LeBon JM, Tanguay RL, Riggs AD. A quantitative HpaII-PCR assay to measure methylation of DNA from a small number of cells. Nucleic Acids Res. 1990;18:687.
147. Feng F, Liu L, Wang S. Fluorescent conjugated polymer-based FRET technique for detection of DNA methylation of cancer cells. Nat Protoc. 2010;5:1255–64.
148. Nygren AOH, Ameziane N, Duarte HMB, Vijzelaar R, Waisfisz Q, Hess CJ, Schouten JP, Errami A. Methylation-specific MLPA (MS-MLPA): simultaneous detection of CpG methylation and copy number changes of up to 40 sequences. Nucleic Acids Res. 2005;33:e128.

149. Wen H, Wang H, Wang H, Yan J, Tian H, Li Z. Ultrasensitive detection of site-specific DNA methylation by loop-mediated isothermal amplification. Anal Methods. 2016;8:5372–7.
150. Pataer P, Zhang P, Li Z. Single methylation sensitive restriction endonuclease-based cascade exponential amplification assay for visual detection of DNA methylation at single-molecule level. Anal Chem. 2024;96:13335–43.
151. Sun Y, Sun Y, Tian W, Liu C, Gao K, Li Z. A novel restriction endonuclease GlaI for rapid and highly sensitive detection of DNA methylation coupled with isothermal exponential amplification reaction. Chem Sci. 2018;9:1344–51.
152. Zeng Y-P, Hu J, Long Y, Zhang C-Y. Sensitive detection of DNA methyltransferase using hairpin probe-based primer generation rolling circle amplification-induced chemiluminescence. Anal Chem. 2013;85:6143–50.
153. Weber M, Davies JJ, Wittig D, Oakeley EJ, Haase M, Lam WL, Schubeler D. Chromosome-wide and promoter-specific analyses identify sites of differential DNA methylation in normal and transformed human cells. Nat Genet. 2005;37:853–62.
154. Down TA, Rakyan VK, Turner DJ, Flicek P, Li H, Kulesha E, Graf S, Johnson N, Herrero J, Tomazou EM, Thorne NP, Backdahl L, Herberth M, Howe KL, Jackson DK, Miretti MM, Marioni JC, Birney E, Hubbard TJP, Durbin R, Tavare S, Beck S. A bayesian deconvolution strategy for immunoprecipitation-based DNA methylome analysis. Nat Biotechnol. 2008;26: 779–85.
155. Zilberman D, Gehring M, Tran RK, Ballinger T, Henikoff S. Genome-wide analysis of arabidopsis thaliana DNA methylation uncovers an interdependence between methylation and transcription. Nat Genet. 2007;39:61–9.
156. Zhang X, Yazaki J, Sundaresan A, Cokus S, Chan SWL, Chen H, Henderson IR, Shinn P, Pellegrini M, Jacobsen SE, Ecker JR. Genome-wide high-resolution mapping and functional analysis of DNA methylation in arabidopsis. Cell. 2006;126:1189–201.
157. Tibor R, Pfeifer PG. Methylated-CpG island recovery assay: a new technique for the rapid detection of methylated-CpG islands in cancer. Lab Investig. 2005;85:1172–80.
158. Kuester D, El-Rifai WE, Peng D, Ruemmele P, Kroeckel I, Peters B, Moskaluk CA, Stolte M, Moenkemueller K, Meyer F, Schulz H-U, Hartmann A, Roessner A, Schneider-Stock R. Silencing of MGMT expression by promoter hypermethylation in the metaplasia-dysplasia-carcinoma sequence of Barrett's esophagus. Cancer Lett. 2009;275:117–26.
159. Kelavkar UP, Harya NS, Hutzley J, Bacich DJ, Monzon FA, Chandran U, Dhir R, Keefe DSO. DNA methylation paradigm shift: 15-lipoxygenase-1 upregulation in prostatic intraepithelial neoplasia and prostate cancer by atypical promoter hypermethylation. Prostaglandins Other Lipid Mediat. 2007;82:185–97.
160. Verma M, Srivastava S. Epigenetics in cancer: implications for early detection and prevention. Lancet Oncol. 2002;3:755–63.
161. Gal-Yam EN, Saito Y, Egger G, Jones PA. Cancer epigenetics: modifications, screening, and therapy. Annu Rev Med. 2008;59:267–80.
162. Hesson LB, Cooper WN, Latif F. The role of RASSF1A methylation in cancer. Dis Markers. 2007;23:73–87.
163. Abe M, Westermann F, Nakagawara A, Takato T, Schwab M, Ushijima T. Marked and independent prognostic significance of the CpG island methylator phenotype in neuroblastomas. Cancer Lett. 2007;247:253–8.
164. Esteller M, Garcia-Foncillas J, Andion E, Goodman SN, Hidalgo OF, Vanaclocha V, Baylin SB, Herman JG. Inactivation of the DNA-repair gene MGMT and the clinical response of gliomas to alkylating agents. New Engl J Med. 2000;343:1350–4.
165. Long NK, Kato K, Yamashita T, Makita H, Toida M, Hatakeyama D, Hara A, Mori H, Shibata T. Hypermethylation of the RECK gene predicts poor prognosis in oral squamous cell carcinomas. Oral Oncol. 2008;44:1052–8.
166. Egger G, Liang GN, Aparicio A, Jones PA. Epigenetics in human disease and prospects for epigenetic therapy. Nature. 2004;429:457–63.

167. Constantinides PG, Jones PA, Gevers W. Functional striated muscle cells from non-myoblast precursors following 5-azacytidine treatment. Nature. 1977;267:364–6.
168. Bender CM, Gonzalgo ML, Gonzales FA, Nguyen CT, Robertson KD, Jones PA. Roles of cell division and gene transcription in the methylation of CpG islands. Mol Cell Biol. 1999;19: 6690–8.
169. Sorm F, Piskala A, Cihak A, Vesely J. 5-azacytidine, a new, highly effective cancerostatic. Experientia. 1964;20:202–3.
170. Veselý J, Čihák A. 5-azacytidine: mechanism of action and biological effects in mammalian cells. Pharmacol Therapeut. 1978;2:813–40.
171. Kantarjian H, Oki Y, Garcia-Manero G, Huang XL, O'Brien S, Cortes J, Faderl S, Bueso-Ramos C, Ravandi F, Estrov Z, Ferrajoli A, Wierda W, Shan JQ, Davis J, Giles F, Saba HI, Issa JPJ. Results of a randomized study of 3 schedules of low-dose decitabine in higher-risk myelodysplastic syndrome and chronic myelomonocytic leukemia. Blood. 2007;109:52–7.
172. Cheng JC, Yoo CB, Weisenberger DJ, Chuang J, Wozniak C, Liang GN, Marquez VE, Greer S, Orntoft TF, Thykjaer T, Jones PA. Preferential response of cancer cells to zebularine. Cancer Cell. 2004;6:151–8.
173. Yoo CB, Jones PA. Epigenetic therapy of cancer: past, present and future. Nat Rev Drug Discov. 2006;5:37–50.
174. Minucci S, Pelicci PG. Histone deacetylase inhibitors and the promise of epigenetic (and more) treatments for cancer. Nat Rev Cancer. 2006;6:38–51.
175. Marks PA, Xu WS. Histone deacetylase inhibitors: potential in cancer therapy. J Cell Biochem. 2009;107:600–8.
176. Cameron EE, Bachman KE, Myohanen S, Herman JG, Baylin SB. Synergy of demethylation and histone deacetylase inhibition in the re-expression of genes silenced in cancer. Nat Genet. 1999;21:103–7.
177. Issa JPJ. DNA methylation as a therapeutic target in cancer. Clin Cancer Res. 2007;13:1634–7.
178. Kriaucionis S, Heintz N. The nuclear DNA base 5-hydroxymethylcytosine is present in purkinje neurons and the brain. Science. 2009;324:929–30.
179. He YF, Li BZ, Li Z, Liu P, Wang Y, Tang Q, Ding J, Jia Y, Chen Z, Li L, Sun Y, Li X, Dai Q, Song CX, Zhang K, He C, Xu GL. Tet-mediated formation of 5-carboxylcytosine and its excision by TDG in mammalian DNA. Science. 2011;333:1303–7.
180. Pfaffeneder T, Hackner B, Truß M, Münzel M, Müller M, Deiml CA, Hagemeier C, Carell T. The discovery of 5-formylcytosine in embryonic stem cell DNA. Angew Chem Int Edit. 2011;50:7008–12.
181. Wu H, Zhang Y. Reversing DNA methylation: mechanisms, genomics, and biological functions. Cell. 2014;156:45–68.
182. Maiti A, Drohat AC. Thymine DNA glycosylase can rapidly excise 5-formylcytosine and 5-carboxylcytosine: potential implications for active demethylation of CpG sites. J Biol Chem. 2011;286:35334–8.
183. Kohli RM, Zhang Y. TET enzymes, TDG and the dynamics of DNA demethylation. Nature. 2013;502:472–9.
184. Munzel M, Globisch D, Bruckl T, Wagner M, Welzmiller V, Michalakis S, Muller M, Biel M, Carell T. Quantification of the sixth DNA base hydroxymethylcytosine in the brain. Angew Chem Int Ed. 2010;49:5375–7.
185. Tang Y, Chu JM, Huang W, Xiong J, Yuan BF. Hydrophilic material for the selective enrichment of 5-hydroxymethylcytosine and its liquid chromatography-tandem mass spectrometry detection. Anal Chem. 2013;85:6129–35.
186. Tang Y, Zheng S, Qi C, Feng Y, Yuan B. Sensitive and simultaneous determination of 5-methylcytosine and its oxidation products in genomic DNA by chemical derivatization coupled with liquid chromatography-tandem mass spectrometry analysis. Anal Chem. 2015;87:3445–52.
187. Liutkeviiūt Z, Kriukien E, Grigaityt I, Maseviius V, Klimaauskas S. Methyltransferase-directed derivatization of 5-hydroxymethylcytosine in DNA. Angew Chem Int Ed. 2011;50: 2090–3.

188. Hong T, Wang T, Guo P, Xing X, Ding F, Chen Y. Fluorescent strategy based on cationic conjugated polymer fluorescence resonance energy transfer for the quantification of 5-(hydroxymethyl)cytosine in genomic DNA. Anal Chem. 2013;85:10797–802.
189. Zhou Y, Yang Z, Li X, Wang Y, Yin H, Ai S. Electrochemical biosensor for detection of DNA hydroxymethylation based on glycosylation and alkaline phosphatase catalytic signal amplification. Electrochim Acta. 2015;174:647–52.
190. Zhang Z, Shan X, Zhang P, Liu W, Yan J, Li Z. Highly sensitive quantification of site-specific 5-hydroxymethylcytosine at singlebase resolution by HpaII-mediated ligation PCR. Org Biomol Chem. 2019;17:9849–53.
191. Hayashi G, Koyama K, Shiota H. Base-resolution analysis of 5-hydroxymethylcytosine by one-pot bisulfite-free chemical conversion with peroxotungstate. J Am Chem Soc. 2016;138: 14178–81.
192. Zhang Z, Yan J, Li Z. Peroxotungstate oxidation-mediated two-phase amplification system (POM-TPAS) for bisulfite-free quantification of locusspecific 5-hydroxymethylcytosine. Chem Commun. 2020;56:3111–4.
193. Jin SG, Wu X, Li A, Pfeifer GP. Genomic mapping of 5-hydroxymethylcytosine in the human brain. Nucleic Acids Res. 2011;39:5015–24.
194. Li X, Shi X, Gong Y, Guo W, Liu Y, Peng C, Xu Y. Selective chemical labeling and sequencing of 5-hydroxymethylcytosine in DNA at single-base resolution. Front Genet. 2021;12:749211.
195. Song CX, Clark TA, Lu X, Kislyuk A, Dai Q, Turner SW. Sensitive and specific single-molecule sequencing of 5-hydroxymethylcytosine. Nat Methods. 2012;9:75–7.
196. Booth MJ, Branco MR, Ficz G, Oxley D, Krueger F, Reik W. Quantitative sequencing of 5-methylcytosine and 5-hydroxymethylcytosine at single-base resolution. Science. 2012;336: 934–7.
197. Miao Y, Hon G, Szulwach K, Song CX, Zhang L, Kim A. Base-resolution analysis of 5-hydroxymethylcytosine in the mammalian genome. Cell. 2012;149:1368–80.
198. Sun X, Chung TH, Tan D, Kim A. Practical guidelines and consideration of using RRHP for 5hmC detection. Epigenomics. 2016;8:225–35.

Chapter 10
Biological Functions and Analytical Methods of MicroRNA

Honghong Wang and Kejian Gao

10.1 Overview of MicroRNA

In 1993, two research groups led by Ambros [1] and Ruvkun [2] independently discovered in *Caenorhabditis elegans* (*C. elegans*) that the *lin-4* gene, which controls larval development, does not encode a protein but instead produces a small RNA of 22 nucleotides (nt) in length, named *lin-4*. In 2000, Ruvkun's group [3] further identified another small RNA molecule, *let-7*, which is 21 nt in length and regulates developmental timing in *C. elegans*. Due to their similar functions in controlling development, *lin-4* and *let-7* were initially referred to as small temporal RNAs (stRNAs). Subsequently, researchers discovered many other small RNA molecules of similar lengths in various organisms, deepening the understanding of *lin-4* and *let-7* functions and revealing that these stRNAs are actually part of a specific class of small RNAs. In October 2001, three research teams led by Tuschl [4], Bartel [5], and Ambros [6] simultaneously published papers in *Science*, formally naming these small RNAs as microRNAs (miRNAs). Since then, with the continuous advancement of molecular biology techniques, thousands of miRNAs have been identified in various species, including humans, mice, rats, zebrafish, *Drosophila*, rice, *Arabidopsis*, and many other plant and animal taxa, opening up a completely new and expansive research field.

Mature miRNAs are a class of single-stranded, noncoding small RNAs ranging from 18 to 25 nucleotides (nt) in length. They regulate gene expression at the posttranscriptional level by pairing with target mRNAs to form the RISC, thereby promoting target mRNA degradation or inhibiting its translation. Numerous studies have shown that miRNAs serve as crucial regulatory molecules in key biological processes such as cell development, proliferation, differentiation, and apoptosis.

H. Wang (✉) · K. Gao
University of Science and Technology Beijing, Beijing, China
e-mail: wanghh@ustb.edu.cn; B20190397@xs.ustb.edu.cn

Z. Li (ed.), *Nucleic Acid Detection*, https://doi.org/10.1007/978-981-95-7295-3_10

Additionally, they play essential roles in regulating stem cell self-renewal and antiviral responses. Moreover, miRNAs are involved in the occurrence and progression of certain cancers. The sequences of miRNAs are highly conserved, and their expression exhibits temporal and tissue-specific characteristics. A single miRNA can regulate multiple target genes, while a single gene may be simultaneously regulated by multiple miRNAs, contributing to a highly complex regulatory network. Typically, commonly used databases can facilitate the systematic analysis of miRNA functions. Table 10.1 lists several commonly used databases for miRNA research.

10.2 Biological Functions of miRNA

10.2.1 miRNA Biogenesis Pathways

10.2.1.1 Biogenesis of miRNA in Animals

MiRNA genes are distributed throughout the genome [32]. Some miRNA genes are located in intergenic regions, far from protein-coding genes, possessing independent promoters and transcribing as separate transcription units. In animals, miRNAs often exist in clusters, meaning that multiple miRNAs are typically derived from the same precursor miRNA. Additionally, some miRNA genes are located within the introns of protein-coding genes. These intronic miRNAs are co-transcribed with their host genes and subsequently excised from the introns. As a result, miRNAs originating from intronic regions of protein-coding genes are often processed during mRNA splicing. Studies have also identified a small number of miRNAs located in exon regions of genes.

In animals, the synthesis pathway of mature miRNAs is illustrated in Fig. 10.1 [33]. Within the nucleus, most miRNA genes are transcribed by RNA polymerase II, while a few are transcribed by RNA polymerase III. The primary miRNA transcript (pri-miRNA) produced by RNA polymerase II transcription is typically several kilonucleotides long and possesses a characteristic 5′ cap structure (m^7G) and a 3′ poly(A) tail, which are hallmarks of RNA polymerase II transcripts. Each pri-miRNA contains at least one region capable of folding into a hairpin structure, which serves as a substrate for the microprocessor complex. The microprocessor complex consists of a Drosha endonuclease (a nuclear RNase III-type endonuclease) and two molecules of the double-stranded RNA-binding protein DGCR8 (known as Pasha in Drosophila and *C. elegans*). With the assistance of DGCR8, Drosha recognizes the hairpin structure of the pri-miRNA and cleaves it at the stem, releasing a hairpin-shaped precursor miRNA (pre-miRNA) approximately 60 nt in length, with a 2-nt 3′ overhang. After nuclear processing, the pre-miRNA is transported to the cytoplasm by the Exportin-5-RAN-GTP transport complex. In the cytoplasm, pre-miRNA undergoes further processing by the Dicer endonuclease, which removes the loop region of the hairpin structure, generating an approximately 22-nt-long miRNA/miRNA* duplex with incomplete base pairing, where both

Table 10.1 Commonly used databases for studying miRNA

Database	Introduction/Description	Version/ Update date	Website	References
miRbase	It is a comprehensive database that provides information including miRNA sequence data, annotations, predicted gene targets, etc. It is one of the most important public databases for storing miRNA information	miRbase 1.0 (2002) miRbase 22.1 (2018)	http://www.mirbase.org/	[7–12]
miRTarBase	The experimentally validated microRNA-target interactions database. It serves as a key resource for understanding the regulatory roles of microRNAs in various biological processes by providing high-confidence, experimentally supported miRNA-target interaction data	miRTarBase (2011) miRTarBase (2020)	http://mirtarbase.mbc.nctu.edu.tw/	[13–16]
miRcode	MiRcode offers "whole transcriptome" microRNA target predictions for human genes, utilizing the comprehensive GENCODE gene annotation. This includes predictions for over 10,000 long non-coding RNA genes, as well as coding genes, covering atypical regions such as 5′ untranslated regions (UTRs) and coding sequences (CDS)	miRcode 11 (2012)	http://www.mircode.org	[17]
ENCORI (StaBase)	ENCORI is an open-source platform designed for studying a wide range of RNA interactions, including miRNA-ncRNA, miRNA-mRNA, ncRNA-RNA, RNA-RNA, RBP-ncRNA, and RBP-mRNA interactions. It integrates data from CLIP-seq, degradome-seq, and RNA–RNA interactome studies. Unlike databases that rely solely on predictions, ENCORI's data is largely supported by experimental evidence, including Co-Immunoprecipitation (Co-IP) data, providing a more reliable and robust resource for RNA interaction research	StarBase (2011) StarBase 2.0 (2014)	http://starbase.sysu.edu.cn/	[18, 19]
TargetScan	TargetScan predicts biological targets of miRNAs by searching for the presence of conserved sites	TargetScan 7.2 (2018)	http://www.targetscan.org	[20, 21]

(continued)

Table 10.1 (continued)

Database	Introduction/Description	Version/ Update date	Website	References
	that match the seed region of each miRNA			
PicTar	PicTar is an algorithm for the identification of microRNA targets, which can provide details of miRNA target predictions in various species, including vertebrates, Drosophila, nematode and human	PicTar (2007)	https://pictar.mdc-berlin.de/	[22, 23]
CoGemiR	It is a publicly available miRNA-centered database whose aim is to offer an overview of the genomic organization of miRNA and of its extent of conservation during evolution in different metazoan species. The database collects information on genomic location, conservation, and expression data of both known and newly predicted miRNAs	CoGemiR (2008)	http://cogemir.tigem.it/	[24]
PMRD	It is a plant microRNA database, which tries to integrate the available large information of plant microRNAs data, consisting of microRNA sequence and their target genes, secondary dimension structure, expression profiling, genome browser, etc.	PMRD (2010) PNRD (2014)	http://bioinformatics.cau.edu.cn/PMRD/	[25, 26]
miRNEST	miRNEST is an integrative collection of animal, plant, and virus microRNA data	miRNEST (2012) miRNEST 2.0 (2014)	http://rhesus.amu.edu.pl/mirnest/copy/	[27, 28]
miRvar	A comprehensive database of miRNA genomic variations	miRvar (2011)	http://genome.igib.res.in/mirlovd/home.php	[29]
miRandola	miRandola is a comprehensive manually curated classification of different extracellular circulating noncoding RNA types	miRandola (2017)	http://mirandola.iit.cnr.it/	[30]
EVmiRNA	EVmiRNA is the Extracellular Vesicles (EV) miRNA Database, collecting data from EV samples including miRNA expression profiles, miRNA related drugs, miRNA regulatory pathways, miRNA functions, and related publications	EVmiRNA (2018)	http://bioinfo.life.hust.edu.cn/EVmiRNA	[31]

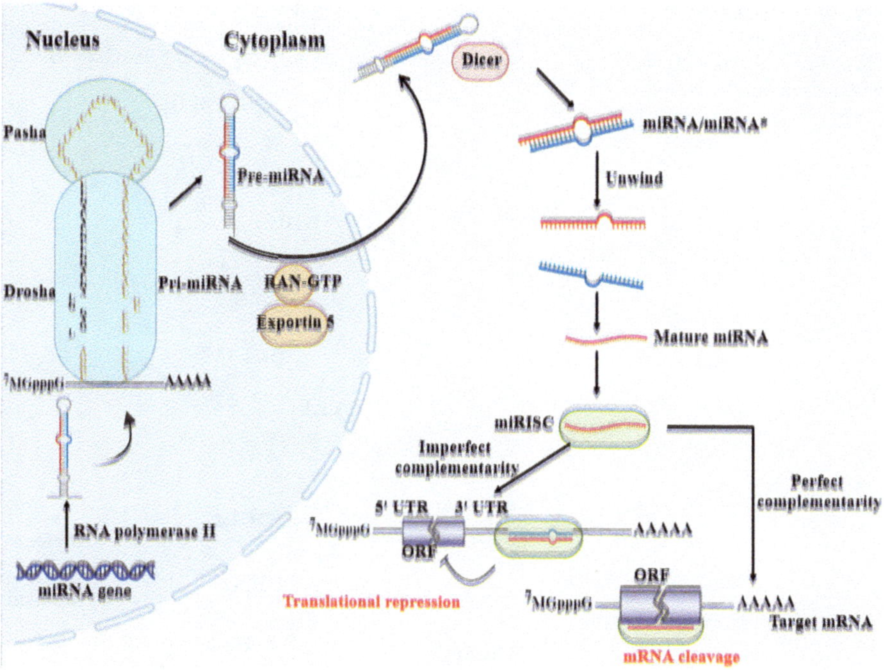

Fig. 10.1 Biogenesis of miRNA in animals [33]. (Revised and reprinted with permission from Ref. [33])

strands exhibit a 2-nt 3′ overhang [34]. Studies suggest that the selection of which strand is retained in the RNA-Induced Silencing Complex (RISC) depends on multiple factors, such as the thermodynamic asymmetry of the duplex and the stability of base pairing at the 5′ end. However, in some cases, both strands of the duplex can be functional, acting as independent miRNAs targeting different mRNAs [35]. The mature miRNA is then loaded into RISC, where it guides the complex to its target mRNA through sequence complementarity, thereby regulating gene expression by either posttranscriptional translational repression or target mRNA degradation.

10.2.1.2 Biogenesis of miRNA in Plants

Due to the lack of significant sequence similarity between animal and plant miRNAs, they are generally considered to have arisen independently during evolution. This evolutionary independence has led to notable differences between plant and animal miRNAs in terms of sequence, precursor structure, and biosynthetic mechanisms [36, 37]. In plants, miRNA-encoding genes are primarily located in intergenic

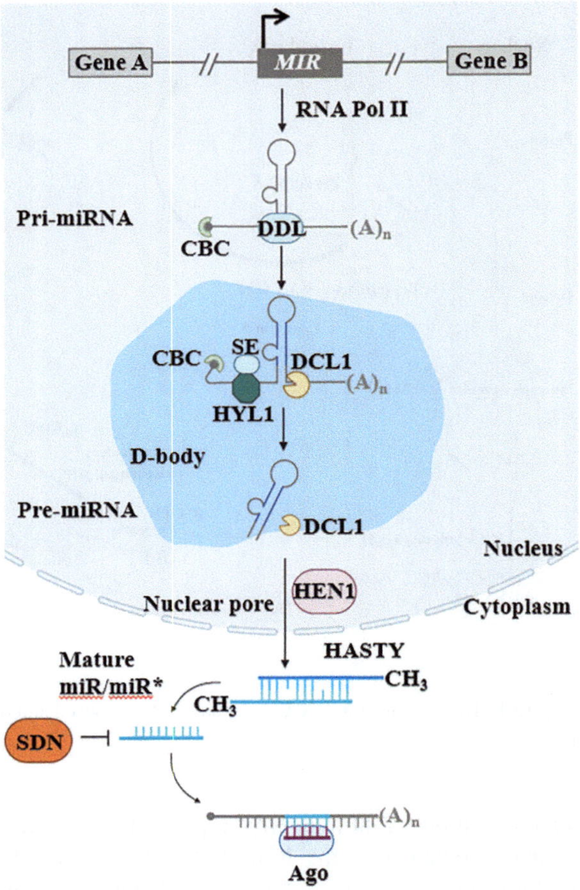

Fig. 10.2 Biogenesis of miRNA in plants [38]. (Revised and reprinted with permission from Ref. [38])

regions of the genome. Most plant miRNA genes have independent transcription units and rarely occur in clusters.

In plants, the synthesis pathway of mature miRNAs is illustrated in Fig. 10.2 [38]. Plant miRNA genes are located between protein-coding genes, and most are transcribed by RNA polymerase II into pri-miRNAs. Similar to animal miRNAs, plant pri-miRNAs possess a 5′ cap structure (m7G) and a 3′ poly(A) tail. However, compared to animal pri-miRNAs, plant pri-miRNAs are typically longer, reaching several hundred nucleotides, and exhibit more complex and variable folded structures. Additionally, the processing of pri-miRNAs into pre-miRNAs differs significantly between plants and animals. In plants, the RNA-binding protein DAWDLE (DDL) stabilizes pri-miRNAs and promotes their processing into stem-loop-structured pre-miRNAs. This processing occurs within nuclear processing centers known as D-body, whereas in animals, this step generally takes place in the cytoplasm. Notably, plants lack the *Drosha* gene responsible for pri-miRNA to pre-miRNA conversion in animals. Instead, this function is carried out by *Dicer-like 1* (DCL1), an RNase III-type endonuclease unique to plants. The conversion of pri-miRNAs relies on the coordinated activity and physical interactions among the acetylene-zinc finger protein SE, the double-stranded RNA-binding protein HYL1, DCL1, and the cap-binding complex (CBC). After DCL1 cleaves pri-miRNA into pre-miRNA, it continues to process pre-miRNA into the miRNA/miRNA* duplex. This duplex undergoes methylation at the 3′ end by the methyltransferase HEN1, which protects it from degradation by small RNA-degrading nucleases (SDNs). Subsequently, the duplex is transported to the cytoplasm via the plant-specific transporter protein HASTY, where it unwinds into a single strand. Finally, the guide strand of the miRNA is incorporated into an AGO protein, forming the RISC, which carries out its biological function.

10.2.2 Mechanism of miRNA Function

MiRNAs guide the RISC to complementary sites on target mRNAs, thereby inhibiting their translation. Understanding how miRNAs recognize and suppress target mRNAs has been a major focus of research [39]. Scientists have proposed various theoretical models to explain the molecular mechanisms of miRNA-mediated gene silencing. This posttranscriptional repression by miRNAs may occur through translational inhibition, leading to reduced translation efficiency, or by inducing target mRNA degradation, thereby lowering its expression level.

In animals, early studies suggested that miRNAs primarily function by inhibiting the translation of target mRNAs, with minimal impact on mRNA abundance. In contrast, plant miRNAs are believed to act almost exclusively by promoting target mRNA degradation. This difference mainly arises from the degree of base pairing between miRNAs and their targets. In animals, the pairing is generally weak, whereas in plants, miRNA-mRNA pairing is typically highly complementary [40]. As illustrated in Fig. 10.3, plant miRNAs recognize binding sites that are completely (or nearly completely) complementary to their sequences, which are usually located within the open reading frame (ORF). Notably, nucleotides 9–12 of the miRNA typically form base pairs with the target, allowing AGO proteins to cleave the target mRNA. AGO proteins cut the mRNA between the 10th and 11th nucleotides of the base-paired region. Additionally, plant miRNAs often have their 3′-end nucleotides modified with a 2′-O-methyl group to prevent degradation. In contrast, animal miRNAs recognize partially complementary binding sites, which are usually located in the 3′ untranslated region (3′ UTR). The key determinant for target recognition is the miRNA's 5′-end complementary sequence, known as the seed sequence (nucleotides 2–7), which is sufficient to trigger gene silencing. For most miRNA binding sites, complementarity is restricted to the seed sequence. However, in some rare cases, complementarity to the miRNA's 3′ region may contribute to target selection, particularly when seed sequence pairing is weak. Even so, nucleotides 9–12 of the miRNA generally protrude, preventing AGO proteins from cleaving the target mRNA.

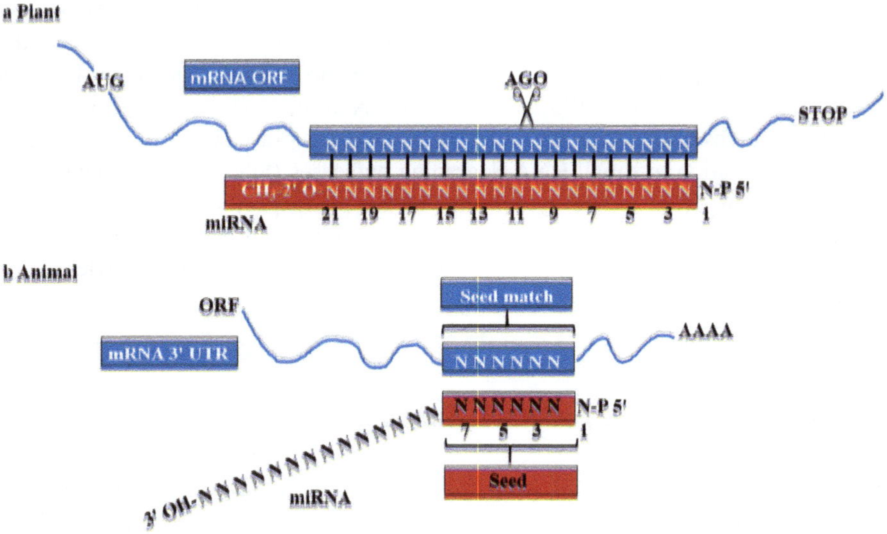

Fig. 10.3 MiRNA-target recognition in plants and animals [40]. (Revised and reprinted with permission from Ref. [40])

Early studies on the posttranscriptional regulatory mechanisms of miRNAs and their target mRNAs proposed two classic models: mRNA degradation and translational repression [41]. The extent of base pairing between the miRNA and its target mRNA is the key factor in determining which posttranscriptional mechanism is employed. When the miRNA and its target site exhibit complete or nearly complete complementarity, the RISC complex specifically cleaves the mRNA, a mechanism that is more common in plants. Conversely, when miRNA-mRNA pairing is incomplete, the RISC complex represses translation without affecting mRNA stability, a mechanism predominantly observed in animals. However, this rule is not without exceptions. For instance, even when the miRNA and its target mRNA have near-perfect complementarity, gene expression regulation may occur through translational inhibition rather than mRNA degradation. A study by Broderson et al. [42] demonstrated that, despite near-perfect complementarity between multiple miRNAs and their target mRNAs in *Arabidopsis thaliana*, these miRNAs primarily function through translational repression, providing strong genetic evidence for miRNA-mediated translational inhibition in plants. Furthermore, studies have revealed that miRNA-mediated mRNA cleavage also occurs in animals. Among the four AGO proteins (AGO1–4) in mammals, only AGO2 possesses cleavage activity. Since miRNAs are equally likely to be loaded into any of the four AGO proteins, two conditions must be met for mRNA cleavage to occur: (1) The miRNA must be loaded into AGO2 rather than other AGO proteins. (2) The miRNA and its target mRNA must exhibit a high degree of base pairing. A classic example is the interaction between miR-196 and HOXB8. The nearly perfect base pairing between miR-196 and HOXB8 leads to mRNA cleavage and degradation [43].

In addition to the classic miRNA-mediated posttranscriptional gene silencing mechanism, increasing research has uncovered various noncanonical miRNA regulatory pathways [44]. These unconventional functions include the following:

1. Expanded target regions: Early studies suggested that miRNAs interact exclusively with the 3′ UTR of mRNAs. However, recent findings indicate that miRNAs can also bind to the 5′ UTR and coding sequence (CDS) regions of mRNAs [45, 46].
2. MiRNA-encoded peptides (miPEPs): Some pri-miRNAs are capable of encoding short peptides, known as miRNA-encoded peptides (miPEPs). Lauressergues et al. [47] first discovered this phenomenon in plants. For example, pri-miR-171b in *Medicago truncatula* and pri-miR-165a in *Arabidopsis thaliana* can produce short peptides.
3. Interactions with non-AGO proteins: Typically, miRNAs are loaded into AGO protein complexes to participate in classic gene silencing mechanisms. However, Eiring et al. [48] found that miRNAs can also interact with non-AGO proteins, expanding their functional scope.
4. Promotion of protein expression: Vasudevan et al. [49] discovered that miRNA regulation of protein expression may depend on different phases of the cell cycle. For example, certain miRNAs can enhance the translation of their target mRNAs during cell cycle arrest but suppress translation during cell proliferation.
5. Targeting mitochondria-related genes: Although miRNAs have not been identified within the mitochondrial genome, Das et al. [50] discovered that miR-181c can translocate to mitochondria, where it suppresses the expression of the mitochondrially encoded MT-COX1 protein while simultaneously increasing the levels of MT-COX2 mRNA and protein.
6. Activating gene transcription: In addition to suppressing gene expression, miRNAs can also directly activate gene transcription [51].
7. Targeting nuclear ncRNA precursor transcripts: Tang et al. [52] found that nuclearly localized miR-709 can target the pri-miRNA precursor transcripts of miR-15a and miR-16-1, thereby inhibiting their maturation.

Additionally, it has long been believed that in the miRNA/miRNA* duplex, only one strand is responsible for gene silencing. However, recent studies indicate that in certain cancers, both strands of the miRNA duplex can act cooperatively to promote cancer cell invasion and growth [53]. These discoveries have significantly expanded our understanding of miRNA regulatory functions, revealing their complex and diverse roles in cellular processes.

In conclusion, while substantial progress has been made in understanding gene silencing mechanisms in animals and plants, some molecular mechanisms remain incompletely elucidated. With ongoing advancements in scientific technology and research, our comprehension of the gene expression regulatory network will continue to deepen, ultimately providing a more comprehensive understanding of miRNA-mediated gene silencing mechanisms.

10.2.3 Biological Functions of miRNAs

It is estimated that the human genome encodes up to 1000 miRNAs, while the number of mRNAs is approximately 30,000. Each miRNA can target multiple different mRNAs, and a single mRNA can be cooperatively inhibited by multiple miRNAs. Additionally, miRNAs sharing the same seed sequence often regulate the same target mRNAs [54], suggesting that miRNA-mediated biological regulation forms a complex and highly coordinated regulatory network. MiRNAs play essential roles in nearly all biological processes, including cell development, differentiation, proliferation, apoptosis, and the cell cycle. Furthermore, miRNAs are involved in various cellular activities, such as immune responses, insulin secretion, neurotransmitter synthesis, circadian rhythm regulation, and viral replication [55].

10.2.3.1 miRNAs in Cell Development Regulation

The role of miRNAs in regulating cellular development was first demonstrated by the discovery of lin-4 and let-7, both of which play critical roles in controlling larval development in *Caenorhabditis elegans*. Lin-4 miRNA regulates development by targeting the genes lin-14 and lin-28. Lin-14 encodes a nuclear protein that controls the transition from the L1 to the L2 larval stage. Lin-28 encodes a cytoplasmic protein that promotes the transition from the L2 to the L3 stage. Loss of lin-4 function results in L1-stage characteristics persisting into later larval stages, leading to the absence of mature structures such as the epidermis and vulva, ultimately disrupting ovulation. Similarly, let-7 also plays a regulatory role in nematode development by targeting lin-41, hbl-1, daf-12, pha-4, and ras. Among these, lin-41 encodes an RBCC (RING-B-box-coiled-coil) protein. Inhibition of lin-41 causes extra molting, leading to the formation of a mature epidermis. Overexpression of lin-41 reduces the number of molts, causing the adult worm to retain larval epidermal characteristics.

Studies across various organisms have further revealed the widespread role of miRNAs in developmental processes. In animals, gene knockout and other experimental techniques have demonstrated that miRNAs play crucial regulatory roles in early embryonic development, muscle development, neural development, and lymphocyte development. For example, in studies of mutant mice, miR-1 has been identified as an early regulatory factor in the transition from proliferation to differentiation during muscle development. Additionally, miR-1 is essential for maintaining physiological functions in the mature heart. In vertebrates, miR-124 is recognized as a key regulator of neural development. Overexpression of miR-124 promotes neural development, while its downregulation inhibits the process. Studies have also shown that miR-124 is highly expressed in both neural progenitor cells and mature neurons, where it plays a critical role in promoting neural development. These findings further underscore the importance of miRNA homeostasis in normal development.

The role of miRNAs in heart, vascular tissues, and blood development has also received extensive attention. Studies have shown that miRNAs are closely associated with various cardiovascular pathological conditions, including arrhythmias (miR-1 and miR-133), fibrosis (miR-21 and miR-29), and metabolic disorders (miR-33). Among these, miR-1 is the most abundant miRNA in cardiomyocytes and was the first miRNA identified to be involved in heart development.

10.2.3.2 miRNAs in Cell Proliferation Regulation

Cell proliferation is one of the core characteristics of cellular life activities and serves as the foundation for biological reproduction. Processes such as wound healing, tissue regeneration, and pathological tissue repair all rely on cell proliferation, which is precisely regulated. Many miRNAs modulate key signaling pathways involved in cell proliferation by directly interacting with crucial regulatory factors. Key regulators in cell proliferation pathways include RAS, RHOA, PI3K/PTEN, and ABL, whose dysregulation plays a critical role in tumor progression. Slack et al. [56] discovered that RAS proteins are regulated by the let-7 family. Overexpression of let-7 inhibits RAS gene expression, whereas reduced let-7 levels lead to an upregulation of RAS protein expression. In addition to RAS, other GTPases involved in cell proliferation and tumorigenesis are also regulated by miRNAs. For example, RHOA, RAC1, and CDC42 are oncogenic GTPases. MiR-124 regulates neurite outgrowth during neuronal differentiation by downregulating RAC1 and CDC42 [57]. MiR-133 controls cardiomyocyte proliferation by suppressing RHOA [58].

The ABL gene is a proto-oncogene commonly implicated in hematologic malignancies, with upregulated expression in certain lymphomas and leukemias. In specific cases, ABL undergoes chromosomal translocation, leading to oncogenic activation. A well-known example is the Philadelphia chromosome, a hallmark genetic rearrangement in chronic myeloid leukemia (CML), which results from a fusion between the BCR and ABL genes. In this process, the ABL gene translocates downstream of BCR, forming the BCR-ABL fusion gene. This fusion gene exhibits antiapoptotic properties and high tyrosine kinase activity, activating multiple signaling pathways that drive excessive cell proliferation, ultimately disrupting normal regulatory mechanisms. Studies have shown that miR-203 binds to the 3′ UTR of ABL mRNA, inhibiting its translation and significantly reducing ABL protein levels. Notably, miR-203 also suppresses BCR-ABL fusion gene expression, as the ABL 3′ UTR remains conserved within the fusion gene, making it susceptible to miR-203-mediated inhibition [59]. In normal hematopoietic cells, miR-203 is naturally expressed, helping to maintain low ABL expression levels. Therefore, miR-203 functions as both an effective antiproliferative miRNA and a key regulator in the malignant progression of tumors.

10.2.3.3 miRNA in Cell Differentiation

Extensive research has demonstrated that miRNAs play a crucial role in regulating stem cell differentiation by targeting specific mRNAs. Let-7 is expressed early in mammalian stem cell differentiation. A study by Melton et al. [60] identified let-7 as a key factor in inducing mouse embryonic stem cell (ESC) differentiation. Let-7 binds to the mRNAs encoding stem cell identity factors, regulating their protein levels. By downregulating these stem cell identity factors, let-7 promotes cellular differentiation. Frank J. Slack proposed a model of miRNA function in embryonic stem cells [61]. In embryonic stem cells, high levels of stem cell identity factors (such as LIN-28 protein, MYC protein, and ESCC miRNAs) are expressed. Studies on mouse embryonic stem cells have shown that let-7 expression levels are low in stem cells, and let-7 plays a critical role in reducing the concentration of stem cell identity factors. When let-7 expression increases, it leads to a decrease in stem cell identity factors, thereby inducing differentiation. Differentiated cells exhibit low levels of stem cell identity factors and high let-7 expression, reinforcing their commitment to differentiation.

Beyond embryonic stem cells, miRNAs also play a crucial role in the differentiation of hematopoietic stem cells, neural stem cells, and other stem cell types. Various miRNAs, including miR-9, miR-222, miR-10a, miR-126, miR-106, miR-10b, miR-17, miR-20, miR-146, miR-155, miR-223, miR-221, miR-92, miR-150, miR-126, and miR-142, have been implicated in hematopoietic stem cell differentiation [62]. MiR-9 promotes myeloid cell generation while inhibiting lymphocyte formation. MiR-150 facilitates megakaryocyte-erythroid progenitor (MEP) differentiation into megakaryocytes, while simultaneously suppressing erythropoiesis, highlighting the critical regulatory role of miRNAs in differentiation. In neural stem cells, miR-138-5p overexpression inhibits neural stem cell proliferation while promoting differentiation. Conversely, downregulating miR-138-5p using the miRNA inhibitor enhances proliferation while reducing differentiation potential [63]. Additionally, miRNAs regulate early osteogenic differentiation in human mesenchymal stem cells (hMSCs). Specifically, miR-148b, miR-27a, and miR-489 play essential roles in osteogenesis [64]. Research has also shown that miRNA depletion significantly impacts human muscle precursor cell differentiation. Several miRNAs, including miR-133, miR-1, miR-20, miR-181, and miR-27a, are essential for terminal differentiation of skeletal muscle cells [65].

10.2.3.4 miRNA in Cell Cycle Regulation

The cell cycle is a series of finely regulated events that occur during cell growth and division. In eukaryotic cells, the cell cycle is typically divided into four phases: G1 phase, S phase, G2 phase, and M phase. G1 phase: The cell synthesizes proteins required for DNA replication and prepares for the next phase. S phase: This is the critical stage of DNA replication, where the genome is duplicated. G2 phase: The cell continues to grow and prepares for mitosis. M phase: The mitotic phase, in which the cell divides into two daughter cells through mitosis. After mitosis, the newly formed daughter cells either enter the G1 phase to begin a new cycle or transition into a quiescent state (G0 phase). The cell cycle is a highly complex process, requiring the precise coordination of multiple regulatory proteins that guide the cell through each stage, ultimately ensuring the successful formation of two daughter cells through mitosis.

In the regulation of the cell cycle, the core regulatory factors are cyclin-dependent kinases (CDKs). These kinases are activated at specific time points in the cell cycle and strictly regulate its progression. In mammals, at least seven types of CDKs exist, and they function by binding to their corresponding cyclins to form cyclin/CDK complexes, which control different stages of the cell cycle. These complexes precisely regulate cellular activities at each phase of the cell cycle and are further modulated by various proteins, including pRB, p53, p21, p19, p16, and CDC25 [66]. In summary, the cell cycle is maintained by a tightly controlled regulatory mechanism. Additionally, research has shown that cell cycle regulators are also subject to regulation by miRNAs.

miRNAs precisely regulate different stages of the cell cycle by targeting cell cycle regulatory factors. Some miRNAs promote cell cycle entry by targeting CDK inhibitors or pRB family transcriptional repressors, while others induce cell cycle arrest by downregulating multiple key cell cycle components [67, 68]. During the G1-to-S phase transition, the pRB family suppresses the transcription of genes required for this transition by binding to E2F transcription factors, thereby preventing S-phase entry and inhibiting cell growth. However, cyclin/CDK complexes can phosphorylate and inactivate pRB proteins, leading to the release of E2F, which drives the G1-to-S transition. MiRNAs regulate multiple key factors in this pathway, controlling the transition from G1 to S phase. Specifically, miR-15, miR-24, miR-34a, miR-124, miR-125b, miR-129, miR-137, miR-195, miR-449, and the let-7 family regulate the cell cycle by targeting CDK2, CDK4, and CDK6. miR-15, miR-17/20, miR-19a, miR-26a, and miR-34a target Cyclin D and Cyclin E, preventing cyclin/CDK complex formation, thereby inducing cell cycle arrest. miR-17-5p, miR-149, miR-330, miR-331-3p, miR-125, miR-210, and miR-195* reduce E2F levels by directly targeting E2F transcription factors, leading to cell cycle arrest. On the other hand, miR-106a and miR-290 target pRB family members, while miR-24 and miR-31 target CDK inhibitors (such as p16 and p19), promoting E2F release and accelerating cell cycle progression. Together, these miRNAs form a complex regulatory network, ensuring the precise control of cell cycle progression.

10.2.3.5 miRNA in the Regulation of Apoptosis

Apoptosis is the primary form of programmed cell death, controlled by genetic mechanisms and initiated through intrinsic cellular pathways to maintain homeostasis. This evolutionarily conserved self-destructive process is characterized by the activation of endonucleases, leading to the degradation of DNA into fragments of

1.8–2.0 kb in size. The regulation of apoptosis is closely linked to various human diseases. In recent years, studies have revealed that miRNAs play a critical role in the regulation of apoptotic pathways, influencing both proapoptotic and antiapoptotic mechanisms.

Studies have shown that a series of apoptosis-related proteases participate in the process of cell apoptosis, among which the caspase family is the most critical member, acting as the core executors of apoptosis [69]. Apoptosis is initiated through a caspase activation cascade, leading to extensive cleavage of intracellular substrates that execute cell death. Initiator caspases (such as caspase-9 in mammals and Dronc in Drosophila) become activated through autocleavage. The apoptosome, a multimeric protein complex, interacts with initiator caspases and significantly accelerates their self-activation process. In living organisms, the inhibitor of apoptosis proteins (IAP) family regulates apoptosis by inhibiting caspase activity. IAP proteins typically bind to caspases, preventing them from interacting with their substrates or even leading to their degradation, thereby suppressing caspase activity. These regulatory mechanisms collectively help maintain the balance of apoptosis and ensure normal physiological function.

MiR-14 and bantam were the first miRNAs identified to be involved in apoptosis regulation in Drosophila [70, 71]. Dark is the apoptosome protein in Drosophila, and its involvement is essential for the autocatalytic activation of the initiator caspase Dronc. Upon receiving apoptotic signals, Dark forms the apoptosome, which is responsible for recruiting and activating Dronc. The activated Dronc subsequently activates the effector caspase Drice, which serves as the executioner of apoptosis. DIAP1, a member of the Drosophila IAP protein family, binds to Dronc and Drice, preventing their interaction with target substrates and thereby inhibiting their activity. Studies on Drosophila apoptotic mutants have also identified several IAP-binding motif proteins, including Reaper, Grim, Hid, and Sickle, which compete with Dronc and Drice for binding to DIAP1, thereby overcoming DIAP1-mediated inhibition and promoting apoptosis. MiRNAs play a crucial role in the posttranscriptional regulation of these proapoptotic proteins. For example, bantam and the miR-2 family have been confirmed to be involved in this regulatory process. This mechanism is not limited to Drosophila, as similar miRNA functions in apoptosis have also been observed in mammals, including miR-21, miR-15, miR-16, and the miR-17 cluster.

10.3 miRNA and Cancer

Cells have developed various safeguard mechanisms to ensure that cell division, differentiation, and apoptosis occur in a precise and coordinated manner during both development and adulthood. Many regulatory factors maintain this balance by activating or repressing genes that control cell proliferation and differentiation. However, when the function of these genes becomes dysregulated, it may lead to cancer initiation. Genes associated with cancer mainly include tumor suppressor genes and oncogenes. Tumor suppressor genes, also known as anti-oncogenes, are genes that inhibit cell growth and have potential tumor-suppressing functions. These genes include those involved in cell cycle regulation, encoding receptors or signaling proteins that suppress proliferation, induce apoptosis, and participate in DNA repair. Proto-oncogenes, on the other hand, are another class of genes closely related to tumorigenesis. Under normal conditions, proto-oncogenes regulate cell proliferation and differentiation. However, when mutations occur, proto-oncogenes can be activated into oncogenes, gaining the ability to induce tumor transformation. The activation of proto-oncogenes can occur through various genetic mechanisms, including mutations, chromosomal translocations, viral insertions, and insertional mutations. This activation allows proto-oncogenes to escape normal regulatory control, leading to uncontrolled cell proliferation. Most tumor suppressor genes and oncogenes exert their functions through the transcription of DNA into RNA, which is then translated into proteins. Since miRNAs play a crucial role in posttranscriptional regulation by binding to the 3′ UTR of target mRNAs to inhibit their translation, dysregulation of miRNAs can disrupt the expression of oncogenes or tumor suppressor genes. This dysregulation is closely linked to cancer pathogenesis. Studies have shown that in various human cancers, miRNA expression levels can be either upregulated or downregulated, significantly influencing cancer initiation and progression. These findings provide important insights into the molecular mechanisms of cancer and offer potential therapeutic strategies for treatment.

Studies have found that human miRNA-encoding genes are often located in genomic regions associated with various cancers or at fragile sites—chromosomal regions susceptible to breakage, DNA replication stress, or fusion with other chromosomes [72]. Additionally, miRNA expression is linked to multiple cancers, and miRNAs may function as tumor suppressors or oncogenes, playing critical roles in cancer initiation and progression. MiRNAs that downregulate oncogenes are classified as tumor suppressor miRNAs (TS-miRs). A reduction or loss of TS-miR expression leads to increased translation of their target oncogenes, promoting the accumulation of oncogenic proteins, which ultimately contributes to tumor formation [73]. TS-miRs are often downregulated or lost in cancers. Conversely, miRNAs that target tumor suppressor genes are classified as oncogenic miRNAs (oncomiRs). Overexpression of oncomiRs inhibits the translation of tumor suppressor genes, reducing the production of tumor-suppressing proteins and ultimately promoting tumorigenesis. As a result, oncomiRs are frequently upregulated in various types of cancer. Table 10.2 lists several cancer-associated oncomiRs and TS-miRs, such as the let-7 family, which acts as a TS-miR, and miR-221/miR-222, which function as oncomiRs [74].

OncomiR-encoding genes are often located in genomic regions with abnormal amplification or increased expression. The upregulation of oncomiRs is typically driven by multiple mechanisms that enhance oncomiR gene transcription, including increased transcription factor accessibility and reduced promoter methylation. The first identified oncomiR is the miR-17~92 cluster (also known as oncomiR-1). OncomiR-1 exerts its function by targeting BIM, a proapoptotic protein, and its overexpression has been shown to induce lymphoproliferative and autoimmune

Table 10.2 miRNAs as oncogenes and oncogenes and their relationship with cancers

miRNA	Function	Cancer type	Dysregulation	References
miR-17~19 clusters	oncomiR	B-cell malignant lymphoma	Upregulation	[75]
miR-21	oncomiR	Glioblastoma, Breast cancer	Upregulation	[76, 77]
BIC/miR-155	oncomiR	Breast cancer, B-cell malignant lymphoma	Upregulation	[77, 78]
miR-221/222	oncomiR	Breast cancer, prostate cancer, liver cancer	Upregulation	[79–81]
miR-372/373	oncomiR	Gastric cancer, colorectal cancer	Upregulation	[82, 83]
miR-15a, miR-16-1	TS-miR	B-cell chronic lymphocytic leukemia	Downregulation	[84]
miR-143, miR-145	TS-miR	Colorectal cancer, breast cancer, prostate cancer, cervical carcinoma	Downregulation	[77, 85]
let-7 family	TS-miR	Lung cancer	Downregulation	[86]
miR-655	TS-miR	Pancreatic cancer	Downregulation	[87]
miR-106a	TS-miR	Kidney cancer	Downregulation	[88]

diseases in mice [89]. Beyond BIM, the miR-17~92 cluster also regulates multiple oncogenic molecules, including PTEN, E2Fs, and MYC, thereby promoting tumorigenesis. Given the crucial roles of miR-21 and miR-155 in various solid tumors and malignant hematologic cancers, these two miRNAs have also been classified as oncomiRs [90–92]. Overexpression of either miR-21 or miR-155 alone is sufficient to induce lymphoproliferative diseases. Although the precise mechanism by which miR-155 drives tumorigenesis remains unclear, studies in mouse leukemia models suggest that miR-155 may promote cancer progression by gradually downregulating its targets SHIP and C/EBP. In miR-dependent mouse models of lung cancer or precursor B-cell lymphoma, the suppression of miR-21 targets PTEN and PDCD4—negative regulators of cell death and cell cycle progression, respectively—contributes to enhanced cancer cell proliferation and growth [93, 94].

Among miRNAs that suppress oncogene expression, approximately 50% are located in or adjacent to genomic fragile sites that are frequently deleted in cancer. Moreover, cancer cells employ various mechanisms to downregulate tumor-suppressor miRNAs (TS-miRs), including loss of heterozygosity (LOH), hypermethylation of promoter regions, and activation of transcriptional repressors, all of which specifically reduce the expression levels of these miRNAs. MiR-15a and miR-16-1 were among the first miRNAs identified to be negatively associated with cancer progression and have been recognized as TS-miRs. The deletion of the miR-15a~16-1 gene cluster is closely linked to the upregulation of BCL-2, a pro-survival factor that prevents apoptosis. In addition to targeting BCL-2, miR-15a~16-1 also regulates several key cellular proteins, including MCL1 (an antiapoptotic protein), CCND1 (a cell cycle regulator), and WNT3A (a protein that induces multiple tumorigenic traits such as survival, proliferation, and invasion). Thus, it is hypothesized that the concurrent overexpression of these oncogenic proteins, resulting from the loss of the miR-15a~16-1 cluster, may synergistically drive cancer progression.

Another well-recognized example of a tumor-suppressor miRNA is let-7. The primary function of the let-7 family is to promote cell differentiation, whereas cancer progression is often associated with the reversal from a differentiated to an undifferentiated state. Consequently, the downregulation or loss of let-7 family members is a common feature in tumorigenesis. In normal, fully differentiated cells, let-7 is highly expressed at multiple genomic loci, effectively suppressing the expression of its target genes. These targets include several key oncogenes, such as KRAS, NRAS, HMGA2, LIN28, and MYC. However, because most let-7 family members are located in genomic regions frequently deleted in cancer, their expression is significantly reduced in various malignancies. This downregulation provides a crucial mechanistic basis for cancer initiation and progression, further highlighting the essential role of let-7 in tumor suppression.

The primary cause of cancer-related mortality is complications arising from tumor metastasis. Studies have shown that certain miRNAs can directly target genes involved in cancer cell migration, invasion, colonization, tumor stemness, and epithelial-mesenchymal transition (EMT), thereby either promoting or inhibiting tumor metastasis. These miRNAs are referred to as metastasis-associated miRNAs (metastamiRs). MiR-335 and miR-126 were among the first metastamiRs identified as metastasis suppressors in human breast cancer, with their expression levels decreasing as the metastatic potential of breast cancer cells increases [95]. Research has demonstrated that restoring the expression of these miRNAs in malignant cells significantly inhibits cancer cell metastasis to the lungs and bones. In the primary tumors of most breast cancer patients who experience recurrence, miR-126 and miR-335 expression is lost. Additionally, miR-146a/b inhibits breast cancer cell migration and invasion by targeting IRAK1 and TRAF6, thereby downregulating the NFκB signaling pathway [96]. Conversely, researchers have also identified metastamiRs that promote metastasis. MiR-10b was the first metastamiR [97] discovered with metastasis-promoting activity and is highly expressed in metastatic breast cancer cells, where it positively regulates cell migration and invasion. Overexpression of miR-10b in nonmetastatic breast tumors induces strong invasive and metastatic behavior. To identify potential metastasis-promoting miRNAs, Huang et al. [98] established a genetic screening mechanism using a nonmetastatic human breast tumor cell line. These cells were transduced with the miRNA expression library and subjected to migration assays, leading to the identification of miR-373 and miR-520c, both of which enhance migration, likely by targeting and suppressing the adhesion molecule CD44, thereby promoting metastasis in vivo. As research continues to uncover the mechanisms by which miRNAs regulate tumor metastasis, more metastasis-associated miRNAs are expected to be identified. This growing knowledge provides a critical theoretical foundation for developing new therapeutic strategies to inhibit cancer cell migration, invasion, and metastasis.

As mentioned earlier, cancer initiation and progression are driven by alterations in oncogenes and tumor suppressor genes, whose expression can be regulated by miRNAs. Consequently, changes in miRNA expression levels play a crucial role in the occurrence, development, and metastasis of various tumors. Different types of cancer exhibit distinct miRNA expression profiles, distinguishing them from normal tissues as well as from other cancer types. More importantly, miRNAs not only remain stable within cells and tissues but also persist in various body fluids, including serum [99], plasma [100], urine [101], saliva [102], and semen [103]. These characteristics make miRNAs promising noninvasive biomarkers for cancer diagnosis and prognosis, with the potential to facilitate early cancer detection and serve as therapeutic targets for cancer treatment.

10.4 RNA Interference (RNAi)

Based on the mechanism by which miRNA inhibits mRNA translation, introducing antisense RNA that hybridizes with mRNA in the cell should be able to suppress mRNA translation. In 1998, Andrew Z. Fire from the Carnegie Institution of Washington and Craig C. Mello from the University of Massachusetts Cancer Research Center [104] conducted experiments on gene regulation in *Caenorhabditis elegans* and made a groundbreaking discovery. They found that antisense RNA strands paired with mRNA exhibited weak inhibitory effects. However, when a double-stranded RNA (dsRNA) molecule formed by both sense and antisense RNA was introduced, it silenced the target gene far more effectively than either strand alone. Based on these observations, Fire and Mello proposed a novel biological phenomenon—RNA interference (RNAi). They hypothesized that dsRNA could silence genes in a sequence-specific manner: only dsRNA molecules that were identical or highly similar to the gene's coding sequence could effectively trigger gene silencing. Furthermore, they noted that RNA interference could spread between cells and even be inherited. Their experiments also demonstrated that injecting a small amount of dsRNA was sufficient to induce gene silencing, suggesting that the RNAi process might be catalytic or involve an amplification mechanism. Their discovery clarified many previously puzzling and contradictory experimental findings and revealed a fundamental mechanism governing the flow of genetic information. This breakthrough marked the beginning of a new field of research.

RNA interference (RNAi) is a double-stranded RNA (dsRNA)-mediated, enzyme-dependent process of sequence-specific gene silencing, which inhibits gene expression by blocking the translation or transcription of target genes. When dsRNA homologous to the coding region of endogenous mRNA is introduced into a cell, the corresponding mRNA is degraded, leading to gene silencing. This process is classified as a posttranscriptional gene silencing (PTGS) mechanism. The discovery, mechanistic elucidation, and application of RNAi have led to groundbreaking advancements. From 2001 to 2003, Science magazine recognized RNAi as one of the top ten scientific breakthroughs each year, and it was also listed among the top

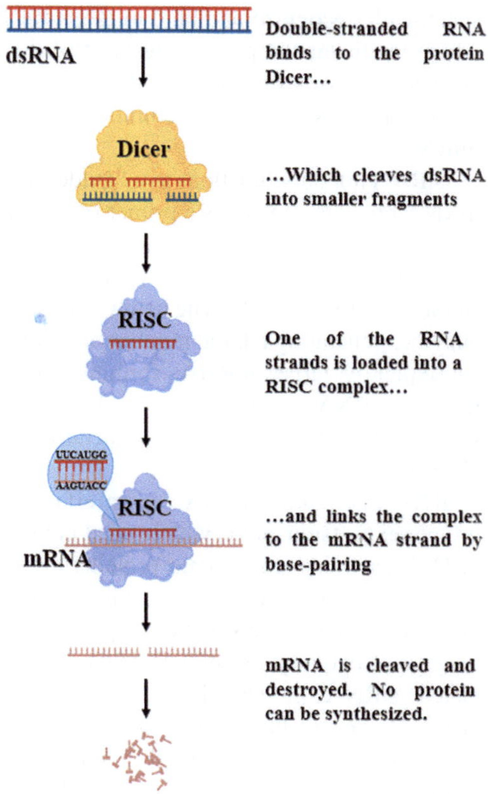

Fig. 10.4 Mechanism of RNAi

ten breakthroughs of 2002. In 2006, Andrew Z. Fire and Craig C. Mello were awarded the Nobel Prize in Physiology or Medicine for their contributions to the study of the RNAi mechanism. RNAi is a highly conserved mechanism found in eukaryotes, playing essential roles in antiviral defense, transposon suppression, and gene regulation. Because RNAi technology enables the targeted elimination or silencing of specific gene expression, it has been widely applied in molecular biology research on gene function as well as in gene therapy for infectious diseases and malignant tumors.

The mechanism of RNA interference (RNAi) is illustrated in Fig. 10.4. Exogenous double-stranded RNA (dsRNA) introduced into the cell binds to the Dicer enzyme in the cytoplasm, where it is cleaved into multiple small RNA fragments with specific lengths and structures, known as small interfering RNA (siRNA). The structural characteristics of siRNA include a length of approximately 21~23 base pairs (bp), 5′ monophosphate and 3′ hydroxyl termini, and a 2~3 nucleotide (nt) overhang at the 3′ end of each complementary strand. Under the action of helicase, siRNA unwinds into a sense strand and an antisense strand. The antisense siRNA then associates with specific proteins in the cell to form the RISC. Once activated, RISC is guided by the unwound siRNA strand (guide strand) to specifically recognize and bind to target mRNA. RISC possesses nuclease activity, which facilitates the precise cleavage of the target mRNA at the binding site. The cleaved mRNA is subsequently degraded, leading to gene silencing. This process is essentially the same as the maturation of pre-miRNA and the formation of RISC by miRNA.

SiRNA not only functions as a "guide strand" to direct RISC in cleaving the target mRNA but can also act as a primer on the target mRNA, similar to primers used in PCR. Under the action of RNA-dependent RNA polymerase (RdRP), an extension reaction occurs, ultimately converting the single-stranded target mRNA into double-stranded RNA (dsRNA). The newly synthesized dsRNA is then cleaved by the Dicer enzyme, generating a large number of secondary siRNAs. These secondary siRNAs subsequently target mRNA, forming an amplification mechanism. This RdRP-mediated RNAi amplification model is referred to as the "random degradative PCR model," enabling RNAi to achieve a highly efficient gene silencing effect [105, 106].

RNA interference (RNAi) is an effective technology for reducing endogenous protein expression and is widely used in biological research, as well as for silencing mRNA encoding pathogenic proteins to achieve therapeutic purposes. However, RNAi therapy faces numerous challenges in clinical translation, and its progress has been significantly slower than expected. The development of RNAi therapy has undergone two major stages. The first stage involved the direct injection of synthetic naked siRNA or chemically modified stable siRNA for systemic delivery [107]. However, due to the negatively charged nature and relatively large size of siRNA molecules, they have difficulty crossing biological barriers and effectively reaching target genes. Additionally, siRNA is rapidly cleared by the kidneys and excreted from the body, significantly reducing its therapeutic efficacy. To address these issues, RNAi therapy entered the second stage, focusing on the design of various delivery vectors (such as lipid nanoparticles, cationic polymers, and viral vectors) or conjugated ligands (such as N-acetylgalactosamine) to significantly enhance the in vivo delivery efficiency of siRNA [108, 109]. These advancements have improved the clinical prospects of RNAi therapy, but further optimization is still required to overcome existing technical and biological challenges. For example, patisiran is a double-stranded siRNA encapsulated in lipid nanoparticles for delivery to hepatocytes. It specifically binds to a conserved genetic sequence in the 3′ untranslated region of transthyretin (TTR) mRNA, utilizing the RNAi mechanism to degrade TTR mRNA, thereby reducing serum TTR protein levels and decreasing TTR protein deposition in tissues. Developed by Alnylam Pharmaceuticals, patisiran was approved by the US Food and Drug Administration (FDA) in 2018 for the treatment of adult polyneuropathy caused by hereditary transthyretin amyloidosis (hATTR). This breakthrough made patisiran the first RNAi-based biopharmaceutical approved worldwide, marking the beginning of a new era in RNAi therapy. Subsequently, in November 2019, the FDA approved a second RNAi drug, Givlaari (givosiran), for the treatment of acute hepatic porphyria (AHP), a rare genetic disorder primarily affecting female patients. The successful development of these

RNAi drugs has laid the foundation for the clinical application of RNAi therapy and has driven further advancements in the field.

SiRNA and miRNA are similar in size, and both function in gene silencing, sharing biological relevance in their biogenesis and mechanisms of action. Both types of small RNA rely on the same two protein families: Dicer (which processes precursor RNA) and Argonaute (Ago) proteins (the core components of RNA silencing). However, there are several differences between miRNA and siRNA. For instance, miRNA is endogenously produced within the organism, whereas siRNA is exogenously introduced. MiRNA is processed from an imperfectly paired double-stranded RNA stem-loop precursor, whereas siRNA originates from a fully complementary double-stranded RNA. SiRNA generally exhibits high complementarity with its target mRNA, leading to its degradation. In contrast, miRNA in mammals typically has lower complementarity with its target mRNA, allowing it to regulate gene expression by inhibiting mRNA translation without affecting mRNA stability.

Given the similarities between siRNA and miRNA, as well as the rapid advancement of RNAi-based drugs, miRNA-based therapeutic strategies have emerged, including miRNA mimics and miRNA inhibitors (also known as antimiRs) [110]. MiRNA mimics are synthetically designed double-stranded small RNA molecules that, once introduced into cells, function similarly to endogenous miRNAs. They are intended to restore tumor-suppressor miRNAs lost in disease conditions. In contrast, antimiRs are synthetically designed single-stranded small RNA molecules that are complementary to the sequence of mature target miRNAs. By binding to and sequestering target miRNAs, antimiRs inhibit their function, typically targeting oncomiRs to block their oncogenic activity. One of the key challenges in miRNA-based therapy is ensuring that introduced miRNA mimics or antimiRs can resist degradation by endogenous endonucleases in the human body. To address this, researchers have developed chemical modifications such as 2′-O-methyl (2′-OMe) modifications or locked nucleic acids (LNAs) to enhance RNA stability. However, while such modifications improve stability, they may also hinder the ability of miRNA mimics to properly associate with the RISC complex, thereby reducing their mRNA silencing efficiency. Encapsulating miRNA mimics in nanoparticles or other delivery carriers increases the stability of therapeutic candidates and improves drug efficacy in vivo. However, the key challenges in translating these delivery systems into clinical applications include avoiding potential immune stimulation and achieving precise targeting of disease sites. Additionally, several other hurdles must be addressed for clinical implementation, including identifying optimal miRNA targets for each disease, ensuring tissue-specific targeting while avoiding potential toxicity and off-target effects. Overcoming these challenges is crucial for advancing miRNA-based therapies toward clinical application.

MiRNA mimics and antimiRs currently in preclinical development have shown promise as novel therapeutic agents. Table 10.3 lists several clinical trials involving miRNA-based therapies. However, there remains a significant gap between basic miRNA research and clinical applications. Extensive preclinical and translational

Table 10.3 MiRNA therapeutics in clinical trials

Name (company)	Therapeutic agent	Delivery system	Target diseases	Trial details	Clinical trials gov identifier
Mirvirasen (Santaris Pharma A/S and Hoffmann-La Roche)	AntimiR-122	LNA-modified anti-sense inhibitor	Hepatitis C (chronic infections included)	Single-center phase I, completed	NCT01646489
				Multicenter phase II, completed	NCT01200420
				Multicenter phase II, ongoing	NCT01872936
				Single-center phase II, ongoing	NCT02031133
				Single-center phase II, ongoing	NCT02508090
RG-125/ AZD4076 (Regulus Therapeutics)	AntimiR-103/107	GalNAc-conjugated antimiR	Patients with type 2 diabetes and nonalcoholic fatty liver diseases	Single-center phase I, ongoing	NCT02612662
				Single-center phase I/IIa, ongoing	NCT02826525
MRG-106 (miRagen Therapeutics)	AntimiR-155	LNA-modified anti-sense inhibitor	Cutaneous T cell lymphoma and mycosis fungoides	Multicenter phase I, ongoing	NCT02580552
MRG-201 (miRagen Therapeutics)	miR-29 mimic	Cholesterolconjugated miRNA duplex	Scleroderma	Single-center phase I, ongoing	NCT02603224
MesomiR-1 (EnGeneIC)	miR-16 mimic	EnGeneIC delivery vehicle	Mesothelioma, non-small cell lung cancer	Multi-center Phase I, ongoing	NCT02369198
MRX34 (Mirna Therapeutics)	miR-34 mimic	LNPs (Smarticles)	Multiple solid tumors	Multicenter phase I, terminated	NCT01829971

studies are necessary to improve the in vivo efficacy of delivered miRNA therapeutics while minimizing side effects.

10.5 Analysis Methods of miRNA

10.5.1 Direct Detection of miRNA Based on Hybridization

10.5.1.1 Northern Blotting

In 1977, Alwine et al. [111] invented the Northern blot hybridization method, similar to the Southern blot hybridization method for DNA detection, for RNA analysis. Northern blotting was one of the earliest methods used for miRNA analysis, playing a significant role in the early detection and study of miRNAs, and has become the standard method for miRNA analysis and identification. First, total RNA samples are extracted from the sample, and the target miRNA is separated from other RNA species by denaturing polyacrylamide gel electrophoresis. Then, a nylon or nitrocellulose membrane is placed on the gel surface, and the target miRNA is transferred to the membrane using salt gradient diffusion or electroblotting methods. Due to the negative charge of miRNA, it binds effectively to the positively charged membrane. The miRNA is then stabilized on the membrane through vacuum high-temperature baking or ultraviolet crosslinking. Subsequently, a radiolabeled complementary probe is added to hybridize with the miRNA sequence on the blotting membrane, forming a radiolabeled band. These bands can be detected using autoradiography or phosphor imaging technology. In the early stages of miRNA research, Tuschl et al. [4] used the Northern blotting method to analyze the differences in miRNA expression at various developmental stages of *C. elegans*, including the embryo stage, larval stages L1/L2/L3, pupal stage, and adult stage, which laid an important foundation for miRNA research.

The traditional Northern blotting method is limited in its widespread application due to several drawbacks, including cumbersome procedures, time consumption, low detection sensitivity, large sample requirements, and the potential hazards of radioactivity. Currently, several modified Northern blotting techniques have been reported, with improvements mainly focused on crosslinking technology, probe design, and labeling methods.

In the traditional Northern blotting method, the key step for fixing RNA onto a nylon or nitrocellulose membrane is ultraviolet (UV) crosslinking. This process aims to increase RNA retention during subsequent steps such as hybridization, washing, and probe elution. However, a study by Pall et al. [112] showed that UV crosslinking somewhat inhibits the hybridization process of miRNA, and simply increasing the UV dose does not improve miRNA detection sensitivity. In contrast, chemical crosslinking mediated by soluble carbodiimide (1-ethyl-3-(3-dimethylaminopropyl) carbodiimide, EDC) significantly enhances miRNA detection capability. Compared to the standard UV crosslinking method, EDC crosslinking can increase miRNA detection sensitivity by 25–50 times. This method does not require special equipment, is cost-effective, and shortens the experimental time. EDC crosslinking achieves stable covalent binding of one end of miRNA to the membrane by promoting the formation of covalent bonds between the 5′ phosphate group of miRNA and the free amino groups on the nylon membrane. Since this crosslinking method depends on the 5′ phosphate group of RNA, it applies not only to miRNA but also to other small RNA classes, such as piRNA and siRNA. The advent of EDC crosslinking technology has not only addressed the limitations of traditional UV crosslinking but also provided a more efficient and reliable tool for small RNA detection.

The traditional Northern blotting method requires a large amount of total RNA when detecting low-abundance miRNAs, which significantly limits its sensitivity. This makes the method unsuitable when cell or tissue samples are limited. Locked nucleic acid (LNA)-modified DNA probes, which enhance the affinity for complementary sequences, significantly improve the stability of duplex formation with target DNA or RNA molecules [113], providing an effective solution to this problem. Válóczi et al. [114] proposed an improved Northern blotting method using LNA-modified oligonucleotide probes for efficient miRNA detection. By modifying every three nucleotides of the DNA probe with LNA, this method greatly enhanced detection performance. Compared to unmodified DNA probes, the detection sensitivity of LNA-modified probes was increased at least tenfold while maintaining high specificity. This improvement not only enhances the ability of Northern blotting to detect low-abundance miRNAs but also provides a more sensitive and reliable solution for studies with limited sample quantities.

Considering the potential health hazards of radioactive labels and the challenges and high costs associated with radioactive waste disposal, Ramkissoon et al. [115] proposed a nonradioactive isotope Northern blotting method that uses 3-′-digoxigenin (DIG)-labeled RNA oligonucleotide probes to detect miRNAs. DIG-labeled RNA probes can hybridize with target miRNAs fixed on the membrane, and then, DIG-labeled antibodies conjugated to alkaline phosphatase (AP) bind to the DIG in the probe through an immune reaction. AP catalyzes a chemiluminescent substrate, producing a chemiluminescent signal that can be detected by X-ray film. After performing Northern blot analysis, RNA probes labeled with DIG can achieve the same sensitivity as 32P-labeled probes. Compared to traditional radioactive labeling techniques, the DIG labeling system offers several advantages: higher sensitivity, shorter exposure times, longer shelf life, and a significant reduction in radiation exposure, thereby improving the safety of the experiment. Additionally, other types of nonradioactive labeling probes, such as biotin [116] and fluorescein [117], have also been successfully applied in Northern blotting for miRNA detection.

Kim et al. [118] combined LNA modification, DIG labeling, and EDC crosslinking technology to develop a highly sensitive nonradioactive Northern blotting method. This method can detect miRNA as low as 0.05 fmol and produce visible signals. Compared to traditional radioactive isotope methods, this approach requires only a few seconds of film exposure to achieve signal strength greater than that obtained from overnight exposure with isotopes, shortening the exposure time by approximately 1000-fold. Additionally, while the harmful probes used in traditional radioactive Northern blotting need to be freshly prepared, the probes used in this method have a longer shelf life and can be stored for at least 6 months, improving experimental efficiency, reducing costs, and being more environmentally friendly.

10.5.1.2 Microarray

miRNA microarray is a high-throughput analytical method based on nucleic acid hybridization between target molecules and their complementary probes. As shown in Fig. 10.5, this method covalently attaches capture probes complementary to miRNA onto the surface of solid-phase supports such as glass slides, nylon membranes, or polymer films, forming high-density DNA microarray chips. These chips can be used for simultaneous analysis of multiple miRNAs. By labeling miRNA molecules with fluorescent dyes, the fluorescently labeled miRNA molecules hybridize with specific capture probes, and the fluorescent signals emitted by the labeled miRNA molecules can be detected at specific locations on the solid-phase

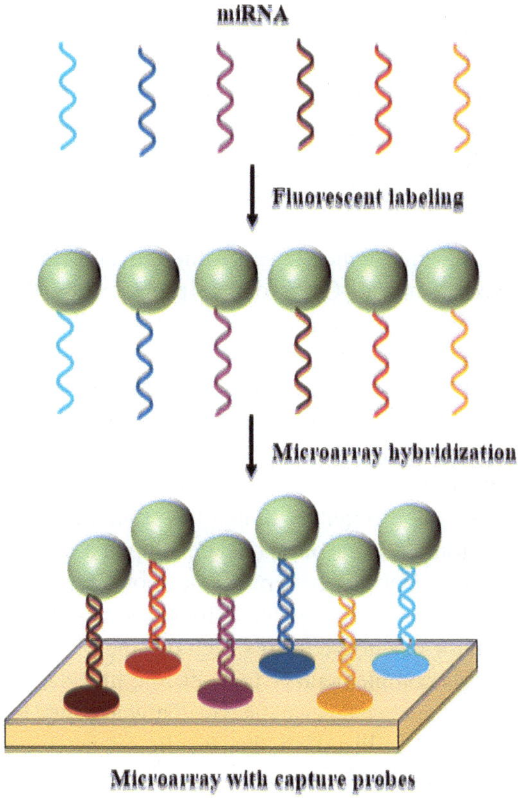

Fig. 10.5 Schematic representation of microarray-based miRNA expression profiling

support. By analyzing these fluorescence signals, the types and expression levels of miRNAs in the sample can be quantitatively evaluated. Typically, the miRNA microarray chip technology has several advantages: first, it enables the global expression profiling of hundreds of genes from the same sample at the same time; second, with carefully designed oligonucleotide probes, it can identify both mature miRNAs and detect the expression of precursor miRNAs; and third, compared to Northern blotting, this method requires a smaller amount of total RNA sample.

Before hybridizing with the capture probes on the microarray, the miRNA to be tested must first be fluorescently labeled. Due to the small size of miRNAs and the limited sequence available for labeling and probe design, their analysis is more complex than that of mRNAs. Moreover, since miRNAs account for only a tiny fraction of the total RNA sample (approximately 0.01%), specificity in labeling becomes especially important. In 2004, Liu et al. [119] developed a fluorescence labeling method based on reverse transcription, applied to miRNA analysis in humans and mice. In this method, the miRNA to be tested serves as the template, and reverse transcription is carried out using random primers labeled with biotin, introducing biotin molecules onto the reverse transcription products. The reverse transcription products then combine with streptavidin-conjugated Alexa647 dye, further incorporating fluorescent molecules. Using this labeling method, the authors developed a chip containing oligonucleotide probes corresponding to 245 miRNAs from the human and mouse genomes, for genome-wide miRNA analysis. The method yielded highly reproducible experimental results and revealed tissue-specific miRNA expression patterns.

In addition to the reverse transcription-based fluorescence labeling method mentioned above, Hammond's group [120] employed a different labeling strategy, using T4 RNA ligase to directly add fluorescence-modified dinucleotides (e.g., pCU-Cy3) to the 3′ end of miRNA. Combined with microarray chip technology, this method enabled the analysis of the expression profiles of 124 miRNAs in adult mouse tissues and at embryonic stages. Although dinucleotides as ligation donors are often in excess, their 3′-OH group is protected by the fluorescent group, preventing the occurrence of tandem side reactions. However, miRNAs in biological samples have a phosphate group at the 5′ end, and under the catalysis of T4 RNA ligase, there is a risk of intramolecular ligation forming a circular structure. To eliminate this risk, a dephosphorylation step is typically required before using T4 RNA ligase to remove the 5′-phosphate group from the miRNA. This method eliminates the potential biases introduced by reverse transcription labeling but cannot avoid biases caused by differences in ligation efficiency.

Mourelatos' group [121] developed a method for labeling miRNA based on the Klenow DNA polymerase extension reaction (RNA-primed, array-based Klenow enzyme, RAKE), which eliminates biases caused by reverse transcription and ligation reactions. In this method, the 3′ end of the DNA oligonucleotide probes used is complementary to a specific miRNA, while the 5′ end contains a universal fixed sequence that can be covalently attached to the microarray chip on a glass slide. As shown in Fig. 10.6, the specific sequence complementary to the miRNA and the universal fixed sequence (purple portion) are separated by three thymine

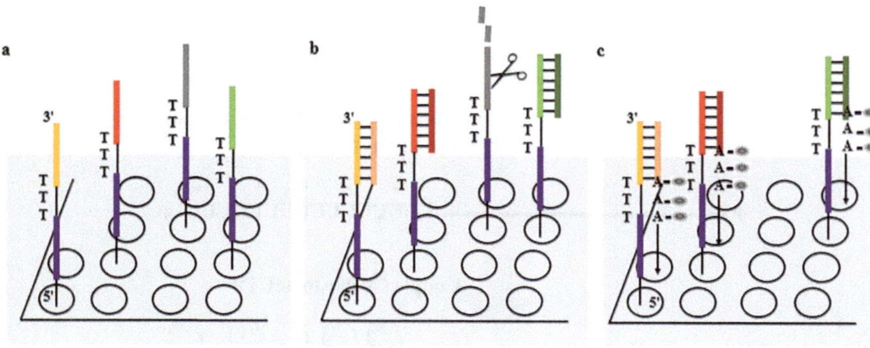

Fig. 10.6 Schematic diagram of RAKE-based miRNA microarray analysis [121]. (Revised and reprinted with permission from Ref. [121])

(T) residues. RNA samples containing miRNA hybridize with the oligonucleotide probes on the chip and are then treated with exonuclease I (Exo I). The oligonucleotide probes bound to miRNA are protected from hydrolysis by Exo I due to their double-stranded structure at the 3′ end, while the unbound single-stranded oligonucleotide probes are specifically hydrolyzed. Subsequently, in the presence of Klenow DNA polymerase and biotin-labeled dATP (B-dATP), the fixed DNA probe serves as the template, and the hybridized miRNA acts as the primer for Klenow DNA polymerase extension, integrating B-dATP into the miRNA. After washing off unreacted B-dATP, fluorescence-labeled streptavidin reacts with the biotin, producing fluorescent spots at the hybridization sites as detection signals. Studies on human cell lines and brain tumor cells using this labeling method with microarray chip technology showed that the method is sensitive and specific, making it highly suitable for rapid expression profiling of all known miRNAs.

Shingara et al. [122] developed the miRNA labeling method based on a poly (A) polymerase (PAP)-tailing reaction. As shown in Fig. 10.7, poly(A) polymerase can repeatedly add 20–50 nucleotides to the 3′ end of miRNA. By incorporating a certain proportion of amino-modified Uridine Triphosphate (UTP) (aaUTP*) into the normal UTP, the 3′ tail of the miRNA is composed of both normal and amino-modified nucleotides. As a result, the tailed miRNA can bind to any dye with amine-reactive groups (e.g., Cy3 and Cy5). In this labeling method, each miRNA can be labeled with multiple fluorescent molecules, greatly enhancing the detection sensitivity. The authors applied this system to analyze 26 human tissue samples, enabling classification of normal tissues based on miRNA expression profiles.

In microarray chips, there are thousands of probe molecules, and their hybridization with target miRNAs occurs over a wide range of melting temperatures (T_m). Since a single hybridization temperature is typically used for the entire chip, probes with T_m values higher than the hybridization temperature will exhibit higher hybridization efficiency with miRNA, while probes with T_m values lower than the hybridization temperature will hybridize with miRNA less efficiently. This variation in hybridization efficiency can distort the final detection signals. If a lower

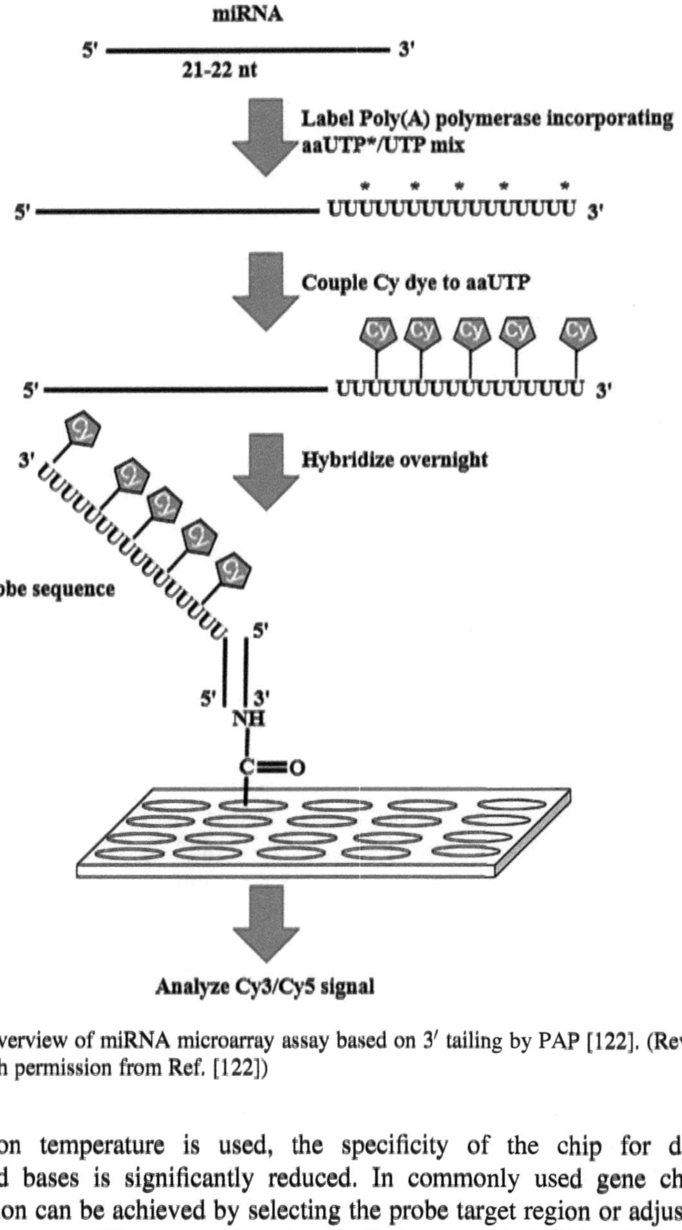

Fig. 10.7 Overview of miRNA microarray assay based on 3′ tailing by PAP [122]. (Revised and reprinted with permission from Ref. [122])

hybridization temperature is used, the specificity of the chip for detecting mismatched bases is significantly reduced. In commonly used gene chips, T_m normalization can be achieved by selecting the probe target region or adjusting the length of the probes. However, due to the size limitations of miRNAs, these methods are not suitable for normalizing the T_m of miRNA probes. To address this issue, Castoldi et al. [123] employed locked nucleic acid (LNA)-modified probes to increase the thermal stability of the miRNA-probe hybrid, thereby raising the T_m value. By adjusting the LNA content and length of the probes, they were able to normalize the T_m of the probes. Compared to unmodified probes, LNA-modified

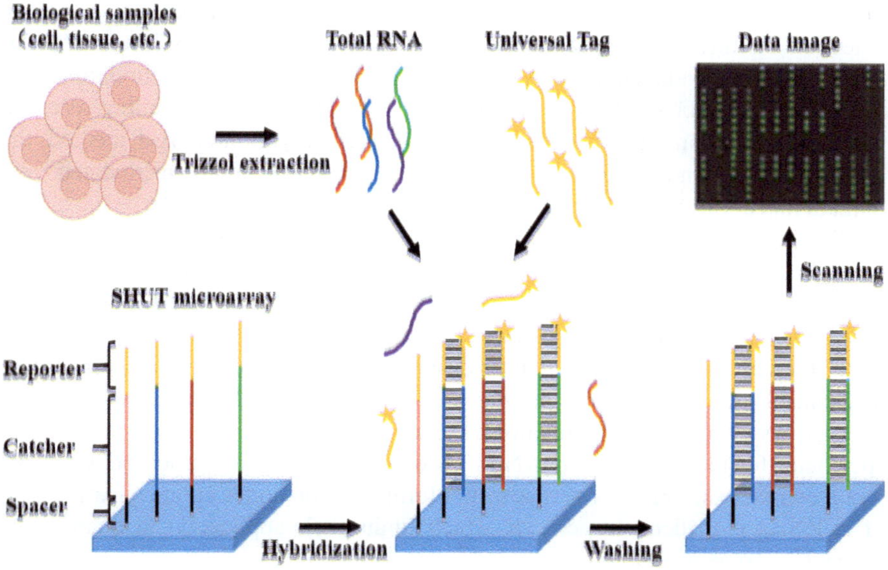

Fig. 10.8 SHUT method for miRNA assay [124]. (Revised and reprinted with permission from Ref. [124])

capture probes not only increase sensitivity but also effectively distinguish single-base differences.

In enzyme- or chemically-labeled, other types of RNA in the sample, besides miRNA, may also be labeled, leading to nonspecific hybridization and background signals. Therefore, RNA purification before labeling is essential. Additionally, the labeling process is not only labor-intensive and expensive but also prone to labeling errors and human mistakes. Currently, commercially available miRNA microarray systems show poor consistency between platforms. Duan et al. [124] proposed a hybridization-triggered fluorescence strategy for label-free, high-throughput miRNA expression profiling, which effectively simplifies the operational process. This method is called the Stacking-Hybridized Universal Tag (SHUT) strategy. Stacking hybridization refers to the increased stability of a DNA hybridization reaction when two or more DNA oligonucleotides are hybridized in a continuous tandem arrangement to a longer complementary single-stranded DNA. As shown in Fig. 10.8, the basic design of the SHUT probe consists of three functional regions: the Reporter sequence, the Catcher sequence, and the Spacer sequence. The Catcher sequence is complementary to the target miRNA, while the Reporter sequence is complementary to a fluorescently labeled universal probe (Universal Tag, UT). When the target miRNA binds to the Catcher sequence in the SHUT probe, the base stacking effect between the UT and miRNA enhances the hybridization stability of the SHUT probe with the UT, enabling stable binding of the UT to the SHUT probe. In contrast, by designing a lower UT T_m value (<35 °C), without the target miRNA, the short UT will have difficulty maintaining stable hybridization with the SHUT probe, thus producing only very weak background fluorescence. This method allows specific detection of miRNA without labeling and can detect miRNA as low as 20 fM. Additionally, when using a polyA tailing-based miRNA analysis method, the 3′-O-methylation modification of plant miRNAs can hinder the polyA tailing reaction, leading to detection bias for plant miRNAs. However, with the SHUT method, the group [125] found consistent results when detecting both methylated and unmethylated miRNAs, indicating that the method is highly versatile.

The main advantage of miRNA microarrays lies in their ability to perform high-throughput analysis of miRNAs. Currently, several companies have launched microarray platforms for miRNA analysis, which are widely used in various research fields, including miRNA biogenesis, tissue distribution, expression differences under normal and abnormal conditions, stem cell development, pathway mapping, mechanisms of action, and tumorigenesis. However, miRNA detection using microarrays also has some limitations, including a narrow linear range and weaker specificity for highly similar miRNAs. Additionally, the consistency between different commercial miRNA microarray platforms is poor, and accurate quantification of miRNAs is difficult to achieve, which limits their application in quantitative research to some extent.

10.5.1.3 Graphene Oxide (GO)-Based miRNA Analysis Method

Geim and his team at the University of Manchester first prepared graphene [126] and revealed its unique structural characteristics. In graphene, the carbon atoms adopt sp^2 hybridized orbitals and form a honeycomb crystal lattice, the same atomic arrangement as in monolayer graphite, creating a typical two-dimensional monolayer crystal. Thanks to this special structure, graphene exhibits a range of exceptional properties, including ultrahigh mechanical strength, excellent thermal and electrical conductivity, strong surface adsorption ability, and good biocompatibility. These characteristics make graphene promising for applications across a variety of fields.

Graphene oxide (GO) can be prepared by treating graphene materials with strong oxidants. Although oxidation does not alter the basic crystal lattice structure of graphene, it introduces a large number of oxygen-containing functional groups on the GO surface, significantly enhancing its hydrophilicity and biocompatibility. Additionally, GO exhibits strong non-covalent interactions with bases and aromatic compounds [127]. Based on this characteristic, a fluorescence resonance energy transfer (FRET) fluorescence reporting system can be constructed between fluorescence-labeled nucleic acid molecules and GO. In 2009, Yang Huanghao's research group [128] developed a GO-based biosensor that enables the detection of DNA. As shown in Fig. 10.9, fluorescence-labeled single-stranded DNA probes strongly adsorb to the GO surface via π-π stacking interactions, resulting in resonance energy transfer and efficient quenching of the fluorescence signal. When the target molecule is present, the fluorescence-labeled single-stranded DNA probe binds to the target molecule to form a double-stranded structure, disrupting the interaction between the fluorescence-labeled DNA probe and GO, causing the

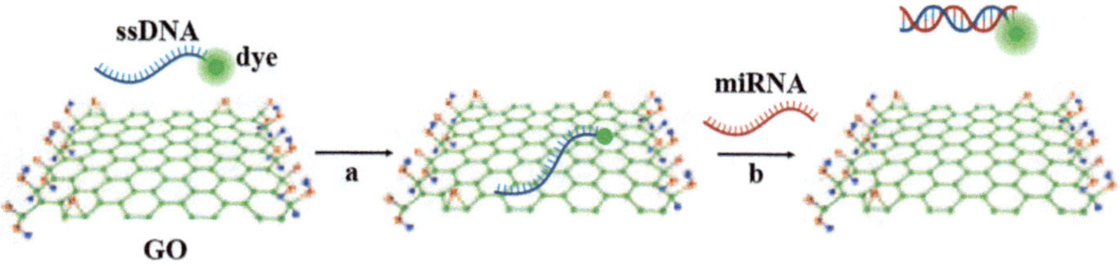

Fig. 10.9 Biosensor platform constructed by fluorescence-labeled nucleic acid molecules and GO [128]. (Revised and reprinted with permission from Ref. [128])

fluorescence-labeled DNA to detach from the GO surface, thereby restoring the fluorescence signal. This design enables specific fluorescence-enhanced detection of target molecules, providing a new technological platform for biosensing and molecular detection. Moreover, research from Fan Chunhai's group indicates that the large surface area of GO can effectively quench different fluorescence signals on DNA probes [129], allowing for the construction of a multicolor sensor that can simultaneously detect multiple target molecules. Lu et al. [130] applied this type of GO sensing platform to miRNA detection, achieving the ability to detect miRNA concentrations as low as 50 nM.

10.5.1.4 Capillary Electrophoresis-Based miRNA Analysis

Capillary electrophoresis (CE) technology, known for its ultrahigh separation efficiency, extremely low sample consumption, and short analysis time, has been widely applied in fields such as food safety, drug screening, environmental monitoring, and life sciences. Wegman et al. [131] developed a method for directly quantifying multiple miRNAs based on capillary electrophoresis hybridization technology (Direct Quantitative Analysis of Multiple miRNAs, DQAMmiR). This method combines two separation-enhancement principles: on one hand, single-stranded DNA-binding protein (SSB) is added to the electrophoresis buffer, and on the other hand, drag tags are modified on DNA probes. By modifying DNA probes for different target sequences with different drag tags (such as biotin, hairpin structures, etc.), the migration rate of hybridized double-stranded molecules is altered, enabling separation analysis. As shown in Fig. 10.10, during the hybridization step, excess DNA probes are mixed with miRNA, ensuring that all miRNA molecules hybridize with DNA probes. SSB can bind to unhybridized single-stranded DNA probes but does not bind to double-stranded miRNA-DNA hybrids. When an electric field is applied, the migration speed of DNA probes bound to SSB is faster than that of double-stranded miRNA-DNA hybrids (SSB acts as a propellant). Due to different drag tags, different target miRNA-DNA hybrids migrate at different speeds, allowing separation by capillary electrophoresis. Finally, the concentration of different miRNAs is calculated from the integral signal (the peak area in the figure). The DQAMmiR method achieved separation and detection of three

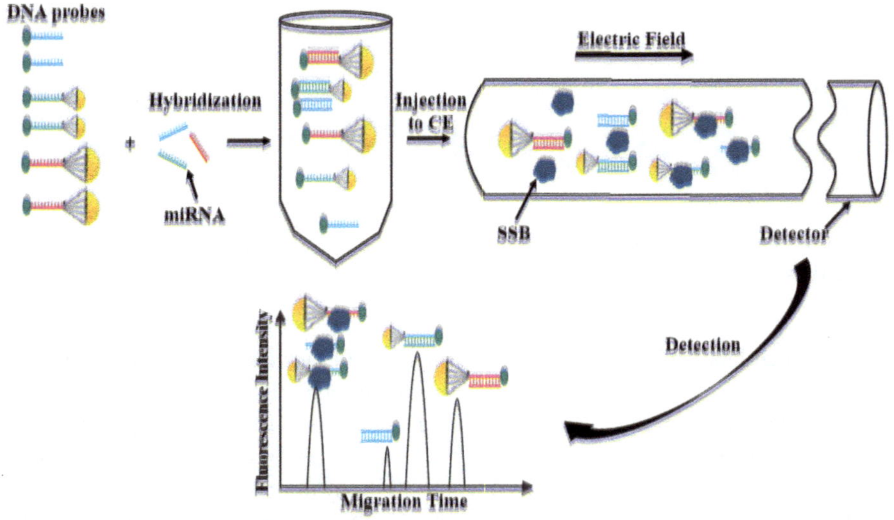

Fig. 10.10 Schematic diagram of multiple miRNA analysis based on DQAMmiR [131]. (Revised and reprinted with permission from Ref. [131])

miRNAs with a sensitivity of 100 pM. Building on this, the research team further improved the DQAMmiR method by using short neutral peptides as universal drag tags to label single-stranded DNA probes. By adjusting the length of the peptide by increasing the number of neutral amino acids, this peptide label did not interfere with the binding of DNA probes to SSB, enabling the simultaneous separation of over ten miRNAs [132].

Khan et al. [133] developed a rapid capillary electrophoresis method to quantitatively measure miRNA expression levels in serum, using SSB and double-stranded RNA-binding protein (p19) as capillary electrophoresis separation enhancers. Unlike the SSB protein, the p19 protein binds specifically to dsRNA of certain sizes, and the length of the dsRNA is critical for high-affinity binding. Specifically, the p19 protein binds tightly to dsRNA of 21–23 bp, but the binding affinity gradually decreases for dsRNA of 24–26 bp, and the binding is weaker for dsRNA of 19 bp or shorter. In this study, the use of SSB and p19 proteins as separation enhancers improved the separation between fluorescence-labeled single-stranded DNA, RNA probes, and miRNA/probe hybrids. The binding of p19 to the miRNA-RNA probe double-stranded complex altered its migration rate, enabling effective separation from excess RNA probes. This method can detect up to 30,000 miRNA molecules per milliliter of serum.

The above hybridization-based direct miRNA detection method enables simple and rapid miRNA analysis and allows for the simultaneous detection of multiple miRNAs. However, due to the lack of nucleic acid amplification and signal amplification mechanisms, its detection sensitivity is generally low.

10.5.2 Analytical Methods for miRNA Based on Nucleic Acid Amplification

10.5.2.1 RT-PCR for miRNA Detection

Reverse transcription-polymerase chain reaction (RT-PCR) is a commonly used technique for RNA analysis, but the length of mature miRNA is similar to that of conventional PCR primers, making the design of specific PCR primers for mature miRNAs challenging. Moreover, the reverse transcription product of mature miRNA cannot be directly used as a template for PCR amplification, preventing subsequent PCR amplification. As shown in Fig. 10.11, Shi et al. [134] used poly(A) polymerase to add a poly(A) tail to the 3′ end of miRNA, thereby elongating the miRNA sequence. They then used primers containing oligo-dT for reverse transcription, followed by PCR amplification of the resulting long-chain cDNA, with SYBR Green I as the fluorescent dye to achieve real-time quantitative analysis of miRNA. This method can detect miRNA in total RNA as low as 100 pg. However, this method cannot effectively distinguish pre-miRNA from mature miRNA. Additionally, the 3′ end of plant miRNAs is 2-O-methylated, preventing the addition of a poly(A) tail using this method, so it cannot be applied to the detection of plant miRNAs.

Chen et al. [135] innovatively designed a stem-loop structure (Stem-loop) reverse transcription primer for high-sensitivity and high-specificity analysis of miRNA. As shown in Fig. 10.12, the 3′ end of the stem-loop structure reverse transcription primer (RT Primer) contains a 6-nucleotide sticky end. Due to base stacking

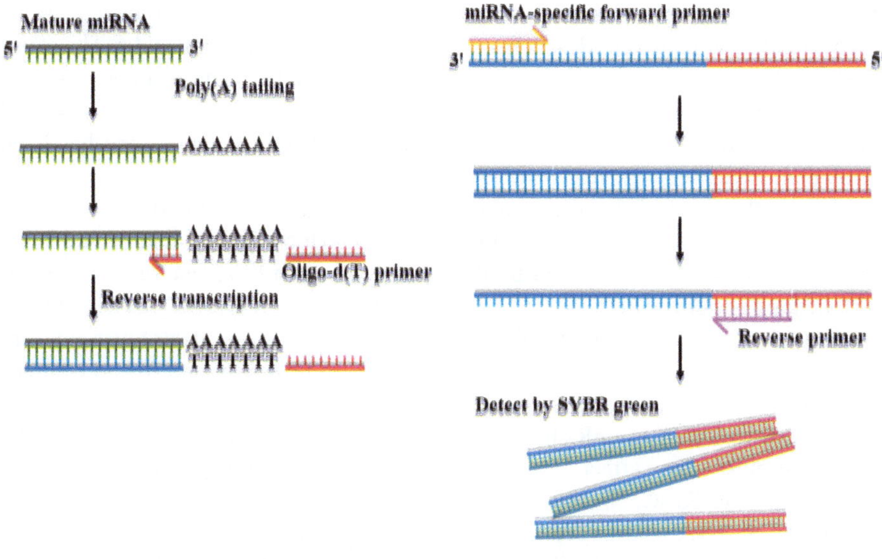

Fig. 10.11 Schematic diagram of RT-PCR method for detecting miRNA based on the 3′ end tailing of miRNA [134]. (Revised and reprinted with permission from Ref. [134])

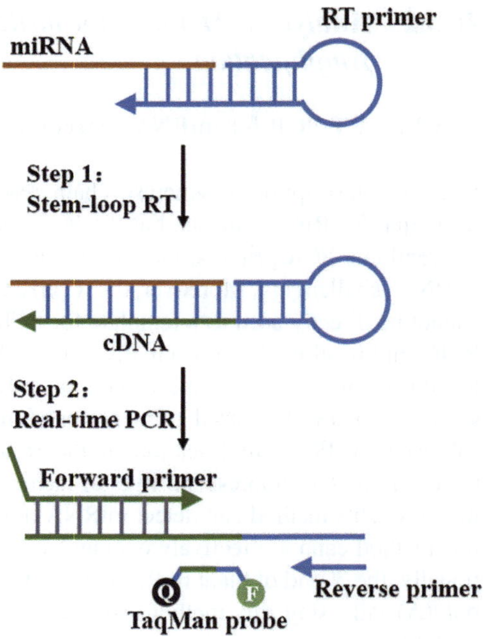

Fig. 10.12 Schematic diagram of RT-PCR for miRNA detection based on stem-loop primer [135]. (Revised and reprinted with permission from Ref. [135])

interactions, it can stably hybridize with miRNA and undergo reverse transcription to generate cDNA under the catalysis of reverse transcriptase. The cDNA then serves as the template for PCR amplification. By combining with TaqMan probes, the PCR amplification process can be monitored in real time. The study results show that this method can accurately detect miRNA with fewer than ten copies and covers a detection range of seven orders of magnitude. Additionally, this method can specifically distinguish mature miRNAs from their corresponding pre-miRNAs. The stem-loop structure reverse transcription primer-mediated RT-PCR method demonstrates higher specificity and sensitivity. Therefore, this method has become the standard technique for miRNA analysis, widely applied in miRNA detection, and has been commercialized.

The development of deep sequencing technology has led to the discovery of an increasing number of miRNAs, many of which exhibit length and/or sequence heterogeneity. MiRNAs encoded by the same gene often have one or more nucleotides added or deleted at the 5′ or 3′ ends, and their length is typically different from the canonical sequence annotated in the miRBase database. These variants are known as isomiRs, with variants resulting from 5′ end changes and 3′ end changes referred to as 5′-isomiRs and 3′-isomiRs, respectively. Many studies have shown that the expression of isomiRs has significant functional implications. For example, certain isomiRs may affect target selection, miRNA stability, and the loading process into the RISC. Specifically, since the seed sequence is located at the 5′ end of the miRNA, the targets recognized by 5′-isomiRs may differ from those recognized by the canonical miRNA. Moreover, changes in the 5′ end sequence may affect the stability of the duplex ends, which could, in turn, influence the selection of

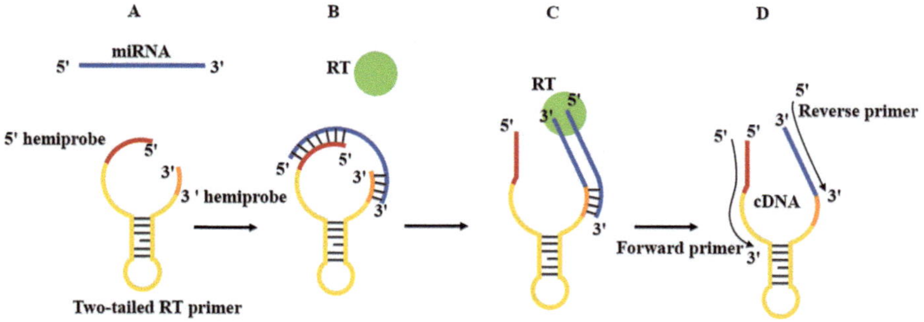

Fig. 10.13 Schematic diagram of two-tailed RT-qPCR for miRNA detection [137]. (Revised and reprinted with permission from Ref. [137])

the miRNA/miRNA* strand for AGO loading, or even the type of AGO protein loaded. Additionally, the posttranscriptional addition of uridine and adenosine to the 3′ end of miRNAs significantly alters miRNA stability, and thus, 3′-isomiRs generated from variations at the miRNA 3′ end play a key role in gene expression regulation [136].

Neither the 3′ tailing RT-PCR nor the stem-loop primer RT-PCR can distinguish 5′-isomiRs of miRNAs. Androvic et al. [137] proposed a novel Two-tailed Reverse Transcription Polymerase Chain Reaction (Two-tailed RT-PCR) method that can simultaneously distinguish 5′-isomiRs and 3′-isomiRs. The design, shown in Fig. 10.13, involves a two-tailed reverse transcription primer composed of two half-probes (hemiprobes) connected by a hairpin folding sequence. The 5′-half probe and 3′-half probe are complementary to the 5′ and 3′ ends of the target miRNA, respectively. By precisely controlling the hybridization temperature, the reverse transcription reaction of the 3′-half probe is triggered only when the 5′-half probe and 3′-half probe match the target miRNA perfectly. If there is a mismatch in the 5′-half probe, the hybridization becomes unstable, preventing the reverse transcription reaction, thus distinguishing highly homologous 5′-isomiRs. Similarly, when there is a mismatch in the 3′-half probe, the reverse transcription reaction cannot occur, allowing for the distinction of highly homologous 3′-isomiRs. This method can detect all terminal variants (isomiRs) of any miRNA, accurately reflecting the true miRNA content in the sample, enabling precise detection of as low as ten copies of miRNA molecules.

10.5.2.2 Ligase-Dependent PCR for miRNA Detection

T4 DNA ligase can repair nicks in the DNA strand of DNA/RNA hybrids, and this specificity has been applied in various RNA detection methods. Based on the activity of T4 DNA ligase, Arefian et al. [138] developed a simple and efficient miRNA analysis method using ligase-dependent PCR, which does not require reverse transcription and enables multiplex miRNA detection with length-encoded probes. The

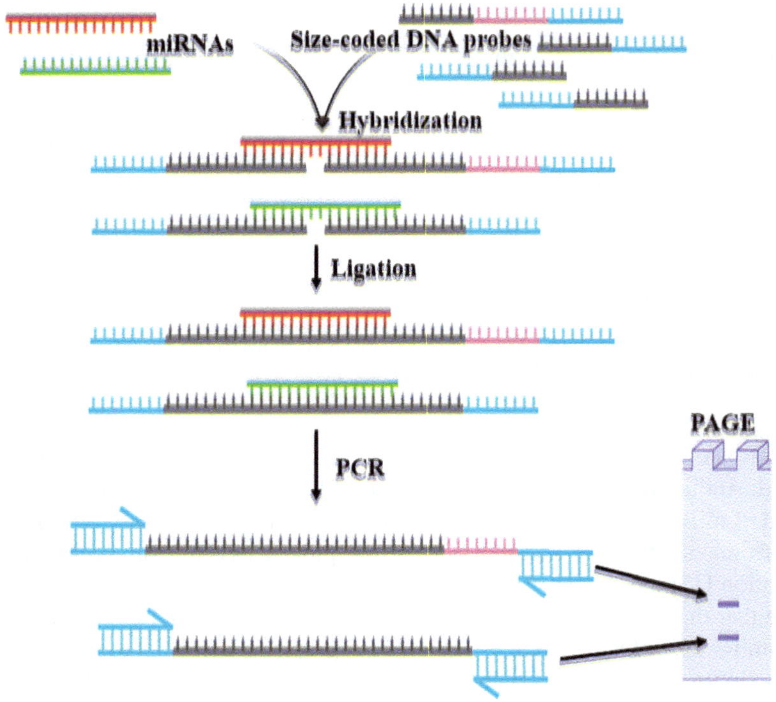

Fig. 10.14 Schematic diagram of T4 DNA ligase-dependent PCR assay for miRNA detection [138]. (Revised and reprinted with permission from Ref. [138])

principle of miRNA detection is shown in Fig. 10.14. First, two DNA probes are designed for each target miRNA sequence. When the target miRNA is present, a pair of DNA probes hybridizes with the target miRNA. Each probe has a predefined length, which serves as the "length code" for the target miRNA (the pink region on the probe). Therefore, multiple probes designed for different miRNAs can be used in the same reaction. Then, under the action of T4 DNA ligase, the DNA probes hybridized with the target miRNA are ligated to form PCR amplification templates, with different lengths of PCR templates formed for different miRNAs. Notably, all probes share the same PCR primer sequence, so the PCR amplification only requires a pair of universal primers. After PCR amplification, the products are analyzed and separated by high-resolution polyacrylamide gel electrophoresis (PAGE). Each electrophoresis band represents a specific miRNA, and the band intensity represents the expression level of the miRNA. By analyzing the number and intensity of the electrophoresis bands, multiplex miRNA detection and quantitative analysis can be achieved. However, T4 DNA ligase has a non-template-dependent, nonspecific ligation activity that can lead to false-positive signals. Additionally, the ligation efficiency of RNA templates to DNA probes catalyzed by T4 DNA ligase is low, and this method can only detect miRNA as low as 1 pM.

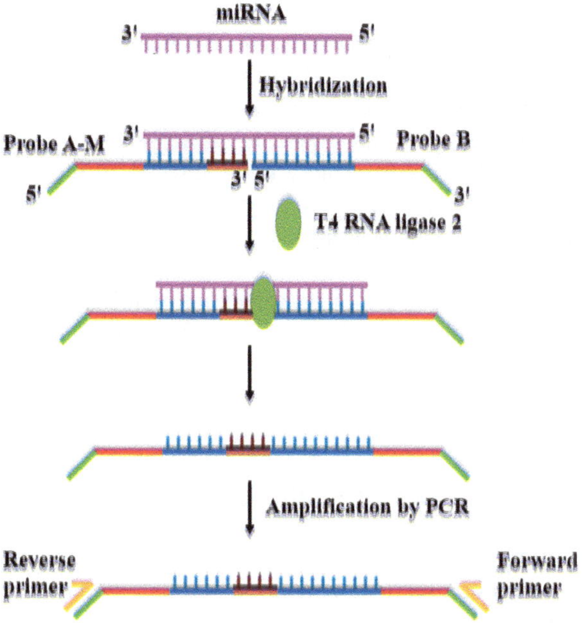

Fig. 10.15 Schematic description of T4 RNA ligase 2-dependent PCR assay for miRNA detection [139]. (Revised and reprinted with permission from Ref. [139])

T4 RNA ligase 2 can efficiently ligate the 3′ hydroxyl group of RNA and the 5′ phosphate group of DNA in DNA/RNA hybrids. Zhang Jiangyan et al. [139] used T4 RNA ligase 2 to establish a ligation-dependent PCR strategy for highly sensitive and quantitative miRNA detection. The principle of miRNA analysis is shown in Fig. 10.15. First, a pair of DNA probes (Probe A-M and Probe B) is designed, each containing a universal PCR primer sequence (green), a complementary sequence to the target miRNA (blue), and a filler sequence (red). The complementary sequence of Probe A-M and Probe B hybridizes with the 3′ and 5′ ends of the target miRNA, respectively. When the target miRNA is present, Probe A-M and Probe B hybridize with the target miRNA, and under the catalysis of T4 RNA ligase 2, the DNA probes are ligated to form a PCR amplification template. Subsequently, PCR amplification can be used to detect and quantitatively analyze the target miRNA. In the study, they found that modifying two RNA bases at the 3′ end of Probe A-M significantly improved the ligation efficiency of the DNA probes using miRNA as the template. Compared to the T4 DNA ligase-based PCR method [138], the detection sensitivity was increased by 5000 times, enabling detection as low as 0.2 fM of let-7a. Additionally, another advantage of ligase-dependent PCR is its high specificity for recognizing single-base mismatched miRNAs. Based on this, combining length-encoded probes and capillary electrophoresis separation techniques allows for high-throughput, highly sensitive quantitative analysis of multiple miRNAs [140]. By combining ligase-dependent PCR with droplet digital PCR (ddPCR) technology [141] and polymerase cloning techniques [142], digital quantitative analysis of miRNA in single cells can be achieved.

In 2014, Lohman et al. [143] discovered SplintR ligase in the Chorella virus, an enzyme that efficiently catalyzes the ligation of adjacent DNA probes using RNA molecules as templates. Based on this property, Jin et al. [144] developed a SplintR ligase-dependent PCR method for simple and sensitive miRNA detection. This method uses SplintR ligase to ligate two DNA oligonucleotide probes that are specifically complementary to the miRNA, forming a PCR amplification template. Combined with TaqMan probes, it enables specific detection and quantitative analysis of target miRNA. This method can detect as low as 4 zmol (approximately 2.5×10^3 molecules) of target miRNA. SplintR ligase can ligate two DNA probes using RNA as a template, achieving high ligation efficiency without the need for 3′-end modified RNA bases. Compared to T4 DNA ligase or T4 RNA ligase 2, SplintR ligase exhibits faster reaction kinetics, significantly reducing the ligation reaction time.

10.5.2.3 Ligase Chain Reaction (LCR) for miRNA Detection

MiRNA sequences are short and cannot be directly used as templates for RT-PCR, but they can serve as templates to ligate two DNA probes. Yan Jingli et al. [145] designed a set of ligase chain reaction (LCR) probes specific to the target miRNA, establishing an LCR-based miRNA analysis method. As shown in Fig. 10.16, using miRNA as the template, T4 RNA ligase 2 catalyzes the ligation of two DNA probes that specifically hybridize to miRNA, forming the initial template for the LCR reaction. This is followed by cyclic ligation of four DNA probes under the catalysis of a heat-stable DNA ligase, triggering an exponential LCR amplification. The method, combined with gel electrophoresis for separation, established a specific miRNA analysis method. However, gel electrophoresis uses grayscale values for quantification, which results in lower accuracy and reliability. Building on this, Zhang et al. [146] used T4 RNA ligase 2 and 3′-end RNA base-modified DNA probes to enhance the ligation efficiency of DNA probes, improving the sensitivity of the method, which can detect miRNA as low as 0.2 fM. Furthermore, by using length-encoded DNA probes, different miRNAs correspond to different lengths of LCR amplification products. This method, combined with capillary electrophoresis for separation and detection, enables multiplex miRNA quantitative analysis based on LCR technology.

10.5.2.4 Hybridization Chain Reaction (HCR) for miRNA Analysis

Hybridization reactions were proposed by Pierce et al. [147] in 2004. It is an isothermal amplification method that does not require nuclease catalysis. The technology consists of two stem-loop DNA probes with sticky ends. In the absence of a target nucleic acid molecule, the two stem-loop DNA probes can stably coexist. However, when the target nucleic acid is present, it triggers a cascading hybridization reaction, forming long double-stranded DNA and amplifying the signal of the

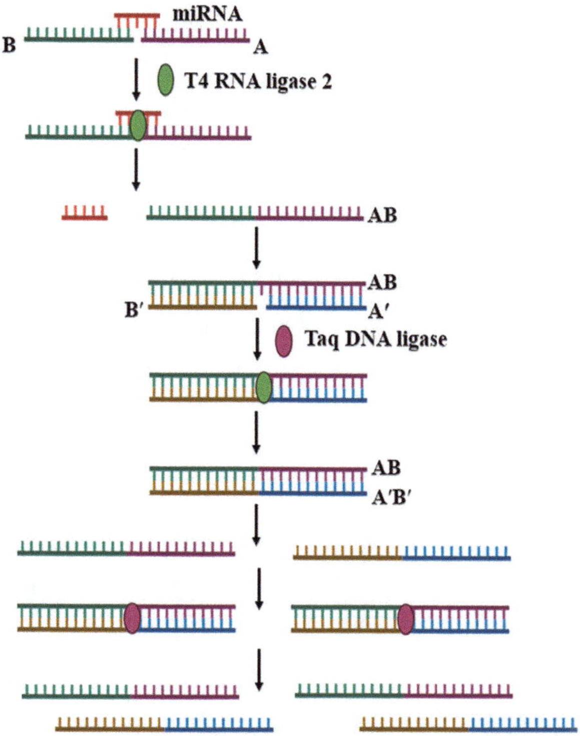

Fig. 10.16 Principle of miRNA detection based on LCR [145]. (Revised and reprinted with permission from Ref. [145])

target sequence under isothermal conditions. The principle of miRNA analysis based on HCR technology is shown in Fig. 10.17. H1* and H2* are a pair of stem-loop DNA probes with sticky ends, and the 5′ end of H1* and the 3′ end of H2* are both modified with fluorescent tags. In the absence of the target miRNA, H1* and H2* stably coexist in solution as stem-loop structures. When the target miRNA is present, the sticky-end sequence (1) of H1* hybridizes with the miRNA, and through a strand displacement reaction, H1* opens its stem-loop structure. The newly exposed sticky end (3–2*) of H1* can further hybridize with the sticky end (3*) of H2*, opening the stem-loop structure of H2* and exposing the sticky end (1*–2*) of H2. The sequence of 1*–2* is identical to the target miRNA, thus triggering another round of hybridization reactions with H1* and H2*. Therefore, each copy of the miRNA can alternately open the stem-loop structures of H1* and H2*, initiating a chain reaction based on target molecule hybridization. The final product is a long, gap-containing double-stranded DNA. HCR enables signal amplification of miRNA without enzymes; however, the fluorescence signals produced by HCR products and by H1* and H2* in homogeneous solutions cannot be distinguished, limiting the widespread application of this method.

Yang et al. [148] established a fluorescence sensing method for miRNA detection in homogeneous solutions, using graphene oxide (GO) to exploit the difference in binding ability of fluorescently labeled stem-loop DNA probes before and after

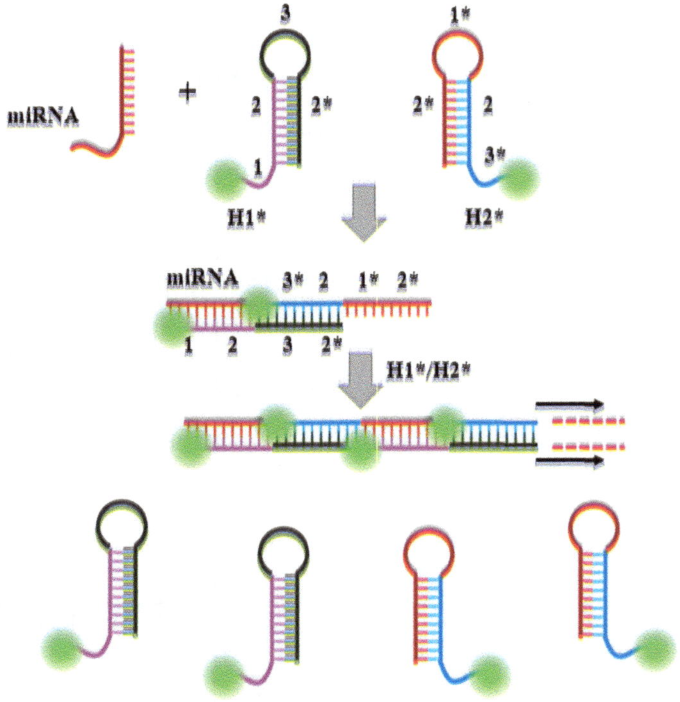

Fig. 10.17 Schematic diagram of HCR-based detection of miRNA [147]. (Revised and reprinted with permission from Ref. [147])

HCR. This method does not require nuclease catalysis and enables sensitive detection of miRNA, with a detection limit of 1 pM.

Miao et al. [149] developed a universal and label-free sensor for the sensitive detection of miRNA based on positively charged gold nanoparticles (AuNPs) combined with HCR. As shown in Fig. 10.18, using miRNA as an initiator, HCR reactions between two stem-loop probes (H1 and H2) produce long-chain double-helix double-stranded DNA polymers. Positively charged AuNPs bind to the negatively charged surface of the HCR amplification products through electrostatic interactions, causing AuNPs to precipitate. The concentration of AuNPs in the supernatant can be easily measured using a simple UV-visible spectrophotometer to detect miRNA. This method can detect miRNA concentrations ranging from 20 pM to 10 nM. Compared to other amplification methods, the linear amplification mechanism of HCR leads to a relatively lower signal amplification, and thus, the detection limit for HCR-based miRNA analysis is generally at the pM level.

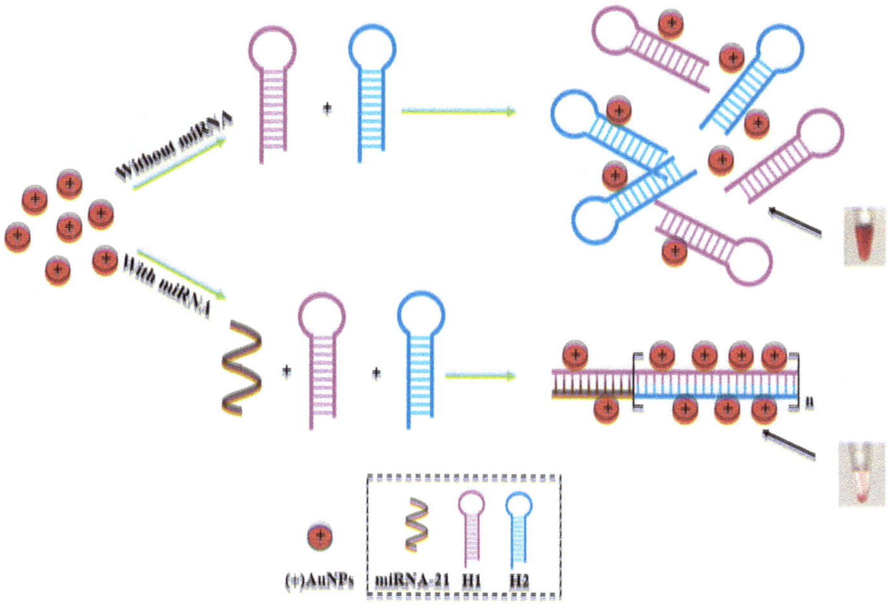

Fig. 10.18 Principle of miRNA analysis based on HCR/AuNPs system [149]. (Revised and reprinted with permission from Ref. [149])

10.5.2.5 Exponential Amplification Reaction (EXPAR) for miRNA Detection

In 2010, Jia Hongxia et al. [150] first discovered that miRNA molecules could trigger the EXPAR reaction, thereby establishing a simple, rapid, and ultrasensitive miRNA analysis method, as shown in Fig. 5.14 of Chap. 5. However, subsequent studies found that background signals generated by nonspecific amplification limited the sensitivity of EXPAR. To effectively reduce the background of nucleic acid amplification and achieve single-molecule and single-cell analysis, in 2015, Zhang Xiaobo et al. [151] developed a single-microsphere fluorescence enrichment platform based on the EXPAR amplification reaction, achieving miRNA detection at the single-molecule level. As shown in Fig. 10.19, through the specific interaction between biotin and streptavidin (STV), the 3′-end biotin-modified EXPAR template XYY is conjugated to the surface of a single magnetic agarose microsphere modified with STV. The XYY captures target miRNA molecules, enriching them on the surface of the microsphere for EXPAR. During the EXPAR amplification process, biotin-modified dATP (Biotin-dATP) is incorporated, and after the reaction, double-stranded DNA containing biotin is formed on the microsphere surface. Then, fluorescently labeled STV (Alexa Fluor 546-STV) is incubated with the microspheres, and biotin and STV specifically bind, highly enriching the fluorescence signal on the surface of the single microsphere. Subsequently, magnetic separation and washing are used to remove unbound Alexa Fluor 546-STV. Finally, the

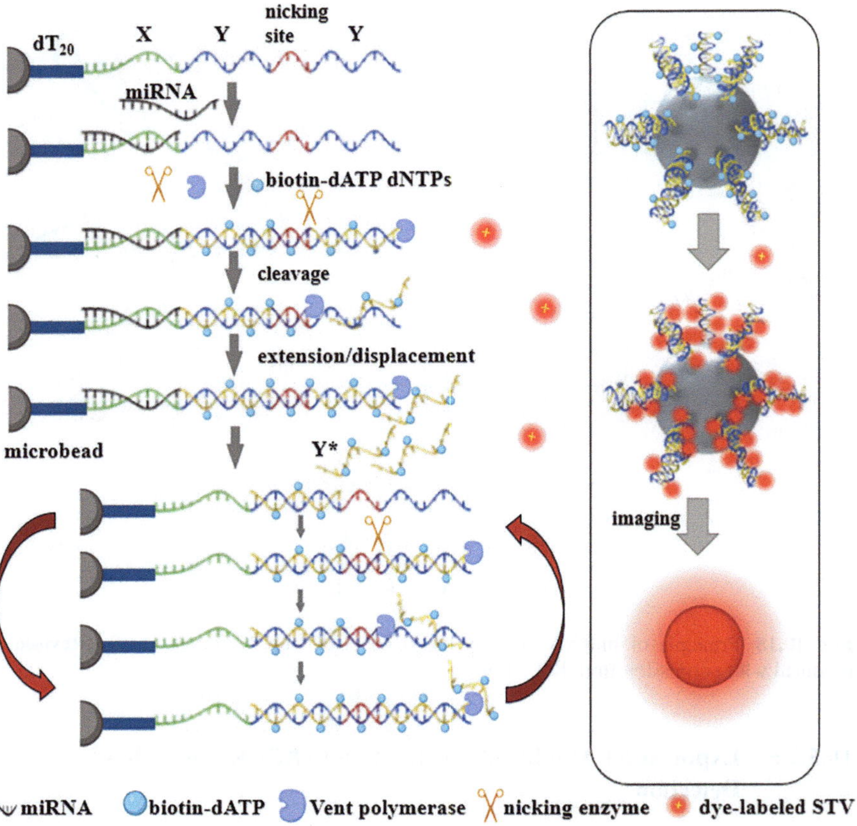

Fig. 10.19 Schematic diagram of EXPAR-based single microbead fluorescence enrichment platform for miRNA detection [151]. (Revised and reprinted with permission from Ref. [151])

microspheres are imaged and analyzed under a confocal fluorescence microscope. By detecting the target miRNA with a single magnetic microsphere, this method not only achieves enrichment of the target but also enriches the fluorescence signal, allowing miRNA analysis to reach a single-molecule level (1 aM in a 5 μL system, equivalent to 3 molecules). The method also exhibits good versatility and selectivity, enabling accurate quantitative detection of miRNA in single cells.

In addition, various specific EXPAR designs combined with multiple detection methods are used for miRNA detection, as detailed in Chap. 5.

10.5.2.6 Rolling Circle Amplification (RCA) for miRNA Detection

Cheng Yongqiang et al. [152] used miRNA as a template, linked the padlock probe with T4 RNA ligase 2, and directly initiated RCA using miRNA as a primer. The design process is shown in Fig. 10.20. By adding a second primer complementary to

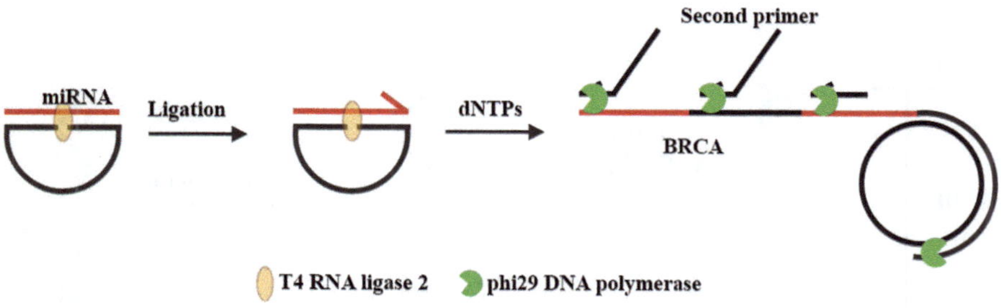

Fig. 10.20 Sensitive detection of miRNA based on BRCA [152]. (Revised and reprinted with permission from Ref. [152])

the RCA product, branched RCA amplification (BRCA) is triggered, generating a large number of amplification products of different lengths. The fluorescent dye SYBR Green I can be used for homogeneous and sensitive detection of miRNA.

Fan Chunhai's research group [153] designed a dumbbell-shaped probe-mediated RCA strategy for highly sensitive and selective miRNA detection. This method utilizes a dumbbell-shaped probe with two stem-loop structures as recognition elements for highly sensitive miRNA-specific detection. As shown in Fig. 10.21, the dumbbell-shaped probe consists of three regions: the miRNA-binding domain (MBD), the SG-binding domain (SGBD), and the loop region. The target miRNA specifically binds to the MBD region of the dumbbell-shaped probe, with two fragments in this region designed to be complementary to the target miRNA. The SGBD is a double-stranded region where SG can bind and produce a strong fluorescent signal. The loop region serves as a spacer. In the presence of miRNA, after the miRNA binds to the MBD region of the dumbbell-shaped probe, it forms a double-strand with a nick, which is then ligated into a closed loop under the action of T4 DNA ligase. Subsequently, using miRNA as a primer, phi29 DNA polymerase initiates linear RCA amplification, producing long DNA chains with repeated SGBD structures. The RCA products containing SGBD can be quantified by the fluorescence intensity of SG. The dumbbell probe is a multifunctional DNA probe that integrates target binding, amplification, and signal detection, enhancing both the sensitivity and specificity of miRNA detection. This method can detect miRNA as low as 1 fM.

Hong et al. [154] combined a GO-based fluorescence analysis method with an RCA-based miRNA amplification strategy for quantitative miRNA detection. The principle is shown in Fig. 10.22. In the presence of target miRNA, a locked probe complementary to the target miRNA is specifically ligated to form a circular DNA, which then serves as a template for RCA. Using miRNA as a primer, phi29 DNA polymerase initiates the RCA reaction, generating a single-stranded DNA containing multiple target miRNA sequences. Fluorescently labeled peptide nucleic acids (F-PNA) are designed to complement the target miRNA sequence and can anneal to multiple sites on the long single-stranded RCA product to form double-stranded structures. These structures rarely adsorb onto the GO monolayer, weakening the fluorescence quenching effect of GO on F-PNA. As a result, the fluorescence is in

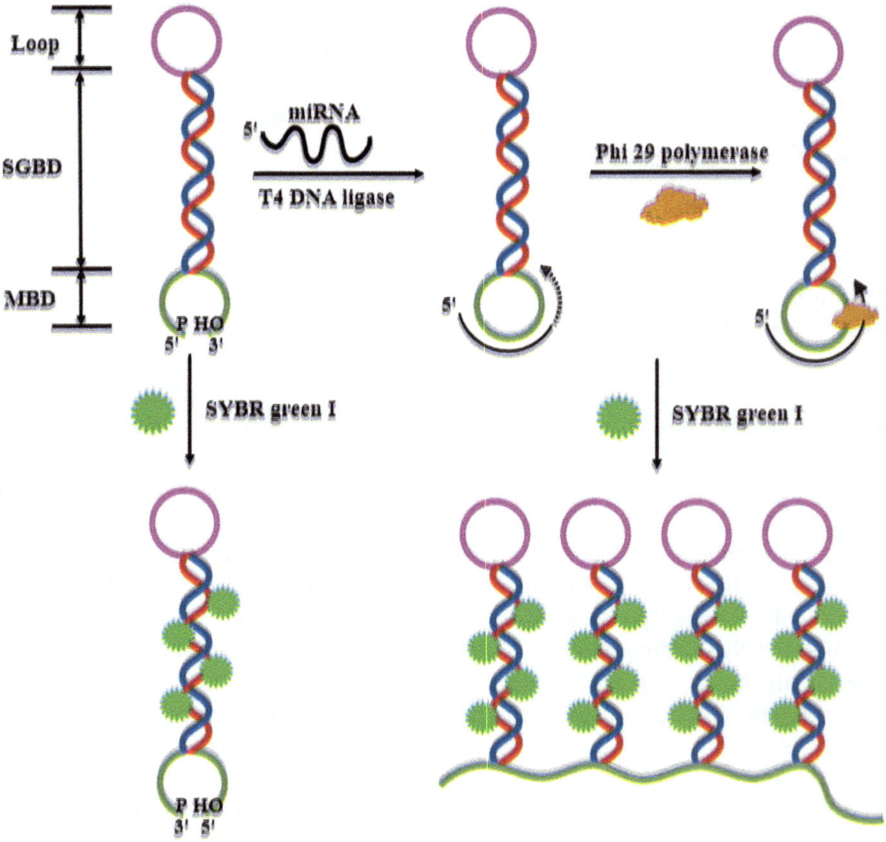

Fig. 10.21 Schematic diagram of dumbbell probe-mediated RCA for miRNA detection [153]. (Revised and reprinted with permission from Ref. [153])

the "on" state, and the fluorescence signal of the system can be detected. In the absence of target miRNA, the locked probe cannot be ligated into a circle, and no corresponding RCA amplification product is generated. The free single-stranded F-PNA probe completely adsorbs to the GO surface, resulting in fluorescence quenching. The fluorescence is in the "off" state, and no fluorescence signal is generated in the system. Therefore, by adding GO to the reaction system, miRNA can be quantitatively detected by simply measuring the fluorescence signal. This method can detect miRNA as low as 0.4 pM.

Additionally, other RCA-based miRNA detection methods have also been reported [155–158]. For example, an RCA-assisted CRISPR/Cas9 cleavage detection platform has been used for the highly specific detection of various extracellular vesicle miRNAs, a locked probe-based exponential RCA for ultrasensitive miRNA detection, and a toehold-mediated strand displacement process triggering RCA for in situ imaging analysis of miRNA in single cells (see Chap. 4 for details).

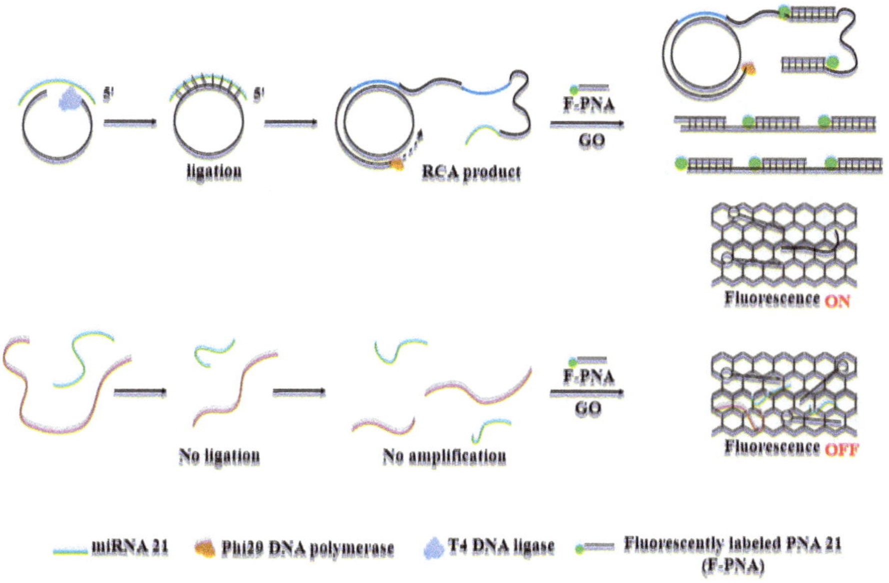

Fig. 10.22 Schematic illustration of the miRNA detection process using the GO-based fluorometric assay combined with RCA-based miRNA amplification [154]. (Revised and reprinted with permission from Ref. [154])

10.5.2.7 Loop-Mediated Isothermal Amplification (LAMP) for miRNA Detection

Li et al. [159] designed miRNA as the initial forward primer for LAMP, allowing it to play the role of F3 and initiating LAMP through the miRNA extension reaction, successfully applying the LAMP method for miRNA detection. The detection principle is shown in Fig. 3.9a of Chap. 3. However, the efficiency of the synergistic hybridization and extension of multiple primers with long templates in this method is low, significantly reducing the efficiency of the LAMP reaction, leading to lower sensitivity of the miRNA-induced LAMP reaction, which fails to meet the detection requirements for low-abundance miRNAs in samples. Sun et al. [160] developed a novel target-triggered loop-mediated isothermal amplification reaction (TT-LAMP) for one-step miRNA analysis and detection, greatly simplifying probe design and significantly improving detection sensitivity. TT-LAMP completes miRNA detection in just one step. For details on the principle, see Chap. 3, Fig. 3.9b.

Gao et al. [161] established a stem-loop primer-based loop-mediated isothermal amplification (LAMP) reaction for detecting miRNA in single cells. The principle is shown in Fig. 10.23. A pair of stem-loop primers, SLRT and SLE, was specifically designed. The 3′ end of SLRT, with protruding single-stranded DNA, can stably hybridize with the 3′ end of miRNA, and the reverse transcription reaction is catalyzed by ProtoScript II reverse transcriptase to generate cDNA. The 3′ end of SLE′s single-stranded DNA can stably hybridize with the 3′ end of the generated

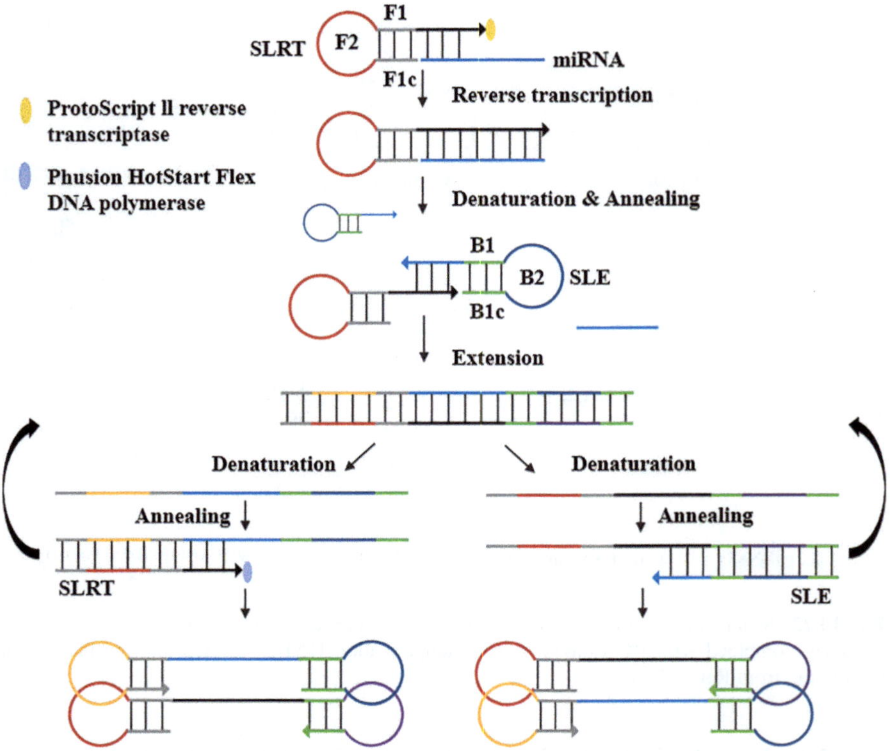

Fig. 10.23 Schematic diagram of stem-loop primers-based LAMP for detection [161]. (Revised and reprinted with permission from Ref. [161])

cDNA. Subsequently, the cDNA and SLE undergo an extension reaction simultaneously in the presence of Phusion HotStart Flex DNA polymerase and dNTPs, forming double-stranded DNA. Then, SLRT and SLE primers are used to perform thermal cycling (denaturation, annealing, and extension), exponentially amplifying and generating more double-stranded DNA. The two strands of this double-stranded DNA can maintain the double stem-loop structure through self-hybridization at their 3′ and 5′ ends. Thus, a target miRNA can generate multiple double stem-loop structure DNAs, and each double stem-loop structure DNA molecule can trigger a LAMP amplification reaction, producing large quantities of double-stranded DNA of varying lengths, with the final product monitored in real-time using SYBR Green I. By combining two-step exponential amplification, this method achieves ultrahigh sensitivity, detecting miRNA as low as 1 aM, and enables simple, rapid, and homogeneous solution-based analysis of miRNA in single cells. Moreover, the dual recognition of miRNA targets by the two stem-loop primers confers good specificity, allowing the method to distinguish single-base differences in miRNA sequences.

10.5.2.8 Nicking Endonuclease-Mediated Amplification (NEMA) for miRNA Detection

Nicking endonucleases are a type of specific nucleic acid endonuclease, similar to restriction endonucleases, that recognize short specific DNA sequences and cleave DNA at fixed positions relative to the recognition sequence. However, unlike restriction endonucleases, nicking endonucleases cleave only one strand of the DNA, resulting in "nicked" DNA molecules rather than complete strand breaks [162, 163]. Shi et al. [164] developed an isothermal exponential amplification method for detecting miRNA with the assistance of a nicking endonuclease. The principle is shown in Fig. 10.24. Both Primer 1 and Primer 2 contain recognition sites for the nicking endonuclease *Nt.AlwI* (GGATC). When miRNA is absent, the recognition sites on Primer 1 and Primer 2 are single-stranded and cannot be cleaved by *Nt.AlwI*. Once the target miRNA hybridizes with the 3′ end of Primer 1, Primer 1 and miRNA each undergo an extension reaction in the presence of Klenow DNA polymerase (exo−), forming double-stranded DNA fragments. The *Nt.AlwI* recognition site on Primer 1 is restored to a double-stranded state, allowing *Nt.AlwI* to cleave at the fourth random nucleotide downstream of the recognition site, creating a nick. The enzyme extends at the nick and displaces the downstream sequence. This repeated cycle of cleavage, polymerization, and displacement forms the first amplification cycle, generating a large amount of single-stranded DNA

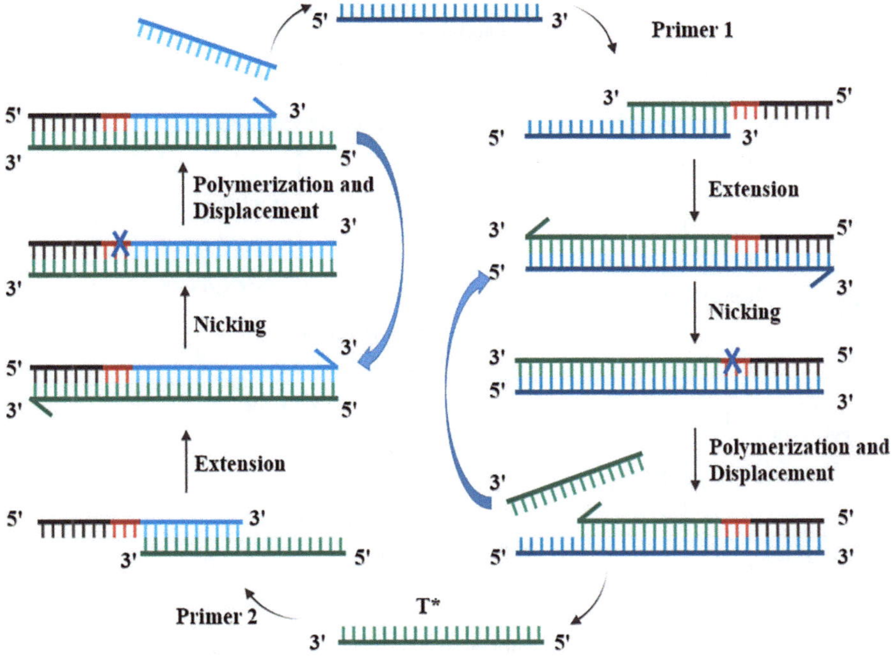

Fig. 10.24 Schematic diagram of NEMA for miRNA detection [164]. (Revised and reprinted with permission from Ref. [164])

(T*) complementary to the target miRNA. T* can hybridize with the 3′ end of Primer 2, and in the presence of DNA polymerase, it extends to form double-stranded DNA. Similarly, the 5′ end of Primer 2 also generates a recognition site for *Nt.AlwI*, triggering a second amplification cycle. The displaced single-stranded DNA (T) has the same sequence as the target miRNA and can serve as the starting template for the first amplification cycle. These two amplification cycles continue to produce T and T*, which hybridize and form double-stranded DNA detected by SYBR Green I. This method can detect as low as 16 zmol of target miRNA.

10.5.2.9 Helicase-Dependent Amplification (HDA) for miRNA Detection

Zhang Chunyang's research group [165] successfully applied HDA (Helicase-Dependent Amplification) to the detection of miRNA. The principle is shown in Fig. 10.25. The target miRNA hybridizes with the 3′ end of the probe to form a DNA/RNA duplex. The probe hybridized with miRNA is protected from digestion by the exonuclease Exo I, while the unhybridized probe is digested by Exo I. After heat-inactivating Exo I, the forward primer hybridizes with the probe, and in the presence of polymerase, it extends to form double-stranded DNA. Helicase can unwind the double-stranded DNA to form a single-stranded template, and SSB

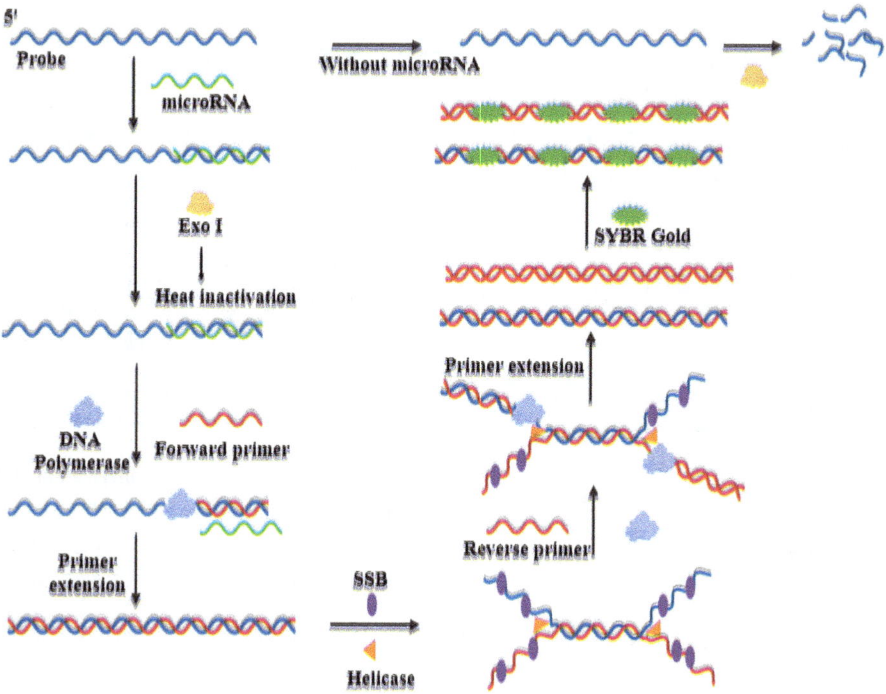

Fig. 10.25 Schematic diagram of HDA for miRNA detection [165]. (Revised and reprinted with permission from Ref. [165])

(single-strand binding protein) stabilizes the single-stranded template, allowing both the forward and reverse primers to hybridize with their respective templates. Under the action of DNA polymerase, both primers can extend to form new double-stranded DNA, which triggers the next round of reaction and leads to exponential amplification. SYBR Gold is used as a fluorescent dye to monitor the HDA process. When the target miRNA is absent, the probe is completely digested by Exo I, preventing the initiation of the HDA reaction and signal amplification. This method can detect miRNA in the range of 100 fM to 10 nM.

10.5.2.10 Enzymatic Signal Amplification Strategy for miRNA Detection

Duan et al. [166] developed an ultrasensitive method for detecting miRNA at the single-cell level by combining a nicking endonuclease with molecular beacon technology, known as hairpin-mediated quadratic enzymatic amplification (HQEA). As shown in Fig. 10.26, when the target miRNA is present, it hybridizes

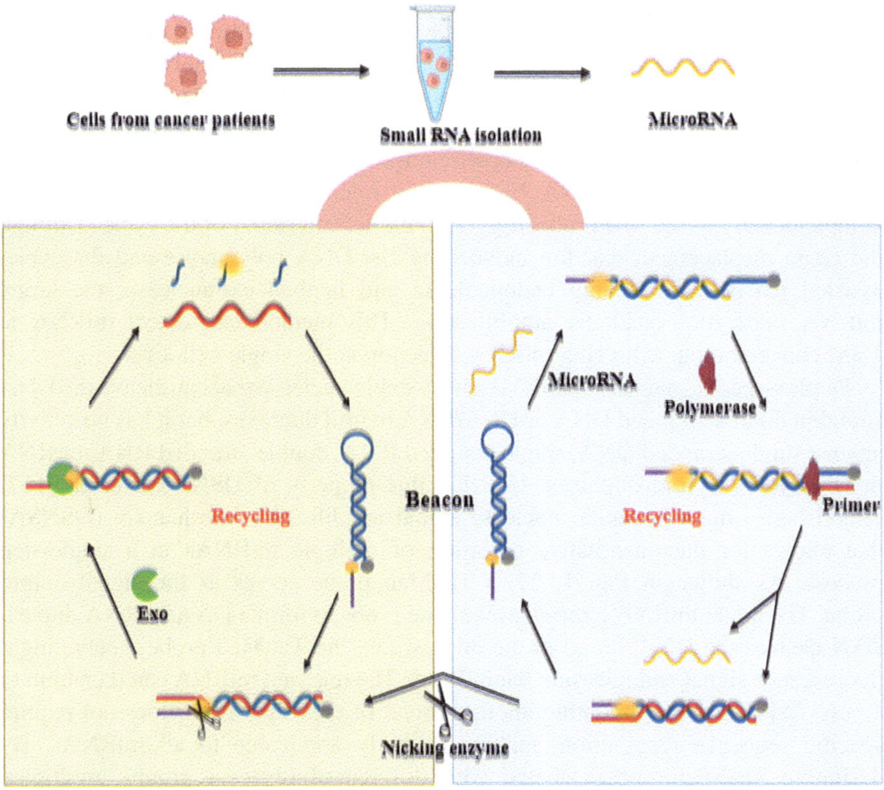

Fig. 10.26 Schematic diagram of HQEA for miRNA detection [166]. (Revised and reprinted with permission from Ref. [166])

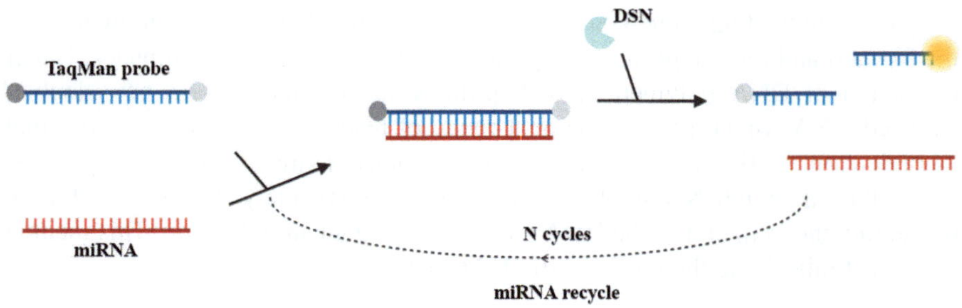

Fig. 10.27 Schematic diagram of miRNA detection based on DSN combined with TaqMan probe [167]. (Revised and reprinted with permission from Ref. [167])

with the loop region of the molecular beacon (MB) probe, opening the MB and causing a conformational change that generates a fluorescence signal. The primer then anneals to the opened stem, and under the action of Bst DNA polymerase, extends using the MB probe as a template to form double-stranded DNA, replacing the target miRNA. At this point, the molecular beacon remains in its activated state, and the released target miRNA can bind with another MB probe, initiating a new cycle. With each cycle, another MB probe is activated, forming a double-stranded structure and releasing the target miRNA again. The new double-stranded MB structure formed in the first round contains a recognition site for the nicking endonuclease. The cleavage of the beacon exposes a recognition site for lambda exonuclease, which catalyzes the stepwise removal of single nucleotides from the 5′ end of the double-stranded DNA. Each reaction releases a new single-stranded DNA complementary to the MB probe, triggering the second round of the cycle. Through the chain displacement reaction induced by Bst DNA polymerase and the cycle-assisted reaction by nicking endonuclease and lambda exonuclease, the target miRNA undergoes quadratic amplification. This method can detect miRNA at 1 aM concentration, achieving miRNA detection at the single-cell level.

Duplex-specific nuclease (DSN) is a heat-stable nuclease that can cleave the DNA strand in double-stranded DNA or DNA/RNA hybrid duplexes, but it has no activity against single-stranded DNA, single-stranded RNA, double-stranded RNA, or RNA in DNA/RNA hybrid duplexes. Utilizing this property of DSN, Yin et al. [167] established a duplex-specific nuclease signal amplification mechanism (DSNSA) that allows for the quantitative detection of multiple miRNAs in a single-step process. As shown in Fig. 10.27, a TaqMan probe serves as the signal output probe. The target miRNA hybridizes with the probe to form a DNA/miRNA duplex. DSN cleaves the DNA strand in the duplex, i.e., the TaqMan probe, generating a fluorescence signal and releasing the miRNA. The released miRNA can then bind to a new TaqMan probe, initiating another round of cleavage. DSN does not require specific sequence recognition, making it easily applicable to all miRNAs. By designing TaqMan probes labeled with different fluorescent groups, multiplex miRNA detection can be achieved. Similarly, Li et al. [168] used molecular beacon

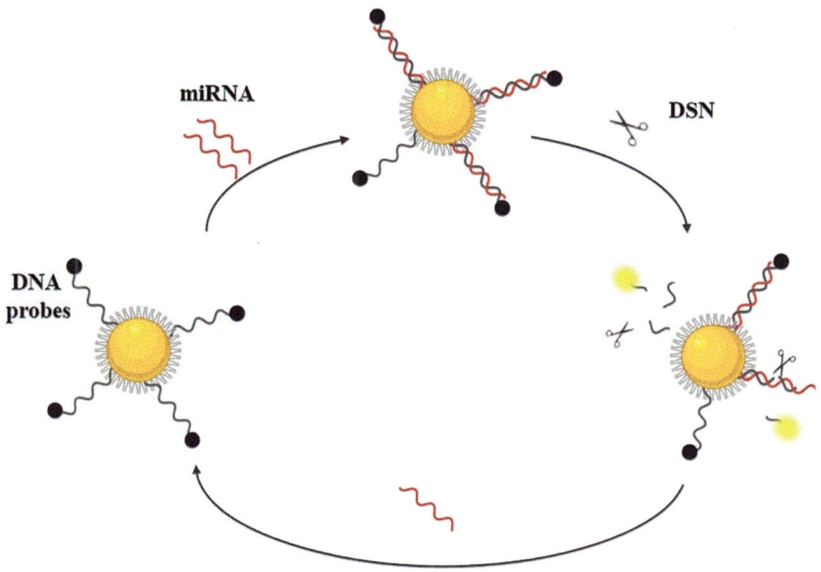

Fig. 10.28 Schematic diagram of miRNA detection based on DSN combined with AuNPs [169]. (Revised and reprinted with permission from Ref. [169])

probes instead of TaqMan probes as the signal output probe for the quantitative detection of miRNAs.

Degliangeli et al. [169] immobilized fluorescently labeled DNA probes onto AuNPs, where the fluorescence group is effectively quenched due to its proximity to the AuNPs. As shown in Fig. 10.28, the target miRNA binds to the DNA probe on the AuNPs, forming a DNA-RNA hybrid duplex. DSN then cleaves the DNA strand in the duplex, releasing the fluorescent group and generating a fluorescence signal. After the cleavage, the target miRNA remains intact, and the released miRNA can bind to a new DNA probe, initiating a cycle of enzymatic cleavage and significantly enhancing the fluorescence signal. By utilizing the strong quenching effect of AuNPs, this method only requires a single probe labeled with a fluorescence group to detect miRNA. This method can detect miRNA as low as 5 pM.

To further reduce detection costs, Ma et al. [170] developed a DSN-assisted target cycling and pyrene molecular switch-based fluorescence method for miRNA detection, as shown in Fig. 10.29. Pyrene molecules are a special type of fluorescence dye with spatial sensitivity: pyrene monomers have two characteristic emission peaks at 375 nm and 400 nm; however, when two pyrene molecules are in close proximity, they can form an excimer complex with a fluorescence emission peak at 485 nm. In this method, a specific linear probe that recognizes the target miRNA and a hairpin probe used for generating fluorescence signals are designed (Fig. 10.29a). The ends of the hairpin probe are labeled with pyrene molecules as reporters. The formation of the hairpin structure brings the two pyrene molecules close together, forming the pyrene excimer complex, which emits strongly at 485 nm. In the absence of the

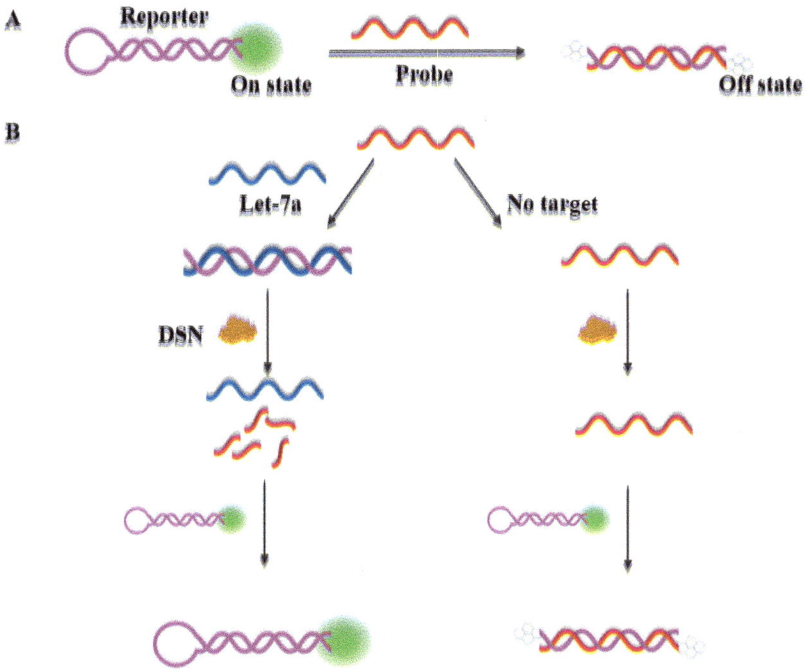

Fig. 10.29 Schematic diagram of miRNA detection based on DSN combined with pyrene excimer [170]. (Revised and reprinted with permission from Ref. [170])

target miRNA, the linear probe remains intact and can hybridize with the loop region of the hairpin probe, opening the hairpin structure and causing the two pyrene molecules to spatially separate, preventing the formation of the pyrene excimer complex and the corresponding fluorescence emission, switching from the "on" to the "off" state. When the target miRNA is present, it hybridizes with the linear probe to form a DNA/RNA duplex. The DNA probe is cleaved by the DSN enzyme, releasing let-7a. The released let-7a can hybridize with another probe, triggering a large cycle of probe cleavage. As a result, the DNA probe cannot open the hairpin structure, keeping the fluorescence in the "on" state. Pyrene molecules have a large Stokes shift (~130 nm), which avoids self-quenching even at higher concentrations. Therefore, compared to fluorescence-labeled probe-based methods, this approach effectively eliminates false positives caused by incomplete quenching.

DNase I is a nonspecific nuclease that can cleave DNA, producing dinucleotide, trinucleotide, and oligonucleotide products with 3′-hydroxyl and 5′-phosphate ends. DNase I acts on single-stranded DNA, double-stranded DNA, chromatin, and RNA: DNA hybrid chains, but does not affect RNA. Cui et al. [171] developed a cyclic enzymatic cleavage signal amplification method for detecting miRNA, based on the characteristics of DNase I combined with GO. As shown in Fig. 10.30, in the absence of miRNA, different fluorescence-labeled probes remain in a single-stranded state and adsorb onto the GO surface. Due to the strong quenching effect

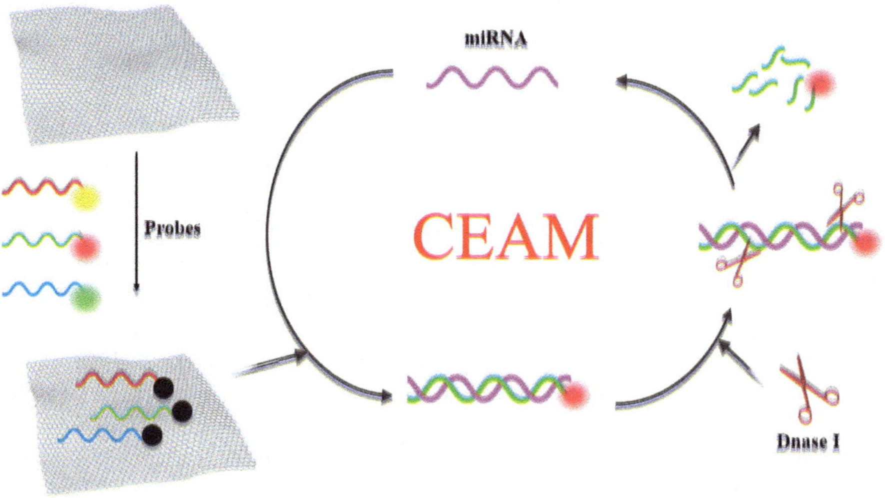

Fig. 10.30 Schematic diagram of DNase I-assisted cyclic enzymatic amplification method for multiplex miRNA analysis based on GO-protected DNA probes [171]. (Revised and reprinted with permission from Ref. [171])

of GO on the dye, the probes do not exhibit fluorescence. When a specific miRNA is added, the miRNA/DNA complex dissociates from the GO surface, and the fluorescence of the probes is restored. More importantly, the released complex immediately becomes a substrate for DNase I, which digests the DNA component, releasing miRNA that can bind to another probe on the GO surface and start the next round of cleavage. This cyclic reaction continues, restoring the fluorescence of the probe's fluorophore, resulting in a significant fluorescence signal amplification. GO is an efficient quencher for almost all fluorophores and can quench various dyes simultaneously. Based on this property, different fluorescence-labeled probes can be designed for multi-analyte miRNA detection. This method can detect miRNA in the range of 20 pM to 1 nM and can differentiate single-base differences. In addition to DNase I, the group also developed miRNA detection methods based on Exo III [172] and T7 exonuclease-assisted [173] cyclic enzymatic cleavage signal amplification.

10.6 Application of miRNA in Disease Diagnosis

Early diagnosis is crucial for many common and major diseases. With the rapid development of molecular biology, numerous biomarkers have been widely applied in disease monitoring, screening, diagnosis, and prognostic analysis. As potential therapeutic targets, highly sensitive detection of biomarkers is of great significance for the advancement of precision medicine. A large body of research has shown that miRNAs play a "hub" role in gene regulatory networks, and their expression reflects information dispersed across hundreds or even thousands of target genes. In mammals, dysregulation of miRNAs is associated with a variety of diseases, ranging from cancer to autoimmune disorders, and plays a significant role in the pathogenesis of these diseases.

The expression patterns of miRNAs are tissue-specific and exhibit differences between disease and normal tissues. Certain miRNAs are specifically upregulated or downregulated in diseases. This specificity, along with their remarkable stability across various sample types, such as formalin-fixed paraffin-embedded tissues (FFPET), plasma, and other body fluids like urine, saliva, semen, amniotic fluid, and pleural effusion, makes miRNAs robust and reproducible biomarkers for disease diagnosis. By assessing the expression levels of multiple miRNAs, accurate predictions can be made regarding the prognosis of various diseases [174–176].

MiRNAs can be secreted by cells and are detectable in a wide range of body fluids, including blood, saliva, and urine. As a result, circulating miRNAs are candidate biomarkers for almost all human diseases, including cancer, in a noninvasive manner. The abnormal expression levels of miRNAs have been evaluated in various cancers. In 2008, Lawrie et al. [177] discovered that serum miR-21 levels were associated with the diagnosis and prognosis of patients with diffuse large B-cell lymphoma (DLBCL). Fayyad-Kazan et al. [178] found that, compared to healthy controls, two important plasma miRNAs (miR-150 and miR-342) were upregulated in patients with acute myeloid leukemia (AML), suggesting that these miRNAs could serve as potential and specific biomarkers for early AML diagnosis. Studies have also indicated that the levels of circulating miR-195, miR-29a, and miR-222 are good indicators for distinguishing chronic lymphocytic leukemia (CLL) patients from healthy controls [179]. Non-small cell lung cancer (NSCLC), the most common type of lung cancer, can be diagnosed at different stages through noninvasive extracellular circulating miRNAs. Two effective serum miRNAs (miR-25 and miR-223) have been reported to be involved in the carcinogenesis process and are significantly elevated in NSCLC patients, suggesting their potential use for early NSCLC diagnosis [100]. In other cancers, circulating miRNAs have proven to be useful diagnostic/prognostic markers. Table 10.4 summarizes several commonly used miRNA databases related to human diseases.

The identification of tumor subtypes is crucial for patient treatment and survival, and the distinct expression patterns of miRNAs across various tumor types can aid in this classification. Breast cancer, being a highly heterogeneous disease with diverse molecular alterations, is classified into several intrinsic subtypes, including Luminal (Luminal A and Luminal B), Basal, HER2, and Normal types. Studies have shown that aberrant miRNA expression is closely associated with these molecular subtypes of breast cancer. In plasma, a combination of four downregulated miRNAs (miR-29a, miR-181a, miR-223, miR-652) and three upregulated miRNAs (miR-29a, miR-181a, miR-652) has been significantly linked to the Luminal A breast cancer subtype [185]. MiRNA-342 is highly expressed in the Luminal B subtype and is considered a highly sensitive and specific biomarker for this breast cancer type [186]. Furthermore, miRNAs play an essential role in the classification

Table 10.4 Several commonly used disease-associated miRNA databases

Database	Description	Version/ Update date	Website	References
dbDEMC	A comprehensive database for storing and displaying differentially expressed miRNAs detected in cancer through both high-throughput and low-throughput methods	dbDEMC 3.0 (2021)	https://www.biosino.org/dbDEMC/index	[180]
HMDD	The Human microRNA Disease Database (HMDD) is a database that compiles experimentally supported evidence of associations between human microRNAs and diseases	HMDD 3.2 (2019)	http://www.cuilab.cn/hmdd	[181]
miR2Disease	A manually curated database, aims at providing a comprehensive resource of microRNA deregulation in various human diseases	miR2Disease (Founded in 2008)	http://www.mir2disease.org/	[182]
miRcancer	A database provides comprehensive collection of miRNA expression profiles in various human cancers, which are automatically extracted from published literature in PubMed. It utilizes text mining techniques for information collection. Manual revision is applied after auto-extraction to provide 100% precision	miRcancer (Founded in 2013, updated in 2020)	http://mircancer.ecu.edu/	[183]
PhenomiR	A database provides information about differentially regulated miRNA expression in diseases and other biological processes	PhenomiR 2.0 (2011)	http://mips.helmholtz-muenchen.de/phenomir	[184]

of other cancers, including renal cancer [187], lung cancer [188], and lymphoma [189].

In addition to cancer, miRNAs are also pivotal in the early diagnosis of cardiovascular diseases (CVDs) [190], including acute myocardial infarction (AMI), hypertrophy, heart failure, arrhythmias, and atherosclerosis. Tijsen et al. [191] confirmed that blood levels of miR-423-5p are elevated in heart failure patients (independent of age and gender), suggesting it as an important and sensitive cardiac biomarker. D′Alessandra et al. [192] found that after the onset of AMI-induced inflammation, plasma levels of miR-1, miR-133a/b were increased, while plasma levels of miR-122 and miR-375 were lower in STEMI patients compared to healthy controls. Takotsubo cardiomyopathy (CTT), also known as "broken heart syndrome," is a nonischemic cardiomyopathy unrelated to AMI, and a severe,

Table 10.5 miRNAs as potential diagnostic biomarkers for diseases

Cancer type	miRNA	Sample source	Expression level	References
Cardiovascular disease	miR-423-5p	Whole blood	Upregulation	[196]
	miR-1, miR-133a/b	Exosomes	Upregulation	[192]
	miR-122, miR-375	Plasma	Upregulation	
	miR-21, miR-1, miR-26, miR-133, miR-30, miR-328, miR-499	Plasma	Upregulation	
	miR-1, miR-16, miR-26a, miR-133a	Serum	Upregulation	[193]
Diabetes	miR-21, miR-210	Urine and plasma	Upregulation	[194]
	miR-29a	Urine	Upregulation	[197]
	miR-223, miR-146a	Plasma, platelets	Downregulation	[198]
	miR-130b	Whole blood	Downregulation	[195]
AIDS	miR-29b-3p, miR-33a-5p, miR-146a-5p	Plasma	Downregulation	[199]
	miR-29b	Plasma	Upregulation	[200]
	miR-29c	Plasma	Downregulation	
	miR-150, miR-146b-5p	Plasma	Differential expression	[201]
	miR-21	Plasma	Upregulation	[202]
	miR-122	Plasma	Upregulation	[203]
Medicolegal expertise	miR-451, miR-16, miR-135b, miR-10b, miR-658, miR-205, miR-124a, miR-372, miR-412	Blood, semen, saliva, vaginal secretions, menstrual blood	Differential expression	[204]

life-threatening condition. Jaguszewski et al. [193] discovered that a unique expression profile consisting of miR-1, miR-16, miR-26a, and miR-133a could not only differentiate CTT patients from healthy individuals but also distinguish them from AMI patients. Research has shown that circulating miRNAs are linked to Type 1 and Type 2 diabetes, as well as gestational diabetes and its complications. Compared to healthy controls, plasma and urine levels of miR-21 and miR-210 were significantly elevated in Type 1 diabetes patients, while miR-126 levels in urine were significantly reduced [194]. Luo et al. [195] confirmed that platelet-derived miR-103b is specifically downregulated in early-stage Type 2 diabetes. MiRNAs are not only associated with the diseases mentioned above but also serve as potential biomarkers in a wide range of other diseases. Table 10.5 summarizes the abnormal expression and characteristics of miRNAs in various diseases.

An ideal biomarker should be sensitive, specific, robust, and noninvasive. Circulating free miRNAs and tissue miRNAs have emerged as typical biomarkers, offering new platforms and approaches for disease diagnosis and therapy. The combination of multiple miRNAs can enhance the sensitivity and specificity of biomarkers, thereby improving the accuracy of disease diagnosis. Despite the encouraging progress made, several significant theoretical and technical challenges remain to be addressed. For instance, more basic research is needed to further explore the roles and functions of miRNAs in cancer, cardiovascular diseases, and other conditions. Furthermore, before being translated into clinical practice, all circulating miRNAs used as biomarkers must undergo additional validation steps, along with precise standardization of pre- and post-analysis procedures to control potential technical biases.

References

1. Lee RC, Feinbaum RL, Ambros V. The *C. elegans* heterochronic gene lin-4 encodes small RNAs with antisense complementarity to lin-14. Cell. 1993;75:843–54.
2. Wightman B, Ha I, Ruvkun G. Posttranscriptional regulation of the heterochronic gene lin-14 by lin-4 mediates temporal pattern formation in *C. elegans*. Cell. 1993;75:855–62.
3. Reinhart BJ, Slack FJ, Basson M, Pasquinelli AE, Bettinger JC, Rougvie AE, Horvitz HR, Ruvkun G. The 21-nucleotide let-7 RNA regulates developmental timing in *Caenorhabditis elegans*. Nature. 2000;403:901–6.
4. Lagos-Quintana M, Rauhut R, Lendeckel W, Tuschl T. Identification of novel genes coding for small expressed RNAs. Science. 2001;294:853–8.
5. Lau NC, Lim LP, Weinstein EG, Bartel DP. An abundant class of tiny RNAs with probable regulatory roles in *Caenorhabditis elegans*. Science. 2001;294:858–62.
6. Lee RC, Ambros V. An extensive class of small RNAs in *Caenorhabditis elegans*. Science. 2001;294:862–4.
7. Griffiths-Jones S. The microRNA registry. Nucleic Acids Res. 2004;32(Database issue): D109–11.
8. Griffiths-Jones S, Grocock RJ, van Dongen S, Bateman A, Enright AJ. miRBase: microRNA sequences, targets and gene nomenclature. Nucleic Acids Res. 2006;34(Database issue): D140–4.
9. Griffiths-Jones S, Saini HK, van Dongen S, Enright AJ. miRBase: tools for microRNA genomics. Nucleic Acids Res. 2008;36(Database issue):D154–8.
10. Kozomara A, Griffiths-Jones S. miRBase: integrating microRNA annotation and deep-sequencing data. Nucleic Acids Res. 2011;39(Database issue):D152–7.
11. Kozomara A, Griffiths-Jones S. miRBase: annotating high confidence microRNAs using deep sequencing data. Nucleic Acids Res. 2014;42(Database issue):D68–73.
12. Kozomara A, Birgaoanu M, Griffiths-Jones S. miRBase: from microRNA sequences to function. Nucleic Acids Res. 2019;47(D1):D155–62.
13. Hsu SD, Lin FM, Wu WY, Liang C, Huang WC, Chan WL, Tsai WT, Chen GZ, Lee CJ, Chiu CM, Chien CH, Wu MC, Huang CY, Tsou AP, Huang HD. miRTarBase: a database curates experimentally validated microRNA-target interactions. Nucleic Acids Res. 2011;39(Database issue):D163–9.
14. Chou CH, Chang NW, Shrestha S, Hsu SD, Lin YL, Lee WH, Yang CD, Hong HC, Wei TY, Tu SJ, Tsai TR, Ho SY, Jian TY, Wu HY, Chen PR, Lin NC, Huang HT, Yang TL, Pai CY, Tai CS, Chen WL, Huang CY, Liu CC, Weng SL, Liao KW, Hsu WL, Huang HD. miRTarBase 2016: updates to the experimentally validated miRNA-target interactions database. Nucleic Acids Res. 2016;44(D1):D239–47.
15. Chou CH, Shrestha S, Yang CD, Chang NW, Lin YL, Liao KW, Huang WC, Sun TH, Tu SJ, Lee WH, Chiew MY, Tai CS, Wei TY, Tsai TR, Huang HT, Wang CY, Wu HY, Ho SY, Chen PR, Chuang CH, Hsieh PJ, Wu YS, Chen WL, Li MJ, Wu YC, Huang XY, Ng FL, Buddhakosai W, Huang PC, Lan KC, Huang CY, Weng SL, Cheng YN, Liang C, Hsu WL,

Huang HD. miRTarBase update 2018: a resource for experimentally validated microRNA-target interactions. Nucleic Acids Res. 2018;46(D1):D296–302.
16. Huang HY, Lin YC, Li J, Huang KY, Shrestha S, Hong HC, Tang Y, Chen YG, Jin CN, Yu Y, Xu JT, Li YM, Cai XX, Zhou ZY, Chen XH, Pei YY, Hu L, Su JJ, Cui SD, Wang F, Xie YY, Ding SY, Luo MF, Chou CH, Chang NW, Chen KW, Cheng YH, Wan XH, Hsu WL, Lee TY, Wei FX, Huang HD. miRTarBase 2020: updates to the experimentally validated microRNA-target interaction database. Nucleic Acids Res. 2020;48(D1):D148–54.
17. Jeggari A, Marks DS, Larsson E. miRcode: a map of putative microRNA target sites in the long non-coding transcriptome. Bioinformatics. 2012;28(15):2062–3.
18. Yang JH, Li JH, Shao P, Zhou H, Chen YQ, Qu LH. starBase: a database for exploring microRNA-mRNA interaction maps from Argonaute CLIP-Seq and Degradome-Seq data. Nucleic Acids Res. 2011;39(Database issue):D202–9.
19. Li JH, Liu S, Zhou H, Qu LH, Yang JH. starBase v2.0: decoding miRNA-ceRNA, miRNA-ncRNA and protein-RNA interaction networks from large-scale CLIP-Seq data. Nucleic Acids Res. 2014;42(Database issue):D92–7.
20. Lewis BP, Burge CB, Bartel DP. Conserved seed pairing, often flanked by adenosines, indicates that thousands of human genes are microRNA targets. Cell. 2005;120:15–20.
21. Agarwal V, Bell GW, Nam JW, Bartel DP. Predicting effective microRNA target sites in mammalian mRNAs. elife. 2015;4:e05005.
22. Krek A, Grün D, Poy MN, Wolf R, Rosenberg L, Epstein EJ, MacMenamin P, da Piedade I, Gunsalus KC, Stoffel M, Rajewsky N. Combinatorial microRNA target predictions. Nat Genet. 2005;37:495–500.
23. Chen K, Rajewsky N. Natural selection on human microRNA binding sites inferred from SNP data. Nat Genet. 2006;38:1452–6.
24. Maselli V, Di Bernardo D, Banfi S. CoGemiR: a comparative genomics microRNA database. BMC Genomics. 2008;9:457.
25. Zhang Z, Yu J, Li D, Zhang Z, Liu F, Zhou X, Wang T, Ling Y, Su Z. PMRD: plant microRNA database. Nucleic Acids Res. 2010;38(Database issue):D806–13.
26. Yi X, Zhang Z, Ling Y, Xu W, Su Z. PNRD: a plant non-coding RNA database. Nucleic Acids Res. 2015;43(Database issue):D982–9.
27. Szcześniak MW, Deorowicz S, Gapski J, Kaczyński Ł, Makalowska I. miRNEST database: an integrative approach in microRNA search and annotation. Nucleic Acids Res. 2012;40(Database issue):D198–204.
28. Szczesniak MW, Makalowska I. miRNEST 2.0: a database of plant and animal microRNAs. Nucleic Acids Res. 2014;42(Database issue):D74–7.
29. Bhartiya D, Laddha SV, Mukhopadhyay A, Scaria V. miRvar: a comprehensive database for genomic variations in microRNAs. Hum Mutat. 2011;32:E2226–45.
30. Russo F, Di Bella S, Vannini F, Berti G, Scoyni F, Cook HV, Santos A, Nigita G, Bonnici V, Laganà A, Geraci F, Pulvirenti A, Giugno R, De Masi F, Belling K, Jensen LJ, Brunak S, Pellegrini M, Ferro A. miRandola 2017: a curated knowledge base of non-invasive biomarkers. Nucleic Acids Res. 2018;46(D1):D354–9.
31. Liu T, Zhang Q, Zhang J, Li C, Miao YR, Lei Q, Li Q, Guo AY. EVmiRNA: a database of miRNA profiling in extracellular vesicles. Nucleic Acids Res. 2019;47(D1):D89–93.
32. Berezikov E. Evolution of microRNA diversity and regulation in animals. Nat Rev Genet. 2011;2:846–60.
33. Esquela-Kerscher A, Slack FJ. Oncomirs—microRNAs with a role in cancer. Nat Rev Cancer. 2006;6:259–69.
34. Guo L, Lu Z. The fate of miRNA* strand through evolutionary analysis: implication for degradation as merely carrier strand or potential regulatory molecule? PLoS One. 2010;5:e11387.
35. Suzuki HI, Katsura A, Yasuda T, Ueno T, Mano H, Sugimoto K, Miyazono K. Small-RNA asymmetry is directly driven by mammalian Argonautes. Nat Struct Mol Biol. 2015;22:512–21.

36. Moran Y, Agron M, Praher D, Technau U. The evolutionary origin of plant and animal microRNAs. Nat Ecol Evol. 2017;1:27.
37. Jones-Rhoades MW, Bartel DP, Bartel B. MicroRNAs and their regulatory roles in plants. Annu Rev Plant Biol. 2006;57:19–53.
38. Mallory AC, Vaucheret H. Functions of microRNAs and related small RNAs in plants. Nat Genet. 2006;38:S31–6.
39. Bartel DP. Metazoan microRNAs. Cell. 2018;173:20–51.
40. Huntzinger E, Izaurralde E. Gene silencing by microRNAs: contributions of translational repression and mRNA decay. Nat Rev Genet. 2011;12:99–110.
41. Bartel DP. MicroRNAs: genomics, biogenesis, mechanism, and function. Cell. 2004;116:281–97.
42. Brodersen P, Sakvarelidze-Achard L, Bruun-Rasmussen M, Dunoyer P, Yamamoto YY, Sieburth L, Voinnet O. Widespread translational inhibition by plant miRNAs and siRNAs. Science. 2008;320:1185–90.
43. Yekta S, Shih IH, Bartel DP. MicroRNA-directed cleavage of HOXB8 mRNA. Science. 2004;304:594–6.
44. Dragomir MP, Knutsen E, Calin GA. SnapShot: unconventional miRNA functions. Cell. 2018;174:1038–1038.e1.
45. Tay Y, Zhang J, Thomson AM, Lim B, Rigoutsos I. MicroRNAs to Nanog, Oct4 and Sox2 coding regions modulate embryonic stem cell differentiation. Nature. 2008;455:1124–8.
46. Fabian MR, Sonenberg N, Filipowicz W. Regulation of mRNA translation and stability by microRNAs. Annu Rev Biochem. 2010;79:351–79.
47. Lauressergues D, Couzigou JM, Clemente HS, Martinez Y, Dunand C, Bécard G, Combier JP. Primary transcripts of microRNAs encode regulatory peptides. Nature. 2015;520:90–3.
48. Eiring AM, Harb JG, Neviani P, Garton C, Oaks JJ, Spizzo R, Liu S, Schwind S, Santhanam R, Hickey CJ, Becker H, Chandler JC, Andino R, Cortes J, Hokland P, Huettner CS, Bhatia R, Roy DC, Liebhaber SA, Caligiuri MA, Marcucci G, Garzon R, Croce CM, Calin GA, Perrotti D. miR-328 functions as an RNA decoy to modulate hnRNP E2 regulation of mRNA translation in leukemic blasts. Cell. 2010;140:652–65.
49. Vasudevan S, Tong Y, Steitz JA. Switching from repression to activation: microRNAs can up-regulate translation. Science. 2007;318:1931–4.
50. Das S, Ferlito M, Kent OA, Fox-Talbot K, Wang R, Liu D, Raghavachari N, Yang Y, Wheelan SJ, Murphy E, Steenbergen C. Nuclear miRNA regulates the mitochondrial genome in the heart. Circ Res. 2012;110:1596–603.
51. Hwang HW, Wentzel EA, Mendell JT. A hexanucleotide element directs microRNA nuclear import. Science. 2007;315:97–100.
52. Tang R, Li L, Zhu D, Hou D, Cao T, Gu H, Zhang J, Chen J, Zhang CY, Zen K. Mouse miRNA-709 directly regulates miRNA-15a/16-1 biogenesis at the posttranscriptional level in the nucleus: evidence for a microRNA hierarchy system. Cell Res. 2012;22:504–15.
53. Mitra R, Adams CM, Jiang W, Greenawalt E, Eischen CM. Pan-cancer analysis reveals cooperativity of both strands of microRNA that regulate tumorigenesis and patient survival. Nat Commun. 2020;11:968.
54. Di Leva G, Garofalo M, Croce CM. MicroRNAs in cancer. Annu Rev Pathol. 2014;9:287–314.
55. Dong H, Lei J, Ding L, Wen Y, Ju H, Zhang X. MicroRNA: function, detection, and bioanalysis. Chem Rev. 2013;113:6207–33.
56. Johnson SM, Grosshans H, Shingara J, Byrom M, Jarvis R, Cheng A, Labourier E, Reinert KL, Brown D, Slack FJ. RAS is regulated by the let-7 microRNA family. Cell. 2005;120:635–47.
57. Yu JY, Chung KH, Deo M, Thompson RC, Turner DL. MicroRNA miR-124 regulates neurite outgrowth during neuronal differentiation. Exp Cell Res. 2008;314(14):2618–33.
58. Carè A, Catalucci D, Felicetti F, Bonci D, Addario A, Gallo P, Bang ML, Segnalini P, Gu Y, Dalton ND, Elia L, Latronico MV, Høydal M, Autore C, Russo MA, Dorn GW, Ellingsen O,

Ruiz-Lozano P, Peterson KL, Croce CM, Peschle C, Condorelli G. MicroRNA-133 controls cardiac hypertrophy. Nat Med. 2007;13:613–8.
59. Bueno MJ, Pérez de Castro I, Gómez de Cedrón M, Santos J, Calin GA, Cigudosa JC, Croce CM, Fernández-Piqueras J, Malumbres M. Genetic and epigenetic silencing of microRNA-203 enhances ABL1 and BCR-ABL1 oncogene expression. Cancer Cells. 2008;13:496–506.
60. Melton C, Judson RL, Blelloch R. Opposing microRNA families regulate self-renewal in mouse embryonic stem cells. Nature. 2010;463:621–6.
61. Slack FJ. Stem cells: big roles for small RNAs. Nature. 2010;463:616.
62. Ghafouri-Fard S, Niazi V, Taheri M. Role of miRNAs and lncRNAs in hematopoietic stem cell differentiation. Non-coding RNA Res. 2020;6:8–14.
63. Wang J, Li J, Yang J, Zhang L, Gao S, Jiao F, Yi M, Xu J. MicroRNA-138-5p regulates neural stem cell proliferation and differentiation in vitro by targeting TRIP6 expression. Mol Med Rep. 2017;16:7261–6.
64. Schoolmeesters A, Eklund T, Leake D, Vermeulen A, Smith Q, Force Aldred S, Fedorov Y. Functional profiling reveals critical role for miRNA in differentiation of human mesenchymal stem cells. PLoS One. 2009;4:e5605.
65. Polesskaya A, Degerny C, Pinna G, Maury Y, Kratassiouk G, Mouly V, Morozova N, Kropp J, Frandsen N, Harel-Bellan A. Genome-wide exploration of miRNA function in mammalian muscle cell differentiation. PLoS One. 2013;8:e71927.
66. Schafer KA. The cell cycle: a review. Vet Pathol. 1998;35:461–78.
67. Bueno MJ, Malumbres M. MicroRNAs and the cell cycle. Biochim Biophys Acta. 2011;1812: 592–601.
68. Mens MMJ, Ghanbari M. Cell cycle regulation of stem cells by microRNAs. Stem Cell Rev Rep. 2018;14:309–22.
69. Shi Y. Mechanisms of caspase activation and inhibition during apoptosis. Mol Cell. 2002;9: 459–70.
70. Xu P, Vernooy SY, Guo M, Hay BA. The Drosophila microRNA Mir-14 suppresses cell death and is required for normal fat metabolism. Curr Biol. 2003;13:790–5.
71. Thompson BJ, Cohen SM. The Hippo pathway regulates the bantam microRNA to control cell proliferation and apoptosis in Drosophila. Cell. 2006;126:767–74.
72. Calin GA, Sevignani C, Dumitru CD, Hyslop T, Noch E, Yendamuri S, Shimizu M, Rattan S, Bullrich F, Negrini M, Croce CM. Human microRNA genes are frequently located at fragile sites and genomic regions involved in cancers. Proc Natl Acad Sci USA. 2004;101:2999–3004.
73. Svoronos AA, Engelman DM, Slack FJ. OncomiR or tumor suppressor? The duplicity of microRNAs in cancer. Cancer Res. 2016;76:3666–70.
74. Lu J, Getz G, Miska EA, Alvarez-Saavedra E, Lamb J, Peck D, Sweet-Cordero A, Ebert BL, Mak RH, Ferrando AA, Downing JR, Jacks T, Horvitz HR, Golub TR. MicroRNA expression profiles classify human cancers. Nature. 2005;435:834–8.
75. He L, Thomson JM, Hemann MT, Hernando-Monge E, Mu D, Goodson S, Powers S, Cordon-Cardo C, Lowe SW, Hannon GJ, Hammond SM. A microRNA polycistron as a potential human oncogene. Nature. 2005;435:828–33.
76. Wu X. Expressions of miR-21 and miR-210 in Breast Cancer and their predictive values for prognosis. Iran J Public Health. 2020;49:21–29.
77. Iorio MV, Ferracin M, Liu CG, Veronese A, Spizzo R, Sabbioni S, Magri E, Pedriali M, Fabbri M, Campiglio M, Ménard S, Palazzo JP, Rosenberg A, Musiani P, Volinia S, Nenci I, Calin GA, Querzoli P, Negrini M, Croce CM. MicroRNA gene expression deregulation in human breast cancer. Cancer Res. 2005;65:7065–70.
78. Eis PS, Tam W, Sun L, Chadburn A, Li Z, Gomez MF, Lund E, Dahlberg JE. Accumulation of miR-155 and BIC RNA in human B cell lymphomas. Proc Natl Acad Sci USA. 2005;102: 3627–32.

79. Miller TE, Ghoshal K, Ramaswamy B, Roy S, Datta J, Shapiro CL, Jacob S, Majumder S. MicroRNA-221/222 confers tamoxifen resistance in breast cancer by targeting p27Kip1. J Biol Chem. 2008;283:29897–903.
80. Fornari F, Gramantieri L, Ferracin M, Veronese A, Sabbioni S, Calin GA, Grazi GL, Giovannini C, Croce CM, Bolondi L, Negrini M. MiR-221 controls CDKN1C/p57 and CDKN1B/p27 expression in human hepatocellular carcinoma. Oncogene. 2008;27:5651–61.
81. Terasawa K, Ichimura A, Sato F, Shimizu K, Tsujimoto G. Sustained activation of ERK1/2 by NGF induces microRNA-221 and 222 in PC12 cells. FEBS J. 2009;276:3269–76.
82. Zhang X, Li X, Tan Z, Liu X, Yang C, Ding X, Hu X, Zhou J, Xiang S, Zhou C, Zhang J. MicroRNA-373 is upregulated and targets TNFAIP1 in human gastric cancer, contributing to tumorigenesis. Oncol Lett. 2013;6(5):1427–34.
83. Wang LQ, Yu P, Li B, Guo YH, Liang ZR, Zheng LL, Yang JH, Xu H, Liu S, Zheng LS, Zhou H, Qu LH. miR-372 and miR-373 enhance the stemness of colorectal cancer cells by repressing differentiation signaling pathways. Mol Oncol. 2018;12:1949–64.
84. Calin GA, Dumitru CD, Shimizu M, Bichi R, Zupo S, Noch E, Aldler H, Rattan S, Keating M, Rai K, Rassenti L, Kipps T, Negrini M, Bullrich F, Croce CM. Frequent deletions and down-regulation of microRNA genes miR15 and miR16 at 13q14 in chronic lymphocytic leukemia. Proc Natl Acad Sci USA. 2002;99:15524–9.
85. Michael MZ, O'Connor SM, van Holst Pellekaan NG, Young GP, James RJ. Reduced accumulation of specific microRNAs in colorectal neoplasia. Mol Cancer Res. 2003;1:882–91.
86. Takamizawa J, Konishi H, Yanagisawa K, Tomida S, Osada H, Endoh H, Harano T, Yatabe Y, Nagino M, Nimura Y, Mitsudomi T, Takahashi T. Reduced expression of the let-7 microRNAs in human lung cancers in association with shortened postoperative survival. Cancer Res. 2004;64:3753–6.
87. Harazono Y, Muramatsu T, Endo H, Uzawa N, Kawano T, Harada K, Inazawa J, Kozaki K. miR-655 Is an EMT-suppressive microRNA targeting ZEB1 and TGFBR2. PLoS One. 2013;8:e62757.
88. Ma J, Wang W, Azhati B, Wang Y, Tusong H. miR-106a-5p functions as a tumor suppressor by targeting VEGFA in renal cell carcinoma. Dis Markers. 2020;2020:8837941.
89. Xiao C, Srinivasan L, Calado DP, Patterson HC, Zhang B, Wang J, Henderson JM, Kutok JL, Rajewsky K. Lymphoproliferative disease and autoimmunity in mice with increased miR-17-92 expression in lymphocytes. Nat Immunol. 2008;9:405–14.
90. Babar IA, Cheng CJ, Booth CJ, Liang X, Weidhaas JB, Saltzman WM, Slack FJ. Nanoparticle-based therapy in an in vivo microRNA-155 (miR-155)-dependent mouse model of lymphoma. Proc Natl Acad Sci USA. 2012;109:E1695–704.
91. Costinean S, Zanesi N, Pekarsky Y, Tili E, Volinia S, Heerema N, Croce CM. Pre-B cell proliferation and lymphoblastic leukemia/high-grade lymphoma in E(mu)-miR155 transgenic mice. Proc Natl Acad Sci USA. 2006;103:7024–9.
92. Medina PP, Nolde M, Slack FJ. OncomiR addiction in an in vivo model of microRNA-21-induced pre-B-cell lymphoma. Nature. 2010;467:86–90.
93. Lu Z, Liu M, Stribinskis V, Klinge CM, Ramos KS, Colburn NH, Li Y. MicroRNA-21 promotes cell transformation by targeting the programmed cell death 4 gene. Oncogene. 2008;27:4373–9.
94. Buscaglia LE, Li Y. Apoptosis and the target genes of microRNA-21. Chin J Cancer. 2011;30:371–80.
95. Tavazoie SF, Alarcón C, Oskarsson T, Padua D, Wang Q, Bos PD, Gerald WL, Massagué J. Endogenous human microRNAs that suppress breast cancer metastasis. Nature. 2008;451:147–52.
96. Bhaumik D, Scott GK, Schokrpur S, Patil CK, Campisi J, Benz CC. Expression of microRNA-146 suppresses NF-kappaB activity with reduction of metastatic potential in breast cancer cells. Oncogene. 2008;27:5643–7.
97. Ma L, Teruya-Feldstein J, Weinberg RA. Tumour invasion and metastasis initiated by microRNA-10b in breast cancer. Nature. 2007;449:682–8.

98. Huang Q, Gumireddy K, Schrier M, le Sage C, Nagel R, Nair S, Egan DA, Li A, Huang G, Klein-Szanto AJ, Gimotty PA, Katsaros D, Coukos G, Zhang L, Puré E, Agami R. The microRNAs miR-373 and miR-520c promote tumour invasion and metastasis. Nat Cell Biol. 2008;10:202–10.
99. Mitchell PS, Parkin RK, Kroh EM, Fritz BR, Wyman SK, Pogosova-Agadjanyan EL, Peterson A, Noteboom J, O'Briant KC, Allen A, Lin DW, Urban N, Drescher CW, Knudsen BS, Stirewalt DL, Gentleman R, Vessella RL, Nelson PS, Martin DB, Tewari M. Circulating microRNAs as stable blood-based markers for cancer detection. Proc Natl Acad Sci USA. 2008;105:10513–8.
100. Chen X, Ba Y, Ma L, Cai X, Yin Y, Wang K, Guo J, Zhang Y, Chen J, Guo X, Li Q, Li X, Wang W, Zhang Y, Wang J, Jiang X, Xiang Y, Xu C, Zheng P, Zhang J, Li R, Zhang H, Shang X, Gong T, Ning G, Wang J, Zen K, Zhang J, Zhang CY. Characterization of microRNAs in serum: a novel class of biomarkers for diagnosis of cancer and other diseases. Cell Res. 2008;18:997–1006.
101. Hanke M, Hoefig K, Merz H, Feller AC, Kausch I, Jocham D, Warnecke JM, Sczakiel G. A robust methodology to study urine microRNA as tumor marker: microRNA-126 and microRNA-182 are related to urinary bladder cancer. Urol Oncol. 2010;28:655–61.
102. Park NJ, Zhou H, Elashoff D, Henson BS, Kastratovic DA, Abemayor E, Wong DT. Salivary microRNA: discovery, characterization, and clinical utility for oral cancer detection. Clin Cancer Res. 2009;15:5473–7.
103. Selth LA, Roberts MJ, Chow CW, Marshall VR, Doi SA, Vincent AD, Butler LM, Lavin MF, Tilley WD, Gardiner RA. Human seminal fluid as a source of prostate cancer-specific microRNA biomarkers. Endocr Relat Cancer. 2014;21:L17–21.
104. Fire A, Xu S, Montgomery MK, Kostas SA, Driver SE, Mello CC. Potent and specific genetic interference by double-stranded RNA in *Caenorhabditis elegans*. Nature. 1998;391:806–11.
105. Ramaswamy G, Slack FJ. siRNA: a guide for RNA silencing. Chem Biol. 2002;9:1053–5.
106. Nishikura K. A short primer on RNAi: RNA-directed RNA polymerase acts as a key catalyst. Cell. 2001;107:415–8.
107. Davidson BL, McCray PB. Current prospects for RNA interference-based therapies. Nat Rev Genet. 2011;12:329–40.
108. Singh A, Trivedi P, Jain NK. Advances in siRNA delivery in cancer therapy. Artif Cells Nanomed Biotechnol. 2018;46:274–83.
109. Nair JK, Willoughby JL, Chan A, Charisse K, Alam MR, Wang Q, Hoekstra M, Kandasamy P, Kel'in AV, Milstein S, Taneja N, O'Shea J, Shaikh S, Zhang L, van der Sluis RJ, Jung ME, Akinc A, Hutabarat R, Kuchimanchi S, Fitzgerald K, Zimmermann T, van Berkel TJ, Maier MA, Rajeev KG, Manoharan M. Multivalent N-acetylgalactosamine-conjugated siRNA localizes in hepatocytes and elicits robust RNAi-mediated gene silencing. J Am Chem Soc. 2014;136:16958–61.
110. Rupaimoole R, Slack FJ. MicroRNA therapeutics: towards a new era for the management of cancer and other diseases. Nat Rev Drug Discov. 2017;16:203–22.
111. Alwine JC, Kemp DJ, Stark GR. Method for detection of specific RNAs in agarose gels by transfer to diazobenzyloxymethyl-paper and hybridization with DNA probes. Proc Natl Acad Sci USA. 1977;74:5350–4.
112. Pall GS, Hamilton AJ. Improved northern blot method for enhanced detection of small RNA. Nat Protoc. 2008;3:1077–84.
113. Kurreck J, Wyszko E, Gillen C, Erdmann VA. Design of antisense oligonucleotides stabilized by locked nucleic acids. Nucleic Acids Res. 2002;30:1911–8.
114. Válóczi A, Hornyik C, Varga N, Burgyán J, Kauppinen S, Havelda Z. Sensitive and specific detection of microRNAs by northern blot analysis using LNA-modified oligonucleotide probes. Nucleic Acids Res. 2004;32:e175.
115. Ramkissoon SH, Mainwaring LA, Sloand EM, Young NS, Kajigaya S. Nonisotopic detection of microRNA using digoxigenin labeled RNA probes. Mol Cell Probes. 2006;20:1–4.

116. Huang Q, Mao Z, Li S, Hu J, Zhu Y. A non-radioactive method for small RNA detection by northern blotting. Rice (N Y). 2014;7:26.
117. Peterhaensel C, Obermaier I, Rueger B. Nonradioactive Northern blot analysis of plant RNA and the application of different haptens for reprobing. Anal Biochem. 1998;264:279–83.
118. Kim SW, Li Z, Moore PS, Monaghan AP, Chang Y, Nichols M, John B. A sensitive non-radioactive northern blot method to detect small RNAs. Nucleic Acids Res. 2010;38:e98.
119. Liu CG, Calin GA, Meloon B, Gamliel N, Sevignani C, Ferracin M, Dumitru CD, Shimizu M, Zupo S, Dono M, Alder H, Bullrich F, Negrini M, Croce CM. An oligonucleotide microchip for genome-wide microRNA profiling in human and mouse tissues. Proc Natl Acad Sci USA. 2004;101:9740–4.
120. Thomson JM, Parker J, Perou CM, Hammond SM. A custom microarray platform for analysis of microRNA gene expression. Nat Methods. 2004;1:47–53.
121. Nelson PT, Baldwin DA, Scearce LM, Oberholtzer JC, Tobias JW, Mourelatos Z. Microarray-based, high-throughput gene expression profiling of microRNAs. Nat Methods. 2004;1:155–61.
122. Shingara J, Keiger K, Shelton J, Laosinchai-Wolf W, Powers P, Conrad R, Brown D, Labourier E. An optimized isolation and labeling platform for accurate microRNA expression profiling. RNA. 2005;11:1461–70.
123. Castoldi M, Schmidt S, Benes V, Hentze MW, Muckenthaler MU. miChip: an array-based method for microRNA expression profiling using locked nucleic acid capture probes. Nat Protoc. 2008;3:321–9.
124. Duan D, Zheng KX, Shen Y, Cao R, Jiang L, Lu Z, Yan X, Li J. Label-free high-throughput microRNA expression profiling from total RNA. Nucleic Acids Res. 2011;39:e154.
125. Shen Y, Zheng KX, Duan D, Jiang L, Li J. Label-free microRNA profiling not biased by 3′ end 2′-O-methylation. Anal Chem. 2012;84:6361–5.
126. Novoselov KS, Geim AK, Morozov SV, Jiang D, Zhang Y, Dubonos SV, Grigorieva IV, Firsov AA. Electric field effect in atomically thin carbon films. Science. 2004;306:666–9.
127. Varghese N, Mogera U, Govindaraj A, Das A, Maiti PK, Sood AK, Rao CN. Binding of DNA nucleobases and nucleosides with graphene. ChemPhysChem. 2009;10:206–10.
128. Lu CH, Yang HH, Zhu CL, Chen X, Chen GN. A graphene platform for sensing biomolecules. Angew Chem Int Ed Eng. 2009;48:4785–7.
129. He S, Song B, Li D, Zhu C, Qi W, Wen Y, Wang L, Song S, Fang H, Fan C. A graphene nanoprobe for rapid, sensitive, and multicolor fluorescent DNA analysis. Adv Funct Mater. 2009;19:1–7.
130. Lu Z, Zhang L, Deng Y, Li S, He N. Graphene oxide for rapid microRNA detection. Nanoscale. 2012;4:5840–2.
131. Wegman DW, Krylov SN. Direct quantitative analysis of multiple miRNAs (DQAMmiR). Angew Chem Int Ed Eng. 2011;50:10335–9.
132. Wegman DW, Cherney LT, Yousef GM, Krylov SN. Universal drag tag for direct quantitative analysis of multiple microRNAs. Anal Chem. 2013;85:6518–23.
133. Khan N, Cheng J, Pezacki JP, Berezovski MV. Quantitative analysis of microRNA in blood serum with protein-facilitated affinity capillary electrophoresis. Anal Chem. 2011;83:6196–201.
134. Shi R, Chiang VL. Facile means for quantifying microRNA expression by real-time PCR. Biotechniques. 2005;39:519–25.
135. Chen C, Ridzon DA, Broomer AJ, Zhou Z, Lee DH, Nguyen JT, Barbisin M, Xu NL, Mahuvakar VR, Andersen MR, Lao KQ, Livak KJ, Guegler KJ. Real-time quantification of microRNAs by stem-loop RT-PCR. Nucleic Acids Res. 2005;33:e179.
136. Neilsen CT, Goodall GJ, Bracken CP. IsomiRs—the overlooked repertoire in the dynamic microRNAome. Trends Genet. 2012;28:544–9.
137. Androvic P, Valihrach L, Elling J, Sjoback R, Kubista M. Two-tailed RT-qPCR: a novel method for highly accurate miRNA quantification. Nucleic Acids Res. 2017;45:e144.

138. Arefian E, Kiani J, Soleimani M, Shariati SA, Aghaee-Bakhtiari SH, Atashi A, Gheisari Y, Ahmadbeigi N, Banaei-Moghaddam AM, Naderi M, Namvarasl N, Good L, Faridani OR. Analysis of microRNA signatures using size-coded ligation-mediated PCR. Nucleic Acids Res. 2011;39:e80.
139. Zhang J, Li Z, Wang H, Wang Y, Jia H, Yan J. Ultrasensitive quantification of mature microRNAs by real-time PCR based on ligation of a ribonucleotide-modified DNA probe. Chem Commun. 2011;47:9465–7.
140. Zhang P, Liu Y, Zhang Y, Liu C, Wang Z, Li Z. Multiplex ligation-dependent probe amplification (MLPA) for ultrasensitive multiplexed microRNA detection using ribonucleotide-modified DNA probes. Chem Commun. 2013;49:10013–5.
141. Tian H, Sun Y, Liu C, Duan X, Tang W, Li Z. Precise quantitation of microRNA in a single cell with droplet digital PCR based on ligation reaction. Anal Chem. 2016;88:11384–9.
142. Wang H, Wang HH, Duan XR, Liu CH, Li ZP. Digital quantitative analysis of microRNA in single cell based on ligation-depended polymerase colony (Polony). Biosens Bioelectron. 2017;95:146–51.
143. Lohman GJ, Zhang Y, Zhelkovsky AM, Cantor EJ, Evans TC. Efficient DNA ligation in DNA-RNA hybrid helices by Chlorella virus DNA ligase. Nucleic Acids Res. 2014;42:1831–44.
144. Jin J, Vaud S, Zhelkovsky AM, Posfai J, McReynolds LA. Sensitive and specific miRNA detection method using SplintR Ligase. Nucleic Acids Res. 2016;44:e116.
145. Yan J, Li Z, Liu C, Cheng Y. Simple and sensitive detection of microRNAs with ligase chain reaction. Chem Commun. 2010;46:2432–4.
146. Zhang P, Zhang J, Wang C, Liu C, Wang H, Li Z. Highly sensitive and specific multiplexed microRNA quantification using size-coded ligation chain reaction. Anal Chem. 2014;86: 1076–82.
147. Dirks RM, Pierce NA. Triggered amplification by hybridization chain reaction. Proc Natl Acad Sci USA. 2004;101:15275–8.
148. Yang L, Liu C, Ren W, Li Z. Graphene surface-anchored fluorescence sensor for sensitive detection of microRNA coupled with enzyme-free signal amplification of hybridization chain reaction. ACS Appl Mater Interfaces. 2012;4:6450–3.
149. Miao X, Ning X, Li Z, Cheng Z. Sensitive detection of miRNA by using hybridization chain reaction coupled with positively charged gold nanoparticles. Sci Rep. 2016;6:32358.
150. Jia H, Li Z, Liu C, Cheng Y. Ultrasensitive detection of microRNAs by exponential isothermal amplification. Angew Chem Int Ed Eng. 2010;49:5498–501.
151. Zhang X, Liu C, Sun L, Duan X, Li Z. Lab on a single microbead: an ultrasensitive detection strategy enabling microRNA analysis at the single-molecule level. Chem Sci. 2015;6:6213–8.
152. Cheng Y, Zhang X, Li Z, Jiao X, Wang Y, Zhang Y. Highly sensitive determination of microRNA using target-primed and branched rolling-circle amplification. Angew Chem Int Ed Eng. 2009;48:3268–72.
153. Zhou Y, Huang Q, Gao J, Lu J, Shen X, Fan C. A dumbbell probe-mediated rolling circle amplification strategy for highly sensitive microRNA detection. Nucleic Acids Res. 2010;38: e156.
154. Hong C, Baek A, Hah SS, Jung W, Kim DE. Fluorometric detection of microRNA using isothermal gene amplification and graphene oxide. Anal Chem. 2016;88:2999–3003.
155. Wang R, Zhao X, Chen X, Qiu X, Qing G, Zhang H, Zhang L, Hu X, He Z, Zhong D, Wang Y, Luo Y. Rolling circular amplification (RCA)-assisted CRISPR/Cas9 cleavage (RACE) for highly specific detection of multiple extracellular vesicle microRNAs. Anal Chem. 2020;92: 2176–85.
156. Liu H, Li L, Duan L, Wang X, Xie Y, Tong L, Wang Q, Tang B. High specific and ultrasensitive isothermal detection of microRNA by padlock probe-based exponential rolling circle amplification. Anal Chem. 2013;85:7941–7.

157. Ali MM, Li F, Zhang Z, Zhang K, Kang DK, Ankrum JA, Le XC, Zhao W. Rolling circle amplification: a versatile tool for chemical biology, materials science and medicine. Chem Soc Rev. 2014;43:3324–41.
158. Deng R, Tang L, Tian Q, Wang Y, Lin L, Li J. Toehold-initiated rolling circle amplification for visualizing individual microRNAs in situ in single cells. Angew Chem Int Ed Eng. 2014;53: 2389–93.
159. Li C, Li Z, Jia H, Yan J. One-step ultrasensitive detection of microRNAs with loop-mediated isothermal amplification (LAMP). Chem Commun. 2011;47:2595–7.
160. Sun Y, Tian H, Liu C, Sun Y, Li Z. One-step detection of microRNA with high sensitivity and specificity via target-triggered loop-mediated isothermal amplification (TT-LAMP). Chem Commun. 2017;53:11040–3.
161. Gao K, Zhang P, Wang H, Wang H, Su F, Li Z. Ultrasensitive homogeneous detection of microRNAs in a single cell with specifically designed exponential amplification. Chem Commun. 2021;57:5570–3.
162. Zheleznaya LA, Kachalova GS, Artyukh RI, Yunusova AK, Perevyazova TA, Matvienko NI. Nicking endonucleases. Biochemistry. 2009;74(13):1457–66.
163. Peng T. Nucleic acid isothermal amplification technology and its applications. Beijing: Science Press; 2009. p. 86–96.
164. Shi C, Liu Q, Ma C, Zhong W. Exponential strand-displacement amplification for detection of microRNAs. Anal Chem. 2014;86:336–9.
165. Ma F, Liu M, Tang B, Zhang CY. Sensitive quantification of MicroRNAs by isothermal helicase-dependent amplification. Anal Chem. 2017;89:6182–7.
166. Duan R, Zuo X, Wang S, Quan X, Chen D, Chen Z, Jiang L, Fan C, Xia F. Lab in a tube: ultrasensitive detection of microRNAs at the single-cell level and in breast cancer patients using quadratic isothermal amplification. J Am Chem Soc. 2013;135:4604–7.
167. Yin BC, Liu YQ, Ye BC. One-step, multiplexed fluorescence detection of microRNAs based on duplex-specific nuclease signal amplification. J Am Chem Soc. 2012;134:5064–7.
168. Li Y, Zhang J, Zhao J, Zhao L, Cheng Y, Li Z. A simple molecular beacon with duplex-specific nuclease amplification for detection of microRNA. Analyst. 2016;141:1071–6.
169. Degliangeli F, Kshirsagar P, Brunetti V, Pompa PP, Fiammengo R. Absolute and direct microRNA quantification using DNA-gold nanoparticle probes. J Am Chem Soc. 2014;136: 2264–7.
170. Ma F, Liu WJ, Zhang Q, Zhang CY. Sensitive detection of microRNAs by duplex specific nuclease-assisted target recycling and pyrene excimer switching. Chem Commun. 2017;53: 10596–9.
171. Cui L, Lin X, Lin N, Song Y, Zhu Z, Chen X, Yang CJ. Graphene oxide-protected DNA probes for multiplex microRNA analysis in complex biological samples based on a cyclic enzymatic amplification method. Chem Commun. 2012;48:194–6.
172. Cui L, Ke G, Lin X, Song Y, Zhang H, Guan Z, Zhu Z, Yang CJ. Cyclic enzymatic amplification method (CEAM) based on exonuclease III for highly sensitive bioanalysis. Methods. 2013;63:202–11.
173. Cui L, Zhu Z, Lin N, Zhang H, Guan Z, Yang CJ. A T7 exonuclease-assisted cyclic enzymatic amplification method coupled with rolling circle amplification: a dual-amplification strategy for sensitive and selective microRNA detection. Chem Commun. 2014;50:1576–8.
174. Schwarzenbach H, Nishida N, Calin GA, Pantel K. Clinical relevance of circulating cell-free microRNAs in cancer. Nat Rev Clin Oncol. 2014;11:145–56.
175. Faruq O, Vecchione A. microRNA: diagnostic perspective. Front Med. 2015;2:51.
176. Cortez MA, Bueso-Ramos C, Ferdin J, Lopez-Berestein G, Sood AK, Calin GA. MicroRNAs in body fluids—the mix of hormones and biomarkers. Nat Rev Clin Oncol. 2011;8:467–77.
177. Lawrie CH, Gal S, Dunlop HM, Pushkaran B, Liggins AP, Pulford K, Banham AH, Pezzella F, Boultwood J, Wainscoat JS, Hatton CS, Harris AL. Detection of elevated levels of tumour-associated microRNAs in serum of patients with diffuse large B-cell lymphoma. Br J Haematol. 2008;141:672–5.

178. Fayyad-Kazan H, Bitar N, Najar M, Lewalle P, Fayyad-Kazan M, Badran R, Hamade E, Daher A, Hussein N, ElDirani R, Berri F, Vanhamme L, Burny A, Martiat P, Rouas R, Badran B. Circulating miR-150 and miR-342 in plasma are novel potential biomarkers for acute myeloid leukemia. J Transl Med. 2013;11:31.
179. Moussay E, Wang K, Cho JH, van Moer K, Pierson S, Paggetti J, Nazarov PV, Palissot V, Hood LE, Berchem G, Galas DJ. MicroRNA as biomarkers and regulators in B-cell chronic lymphocytic leukemia. Proc Natl Acad Sci USA. 2011;108:6573–8.
180. Yang Z, Wu L, Wang A, Tang W, Zhao Y, Zhao H, Teschendorff AE. dbDEMC 2.0: updated database of differentially expressed miRNAs in human cancers. Nucleic Acids Res. 2017;45 (D1):D812–8.
181. Huang Z, Shi J, Gao Y, Cui C, Zhang S, Li J, Zhou Y, Cui Q. HMDD v3.0: a database for experimentally supported human microRNA-disease associations. Nucleic Acids Res. 2019;47(D1):D1013–7.
182. Jiang Q, Wang Y, Hao Y, Juan L, Teng M, Zhang X, Li M, Wang G, Liu Y. miR2Disease: a manually curated database for microRNA deregulation in human disease. Nucleic Acids Res. 2009;37(Database issue):D98–104.
183. Xie B, Ding Q, Han H, Wu D. miRCancer: a microRNA-cancer association database constructed by text mining on literature. Bioinformatics. 2013;29:638–44.
184. Ruepp A, Kowarsch A, Schmidl D, Buggenthin F, Brauner B, Dunger I, Fobo G, Frishman G, Montrone C, Theis FJ. PhenomiR: a knowledgebase for microRNA expression in diseases and biological processes. Genome Biol. 2010;11:R6.
185. McDermott AM, Miller N, Wall D, Martyn LM, Ball G, Sweeney KJ, Kerin MJ. Identification and validation of oncologic miRNA biomarkers for luminal A-like breast cancer. PLoS One. 2014;9:e87032.
186. Lowery AJ, Miller N, Devaney A, McNeill RE, Davoren PA, Lemetre C, Benes V, Schmidt S, Blake J, Ball G, Kerin MJ. MicroRNA signatures predict oestrogen receptor, progesterone receptor and HER2/neu receptor status in breast cancer. Breast Cancer Res. 2009;11:R27.
187. Youssef YM, White NM, Grigull J, Krizova A, Samy C, Mejia-Guerrero S, Evans A, Yousef GM. Accurate molecular classification of kidney cancer subtypes using microRNA signature. Eur Urol. 2011;59:721–30.
188. Gilad S, Lithwick-Yanai G, Barshack I, Benjamin S, Krivitsky I, Edmonston TB, Bibbo M, Thurm C, Horowitz L, Huang Y, Feinmesser M, Hou JS, St Cyr B, Burnstein I, Gibori H, Dromi N, Sanden M, Kushnir M, Aharonov R. Classification of the four main types of lung cancer using a microRNA-based diagnostic assay. J Mol Diagn. 2012;14:510–7.
189. Lawrie CH, Soneji S, Marafioti T, Cooper CD, Palazzo S, Paterson JC, Cattan H, Enver T, Mager R, Boultwood J, Wainscoat JS, Hatton CS. MicroRNA expression distinguishes between germinal center B cell-like and activated B cell-like subtypes of diffuse large B cell lymphoma. Int J Cancer. 2007;121:1156–61.
190. Small EM, Olson EN. Pervasive roles of microRNAs in cardiovascular biology. Nature. 2011;469:336–42.
191. Tijsen AJ, Creemers EE, Moerland PD, de Windt LJ, van der Wal AC, Kok WE, Pinto YM. MiR423-5p as a circulating biomarker for heart failure. Circ Res. 2010;106:1035–9.
192. D'Alessandra Y, Devanna P, Limana F, Straino S, Di Carlo A, Brambilla PG, Rubino M, Carena MC, Spazzafumo L, De Simone M, Micheli B, Biglioli P, Achilli F, Martelli F, Maggiolini S, Marenzi G, Pompilio G, Capogrossi MC. Circulating microRNAs are new and sensitive biomarkers of myocardial infarction. Eur Heart J. 2010;31:2765–73.
193. Jaguszewski M, Osipova J, Ghadri JR, Napp LC, Widera C, Franke J, Fijalkowski M, Nowak R, Fijalkowska M, Volkmann I, Katus HA, Wollert KC, Bauersachs J, Erne P, Lüscher TF, Thum T, Templin C. A signature of circulating microRNAs differentiates takotsubo cardiomyopathy from acute myocardial infarction. Eur Heart J. 2014;35:999–1006.
194. Osipova J, Fischer DC, Dangwal S, Volkmann I, Widera C, Schwarz K, Lorenzen JM, Schreiver C, Jacoby U, Heimhalt M, Thum T, Haffner D. Diabetes-associated microRNAs

in pediatric patients with type 1 diabetes mellitus: a cross-sectional cohort study. J Clin Endocrinol Metab. 2014;99:E1661–5.
195. Luo M, Li R, Deng X, Ren M, Chen N, Zeng M, Yan K, Xia J, Liu F, Ma W, Yang Y, Wan Q, Wu J. Platelet-derived miR-103b as a novel biomarker for the early diagnosis of type 2 diabetes. Acta Diabetol. 2015;52:943–9.
196. Lloyd-Jones D, Adams RJ, Brown TM, Carnethon M, Dai S, De Simone G, Ferguson TB, Ford E, Furie K, Gillespie C, Go A, Greenlund K, Haase N, Hailpern S, Ho PM, Howard V, Kissela B, Kittner S, Lackland D, Lisabeth L, Marelli A, McDermott MM, Meigs J, Mozaffarian D, Mussolino M, Nichol G, Roger VL, Rosamond W, Sacco R, Sorlie P, Stafford R, Thom T, Wasserthiel-Smoller S, Wong ND, Wylie-Rosett J. Executive summary: heart disease and stroke statistics—2010 update: a report from the American Heart Association. Circulation. 2010;121:948–54.
197. Fan KL, Zhang HF, Shen J, Zhang Q, Li XL. Circulating microRNAs levels in Chinese heart failure patients caused by dilated cardiomyopathy. Indian Heart J. 2013;65:12–6.
198. Duan X, Zhan Q, Song B, Zeng S, Zhou J, Long Y, Lu J, Li Z, Yuan M, Chen X, Yang Q, Xia J. Detection of platelet microRNA expression in patients with diabetes mellitus with or without ischemic stroke. J Diabetes Complicat. 2014;28:705–10.
199. Reynoso R, Laufer N, Hackl M, Skalicky S, Monteforte R, Turk G, Carobene M, Quarleri J, Cahn P, Werner R, Stoiber H, Grillari-Voglauer R, Grillari J. MicroRNAs differentially present in the plasma of HIV elite controllers reduce HIV infection in vitro. Sci Rep. 2014;4:5915.
200. Monteleone K, Selvaggi C, Cacciotti G, Falasca F, Mezzaroma I, D'Ettorre G, Turriziani O, Vullo V, Antonelli G, Scagnolari C. MicroRNA-29 family expression and its relation to antiviral immune response and viro-immunological markers in HIV-1-infected patients. BMC Infect Dis. 2015;15:51.
201. Munshi SU, Panda H, Holla P, Rewari BB, Jameel S. MicroRNA-150 is a potential biomarker of HIV/AIDS disease progression and therapy. PLoS One. 2014;9:e95920.
202. Sekar D, Hairul Islam VI, Thirugnanasambantham K, Saravanan S. Relevance of miR-21 in HIV and non-HIV-related lymphomas. Tumour Biol. 2014;35:8387–93.
203. Jansen C, Reiberger T, Huang J, Eischeid H, Schierwagen R, Mandorfer M, Anadol E, Schwabl P, Schwarze-Zander C, Warnecke-Eberz U, Strassburg CP, Rockstroh JK, Peck-Radosavljevic M, Odenthal M, Trebicka J. Circulating miRNA-122 levels are associated with hepatic necroinflammation and portal hypertension in HIV/HCV coinfection. PLoS One. 2015;10:e0116768.
204. Hanson EK, Lubenow H, Ballantyne J. Identification of forensically relevant body fluids using a panel of differentially expressed microRNAs. Anal Biochem. 2009;387:303–14.

Chapter 11
mRNA Biomarkers and Their Analytical Methods

Hui Wang and Weiliang Liu

11.1 Alternative Splice Variants

11.1.1 RNA Splicing and Alternative Splicing Variants

In most higher eukaryotes, the coding regions (exons) of multi-exon genes are separated by noncoding regions (introns). RNA polymerase II cannot distinguish between coding and noncoding regions during transcription. Instead, it transcribes the entire sequence of the target gene into a precursor messenger RNA (pre-mRNA), which lacks translational functionality. The cell must remove the noncoding introns through RNA splicing and connect the coding exons to produce a mature mRNA molecule capable of directing protein synthesis. During RNA splicing, the pre-mRNA transcribed from the same gene can undergo different splicing mechanisms and exon combinations, resulting in mRNA isoforms that encode proteins with distinct functions. These isoforms are referred to as alternative splicing variants (ASVs) or splice isoforms, and the process is known as alternative splicing (AS) [1, 2].

11.1.2 Patterns of Alternative Splicing

The main patterns of alternative splicing are shown in Fig. 11.1, including exon skipping (cassette exon), alternative 5′ splice site (donor site), alternative 3′ splice site (acceptor site), mutually exclusive alternative exons, alternative promoters, alternative poly(A), and intron retention [1–7]. Exon skipping or cassette alternative

H. Wang (✉) · W. Liu
University of Science and Technology Beijing, Beijing, China
e-mail: winscavin@ustb.edu.cn; lwliang@tzc.edu.cn

Z. Li (ed.), *Nucleic Acid Detection*, https://doi.org/10.1007/978-981-95-7295-3_11

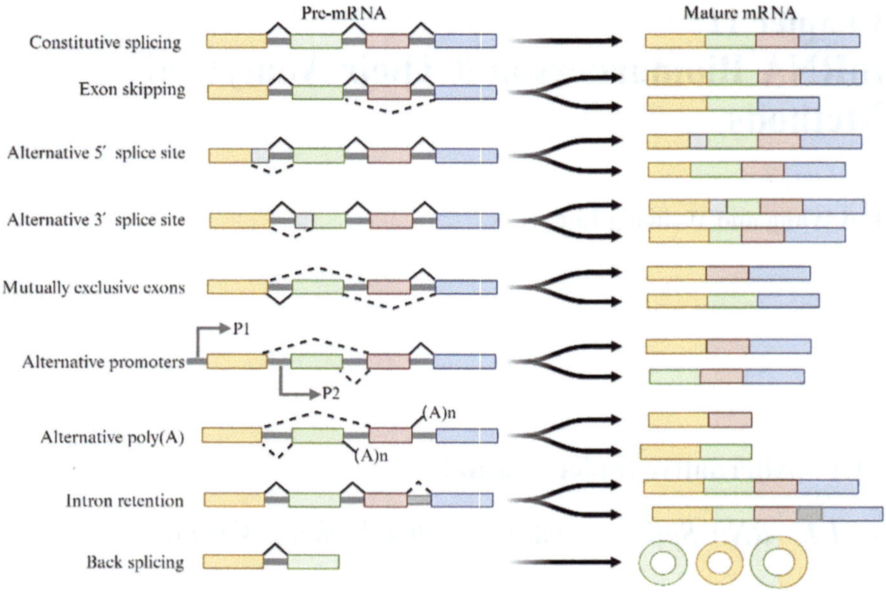

Fig. 11.1 Patterns of pre-mRNA alternative splicing

exon, in which an exon may be spliced out of the pre-mRNA or completely retained, is the predominant splicing mode of mammalian pre-mRNAs [6, 8, 9]. Nearly 38% of mature mRNAs are formed by this splicing mode. The second and third modes are alternative 5′ splice site (alternative 5′ SS) and alternative 3′ splice site (alternative 3′ SS). These modes of alternative splicing occur when two or more donor sites or acceptor junctions are selected, changing the 3′ boundary of the upstream exon or the 5′ boundary of the downstream exon. About 18% and 8% of mature mRNA are generated by alternative 5′ SS and alternative 3′ SS, respectively. Mutually exclusive exons mode, in which one of the two exons but not both are retained during RNA splicing, yields about 33% of mature mRNAs. For instance, the Dascm gene generates over 38,000 splice variants primarily through this splicing mode. Alternative promoters, also known as alternative 5′ untranslated regions (5′ UTRs), regulate the selection of alternative transcription start sites, leading to mutually exclusive splicing of the first exon. Similarly, alternative polyadenylation, also referred to as alternative 5′ UTRs, regulates polyadenylation sites, resulting in mutually exclusive splicing of the final exon. Intron retention, in which introns remain in the mature mRNA transcripts, accounts for approximately 3% of all splicing events. If the retained intron is in the coding region, the intron must encode amino acids in the frame with the neighboring exons, or a stop codon or a shift in the reading frame, producing nonfunctional proteins. This is the rarest way in mammals but the most prevalent type in short intron genes, unicellular eukaryotes, lower multicellular animals, and plants [7]. Alternative promoters, also known as alternative 5′ noncoding regions, refer to the first exon that generates a mutually exclusive splicing event through the regulation of the coding start site. Recent studies have revealed that, in addition to the aforementioned splicing modes, pre-mRNA can also undergo back-splicing to form circular RNA (circRNA). Circular RNA regulates gene expression by acting as a miRNA sponge, interfering with alternative splicing, or binding proteins [10, 11].

11.1.3 Alternative Splicing Mechanism

The pre-mRNA, transcribed from DNA, typically contains multiple introns and exons. If the introns are retained during the splicing process, the mature mRNA transcript will be spaced by noncoding sequences. If the exons are removed from the pre-mRNA, the mature mRNA will lose important sequences, which always guide and translate inactive or abnormal function proteins. Therefore, precise regulation and selection of splice sites on pre-mRNA are essential for the formation of mature mRNAs.

The nucleotide sequences at the boundaries of the exons and introns are called splice sites or splice junctions, which can be recognized by the spliceosome. Comparative analyses of splice sites have revealed that nearly all splice sites possess highly conserved sequence characteristics: the first two bases at the 5′ end of an intron are typically GU (referred to as the 5′ splice site, 5′-SS), while the last two bases at the 3′ end are almost always AG (referred to as the 3′ splice site, 3′-SS). This splicing pattern is known as the GU-AG rule and serves as a splicing signal recognized by spliceosome-related protein complexes. However, because introns often contain multiple GU-AG sequences, this signal alone is insufficient to ensure accurate intron removal. Therefore, besides the GU-AG signal, other conserved sequences contribute to splicing signals. For example, a conserved polypyrimidine tract (PPT), composed mainly of pyrimidine bases (U or C), is located upstream of the 3′-SS. Additionally, a conserved sequence within the intron, known as the branch point, plays a critical role. These elements collectively ensure the precise splicing of pre-mRNA [1, 7, 12, 13]. The conserved sequences at exon–intron boundaries in mammals are illustrated in Fig. 11.2. In this context, "Y" represents pyrimidine bases (U or C), "N" denotes any nucleotide, "R" indicates purine bases (A or G), "(Py)n" refers to a sequence of multiple pyrimidine nucleotides, and "A" denotes a specific adenosine nucleotide involved in forming the lariat intermediate during splicing. The uppercase bases indicate highly conserved nucleotides. While the conserved

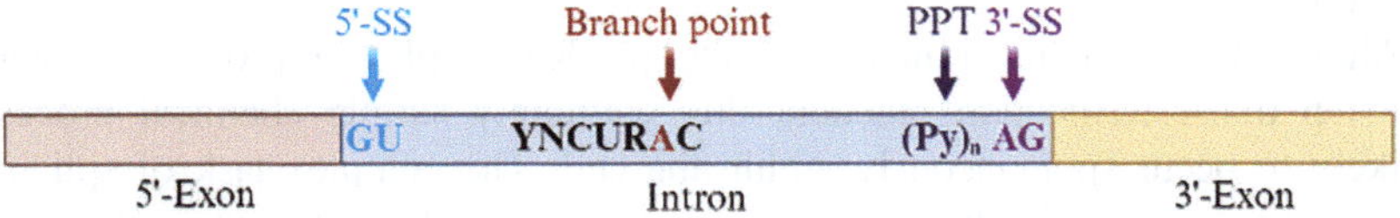

Fig. 11.2 Schematic representation of conserved sequences (splice signals) at mammalian exon and intron boundaries

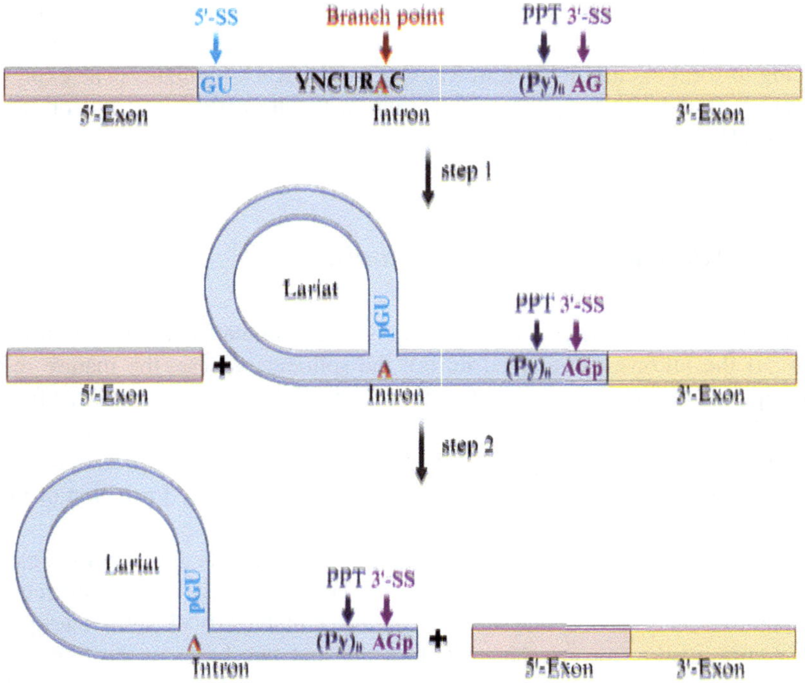

Fig. 11.3 Pre-mRNA splicing reaction: two-step transesterification reaction mediated by a lariat intermediate

sequences at exon–intron boundaries vary across different eukaryotes, the fundamental splicing principles remain consistent.

In the nucleus, the chemical nature of the alternative splicing of pre-mRNA is a two-step transesterification reaction [1, 5, 13, 14]. The reaction processes are displayed in Fig. 11.3. First step (Branching Reaction): At the branch point upstream of the 3′ splice site (3′-SS), the 2′-OH group of an adenine nucleotide (A) acts as a nucleophile, attacking the phosphodiester bond at the junction between the first exon (Exon 1) and the intron, specifically at the 5′ splice site (5′-SS). This reaction forms a lariat-like circular structure, simultaneously excising the first exon. Second step (Exon Ligation Reaction): The 3′-OH group of the excised exon nucleophilically attacks the phosphodiester bond at the junction between the intron and the second exon (Exon 2), specifically at the 3′ splice site. This step links the two exons via a 5′-3′ phosphodiester bond, while the intron is released in the form of a lariat structure.

Although the chemical nature of the pre-RNA splicing process involves a two-step transesterification reaction, this seemingly simple chemical process is unlikely to occur spontaneously within the cell. The complex task of splicing is carried out by a highly dynamic and intricate molecular machine in the nucleus spliceosome. The spliceosome is a massive RNA-protein complex comprising five small nuclear RNAs (snRNAs) that are rich in uridine and range in length from

106 to 185 nucleotides. Each snRNA associates with specific proteins to form small nuclear ribonucleoproteins (snRNPs), which are designated as U1, U2, U4, U5, and U6. Additionally, several auxiliary proteins contribute to the assembly and function of the spliceosome, such as the U2 snRNA auxiliary factor (U2AF) and the splicing factor SF1 [1, 4, 14–16].

Since the first discovery of RNA splicing in 1977, scientists have utilized various research methods, including immunoprecipitation, gene knockout, crosslinking mass spectrometry, and in vitro splicing systems, to establish an initial understanding of the assembly, activation, and disassembly of the spliceosome. These studies have also shed light on the complex RNA splicing regulatory network, encompassing protein–protein, protein–nucleic acid interactions, and their mutual regulation. During the splicing reaction, numerous protein–nucleic acid complexes and splicing factors participate in the spliceosome assembly, activation, splicing reaction, and disassembly (Disassembly) in a highly precise sequence. This process gives rise to various spliceosomal complexes, including the preassembled U4/U6.U5 Tri-snRNP complex (U4/U6.U5 tri-small nuclear ribonucleoprotein complex) and spliceosomes in different states of assembly, such as the pre-catalytic spliceosome precursor (pre-B complex), the pre-catalytic spliceosome (B complex), the activated spliceosome (B^{act} complex), the catalytically activated spliceosome (B* complex), the post-first-step spliceosome (C complex), the second-step catalytically activated spliceosome (C* complex), the spliceosome completing two transesterification steps (P complex), and the intron lariat spliceosome (ILS complex) [14, 17].

The process of pre-mRNA splicing in eukaryotes (yeast) is described in Fig. 11.4 [14]. Initially, SF1 and U2AF proteins recognize and bind to the branch point (BP) and 3′ splice site (3′ SS), respectively. Simultaneously, the snRNA in the U1 snRNP complex (a combination of U1 snRNA and proteins) pairs complementarily with the 5′ splice site (5′-SS), forming the early spliceosome E complex (E complex). Subsequently, U2 snRNP replaces SF1 and base pairs with the conserved sequence at the BP, creating an RNA-RNA double helix. This process requires the participation of U2 auxiliary factors (U2AF) and is dependent on ATP hydrolysis, representing the rate-limiting step of the entire splicing process. During this pairing, the A base at the BP remains unpaired and protrudes from the double helix, enabling the first transesterification reaction and forming the pre-spliceosome, known as the A complex (A complex). Next, the U4/U6.U5 tri-snRNP enters the A complex to form the catalytic spliceosome precursor complex (B complex). U6 snRNP binds to U4 snRNP through snRNA before entering the A complex, and after entering the A complex, it separates from U4 snRNP and then binds to U2 snRNP, replacing U1 snRNP and binding to 5′-SS. With the cooperative action of auxiliary proteins (e.g., yeast 19 complex protein NTC and NTC-related protein NTR; mammalian PRP19-CDC5L complex protein, etc.), the spliceosome assembles into the activated spliceosome (B^{act} complex). With the aid of ATP and auxiliary proteins, the B^{act} complex is converted into a catalytically active spliceosome (B* complex). In the B* complex, U6 snRNP acts as a bridge, bringing the 2′-OH group of the prominent A at the branch point site adjacent to the 5′-SS. Then, the 2′-OH nucleophilically attacks the phosphodiester bond of 5′-SS,

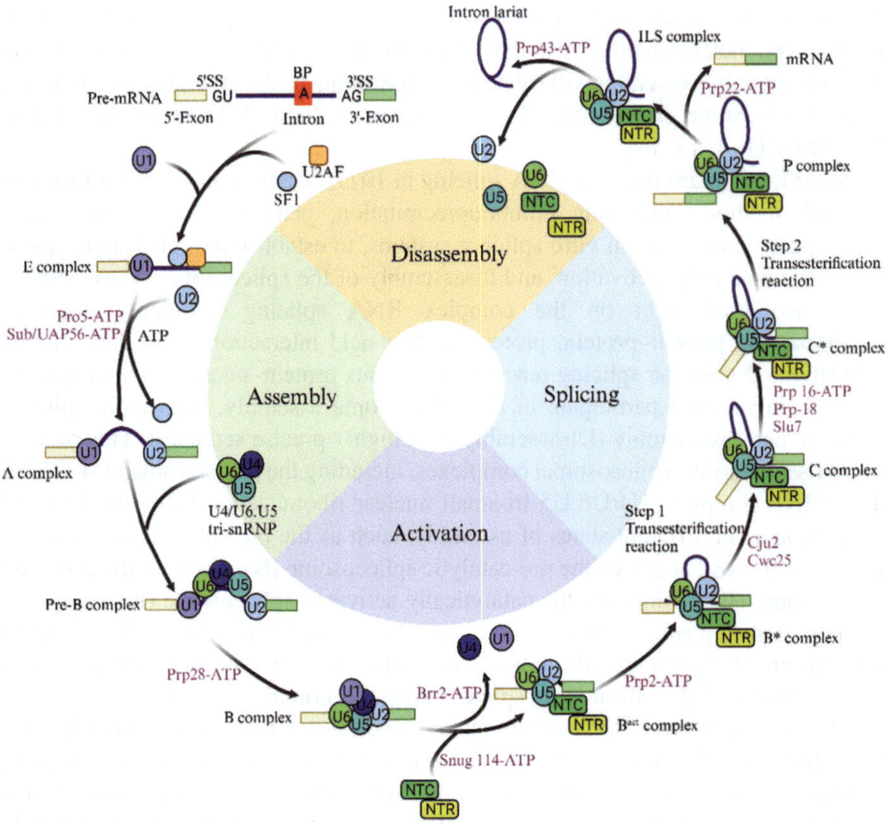

Fig. 11.4 Schematic diagram of eukaryotic (yeast) spliceosome assembly and the complete splicing reaction processes [14]. (Revised and reprinted with permission from Ref. [14])

breaking the 3′,5′-phosphodiester bond at the 5′-SS site. Simultaneously, a 2′,5′ phosphodiester bond forms between the BP A nucleotide's 2′-OH and the 5-′-phosphate of the intron, completing the first transesterification reaction. The spliceosome resulting from this reaction is referred to as the C complex (C complex). With the aid of ATP and auxiliary proteins, U5 snRNP in the C complex rearranges and brings the adjacent exons close to each other, providing the conditions for the second step of the transesterification reaction. Subsequently, the C complex then transitions into the catalytically activated spliceosome (C* complex), catalytically active for the second transesterification reaction. In the C* complex, the 3′-OH of the exon released during the first transesterification reaction nucleophilically attacks the phosphodiester bond at the 5′-SS. This action forms a lariat structure and joins the exons, completing the second transesterification reaction. After the second transesterification reaction, the spliceosome releases the mature mRNA molecule and the intron lariat structure via the post-spliceosome (P complex) and intron lariat spliceosome (ILS complex), respectively. The mature mRNA molecule proceeds with its role in the cellular processes, while the intron lariat structure is quickly degraded. Concurrently, all snRNPs and auxiliary proteins dissociate from the spliceosome, entering a new round of pre-mRNA splicing.

Over nearly a decade of research, the team of structural biologist Shiyi Gong has resolved the atomic-level, three-dimensional structures of spliceosome complexes in various states for yeast and humans using single-particle cryo-electron microscopy, including the preassembled U4/U6.U5 tri-small nuclear ribonucleoprotein complex (U4/U6.U5 Tri-snRNP), the pre-B complex, B complex, Bact complex, B* complex, C complex, C* complex, P complex, and ILS complex [18–28]. These structures comprehensively cover the whole RNA splicing cycle, from spliceosome assembly and activation to the splicing reaction and disassembly, and provide the outline of the spliceosome-mediated pre-mRNA splicing process. These studies have given the clearest structural information of spliceosome complexes at different stages of the splicing cycle, elevating our understanding of the molecular mechanism of RNA splicing. These also offer an important foundation for understanding the pre-RNA splicing process in higher organisms and provide a molecular basis for further exploration of the pathogenesis and drug development of diseases associated with splicing abnormalities.

11.1.4 Biological Function of Alternative Splicing

In 1980, Leroy Hood and colleagues first identified an alternative splicing event in the μ heavy chain gene of mouse immunoglobulin (IgM). They demonstrated that alternative splicing can determine the properties of a gene's protein products, playing a critical regulatory role in gene expression and influencing protein function [29]. There are two variants of immunoglobin μ heavy chain in mice, secretory variant (μ_s) and membrane-bound variant (μ_m). The C-terminus of the μ_m protein can be anchored to a hydrophobic region of the membrane, whereas the C-terminus of the μ_s protein does not have this hydrophobic functional region. As illustrated in Fig. 11.5, the pre-mRNA of the μ heavy chain gene includes several exons. The exon S encodes the signal peptide that directs the protein product to the plasma membrane

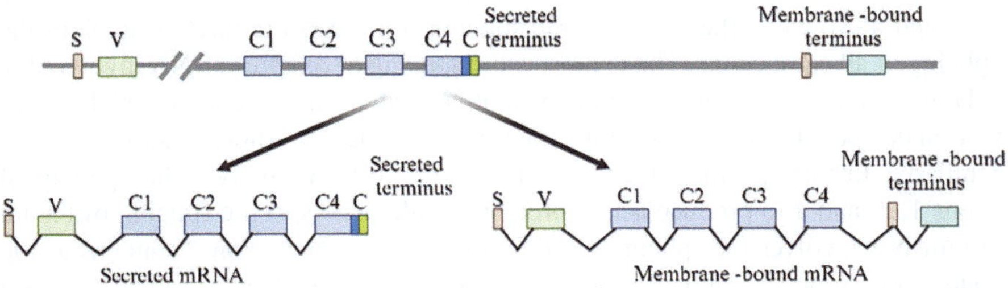

Fig. 11.5 Schematic diagram of alternative splicing of the mouse immunoglobulin μ heavy chain gene. The C_μ gene structure is shown at the top of the figure, and the bars represent exons

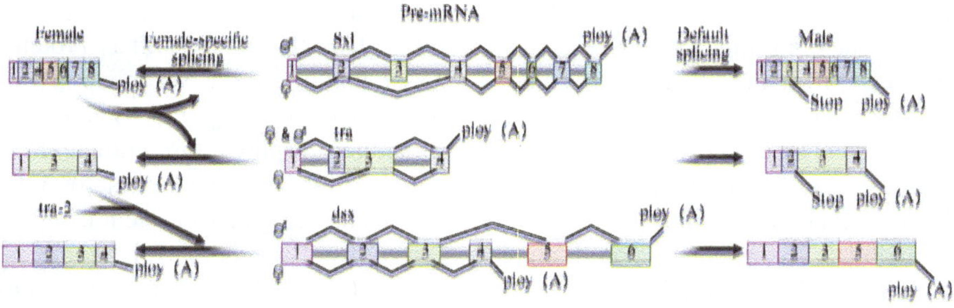

Fig. 11.6 Schematic diagram of the alternative splicing process for the determination of Drosophila sex

or extracellular space, exon V encodes the variable region of the protein, and exons C1–C4 encode the constant regions of the protein. The μ_s protein's secreted-form terminal sequence is located near the end of the fourth constant exon (C4), followed by a noncoding region, a long intron, and two additional exon sequences. The first of these exons encodes the membrane-anchoring region of the μ_m mRNA, while the second exon is a noncoding region of the μ_m mRNA. Hood and colleagues used nucleic acid hybridization to identify the germline gene encoding the constant regions of these two protein forms (the C_μ gene). This gene encodes the 3′ regions of both the μ_s and μ_m, but these sequences belong to two separate exons. This indicates that a single pre-mRNA molecule can generate mature mRNA variants through different splicing and joining mechanisms, resulting in μ_s mRNA and μ_m mRNA, which in turn produce the μ_s and μ_m proteins, respectively.

In 1989, Bruce Baker found that Drosophila sex is determined by alternative splicing of pre-mRNA during gene expression, which further verified that alternative splicing has an important biological function [30]. Sex determination in *Drosophila* involves the alternative splicing of pre-mRNAs from three key genes, Sex-lethal gene (sxl), transformer gene (tra), and double-sex gene (dsx). The female-specific splicing of these pre-mRNAs will lead to female development, on the contrary, the male-specific splicing of these transcription products will cause male development. As exhibited in Fig. 11.6, the structures of the Sxl, tra, and dsx gene precursor mRNA molecules are identical in male and female individuals. The female-specific splicing of the Sxl pre-mRNA involves the splicing of seven exons (including exons 1, 2, and 4–8), while the male-specific splicing of the Sxl pre-mRNA involves the splicing of all eight exons. The male-specific splicing of Sxl pre-mRNA involves the splicing of all eight exons, of which exon three contains a stop codon, which means that male-specific splicing of this transcript produces a short inactive protein. Similarly, female-specific splicing of the tra pre-mRNA involves the splicing of exons 1, 3, and 4 to produce active proteins, while male-specific splicing of the tra pre-mRNA involves the splicing of all four exons, in which exon 2 contains a stop codon, to produce inactive protein. In contrast, male-specific splicing of Sxl pre-mRNA produces nonfunctionally active proteins because one exon contains a stop codon. Then the pre-mRNA can be spliced by the default pattern (male-specific splicing), and the mature mRNA cannot translate a function-activated tra protein because it contains a stop codon in the mature mRNA. Then cells will splice the dsx pre-mRNA in the default male-specific splicing pattern, and the encoding protein will inactivate the female-specific gene, resulting in the development of male individuals. In addition, the female-specific splicing of the Sxl pre-mRNA yields the function-activated Sxl protein, which further enhances the female-specific splicing of the Sxl gene, and also triggers the female-specific splicing of the tra pre-mRNA encoding the active tra protein. The active tra protein acts synergistically with the tra-2 protein to trigger the dsx pre-mRNA to repress the male-specific splicing, and its encoding protein represses the expression of male-specific genes, resulting in the development of female individuals.

More and more studies have found that alternative splicing of pre-mRNAs is prevalent in eukaryotes [1]. Dietmar Schmucker et al. found that the Drosophila Down syndrome cell adhesion molecule (Dscam) gene may generate more than 38,000 splice isoforms through alternative splicing by cDNA and comparative genomics analysis [31]. In 2011, by combining RNA physical competition principle, mathematical models, and molecular biology approaches, Yongfeng Jin's research team cracked the veil of the RNA mutually exclusive control mechanism and clarified the process of alternative splicing of the Dscam gene [32]. The Dscam gene can generate 38,016 different splice isoforms through alternative splicing, each of which can encode different protein receptors, and each protein receptor molecule can recognize different directional signals, thus accurately guiding the axon direction of neurons. This gene is the "record holder" for the largest number of splice variants and is known as the variable gene that connects nerves.

It can be seen that through the process of alternative splicing, multiple transcript isoforms are generated from the same gene. The structure and function of the proteins encoded by each transcript isoform are different, and these proteins play different functions at different stages of the organism's development. Alternative splicing is not only an important mechanism for regulating gene expression at the posttranscriptional RNA level. It is also an important mechanism for the formation of protein biodiversity and greatly expands the varying scope of species [1–3, 33, 34]. More than 95% of human multi-exon genes have undergone alternative splicing at least once [8, 35]. In addition, misregulation of splicing and aberrant spliceosome expression are closely associated with the development of many diseases and cancers [36, 37]. Therefore, mRNA alternative splicing variants can be used as important biomarkers for disease diagnosis and treatment [6, 38–41].

11.1.5 Analysis Methods of Alternative Splicing

Alternative splicing of pre-mRNA not only plays an important regulatory role in gene expression, but also is an important mechanism for the formation of proteins and biodiversity, and is also closely related to the occurrence and development of various diseases and even cancers. Therefore, the identification and analysis of alternative splicing events and their variants are crucial to the in-depth study of alternative splicing regulatory mechanisms, as well as the diagnosis and treatment of diseases. Currently, the analysis methods of alternative splicing mainly rely on hybridization techniques, PCR, isothermal nucleic acid amplification methods, and high-throughput sequencing.

11.1.5.1 Hybridization-Based Methods

1. Exon DNA microarray
   There are two types of microarrays used for alternative splicing analysis, the splicing site microarray and the exon DNA microarray (ExonChip). The splicing site microarray is constructed by immobilizing a large number of oligonucleotide probes on a solid phase carrier. These probes are divided into multiple groups, each group corresponding to a specific exon or splice site. Splice site microarrays enable efficient detection of alternative splicing events. Due to their high-throughput nature, this technology is well-suited for genome-wide analysis of alternative splicing [2]. ExonChip is another important tool for detecting alternative splicing events and discovering new splice isoforms. The widely used one is the Human ExonChip developed by Affymetrix (e.g., GeneChip Human Exon 1.0 ST Array). This chip has 5.4 million probe sites, which contain all the annotated exons, predicted exons, and some intron regions in the whole genome, covering classical databases (such as NCBI and Ensembl) and more than 1 million exon clusters. The microarray can be used for both alternative splicing analysis and the discovery of novel splice variants, as well as genome-wide gene expression analysis [42]. The probe design of the exon microarray differs from the conventional microarray. As described in Fig. 11.7, the conventional gene expression microarray only designs a set of probes at the 3′ ends of the mRNA transcripts, the exon microarray designs four oligonucleotide probes in each exon region. These four probes together constitute Probe Selection Region (PSR), which usually

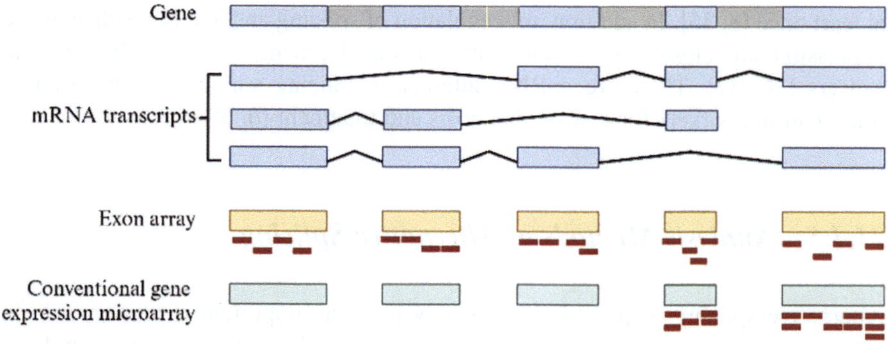

Fig. 11.7 Schematic design of conventional gene expression microarrays and exon microarray probes

corresponds to a single exon with an average length of about 120 bp. Exon microarrays not only design PSRs on each exon but also have about four times more probes than conventional microarrays, thus outperforming conventional gene expression microarrays in terms of sensitivity and accuracy.

2. Surface-enhanced Raman scattering (SERS)

   Breast cancer susceptibility 1 (BRCA1) is the earliest identified breast cancer susceptibility gene. It is located on human chromosome 17q21 (chromosome 17, long arm, region 2, band 1), containing 22 introns, 22 coding exons, and 2 non-coding exons, with a total length of about 100 kb. The transcription mRNA of BRCA1 is about 7.8 kb, in which exon 11 is the core exon (3758 bases) and encodes more than 60% of the amino acid residues. It was found that the BRCA1 gene can generate at least 11 splice isoforms through alternative splicing, which can encode different proteins with different biological functions [43, 44]. Additionally, the types and expression levels of these isoforms vary depending on cell type and the stage of cell growth [45]. Sun et al. developed a nonfluorescent Surface-enhanced Raman scattering (SERS) platform for multiplex detection of alternative splice variants of BRCA1 by using DNA-AuNP-Raman Tag probes [46]. As shown in Fig. 11.8, four alternative variants of BCRA1, △(11q) (the last 3309 nt deleted from exon 11), △(9, 10) (exon 9 and exon 10 are deleted), △(5) (exon 5 deleted), and △(5q, 6) (the last 22 nt of exon 5 and the entire exon 6 deleted) were selected as model targets. The target strand (TS) is a single-stranded DNA (ssDNA) sequence specific to the exon/exon splice site. The probe strand (PS) used on the DNA-AuNP-Roman Tag is ssDNA complementary to the downstream half of the TS, its 5′ end is modified with sulfhydryl groups. The capture strand (CS) is ssDNA complementary to the upstream half of the TS, and its 3′ end is modified with a sulfhydryl group. T_{10} is used as a spacer sequence for PS and CS to reduce nonspecific binding and improve hybridization efficiency because thymidine has the lowest affinity to gold compared to other nucleotides [47]. In the presence of the TS, the TS hybridizes with PS and CS, forming a sandwich complex on the gold-plated glass slide. When the sandwich structure is formed, the TS-dependent Raman signal of the nonfluorescent RTag can be observed. To achieve multiplex analysis of the four splice variants, the researchers conjugated four distinct Raman signal molecules (4-mercaptopurine, 2-thiazoline-2-thiol, 4,6-dimethyl-2-pyrimidinethiol, and 2-thiouracil) to gold nanoparticles, serving as specific labels (RTag 1–4) for each variant [46]. Building on this study, they also demonstrated a SERS-based array platform to monitor alternative splicing variants in cancer cells in a multiplex and quantitative format without nucleic acid amplification steps [48].

3. In Situ Hybridization Plasmon Resonance Imaging

   Gold nanoparticles (AuNPs) exhibit localized surface plasmon resonance (LSPR), which produces plasmonic peaks at specific wavelengths and is highly sensitive to nanoscale interparticle distances [49]. The plasmonic peak of adjacent localized AuNPs undergoes a redshift as the interparticle distance decreases, with the magnitude of the redshift being inversely proportional to the distance. Moreover, the localized field enhancement effect of AuNPs results in additional signal

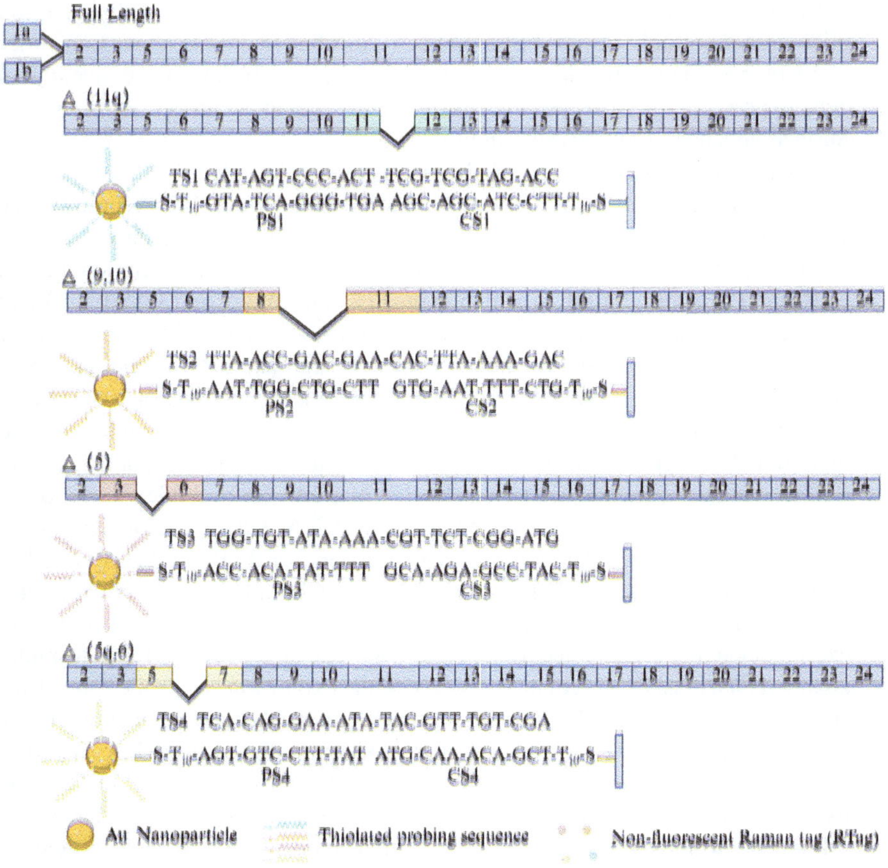

Fig. 11.8 Surface-enhanced Raman scattering (SERS) platform for the detection of alternative splice variants [46]. The numbers indicate the different exons, and the deleted exons in each variant are indicated by connecting lines; the exons are not drawn to scale. (Revised and reprinted with permission from Ref. [46])

amplification due to plasmonic coupling between nanoparticles [50]. Plasmonic in situ hybridization (Plasmonic ISH) probes are created by functionalizing AuNPs with oligonucleotides complementary to the target RNA sequences. When these probes hybridize with the target RNA, gold nanoparticle dimers are formed, exhibiting distinctive spectral shifts due to plasmonic coupling. Irudayaraj et al. developed a plasmonic nanoparticle resonance imaging technique for analyzing RNA splicing variants at the single-cell level [51]. As illustrated in Fig. 11.9, AuNP-DNA probes (AuNP-DNA1 and AuNP-DNA2) hybridize with different splice transcripts, forming gold nanoparticle dimers. These dimers exhibit strong plasmonic coupling, resulting in resonance spectral shifts that enable resonance spectral imaging analysis of single-molecule mRNA splicing variants within single cells. This study represents the first report using a

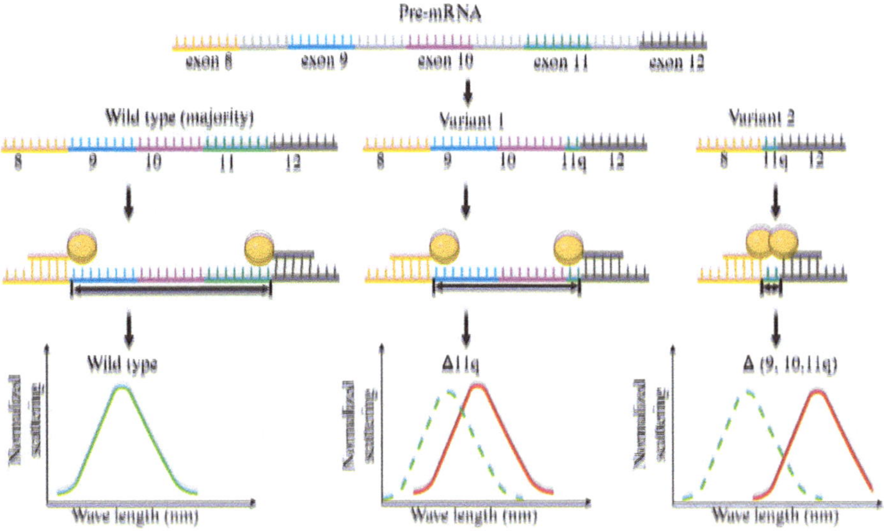

Fig. 11.9 Schematic diagram of plasmonic nanoparticle resonance imaging technique for analyzing RNA splicing variants at the single-cell level [51]. (Revised and reprinted with permission from Ref. [51])

nonfluorescent method to analyze single-molecule pre-mRNA alternative splicing in living cells. It provides a novel approach for studying the dynamic processes of RNA splicing and other real-time molecular interactions.

11.1.5.2 PCR-Based Strategies

1. Reverse Transcription Polymerase Chain Reaction (RT-PCR)
   So far, reverse transcription polymerase chain reaction (RT-PCR) is the most commonly used method for quantitative detection of mRNA splice variants [51–57]. However, these PCR-based methods have some inherent limitations. As shown in Fig. 11.10, a pre-mRNA consisting of exons a, b, and c, for example, can be spliced into two mRNA splice variants: ac and abc. The detection of the ac splice variant using RT-PCR requires the design of two amplification primers: a boundary-spanning primer (BSP, red) and a reverse primer (black). BSP can hybridize to the corresponding junction site of exon a* and exon c* in cDNA, and then the cDNA can be amplified by PCR. However, mispriming often results in undesired nonspecific amplification owing to the partial hybridization between either the 5′- or 3′-end of the BSP and other homologous splicing isoforms, which greatly challenges the specificity and sensitivity for PCR-based splicing variant assay. To address this problem, Wang et al. [58] designed two DNA probes at the splice junction of exons a and c. In the presence of an ac isoform, two probes can adjacently hybridize with the target ac isoform and are directly ligated with the

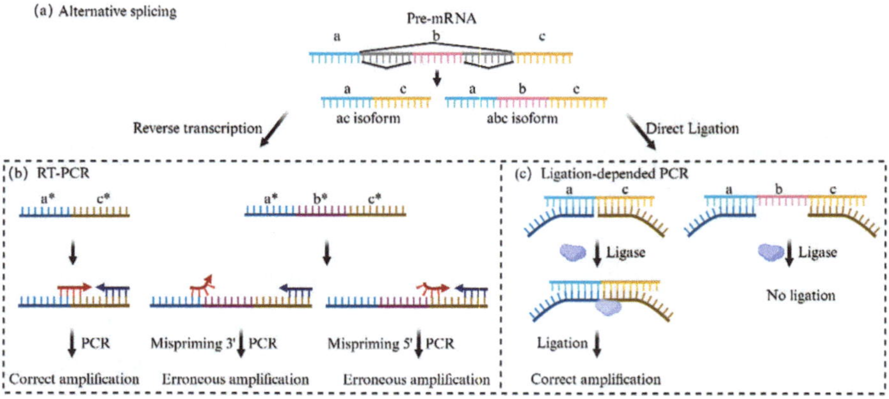

Fig. 11.10 (**a**) Alternative splicing of pre-mRNA. (**b**) Principle of mRNA splice variants detection by RT-PCR and primer-induced nonspecific amplification mechanism; a*, b*, c* are corresponding cDNAs after reverse transcription. (**c**) Principle of mRNA splice variants detection by ligation-PCR

catalysis of T4 RNA ligase 2. Then, ligated probes can be used as the PCR template to initiate a universal PCR reaction. As such, when there is no ac isoform, the two probes can also hybridize with the abc isoform. But because of two probes are separated from the exon b, they cannot be ligated together, and the PCR reaction won't be initiated. So, compared to the RT-PCR method, direct ligation of DNA probes at exon junction achieves high specificity for mRNA splice variants assay.

2. Ligation-PCR

By combining the direction ligation of two exon–exon junction-specific DNA probes and universal PCR amplification, Wang et al. developed a highly selective and sensitive method for the quantification of mRNA splice variants without reverse transcription [58]. In this work, three splicing variants of the BRCA1 gene, including Δ(11q, 3642-), Δ(11q, 333-), and Δ(9, 10), were chosen as the model target (Fig. 11.11). Firstly, two splicing junction site-specific DNA probes (Probe-PO_4 and Probe-RR) were designed according to the junction sequences of different mRNA splice isoforms. Each probe contains an anti-target sequence (ATS), a coding sequence of different lengths (CS), and a universal PCR primer sequence (UFP and URP). Probe-PO_4 was modified with a phosphate group at its 5′-terminus for the ligation reaction. To improve the ligation efficiency of T4 RNA ligase, two ribonucleotides were modified at the 3′-terminus of Probe-RR. The coding sequence of DNA probes has a unique length that acts as a "size code" for different variants. The universal PCR primer sequences have identical sequences at their 5′- and 3′-terminus, which permit simultaneous amplification in the one-tube PCR with uniform PCR primers. In the presence of the target variants, the two DNA probes hybridize with the target mRNA at the splicing junction and are ligated by T4 RNA ligase. The ligated DNA probes with different lengths, as the PCR templates, can be amplified by PCR with universal primers. Furthermore, PCR products can be analyzed by polyacrylamide gel

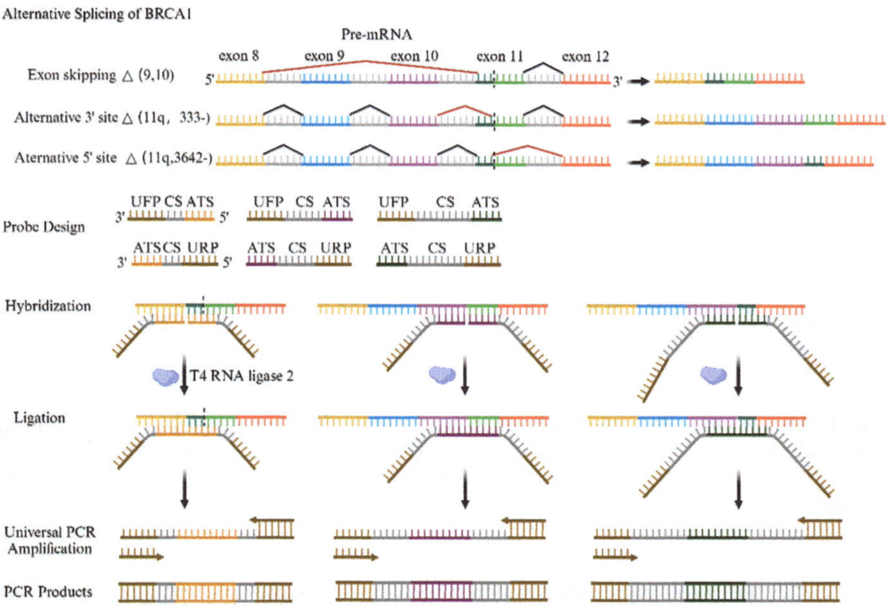

Fig. 11.11 Schematic diagram of the direct ligation of DNA probes at the exon junction and universal PCR amplification for the detection of mRNA splice variants from the BRCA1 gene [58]. (Revised and reprinted with permission from Ref. [58])

electrophoresis (PAGE), and multiple splice variants can be detected simultaneously. The ligation-based PCR assay has been successfully applied to the quantification of splicing isoforms in total RNA samples from different cell lines. The results showed that the mRNA splice variants of the BRCA1 gene were differentially expressed in different cell lines, which confirmed the existence of diversity in the expression of different mRNA splice variants of the same gene. Nevertheless, this strategy requires ribonucleotides to modify the DNA probe at its 3′-terminus to improve the ligation efficiency, reducing the stability of the probe and increasing the cost. To address this shortage, they recruit SplintR ligase for further research. The SplintR ligase can efficiently connect the splice-junction anchored DNA probes without modification. Moreover, by the design of TaqMan probes and encoding them with different fluorophores for different splicing variants, the developed method can simultaneously detect multiple splicing variants in a one-tube reaction and greatly improve the sensitivity down to 10 aM [59].

11.1.5.3 Isothermal Nucleic Acid Amplification Strategies

1. Rolling circle amplification (RCA)

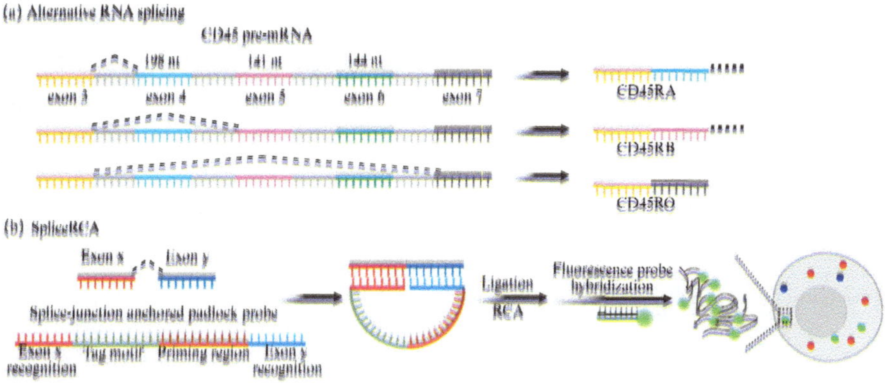

Fig. 11.12 Schematic diagram of SpliceRCA for multiplex detection of mRNA splice variants in single cells [64]. (**a**) Alternative splicing modes of CD45 during T-cell activation. (**b**) The steps of SpliceRCA for the detection of splicing variants in single cells. (Revised and reprinted with permission from Ref. [64])

CD45, the prototypic receptor-like protein tyrosine phosphatase gene, is expressed in all leukocytes and plays an important regulatory role in signaling transduction pathways of immune cells [60]. CD45 pre-mRNA alternative splicing marks the transition from primitive T cells (lymphocytes) to activated T cells. Abrant CD45 splicing isoform expressions are associated with autoimmune and infectious diseases [61–63].

Li's group designed a high-specificity method termed splice-junction anchored padlock probe-mediated rolling circle amplification (SpliceRCA) for single-cell imaging of CD45 splicing isoforms [64]. As displayed in Fig. 11.12, alternative splicing of exons 4 (198 nt), 5 (141 nt), and 6 (144 nt) in CD45 pre-mRNA are strictly regulated in T cell activation: primitive T cells express various larger variants, while memory T cells tend to produce smaller variants. Three key alternative splicing isoforms of the CD45 referred to the T cell activation process, including the biggest variant CD45RA (all exons are retained), the middle isoform CD45RB (exon skipping, the exon 4 is deleted), and the smallest variant CD45RO (exon skipping, the exons 4, 5, and 6 are deleted) (Fig. 11.15a). The splicing site-anchored padlock probe consists of four elements: exon–exon junction recognition elements (Exon x recognition-Rx and Exon y recognition-Ry), universal priming sequence (P), and tag motif (T) (Fig. 11.15b). The splicing sites of the target splicing variants can hybridize with the padlock probes of Rx and Ry, thereby closing the Rx and Ry and being circularized by the ligase. Then the amplification primer hybridizes with the P region to trigger the in situ RCA amplification, and the one-target-one-amplicon amplification will generate a long DNA amplicon with hundreds of copies of the padlock probe. By adding to a fluorescently labeled DNA probe complementary to the T region, the RCA amplicon can be visualized by in situ imaging. By coding the T sequence corresponding to different fluorophores, three splicing isoforms can be

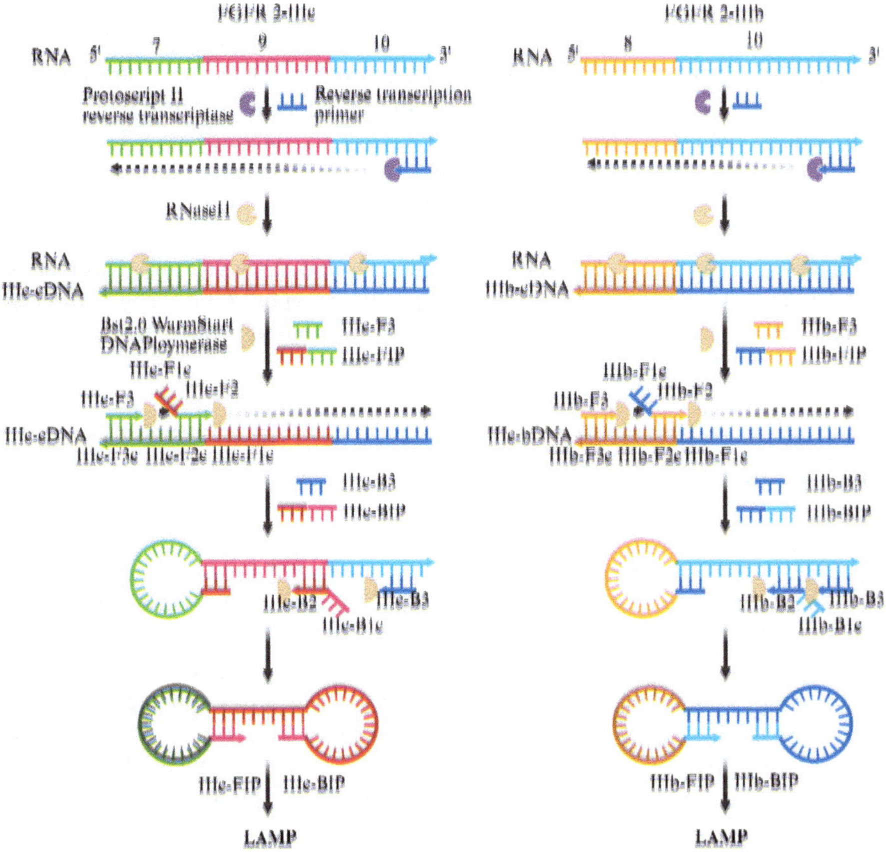

Fig. 11.13 Schematic diagram of LAMP-based detection of mRNA splice variants [67]. (Revised and reprinted with permission from Ref. [67])

simultaneously distinguished and detected in single cells with single-molecule resolution. Furthermore, by using the padlock probe and RCA amplification, Li's group established a padlock probe-based isothermal amplification assay to quantify the specific products in the two splicing steps [65].

2. Loop-mediated isothermal nucleic acid amplification (LAMP)

   Splicing variants of fibroblast growth factor receptor 2 (FGFR 2) are prevalent in malignant tumors such as gastric, lung, breast, ovarian, and endometrial cancers [66]. By designing splicing junction-specific primers, Su et al. developed a reverse transcription-loop-mediated isothermal amplification (RT-LAMP) for splicing isoforms assay [67]. As illustrated in Fig. 11.13, FGFR 2 pre-mRNA can generate two distinct splice variants: FGFR 2-IIIc, which includes exons 7, 9, and 10, and FGFR 2-IIIb, which includes exons 8 and 10. For detection, a reverse transcription primer hybridizes to exon 10 of either FGFR 2-IIIc or FGFR 2-IIIb transcripts. Under the catalysis of reverse transcriptase, complementary DNA (cDNA) strands, IIIc-cDNA and IIIb-cDNA, are synthesized. RNase H is then used to specifically hydrolyze the RNA strand in RNA-DNA hybrids, leaving the single-stranded cDNA. For FGFR 2-IIIc detection, four specific LAMP primers (IIIc-FIP, IIIc-F3, IIIc-BIP, IIIc-B3) are designed to recognize six distinct regions of the FGFR 2-IIIc splice variant. These primers initiate the formation of a LAMP reaction structure exclusively when FGFR 2-IIIc (containing exons 7, 9, and 10) is present. Similarly, to detect the FGFR 2-IIIb variant, LAMP primers specific to FGFR 2-IIIb are designed (Fig. 11.13b). The LAMP amplification process is monitored in real time using double-strand-specific fluorescent dyes, enabling high-sensitivity and high-specificity detection of the splice variants. This method provides a powerful tool for the precise quantification of FGFR 2 splice variants, facilitating further research on their role in cancer biology and their potential as diagnostic markers.

11.1.5.4 High-Throughput Sequencing Technology

Next-generation high-throughput sequencing technology can be used not only to analyze genome-wide DNA changes but also as a powerful tool to detect and analyze alternative splicing events in pre-mRNA at the whole transcriptome level [35, 68–70]. Using high-throughput sequencing technology, Blencower et al. systematically investigated splicing diversity in human tissues (including brain, heart, cerebral cortex, lung, skeletal muscle, and liver) and identified novel splicing variants in 20% of multi-exon genes, and most of these splice variants were tissue-specific; comparison of sequencing data with EST-cDNA data revealed that 95% of human multi-exon genes are undergoing alternative splicing [35]. Because of its high throughput and single-base resolution, high-throughput sequencing technology has been considered the core tool for transcriptome-wide alternative splicing analysis, discovering new splice sites and mRNA splice variants [71, 72]. The development of new personal genome sequencing (PGS) platforms will accelerate the development of transcriptome sequencing, making it a routine method for alternative splicing analysis in the future. Nevertheless, high-throughput sequencing requires a series of experimental steps, such as fragmentation, reverse transcription, adaptor ligation, PCR amplification, purification, etc., to construct sequencing libraries for sequencing, and each experimental step may introduce certain bias; moreover, the analysis cost of high-throughput sequencing technology is still high, and there are still various challenges in data analysis [5, 70, 73, 74].

11.2 Fusion Gene

11.2.1 Overview of Fusion Genes

Fusion genes are chimeric genes formed by chromosomal rearrangements (including translocations, deletions, inversions, and insertions) that link two or more gene coding regions under the same set of regulatory sequences (including promoters, enhancers, terminators, and ribosomal-binding sequences). The fusion genes generate fusion transcripts encoding fusion proteins with abnormal activity or function during gene expression, which play a causal role in tumorigenesis and development [75–77]. For example, the translocation of the ABL proto-oncogene on human chromosome 9 to the BCR gene on chromosome 22 forms the Philadelphia chromosome (Ph chromosome), also known as the BCR-ABL fusion gene. The resulting BCR-ABL fusion protein exhibits abnormally high tyrosine kinase activity, disrupting cellular regulatory systems and inhibiting apoptosis, ultimately leading to chronic myelogenous leukemia (CML) [78–80]. Similarly, an intrachromosomal deletion on chromosome 21 leads to low-level expression of the ERG gene being hijacked by the highly active TMPRSS2 gene promoter, forming the TMPRSS2-ERG fusion gene. This gene drives the overexpression of the oncogenic ERG protein, which activates downstream oncogenic genes as a malignant pro-cancer transcription factor, promoting the onset, progression, invasion, and metastasis of prostate cancer (PCa) [81–84]. Another example is the inversion and insertion of the EML4 gene on chromosome 2 into exon 20 of the ALK gene, forming the EML4-ALK fusion gene. The encoded EML4-ALK fusion protein can activate downstream signaling pathways without external signals or ligands, leading to tumor suppressor gene inactivation and promoting continuous cell proliferation, ultimately causing non-small cell lung cancer (NSCLC) [85–87].

With the completion of the Human Genome Project, advancements in post-genomic research, and the rapid development of high-throughput sequencing technologies, the relationships between fusion genes and various diseases have gradually been uncovered. Studies have shown that fusion genes are present in nearly all malignant tumors [75, 88–90]. According to the Mitelman Database of Chromosome Aberrations and Gene Fusions in Cancer, 14,078 genes associated with cancer have been identified to form fusions, resulting in 34,096 fusion genes [91]. Increasing numbers of fusion genes have been incorporated into diagnostic guidelines for malignant tumors by the World Health Organization (WHO) and the National Comprehensive Cancer Network (NCCN), serving as standards for classification, clinical diagnosis, and therapeutic monitoring. In the diagnosis, prognosis, and treatment of malignant tumors, detection and expression analysis of fusion gene transcripts have become essential indicators [76, 89].

11.2.2 Methods for Fusion Gene Analysis

Conventional methods for detecting fusion genes include traditional cytogenetic analysis (conventional cytogenetics, CC), also known as karyotyping, fluorescence in situ hybridization (FISH), and reverse transcription-polymerase chain reaction (RT-PCR)-based fusion gene analysis [92–94]. CC requires fresh, viable cells for short-term culture to obtain metaphase chromosomes, which can be challenging for solid tumors. Furthermore, this technique is time-consuming and requires highly trained personnel to accurately interpret karyotypes and identify potential chromosomal rearrangements. Although CC has low sensitivity (requiring a positive cell rate >5%), its advantage lies in the unbiased screening of all chromosomal rearrangements at the resolution of chromosomes. FISH, which uses site-specific fluorescently labeled probes, can detect chromosomal rearrangements at the DNA level. Despite being a standard reference for fusion gene detection, FISH is labor-intensive and time-consuming. RT-PCR, on the other hand, is widely used in fusion gene detection and clinical diagnostics due to its speed, sensitivity, and specificity. However, both FISH and RT-PCR are limited to detecting one or a few known candidate fusion genes due to restrictions in fluorescent channels, making them unsuitable for large-scale screening [92, 93].

Recent advancements in bioinformatics, gene expression microarrays, and high-throughput sequencing technologies have enabled high-throughput analysis of fusion genes at the whole genome and transcriptome levels [92, 93, 95–102]. Fusion transcripts encoding fusion proteins are typically formed by fusing two or more genes, each with its promoter. Consequently, RNA expression levels of 5′ or 3′ exons often differ in fusion events, providing the basis for identifying fusion genes using gene expression profiling with microarray technology. Exon DNA microarrays can not only identify splice variants at high throughput but also effectively detect and characterize fusion transcripts [102, 103]. With the rapid development of next-generation sequencing (NGS) technologies, along with improvements in bioinformatics databases and fusion gene analysis tools, RNA sequencing (RNA-Seq) based on high-throughput sequencing has become the most powerful tool for fusion gene identification and analysis. RNA-Seq enables genome-wide transcriptome analysis of gene fusions and efficient identification of novel fusion transcripts. However, complex and time-consuming experimental procedures and data analysis remain the major bottlenecks for both microarray and sequencing technologies. Additionally, these techniques require high-quality RNA samples to ensure reliable results [70, 73, 74, 92].

Beyond these methods, several emerging technologies have been applied for fusion gene detection and quantitative analysis. These include next-generation RNA in situ hybridization (RNAScope), single-molecule detection using fluorescent barcodes (NanoString nCounter), and isothermal nucleic acid amplification technologies (e.g., RCA, LAMP), which have been successfully utilized in fusion gene analysis.

11.2.2.1 RNAScope Technology

RNAScope, developed in 2011 by Advanced Cell Diagnostics, Inc. (California, USA), is an innovative quantitative RNA fluorescent in situ hybridization (ISH) method. The core of RNAScope lies in the design of its dual "ZZ" probes (Target Probes) and a signal amplification mechanism resembling branched DNA (bDNA) hybridization technology. As shown in Fig. 11.14, each "Z" probe consists of three parts: a region complementary to the target mRNA sequence (18–25 bases), a spacer sequence, and a tail sequence (14 bases). Two "Z" probes, each with distinct tail sequences, hybridize sequentially to a target mRNA region (50 bases). Together, their tail sequences form a 28-base preamplifier probe hybridization site. The preamplifier probe includes 20 hybridization-binding sites for amplifier probes, each of which contains 20 binding sites for fluorescently labeled reporter probes. After hybridizing with the Preamplifier, Amplifier, and Label Probes, each target RNA molecule theoretically can yield up to 400 labels, which allows the detection of RNA targets with single-molecule sensitivity. The Label probe can be either modified by fluorescent labels for visualization under a fluorescent microscope or conjugated to an alkaline phosphatase or horseradish peroxidase (HRP) molecule for chromogenic reactions. The alkaline phosphatase or HRP-labeled probes can be viewed under a standard bright-field microscope similar to IHC procedures, making RNAscope assay results easier to read and archive in a clinical setting [103].

The workflow of mRNA detection based on the RNAScope technique is depicted in Fig. 11.15. First step (Sample Preparation): Fix the target cells onto the surface of a glass slide and alter their permeability. Second step (Target Hybridization): Incubate the target-specific "ZZ" probes with the cells fixed on the glass slide, allowing the "ZZ" probes to hybridize with the target mRNA molecules. Wash away unhybridized "ZZ" probes. Third step (Signal Amplification): preamplified probes, amplified probes, and fluorescently labeled reporter probes are bound to the target mRNA molecules in an orderly manner, and the unhybridized "ZZ" probes are

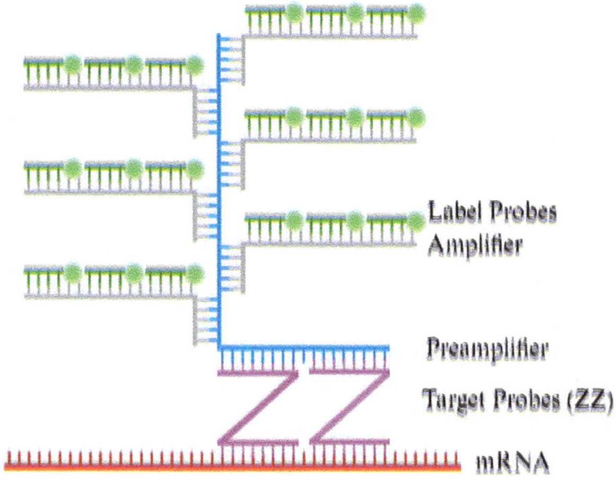

Fig. 11.14 Schematic diagram of double "ZZ" probes and branching DNA (bDNA) signal amplification of the RNAScope technique for mRNA analysis

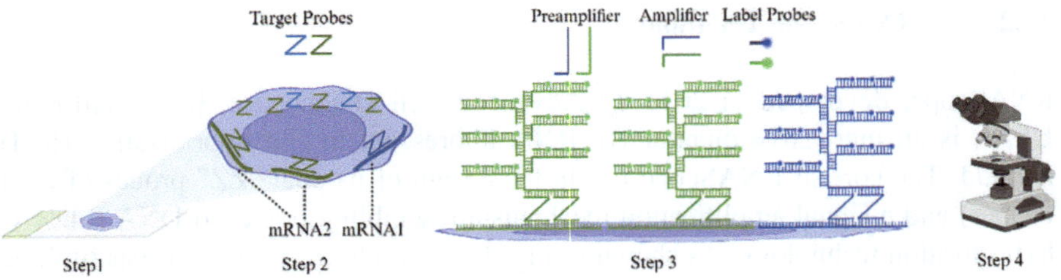

Fig. 11.15 Overview of mRNA analysis based on RNAScope technology

removed. Fourth step (Signal acquisition and quantitative analysis): Observe the signals using a fluorescence microscope or multispectral imaging system. Each target mRNA molecule appears as a dot-like fluorescent signal. Utilize image analysis software to identify and quantify individual mRNA molecules.

The unique dual "ZZ" probe design of RNAScope technology effectively prevents nonspecific probe binding, thereby reducing background noise interference. Since a single "Z" probe cannot form a complete signal amplification molecule–binding site, any unbound probes are washed away during the hybridization process, avoiding the amplification of nonspecific signals and ensuring high specificity. Additionally, this technology incorporates a signal amplification mechanism similar to the branched DNA (bDNA) hybridization technique, significantly improving the signal-to-noise ratio. It enables the localization and quantitative analysis of various RNA molecules within cells (including fusion gene transcripts, splice variants, lncRNA, circRNA, miRNA, etc.) at the single-cell and single-molecule levels. The technology also supports the simultaneous detection of multiple RNA molecules (https://acdbio.com/science/applications). RNAScope is currently one of the most sensitive methods for in situ RNA detection [104–106].

11.2.2.2 NanoString nCounter Technique

The fluorescent barcode-labeled single molecule detection technology, NanoString nCounter, was first developed at the Institute for Systems Biology (ISB), founded by Dr. Leroy Hood. It was featured in a cover article in *Nature Biotechnology* in 2008 [107]. NanoString nCounter is a novel high-throughput nucleic acid molecular detection technology, which can be used for the analysis of gene expression profiles, miRNAs, and more. Compared to chip technologies and real-time quantitative PCR, NanoString nCounter offers higher sensitivity, with its quantitative analysis accuracy and sensitivity comparable to that of real-time quantitative PCR [107–109].

NanoString nCounter technology is based on the hybridization reaction between target nucleic acids and a pair of DNA probes. The probe design is illustrated in Fig. 11.16. For each target nucleic acid molecule, a pair of DNA probes is designed and synthesized: a Capture Probe and a Reporter Probe, both consisting of 50-base DNA sequences. The Capture Probe is labeled with biotin at its 3′ end, which allows the hybridized complex to be immobilized on the surface of a reaction cartridge

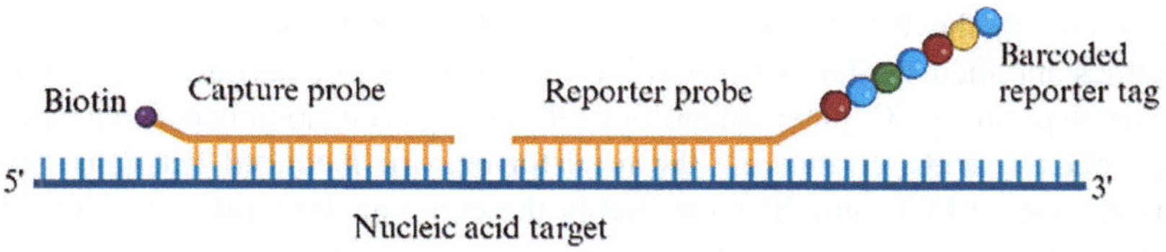

Fig. 11.16 NanoString nCounter technology probe design schematic

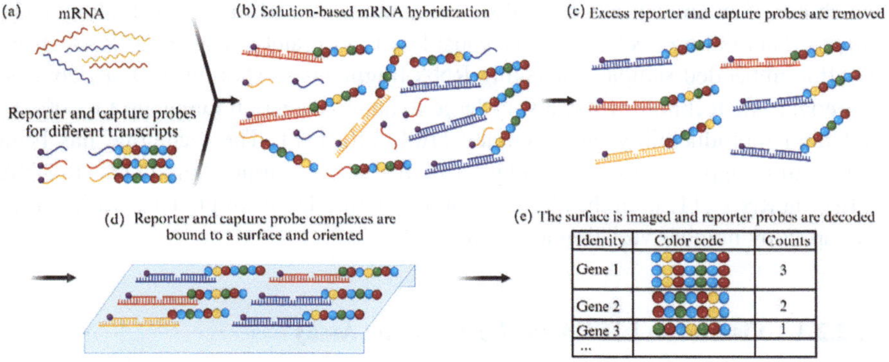

Fig. 11.17 Overview of nucleic acid analysis based on NanoString nCounter technology [107]. (Revised and reprinted with permission from Ref. [107])

modified with streptavidin through biotin-streptavidin interaction. The Reporter Probe carries a fluorescent barcode at its 5′ end, serving as the primary signal source. The fluorescent barcode (Barcoded reporter tag) is composed of six fluorescent microspheres (200 nm in diameter), each of which can be one of four colors: red, yellow, blue, or green. These colors are arranged in fixed combinations, enabling a one-to-one correspondence between the barcode and the target nucleic acid molecule. Theoretically, the four-color arrangement of six microspheres can generate $4^6 = 4096$ unique fluorescent barcodes. However, to facilitate software recognition of microsphere colors, adjacent microspheres cannot be the same color. Additionally, some built-in positive controls are designed. This technology supports the simultaneous analysis of up to 800 target nucleic acid molecules.

The overview of nucleic acid analysis based on NanoString nCounter technology is described in Fig. 11.17. The sample containing the different target nucleic acid molecules (mRNAs) is incubated with the capture and reporter probes, and then the capture and reporter probes hybridize with the mRNA molecule (Fig. 11.17a, b). Next, the excess free probes are removed through a series of hybridization and elution processes (Fig. 11.17c). Subsequently, the purified hybridization complexes are immobilized on the streptavidin-coated chip surface by the streptavidin–biotin interaction. Subsequently, the hybridization complexes are elongated, oriented, and aligned by adding an electric field to the ends of the chip (Fig. 11.17d). Finally, after chip imaging, the target nucleic acid molecule information, including the gene identities and their digital expression level (counts), can be decoded (Fig. 11.17e).

The NanoString nCounter technology relies on hybridization between probes and target nucleic acids, eliminating the need for enzyme-mediated reactions, reverse transcription, or PCR amplification during the entire analysis process. This allows for direct analysis of nucleic acids in lysates from formalin-fixed, paraffin-embedded (FFPE) samples, cell lysates, or blood samples. It is particularly well suited for severely degraded RNA samples, such as those derived from FFPE specimens, which are often challenging to analyze using conventional molecular biology techniques. NanoString technology is capable of accurately analyzing RNA from paraffin-embedded samples, detecting RNA fragments as short as 100 nucleotides. Moreover, the technology has a very large dynamic detection range, up to $7.0*10^7$, and high reproducibility and accuracy [108, 110–112]. The technique has been successfully applied to the quantitation of fusion gene transcripts [113], miRNA [114], lncRNA [115], circRNA [116], and genomic DNA single base variation as well as copy number variation analysis [117].

11.2.2.3 Vis-Fusion LIEXA for Fusion Gene Assay

Chronic granulocytic leukemia (CML) is a clonal malignant proliferative disease of hematopoietic stem cells. The Ph chromosome is the most important marker of CML. In CML, 90–95% of patients have the BCR-ABL fusion transcripts from the Ph chromosome. The BCR gene has several breakpoints (Fig. 11.18a). As shown in Fig. 11.18b, the BCR gene break occurs in the minor breakpoint cluster region, e1a2 (exon 1 from the ABL gene and exons 2–13 from the BCR gene) fusion transcript is generated, encoding $\text{BCR-ABL}^{\text{p190}}$ fusion protein with a molecular weight of 190 kD. When the gene break occurs in the major breakpoint cluster region (M-BCR), e13a2 (exons 1–13 from the ABL gene and exons 2–13 from the BCR gene) and e14a2 (exons 1–14 from the ABL gene and exons 2–13 from BCR gene) fusion transcripts are produced, encoding $\text{BCR-ABL}^{\text{p210}}$ fusion proteins with a molecular weight of 210 kD. If the gene break occurs in the micro breakpoint cluster region (μ-BCR), the e19a2 (exons 1–19 from ABL gene and exons 2–13 from BCR gene) fusion transcript is generated, encoding a 230 kD $\text{BCR-ABL}^{\text{p230}}$ fusion protein (Fig. 11.18b) [80]. These fusion proteins greatly enhance tyrosine kinase activity, alter the phosphorylation levels of various proteins, disrupt normal intracellular signaling pathways, inhibit apoptosis, and thus promote the growth of leukemic cells [118]. Quantitative detection of BCR-ABL fusion transcripts is essential in CML diagnosis, treatment, and prognosis.

Combining ligation reaction with excellent specificity, isothermal exponential amplification (IEXA) with high amplified efficiency, and a pH indicator, Wang et al. developed a sensitive and specific method for Visual detection of Fusion genes by Ligation-triggered Isothermal Exponential Amplification (termed Vis-Fusion LIEXA) [119]. Herein, the e13a2 transcript of the BCR-ABL fusion gene was used as the target model. A pair of fusion junction-specific stem-loop DNA probes

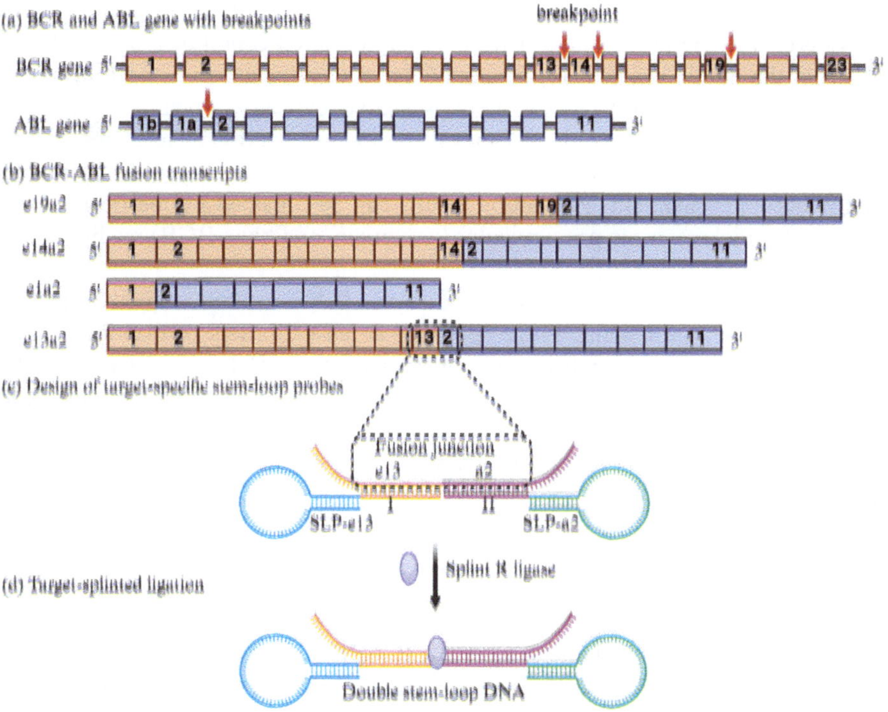

Fig. 11.18 Overview of the Vis-Fusion LIEXA method for fusion gene assay [119]. (**a**) BCR and ABL genes with breakpoints. (**b**) BCR-ABL fusion transcripts. (**c**) Design of fusion transcript (e13a2 as a model in this work)-specific stem-loop DNA probes (SLP-e13 and SLP-a2). (**d**) Target-splinted ligation of two stem-loop probes by SplintR ligase. (Revised and reprinted with permission from Ref. [119])

(SLP-e13 and SLP-a2) was first designed. Each probe contains a stem-loop structure and a 20-nucleotide anti-target-specific sequence (I and II) complementary to the 3′-terminus of BCR exon 13 and the 5′-terminus of ABL exon 2, respectively (Fig. 11.18c). When e13a2 is present, the SLP-e13 and SLP-a2 hybridize adjacent to the e13a2 transcript at the fusion junction, and the two DNA probes are ligated by SplintR ligase to form a double stem-loop DNA template (Fig. 11.18d), a unique DNA molecule, which can initiate the rapid and efficient IEXA amplification by DNA polymerase under isothermal conditions; the amplification processes are the most similar to the recycling amplification steps of loop-mediated isothermal amplification (LAMP) [120, 121]. When other fusion transcripts (e.g., e1a2, e14a2, and e19a2) are present, SLP-e13 and SLP-a2 cannot be ligated, and no subsequent amplification occurs due to the absence of the double stem-loop DNA template. The superior specificity of the ligase ensures that the method can effectively avoid the interference of other fusion transcripts. The fast and efficient amplification property of IEXA ensures ultrahigh sensitivity for the quantitative detection of low-abundance fusion transcripts. During the IEXA amplification process, a pyrophosphate (PPi) and a hydrogen ion (H^+) are released when each dNTP is polymerized to the DNA strand. Under weakly buffered conditions, the release of H^+ from the highly efficient IEXA amplification is sufficient to cause a pH drop, which results in a color change in pH-sensitive indicators such as cresol red, generating a color shift from red to yellow in the neutral range. Thus, visual detection of fusion gene transcripts is possible using the naked eye. More importantly, accurate quantitative detection results can be obtained by the open-source App (Color Picker), triple primary color (red, green, and blue, RGB) reading mode on smartphones.

They have demonstrated that the Vis-Fusion LIEXA method is highly sensitive and specific, with a dynamic range spanning at least 6 orders of magnitude. The proposed method can be applied to visualize fusion transcripts in real samples with the naked eye or to accurately quantify fusion gene transcripts without other expensive instruments using a smartphone-assisted RGB reading mode. The Vis-Fusion LIEXA method provides a new platform for rapid and sensitive clinical detection of fusion genes.

11.2.2.4 RCA-Based Fusion Gene Assay

More than 95% of CML patients have BCR-ABL^{p210} fusion protein [122]. This fusion protein has two distinct transcript isoforms, e13a2 (40%) and e14a2 (55%) [123], which are formed by fusing exon 2 of the ABL gene with exon 13 or 14 of the BCR gene, respectively. The fusion transcript e14a2 isoform has 75 bases (exon 14) more than e13a2, resulting in a protein with only 25 amino acid differences in their proteins. Although the differences between the two variants are small, they lead to significant differences in the pathological features and the therapeutic effect of the targeted drug imatinib [124]. Studies found that the CML patients carrying the e14a2 isoform showed higher platelet counts [125, 126] and lower white blood cells [127] than the patients with the e13a2 isoform. In addition, the CML patients with the e14a2 subtype have a faster and deeper response to imatinib, a better prognosis, and a lower relapse rate [128, 129]. Therefore, the accurate identification and detection of different BCR-ABL fusion gene transcript subtypes are very important for the typing of CML, disease risk classification, and the assessment of CML patients' effect on imatinib treatment and prognosis.

In 2019, Luo et al. reported an RNA fusion-triggered rolling circle amplification (RF-RCA) strategy for distinguishing and detecting the e13a2 and e14a2 transcripts of BCR/ABL^{p210} [130]. The principle of this method is shown in Fig. 11.19. It involves the design of a universal dumbbell-shaped circular DNA template (Dumbbell Template, DT), which consists of three sequence parts: primer-binding sequences (5 nt, which initiate the RCA amplification reaction upon binding with the DT template), a pyrimidine-rich sequence (whose complementary strand DNA formed after amplification can create a G-4 quadruplex structure), and an ABL gene fusion site-specific recognition sequence (16 nt, complementary to the 5′-end of exon 2 of the ABL gene). Furthermore, probes P13 and P14 were designed to

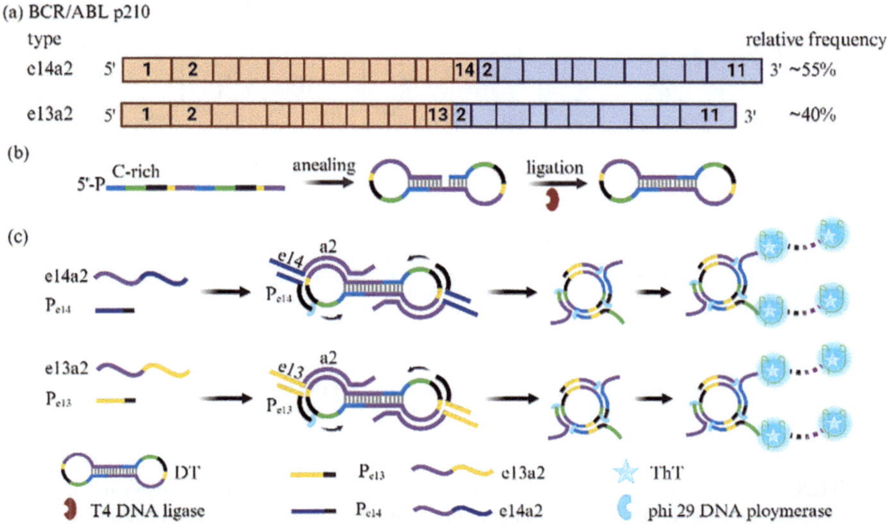

Fig. 11.19 Schematic diagram of RF-RCA-based BCR-ABL^{P210} fusion transcript isoform analysis [130]. (**a**) Exon composition of e13a2 and e14a2 fusion transcripts. (**b**) Dumbbell-shaped circular template preparation. (**c**) BCR-ABL^{P210} fusion transcript isoform analysis. (Revised and reprinted with permission from Ref. [130])

specifically recognize the BCR exon 13 and exon 14 in the e13a2 and e14a2 transcript fusion sites, respectively. These two probes were used to prepare amplification systems, Mix e13a2 and Mix e14a2, which specifically respond to e13a2 and e14a2. In the presence of a specific fusion transcript, the DT recognizes and binds to the ABL gene 5′-end sequence at exon 2 of the fusion site, while the BCR gene's exon 14 (or exon 13) specifically binds, bringing the primer sequence close to the DT template. This interaction forms a "Y"-shaped structure through base stacking, triggering the RCA amplification reaction and producing numerous tandem G-4 quadruplex structures. Thioflavin T (ThT) can intercalate into the G-4 quadruplex structure, generating a fluorescent signal. The study confirmed that this method can sensitively detect a few copies of fusion gene transcripts, enabling accurate differentiation of BCR-ABLp210 fusion transcript subtypes in clinical samples and dynamic monitoring of transcript subtype changes during CML treatment.

11.3 Methylation Modifications of mRNA (m^{6}A)

11.3.1 Discovery and Distribution of m^{6}A

Epigenetics has become a pivotal field in life sciences research. While significant progress has been made in understanding the epigenetic modifications of DNA and proteins, RNA modifications have also garnered increasing attention from scientists,

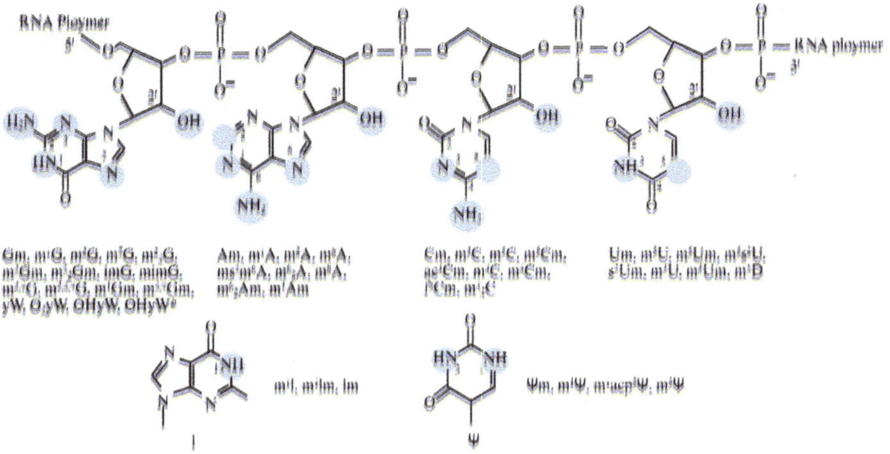

Fig. 11.20 Chemical structures of the four major ribonucleosides (adenosine, guanosine, cytidine, and uridine), along with inosine and pseudouridine, and their methylation modification sites

emerging as a new focus in epigenetic studies. To date, 171 types of RNA modifications have been identified, with methylation being the most prevalent class, encompassing 72 distinct methylation modifications [131]. RNA methylation modifications predominantly occur on the four canonical ribonucleotides (A, C, G, U), as well as on inosine (I) and pseudouridine (Ψ), with the specific modification sites illustrated in Fig. 11.20. Based on the location of the modification, RNA methylation can be categorized into three major types: methylation of the hydroxyl oxygen atom at the 2′ position of the ribose, methylation of the nitrogen or carbon atoms in the base ring, and methylation of the exocyclic amino groups [132]. Notably, methylation on the nitrogen atom in the glycosidic bond or on the third and seventh nitrogen atoms of adenosine has not been observed.

Methylation modifications of RNA are widespread across various organisms and are found in multiple types of RNA molecules, including messenger RNA (mRNA), transfer RNA (tRNA), ribosomal RNA (rRNA), long noncoding RNA (lncRNA), and various small nuclear RNAs (snRNAs). However, the distribution of RNA methylation modifications exhibits significant variation among different organisms [132]. As illustrated in Fig. 11.21, modifications such as m^6Am, m^6A, m^1A, Cm, m^5C, Gm, m^1G, m^7G, and m^5U have been identified in archaea, eukaryotes, and prokaryotes. In contrast, m^3C appears to be primarily distributed in eukaryotes and prokaryotes, with no reports of its presence in archaea to date. Among these RNA methylation modifications, N^6-methyladenosine (m^6A) is the most common and abundant type in eukaryotic mRNA, highlighting its significance as a prominent posttranscriptional modification in gene regulation.

N^6-methyladenosine (m^6A) is a methylation modification of the nitrogen atom at the sixth position of adenosine (Fig. 11.22). Initially discovered in archaea in 1955 [133], subsequent studies have demonstrated that m^6A is widely distributed in the RNA of animals, plants, yeast, and viruses [134–137]. In eukaryotes, m^6A is present

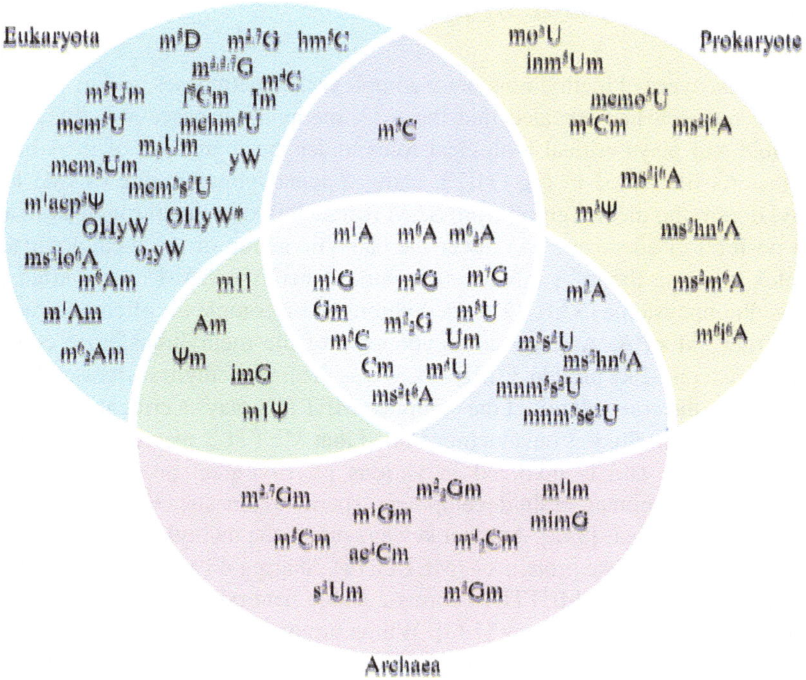

Fig. 11.21 Distribution of methylated RNA nucleosides across different organisms [132]. (Revised and reprinted with permission from Ref. [132])

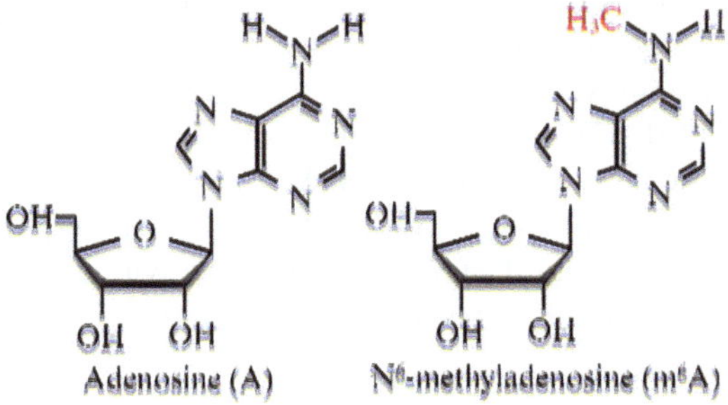

Fig. 11.22 Structural representation of adenosine (A) and N^6-methyladenosine (m^6A)

not only in mRNA but also in various RNA species, including tRNA, rRNA, snRNA, miRNA, and lncRNA [138–140]. Furthermore, studies have revealed that the distribution of m^6A exhibits sequence conservation and tissue specificity, predominantly occurring in the conserved sequence motif RRACH (R = A/G, A = m^6A, H = A/C/U) [141]. In mammals, m^6A is enriched in the brain, liver, and kidneys, whereas in plants, it is primarily concentrated in roots, leaves, and flowers [140]. Due to its ubiquitous presence and critical role in regulating biological processes, m^6A has rapidly emerged as a focal point in epigenetic research.

11.3.2 Process of m⁶A Methylation

With the discovery of methyltransferase complexes, demethylases, and m^6A-binding proteins, research has revealed that the m^6A methylation process is dynamically reversible and plays critical biological roles under the regulation of m^6A-binding proteins. As illustrated in Fig. 11.23, using *S*-adenosylmethionine (SAM) as the methyl donor, the methyl group from SAM is transferred to the nitrogen atom at the sixth position of adenosine (A) under the catalytic action of the methyltransferase complex, forming the m^6A modification. Simultaneously, SAM is converted to *S*-adenosylhomocysteine (SAH). The methyltransferase complex is often referred to as the "Writer" of m^6A. Among the components of the methyltransferase complex, methyltransferase-like protein 3 (METTL3) was the first methyltransferase identified and serves as the core subunit of the complex. METTL3 plays a critical catalytic role in m^6A formation. Studies have demonstrated that METTL3 specifically methylates target transcripts and is involved in various physiological processes, including embryonic development, brain development, spermatogenesis, cell reprogramming, and T-cell homeostasis [142]. Another key enzyme in the methyltransferase complex is methyltransferase-like protein 14 (METTL14). Sharing 43% amino acid sequence identity with METTL3, METTL14 forms a stable heterodimer with METTL3 and enhances methylation efficiency [143]. Wilms tumor 1-associated protein (WTAP) is another essential component of the methyltransferase complex. Although WTAP alone lacks methyltransferase activity in vitro, it forms a complex with METTL3 and METTL14, enhancing the complex's RNA-binding capacity. WTAP depletion significantly reduces the RNA-binding ability of the methyltransferase complex [144]. Additionally, other proteins have been identified as contributors to the RNA methylation process. For instance, methyltransferase-like protein 16 (METTL16) functions as the methyltransferase for U6 spliceosomal small nuclear RNA [145]. The KIAA1429 protein recruits and directs the METTL3/METTL14/WTAP complex to specific RNA sequences, accelerating the methylation reaction [146]. Zinc finger CCCH domain-containing protein 13 (ZC3H13), a newly identified auxiliary factor in the m^6A methyltransferase complex, plays a crucial role in regulating m^6A levels and sex determination in Drosophila [147].

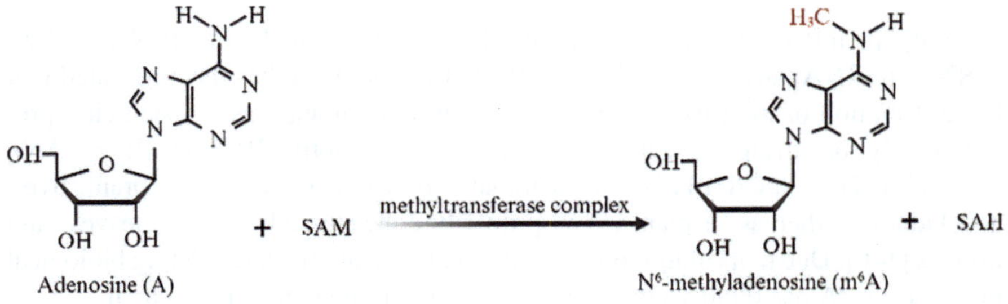

Fig. 11.23 The process of m^6A methylation

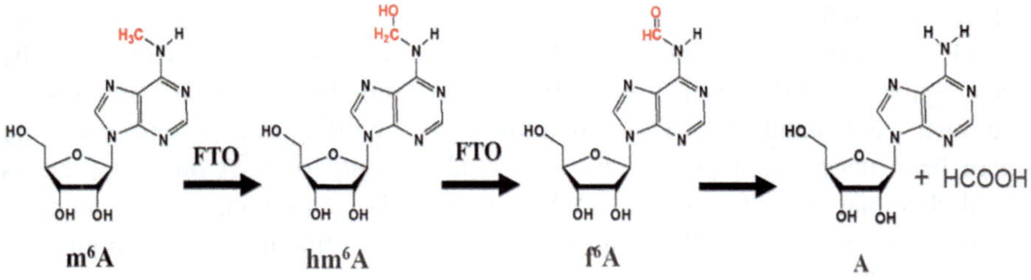

Fig. 11.24 Schematic representation of the FTO demethylation process

The methylation process of RNA m^6A modification is dynamic and reversible. Under the catalytic action of demethylases, m^6A is demethylated to restore adenosine (A). Hence, demethylases are also referred to as "Erasers" of m^6A. To date, two m^6A demethylases have been identified: Fat mass and obesity-associated protein (FTO) and AlkB homolog 5 (ALKBH5), both of which belong to the AlkB family of dioxygenases. FTO was the first demethylase discovered to remove m^6A modifications, and its demethylation mechanism is illustrated in Fig. 11.24. Initially, FTO oxidizes m^6A to N^6-hydroxymethyl adenosine (hm^6A), which is subsequently further oxidized to N^6-formyl adenosine (f^6A). Due to the instability of f^6A, it ultimately decomposes into adenosine (A) and formic acid [148]. ALKBH5, the second identified m^6A demethylase, is capable of removing m^6A modifications both in vivo and in vitro. However, its demethylation mechanism differs from that of FTO. ALKBH5 directly catalyzes the conversion of m^6A to adenosine (A) without producing detectable intermediates [149]. The discovery of m^6A demethylases has significantly expanded our understanding of the m^6A methylation process, unveiling its dynamic regulatory mechanisms. This represents a crucial advancement in epigenetics, highlighting the intricate control of RNA modifications and their roles in cellular processes.

The m^6A-binding proteins, also known as "readers" of m^6A methylation, specifically recognize and bind to transcripts containing m^6A modifications. Identified m^6A-binding proteins include members of the YTH family (YTHDC1/2, YTHDF1/2/3), the HNRNP family (HNRNPA2B1, HNRNPC, HNRNPG), and IGF2BP1/2/3, among others. These proteins interact with m^6A-modified RNA to regulate various processes, such as mRNA splicing, nuclear export, translation, and degradation [150–154]. For instance, YTHDC1 binds to m^6A-modified mRNA and selectively recruits serine/arginine-rich splicing factor 3 (SRSF3) while blocking the binding of serine/arginine-rich splicing factor 10 (SRSF10) [151]. This selective interaction protects m^6A-modified exons from being spliced out. Additionally, YTHDC1 facilitates the nuclear export of m^6A-containing transcripts by interacting with nuclear transport receptors [153]. HNRNPA2B1 interacts with the microRNA microprocessor complex protein DGCR8, directly binding to and regulating the maturation of pre-miRNAs containing m^6A modifications [155]. Meanwhile, YTHDF proteins display diverse roles: YTHDF1 enhances mRNA translation by interacting with translation initiation factors and ribosomes to accelerate ribosome recruitment, whereas YTHDF2 promotes mRNA decay by recruiting RNA degradation machinery [154]. YTHDF3 cooperates with YTHDF1 to facilitate protein synthesis while also regulating YTHDF2-mediated mRNA degradation [156]. IGF2BP proteins, on the other hand, stabilize m^6A-modified mRNAs by binding to them and protecting them from degradation [157]. The functional diversity of m^6A-binding proteins highlights their complex roles in RNA regulation. These proteins can act cooperatively or antagonistically, depending on the context, to fine-tune the posttranscriptional fate of m^6A-modified transcripts.

11.3.3 Biological Functions of m^6A

The discovery of methyltransferases, demethylases, and m^6A-binding proteins has significantly advanced the study of the biological functions of m^6A methylation in RNA. Researchers primarily investigate the roles of m^6A modification by selectively knocking out genes such as METTL3, ALKBH5, or YTHDF2 in various species.

1. Regulation of Hematopoietic Stem Cell Development: The hematopoietic system provides blood cells for the body throughout life, originating from multipotent hematopoietic stem cells (HSCs). Studies have shown that m^6A modifications play a critical role in HSC development and are essential for the maturation of the hematopoietic system [158]. Zebrafish research has revealed that m^6A methylation determines the transition from hemogenic endothelial cells to HSCs. Knocking out METTL3 in zebrafish embryos significantly reduces m^6A methylation levels in RNA and inhibits this transition [159]. In mice, METTL3 deletion in embryonic stem cells promotes the formation of endogenous double-stranded RNA (dsRNA), activating the MDA5-RIG-I, PKR-eIF2α, and OAS-RNase L signaling pathways in HSCs, ultimately causing hematopoietic failure and death [160]. Additionally, METTL3 promotes the proliferation of HSCs by increasing mRNA methylation levels of genes such as c-Myc, PTEN, and BCL2. In contrast, YTHDF2 suppresses HSC proliferation and differentiation by degrading the mRNA of genes such as ccnd1, c-Myc, and Axin2, thereby inhibiting the Wnt signaling pathway [161].
2. Regulation of Central Nervous System Development: The m^6A methylation level in RNA from the central nervous system (CNS) is higher than in other organs and increases from embryonic development to adulthood, suggesting its critical role in brain development [150]. Studies have demonstrated that m^6A modifications regulate Drosophila brain neurodevelopment [162], learning, and memory [163]. For instance, the deletion of METTL3 or METTL14 in the embryonic mouse brain prolongs the cell cycle of radial glial cells. In the hippocampus, METTL3 knockout impairs memory consolidation, while its overexpression significantly enhances long-term memory [164]. Furthermore, m^6A methylation regulates cortical neuron generation in the forebrain, influencing CNS development. Deletion of METTL14 reduces oligodendrocyte numbers and results in

CNS hypomyelination, disrupting postmitotic oligodendrocyte maturation [165]. FTO, highly expressed in adult neural stem cells and neurons, promotes neural stem cell proliferation and neuron differentiation. Its absence leads to reduced brain volume, learning, and memory deficits [163].

3. Regulation of Reproductive System Development: Primordial germ cells proliferate and differentiate into spermatogonia and oogonia in embryonic testes and ovaries, respectively. These cells develop into mature sperm and oocytes after meiotic division. Studies indicate that METTL3 and ALKBH5 promote sperm cell proliferation and viability by modulating mRNA m^6A levels. YTHDC1 and YTHDF2 regulate the maturation and translation of CPSF6 and Trcp5 transcripts, promoting oocyte proliferation and maturation. For example, YTHDF2 knockout in female mice results in immature oocytes and infertility [166]. In male mice, high ALKBH5 expression maintains normal reproductive function, while its deletion increases mRNA m^6A levels, causing seminiferous tubule abnormalities, testicular atrophy, and reduced fertility [167]. In zebrafish, METTL3 mutations lower m^6A methylation levels, disrupt gamete maturation, and significantly reduce fertility [168].
4. Regulation of Other Life Processes: Beyond hematopoietic, nervous, and reproductive system regulation, m^6A methylation plays vital roles in various biological processes. For instance, FTO influences adipogenesis by regulating alternative splicing of growth-related transcription factors via m^6A methylation [150]. Additionally, m^6A modifications participate in circadian rhythm regulation. The circadian clock is governed by the expression rates of clock genes, many of which exhibit m^6A methylation [169, 170]. Deletion of METTL3, which reduces m^6A levels, leads to lengthened circadian rhythms.

   Abnormal m^6A methylation is closely associated with numerous diseases, including psychiatric disorders [171–173], metabolic syndromes [174, 175], cardiovascular diseases [176, 177], and cancers such as acute myeloid leukemia [178, 179], brain tumors [180–183], and reproductive system-related cancers [184–188]. Therefore, quantitative detection of m^6A methylation in RNA is crucial for understanding its biological functions and its role in disease diagnosis and research.

11.3.4 Methods for m^6A Analysis

Early studies on RNA m^6A methylation primarily relied on the use of methylation inhibitors (e.g., adenosine analogs, cycloleucine, homocysteine) or the silencing of methyltransferases to observe phenotypic changes, thereby inferring the functional roles of m^6A modifications [189–192]. However, these approaches cannot precisely identify which gene regions are methylated and responsible for the observed phenotypic changes. Therefore, determining the exact locations and modification levels of m^6A on RNA is essential for accurately elucidating its biological significance. Currently, the detection methods for RNA m^6A methylation can be categorized into three main approaches: Single-nucleotide m^6A analysis, High-throughput sequencing of m^6A modifications, and Quantitative detection of m^6A at specific sites.

11.3.4.1 Single Nucleotide Analysis Method

The single nucleotide analysis of RNA m^6A methylation involves the complete hydrolysis of RNA fragments into free nucleotides, followed by detection using chemical or physical methods. This approach is suitable for quantifying the overall m^6A content, but cannot determine the specific distribution or percentage of m^6A at particular sites within the RNA. The primary techniques include liquid chromatography (LC), liquid chromatography-mass spectrometry (LC-MS), and thin-layer chromatography (TLC). Among these, LC and LC-MS offer high sensitivity and accuracy, with LC-MS particularly suited for precise quantification of overall m^6A percentages. In contrast, TLC, especially two-dimensional TLC, allows for rapid separation and qualitative analysis but has lower resolution and sensitivity. While single nucleotide analysis cannot provide information on m^6A's specific distribution in RNA, it is useful for preliminary screening or studies on total m^6A content variation.

LC was one of the earliest methods used to analyze m^6A methylation in RNA [193, 194]. This technique employs ribonuclease A, phosphodiesterase, and alkaline phosphatase to hydrolyze RNA into single nucleotides completely. The nucleotides are then separated using liquid chromatography, where different nucleotides exhibit distinct retention times on the chromatographic column. Quantification of m^6A is achieved by comparing peak areas to a standard curve. To improve accuracy, researchers have developed LC-MS, which combines the high separation efficiency of LC with the high sensitivity of mass spectrometry, enabling precise measurement of m^6A's overall percentage in RNA.

TLC includes one-dimensional TLC and two-dimensional TLC, with the latter offering higher resolution. Sample preparation is identical for both methods. Magnetic beads modified with oligo dT are used to enrich and isolate target mRNA, followed by ribonuclease T1 digestion to cleave phosphodiester bonds at guanosine residues in mRNA, resulting in fragmented mRNA. After enzymatic digestion, T4 polynucleotide kinase is used to label the 5′ ends of mRNA fragments with radioactive phosphate groups, and ribonuclease P1 is applied to digest the fragments into single nucleotides. These nucleotides are then separated and analyzed using TLC. One-dimensional TLC separates samples in a single direction and is simple to perform, but has low resolution, while two-dimensional TLC separates samples in two directions, significantly improving the resolution of m^6A from other nucleotides [137, 195, 196].

11.3.4.2 High-Throughput Sequencing

High-throughput sequencing techniques for analyzing m^{6}A methylation include immunoprecipitation-based sequencing methods (e.g., m^{6}A-seq [197] and MeRIP-seq [198]), photo-crosslinking-assisted sequencing methods [199, 200], and antibody-independent approaches [201–203]. m^{6}A-seq and MeRIP-seq were the first methods to map m^{6}A distribution and have significantly advanced our understanding of m^{6}A functions. However, their resolution is limited to approximately 100–200 nucleotides, and they require large amounts of RNA. Photo-crosslinking-assisted techniques (e.g., PAm^{6}A-seq) and crosslinking immunoprecipitation sequencing (miCLIP) improved resolution to the single-nucleotide level but are experimentally complex and rely heavily on antibody quality or specific RNA labeling.

Antibody-independent methods such as MazF digestion sequencing (m^{6}A-REF-seq) and sodium nitrite sequencing offer high specificity for detecting m^{6}A at specific sequences. Meanwhile, DART-seq, based on YTH family proteins and APOBEC1, stands out for its low sample requirement (10 ng) and suitability for single-cell analyses. Additionally, FTO-assisted m^{6}A-SEAL achieves high sensitivity and reliability by chemically labeling m^{6}A, making it well-suited for transcriptome-wide applications. These methods provide powerful tools for exploring m^{6}A distribution and functional mechanisms.

1. m^{6}A-seq and MeRIP-seq

   M^{6}A-seq and MeRIP-seq are the earliest reported transcriptome-wide m^{6}A detection methods and remain the most widely used techniques for m^{6}A high-throughput analysis [197, 198]. Both methods share the same principles, as illustrated in Fig. 11.25. RNA samples are fragmented into ~100 nt segments and divided into two portions: one serves as a reference sample, directly undergoing adapter ligation and cDNA library construction for sequencing, while the other is incubated with m^{6}A antibodies conjugated to magnetic beads for immunoprecipitation. RNA fragments containing m^{6}A modifications are eluted, converted into cDNA libraries, and sequenced. Comparative analysis of the sequencing results from the reference and immunoprecipitated samples allows the identification of m^{6}A-modified regions. These methods enabled the first transcriptome-wide mapping of m^{6}A, driving research on its functions and mechanisms. However, limitations include low resolution due to fragment size (~100–200 nt), lack of quantitative capability, and high RNA input requirements, making them unsuitable for rare samples.

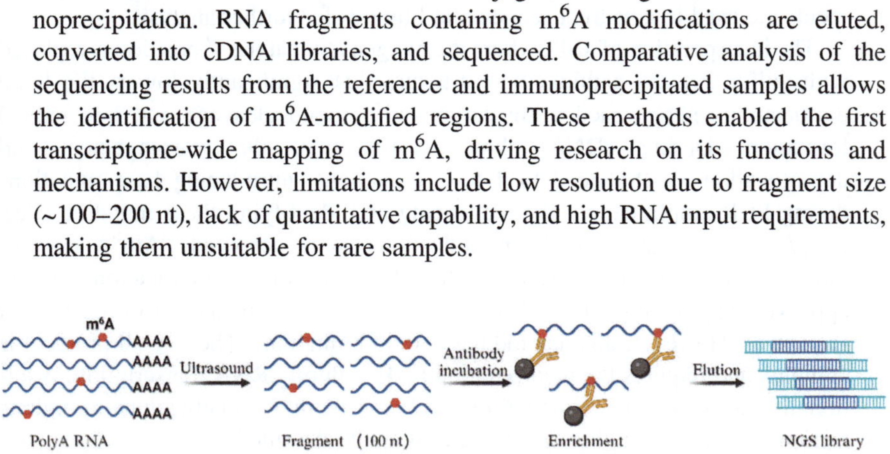

Fig. 11.25 Workflow of m^{6}A-seq [197]. (Revised and reprinted with permission from Ref. [197])

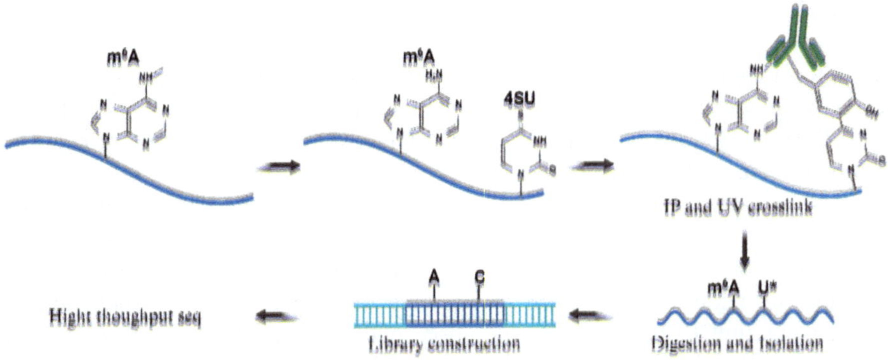

Fig. 11.26 Workflow of PAm^{6}A-seq [199]. (Revised and reprinted with permission from Ref. [199])

2. Photo-Crosslinking-Assisted m^{6}A Sequencing (PAm^{6}A-seq)

   To enhance the resolution of immunoprecipitation-based sequencing for m^{6}A detection, He et al. introduced a photo-crosslinking-assisted m^{6}A sequencing technology (PAm^{6}A-seq) in 2015, enabling more precise localization of m^{6}A sites within RNA [199]. The principle of this method is illustrated in Fig. 11.26. First, cells are cultured in a medium containing 4-thiouridine (4SU), which is randomly incorporated into newly synthesized mRNA sequences. After sufficient incubation, mRNA is extracted from the cells and incubated with m^{6}A-specific antibodies. Simultaneously, the samples are exposed to 365 nm UV light to induce crosslinking between the antibodies and the m^{6}A sites, as well as any nearby 4SU residues. Subsequently, the RNA is digested with RNase T1, which cleaves the mRNA into ~30 nt fragments. The cross-linked antibodies protect the mRNA fragments near the m^{6}A sites from degradation. Finally, the crosslinked antibodies are removed by proteinase K digestion, and the resulting RNA fragments are used to construct sequencing libraries for further analysis.

   The incorporation of 4SU, where the oxygen atom at the 4′ position is replaced with sulfur, creates a thioketone structure that significantly lowers the bond dissociation energy, facilitating carbon-sulfur bond cleavage and free radical formation. During cDNA library construction, the rearrangement and deprotonation of 4SU lead to T-to-C transitions, introducing base mutations during PCR amplification. By sequencing and identifying these mutation sites, the precise localization of m^{6}A modifications can be achieved. PAm^{6}A-seq provides single-nucleotide resolution for detecting m^{6}A modifications within approximately 30 nt regions, allowing for high-resolution analysis of methylated sequences. However, this method has certain limitations. The overall protocol is complex and requires the incorporation of 4SU during RNA transcription. Additionally, its detection is confined to specific sites; m^{6}A modifications in regions lacking 4SU incorporation cannot be accurately identified.
3. Crosslinking Immunoprecipitation Sequencing (miCLIP)

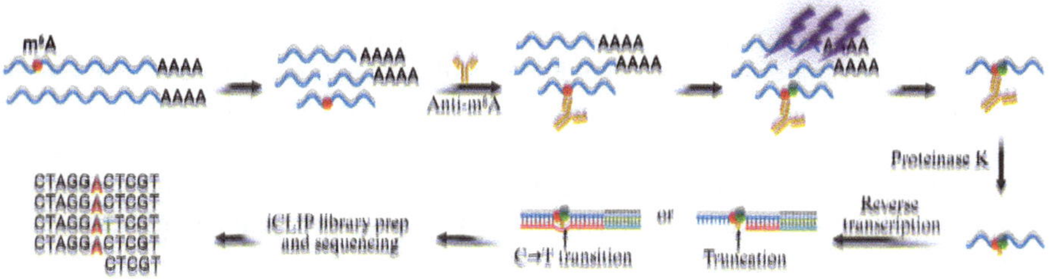

Fig. 11.27 Workflow of miCLIP [200]. (Revised and reprinted with permission from Ref. [200])

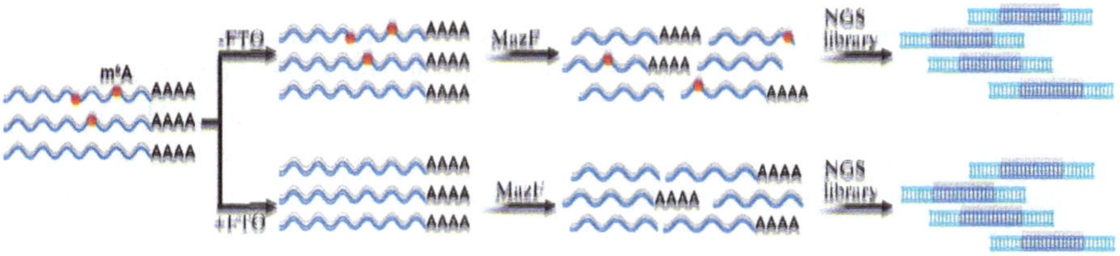

Fig. 11.28 Workflow of the m^{6}A-REF-seq [201]. (Revised and reprinted with permission from Ref. [201])

Jaffrey et al. developed and established the crosslinking immunoprecipitation sequencing method, known as miCLIP, which enables single-nucleotide resolution identification of m^{6}A modifications [200]. The principle of miCLIP is illustrated in Fig. 11.27: First, polyA RNA samples are fragmented and incubated with an m^{6}A-specific antibody. During the incubation, ultraviolet (UV) light at 254 nm is used to induce covalent crosslinking between the m^{6}A antibody and the RNA fragments. Subsequently, proteinase K is applied to partially digest the crosslinked antibody. During the reverse transcription process for cDNA library construction, residual antibody fragments at m^{6}A sites may cause reverse transcription termination or result in T-to-C base mutations. By sequencing and detecting these mutation sites, the precise locations of m^{6}A modifications on RNA molecules can be identified.

4. MazF Cleavage-Based Sequencing Method (m^{6}A-REF-seq)

The MazF nuclease specifically recognizes and cleaves the 5′ end of "ACA" sequences in RNA. However, when the "ACA" sequence is modified with m^{6}A (m^{6}ACA), the m^{6}A modification protects the site from cleavage. Leveraging this unique property of MazF, He et al. [201] developed an antibody-independent sequencing method capable of single-nucleotide resolution for m^{6}A detection, termed m^{6}A-REF-seq. The workflow of this method is illustrated in Fig. 11.28. First, Poly(A) RNA samples are treated with the MazF nuclease to cleave unmodified "ACA" sequences. Next, the resulting RNA fragments are repaired and ligated with sequencing adapters to construct a cDNA library for high-throughput sequencing. Finally, the location of "ACA" sequences in the sequenced RNA fragments—either at the termini or internal regions—is analyzed to identify potential m^6A sites. To reduce false positives, a negative control is used. In this control, poly(A) RNA is pretreated with the demethylase FTO to remove m^6A modifications. The RNA is then subjected to MazF digestion, ensuring that both m^6ACA and ACA sites are cleaved. The resulting RNA fragments are sequenced, and a cDNA library is constructed. By comparing sequencing results between untreated and FTO-treated samples, m^6A sites that show reduced methylation levels in the FTO-treated group are identified as true m^6A modification sites.

Using m^6A-REF-seq, researchers explored m^6A modifications in the mRNA of the HEK293T cell line at the transcriptome-wide level. High-throughput sequencing and data analysis identified over 4000 high-confidence m^6A sites, with a distribution pattern consistent with results from MeRIP-seq. The m^6A-REF-seq method offers precise m^6A site localization without relying on antibodies, making it a valuable tool for m^6A methylation research. However, its primary limitation is that it only detects m^6A modifications in the specific "ACA" sequence, leaving m^6A modifications in other sequences (e.g., "m^6ACU") undetectable.

5. Sodium Nitrite Sequencing

   Ryan Hili et al. [202] discovered that sodium nitrite, under acidic conditions, selectively deaminates unmethylated adenosine (A) into inosine (I). During reverse transcription, inosine pairs with cytosine (C), leading to the introduction of A-to-G base mutations. In contrast, methylated m^6A resists deamination and is still recognized as A during reverse transcription. Based on this principle, Ryan Hili developed a sodium nitrite sequencing method. In this method, the RNA sample to be analyzed is divided into two groups. One group serves as a reference sample, directly used to construct a cDNA library for high-throughput sequencing. The other group, serving as the experimental sample, is treated with sodium nitrite prior to constructing the cDNA library and subjected to high-throughput sequencing. By comparing the sequencing results of the experimental and reference samples, sites without "G" mutations are identified as m^6A sites. This method achieves single-base resolution and can also be applied to detect m^6A methylation in DNA. However, sodium nitrite sequencing has certain limitations. Since m^6Am (N^6, 2-O-dimethyladenosine) also resists deamination by sodium nitrite, the method cannot distinguish between m^6A and m^6Am. This limitation may lead to overestimation of m^6A levels in regions enriched with m^6A. Therefore, additional data processing and comparison are required to improve the accuracy of the results.

6. Binding Proteins (YTH) and Cytidine Deaminase (APOBEC1)-based m^6A Sequencing Technology (DART-seq)

   DART-seq is an innovative m^6A sequencing method leveraging the m^6A-binding specificity of YTH proteins and the enzymatic activity of cytidine deaminase APOBEC1. The m^6A modification typically occurs within the conserved sequence Gm^6AC or Am^6AC, where the methylated adenosine (m^6A) is immediately followed by a cytosine (C). APOBEC1 can deaminate cytosine (C) to uracil (U), and by detecting C-to-U conversion events adjacent to m^6A

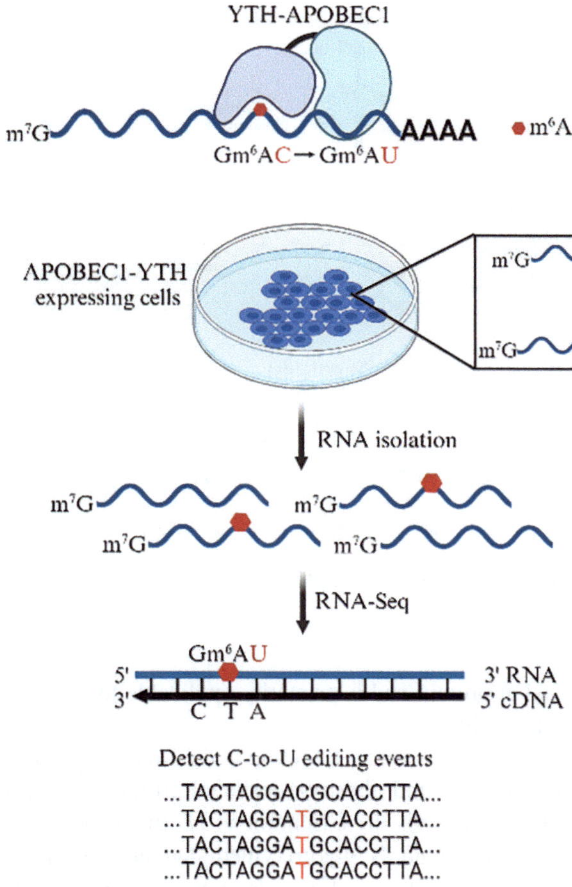

Fig. 11.29 Overview of DART-seq for m^{6}A assay [203]. (Revised and reprinted with permission from Ref. [203])

sites, the location of m^{6}A modifications can be inferred. However, APOBEC1 lacks intrinsic m^{6}A recognition capability, whereas YTH family proteins specifically bind to m^{6}A-modified RNA. In 2019, Meyer et al. [203] engineered a fusion protein combining APOBEC1 with YTH (APOBEC1-YTH), establishing the DART-seq technique. The principle of this method is illustrated in Fig. 11.29. First, the APOBEC1-YTH fusion protein is expressed in cells, where it specifically binds to m^{6}A-modified RNA and catalyzes the deamination of the cytosine immediately following the m^{6}A site, converting it to uracil. The RNA is then extracted and sequenced. By aligning the sequencing data to the reference genome, C-to-U conversions downstream of adenosines indicate m^{6}A modification sites. DART-seq requires as little as 10 ng of total RNA from cells, significantly reducing the RNA sample requirement and enabling single-cell transcriptome-wide m^{6}A analysis. However, the method relies on efficient

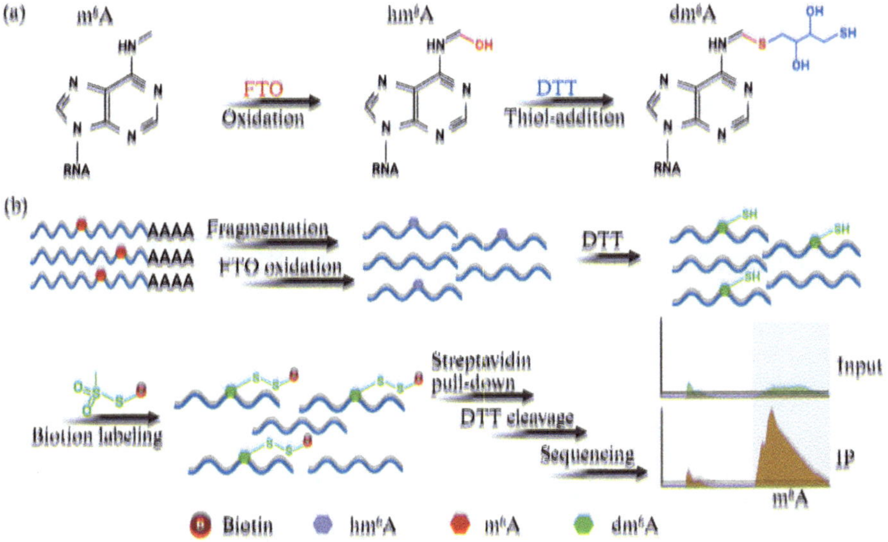

Fig. 11.30 Overview of m^{6}A-SEAL for m^{6}A assay [204]. (Revised and reprinted with permission from Ref. [204])

transfection of the fusion protein and is primarily applicable to in vitro samples, limiting its broader applicability.

7. FTO-Assisted m^{6}A Selective Chemical Labeling and Sequencing (m^{6}A-SEAL)

The enzyme FTO (fat mass and obesity-associated protein), which functions as an m^{6}A demethylase, catalyzes the oxidation of N^6-methyladenosine (m^{6}A) into a highly reactive intermediate, N^6-hydroxymethyladenosine (hm^{6}A). Building on this catalytic property, Jia et al. [204] developed the FTO-assisted m^{6}A selective chemical labeling and sequencing method, termed m^{6}A-SEAL. The overview of m^{6}A-SEAL is illustrated in Fig. 11.30. First, FTO catalyzes the oxidation of m^{6}A to generate hm^{6}A, a highly reactive intermediate. Next, the thiol (-SH) groups in dithiothreitol (DTT) undergo a 1,2-addition reaction with the imine group in hm^{6}A, stabilizing it into a thiol adduct product (dm^{6}A). The free -SH group on dm^{6}A subsequently reacts with methanethiosulfonate (MTSEA) to chemically attach biotin to the m^{6}A methylation site on the mRNA. Biotin-labeled RNA is then efficiently enriched using streptavidin-functionalized magnetic beads. The enriched RNA samples can be subjected to high-throughput sequencing, enabling precise identification of m^{6}A methylation sites. Using this approach, researchers conducted transcriptome-wide sequencing of human HEK293T cells and rice. The results demonstrated that m^{6}A-SEAL offers higher reliability and sensitivity compared to antibody-based sequencing methods or DART-seq. This advancement highlights the potential of m^{6}A-SEAL for robust and accurate mapping of m^{6}A methylation modifications across diverse transcriptomes.

11.3.4.3 Quantitative Detection of Site-Specific m^{6}A Modifications

While sequencing technologies have provided a comprehensive understanding of m^{6}A methylation distribution at the transcriptome level, detailed information on the modification levels of individual methylation sites remains limited. Thus, developing precise methods for site-specific localization and quantitative analysis of m^{6}A is essential. Currently, site-specific m^{6}A detection methods include several advanced techniques: site-specific cleavage and radioactive labeling followed by thin-layer chromatography (SCARLET), single-base extension assays utilizing m^{6}A-sensitive enzymes, ligation-PCR, and ligation-PCR based on single-base extension.

1. SCARLET

   SCARLET achieves high precision and quantitative analysis of m^{6}A methylation at specific sites through selective enzymatic digestion and radioactive labeling. However, its complexity and the use of radioactive reagents limit its practical applications. The single-base extension (SBE) method leverages the property of m^{6}A to block polymerase extension, offering simplicity and compatibility with various RNA types. Similarly, the ligation-PCR method combines ligase activity with amplification technology to enhance detection sensitivity via real-time fluorescence monitoring. The SELECT method integrates single-base extension with ligase-mediated selectivity, enabling highly specific and sensitive site-specific detection. This method has been commercialized, facilitating rapid and quantitative m^{6}A detection.

   Pan et al. [205] developed a novel thin-layer chromatography-based method, SCARLET, leveraging specific enzymatic cleavage, splint ligation reactions, and radioactive labeling to accurately identify m^{6}A methylation sites on mRNA and lncRNA and quantitatively determine the methylation levels at these sites. The principle and overview of the SCARLET are illustrated in Fig. 11.31. A target site (X) is selected on the RNA molecule (mRNA or lncRNA). The site X can be either adenosine (A) or methylated adenosine (m^{6}A). A complementary DNA probe (Probe 1) is designed to partially hybridize with the target RNA near site X. The portion of Probe 1 complementary to the 5′ end of site X is a standard DNA sequence, while the remaining regions are modified with 2′-O-methylation (2′-OMe). Probe 1 hybridizes with the target RNA molecule. Under the action of RNase H, the RNA strand complementary to the standard DNA sequence of

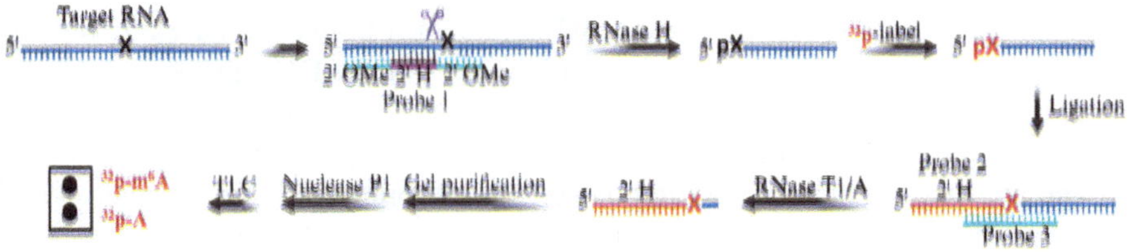

Fig. 11.31 The principle and overview of the SCARLET [205]. (Revised and reprinted with permission from Ref. [205])

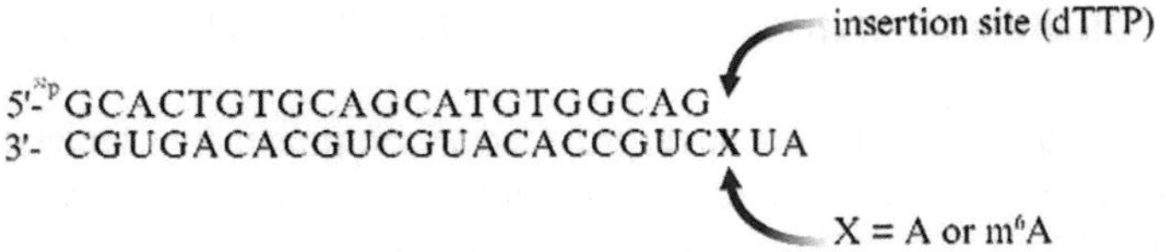

Fig. 11.32 Schematic representation of m^{6}A methylation detection using Tth DNA polymerase single-nucleotide extension

Probe 1 is hydrolyzed, exposing site X at the RNA's 2′ end (pX-RNA). The RNA molecule is labeled with radioactive isotope ^{32}P, converting pX-RNA into ^{32}pX-RNA. Probe 2 hybridizes to the target molecule ^{32}pX-RNA, and Probe 3 forms a splint structure. In the presence of T4 DNA ligase, Probe 3 is ligated to ^{32}pX-RNA, forming DNA-^{32}pX-RNA. Then, RNases T1 or A are used to digest the RNA portion, leaving DNA-^{32}pX-p or DNA-^{32}pX-Cp. The resulting products are gel-purified. Next, the purified DNA is hydrolyzed into individual 5-′-nucleotides using nuclease P1. The nucleotide mixture is applied to a TLC plate, separated, and visualized. Finally, by analyzing the position of the radioactive signal on the TLC plate, the presence of m^{6}A at site X can be determined. The relative signal intensities of ^{32}P-m^{6}A and ^{32}P-A allow for quantification of the proportion of m^{6}A methylation at the target site.

SCARLET enables precise identification of m^{6}A sites and accurate quantification of methylation levels, making it suitable for studying m^{6}A modifications at multiple sites in mRNA and lncRNA across different cell types. However, the method is technically complex and involves radioactive materials, limiting its broader applicability due to operational challenges and safety concerns.

2. DNA Polymerase Single-Nucleotide Extension for the Detection of m^{6}A Methylation

The Kool team discovered that Thermus thermophilus (Tth) DNA polymerase exhibits remarkable selectivity toward m^{6}A methylation. Specifically, Tth DNA polymerase is inhibited by m^{6}A methylation on RNA templates during single-nucleotide extension (SNE) reactions, preventing the extension of DNA primers. Leveraging this property, the team developed a method for detecting m^{6}A methylation sites on RNA using Tth DNA polymerase in SNE assays [206]. The detection principle is illustrated in Fig. 11.32. First, a target site (X) on the RNA is selected, and a radiolabeled DNA primer is designed to hybridize with the RNA template at the region of interest. During the SNE reaction, Tth DNA polymerase and dTTP are used to extend the DNA primer. If the target site X is an unmethylated adenosine (A), the DNA primer extends by one nucleotide (T). However, if the target site X is methylated (m^{6}A), the m^{6}A modification inhibits the extension reaction. The change in primer length before and after the reaction is analyzed using gel electrophoresis, and the intensity of the gel bands is quantified to determine the presence and degree of m^{6}A methylation at the target site. In 2015, Zhou et al. [207] further demonstrated that *Bst* DNA polymerase also exhibits high selectivity for m^{6}A. They developed a similar

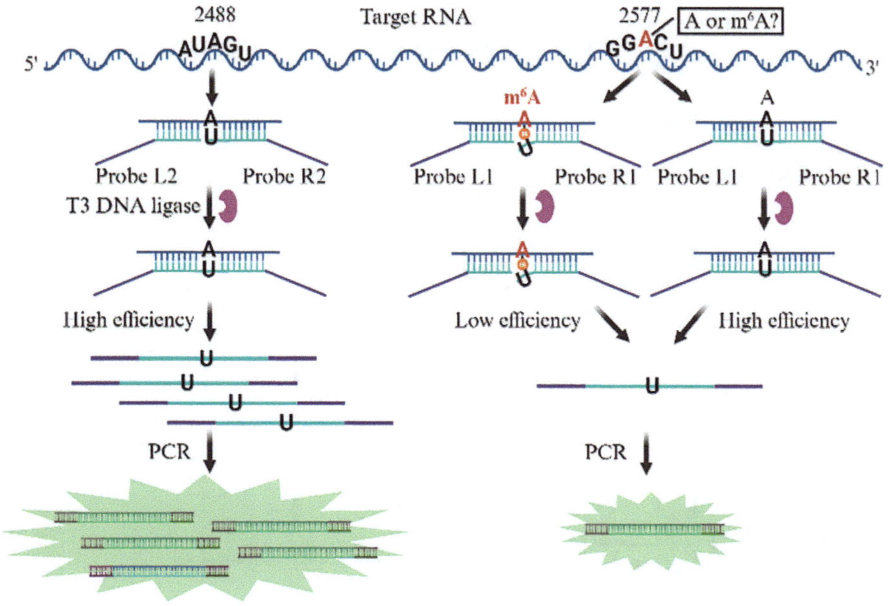

Fig. 11.33 Schematic representation of the ligation-PCR-based method for quantitative analysis of m^6A methylation [208]. (Revised and reprinted with permission from Ref. [208])

m^6A detection method based on *Bst* DNA polymerase SNE reactions, successfully applying it to analyze m^6A modifications on rRNA in HeLa cells.

3. Ligation-PCR for m^6A Assay

   Li et al. [208] found that m^6A methylation modification inhibits the ligation activity of T3 DNA ligase. Based on this finding, they developed a ligation-PCR method for the quantitative analysis of m^6A methylation in RNA (Fig. 11.33). The approach involves designing two DNA probes, L1 and R1, complementary to the sequences upstream and downstream of the target site. When the target site contains adenosine (A), probes L1 and R1 hybridize to the RNA and are ligated by T3 DNA ligase. The ligation product can then be amplified by PCR, and the fluorescence signal generated during the amplification process is monitored in real time using SYBR Green I dye. However, if the target site contains m^6A instead of A, the m^6A modification inhibits the ligation activity of T3 DNA ligase, resulting in reduced ligation product and, consequently, significantly decreased PCR amplification. By monitoring the fluorescence signals in real time and analyzing the cycle threshold (C_T) values from the fluorescence curves, the amount of adenosine (A) at the target site can be quantified. The m^6A content is calculated as the difference between the total RNA target molecule quantity and the adenosine (A) quantity. To determine the total RNA target molecule quantity, a reference site without m^6A modification is selected. A separate pair of probes (L2 and R2) is used for the ligation-PCR reaction at the reference site. The C_T values from the fluorescence curves are then used to calculate the total RNA

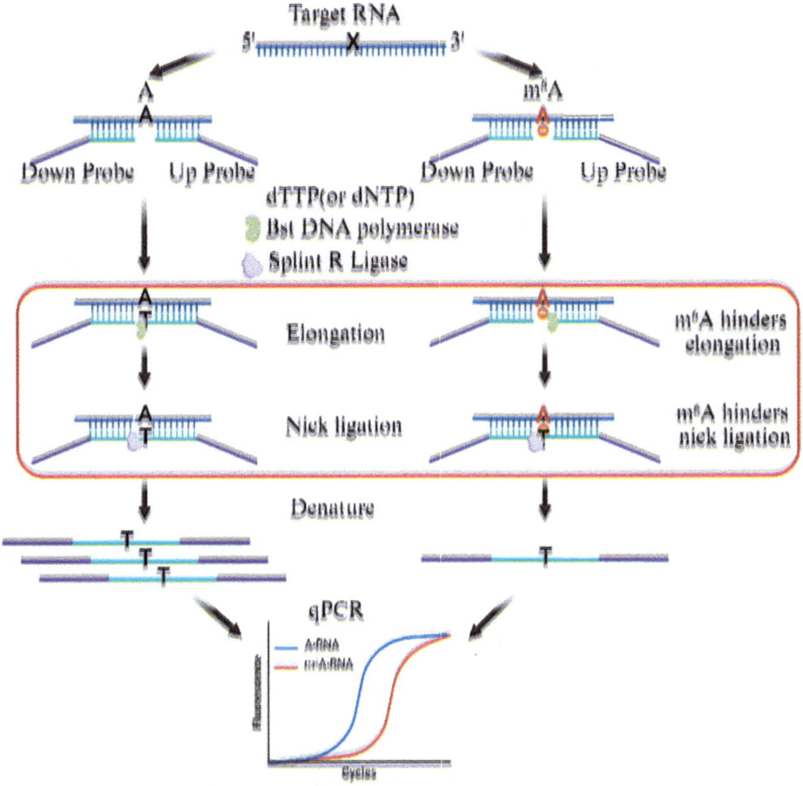

Fig. 11.34 Principle and overview of SELECT [209]. (Revised and reprinted with permission from Ref. [209])

concentration and the percentage of m^6A at the target site. This method holds the high selectivity of T3 DNA ligase for m^6A and combines it with PCR amplification to achieve highly sensitive detection of RNA m^6A methylation. For different m^6A methylation sites, corresponding DNA probes can be designed. Furthermore, the ligation products from different sites can be amplified using the same pair of PCR primers. For multiplex detection, TaqMan probes labeled with different fluorescent groups can be employed to simultaneously detect various amplification products. This expands the method's potential for multiplex m^6A methylation analysis.

4. Single-Base Extension and Ligation-PCR Method (SELECT)

Jia et al. [209] developed a single-base extension and ligation-PCR method (SELECT) to quantitatively detect site-specific m^6A methylation, coupling the inhibitory effects of m^6A on both *Bst* DNA polymerase extension activity and SplintR ligase ligation activity. The principle and overview of SELECT are illustrated in Fig. 11.34. First, a pair of DNA probes (Up Probe and Down Probe) is designed to hybridize with the target RNA molecule, leaving a single-nucleotide gap at the site of interest. If the target site contains adenosine (A), the downstream probe is extended by one thymidine (T) base via *Bst* DNA polymerase to fill the gap. Subsequently, SplintR ligase catalyzes the ligation of the upstream and downstream probes. The resulting ligation product serves as a template for amplification using PCR primers. However, when the target site contains m^6A, the methyl group on m^6A inhibits the single-base extension of the downstream probe. This prevents the ligation of the upstream and downstream probes by SplintR ligase due to the remaining single-nucleotide gap. Even if a partial extension of the downstream probe occurs, m^6A further hinders the ligation activity of SplintR ligase, resulting in significantly reduced ligation and PCR amplification products. Under conditions where the total amount of target RNA is constant, the 2^{CT} value (cycle threshold from the PCR-generated fluorescence curve) is proportional to the proportion of m^6A modifications at the site of interest. By analyzing the C_T value of the fluorescence curve, the degree of m^6A methylation at the target site can be determined. To measure the total RNA amount, a reference site without m^6A modification is included in the experiment. Single-base extension and ligation-PCR reactions are performed at this reference site, and the total RNA quantity is calculated from the corresponding C_T values of the real-time fluorescence curve. The SELECT method employs two layers of selection—single-base extension by polymerase and specific ligation by ligase—significantly enhancing its selectivity for m^6A methylation. Coupled with PCR amplification, the method also ensures high sensitivity. This technique has been successfully commercialized as a rapid and quantitative m^6A detection kit, providing a robust tool for site-specific m^6A methylation analysis.

Pian et al. [210] reported a single-base extension and ligation-loop mediated isothermal strategy (GEXL-LAMP) to quantitatively detect site-specific m^6A methylation, combining the inhibitory effects of m^6A on both *Bst* DNA polymerase extension activity and SplintR ligase ligation activity. The outline of GEXL-LAMP is displayed in Fig. 11.35. A pair of DNA probes (Probe A and Probe B) containing a stem-loop structure is designed to target the RNA. Probe A and probe B hybridize with the target RNA, leaving a single nucleotide gap at the site to be detected. When the target site is A, *Bst* DNA polymerase catalyzes the single-base extension of dUTP at the 3′ end of Probe B. Subsequently, under the action of SplintR ligase, Probe A and Probe B are ligated to form a double dumbbell DNA template, which can initiate rapid and efficient LAMP amplification. On the contrary, when the target site is modified with m^6A, the single-base extension reaction of dUTP catalyzed by *Bst* DNA polymerase is significantly hindered, leaving a gap between Probe A and Probe B. This gap inhibits the ligation reaction between the two probes. Even if minimal extension occurs, the m^6A modification further suppresses the activity of SplintR ligase, thereby impeding the ligation reaction and significantly reducing the ligation product. Consequently, when the target RNA contains m^6A modification, the LAMP product is significantly reduced, enabling the detection of the m^6A modification at the specific site. To quantitatively analyze the m^6A methylation level at this site, the GEXL-LAMP method selects a reference site on the target sequence that does not contain m^6A modification. Single-base

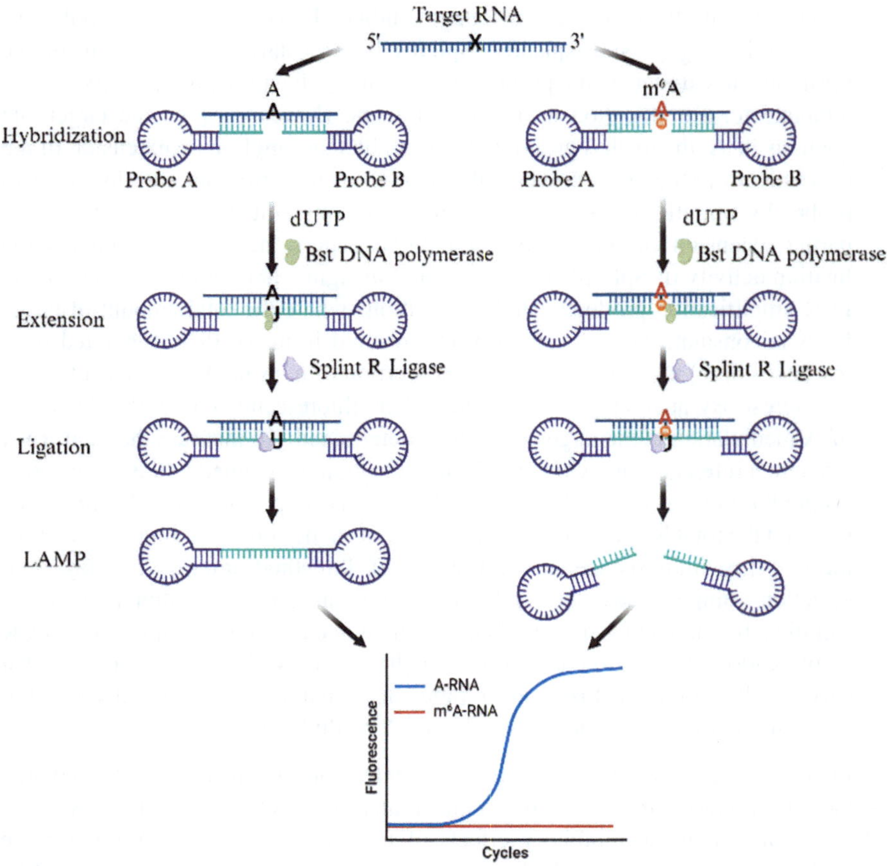

Fig. 11.35 Principle and overview of GEXL-LAMP [210]. (Revised and reprinted with permission from Ref. [210])

extension ligation LAMP reactions are performed, and the total amount of target RNA is determined from the real-time fluorescence curve. The difference between the total amount of target RNA and the unmodified RNA at the site to be measured allows the calculation of the m^{6}A modification level at that site. The results demonstrate that the GEXL-LAMP method can accurately quantify the m^{6}A methylation level in actual biological samples, showing its strong practical applicability.

In summary, different methods exhibit distinct strengths in terms of resolution, sensitivity, and applicability. Single-nucleotide analysis is well-suited for total quantification, high-throughput sequencing excels in exploring distribution patterns, and site-specific quantitative detection is ideal for functional studies and in-depth mechanistic investigations. The continuous advancement of these methodologies provides comprehensive technical support for m^{6}A modification research and lays a solid foundation for the development of more efficient and convenient m^{6}A detection technologies in the future.

References

1. Black DL. Mechanisms of alternative pre-messenger RNA splicing. Annu Rev Biochem. 2003;72:291–336.
2. Blencowe BJ. Alternative splicing: new insights from global analyses. Cell. 2006;126:37–47.
3. Ast G. How did alternative splicing evolve? Nat Rev Genet. 2004;5:773–82.
4. Matlin AJ, Clark F, Smith CWJ. Understanding alternative splicing: towards a cellular code. Nat Rev Mol Cell Biol. 2005;6:386–98.
5. Ren XJ, Zhang KX, Deng RJ, Li JH. RNA splicing analysis: from in vitro testing to single-cell imaging. Chem. 2019;5:2571–92.
6. Le KQ, Prabhakar BS, Hong WJ, Li LC. Alternative splicing as a biomarker and potential target for drug discovery. Acta Pharmacol Sin. 2015;36:1212–8.
7. Keren H, Lev-Maor G, Ast G. Alternative splicing and evolution: diversification, exon definition and function. Nat Rev Genet. 2010;11:345–55.
8. Wang ET, Sandberg R, Luo SJ, Khrebtukova I, Zhang L, Mayr C, Kingsmore SF, Schroth GP, Burge CB. Alternative isoform regulation in human tissue transcriptomes. Nature. 2008;456: 470–6.
9. Sammeth M, Foissac S, Guigó R. A general definition and nomenclature for alternative splicing events. PLoS Comput Biol. 2008;4:e1000147.
10. Chen LL. The biogenesis and emerging roles of circular RNAs. Nat Rev Mol Cell Biol. 2016;17:205–11.
11. Li X, Yang L, Chen LL. The biogenesis, functions, and challenges of circular RNAs. Mol Cell. 2018;71:428–42.
12. Bentley DL. Coupling mRNA processing with transcription in time and space. Nat Rev Genet. 2014;15:163–75.
13. Pagani F, Baralle FE. Genomic variants in exons and introns: identifying the splicing spoilers. Nat Rev Genet. 2004;5:389–96.
14. Shi YG. Mechanistic insights into precursor messenger RNA splicing by the spliceosome. Nat Rev Mol Cell Biol. 2017;18:655–70.
15. Graveley BR, Hertel KJ, Maniatis T. The role of U2AF35 and U2AF65 in enhancer-dependent splicing. RNA. 2001;7:806–18.
16. Jurica MS, Moore MJ. Pre-mRNA splicing: awash in a sea of proteins. Mol Cell. 2003;12:5–14.
17. Will CL, Lührmann R. Spliceosome structure and function. Cold Spring Harb Perspect Biol. 2011;3:a003707.
18. Wan RX, Bai R, Yan CY, Lei JL, Shi YG. Structures of the catalytically activated yeast spliceosome reveal the mechanism of branching. Cell. 2019;177:339–351.e313.
19. Wan RX, Yan CY, Bai R, Lei JL, Shi YG. Structure of an intron lariat spliceosome from Saccharomyces cerevisiae. Cell. 2017;171:120–132.e112.
20. Zhang XF, Yan CY, Hang J, Finci LI, Lei JL, Shi YG. An atomic structure of the human spliceosome. Cell. 2017;169:918–92.e914.
21. Bai R, Wan RX, Yan CY, Lei JL, Shi YG. Structures of the fully assembled Saccharomyces cerevisiae spliceosome before activation. Science. 2018;360:1423–8.
22. Yan CY, Wan RX, Bai R, Huang GXY, Shi YG. Structure of a yeast step II catalytically activated spliceosome. Science. 2017;355:149–55.
23. Wan RX, Yan CY, Bai R, Huang GXY, Shi YG. Structure of a yeast catalytic step I spliceosome at 3.4 Å resolution. Science. 2016;353:895–904.
24. Yan CY, Wan RX, Bai R, Huang GXY, Shi YG. Structure of a yeast activated spliceosome at 3.5 Å resolution. Science. 2016;353:904–11.
25. Zhan XC, Yan CY, Zhang XF, Lei JL, Shi YG. Structure of a human catalytic step I spliceosome. Science. 2018;359:537–44.
26. Hang J, Wan RX, Yan CY, Shi YG. Structural basis of pre-mRNA splicing. Science. 2015;349:1191–8.

27. Yan CY, Hang J, Wan RX, Huang M, Wong CCL, Shi YG. Structure of a yeast spliceosome at 3.6-angstrom resolution. Science. 2015;349:1182–91.
28. Wan RX, Yan CY, Bai R, Wang L, Huang M, Wong CCL, Shi YG. The 3.8 Å structure of the U4/U6.U5 tri-snRNP: insights into spliceosome assembly and catalysis. Science. 2016;351: 466–75.
29. Early P, Rogers J, Davis M, Calame K, Bond M, Wall R, Hood L. Two mRNAs can be produced from a single immunoglobulin μ gene by alternative RNA processing pathways. Cell. 1980;20:313–9.
30. Baker BS. Sex in flies: the splice of life. Nature. 1989;340:521–4.
31. Schmucker D, Clemens JC, Shu H, Worby CA, Xiao J, Muda M, Dixon JE, Zipursky SL. Drosophila Dscam is an axon guidance receptor exhibiting extraordinary molecular diversity. Cell. 2000;101:671–84.
32. Yang Y, Zhan LL, Zhang WJ, Sun F, Wang WF, Tian N, Bi JP, Wang HT, Shi DK, Jiang YJ, Zhang YZ, Jin YF. RNA secondary structure in mutually exclusive splicing. Nat Struct Mol Biol. 2011;18:159–68.
33. Black DL. Protein diversity from alternative splicing: a challenge for bioinformatics and post-genome biology. Cell. 2000;103:367–70.
34. Roy B, Haupt LM, Griffiths LR. Review: alternative splicing (AS) of genes as an approach for generating protein complexity. Curr Genomics. 2013;14:182–94.
35. Pan Q, Shai O, Lee LJ, Frey BJ, Blencowe BJ. Deep surveying of alternative splicing complexity in the human transcriptome by high-throughput sequencing. Nat Genet. 2009;41:762.
36. He CJ, Zhou F, Zuo ZX, Cheng HH, Zhou RJ. A global view of cancer-specific transcript variants by subtractive transcriptome-wide analysis. PLoS One. 2009;4:e4732.
37. Sveen A, Kilpinen S, Ruusulehto A, Lothe RA, Skotheim RI. Aberrant RNA splicing in cancer; expression changes and driver mutations of splicing factor genes. Oncogene. 2016;35: 2413–27.
38. Fackenthal JD, Godley LA. Aberrant RNA splicing and its functional consequences in cancer cells. Dis Model Mech. 2008;1:37–42.
39. Chen J, Weiss WA. Alternative splicing in cancer: implications for biology and therapy. Oncogene. 2015;34:1–14.
40. Skotheim RI, Nees M. Alternative splicing in cancer: noise, functional, or systematic? Int J Biochem Cell Biol. 2007;39:1432–49.
41. Lee SCW, Abdel-Wahab O. Therapeutic targeting of splicing in cancer. Nat Med. 2016;22: 976–86.
42. Gardina PJ, Clark TA, Shimada B, Staples MK, Yang Q, Veitch J, Schweitzer A, Awad T, Sugnet C, Dee S, Davies C, Williams A, Turpaz Y. Alternative splicing and differential gene expression in colon cancer detected by a whole genome exon array. BMC Genomics. 2006;7: 325.
43. Feng L, Huang J, Chen JJ. MERIT40 facilitates BRCA1 localization and DNA damage repair. Genes Dev. 2009;23:719–28.
44. Wang B, Hurov K, Hofmann K, Elledge SJ. NBA1, a new player in the Brca1 A complex, is required for DNA damage resistance and checkpoint control. Genes Dev. 2009;23:729–39.
45. Orban TI, Olah E. Expression profiles of BRCA1 splice variants in asynchronous and in G1/S synchronized tumor cell lines. Biochem Biophys Res Commun. 2001;280:32–8.
46. Sun L, Yu CX, Irudayaraj J. Raman multiplexers for alternative gene splicing. Anal Chem. 2008;80:3342–9.
47. Kimura-Suda H, Petrovykh DY, Tarlov MJ, Whitman LJ. Base-dependent competitive adsorption of single-stranded DNA on gold. J Am Chem Soc. 2003;125:9014–5.
48. Sun L, Irudayaraj J. PCR-free quantification of multiple splice variants in a cancer gene by surface-enhanced Raman spectroscopy. J Phys Chem B. 2009;113:14021–5.
49. Lee K, Drachev VP, Irudayaraj J. DNA-gold nanoparticle reversible networks grown on cell surface marker sites: application in diagnostics. ACS Nano. 2011;5:2109–17.

50. Reinhard BM, Siu M, Agarwal H, Alivisatos AP, Liphardt J. Calibration of dynamic molecular rule based on plasmon coupling between gold nanoparticles. Nano Lett. 2005;5:2246–52.
51. Lee K, Cui Y, Lee LP, Irudayaraj J. Quantitative imaging of single mRNA splice variants in living cells. Nat Nanotechnol. 2014;9:474–80.
52. Yang D, Le J. Targeted amplification of alternatively spliced transcripts of major histocompatibility complex class I heavy chain. J Immunol Methods. 1994;176:265–70.
53. Wang F, Zhao Y, Hao YH, Tan Z. Identification of low-abundance alternatively spliced mRNA variants by exon exclusive reverse transcriptase polymerase chain reaction. Anal Biochem. 2008;383:307–10.
54. Sun B, Tao L, Zheng YL. Simultaneous quantification of alternatively spliced transcripts in a single droplet digital PCR reaction. BioTechniques. 2014;56:319–25.
55. Vargas IME, Vivas-Mejia PE. Assessment of mRNA splice variants by qRT-PCR. Methods Mol Biol. 2013;1049:171–86.
56. Vandenbroucke II, Vandesompele J, Paepe AD, Messiaen L. Quantification of splice variants using real-time PCR. Nucleic Acids Res. 2001;29:e68.
57. Brosseau JP, Lucier JF, Lapointe E, Durand M, Gendron D, Gervais-Bird J, Tremblay K, Perreault JP, Abou Elela S. High-throughput quantification of splicing isoforms. RNA. 2010;16:442–9.
58. Wang HH, Wang H, Duan XR, Sun YY, Wang XD, Li ZP. Highly sensitive and multiplexed quantification of mRNA splice variants by the direct ligation of DNA probes at the exon junction and universal PCR amplification. Chem Sci. 2017;8:3635–40.
59. Jia YT, Han J, Wang H, Hong WX, Wang HH, Zhang M, Li ZP. Ultrasensitive quantification of multiplexed mRNA variants via splice-junction anchored DNA probes and SplintR ligase-initiated PCR. Chem Commun. 2021;57:10011–4.
60. Hermiston ML, Xu Z, Weiss A. CD45: a critical regulator of signaling thresholds in immune cells. Annu Rev Immunol. 2003;21:107–37.
61. Machura E, Mazur B, Pieniazek W, Karczewska K. Expression of naive/memory (CD45RA/CD45RO) markers by peripheral blood $CD4^+$ and $CD8^+$ T cells in children with asthma. Arch Immunol Ther Exp. 2008;56:55–62.
62. Tchilian EZ, Beverley PCL. Altered CD45 expression and disease. Trends Immunol. 2006;27:146–53.
63. Zhu J, Yamane H, Paul WE. Differentiation of effector CD4 T cell populations. Annu Rev Immunol. 2009;28:445–89.
64. Ren XJ, Deng RJ, Zhang KX, Sun YP, Teng XC, Li JH. SpliceRCA: in situ single-cell analysis of mRNA splicing variants. Acs Central Sci. 2018;4:680–7.
65. Ren XJ, Deng RJ, Wang LD, Zhang KX, Li JH. RNA splicing process analysis for identifying antisense oligonucleotide inhibitors with padlock probe-based isothermal amplification. Chem Sci. 2017;8:5692–8.
66. Katoh Y, Katoh M. FGFR2-related pathogenesis and FGFR2-targeted therapeutics. Int J Mol Med. 2009;23:307–11.
67. Su FX, Wang GH, Ji JN, Zhang PB, Wang FF, Li ZP. Real-time detection of mRNA splicing variants with specifically designed reverse-transcription loop-mediated isothermal amplification. RSC Adv. 2020;10:6271–6.
68. Bryant DW Jr, Priest HD, Mockler TC. Detection and quantification of alternative splicing variants using RNA-seq. Methods Mol Biol (Clifton, NJ). 2012;883:97–110.
69. Martin JA, Wang Z. Next-generation transcriptome assembly. Nat Rev Genet. 2011;12:671–82.
70. Ozsolak F, Milos PM. RNA sequencing: advances, challenges and opportunities. Nat Rev Genet. 2011;12:87–98.
71. Sultan M, Schulz MH, Richard H, Magen A, Klingenhoff A, Scherf M, Seifert M, Borodina T, Soldatov A, Parkhomchuk D, Schmidt D, O'Keeffe S, Haas S, Vingron M, Lehrach H, Yaspo ML. A global view of gene activity and alternative splicing by deep sequencing of the human transcriptome. Science. 2008;321:956–60.

72. Hu Y, Huang Y, Du Y, Orellana CF, Singh D, Johnson AR, Monroy A, Kuan PF, Hammond SM, Makowski L, Randell SH, Chiang DY, Hayes DN, Jones C, Liu YF, Prins JF, Liu JZ. DiffSplice: the genome-wide detection of differential splicing events with RNA-seq. Nucleic Acids Res. 2013;41:e39.
73. Goodwin S, McPherson JD, McCombie WR. Coming of age: ten years of next-generation sequencing technologies. Nat Rev Genet. 2016;17:333–51.
74. Stark R, Grzelak M, Hadfield J. RNA sequencing: the teenage years. Nat Rev Genet. 2019;20:631–56.
75. Mitelman F, Johansson B, Mertens F. The impact of translocations and gene fusions on cancer causation. Nat Rev Cancer. 2007;7:233–45.
76. Teixeira MR. Recurrent fusion oncogenes in carcinomas. Crit Rev Oncog. 2006;12:257–71.
77. Weckselblatt B, Rude MK. Human structural variation: mechanisms of chromosome rearrangements. Trends Genet. 2015;31:587–99.
78. Kurzrock R, Kantarjian HM, Druker BJ, Talpaz M. Philadelphia chromosome-positive leukemias: from basic mechanisms to molecular therapeutics. Ann Intern Med. 2003;138:819–30.
79. Druker BJ, Tamura S, Buchdunger E, Ohno S, Segal GM, Fanning S, Zimmermann J, Lydon NB. Effects of a selective inhibitor of the Abl tyrosine kinase on the growth of Bcr-Abl positive cells. Nat Med. 1996;2:561–6.
80. Melo JV. The diversity of BCR-ABL fusion proteins and their relationship to leukemia phenotype. Blood. 1996;88:2375–84.
81. Hong Z, Zhang W, Ding DL, Huang ZL, Yan YQ, Cao W, Pan YQ, Hou XN, Weroha SJ, Karnes RJ, Wang DJ, Wu Q, Wu DL, Huang HJ. DNA damage promotes TMPRSS2-ERG oncoprotein destruction and prostate cancer suppression via signaling converged by GSK3β and WEE1. Mol Cell. 2020;79:1008–1032.e1004.
82. Tomlins SA, Rhodes DR, Perner S, Dhanasekaran SM, Mehra R, Sun XW, Varambally S, Cao XH, Tchinda J, Kuefer R, Lee C, Montie JE, Shah RB, Pienta KJ, Rubin MA, Chinnaiyan AM. Recurrent fusion of TMPRSS2 and ETS transcription factor genes in prostate cancer. Science. 2005;310:644–8.
83. Kumar-Sinha C, Tomlins SA, Chinnaiyan AM. Recurrent gene fusions in prostate cancer. Nat Rev Cancer. 2008;8:497–511.
84. Pederzoli F, Bandini M, Marandino L, Ali SM, Madison R, Chung J, Ross JS, Necchi A. Targetable gene fusions and aberrations in genitourinary oncology. Nat Rev Urol. 2020;17:613–25.
85. Soda M, Choi YL, Enomoto M, Takada S, Yamashita Y, Ishikawa S, Fujiwara SI, Watanabe H, Kurashina K, Hatanaka H, Bando M, Ohno S, Ishikawa Y, Aburatani H, Niki T, Sohara Y, Sugiyama Y, Mano H. Identification of the transforming EML4-ALK fusion gene in non-small-cell lung cancer. Nature. 2007;448:561–6.
86. Shaw AT, Yeap BY, Solomon BJ, Riely GJ, Gainor J, Engelman JA, Shapiro GI, Costa DB, Ou SHI, Butaney M, Salgia R, Maki RG, Varella-Garcia M, Doebele RC, Bang YJ, Kulig K, Selaru P, Tang YY, Wilner KD, Kwak EL, Clark JW, Iafrate AJ, Camidge DR. Effect of crizotinib on overall survival in patients with advanced non-small-cell lung cancer harbouring ALK gene rearrangement: a retrospective analysis. Lancet Oncol. 2011;12:1004–12.
87. Takahashi T, Sonobe M, Kobayashi M, Yoshizawa A, Menju T, Nakayama E, Mino N, Iwakiri S, Sato K, Miyahara R, Okubo K, Manabe T, Date H. Clinicopathologic features of non-small-cell lung cancer with EML4-ALK fusion gene. Ann Surg Oncol. 2010;17:889–97.
88. Mertens F, Johansson B, Fioretos T, Mitelman F. The emerging complexity of gene fusions in cancer. Nat Rev Cancer. 2015;15:371–81.
89. Schram AM, Chang MT, Jonsson P, Drilon A. Fusions in solid tumours: diagnostic strategies, targeted therapy, and acquired resistance. Nat Rev Clin Oncol. 2017;14:735–48.
90. Brien GL, Stegmaier K, Armstrong SA. Targeting chromatin complexes in fusion protein-driven malignancies. Nat Rev Cancer. 2019;19:255–69.
91. Mitelman FJBaMF. Mitelman database of chromosome aberrations and gene fusions in cancer 2021 [updated October 15, 2024]. Available from: https://mitelmandatabase.isb-cgc.org

92. Wang L. Identification of cancer gene fusions based on advanced analysis of the human genome or transcriptome. Front Med. 2013;7:280–9.
93. Skotheim RI, Eken M, Thomassen G, Lind GE, Lothe RA. A universal assay for detection of oncogenic fusion transcripts by oligo microarray analysis. EJC Suppl. 2009;7:147–8.
94. van Dongen JJ, Macintyre EA, Gabert JA, Delabesse E, Rossi V, Saglio G, Gottardi E, Rambaldi A, Dotti G, Griesinger F, Parreira A, Gameiro P, Diaz MG, Malec M, Langerak AW, San Miguel JF, Biondi A. Standardized RT-PCR analysis of fusion gene transcripts from chromosome aberrations in acute leukemia for detection of minimal residual disease. Report of the BIOMED-1 Concerted Action: investigation of minimal residual disease in acute leukemia. Leukemia. 1999;13:1901–28.
95. Maher CA, Kumar-Sinha C, Cao XH, Kalyana-Sundaram S, Han B, Jing XJ, Sam L, Barrette T, Palanisamy N, Chinnaiyan AM. Transcriptome sequencing to detect gene fusions in cancer. Nature. 2009;458:97–U99.
96. Wang L, Motoi T, Khanin R, Olshen A, Mertens F, Bridge J, Dal Cin P, Antonescu CR, Singer S, Hameed M, Bovee J, Hogendoorn PCW, Socci N, Ladanyi M. Identification of a novel, recurrent HEY1-NCOA2 fusion in mesenchymal chondrosarcoma based on a genome-wide screen of exon-level expression data. Genes Chromosom Cancer. 2012;51:127–39.
97. Wang QG, Xia JF, Jia PL, Pao W, Zhao ZM. Application of next generation sequencing to human gene fusion detection: computational tools, features and perspectives. Brief Bioinform. 2013;14:506–19.
98. Hu X, Wang QH, Tang M, Barthel F, Amin S, Yoshihara K, Lang FM, Martinez-Ledesma E, Lee SH, Zheng SY, Verhaak RGW. TumorFusions: an integrative resource for cancer-associated transcript fusions. Nucleic Acids Res. 2018;46:D1144–9.
99. Kim P, Zhou XB. FusionGDB: fusion gene annotation DataBase. Nucleic Acids Res. 2019;47: D994–D1004.
100. Hsieh G, Bierman R, Szabo L, Lee AG, Freeman DE, Watson N, Sweet-Cordero EA, Salzman J. Statistical algorithms improve accuracy of gene fusion detection. Nucleic Acids Res. 2017;45:e126.
101. Heyer EE, Blackburn J. Sequencing strategies for fusion gene detection. BioEssays. 2020;42: e2000016.
102. Jhavar S, Reid A, Clark J, Kote-Jarai Z, Christmas T, Thompson A, Woodhouse C, Ogden C, Fisher C, Corbishley C, De-Bono J, Eeles R, Brewer D, Cooper C. Detection of TMPRSS2-ERG translocations in human prostate cancer by expression profiling using GeneChip Human Exon 1.0 ST arrays. J Mol Diagn. 2008;10:50–7.
103. Lin E, Li L, Guan YH, Soriano R, Rivers CS, Mohan S, Pandita A, Tang J, Modrusan Z. Exon array profiling detects EML4-ALK fusion in breast, colorectal, and non-small cell lung cancers. Mol Cancer Res. 2009;7:1466–76.
104. Wang F, Flanagan J, Su N, Wang LC, Bui S, Nielson A, Wu XY, Vo HT, Ma XJ, Luo YL. RNAscope a novel in situ RNA analysis platform for formalin-fixed, paraffin-embedded tissues. J Mol Diagn. 2012;14:22–9.
105. Wang H, Su N, Wang L-C, Wu X, Bui S, Nielsen A, Vo H-T, Luo Y, Ma X-J. Quantitative ultrasensitive bright-field RNA in situ hybridization with RNAscope. Methods Mol Biol (Clifton, NJ). 2014;1211:201–12.
106. Wang H, Su N, Wang L-C, Wu X, Bui S, Nielsen A, Vo H-T, Luo Y, Ma X-J. Dual-color ultrasensitive bright-field RNA in situ hybridization with RNAscope. Methods Mol Biol (Clifton, NJ). 2014;1211:139–49.
107. Geiss GK, Bumgarner RE, Birditt B, Dahl T, Dowidar N, Dunaway DL, Fell HP, Ferree S, George RD, Grogan T, James JJ, Maysuria M, Mitton JD, Oliveri P, Osborn JL, Peng T, Ratcliffe AL, Webster PJ, Davidson EH, Hood L. Direct multiplexed measurement of gene expression with color-coded probe pairs. Nat Biotechnol. 2008;26:317–25.
108. Godoy PM, Barczak AJ, DeHoff P, Srinivasan S, Etheridge A, Galas D, Das S, Erle DJ, Laurent LC. Comparison of reproducibility, accuracy, sensitivity, and specificity of miRNA quantification platforms. Cell Rep. 2019;29:4212–4222.e4215.

109. Lira ME, Choi YL, Lim SM, Deng SB, Huang DH, Ozeck M, Han JH, Eong JY, Shim HS, Cho BC, Kim J, Ahn MJ, Mao M. A single-tube multiplexed assay for detecting ALK, ROS1, and RET fusions in lung cancer. J Mol Diagn. 2014;16:229–43.
110. Veldman-Jones MH, Brant R, Rooney C, Geh C, Emery H, Harbron CG, Wappett M, Sharpe A, Dymond M, Barrett JC, Harrington EA, Marshall G. Evaluating robustness and sensitivity of the NanoString technologies nCounter platform to enable multiplexed gene expression analysis of clinical samples. Cancer Res. 2015;75:2587–93.
111. Chen X, Deane NG, Lewis KB, Li J, Zhu J, Washington MK, Beauchamp RD. Comparison of Nanostring nCounter® data on FFPE colon cancer samples and Affymetrix microarray data on matched frozen tissues. PLoS One. 2016;11:e0153784.
112. Kulkarni MM. Chapter 25: digital multiplexed gene expression analysis using the NanoString nCounter system. In: Current protocols in molecular biology; 2011. p. Unit25B.10. https://doi.org/10.1002/0471142727.mb25b10s94.
113. Chang KTE, Goytain A, Tucker T, Karsan A, Lee CH, Nielsen TO, Ng TL. Development and evaluation of a Pan-Sarcoma fusion gene detection assay using the NanoString nCounter platform. J Mol Diagn. 2018;20:63–77.
114. Armstrong DA, Green BB, Seigne JD, Schned AR, Marsit CJ. MicroRNA molecular profiling from matched tumor and bio-fluids in bladder cancer. Mol Cancer. 2015;14:194.
115. Permuth JB, Chen DT, Yoder SJ, Li JN, Smith AT, Choi JW, Kim J, Balagurunathan Y, Jiang K, Coppola D, Centeno BA, Klapman J, Hodul P, Karreth FA, Trevino JG, Merchant N, Magliocco A, Malafa MP, Gillies R. Linc-ing circulating long non-coding RNAs to the diagnosis and malignant prediction of intraductal papillary mucinous neoplasms of the pancreas. Sci Rep. 2017;7:10484.
116. Oikonomopoulos A, Polytarchou C, Joshi S, Hommes DW, Iliopoulos D. Identification of circulating microRNA signatures in Crohn's disease using the nanostring nCounter technology. Inflamm Bowel Dis. 2016;22:2063–9.
117. Sendorek DH, Lalonde E, Yao CQ, Sabelnykova VY, Bristow RG, Boutros PC. NanoStringNormCNV: pre-processing of NanoString CNV data. Bioinformatics. 2018;34:1034–6.
118. Quintas-Cardama A, Cortes J. Molecular biology of bcr-abl1-positive chronic myeloid leukemia. Blood. 2009;113:1619–30.
119. Wang H, Wang HH, Jia YT, Sun RY, Hong WX, Zhang M, Li ZP. Visual detection of fusion genes by ligation-triggered isothermal exponential amplification: a point-of-care testing method for highly specific and sensitive quantitation of fusion genes with a smartphone. Anal Chem. 2019;91:12428–34.
120. Wang HH, Wang H, Liu CH, Duan XR, Li ZP. Ultrasensitive detection of telomerase activity in a single cell using stem-loop primer-mediated exponential amplification (SPEA) with near zero nonspecific signal. Chem Sci. 2016;7:4945–50.
121. Notomi T, Okayama H, Masubuchi H, Yonekawa T, Watanabe K, Amino N, Hase T. Loop-mediated isothermal amplification of DNA. Nucleic Acids Res. 2000;28:e63.
122. Burmeister T, Reinhardt R. A multiplex PCR for improved detection of typical and atypical BCR-ABL fusion transcripts. Leuk Res. 2008;32:579–85.
123. Goh HG, Hwang JY, Kim SH, Lee YH, Kim YL, Kim DW. Comprehensive analysis of BCR-ABL transcript types in Korean CML patients using a newly developed multiplex RT-PCR. Transl Res. 2006;148:249–56.
124. Hai A, Kizilbash NA, Zaidi SHH, Alruwaili J, Shahzad K. Differences in structural elements of Bcr-Abl oncoprotein isoforms in Chronic Myelogenous Leukemia. Bioinformation. 2014;10:108–14.
125. Perego RA, Costantini M, Cornacchini G, Gargantini L, Bianchi C, Pungolino E, Rovida E, Morra E. The possible influences of B2A2 and B3A2 BCR/ABL protein structure on thrombopoiesis in chronic myeloid leukaemia. Eur J Cancer. 2000;36:1395–401.
126. Lucas CM, Harris RJ, Giannoudis A, Davies A, Knight K, Watmough SJ, Wang LH, Clark RE. Chronic myeloid leukemia patients with the e13a2 BCR-ABL fusion transcript have

inferior responses to imatinib compared to patients with the e14a2 transcript. Haematologica. 2009;94:1362–7.
127. Hanfstein B, Lauseker M, Hehlmann R, Saussele S, Erben P, Dietz C, Fabarius A, Proetel U, Schnittger S, Haferlach C, Krause SW, Schubert J, Einsele H, Hänel M, Dengler J, Falge C, Kanz L, Neubauer A, Kneba M, Stegelmann F, Pfreundschuh M, Waller CF, Spiekermann K, Baerlocher GM, Pfirrmann M, Hasford J, Hofmann WK, Hochhaus A, Müller MC, Grp SGCS. Distinct characteristics of e13a2 versus e14a2 BCR-ABL1 driven chronic myeloid leukemia under first-line therapy with imatinib. Haematologica. 2014;99:1441–7.
128. Castagnetti F, Gugliotta G, Breccia M, Iurlo A, Levato L, Albano F, Vigneri P, Abruzzese E, Rossi G, Rupoli S, Cavazzini F, Martino B, Orlandi E, Pregno P, Annunziata M, Usala E, Tiribelli M, Sica S, Bonifacio M, Fava C, Gherlinzoni F, Bocchia M, Soverini S, Bochicchio MT, Cavo M, Giovanni M, Saglio G, Pane F, Baccarani M, Rosti G, Party GCW. The BCR-ABL1 transcript type influences response and outcome in Philadelphia chromosome-positive chronic myeloid leukemia patients treated frontline with imatinib. Am J Hematol. 2017;92:797–805.
129. Jain P, Kantarjian H, Patel KP, Gonzalez GN, Luthra R, Shamanna RK, Sasaki K, Jabbour E, Romo CG, Kadia TM, Pemmaraju N, Daver N, Borthakur G, Estrov Z, Ravandi F, O'Brien S, Cortes J. Impact of BCR-ABL transcript type on outcome in patients with chronic-phase CML treated with tyrosine kinase inhibitors. Blood. 2016;127:1269–75.
130. Luo NN, Xia QF, Zhang LT, Zhang YH, Huang LZ, Zhang YL, Zhao J, Ding SJ, Cheng W. One-step discrimination of BCR/ABL^{p210} transcript isoforms directly from RNA extraction with fusion-triggered rolling circle amplification. Anal Chim Acta. 2019;1067:129–36.
131. Ovcharenko A, Rentmeister A. Emerging approaches for detection of methylation sites in RNA. Open Biol. 2018;8:180121.
132. Motorin Y, Helm M. RNA nucleotide methylation. Wiley Interdiscip Rev RNA. 2011;2:611–31.
133. Dunn DB, Smith JD. Occurrence of a new base in the deoxyribonucleic acid of a strain of Bacterium coli. Nature. 1955;175:336–7.
134. Littlefield JW, Dunn DB. Natural occurrence of thymine and three methylated adenine bases in several ribonucleic acids. Nature. 1958;181:254–5.
135. Adams JM, Cory S. Modified nucleosides and bizarre 5′-termini in mouse myeloma mRNA. Nature. 1975;255:28–33.
136. Dubin DT, Taylor RH. The methylation state of poly A-containing messenger RNA from cultured hamster cells. Nucleic Acids Res. 1975;2:1653–68.
137. Zhong SL, Li HY, Bodi Z, Button J, Vespa L, Herzog M, Fray RG. MTA is an Arabidopsis messenger RNA adenosine methylase and interacts with a homolog of a sex-specific splicing factor. Plant Cell. 2008;20:1278–88.
138. Agris PF, Vendeix FAP, Graham WD. tRNA's wobble decoding of the genome: 40 years of modification. J Mol Biol. 2007;366:1–13.
139. Gu J, Patton JR, Shimba S, Reddy R. Localization of modified nucleotides in Schizosaccharomyces pombe spliceosomal small nuclear RNAs: modified nucleotides are clustered in functionally important regions. RNA (New York, NY). 1996;2:909–18.
140. Choi YC, Busch H. Modified nucleotides in T1 RNase oligonucleotides of 18S ribosomal RNA of the Novikoff hepatoma. Biochemistry. 1978;17:2551–60.
141. Yue YN, Liu JZ, He C. RNA N^6-methyladenosine methylation in post-transcriptional gene expression regulation. Genes Dev. 2015;29:1343–55.
142. Xu K, Yang Y, Feng GH, Sun BF, Chen JQ, Li YF, Chen YS, Zhang XX, Wang CX, Jiang LY, Liu C, Zhang ZY, Wang XJ, Zhou Q, Yang YG, Li W. Mettl3-mediated m^6A regulates spermatogonial differentiation and meiosis initiation. Cell Res. 2017;27:1100–14.
143. Wang Y, Li Y, Toth JI, Petroski MD, Zhang ZL, Zhao JC. N^6-methyladenosine modification destabilizes developmental regulators in embryonic stem cells. Nat Cell Biol. 2014;16:191–8.
144. Ping XL, Sun BF, Wang L, Xiao W, Yang X, Wang WJ, Adhikari S, Shi Y, Lv Y, Chen YS, Zhao X, Li A, Yang Y, Dahal U, Lou XM, Liu X, Huang J, Yuan WP, Zhu XF, Cheng T, Zhao

YL, Wang XQ, Danielsen JMR, Liu F, Yang YG. Mammalian WTAP is a regulatory subunit of the RNA N6-methyladenosine methyltransferase. Cell Res. 2014;24:177–89.

145. Pendleton KE, Chen BB, Liu KQ, Hunter OV, Xie Y, Tu BP, Conrad NK. The U6 snRNA m^6A methyltransferase METTL16 regulates SAM synthetase intron retention. Cell. 2017;169: 824–35.
146. Yue YA, Liu J, Cui XL, Cao J, Luo GZ, Zhang ZZ, Cheng T, Gao MS, Shu X, Ma HH, Wang FQ, Wang XX, Shen B, Wang YZ, Feng XH, He C, Liu JZ. VIRMA mediates preferential m^6A mRNA methylation in 3′UTR and near stop codon and associates with alternative polyadenylation. Cell Discov. 2018;4:10.
147. Wen J, Lv RT, Ma HH, Shen HJ, He CX, Wang JH, Jiao FF, Liu H, Yang PY, Tan L, Lan F, Shi Y, He C, Shi Y, Diao JB. Zc3h13 regulates nuclear RNA m^6A methylation and mouse embryonic stem cell self-renewal. Mol Cell. 2018;69:1028–38.
148. Fu Y, Jia GF, Pang XQ, Wang RN, Wang X, Li CJ, Smemo S, Dai Q, Bailey KA, Nobrega MA, Han KL, Cui Q, He C. FTO-mediated formation of N^6-hydroxymethyladenosine and N^6-formyladenosine in mammalian RNA. Nat Commun. 2013;4:1798.
149. Yu F, Zhu AC, Liu S, Gao BY, Wang YZ, Khudaverdyan N, Yu CJ, Wu Q, Jiang YH, Song JK, Jin LT, He C, Qian ZJ. RBM33 is a unique m6A RNA-binding protein that regulates ALKBH5 demethylase activity and substrate selectivity. Mol Cell. 2023;83:2003–19.
150. Hess ME, Hess S, Meyer KD, Verhagen LAW, Koch L, Brönneke HS, Dietrich MO, Jordan SD, Saletore Y, Elemento O, Belgardt BF, Franz T, Horvath TL, Rüther U, Jaffrey SR, Kloppenburg P, Brüning JC. The fat mass and obesity associated gene (Fto) regulates activity of the dopaminergic midbrain circuitry. Nat Neurosci. 2013;16:1042–8.
151. Roundtree IA, He C. Nuclear m^6A reader YTHDC1 regulates mRNA splicing. Trends Genet. 2016;32:320–1.
152. Spitale RC, Flynn RA, Zhang QC, Crisalli P, Lee B, Jung JW, Kuchelmeister HY, Batista PJ, Torre EA, Kool ET, Chang HY. Structural imprints in vivo decode RNA regulatory mechanisms. Nature. 2015;519:486–90.
153. Roundtree IA, Luo GZ, Zhang ZJ, Wang X, Zhou T, Cui YQ, Sha JH, Huang XX, Guerrero L, Xie P, He E, Shen B, He C. YTHDC1 mediates nuclear export of N^6-methyladenosine methylated mRNAs. elife. 2017;6:e31311.
154. Wang X, Zhao BS, Roundtree IA, Lu ZK, Han DL, Ma HH, Weng XC, Chen K, Shi HL, He C. N^6-methyladenosine modulates messenger RNA translation efficiency. Cell. 2015;161: 1388–99.
155. Alarcón CR, Goodarzi H, Lee H, Liu XH, Tavazoie S, Tavazoie SF. HNRNPA2B1 is a mediator of m^6A-dependent nuclear RNA processing events. Cell. 2015;162:1299–308.
156. Shi HL, Wang X, Lu ZK, Zhao BXS, Ma HH, Hsu PJ, Liu C, He C. YTHDF3 facilitates translation and decay of N^6-methyladenosine-modified RNA. Cell Res. 2017;27:315–28.
157. Huang HL, Weng HY, Sun WJ, Qin X, Shi HL, Wu HZ, Zhao BS, Mesquita A, Liu C, Yuan CL, Hu YC, Hüttelmaier S, Skibbe JR, Su R, Deng XL, Dong L, Sun M, Li CY, Nachtergaele S, Wang YG, Hu C, Ferchen K, Greis KD, Jiang X, Wei MJ, Qu LH, Guan JL, He C, Yang JH, Chen JJ. Recognition of RNA N^6-methyladenosine by IGF2BP proteins enhances mRNA stability and translation. Nat Cell Biol. 2018;20:285–95.
158. Hart SM, Foroni L. Core binding factor genes and human leukemia. Haematologica. 2002;87: 1307–23.
159. Zhang CX, Chen YS, Sun BF, Wang L, Yang Y, Ma DY, Lv JH, Heng J, Ding YY, Xue YY, Lu XY, Xiao W, Yang YG, Liu F. m^6A modulates haematopoietic stem and progenitor cell specification. Nature. 2017;549:273–6.
160. Gao YM, Vasic R, Song YB, Teng R, Liu CY, Gbyli R, Biancon G, Nelakanti R, Lobben K, Kudo E, Liu W, Ardasheva A, Fu XY, Wang XM, Joshi P, Lee V, Dura B, Viero G, Iwasaki A, Fan R, Xiao A, Flavell RA, Li HB, Tebaldi T, Halene S. m^6A modification prevents formation of endogenous double-stranded RNAs and deleterious innate immune responses during hematopoietic development. Immunity. 2020;52:1007–21.

161. Wang H, Zuo H, Liu J, Wen F, Gao Y, Zhu X, Liu B, Xiao F, Wang W, Huang G, Shen B, Ju Z. Loss of YTHDF2-mediated m^6A-dependent mRNA clearance facilitates hematopoietic stem cell regeneration. Cell Res. 2018;28:1035–8.
162. Lence T, Akhtar J, Bayer M, Schmid K, Spindler L, Ho CH, Kreim N, Andrade-Navarro MA, Poeck B, Helm M, Roignant JY. m^6A modulates neuronal functions and sex determination in Drosophila. Nature. 2016;540:242–7.
163. Li LP, Zang LQ, Zhang FR, Chen JC, Shen H, Shu LQ, Liang F, Feng CY, Chen D, Tao HK, Xu TL, Li ZY, Kang YH, Wu H, Tang LC, Zhang PM, Jin P, Shu QA, Li XK. Fat mass and obesity-associated (FTO) protein regulates adult neurogenesis. Hum Mol Genet. 2017;26: 2398–411.
164. Zhang ZY, Wang M, Xie DF, Huang ZH, Zhang LS, Yang Y, Ma DX, Li WG, Zhou Q, Yang YG, Wang XJ. METTL3-mediated N^6-methyladenosine mRNA modification enhances long-term memory consolidation. Cell Res. 2018;28:1050–61.
165. Yoon KJ, Ringeling FR, Vissers C, Jacob F, Pokrass M, Jimenez-Cyrus D, Su YJ, Kim NS, Zhu YH, Zheng LL, Kim S, Wang XY, Doré LC, Jin P, Regot S, Zhuang XX, Canzar S, He C, Ming GL, Song HJ. Temporal control of mammalian cortical neurogenesis by m^6A methylation. Cell. 2017;171:877–89.
166. Ivanova I, Much C, Di Giacomo M, Azzi C, Morgan M, Moreira PN, Monahan J, Carrieri C, Enright AJ, O'Carroll D. The RNA m^6A reader YTHDF2 is essential for the post-transcriptional regulation of the maternal transcriptome and oocyte competence. Mol Cell. 2017;67:1059–67.
167. Zheng GQ, Dahl JA, Niu YM, Fedorcsak P, Huang CM, Li CJ, Vågbo CB, Shi Y, Wang WL, Song SH, Lu ZK, Bosmans RPG, Dai Q, Hao YJ, Yang X, Zhao WM, Tong WM, Wang XJ, Bogdan F, Furu K, Fu Y, Jia GF, Zhao X, Liu J, Krokan HE, Klungland A, Yang YG, He C. ALKBH5 is a mammalian RNA demethylase that impacts RNA metabolism and mouse fertility. Mol Cell. 2013;49:18–29.
168. Xia H, Zhong CR, Wu XX, Chen J, Tao BB, Xia XQ, Shi MJ, Zhu ZY, Trudeau VL, Hu W. Mettl3 mutation disrupts gamete maturation and reduces fertility in zebrafish. Genetics. 2018;208:729–43.
169. Siwicki KK, Eastman C, Petersen G, Rosbash M, Hall JC. Antibodies to the period gene product of Drosophila reveal diverse tissue distribution and rhythmic changes in the visual system. Neuron. 1988;1:141–50.
170. Fustin JM, Doi M, Yamaguchi Y, Hida H, Nishimura S, Yoshida M, Isagawa T, Morioka MS, Kakeya H, Manabe I, Okamura H. RNA-methylation-dependent RNA processing controls the speed of the circadian clock. Cell. 2013;155:793–806.
171. Engel M, Eggert C, Kaplick PM, Eder M, Röh S, Tietze L, Namendorf C, Arloth J, Weber P, Rex-Haffner M, Geula S, Jakovcevski M, Hanna JH, Leshkowitz D, Uhr M, Wotjak CT, Schmidt MV, Deussing JM, Binder EB, Chen A. The role of m^6A/m-RNA methylation in stress response regulation. Neuron. 2018;99:389–403.
172. Liu XX, Shimada T, Otowa T, Wu YY, Kawamura Y, Tochigi M, Iwata Y, Umekage T, Toyota T, Maekawa M, Iwayama Y, Suzuki K, Kakiuchi C, Kuwabara H, Kano Y, Nishida H, Sugiyama T, Kato N, Chen CH, Mori N, Yamada K, Yoshikawa T, Kasai K, Tokunaga K, Sasaki T, Gau SSF. Genome-wide association study of autism spectrum disorder in the East Asian populations. Autism Res. 2016;9:340–9.
173. Oldmeadow C, Mossman D, Evans TJ, Holliday EG, Tooney PA, Cairns MJ, Wu JQ, Carr V, Attia JR, Scott RJ. Combined analysis of exon splicing and genome wide polymorphism data predict schizophrenia risk loci. J Psychiatr Res. 2014;52:44–9.
174. Xie W, Ma LL, Xu YQ, Wang BH, Li SM. METTL3 inhibits hepatic insulin sensitivity via N^6-methyladenosine modification of Fasn mRNA and promoting fatty acid metabolism. Biochem Biophys Res Commun. 2019;518:120–6.
175. Li YH, Zhang QY, Cui GS, Zhao F, Tian X, Sun BF, Yang Y, Li W. m^6A regulates liver metabolic disorders and hepatogenous diabetes. Genomics Proteomics Bioinformatics. 2020;18:371–83.

176. Qin YH, Li LQ, Luo E, Hou JT, Yan GL, Wang D, Qiao Y, Tang CC. Role of m^6A RNA methylation in cardiovascular disease (review). Int J Mol Med. 2020;46:1958–72.
177. Zhao K, Yang CX, Li P, Sun W, Kong XQ. Epigenetic role of N^6-methyladenosine (m^6A) RNA methylation in the cardiovascular system. J Zhejiang Univ Sci B. 2020;21:509–23.
178. Vu LP, Pickering BF, Cheng YM, Zaccara S, Nguyen D, Minuesa G, Chou T, Chow A, Saletore Y, MacKay M, Schulman J, Famulare C, Patel M, Klimek VM, Garrett-Bakelman FE, Melnick A, Carroll M, Mason CE, Jaffrey SR, Kharas MG. The N^6-methyladenosine (m^6A)-forming enzyme METTL3 controls myeloid differentiation of normal hematopoietic and leukemia cells. Nat Med. 2017;23:1369–76.
179. Yan F, Al-Kali A, Zhang ZJ, Liu J, Pang JX, Zhao N, He C, Litzow MR, Liu SJ. A dynamic N^6-methyladenosine methylome regulates intrinsic and acquired resistance to tyrosine kinase inhibitors. Cell Res. 2018;28:1062–76.
180. Chai RC, Wu F, Wang QX, Zhang S, Zhang KN, Liu YQ, Zhao Z, Jiang T, Wang YZ, Kang CS. m^6A RNA methylation regulators contribute to malignant progression and have clinical prognostic impact in gliomas. Aging-Us. 2019;11:1204–25.
181. Cui Q, Shi HL, Ye P, Li L, Qu QH, Sun GQ, Sun GH, Lu ZK, Huang Y, Yang CG, Riggs AD, He C, Shi YH. m^6A RNA methylation regulates the self-renewal and tumorigenesis of glioblastoma stem cells. Cell Rep. 2017;18:2622–34.
182. Zhang SC, Zhao BS, Zhou AD, Lin KY, Zheng SP, Lu ZK, Chen YH, Sulman EP, Xie KP, Bögler O, Majumder S, He C, Huang SY. m^6A demethylase ALKBH5 maintains tumorigenicity of glioblastoma stem-like cells by sustaining FOXM1 expression and cell proliferation program. Cancer Cell. 2017;31:591–606.
183. Deng JM, Chen S, Wang F, Zhao HX, Xie ZY, Xu ZY, Zhang QT, Liang P, Zhai X, Cheng Y. Effects of hnRNP A2/B1 knockdown on inhibition of glioblastoma cell invasion, growth and survival. Mol Neurobiol. 2016;53:1132–44.
184. Han J, Wang JZ, Yang X, Yu H, Zhou R, Lu HC, Yuan WB, Lu JC, Zhou ZJ, Lu Q, Wei JF, Yang HW. METTL3 promote tumor proliferation of bladder cancer by accelerating pri-miR221/222 maturation in m6A-dependent manner. Mol Cancer. 2019;18:110.
185. Xie HY, Li JF, Ying YF, Yan HQ, Jin K, Ma XY, He LJ, Xu X, Liu B, Wang X, Zheng XY, Xie LP. METTL3/YTHDF2 m^6A axis promotes tumorigenesis by degrading SETD7 and KLF4 mRNAs in bladder cancer. J Cell Mol Med. 2020;24:4092–104.
186. Zhu HT, Gan XL, Jiang XW, Diao S, Wu H, Hu JG. ALKBH5 inhibited autophagy of epithelial ovarian cancer through miR-7 and BCL-2. J Exp Clin Cancer Res. 2019;38:163.
187. Ma XX, Cao ZG, Zhao SL. m^6A methyltransferase METTL3 promotes the progression of prostate cancer via m^6A-modified LEF1. Eur Rev Med Pharmacol Sci. 2020;24:3565–71.
188. Yuan Y, Du Y, Wang L, Liu XH. The M6A methyltransferase METTL3 promotes the development and progression of prostate carcinoma via mediating MYC methylation. J Cancer. 2020;11:3588–95.
189. Zehring WA, Wheeler DA, Reddy P, Konopka RJ, Kyriacou CP, Rosbash M, Hall JC. P-element transformation with period locus DNA restores rhythmicity to mutant, arrhythmic Drosophila melanogaster. Cell. 1984;39:369–76.
190. Kane SE, Beemon K. Inhibition of methylation at two internal N6-methyladenosine sites caused by GAC to GAU mutations. J Biol Chem. 1987;262:3422–7.
191. Tuck MT, Wiehl PE, Pan T. Inhibition of 6-methyladenine formation decreases the translation efficiency of dihydrofolate reductase transcripts. Int J Biochem Cell Biol. 1999;31:837–51.
192. Bachellerie JP, Amalric F, Caboche M. Biosynthesis and utilization of extensively undermethylated poly(A)+ RNA in CHO cells during a cycloleucine treatment. Nucleic Acids Res. 1978;5:2927–43.
193. Desrosiers R, Friderici K, Rottman F. Identification of methylated nucleosides in messenger RNA from Novikoff hepatoma cells. Proc Natl Acad Sci USA. 1974;71:3971–5.
194. Jia GF, Fu Y, Zhao X, Dai Q, Zheng GQ, Yang Y, Yi CQ, Lindahl T, Pan T, Yang YG, He C. N6-methyladenosine in nuclear RNA is a major substrate of the obesity-associated FTO. Nat Chem Biol. 2011;7:885–7.

195. Grosjean H, Keith G, Droogmans L. Detection and quantification of modified nucleotides in RNA using thin-layer chromatography. Methods Mol Biol (Clifton, NJ). 2004;265:357–91.
196. Schwartz S, Agarwala SD, Mumbach MR, Jovanovic M, Mertins P, Shishkin A, Tabach Y, Mikkelsen TS, Satija R, Ruvkun G, Carr SA, Lander ES, Fink GR, Regev A. High-resolution mapping reveals a conserved, widespread, dynamic mRNA methylation program in yeast meiosis. Cell. 2013;155:1409–21.
197. Dominissini D, Moshitch-Moshkovitz S, Schwartz S, Salmon-Divon M, Ungar L, Osenberg S, Cesarkas K, Jacob-Hirsch J, Amariglio N, Kupiec M, Sorek R, Rechavi G. Topology of the human and mouse m^6A RNA methylomes revealed by m^6A-seq. Nature. 2012;485:201–6.
198. Meyer KD, Saletore Y, Zumbo P, Elemento O, Mason CE, Jaffrey SR. Comprehensive analysis of mRNA methylation reveals enrichment in 3′ UTRs and near stop codons. Cell. 2012;149:1635–46.
199. Chen K, Lu ZK, Wang X, Fu Y, Luo GZ, Liu N, Han DL, Dominissini D, Dai Q, Pan T, He C. High-resolution N^6-methyladenosine (m^6A) map using photo-crosslinking-assisted m^6A sequencing. Angew Chem Int Ed. 2015;54:1587–90.
200. Linder B, Grozhik AV, Olarerin-George AO, Meydan C, Mason CE, Jaffrey SR. Single-nucleotide-resolution mapping of m6A and m6Am throughout the transcriptome. Nat Methods. 2015;12:767–72.
201. Zhang Z, Chen LQ, Zhao YL, Yang CG, Roundtree IA, Zhang ZJ, Ren J, Xie W, He C, Luo GZ. Single-base mapping of m^6A by an antibody-independent method. Sci Adv. 2019;5: eaax0250.
202. Mahdavi-Amiri Y, Chung KCK, Hili R. Single-nucleotide resolution of N^6-adenine methylation sites in DNA and RNA by nitrite sequencing. Chem Sci. 2021;12:606–12.
203. Meyer KD. DART-seq: an antibody-free method for global m^6A detection. Nat Methods. 2019;16:1275–80.
204. Wang Y, Xiao Y, Dong SQ, Yu Q, Jia GF. Antibody-free enzyme-assisted chemical approach for detection of N^6-methyladenosine. Nat Chem Biol. 2020;16:896–903.
205. Liu N, Parisien M, Dai Q, Zheng GQ, He C, Pan T. Probing N^6-methyladenosine RNA modification status at single nucleotide resolution in mRNA and long noncoding RNA. RNA. 2013;19:1848–56.
206. Harcourt EM, Ehrenschwender T, Batista PJ, Chang HY, Kool ET. Identification of a selective polymerase enables detection of N^6-methyladenosine in RNA. J Am Chem Soc. 2013;135: 19079–82.
207. Wang SR, Wang JQ, Zhang XE, Fu BS, Song YY, Ma P, Gu K, Zhou X, Zhang XL, Tian T, Zhou X. N^6-methyladenine hinders RNA- and DNA-directed DNA synthesis: application in human rRNA methylation analysis of clinical specimens. Chem Sci. 2016;7:1440–6.
208. Liu WL, Yan JL, Zhang ZH, Pian HR, Liu CH, Li ZP. Identification of a selective DNA ligase for accurate recognition and ultrasensitive quantification of N^6-methyladenosine in RNA at one-nucleotide resolution. Chem Sci. 2018;9:3354–9.
209. Xiao Y, Wang Y, Tang Q, Wei LH, Zhang X, Jia GF. An elongation- and ligation-based qPCR amplification method for the radiolabeling-free detection of locus-specific N^6-methyladenosine modification. Angew Chem Int Ed. 2018;57:15995–6000.
210. Pian H, Yan J, Liu W, Wang H, Li Z. Highly sensitive quantitation of N6-methyladenosine in long noncoding RNA at precise location with gap extension- and ligation-based loop-mediated isothermal amplification (GEXL-LAMP). Talanta. 2025;284:127276.

Chapter 12
Nucleic Acid Analysis-Based Methods for SARS-CoV-2 Detection

Yang Sun and Weixiang Hong

At the end of 2019, a series of unknown cases of pneumonia were reported worldwide [1]. Through sequencing the samples from pneumonia patients, it was determined that the severe acute respiratory syndrome is caused by a novel coronavirus (SARS-CoV-2). Later, this unknown pneumonia was named COVID-19 by World Health Organization (WHO) in January 2020. Considering its rapid spread, the WHO classified COVID-19 as the pandemic level on March 12, 2020. The outbreak of COVID-19 presented a significant challenge to the security of global public health and severely impeded global economic growth. Furthermore, the asymptomatic/presymptomatic infected person that could also carry and spread SARS-CoV-2 [2, 3] makes timely and effective screening and isolation difficult, exacerbating the escalation of the pandemic. The control strategies based on vaccine-induced preimmunization and pharmacological treatment have demonstrated limited efficacy. So, screening and isolating the infected samples by developing rapid and specific detection methods is meaningful for blocking the source of the epidemic and effectively preventing the spreading of SARS-CoV-2.

12.1 RT-PCR-Based Method for SARS-CoV-2 Detection

With the promotion of governments and the efforts of researchers, the full sequence of the SARS-CoV-2 reference genome (Wuhan-Hu-1) was released by the RNA metagenomic sequencing [4]. As shown in Fig. 12.1, the total length of the SARS-CoV-2 reference genome is 29903nt, and the genome is mainly composed of open reading frame 1a (ORF1a), open reading frame 1ab (ORF1ab), spike protein gene

Y. Sun (✉) · W. Hong
University of Science and Technology Beijing, Beijing, China
e-mail: yangsunbio@ustb.edu.cn

Z. Li (ed.), *Nucleic Acid Detection*, https://doi.org/10.1007/978-981-95-7295-3_12

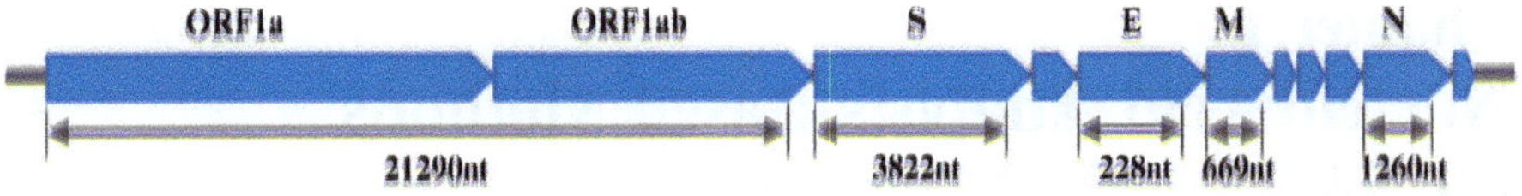

Fig. 12.1 Distribution of major genes in the SARS-CoV-2 reference genome. (Revised and reprinted with permission from Ref. [4])

(S), envelope protein gene (E), membrane protein gene (M), and nucleocapsid protein gene (N).

Based on the released sequence information, a variety of technologies based on nucleic analysis have been developed for SARS-CoV-2 detection. Currently, the international gold standard for molecular diagnosis of COVID-19 is reverse transcription quantitative PCR (RT-qPCR), which could amplify specific sequences in the SARS-CoV-2 genome. Designing the specific probes and primers is the key in SARS-CoV-2 detection, which determines whether the RT-PCR reaction could be successfully completed under the action of reverse transcriptase and DNA polymerase (see Chap. 1 for specific principles).

Based on this, the Chinese Center for Disease Control and Prevention recommends partial sequences of ORF1a and N genes as templates for designing specific primers and probes (Table 12.1). In this system, when the Ct value is less than 37, the tested samples will be judged as positive [5]. The US Centers for Disease Control and Prevention recommends three distinct conserved fragments of the N gene (N1, N2, and N3) as templates for designing specific primers and probes, while using the human RNase P gene as a negative control. In this system, only a result that is positive for the RNase P gene and negative for the SARS-CoV-2 gene can prove that the tested sample is negative [6]. However, this protocol requires multiple tests for each patient sample, which not only consumes a lot of resources but also increases the risk of errors. Based on this, the research team of Yale University designed a multi-RT-QPCR method for the detection of SARS-CoV-2 [7]. Primers to N1, N2, and human RNase P genes released by the Centers for Disease Control and Prevention of the United States were adopted, and multiplex fluorescent probes were designed for simultaneous detection, realizing multichannel detection (Fig. 12.2). The method showed similar accuracy and specificity as the single RT-qPCR method, while effectively saving time and reagent cost. In addition, probes and primers for real-time quantitative detection of SARS-CoV-2 have been continuously developed and optimized by relevant research institutions, which have been strongly enabling epidemic control and the development of nucleic acid kits.

Since the sequence of SARS-CoV-2 is highly similar to that of the SARS virus, the primers and TaqMan probes of these two viruses may have cross-reactions, and the detection capabilities of different primers and probes are also significantly different. Therefore, comparing the detection capabilities of primers and probes is necessary to guide and improve the detection program. Jung's research team [8] systematically compared the performance of three prime-probes acting on the RdRp/ORF1 region and seven prime-probes acting on the N region in detecting SARS-CoV-2. The compared prime-probes were provided by the China Center for Disease

Table 12.1 Primers and probe sequence information of RT-PCR

Supplier	Region	Name	Sequence (5′ > 3′)
The Chinese Center for Disease Control and Prevention	ORF1ab	Forward primer	CCCTGTGGGTTTTACACTTAA
		Reverse primer	ACGATTGTGCATCAGCTGA
		TaqMan probe	FAM-CCGTCTGCGGTATGTGGAAAGGTTATGG-BHQ1
	N	Forward primer	GGGGAACTTCTCCTGCTAGAAT
		Reverse primer	CAGACATTTTGCTCTCAAGCTG
		TaqMan probe	FAM- TTGCTGCTGCTTGACAGATT-TAMRA
The US Centers for Disease Control and Prevention	N1	Forward primer	GACCCCAAAATCAGCGAAAT
		Reverse primer	TCTGGTTACTGCCAGTTGAATCTG
		TaqMan probe	FAM- ACCCCGCATTACGTTTGGTGGACC-BHQ1
	N2	Forward primer	TTACAAACATTGGCCGCAAA
		Reverse primer	GCGCGACATTCCGAAGAA
		TaqMan probe	FAM- ACAATTTGCCCCCAGCGCTTCAG-BHQ1
	N3	Forward primer	GGGAGCCTTGAATACACCAAAA
		Reverse primer	TGTAGCACGATTGCAGCATTG
		TaqMan probe	FAM- AYCACATTGGCACCCGCAATCCTG-BHQ1
	RNase P	Forward primer	AGATTTGGACCTGCGAGCG
		Reverse primer	GAGCGGCTGTCTCCACAAGT
		TaqMan probe	FAM- TTCTGACCTGAAGGCTCTGCGCG-BHQ1

Revised and reprinted with permission from Ref. [4]

Control and Prevention, the US Center for Disease Control, NIID of Japan and Charité of Germany, etc. The results showed that specific primers and probes to "ORF1ab" from the Chinese Center for Disease Control and Prevention and to "NIID_2019-nCOV_N" from the Japanese NIID had better detection performance for RT-qPCR amplification of SARS-CoV-2, and these two sets of probes and primers did not produce nonspecific amplification and cross-reaction to the RNA of other coronaviruses (such as hCoV-229E, hCoV-OC43) and the middle east

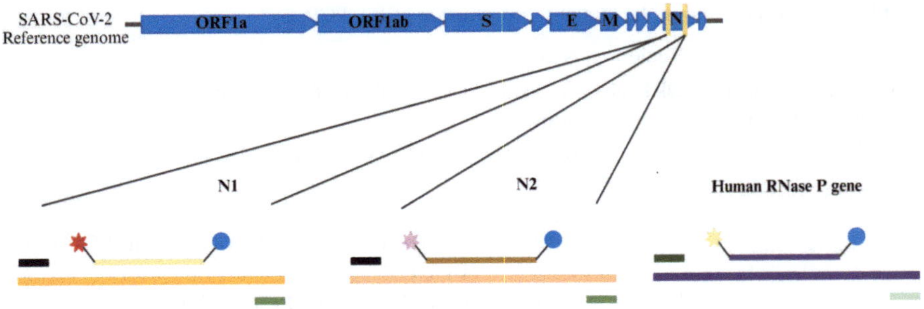

Fig. 12.2 Schematic diagram of multiplex RT-qPCR detection of SARS-CoV-2. (Revised and reprinted with permission from Ref. [7])

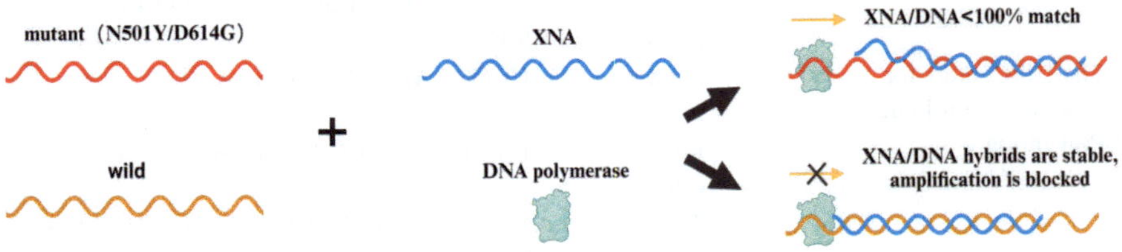

Fig. 12.3 Molecular clamp technology using XNA for detecting SARS-CoV-2 variants. (Revised and reprinted with permission from Ref. [9])

respiratory syndrome virus. Based on this, the authors suggest using the appropriate combination of the two probe-primer sets for reliable detecting SARS-CoV-2.

With the increasing variants emerging, Shen et al. [9] reported a multiplex RT-qPCR detection method based on xenonucleic acids (XNA) molecular clamp technology for typing SARS-CoV-2. XNA is a synthetic polymer that retains the Watson–Crick base-pairing capability. When a single-base mismatch occurs within an XNA/DNA duplex, its melting temperature (Tm) decreases by approximately 10–18 °C. Therefore, when XNA is perfectly complementary to wild-type DNA, the XNA/DNA (wild-type) hybrid exhibits greater stability compared to the XNA/DNA (mutant) hybrid. This property enables XNA to function as a molecular clamp in RT-qPCR reactions or as a highly specific molecular probe for detecting nucleic acid [10]. This study initially designed primer-probe sets that are capable of performing RT-qPCR amplification reactions based on the conserved regions of the E gene and the ORF1ab gene. Additionally, XNA were designed as sequence-complementary to the target sequences of the wild-type virus. As illustrated in Fig. 12.3, when detecting mutant variants, the high affinity between the XNAs and the wild target sequences effectively blocks the qPCR amplification, resulting in negligible amplification signals. Conversely, since the XNAs do not perfectly match the target sequences of the mutant viruses (N501Y and D614G), their affinity is significantly reduced, allowing the qPCR amplification to proceed and generate robust amplification signals. This approach allows for the highly sensitive genotyping of viral variants, with sensitivity of 100 copies/mL and specificity of 100%.

12.2 RT-LAMP-Based Method for SARS-CoV-2 Detection

Although RT-qPCR is widely viewed as the "gold standard" technique for SARS-CoV-2 detection, it relies on sophisticated thermal cyclers and lengthy detection times, which are unfavorable for the convenient deployment and rapid screening of COVID-19. Compared with qPCR, loop-mediated isothermal amplification (LAMP) can recognize six different sequences of the target by using four specific primers to achieve rapid and efficient amplification under a constant temperature. In addition, this method can greatly shorten the reaction time by adding two additionally designed loop primers (LF and LB) to accelerate the amplification reaction. Therefore, the reverse transcription loop-mediated isothermal exponential amplification method (RT-LAMP) has been widely reported as an alternative to the RT-qPCR method [11–13] (see Chap. 3 for specific principles).

Lu et al. [14] designed six sets of LAMP primers acting on different regions of SARS-CoV-2 genome. Firstly, they used the online software for LAMP primer design (http://primerexplorer.jp/lampv5e/index.html) to exclude nonspecific primer sets. Then, the primer set with the highest amplification efficiency will be screened out through the conventional RT-LAMP operating procedure. Finally, as illustrated in Fig. 12.4, the LAMP primer set targeting the RdRp gene region was selected for the amplification reaction. The amplification was carried out using the WarmStart colorimetric LAMP buffer system, which contains the pH-sensitive indicator dye phenol red. The reaction mixture also included WarmStart Bst 3.0 DNA polymerase, WarmStartR reverse transcriptase, and Q5 high-fidelity DNA polymerase to facilitate LAMP amplification. During the amplification process, hydrogen ions produced continuously, leading to a color change in the pH indicator from red to yellow. Experimental results indicated that the sensitivity of this method reaches 30 copies/reaction.

To evaluate amplification capabilities of various reported RT-LAMP primer sets and identify the optimal RT-LAMP condition, Zhang et al. [15] systematically compared the specificity and sensitivity of 19 reported RT-LAMP primer sets [14, 16–24]. Among these 19 primers sets, 15 primer sets target four key genomic regions: Nsp3, RdRp, E, and N, while the remaining 4 primers sets target other gene regions. To minimize the consumption of clinical samples and enhance evaluation

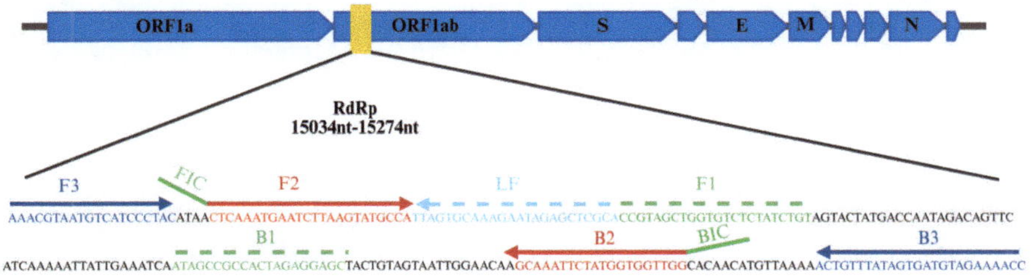

Fig. 12.4 Primer design and targeting for RT-LAMP detection of RdRp. (Revised and reprinted with permission from Ref. [14])

efficiency, the study initially used cDNA derived from reverse-transcription RNA to assess the 15 primer sets. The remaining 4 primer sets, along with those initially selected from the 15, were further evaluated using clinical samples. Following a comprehensive evaluation, the results revealed that the primer set developed by Park et al. [20], targeting the Nsp3 region, demonstrated high accuracy and sensitivity, making it a preferable choice for SARS-CoV-2 detection. Notably, the researchers further enhanced the detection sensitivity of RT-LAMP by incorporating high-fidelity DNA polymerase into the reaction process, effectively eliminating mismatched bases at the 3′ end of primers [25, 26].

With the development of RT-LAMP-based technologies, optimizing detection strategies to further enhance the sensitivity and specificity of RT-LAMP assays, as well as adapting this technology for large-scale, real-time diagnostics, has become increasingly critical. Wang et al. [27] developed a nanoparticle-based lateral flow biosensor (LFB) integrated with multiplex reverse transcription loop-mediated isothermal exponential amplification (mRT-LAMP-LFB) for the visual detection of SARS-CoV-2. As illustrated in Fig. 12.5a, the researchers first designed RT-LAMP primers targeting the ORF1ab and N genes. The forward loop primers were modified at the 5′ end with hapten fluorescein (FITC) and digoxigenin (Dig), respectively, while the backward loop primers were uniformly modified with biotin at the 5′ end. As shown in Fig. 12.5b, following the RT-LAMP reaction, the ORF1ab-RT-LAMP products were labeled with FITC and biotin at their respective ends, and the N-RT-LAMP products were labeled with Dig and biotin. The RT-LAMP products were

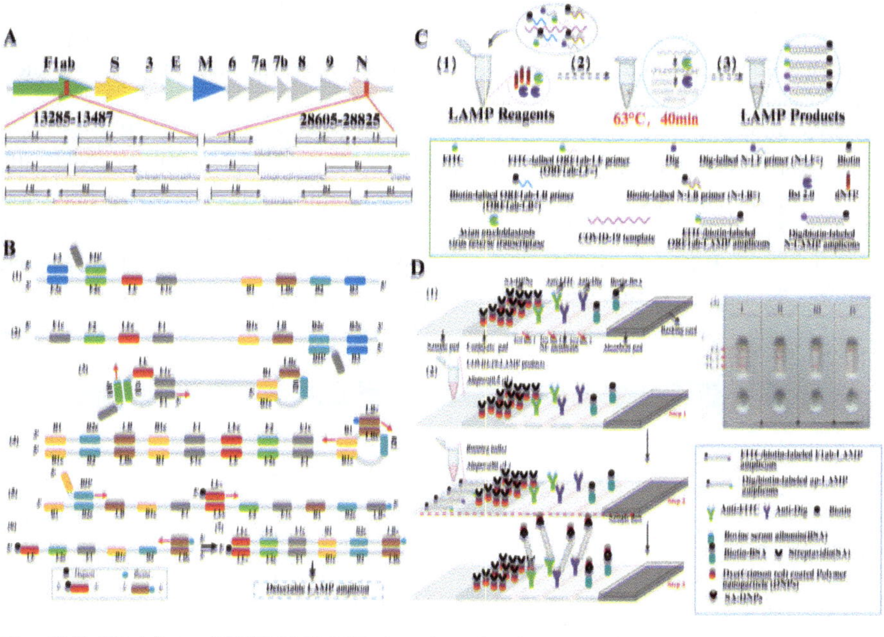

Fig. 12.5 Workflow of RT-LAMP-LFB detection. (Revised and reprinted with permission from Ref. [27])

then subjected to lateral flow biosensor. As depicted in Fig. 12.5c, the FITC-labeled ORF1ab-RT-LAMP products were captured by anti-FITC antibodies immobilized at test line 1, while the Dig-labeled N-RT-LAMP products were captured by anti-Dig antibodies at test line 2. Excess streptavidin-conjugated colored nanoparticles were captured at the control line by biotinylated bovine serum albumin (BSA), serving as an internal quality control. The presence and intensity of color changes at test lines 1 and 2, along with the control line, were used to analyze the RT-LAMP products. Experimental results demonstrated that this method achieved a limit of detection down to 12 copies/reaction for both ORF1ab and N genes, with no cross-reactivity.

Li et al. [28] developed a nucleic acid detection method named Dual Hairpin Ligation-Induced Isothermal Amplification Pro (DHLApro), which has demonstrated significant advantages in SARS-CoV-2 detection. As shown in Fig. 12.6a, a single hairpin probe (SHP) composed of Chain A and Chain B was first constructed. Chain A of the SHP contains a segment β1 that is complementary to the target sequence β*, while Chain B contains a segment β2 that is also complementary to β*. As illustrated in Fig. 12.6b, in the presence of the target nucleic acid sequence and ligase, the β1 segment in Chain A and the β2 segment in Chain B are ligated, forming a dual hairpin probe (DHP). As shown in Fig. 12.6c, the DHP serves as a template for LAMP amplification, generating multiple hairpin loop structures. As depicted in Fig. 12.6d, the authors previously designed a double-stranded fluorescent probe to detect the hairpin loop products. In this fluorescent probe, the

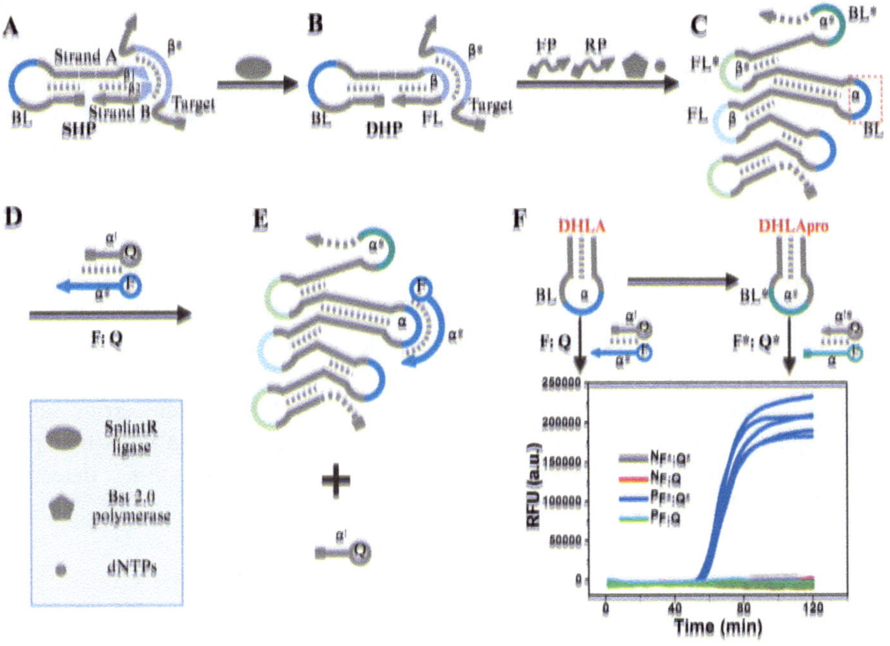

Fig. 12.6 Schematic diagram of DHLApro detection of SARS-CoV-2 RNA. (Revised and reprinted with permission from Ref. [28])

fluorophore and quencher are attached to two hybridized strands, α' and α*, respectively. The single strand α* containing the fluorophore can hybridize with the α segment in the BL region of the SHP. As shown in Fig. 12.6e, the α segment could displace the α' strand to hybridize with the α* strand, producing fluorescence signal. This detection method was named DHLA by the authors. However, since the α segment is initially present in the SHP, fluorescence signals can be generated even in negative samples. Therefore, as shown in Fig. 12.6f, the authors redesigned the fluorescent probe by modifying the sequence of the fluorophore-containing strand to be complementary to the α* segment instead of the α segment. Since the α* segment is only produced through target-triggered specific amplification, the specificity of the fluorescence signal is significantly enhanced. Consequently, this improved method was named DHLApro. By optimizing the signal probe, sequence design, and reaction conditions, DHLApro achieves a detection sensitivity of zM level.

Bokelmann et al. [29] integrated RT-LAMP technology with colorimetric sensing to develop a smartphone-based visual platform for SARS-CoV-2 detection, termed Cap-iLAMP. The RT-LAMP primer is designed to specifically amplify target sequences within the ORF1a and N genes. A corresponding biotin-labeled probe is designed based on the B3 primer-binding site and its adjacent sequences. The detailed detection process is as follows: firstly, the streptavidin C1 bead suspension is added to the sample lysate and the mixture was incubated at 55 °C for 10 min to capture the target. Then the RT-LAMP reagent is mixed with the eluate of the mixture and incubated at 65 °C for 40 min to perform the amplification reaction. Upon completion of the amplification reaction, the reaction tube is shaken to release 10,000 × SYBR Green I dye, which is pre-affixed to the inner wall of the tube cap, into the reaction mixture for color development. Traditional visual assessment is susceptible to subjective interpretation due to the limited color resolution of the human eye. This approach utilizes a smartphone application to quantitatively convert color intensity into a single digital hue value, significantly improving the accuracy and reliability of result interpretation. Moreover, the incorporation of a hybridization capture step to eliminate nontarget nucleic acids, along with the introduction of UvrD helicase to prevent nonspecific late-stage amplification caused by primer interactions, further enhances the specificity of the assay. As a result, Cap-iLAMP demonstrated 100% reliability in detecting at least one of the ORF1a or N genes and achieved a 91% success rate in simultaneously detecting both genes within a single sample. Furthermore, researchers have developed an RT-LAMP-based detection method that enables direct analysis without RNA extraction by transferring saliva or nasopharyngeal swab samples from COVID-19 patients into a virus transport medium. These innovations have significantly facilitated the rapid, on-site detection of SARS-CoV-2 [30–32].

12.3 CRISR-Cas-Based Method for SARS-CoV-2 Detection

In recent years, new gene editing technologies represented by CRISPR-Cas9 have developed rapidly. As more and more Cas proteins are discovered, the CRISPR-Cas family continues to grow. Cas9, Cas12a, and Cas13a systems are the typifiers, which can recognize target nucleic acid sequences with high specificity [33]. The collateral RNase activity of CRISPR-Cas12a and CRISPR-Cas13a enables the nonspecific cleavage of surrounding single-stranded DNA (ssDNA) or single-stranded RNA (ssRNA). Leveraging this property, fluorescence/quencher-labeled ssDNA or ssRNA probes have been designed to facilitate the detection and analysis of nucleic acid biomarkers. Capitalizing on the high specificity of CRISPR-Cas12a and CRISPR-Cas13a in nucleic acid sequence recognition, as well as the fluorescence signal amplification effect arising from their collateral nuclease activity-mediated cleavage of fluorescent probes (Detailed information is provided in Fig. 7.15, Chap. 7), researchers have developed a series of SARS-CoV-2 detection strategies.

Ning et al. [34] developed a novel SARS-CoV-2 detection method, termed COVID-19 CRISPR-FDS, which integrates CRISPR-Cas12a with RT-PCR technologies. As illustrated in Fig. 12.7a, specific RT-PCR primers and gRNA sequences for the CRISPR-Cas12a system were designed based on the sequences of the SARS-CoV-2 ORF1ab and N genes. As shown in Fig. 12.7b, the assay begins with RT-PCR amplification of SARS-CoV-2 RNA extracted from nasopharyngeal swabs of patients. The obtained amplified products are then mixed with

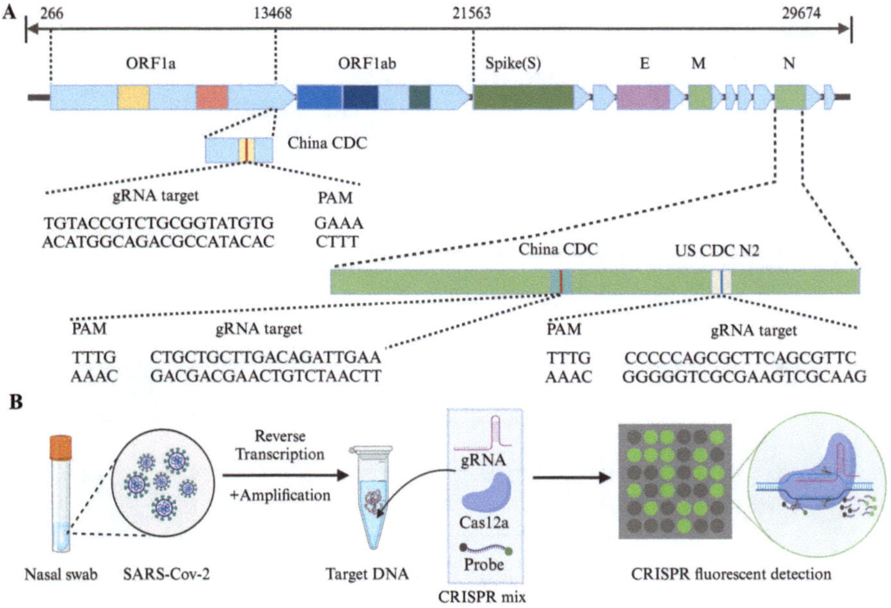

Fig. 12.7 Schematic diagram of CRISPR-FDS detection. (Revised and reprinted with permission from Ref. [34])

CRISPR-Cas12a reagents. Upon the specific recognition of gRNAs to amplified product, the DNase activity of Cas12a is activated. The activated Cas12a protein indiscriminately cleaves a large number of fluorophore/quencher-labeled ssDNA probes, thereby generating a detectable fluorescence signal. Comparative analysis of identical samples demonstrated that the detection results of this method were consistent with those obtained via RT-qPCR, with no ambiguous or invalid outcomes. Moreover, this approach achieved a nice detection limit as low as two copies/20 μL. Compared to conventional RT-PCR, this strategy that combined RT-PCR and CRISPR-Cas12a technologies significantly enhances both the sensitivity and specificity by leveraging dual signal amplification and dual molecular recognition processes.

Broughton et al. [21] reported a visual detection method for SARS-CoV-2 based on a lateral flow strip platform, termed DETECTR, which integrates CRISPR-Cas12a with RT-LAMP. As illustrated in Fig. 12.8, the study designed specific RT-LAMP primer sets and CRISPR-Cas12a guide RNAs (gRNAs) targeting the SARS-CoV-2 E and N genes, as well as the human RNase P gene.

As shown in Fig. 12.8a, during the detection process, the RNA extracted from nasopharyngeal swab samples undergoes simultaneous reverse transcription and loop-mediated isothermal amplification (LAMP), generating amplification products of the E and N genes of SARS-CoV-2 and the RNase P gene of the human host. These amplification products are then added to the CRISPR-Cas12a reaction system.

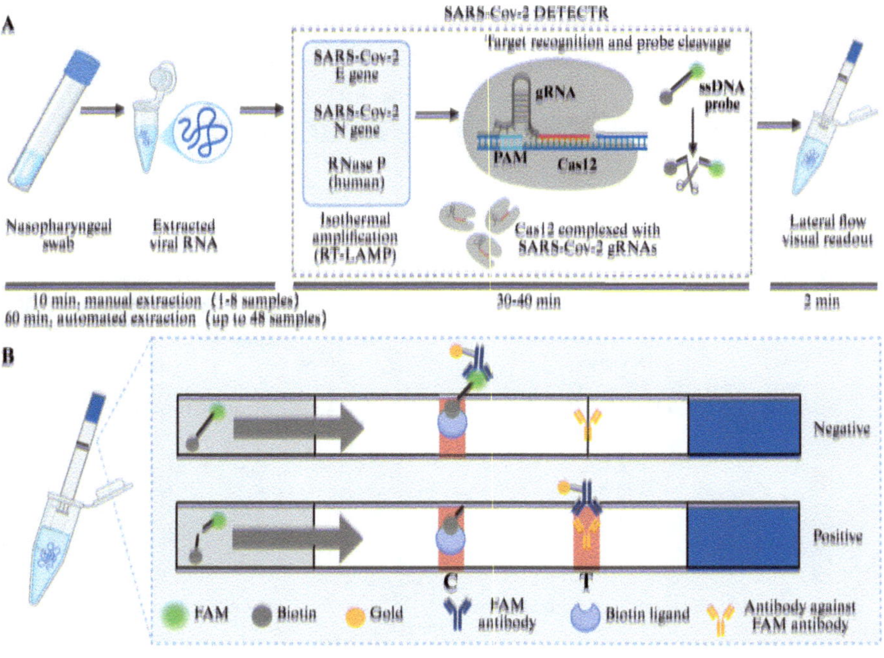

Fig. 12.8 Workflow of SARS-CoV-2 DETECTR. (Revised and reprinted with permission from Ref. [21])

Upon specific recognition of the target sequence, Cas12a is activated and exhibits nonspecific ssDNA cleavage activity, leading to the extensive degradation of a reporter ssDNA labeled with both FAM and biotin.

As depicted in Fig. 12.8b, the lateral flow strip used for visualization consists of two main components: a control line modified with biotin ligands and a test line coated with anti-FAM antibodies. Gold nanoparticles conjugated with anti-FAM antibodies are immobilized on the conjugate pad of the strip. In the absence of target sequences, the sample solution drives the anti-FAM gold nanoparticles to the control line, where capture is mediated by intact FAM-biotin-labeled ssDNA: the biotin group binds to the biotin ligand on the control line, while the FAM group interacts with the gold nanoparticles, resulting in visible accumulation at the control line.

In contrast, when target sequences are present, the FAM-biotin-labeled ssDNA is degraded by the activated Cas12a, preventing capture at the control line. The gold nanoparticles instead continue migrating forward and are eventually captured by the anti-FAM antibodies immobilized on the test line, leading to a color change at the test line due to gold particle accumulation.

In clinical validation studies involving 36 COVID-19 patients and 42 individuals with other viral respiratory infections, the DETECTR assay demonstrated a 95% positive agreement and 100% negative agreement with real-time RT-PCR results reported by the US Centers for Disease Control and Prevention (CDC).

Since the Cas12a protein undergoes denaturation and inactivation at the 62 °C, which is the required condition for the RT-LAMP reaction, the DETECTR necessitates the RT-LAMP amplification and the CRISPR-Cas12a reactions in two separated reaction tubes. Specifically, RT-LAMP was first completed in Tube A. When the products cooled to room temperature, it is transferred to Tube B for the subsequent CRISPR-Cas12a reaction. This additional transfer step for products not only increases procedural complexity but also significantly heightens the risk of cross-contamination. Therefore, despite the excellent analytical performance of the DETECTR, it remains imperfect.

Based on this, Zhang et al. [35] developed a method that enables SARS-CoV-2 RNA detection by combining RT-LAMP and CRISPR-Cas12a reactions within a single reaction tube. The detection principle is relatively conventional, similar to traditional strategies that integrate RT-LAMP amplification with CRISPR-Cas12a-mediated detection. Specifically, a target region of the SARS-CoV-2 genome is first amplified, followed by signal generation through the Cas12a-mediated cleavage of a fluorescently labeled reporter probe. However, it is worth noting that the researchers implemented a remarkably simple approach to achieve the single-tube reaction. During the assay setup, CRISPR-Cas12a reagents were placed in the cap of the PCR tube, while the RT-LAMP reaction mixture was added to the bottom of the tube. After incubating the RT-LAMP reaction at 62 °C for 30 min, the tube was cooled to room temperature. A simple inversion and gentle tapping of the tube allowed the CRISPR-Cas12a reagents to mix with the amplified products, initiating the detection step. Using this method, the researchers tested 50 clinical respiratory swab samples and observed a 94% concordance with RT-qPCR results.

Similar to loop-mediated isothermal amplification, recombinase polymerase amplification (RPA) is an isothermal nucleic acid amplification technique that enables exponential amplification under a constant temperature with high sensitivity and specificity. Due to these advantages, RPA has been widely employed in amplification-based nucleic acid detection. The reaction temperature for RPA ranges between 37 and 42 °C, making it highly compatible with CRISPR-based detection methods in terms of thermal conditions.

Zhang Feng et al. [36] developed a Specific High-sensitivity Enzymatic Reporter Unlocking (SHERLOCK) platform for Zika and dengue viruses detection. Later, the SHERLOCK platform was modified for the detection of SARS-CoV-2 [37–39]. (Detailed information is provided in Chap. 7, Fig. 7.15a). However, this approach requires prior RNA extraction from clinical samples before detection, which adds procedural complexity and increases the risk of contamination. To address this limitation, Cameron et al. [37] developed SHINE (Streamlined Highlight of Infections to Navigate Epidemiology), an optimized sample preprocessing technique. This method integrates RNase inhibitors with a heat- and chemically mediated reduction process to efficiently lyse SARS-CoV-2 to release viral RNA and simultaneously inhibit endogenous ribonuclease activity in biological fluids [40]. This streamlined approach eliminates the labor-intensive and time-consuming nucleic acid extraction steps and enables rapid sample preparation within 10 min. As illustrated in Fig. 12.9, SHINE employs a single-step SHERLOCK assay, in which RT-RPA amplification, T7 in vitro transcription, and Cas13a-mediated target recognition and cleavage occur within a single reaction tube. The Cas13a-mediated cleavage to single-stranded RNA probes could be detected by either lateral flow-based colorimetric assays or fluorescence-based readouts. The detection result obtained by this method demonstrated a 94% concordance rate with that of RT-qPCR for positive detections.

Since the CRISPR-Cas13a system specifically recognizes ssRNA targets, while the RT-RPA amplification product is dsDNA, an additional in vitro transcription step is required to convert dsDNA into ssRNA to activate Cas13a in SHERLOCK assay. For CRISPR-Cas12a system, gRNA can directly recognize dsDNA amplicons and activate Cas12a protein's DNase activity. Su et al. [41] developed a strategy

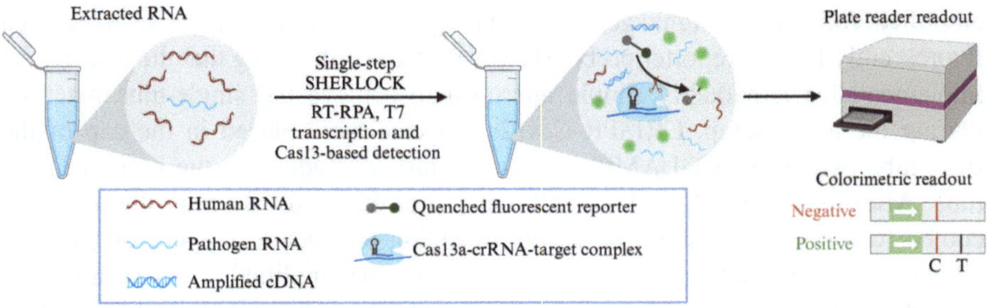

Fig. 12.9 Schematic diagram of one-step SHERLOCK detection for SARS-CoV-2. (Revised and reprinted with permission from Ref. [40])

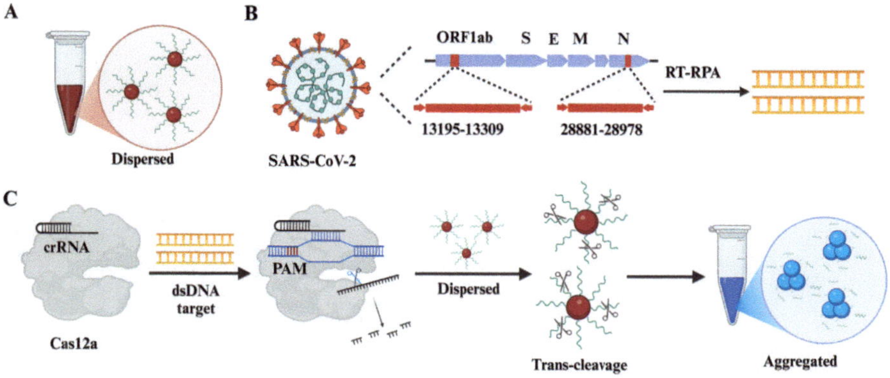

Fig. 12.10 Schematic diagram of one-step RT-RPA coupled with CRISPR/Cas12a-AuNPs colorimetric detection. (Revised and reprinted with permission from Ref. [41]. Copyright 2021 American Chemical Society)

where CRISPR-Cas12a coupled with RT-RPA for the visual SARS-CoV-2 detection. Firstly, as illustrated in Fig. 12.10a, the researchers prepared the gold nanoparticle modified with ssDNA. Due to electrostatic repulsion between ssDNA, the ssDNA-modified AuNPs remained well-dispersed in the solution, imparting a red coloration. Subsequently, as depicted in Fig. 12.10b, the authors employed RT-RPA to amplify the specific sequences of the ORF1ab and N genes, yielding a high concentration of dsDNA amplicons. As shown in Fig. 12.10c, the Cas12a protein is activated in the presence of these dsDNA, leading to extensive degradation of the ssDNA probes on the surface of the gold nanoparticles. This cleavage disrupted the electrostatic repulsion between nanoparticles, inducing their aggregation and resulting in a colorimetric shift from red to blue. The color change in the solution could be quantitatively analyzed via UV-Vis absorption spectroscopy or qualitatively assessed by the naked eye. This work presented a LOD down to 1 copy/reaction.

Liu et al. [42] reported a dual CRISPR-Cas12a-based detection method, termed AIOD-CRISPR (All-In-One Dual CRISPR-Cas12a), which integrates RT-RPA for ultrasensitive and visual detection of SARS-CoV-2. This method targets the N gene of SARS-CoV-2 and employs a pair of specifically designed crRNAs (gRNAs) that are not constrained by the presence of a protospacer adjacent motif (PAM) sequence. This design enables a single target molecule to activate two Cas12a proteins simultaneously, thereby enhancing signal amplification. As illustrated in Fig. 12.11, reverse transcription generates a substantial quantity of dsDNA, which subsequently initiates recombinase polymerase amplification under the action of recombinase, strand-displacing DNA polymerase, and single-strand DNA-binding proteins. This process results in a large number of activation fragments, leading to the robust activation of Cas12a proteins. The fluorescence signal generated by CRISPR-Cas12a-mediated cleavage of ssDNA probes can be observed under ultraviolet (UV) light. Additionally, the color change within the reaction system can be

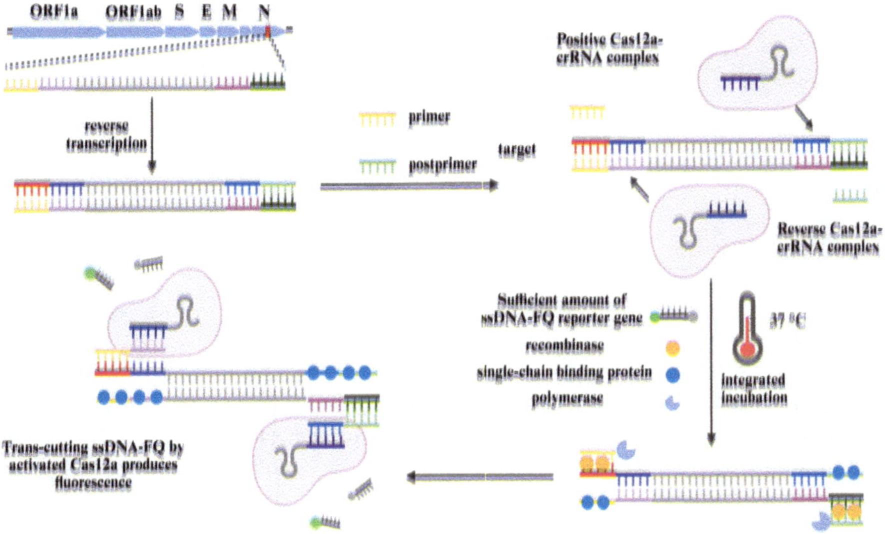

Fig. 12.11 Schematic diagram of AIOD-CRISPR analysis detection. (Revised and reprinted with permission from Ref. [42])

visually assessed, enabling straightforward signal readout. The results obtained by this method presented high concordance with that of RT-qPCR, in testing 28 clinical swab samples. Although the CRISPR-Cas12a system can be activated without the requirement of a PAM sequence, the DNase activity is relatively lower compared to the PAM-dependent CRISPR-Cas12a system [43]. Notably, the ingeniously designed dual CRISPR-Cas12a detection strategy achieved single-molecule sensitivity. This heightened sensitivity may be attributed to the fact that, although the PAM-dependent CRISPR-Cas12a system enables rapid target cleavage, it paradoxically limits detection sensitivity when target molecule concentrations are low [42].

Wang et al. [44] reported a SARS-CoV-2 detection strategy that couples ligation-initiated transcriptional amplification with CRISPR-Cas13a. This method employs fluorescence intensity variations of luminescent RNA aptamers, which selectively bind to dyes to form aptamer-dye complexes [45], as the detection signal. As illustrated in Fig. 12.12, the assay begins with the design of a highly specific DNA probe pair targeting SARS-CoV-2 RNA, wherein Probe A contains a PAM sequence, while Probe B harbors a T7 promoter sequence. Under the catalysis of T4 RNA ligase 2, these probes undergo a ligation reaction. Subsequently, T7 RNA polymerase utilizes the ligation product as a template to initiate in vitro transcription at 37 °C, generating abundant single-stranded RNA molecules carrying the target sequence. These RNA products are then specifically recognized by the CRISPR-Cas13a complex via complementary crRNA binding, which activates the nonspecific endonuclease activity of Cas13a. The activated enzyme cleaves the luminescent RNA aptamer-dye complex, leading to a marked reduction in fluorescence intensity. Compared to other detection approaches, this method capitalizes on

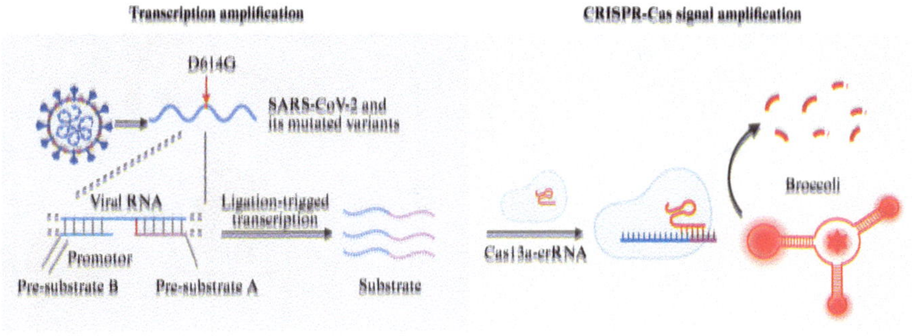

Fig. 12.12 Schematic diagram of the principle of CRISPR/Cas13a-coupled connected reaction for the detection of SARS-CoV-2 and its mutants. (Revised and reprinted with permission from Ref. [44]. Copyright 2021 American Chemical Society)

the high specificity of the ligation reaction toward single-nucleotide variations, enabling precise discrimination between SARS-CoV-2 and its critical D614G variant.

Developing the strategy that could make CRISPR-Cas13a system perform high-sensitive SARS-CoV-2 RNA detection without the need if nucleic acid amplification has grown as a research "hot-spot." Jennifer et al. [46] reported multi-crRNA based CRISPR-Cas13 system for sensitive SARS-CoV-2 RNA detection. This approach designed 12 crRNA targeting different sites of the N gene and 4 crRNA targeting different sites of the E gene. Through experimental screening, two efficient crRNA for the N gene site and one nice crRNA for E gene were identified. By employing the combination of these three crRNAs, a single SARS-CoV-2 RNA molecule could activate at least three Cas13a proteins, significantly enhancing the signal output triggered by per target molecule. This strategy enabled 2 h response time and 30 copies/μL sensitivity in SARS-CoV-2 RNA detection.

Subsequently, Jennifer's research team further optimized the above method and developed a Fast Integrated Nuclease Detection In Tandem (FIND-IT) technique for SARS-CoV-2 RNA detection. FIND-IT integrate Csm6 with multiple CRISPR-Cas13a systems to perform an enzymatic cascade effect [47]. In the FIND-IT strategy, Csm6 is a dimeric RNA endonuclease derived from the Type III CRISPR-Cas system [48, 49]. As illustrated in Fig. 12.13, the researchers designed eight crRNAs targeting different SARS-CoV-2 RNA site. Upon specific recognition, viral RNA triggers Cas13a's RNase activity. Activated Cas13a cleaves both the fluorescent probe and the activator-oligonucleotide conjugate (where the oligonucleotide sequence is 5′-A4-U6–3′). The cleaved conjugate releases a specific Csm6 activator (tetraadenylate with a 2′,3′-cyclic phosphate group), which binds to Csm6 and activates its RNA-cleaving activity. Activated Csm6 further cleaves the fluorescent reporter molecule, releasing a fluorescent signal. Additionally, chemical modifications to the activator (such as the introduction of 2′-fluoro or 2′-deoxy modifications) can prevent Csm6 from degrading the activator, thereby extending its activation time. This method combines Csm6 protein with CRISPR-Cas13a in a tandem system, achieving a detection limit as low as 31 copies/μL within 20 min.

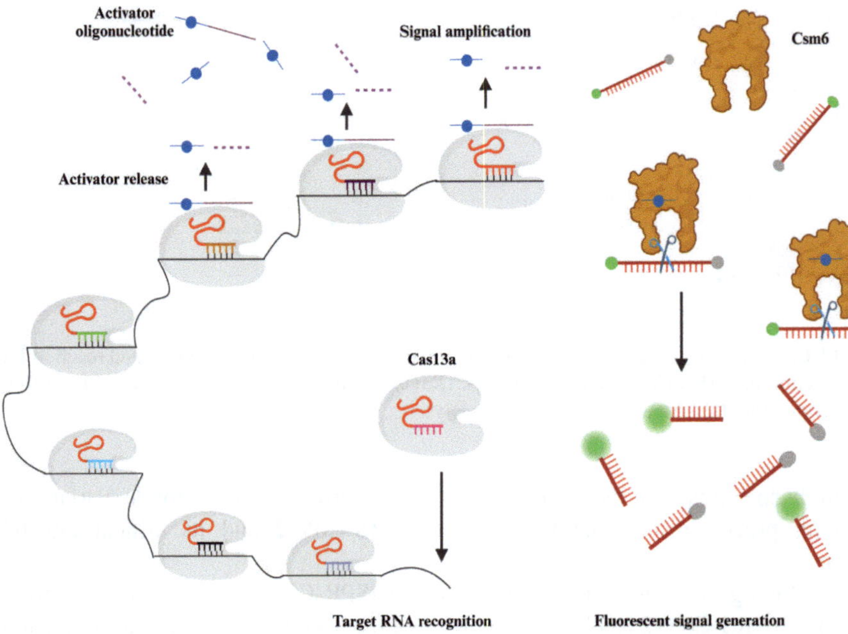

Fig. 12.13 Schematic diagram of FIND-IT detection for SARS-CoV-2. (Revised and reprinted with permission from Ref. [47])

12.4 HTS-Based Method for SARS-CoV-2 Detection

Current methods for detecting SARS-CoV-2 RNA rely on monitoring changes in fluorescence, color, pH, or turbidity signals to perform quantitative analysis. However, these detection techniques are limited to identifying viral RNA with known sequences and are incapable of obtaining sequence information. Consequently, they fail to provide critical insights for tracing and tracking the continuously mutating SARS-CoV-2 variants.

Over the past two decades, next-generation sequencing (NGS) technology has undergone rapid advancements and was increasingly applied to the detection and sequencing of various pathogens. However, despite its high-throughput capabilities, NGS-based sequencing still faces several limitations, including the high cost of instrumentation, operational complexity, significant reagent consumption, and the requirement for highly specialized personnel. To address these challenges, large-scale parallel detection techniques for SARS-CoV-2 have been developed based on NGS, leveraging barcode (Barcodes) technology to directly label different samples or genetic fragments for multiplexed sequencing. By tagging viral RNA from different samples with unique barcodes, this approach enables the simultaneous analysis of thousands or even tens of thousands of specimens, providing crucial and robust support for viral transmission monitoring and genomic epidemiology studies.

For example, Yángüez et al. [50] developed a HiDRA-seq technology. It uses barcode-labeled oligo-dT primers to perform reverse transcription amplification on the highly conserved ORF10 region of the SARS-CoV-2 isolate strain and generates an Illumina sequencing library by PCR. Since the target region is located near the 3′ end, HiDRA-seq can capture both viral genomic RNA and subgenomic RNA transcripts in infected cells. Next-generation sequencing implemented by the reported SwabSeq [51] protocol can detect SARS-CoV-2 RNA in thousands of samples at a time without the need for RNA extraction by labeling mixed samples with specific molecular barcodes. In this protocol, RNA synthesis in vitro is added to each natural sample. Its sequence is the same as the amplicon of the target virus sequence except for the upstream 6 bp. The two RNAs can be distinguished by sequencing. The ratio of the number of natural RNA reads to the number of RNA reads of artificially synthesized in vitro allows for a more accurate measurement of viral genomes compared to natural RNA read counts alone.

The research team of Andrew R. Bassett reported a high-throughput sequencing technology (INSIGHT) based on the isothermal NASBA strategy (nucleic acid sequence-based amplification) for SARS-CoV-2 detection [52]. As shown in Fig. 12.14, the protocol first uses specific primers (including T7 promoter sequence and specific barcode) to reverse-transcribe SARS-CoV-2 RNA to generate single-stranded cDNAs. Then, the SARS-CoV-2 RNA was digested and cleaved by the RNase H. Finally, complementary double-stranded cDNA is generated using second-strand synthesis primers (containing sequencing adapter sequences and specific barcodes). The generated double-stranded cDNA could not only be further amplified to generate a next-generation sequencing library under the action of sequencing primers, but also can be transcribed to generate RNA under the action of the T7 promoter. The synthesized RNA can hybridize with specific primers containing a T7 promoter, thereby facilitating the reiteration of the aforementioned process of double-stranded synthesis from reverse transcription.

Actually, the INSIGHT technology could be viewed as a module package system, the first package includes two rapid test modules (fluorescent molecular beacon detection and test paper-based lateral flow technology), and the second package is a sequencing module. These modules can be used alone or combined in different ways, giving INSIGHT a high degree of flexibility in adapting different testing needs and resource availability. In addition, C19-SPAR-Seq [53] and SARSeq [54] technologies achieve direct amplification of viral RNA also through multiple RT-PCR. Parallel sequencing was achieved by the respective double index combinations from barcode collection and the quality control strategies. Research for SARS-CoV-2 detection using second-generation sequencing technology combined with LAMP amplification technology likewise have been reported, such as LAMp-Seq [55] and COVID [56].

In recent years, the third-generation gene sequencing technology represented by Oxford Nanopore Sequencing (ONT) has been widely used and researched. Compared with the second-generation sequencing instruments represented by Illumina, ONT's equipment has the great advantages of portability, high speed, and scalability [69]. Given ONT technology had been successfully applied to detecting the Ebola

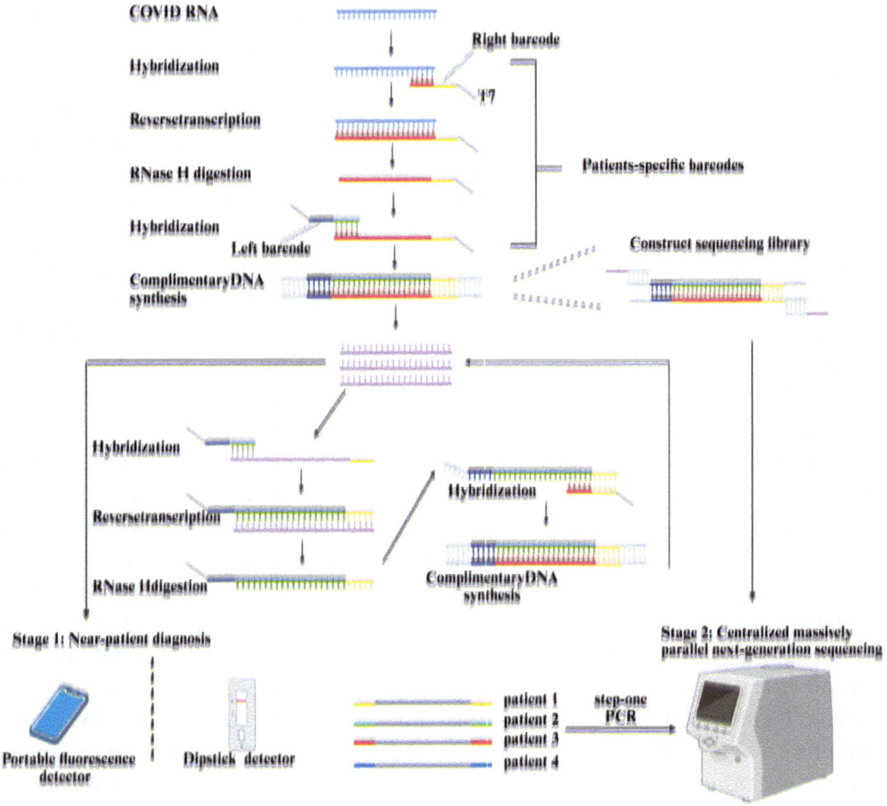

Fig. 12.14 Schematic diagram of INSIGHI detection. (Revised and reprinted with permission from Ref. [52])

virus, Zika virus, and other viruses [70, 71], it is also highly promising for detecting SARS-CoV-2. Based on this, the research team of Oxford Nanopore Sequencing Company developed the LamPORE technology [72]. The technology combines advantages such as the specific amplification of LAMP to multiple targets and the rapid barcode library preparation based on transposon to achieve real-time nanopore sequencing, enabling quick screening of a large number of SARS-CoV-2 samples.

The Wuhan Hope Genomics Medical Laboratory has developed a rapid detection workflow for SARS-CoV-2 RNA using the Nanopore Flongle platform, encompassing RNA extraction, library preparation, sequencing, data analysis, and report generation [73]. This approach targets the Orf1ab and N gene regions of SARS-CoV-2, as well as the HA and M gene regions of influenza B virus, employing specifically designed LAMP primers. By integrating LAMP amplification with Nanopore's rapid library preparation kit, sequencing adapters can be directly ligated for streamlined library construction. This method enables simultaneous, high-throughput pathogen detection, including both SARS-CoV-2 and

Table 12.2 Major mutant strains of the SARS-CoV-2

Mutations	VOC	Changed amino acid	References
Global dominant mutant	A23403G	D614G	[57–61]
British mutant	A23063T	N501Y	[62, 63]
	A23403G	D614G	
South African mutant	G22813T	K417N	[64]
	G23012A	E484K	
	A23063T	N501Y	
Brazilian mutant	A22812C	K417T	[65]
	G23012A	E484K	
	A23063T	N501Y	
New York mutant	G23012A	E484K	[66, 67]
	A23403G	D614G	
Indian mutant	T22917G	L452R	[68]
	G23012C	E484Q	
	A23403G	D614G	

influenza viruses, on a single Flongle ONT sequencing chip, while supporting real-time sequencing data analysis for immediate interpretation of results.

With the continuous spread of the virus and the development of large-scale vaccination work around the world, the virus has undergone a series of mutations under the induction of complex environmental factors, especially the variants in the S gene region (see Table 12.2). These viral variants exhibit enhanced binding affinity to the host cell receptor ACE2 and increased replication efficiency, while simultaneously reducing antibody recognition. This, in turn, strengthens the virus's immune evasion capability, as it adapts more effectively to its surrounding environment.

At present, SARS-CoV-2 genome-wide monitoring has become a key means to rapidly monitor the spread of VOCs. It is worth noting that these VOCs may exhibit hyposensitivity to existing vaccines. Whole-genome sequencing of the SARS-CoV-2 pathogen can inform public health decision-making and provide critical clinical data for tracking epidemics and vaccine exploration [74–77].

12.5 Microfluidic-Based Devices for POCT of the SARS-CoV-2

Microfluidic devices, an umbrella term for miniaturized and systematically integrated detection equipment, are characterized by rapid detection, low cost, portability, and operational simplicity [57, 78–80]. In recent years, propelled by the COVID-19 pandemic, POCT (Point-of-Care Testing) technologies for SARS-CoV-2 detection based on microfluidic devices have rapidly advanced [81].

12.5.1 Test Strip Technology for SARS-CoV-2 Detection

As one of the microfluidic device technologies, the test strip technology has been widely applied for the SARS-CoV-2 detection. This approach offers several advantages, including simplicity in operation, rapid result, compact platform size, and cost-effectiveness [82]. The test strip is capable of swiftly identifying specific biomarkers of SARS-CoV-2, enabling not only antigen and antibody detection, but also facilitating the integration with techniques such as polymerase chain reaction (PCR), enzyme-linked immunosorbent assay (ELISA), and CRISPR/Cas systems to perform nucleic acid detection [83–85].

Currently, the test strip technology that has been reported is primarily categorized into two types: lateral flow assay strips and vertical flow assay strips. As illustrated in Fig. 12.15, the lateral flow assay strip typically consists of a sample pad, conjugate pad, chromatographic membrane, and absorbent pad [86, 87]. The sample pad is a fiber membrane or glass wool that has been treated to rapidly absorb the sample to be tested, while utilizing capillary action to drive the sample toward the conjugate pad. The conjugate pad, made of the same material as the sample pad, serves to adsorb labeled bioactive materials that can form complexes with the target analytes present in the sample solution. The chromatographic membrane, a critical component of the lateral flow assay strip, is primarily made of nitrocellulose. It is immobilized with two reaction zones containing different bioactive substances: the test line (T line) and the control line (C line). The T line (test line) is used to capture the specific complex formed by bioactive materials and the target, displaying the detection results through a visible color change. The C line (control line) is used to verify that the lateral flow test strip can properly drive the sample to flow through the T line via capillary action. Regardless of whether the sample is positive or negative, the C line should always show a color change. The absorbent pad, composed of absorbent paperboard, is designed to absorb the sample solution that flows through the chromatographic membrane. This balances the pressure difference on both sides of the membrane, ensuring that more of the sample can flow across the chromatographic membrane.

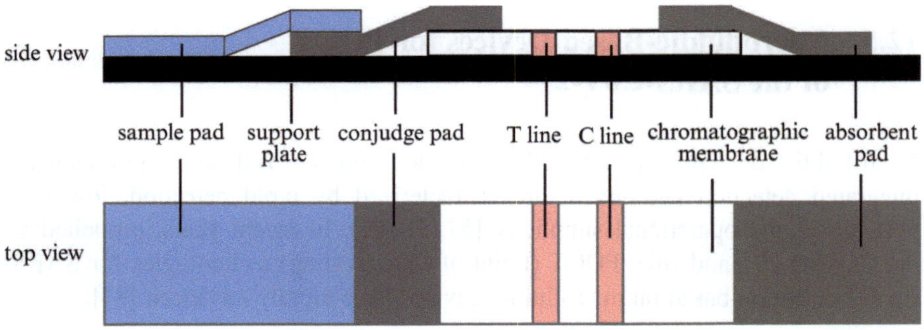

Fig. 12.15 Schematic diagram of the structure of a lateral flow assay strip

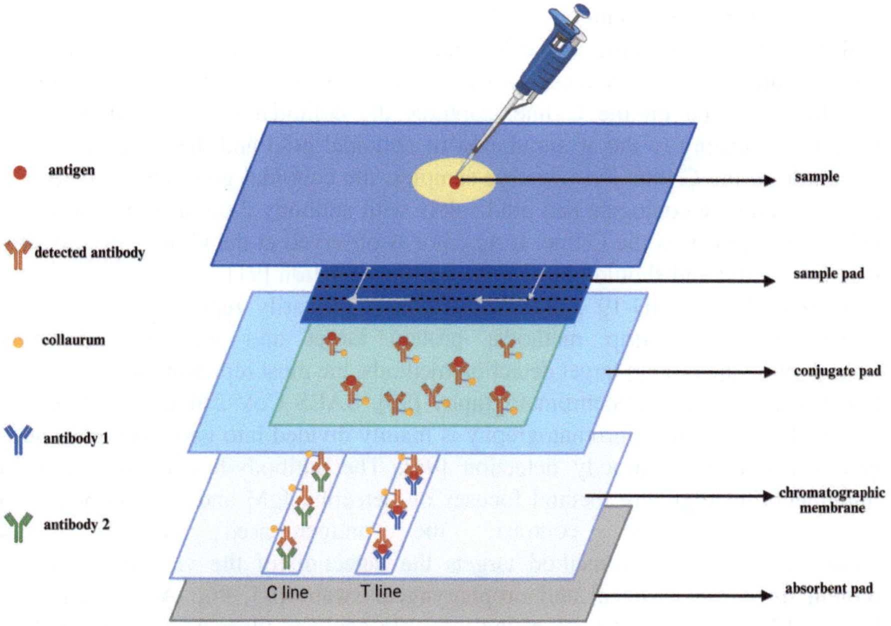

Fig. 12.16 Schematic diagram of the structure of a vertical flow assay strip. (Revised and reprinted with permission from Ref. [90])

Vertical flow assay strip technology is an improvement and advancement over lateral flow test strip technology. Compared to lateral flow detection, vertical flow analysis increases the sample reaction time and enhances the capture efficiency of analytes, contributing to further improvement in detection sensitivity [83, 88, 89]. As illustrated in Fig. 12.16, the structural components of the vertical flow assay strip are identical to those of the lateral flow assay strip, comprising a sample pad, conjugate pad, chromatographic membrane, and absorbent pad [90]. The sample pad constitutes the initial section of the test strip, where the sample is applied. It is typically made of highly absorbent porous material to ensure the sample evenly penetrates the rest of the strip. The conjugate pad contains labeled markers, such as colloidal gold or other tracers, which bind to the target substance and form a complex. Similarly, the chromatographic membrane is made of nitrocellulose, providing a separation and screening function that allows for the specific separation of substances based on their distinct physicochemical properties as they migrate through the strip. Positioned at the bottom of the test strip, the absorbent pad serves to absorb excess liquid, ensuring smooth fluid flow and facilitating the discharge of any surplus liquid.

For positive samples, once the sample is applied to the sample pad, it flows downward under capillary action, uniformly spreading across the sample pad. As the sample migrates through the conjugate pad, the antigen can bind with the detection antibody labeled with colloidal gold, forming a complex. This complex continues to flow downward to the detection pad. At the T line of the chromatographic membrane, antibodies (antibody 1) that are fixed to the membrane specifically bind with the antigen, resulting in the formation of a "sandwich" structure (antibody-antigen-antibody) that produces a visible color change. Simultaneously, antibody 2, which is fixed on the C line, captures the colloidal gold-labeled detection antibodies, leading to the accumulation of colloidal gold and the appearance of a red color on the C line. For negative samples, the colloidal gold-labeled detection antibodies on the conjugate pad bind solely with antibody 2 on the C line, causing color development at the C line. If no color is observed at the C line, the test strip result is invalid and should not be used for interpretation [91].

During the COVID-19 pandemic, scientists primarily reported two types of SARS-CoV-2 test strip methods: protein target and nucleic acid target [92]. Among the protein target detection methods, the most representative technique is colloidal gold immunochromatography [93]. SARS-CoV-2 detection based on colloidal gold immunochromatography is mainly divided into two categories: antigen detection and antibody detection [94]. The antibody-based colloidal gold immunochromatography method focuses on detecting IgM and IgG antibodies in blood samples. In contrast, the antigen-based colloidal gold immunochromatography method targets the detection of the viral nucleocapsid protein on nasopharyngeal and oropharyngeal swabs [95, 96]. As illustrated in Fig. 12.17, for antigen detection: the test strip contains three types of antibodies, categorized by function as labeled antibodies, detection antibodies, and control antibodies. The labeled antibodies, located on the conjugate pad, are conjugated with gold nanoparticles and can specifically bind to the surface protein of the SARS-CoV-2 virus. The detection antibodies, positioned on the test line of the nitrocellulose membrane, are capable of specifically recognizing the SARS-CoV-2 surface protein. The control antibodies, fixed on the control line of the nitrocellulose membrane, bind directly with the labeled antibodies, without requiring mediation by the SARS-CoV-2 surface protein.

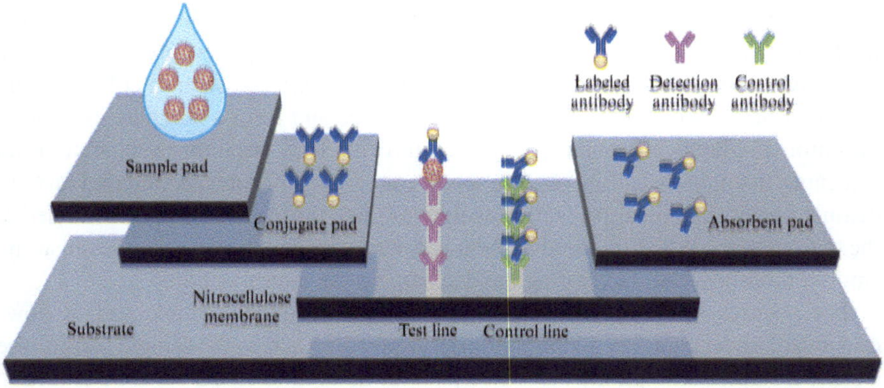

Fig. 12.17 Schematic illustration of colloidal gold immunochromatography for SARS-CoV-2 detection

When the sample is applied to the sample pad, it flows to the conjugate pad under the influence of capillary action. The SARS-CoV-2 particles will form a binary complex with the gold labeled with antibodies on the conjugate pad. As the sample reaches the test line, the binary complex is captured by the detection antibodies on the test line, resulting in the aggregation of gold nanoparticles. The more SARS-CoV-2 particles in the sample, the more gold particles accumulate on the test line, producing a more intense red color. As the liquid continues to flow toward the control line, the gold labeled with antibodies are captured by the control antibodies regardless of the presence of virus particles, causing a significant accumulation of gold particles and forming a distinct red quality control line. For antibody detection, the principle is similar to that of antigen detection. However, as the target analytes are IgM and IgG antibodies, the three antibodies used in the antigen detection process are replaced with specific antibodies targeting IgM and IgG antibodies.

Although the colloidal gold immunochromatographic test strip has played a significant role in COVID-19 testing, several limitations hinder its application in the early detection of SARS-CoV-2 infections. These limitations include the relatively low viral load in nasopharyngeal swabs, the delayed immune response required for antibody production, and the necessity for a substantial accumulation of virus-mediated gold nanoparticles at the test (T) line to produce a detectable colorimetric signal. Consequently, immunoassay-based lateral flow test strips exhibit low sensitivity, rendering them less suitable for the early detection of SARS-CoV-2 infections. In response to these challenges, scientists have developed a series of SARS-CoV-2 lateral flow assay strip technologies based on nucleic acid targets [97–100]. By integrating various nucleic acid amplification techniques, the sensitivity has been significantly improved.

Currently, nucleic acid target-based SARS-CoV-2 lateral flow assay strip technologies can be primarily classified into two types: antibody-dependent (nucleic acid lateral flow immunoassay assay, NALFIA) [101] and antibody-independent (nucleic acid lateral test strip, NALTS) [102]. The antibody-dependent lateral flow strip technology utilizes nucleic acid–antibody interactions and the specific labeling of double-stranded DNA to combine amplification technology with enzyme-linked immunosorbent assay (ELISA) techniques to realize target nucleic acids detection [103]. Based on this principle, scientists have developed various strategies for SARS-CoV-2 RNA detection [104–106].

For example, Liu et al. [104] combined reverse transcription recombinase polymerase amplification (RT-RPA) technology with enzyme-linked immunosorbent assay (ELISA) to develop a sensing chip for rapid on-site detection of SARS-CoV-2. First, a pair of RT-RPA primers targeting a specific sequence of the SARS-CoV-2 genome was designed, with distinct modifications: one primer was labeled with FAM, and the other with biotin. The RT-RPA reaction generates double-stranded amplification products labeled at both ends with FAM and biotin. The detection chip comprises four functional zones: Zone I for RT-RPA amplification of the target sequence, Zone II for storing the buffer required for lateral flow detection, Zone III for mixing the RT-RPA products with the lateral flow buffer, and Zone IV for housing the lateral flow test strip. The conjugate pad of the test strip is coated with colloidal gold which was modified with FAM antibodies, the test line (T line) is coated with biotin-specific antibodies, and the control line (C line) is coated with anti-FAM antibodies. During the detection process, RNA extracted from the clinical sample is added to Zone I, which contains all the RT-RPA reaction components. Detection buffer is added to Zone II. The chip is then sealed and incubated at 42 °C for 15 min. After incubation, the device is placed upright, allowing the reaction mixture to flow into Zone III where it mixes with the detection buffer, and subsequently migrates to the lateral flow test strip in Zone IV. For positive samples, the RT-RPA amplification generates a high yield of dual-labeled amplicons. As the solution reaches the conjugate pad, the FAM label at one end of the product binds to the colloidal gold particles conjugated with anti-FAM antibodies, forming a complex. Through capillary action, the complex migrates to the T line, where the biotin at the other end binds to the biotin-specific antibodies, resulting in the accumulation of gold nanoparticles and the appearance of a red band at the T line. Simultaneously, excess gold-labeled anti-FAM antibodies are captured at the C line by anti-anti-FAM antibodies, ensuring the appearance of a red signal at the control line. For negative samples, failed RT-RPA amplification results in the absence of dual-labeled amplicons. Consequently, the T line cannot capture the gold-labeled complexes and remains colorless. However, since the C line contains anti-anti-FAM antibodies, the FAM antibody–labeled gold particles are still captured, generating a red band at the C line. The absence of a red signal at the C line indicates a failure of the test strip. By leveraging the signal amplification from the RT-RPA process, this strategy enables the successful detection of SARS-CoV-2 in clinical samples with a viral load as low as 1 copy/μL.

The antibody-independent lateral flow test strip is a technology that utilizes nucleic acid hybridization for target detection [82, 107–109]. Taking the colloidal gold-based antibody-independent lateral flow nucleic acid test strip as an example [110], as shown in Fig. 12.18, the conjugate pad of the test strip is fixed with

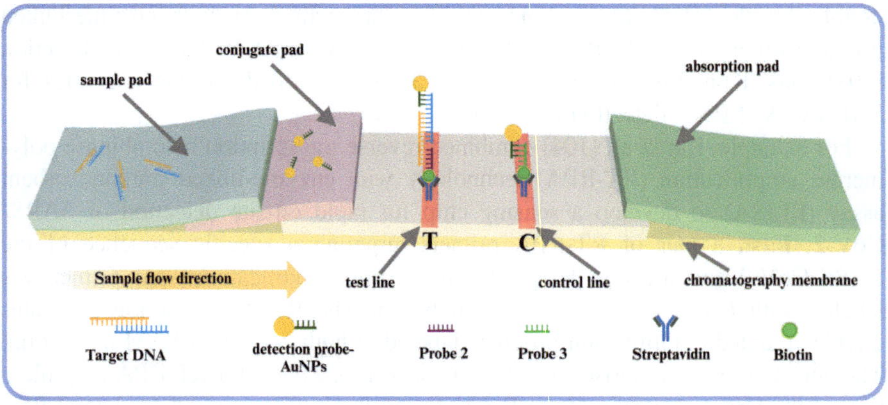

Fig. 12.18 Detection principle of colloidal gold NALTS test strip. (Revised and reprinted with permission from Ref. [107])

colloidal gold conjugated to a target nucleic acid capture probe (Probe 1). The T line contains Probe 2, which is designed to specifically hybridize with the target nucleic acid, facilitating its capture. The C line is fixed with Probe 3, which can specifically hybridize with Probe 1. The T line contains Probe 2, which is designed to specifically hybridize with the target nucleic acid, facilitating its capture. The C-line is fixed with Probe 3, which can specifically hybridize with Probe 1.

In the presence of a positive sample, the asymmetrical double-stranded amplification product formed during amplification hybridizes with the capture probe (Probe 1) on the conjugate pad, forming a complex. This complex is driven by capillary action to the T line, where it is captured by Probe 2, leading to the continuous accumulation of gold particles at the T line, thus displaying a red band. Excess probe 1 conjugated with colloidal gold will migrate past the T line and continue to flow toward the C line, where they are captured by Probe 3, resulting in the accumulation of the colloidal gold at the C line and the appearance of a red band there as well. For negative sample, since Probe 2 on the T line cannot directly capture the colloidal gold conjugated to Probe 1, no color develops at the T line. However, Probe 3 on the C line can directly capture the colloidal gold conjugated to Probe 1, meaning that regardless of the presence of the target nucleic acid, a red band will always appear at the C line. If no red signal appears at the C line, it indicates that the test strip has failed.

Based on this principle, scientists have developed numerous antibody-independent nucleic acid detection technologies [111–113]. For example, Almeida [109] developed an antibody-independent lateral flow test strip for SARS-CoV-2 RNA detection based on reverse transcription-recombinase polymerase amplification (RT-RPA). As illustrated in Fig. 12.19a, this method ingeniously incorporates a set of amplification primers with tracer functionality to amplify the reverse transcription products. These primers consist of two parts: one part targets the predetermined amplification region and guides the polymerase to amplify the target region under the action of recombinase, while the other part does not participate in the amplification process but instead labels each set of amplification products (Label 1 and Label 2). As a result, after the amplification reaction, a large number of amplification products labeled at the ends are generated. As shown in Fig. 12.19b, the conjugate pad of the test strip is coated with a substantial amount of functionalized colloidal gold whose surface is labeled with Probe 1, which can specifically bind to Label 1 at the end of the amplification products. On the T line, Probe 2 is fixed through biotin-streptavidin affinity interactions and is capable of specifically hybridizing with Label 2 at the end of the amplification products. The control line (C line) is immobilized with Probe 3, which can specifically hybridize with Probe 1 on the gold surface. During detection, if the sample is positive, the gold modified with Probe 1 will specifically bind with the amplification product that carries Label 1, forming a complex that will be driven by capillary action to the T line. Label 2 on the amplification product will be specifically captured by Probe 2 on the T line, causing gold particles to accumulate at the T line and resulting in a red band. The detection fluid then continues to flow via capillary action to the C line,

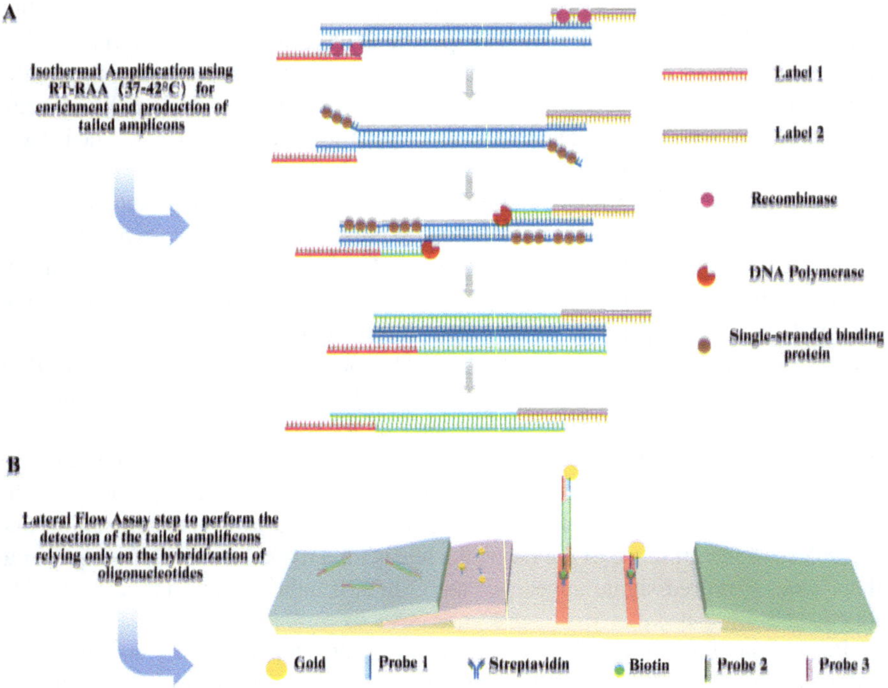

Fig. 12.19 Rapid detection of SARS-CoV-2 nucleic acids using antibody-independent test strips. (Revised and reprinted with permission from Ref. [109])

where Probe 3 can directly capture the colloidal gold labeled with Probe 1, presenting a red band at the C line. This method successfully detects SARS-CoV-2 at a concentration as low as 12 copies/μL.

Academician Li et al. [114] developed a colorimetric test strip (MARVE: Multiplexed Analysis, Nucleic Acid Amplification-Free, Single-Nucleotide-Resolved Viral Evolution) for detecting SARS-CoV-2 and its variants. The core principle of MARVE is based on nucleic acid strand displacement reactions and metal ion-controlled enzymatic reactions. As shown in Fig. 12.20a, this method can specifically identify single-nucleotide mutations in the SARS-CoV-2 genome, such as key mutations in the Alpha, Beta, Gamma, and Delta variants. As illustrated in Fig. 12.20b, the authors designed a double-stranded DNA probe (TEProbe) containing a forward toehold and a reverse toehold. The reverse toehold incorporates a C-Ag(I)-C artificial base pair. When the target RNA binds to the TEProbe, Ag (I) ions are released through a strand displacement reaction. The Ag(I) ions inhibit the catalytic activity of urease, reducing the production of NH_4^+ from urea hydrolysis, thereby altering the pH of the solution and the color of the pH indicator. As shown in Fig. 12.20c, the detection results can be rapidly visualized either by the naked eye or using a smartphone. MARVE does not require nucleic acid amplification and can achieve specific detection of different SARS-CoV-2 variants within 30 min by designing different TEProbes. The method is cost-effective (approximately $0.3 per test) and achieves a detection sensitivity of 400 copies/μL.

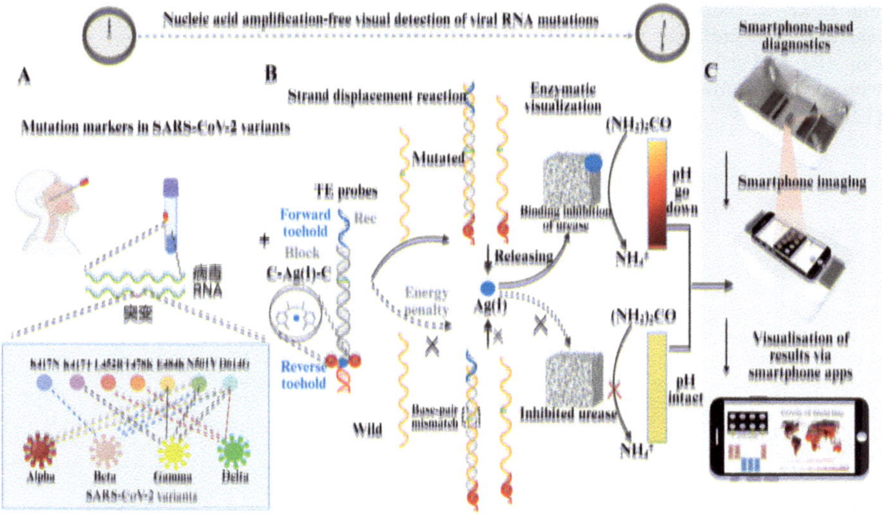

Fig. 12.20 MARVE detect SARS-CoV-2 and its variant. (Revised and reprinted with permission from Ref. [114])

12.5.2 Microfluidic Devices for SARS-CoV-2 Detection

Microfluidic technology is a technique that utilizes channels ranging from tens to hundreds of micrometers in size to process small quantities of fluids (on the scale of 10^{-9} to 10^{-18} liters) [115, 116]. These miniature channels enable precise control over minute volumes of fluid, allowing the fluids to exhibit specific flow behaviors and physicochemical properties [117]. Microfluidic chips are capable of rapidly processing small-volume samples, enabling the acquisition of analytical signals within minutes. This characteristic makes them highly promising for applying in disease diagnostics, particularly in nucleic acid-based biomarker detection.

Microfluidic chips are typically fabricated by the materials that exhibit excellent surface modifiability, biocompatibility, thermal and electrical insulation properties, optical transparency, and cost-effectiveness, such as polymethyl methacrylate (PMMA) [118], polydimethylsiloxane (PDMS) [119], and glass [120]. As illustrated in Fig. 12.21, the structural components of a microfluidic chip consist of microchannels, microstructures, sample inlets, detection windows, and sample outlets [121]. Peripheral equipment includes driving devices such as microinjection pumps, temperature control systems, and detection systems including fluorescence, electrochemical, and chromatographic detection instruments. Samples and reagents are introduced into the sample inlet of the microfluidic chip, and the driving mechanism is activated. Under the influence of the driving device, the sample and reagents flow slowly through the chip, forming a microfluidic path, where one or multiple reactions can occur sequentially. The detection window on the chip is then analyzed using fluorescence, electrochemical, and chemical detection techniques for

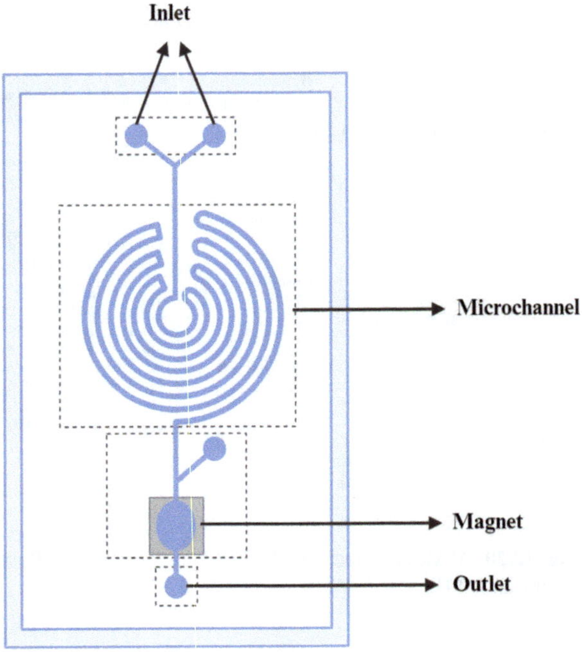

Fig. 12.21 Schematic diagram of a microfluidic chip. (Revised and reprinted with permission from Ref. [121])

rapid, accurate, and high-throughput analysis [122–124]. After detection, the liquid can be continuously expelled through the sample outlet. By repeatedly introducing and expelling samples within the microfluidic chip, continuous monitoring and analysis of the sample are performed, yielding corresponding detection results.

During the COVID-19 pandemic, scientists developed a series of nucleic acid detection methods for the SARS-CoV-2 virus based on microfluidic detection platforms [125, 126]. For instance, Tian et al. [127] combined reverse transcription loop-mediated isothermal amplification (RT-LAMP) with microfluidic technology to develop a microfluidic chip for SARS-CoV-2 RNA detection. As depicted in Fig. 12.22a, the centrifugal microfluidic disc comprises three chambers: Chamber I for loading reagents, Chamber II for loading samples, and Chamber III for the reaction.

As depicted in Fig. 12.22b, c, the detection procedure involves several steps. First, viral lysis buffer and the sample to be tested are injected into Chambers I and II, respectively. Centrifugal force is used to mix the sample and lysis buffer in Chamber III, where the buffer lyses the SARS-CoV-2 virus and releases RNA. In the second step, the fluorescent indicator and the RT-LAMP reaction mixture are added to Chamber I to form the RT-LAMP reaction mixture. Centrifugation then transfers this mixture into Chamber III, where it is combined with the viral RNA. The third step involves adding mineral oil to Chamber I, where centrifugal force mixes the oil into Chamber III, forming an oil layer on top of the reaction mixture. This oil layer serves as a sealing mechanism to prevent evaporation of moisture during the RT-LAMP reaction, thereby ensuring the accuracy of the detection process. In the

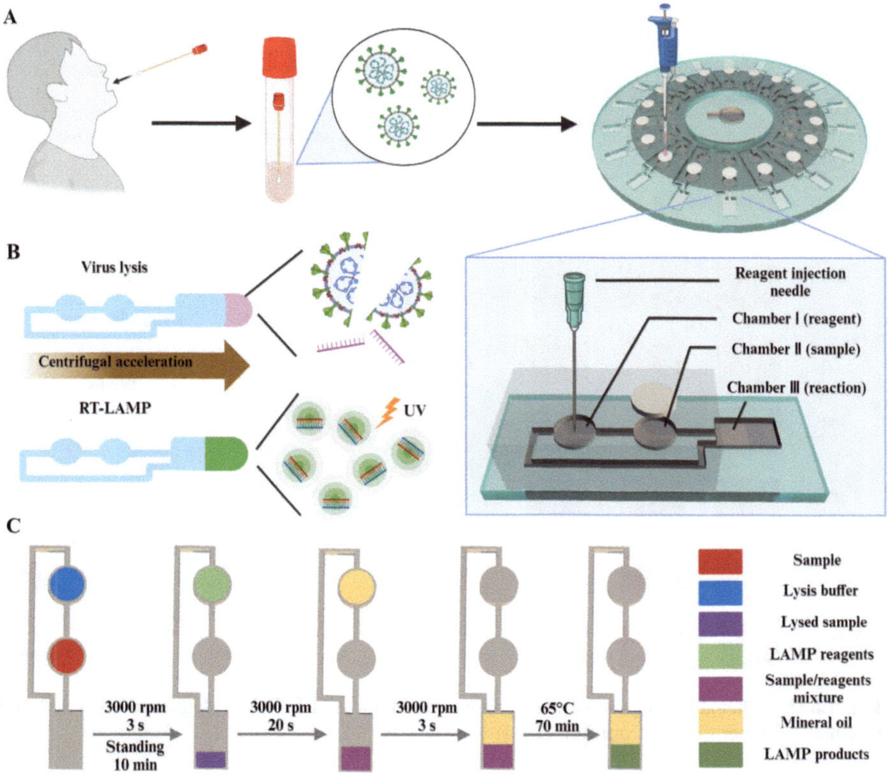

Fig. 12.22 Detection of SARS-CoV-2 using centrifugal microfluidic technology. (Revised and reprinted with permission from Ref. [127])

fourth step, the chip is heated to 65 °C and incubated for 70 min to carry out the RT-LAMP amplification reaction. Finally, after the amplification reaction, the fluorescent signal in the solution is detected. The strategy successfully enables the detection of SARS-CoV-2 RNA with the limits of detection as low as two copies per reaction.

Paper-based microfluidics is a significant branch of microfluidic technology, offering advantages such as low cost, ease of fabrication, and convenience for use and portability [128]. Paper-based microfluidic chips employ inexpensive paper as the substrate, replacing materials such as silicon, glass, and polymers. By utilizing fabrication methods such as physical barrier formation, physical deposition, and chemical modification [129], specific hydrophilic and hydrophobic channel networks could be created on the paper surface, ultimately successfully obtaining paper-based microfluidic diagnostic devices [130].

Utilizing the paper-based microfluidic platform, scientists have developed a variety of methods for SARS-CoV-2 RNA detection [132]. For instance, Davidson et al. [131] proposed a paper-based sensor for detecting SARS-CoV-2 RNA in human saliva. As shown in Fig. 12.23a, the paper-based detection chip consists of

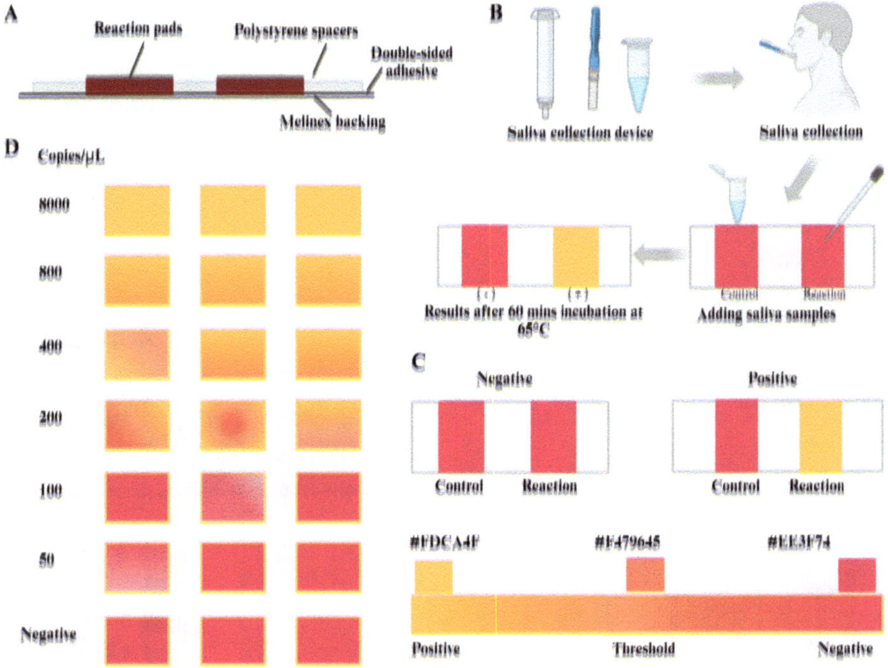

Fig. 12.23 Schematic diagram of paper-based sensor for SARS-CoV-2 detection. (Revised and reprinted with permission from Ref. [131])

a support pad, two reaction pads, and three anti-interference pads. The support pad is made of polyester film, the reaction pads are made of chromatographic paper treated with phenol red, and the anti-interference pads are made of polystyrene. The front side of the support pad is coated with double-sided adhesive tape, allowing the reaction pads and anti-interference pads to be alternately attached to the support pad, enabling simple fabrication of the paper-based sensor. The reaction pads are pre-deposited with reverse transcription loop-mediated isothermal amplification (RT-LAMP) reaction components. As illustrated in Fig. 12.23b, during sample testing, saliva samples collected from the human oral cavity are first heat-inactivated. The inactivated sample is then applied to one of the reaction zones on the paper-based chip and incubated at 65 °C for 60 min. The addition of the sample rehydrates the RT-LAMP components on the reaction pad, ensuring the amplification reaction proceeds efficiently. As shown in Fig. 12.23c, for positive samples, the large amount of H^+ generated by the RT-LAMP reaction causes a decrease in the solution's pH, leading to a color change of phenol red on the reaction pad from red to yellow. In contrast, for negative samples, no amplification occurs, and therefore, the pH of the testing solution remains unchanged, preventing any color alteration of the reaction pad. As shown in Fig. 12.23d, the limit of detection of this paper-based sensor for SARS-CoV-2 RNA in saliva is 200 copies/μL.

12.6 Electrochemical Detection of SARS-CoV-2

In recent years, electrochemistry technology has achieved remarkable advancements in molecular diagnostics, growing as a representative technique in health monitoring. Compared to traditional detection methods, electrochemical sensing offers several significant advantages, including lower cost, higher sensitivity, portability, reduced detection time, and smaller sample requirement. These characteristics make it particularly well-suited for deployment in resource-limited settings as well as for POCT applications [133, 134].

12.6.1 Conventional Electrochemical Detection Technology

Traditional electrochemical sensor consists of a three electrodes system: a working electrode, a reference electrode, and a counter electrode [135–138]. On the surface of working electrode, specific biomolecules, such as antibodies, nucleic acid probes, or antigens, are immobilized, which are capable of selectively capturing the SARS-CoV-2 virus or its specific RNA sequences [139–142]. When the sample containing SARS-CoV-2 were added into a electrochemical sensor, the viral antigens or RNA will be captured by the specific biomolecules immobilized on electrode surface. If the sensor surface is modified with antibodies targeting SARS-CoV-2, the antibodies will capture the virus by recognizing the surface antigens (such as the Spike protein) [143, 144]. Alternatively, if nucleic acid probes are employed, they will undergo a specific hybridization with the viral RNA sequence [145–147]. Once the target molecules (either antigens or RNA) are captured, it will induce an electrochemical change at the electrode surface [148–151]. This alteration will manifest as a variation in current, voltage, or impedance, which can be monitored in real time through the electrochemical instrumentation connected to the sensor.

During COVID-19, researchers developed various electrochemical detection methods, including electrochemical impedance spectroscopy [152], square wave voltammetry [153], differential pulse voltammetry [154], and cyclic voltammetry [155]. Each of these techniques exhibits unique characteristics and provides specific advantages [156]. Moreover, scientists have combined signal amplification strategies, such as rolling circle amplification, catalytic hairpin assembly, and the CRISPR-Cas system with electrochemical technologies, significantly improving the detection performance [157–159]. In summary, the continuously evolving electrochemical technology has played a pivotal role in the diagnosis of COVID-19, owing the unique technical advantages [160, 161]. With the continuous advancement of nanomaterials and biomolecular technologies, electrochemical detection is expected to provide more efficient and reliable solutions for addressing future public health crises [162–165].

Huang et al. [166] proposed an electrochemical impedance spectroscopy-based SARS-CoV-2 RNA detection technique. This technique significantly enhanced the sensitivity and specificity by using a porous electrode and a DNA nanolock strategy. As shown in Fig. 12.24a, a porous silicon electrode was first fabricated using electrochemical etching technology. In Fig. 12.24b, a gold film was deposited on the surface of the porous silicon electrode via electrochemical techniques. As illustrated in Fig. 12.24c, thiol-modified single-stranded DNA probes were immobilized on the gold film surface through Au-S bonds. The large specific surface area of the porous silicon working electrode allows for the modification of a high density of capture probes, thereby enhancing the capture efficiency of target nucleic acid molecules and amplifying the response signal. When the target nucleic acid hybridizes with the capture probes, the resulting rigid double-stranded DNA/RNA physically blocks the nanopores, effectively hindering electron transfer. As shown in Fig. 12.24d, e, this blocking effect significantly increases the impedance change compared to traditional methods, which only create electron barriers on the electrode surface. This amplification of the detection signal improves the sensitivity of the assay. As shown in Fig. 12.24f, this method enables real-time testing by combining a portable wireless detection device, providing an efficient and low-cost solution for rapid screening of SARS-CoV-2 and other infectious diseases.

Experimental results demonstrate that the method achieved a detection limit as low as 0.03 fM, which is one order of magnitude more sensitive than that of qPCR. In clinical validation using 45 saliva samples, the method exhibited a sensitivity of 100% and a specificity of 93% for SARS-CoV-2 detection, showing high consistency with RT-qPCR results.

Naorungroj et al. [167] proposed a technique for the detection of SARS-CoV-2 nucleic acids, based on square wave voltammetry, This technique significantly enhances operational convenience by utilizing a G-quadruplex DNAzyme and a

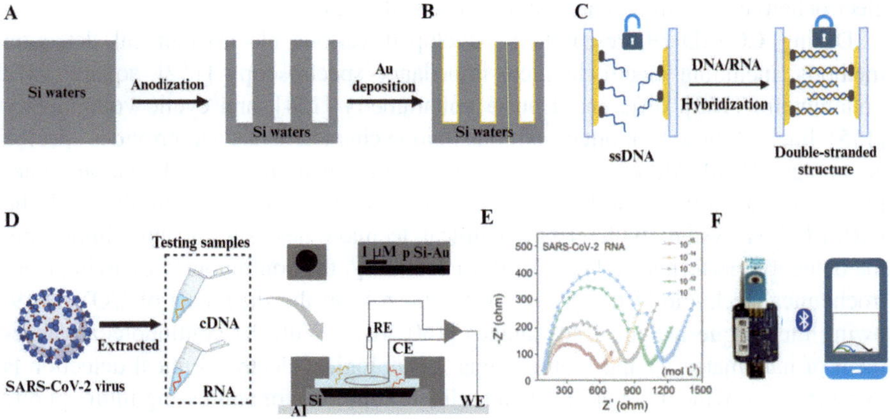

Fig. 12.24 Schematic diagram of electrochemical impedance method for detection of SARS-COV-2. (Revised and reprinted with permission from Ref. [166]. Copyright 2023 American Chemical Society)

controllable sequential flow microfluidic device. The detection mechanism of this method is based on the measurement of redox reactions on the electrode surface using square wave voltammetry. Square wave voltammetry applies a specific potential waveform to induce redox reactions in the target molecules, and the resulting current changes. The current change can be used to quantitatively analyze the concentration of the target nucleic acids. G-quadruplexes (GQ) are specific secondary structures formed by guanine-rich (G-rich) DNA sequences that fold in the presence of cations such as Na^+, K^+, or NH_4^+. When GQ bind to hemin to form a complex, they exhibit peroxidase-like catalytic activity under specific conditions. As shown in Fig. 12.25a, in the detection system, pyrrolidinyl peptide nucleic acid probes (acpcPNA) are pre-immobilized on the surface of a screen-printed graphene working electrode (SPGE) to capture SARS-CoV-2 cDNA. As illustrated in Fig. 12.25b, in the presence of SARS-CoV-2 cDNA, the cDNA is first captured and immobilized on the electrode surface by acpcPNA. Subsequently, the GQ-DNAzyme is anchored to the electrode surface through hybridization and complementary pairing. As shown in Fig. 12.25c, the GQ-DNAzyme can catalyze the oxidation of hydroquinone to benzoquinone in the reaction system. This reaction generates a redox current, the intensity of which is proportional to the concentration of cDNA. As depicted in Fig. 12.25d, the detection of COVID-19 can be achieved by measuring the redox current using square wave voltammetry. Additionally, the study effectively simplified the detection process and enhanced automation by integrating a controllable sequential flow microfluidic device. This device can efficiently perform SARS-CoV-2 RNA detection within 20 min, achieving a detection limit of

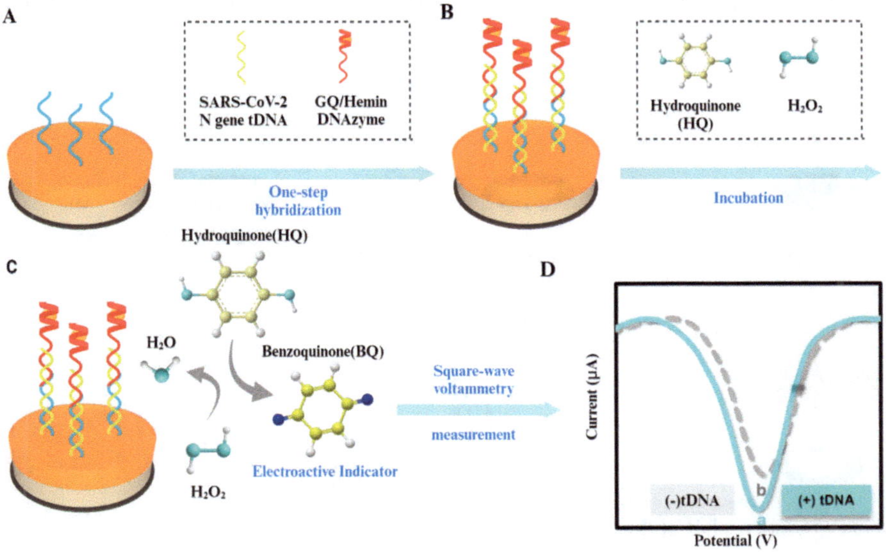

Fig. 12.25 Schematic diagram of square wave voltammetry for the detection of SARS-CoV-2. (Revised and reprinted with permission from Ref. [167]. Copyright 2023 American Chemical Society)

30 pM while being faster than RT-PCR. In tests conducted on five nasopharyngeal swab samples, the detection results of this method were consistent with those of RT-PCR.

Farzin et al. [168] proposed an electrochemical nucleic acid sensor based on differential pulse voltammetry and cyclic voltammetry for the detection of SARS-CoV-2 RNA. As illustrated in Fig. 12.26, this method successfully achieved highly sensitive detection of target nucleic acids by constructing a functionalized carbon paste electrode with high conductivity and selectivity, combined with the electron transfer blocking effect induced by DNA hybridization. The method initially prepared a functionalized carbon paste electrode as the detection platform. Specifically, 55% graphite powder, 15% silver ion-hexathia-18-crown-6 (HT18C6(Ag)) complex, and 30% paraffin oil were mixed and thoroughly ground. The mixture was then packed into an electrode tube to fabricate the carbon paste electrode (Fig. 12.26a). The introduction of the HT18C6(Ag) complex not only enhanced the conductivity of the electrode but also improved the selectivity for target nucleic acids through the specific coordination of silver ions (Ag^{+}), thereby increasing the stability and sensitivity of the sensor. Additionally, to enhance the density of DNA probe modification on the electrode surface, the study functionalized the carbon paste electrode surface with positively charged chitosan (Fig. 12.26b). This modification utilized the electrostatic interactions between chitosan and the DNA probes to immobilize the nucleic acid probes. Furthermore, the electrode surface was further modified with polyamidoamine dendrimer-functionalized silicon quantum dots (PAMAM@SiQDs) (Fig. 12.26c). This nanocomposite material possesses a large specific surface area and abundant amino functional groups, which provide additional binding sites and enhance the DNA capture capability. Finally, using

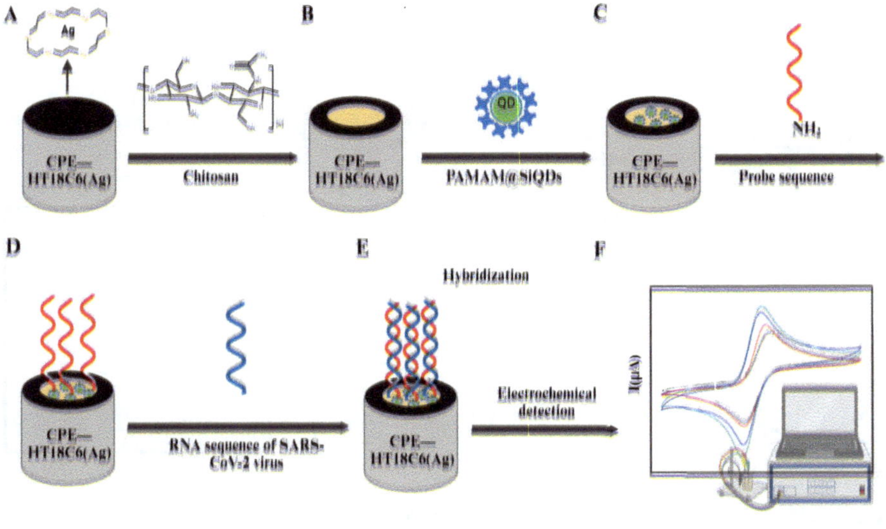

Fig. 12.26 Schematic diagram of differential pulse voltammetry for the detection of SARS-CoV-2. (Revised and reprinted with permission from Ref. [168])

glutaraldehyde as a crosslinking agent, the amino-modified single-stranded DNA capture probes were stably immobilized on the electrode surface, forming an efficient nucleic acid detection interface (Fig. 12.26d). During the detection process, the SARS-CoV-2 target sequence hybridizes with the probe DNA, forming a stable double-stranded structure (Fig. 12.26e). Due to the higher negative charge density of double-stranded DNA, the electrostatic repulsion between the DNA and the HT18C6(Ag) complex is enhanced, inhibiting the electron transfer process of silver ions and resulting in a significant reduction in the differential pulse voltammetry signal (Fig. 12.26f). By monitoring the degree of decrease in the differential pulse voltammetry signal, the concentration of SARS-CoV-2 nucleic acids can be quantitatively analyzed. This method achieves a detection limit as low as 0.3 pM, with a recovery rate exceeding 95% in human sputum samples, demonstrating its suitability for clinical sample detection.

Heo et al. [169] proposed an electrochemical sensor based on the CRISPR/Cas13a system for signal amplification, with the detection signal for SARS-CoV-2 RNA read using cyclic voltammetry. As shown in Fig. 12.27, the authors first fabricated a working electrode with excellent electrochemical activity. Initially, molybdenum disulfide (MoS_2) nanosheets were modified onto the electrode surface. Due to its layered structure and high conductivity, MoS_2 facilitates electron transfer (Fig. 12.27a). Subsequently, flower-like gold nanostructures were deposited onto the MoS_2-modified electrode using electrochemical deposition (Fig. 12.27b) to further enhance the biomolecule immobilization capacity and improve the surface activity of the electrode. 3-mercaptopropionic acid was then linked to the gold nanostructures via gold-thiol bonds, introducing carboxyl functional groups onto the electrode surface (Fig. 12.27c). Streptavidin is a protein containing a large number of amino groups in its molecular structure. These amino groups can react with carboxyl groups under the activation of EDC/NHS, forming amide bonds. As a result, streptavidin is introduced onto the electrode surface (Fig. 12.27d). An RNA reporter probe modified with biotin at one end can be immobilized on the electrode surface through its interaction with streptavidin (Fig. 12.27e). When the sample

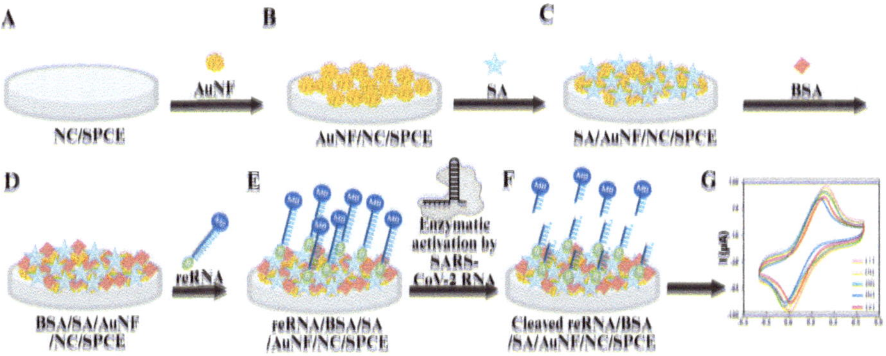

Fig. 12.27 Schematic diagram of cyclic voltammetry combined with CRISPR for the detection of SARS-CoV-2. (Revised and reprinted with permission from Ref. [169])

contains SARS-CoV-2 RNA, the CRISPR/Cas13a system specifically recognizes the relevant gene fragment and activates the RNase activity of Cas13a. This results in the extensive cleavage of the RNA reporter probes immobilized on the electrode surface (Fig. 12.27f). The RNA reporter probes are modified with methylene blue at the other end, which exhibits excellent electrochemical activity. Methylene blue can undergo redox reactions on the electrode surface, generating a significant electrochemical signal. The redox signal on the electrode surface significantly decreases due to the cleavage of the RNA reporter probes, leading to a reduction in the redox signal (Fig. 12.27g). Experimental results demonstrate that this biosensor can achieve rapid detection within 30 min, with a detection limit as low as 0.044 fg/mL. By combining high-performance nanomaterial electrodes with the CRISPR/Cas13a system, this method achieves highly sensitive and rapid detection of SARS-CoV-2 RNA. In the future, it holds promise for further development into POCT devices for on-site nucleic acid detection, offering a supplementary solution for public health monitoring.

Chaibun et al. [170] proposed an electrochemical biosensor based on multiplex RCA for signal amplification, enabling rapid detection of SARS-CoV-2 S and N genes using differential pulse voltammetry. As shown in Fig. 12.28a, the scheme first designs a circular template for RCA amplification based on the target sequence. When the target nucleic acid hybridizes with the circular template, the linear

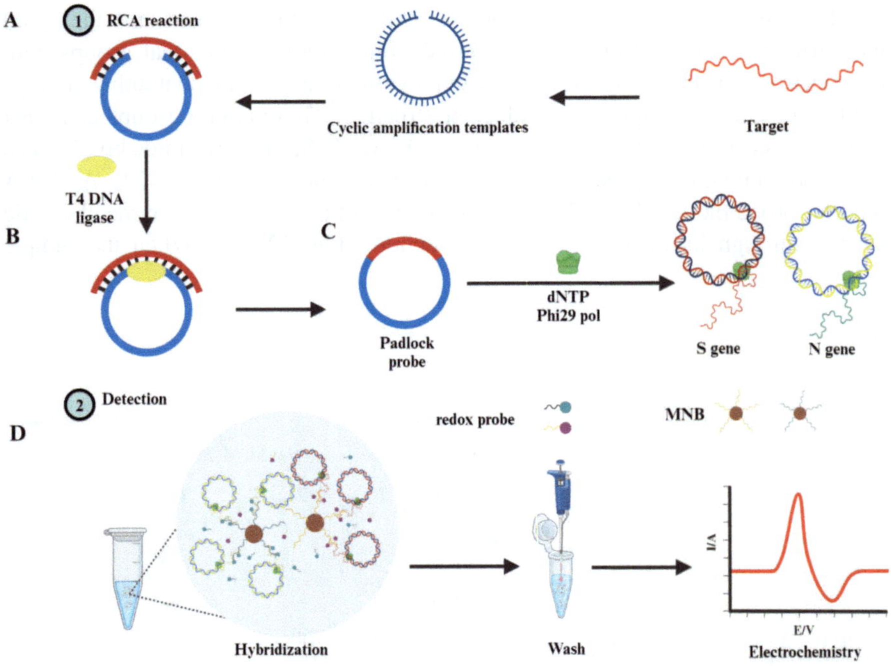

Fig. 12.28 Schematic diagram of electrochemical combined with RCA detection of SARS-CoV-2. (Revised and reprinted with permission from Ref. [170])

template is pulled into a circular form. As illustrated in Fig. 12.28b, the circular template is then closed into a locked DNA structure under the action of T4 DNA ligase. As shown in Fig. 12.28c, the locked DNA bound to the target nucleic acid undergoes continuous amplification of the target sequence under the action of phi29 DNA polymerase, significantly increasing the copy number of each target sequence and greatly enhancing the detection signal. As illustrated in Fig. 12.28d, magnetic beads modified with probes specifically targeting the N and S genes are used to wash and enrich the RCA products. A large number of redox labels hybridize with the RCA products through their surface-modified probes, accumulating on the surface of the RCA products to form massive electrochemically active clusters. After further collection and washing using magnetic beads, the RCA products are collected and introduced into the electrochemical sensor. Quantitative analysis of the target genes is achieved by reading the redox signals using differential pulse voltammetry. The entire detection process, from sample input to result output, can be completed in less than 2 h without the need for a thermal cycler. This method achieves a detection limit of 1 copy/μL for SARS-CoV-2 nucleic acids, demonstrating higher sensitivity compared to RT-qPCR. Furthermore, the detection results of this method showed 100% consistency when compared with RT-qPCR results from 106 clinical samples.

Ji et al. [171] proposed an electrochemical sensor for SARS-CoV-2 nucleic acid detection based on a cantilever molecular recognition system constructed on a graphene working electrode. As shown in Fig. 12.29, the molecular recognition

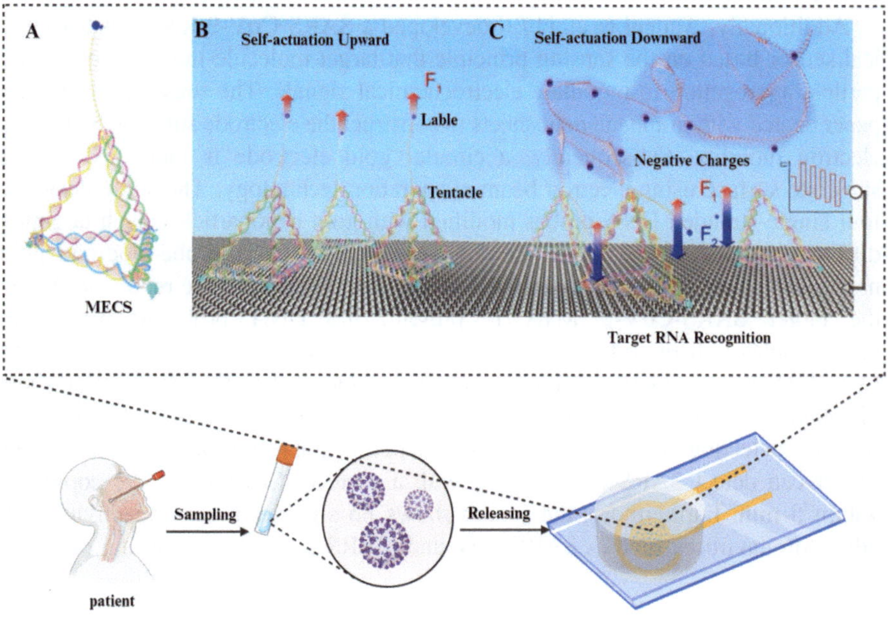

Fig. 12.29 Schematic diagram of a self-powered molecular electrochemical system for the detection of SARS-CoV-2. (Revised and reprinted with permission from Ref. [171]. Copyright 2022 American Chemical Society)

system of this scheme consists of a tetrahedral nucleic acid framework serving as a supporting base and a single-stranded DNA probe at the top acting as a sensing arm (Fig. 12.29a). This design reduces the likelihood of probe entanglement and non-specific adsorption. The single-stranded DNA sensing arm includes a sequence that can hybridize with a fragment of the ORF1ab gene, enabling specific capture of the target nucleic acid. Additionally, the end of the sensing arm is modified with methylene blue, which serves as a redox electrochemical label to generate the electrochemical signal. A sequence of five thymine bases is incorporated between the sensing arm and the tetrahedral base to act as a rotational pivot. For the detection system without target, applying a negative voltage to the graphene working electrode causes the negatively charged DNA sensing arm to stand upright due to electrostatic repulsion, moving away from the electrode surface and resulting in a weak electrochemical signal (Fig. 12.29b). For the detection system with target, the captured SARS-CoV-2 RNA, with its nearly 30,000 bases, exerts a strong electrostatic repulsion on the sensing arm. This causes the electrochemical label to spontaneously move closer to the electrode surface, resulting in a significant electrochemical response (Fig. 12.29c). The changes in electrochemical response caused by the positional shift of the DNA sensing arm can be measured using square wave voltammetry or differential pulse voltammetry. By regulating the electrochemical sensing process at the molecular scale, this method achieves detection sensitivity close to the physical limit, with a sensitivity of four copies/80 μL. Additionally, this strategy can identify ten-pooled clinical positive samples within 1 min, and the double-blind test results show high consistency with RT-PCR results.

Additionally, Alafeef et al. [172] developed a SARS-CoV-2 RNA electrochemical sensor based on the sensing principle that target molecule-induced gold nanoparticle aggregation to modulate electrochemical signals. The sensor utilizes filter paper coated with graphene nanosheets to construct the electrode substrate with high electron mobility. Subsequently, a circular gold electrode is fabricated on the substrate surface using electron beam evaporation technology. The study designed four single-stranded DNA probes modified with gold nanoparticles, each targeting different regions of the SARS-CoV-2 N gene. The DNA probe-modified gold nanoparticles are uniformly distributed due to mutual electrostatic repulsion. When the target SARS-CoV-2 RNA is present, the DNA probe-modified gold nanoparticles will hybridize with the complementary target sequences of SARS-CoV-2 RNA, inducing the aggregation of gold nanoparticles. This aggregation alters the charge distribution and electron mobility of electrode surface, enabling highly sensitive detection of SARS-CoV-2 RNA. Experimental results demonstrate that this sensor can detect SARS-CoV-2 RNA with a viral load as low as 231 copies/μL within 5 min. Furthermore, the sensor shows no significant cross-reactivity with other coronaviruses such as SARS-CoV and MERS-CoV, demonstrating excellent specificity.

12.6.2 *Field-Effect Transistor Electrochemical Detection Technology*

Among the electrochemical sensing technologies, field-effect transistor (FET) sensors offer superior sensitivity compared to traditional electrochemical methods based on redox reactions. This is achieved by leveraging semiconductor materials with excellent electrical properties to detect conductivity changes induced by single or few nucleic acid molecules. Furthermore, FET technology enables direct detection of nucleic acid sequences through conductivity alterations, circumventing the need for labels (such as fluorescent, enzyme, or redox markers) typically required in conventional methods. This reduces both the cost of detection and the potential interference caused by the labeling process. During the COVID-19 pandemic, scientists developed a series of SARS-CoV-2 nucleic acid detection methods based on graphene field-effect transistors.

Kong et al. [173] developed a graphene FET (GFET) with a Y-shaped dual-probe for SARS-CoV-2 nucleic acid detection. As shown in Fig. 12.30a, the Y-shaped DNA probe is assembled from three single-stranded DNA sequences (S1, S2, and S3) and includes two built-in capture probes extending from S1 and S3. These probes target specific fragments of the ORF1ab gene and the N gene, respectively, significantly enhancing the capture efficiency and binding strength of the target genes. As shown in Fig. 12.30b, when the Y-shaped DNA dual-probe specifically

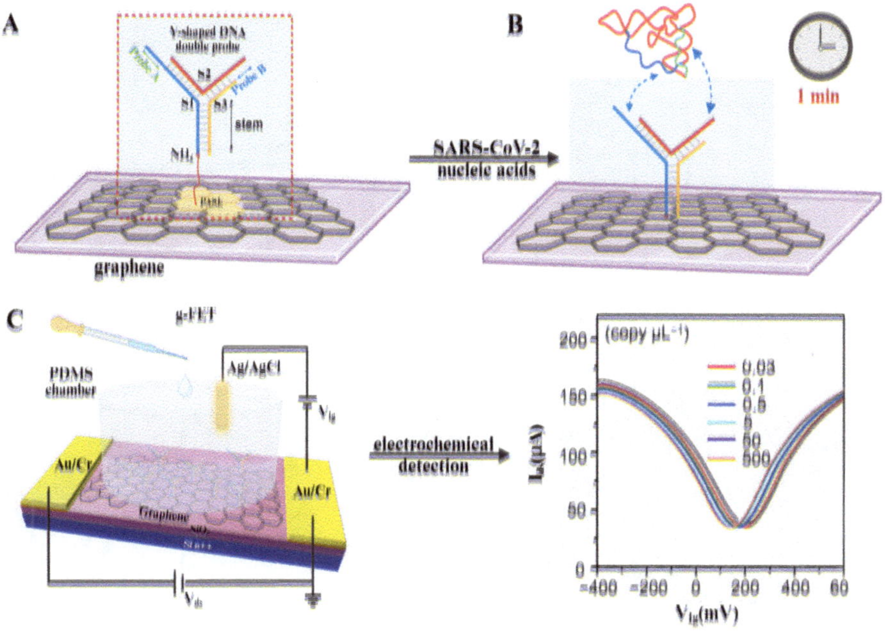

Fig. 12.30 Schematic diagram of Y-shaped DNA dual-probe transistor detection. (Revised and reprinted with permission from Ref. [173]. Copyright 2021 American Chemical Society)

binds to the target nucleic acid sequence, the highly negatively charged SARS-CoV-2 RNA (approximately 30,000 nucleotides in length) induces strong n-type doping in the p-type graphene. This results in a significant leftward shift of the graphene Dirac point and a noticeable reduction in the source-drain current. As shown in Fig. 12.30c, the real-time changes in the Dirac point and source-drain current can be recorded by measuring the transfer characteristic curve and time-current response of the GFET, thereby obtaining the detection signal. Benefitting from the designed Y-shaped probe, this method achieves a high sensitivity as low as 0.03 copies/μL within 1 min. Furthermore, this method enables direct 5-in-1 pooled testing, improving detection efficiency and reducing reagent consumption, which is importance for large-scale population screening.

Wang et al. [174] developed a DNA origami probe-based GFET for SARS-CoV-2 RNA detection. As shown in Fig. 12.31a, the DNA origami probe consists of a flexible single-stranded DNA cantilever and a rigid DNA tetrahedral base. The end of the DNA cantilever is one segment that is capable of recognizing specific fragments of SARS-CoV-2 RNA. The tetrahedral rigid bases allow the probes to stand upright on different bases at a certain spacing, avoiding probe entanglement. Additionally, the DNA tetrahedral base could serve as an anti-contamination layer to resist nonspecific adsorption, maintaining the long-term stability of the sensor. The DNA origami probe is ingeniously modified onto the gate surface made by graphene

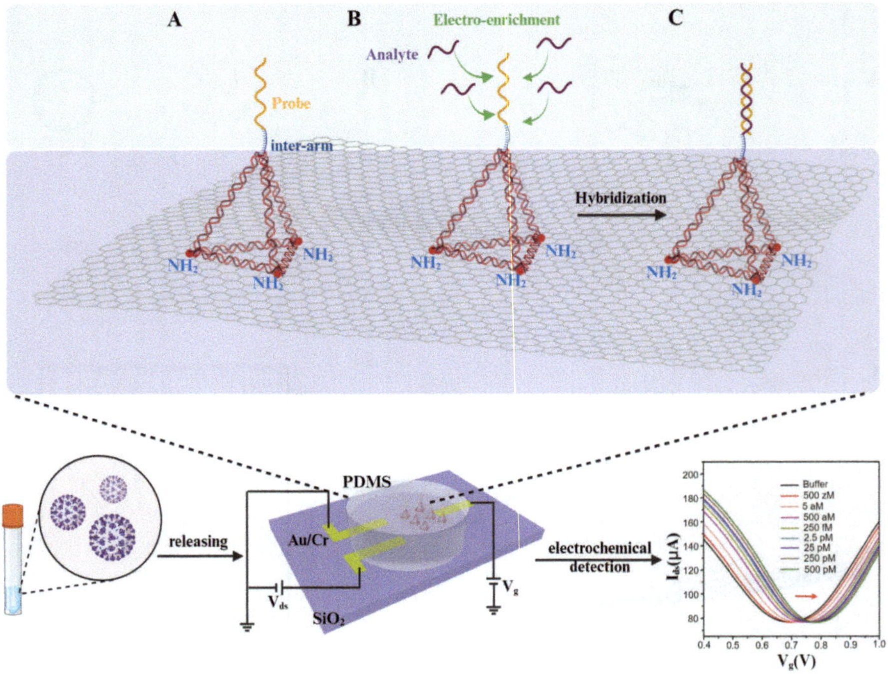

Fig. 12.31 Schematic diagram of liquid-gated FET sensor modified with DNA origami probes. (Revised and reprinted with permission from Ref. [174])

via 1-pyrenebutyric acid N-hydroxysuccinimide ester. As shown in Fig. 12.31b, the target nucleic acids in the test solution can be rapidly enriched onto gate surface through electrostatic attraction by applying a positive voltage to the gate. This enrichment significantly enhances the capture efficiency of the target molecules. As shown in Fig. 12.31c, when SARS-CoV-2 RNA is captured by the DNA origami probe, it affects the electric double layer formed at the gate/electrolyte interface and the electrolyte/graphene interface. This modulates the doping intensity of the gate on graphene, influencing the source-drain current and thereby generating the detection signal. The test results demonstrate that this biosensor can achieve highly sensitive (1–2 copies/100 μL) and rapid (average diagnosis time of about 80 s) detection of SARS-CoV-2.

Furthermore, Wu et al. [175] developed a multi-probe DNA origami-based GFET for SARS-CoV-2 RNA detection. As shown in Fig. 12.32a, the base of this DNA origami is a dimer formed by two DNA tetrahedrons. As shown in Fig. 12.32b, compared to the DNA tetrahedral probe in the work of Wang et al. [176], the lower base of the dimeric base retains its original function, while the upper base replaces the previous cantilever structure. The upper base is modified with three different probes designed to capture specific gene fragments from the ORF1ab, RdRp, and E gene of SARS-CoV-2 RNA. Compared to the previously mentioned Y-shaped DNA dual-probe and the DNA tetrahedral single-probe system in the work of Wang et al., the synergistic effect of the three-probe system not only enhances the capture probability of the target nucleic acids and achieves higher binding affinity but also avoids entanglement between the probes, as shown in Fig. 12.32c. Therefore, this sensor achieves higher sensitivity, faster response times, and improved specificity. In the early or late stages of SARS-CoV-2 infection, the viral load in clinical samples is often low. If the detection platform lacks sufficient sensitivity, this may readily lead

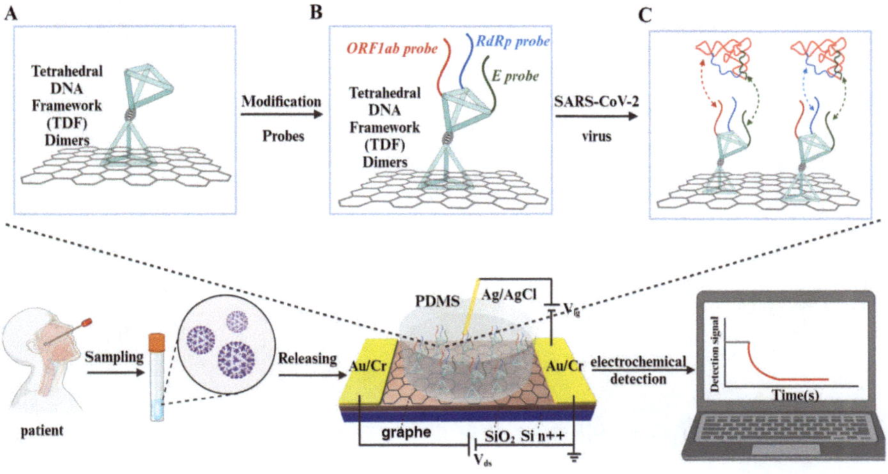

Fig. 12.32 Schematic diagram of g-FET sensor modified with tetrahedral DNA dimer. (Revised and reprinted with permission from Ref. [175]. Copyright 2022 American Chemical Society)

to false-negative results. Meanwhile, in practical large-scale screening, pooled testing strategies are commonly adopted to increase throughput, reduce costs, and improve operational efficiency. However, in pooled testing, dilution of individual samples results in a decrease in the effective viral load of positive specimens relative to the total sample volume, thereby further increasing the risk of false-negative outcomes. Therefore, sensitivity and accuracy are critical requirements for pooled testing. This sensor can directly detect SARS-CoV-2 RNA at concentrations as low as 0.025–0.05 copies/μL and completes the diagnosis of 30 nasopharyngeal swab samples within an average of 74 s, performing 100% consistency with those of RT-qPCR.

Zhengping Li's research group [177] reported a collaborative system combining CRISPR-Cas13a with GFET to generate high-intensity vector signals for SARS-CoV-2 RNA detection. The CRISPR-Cas13a system enhances detection signals for high-intensity, while the GFET imparts directionality to the detection signals. As shown in Fig. 12.33, when the CRISPR-Cas13a system recognizes the target site of SARS-CoV-2 RNA, the RNA-cleaving activity of the Cas13a protein is activated (Fig. 12.33a–c). A single activated Cas13a can nonspecifically degrade tens of thousands of nearby RNA molecules. In this system, RNA-DNA hybrid reporter probes are densely bound to the graphene surface (Fig. 12.33d, e). The reporter probes induce electron doping in graphene (GI-GII), causing a leftward shift in the Dirac point (signal change 1–2). The CRISPR-Cas13a activated by SARS-CoV-2 RNA can extensively cleave the reporter probes (Fig. 12.33e, f), drastically reducing

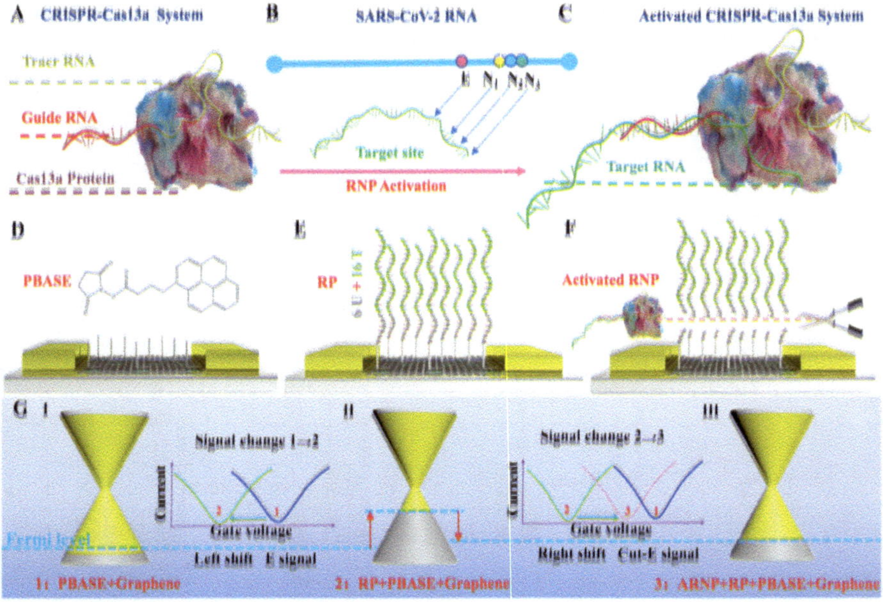

Fig. 12.33 Working principle of the collaborative system for detecting SARS-CoV-2 RNA. (Revised and reprinted with permission from Ref. [177])

the electron doping effect of the reporter probes on graphene (GII-GIII). This results in a significant rightward shift of the Dirac point, generating a large subtraction signal (signal change 2–3). Other nontarget substances, unable to activate CRISPR-Cas13a and lacking nonspecific adsorption to graphene, function similarly to the reporter probes by inducing electron doping in graphene, further shifting the Dirac point to the left. This synergistic system demonstrates high sensitivity in detecting SARS-CoV-2 RNA, with LOD of three copies/50 μL. Repetitive experiments on six SARS-CoV-2 positive patient samples confirmed the reliability and significant potential of this synergistic system for clinical applications. Additionally, by hydrophobically treating the graphene field-effect transistor substrate, this work greatly enhanced the stability of graphene in liquid environments, completely eliminating the nonspecific signal of the graphene Dirac point shifting leftward over time in the analytical solution. The instability of graphene in analytical solutions has been widely noted in the GFET field but has remained unresolved [178]. The hydrophobic treatment of the graphene substrate is of significant importance in advancing the reliable application of GFETs for clinical SARS-CoV-2 RNA detection. This highly stable collaborative system, characterized by high-intensity vector signals, holds promise for advancing SARS-CoV-2 RNA detection toward single-molecule sensitivity and practical clinical applications.

In recent years, driven by the COVID-19 pandemic, a series of high-level studies have demonstrated that FET electrochemical sensing platforms exhibit advantages in sensitivity, specificity, speed, cost-effectiveness, ease of use, and high-throughput analysis. These platforms hold promise as a more commercially attractive and feasible technology for virus detection [179]. Most of the technologies in this method are constructed using standard, commercialized, and mass-produced equipment, which is one of the reasons they have become highly popular and successful. The continuous advancement of nanotechnology and improvements in readout systems have further enhanced biosensing platforms. These platforms facilitate high-throughput screening of biomolecular analysis and meet the demand for rapid, robust, and cost-effective POCT [180]. However, there are still significant challenges to address on the path to commercialization. For instance, biological detection elements must be immobilized on the functionalized FET channel in a robust and reproducible manner, and biochemical reagents should remain effective under natural conditions for extended periods without requiring complex and precise storage and transportation conditions. Additionally, taking classic graphene as an example, the large-scale production and homogeneous preparation of sensing materials are also crucial. The successful translation of emerging technologies from the laboratory to the factory still requires the joint efforts of researchers, engineers, and enterprises.

12.6.3 Detection Technology Based on the Mechanism of Surface Wettability Switching

Diseases caused by pathogenic infections—and, in extreme cases, pandemics—pose a profound threat to human health. Efficient nucleic acid detection of pathogens is essential for clinical management and public-health responses to major outbreaks. Experience from the COVID-19 pandemic has highlighted a persistent "sensitivity-cost-convenience" trade-off in conventional nucleic acid diagnostics: achieving ultrahigh sensitivity typically comes at the expense of affordability and ease of use. Hence, the development of disruptive sensing principles and novel readout paradigms that break this compromise and enable simultaneously low cost, high sensitivity, and operational simplicity would be of considerable strategic value.

To this end, Zhengping Li's research group [181] devised a new sensing mechanism that generates and reads out detection signals by coupling target-triggered interfacial wetting-property switching. Specifically, they proposed a sensor concept in which target nucleic acid recognition triggers a conformational change at the test-tube surface from hydrophobic to hydrophilic (or vice versa), and the resulting alteration in liquid behavior under tube inversion is read out by the naked eye. Implemented on a single glass tube as the sensing platform, this approach yielded an ultrasensitive single-tube biosensor (USTB) that reports nucleic-acid presence by observing whether the liquid droplet detaches from or remains adhered to the tube bottom when the tube is inverted. Two operational modes of the USTB were established. In Mode 1, when the sample contains the target nucleic acid, the liquid remains stably attached to the tube bottom in the inverted state, whereas in the absence of target the liquid falls away from the tube bottom (Fig. 12.34a). In Mode 2, the converse behavior is observed: the liquid falls away from the tube bottom in the presence of target, while in the absence of target the liquid remains adhered in the inverted tube (Fig. 12.34b).

Taking Mode 1 as an example, the sensor architecture was realized by rational design of a three-segment probe possessing a hydrophilic moiety, a central responsive group, and a hydrophobic tail. The probe was immobilized on the tube surface with the hydrophilic domain positioned proximal to the surface and the hydrophobic tail distal, rendering the tube surface overall hydrophobic and causing the liquid to drip when the tube is inverted. For SARS-CoV-2 RNA detection, the central responsive group of the three-segment probe was designed as a ssRNA. When target RNA is present in the test sample, it will be specifically recognized by the CRISPR-Cas13a system, thereby activating the RNase activity of Cas13a (Fig. 12.34cI). Activated Cas13a mediates effective cleavage of the ssRNA responsive segments within the immobilized three-segment probes, leading to dissociation of the distal hydrophobic tail from the tube surface and leaving behind the proximal hydrophilic groups (Fig. 12.34cII). This molecular-scale event, in which CRISPR-Cas mediated removal of hydrophobic tails, induces a wettability transition from a hydrophobic to a hydrophilic surface, which is transduced into an easily observed change in the liquid's behavior upon tube inversion. For DNA pathogen detection, the central

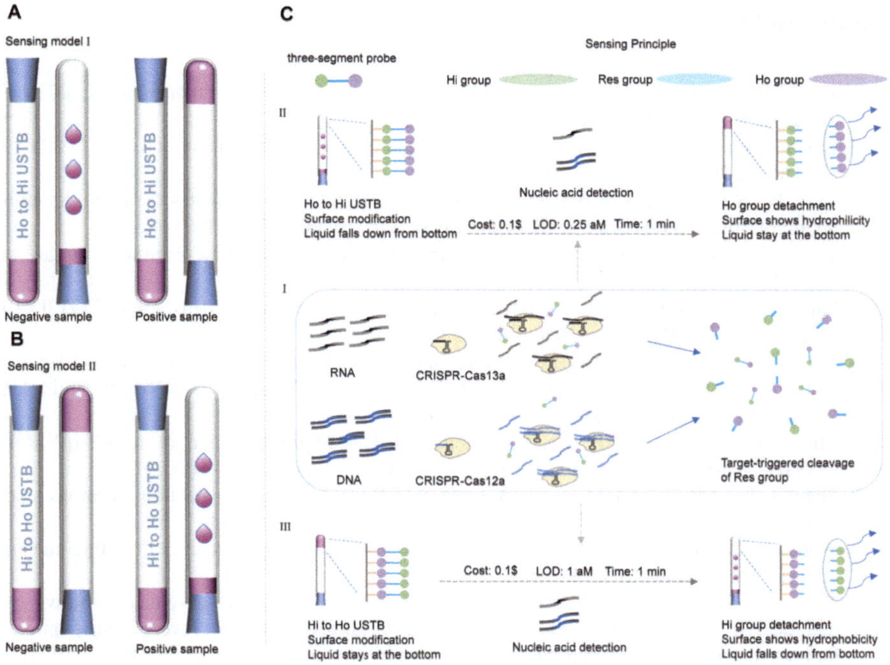

Fig. 12.34 USTB sensing mechanism and sensing modes. (Revised and reprinted with permission from Ref. [181])

responsive group can be replaced by ssDNA and the CRISPR-Cas13a system can be substituted by a CRISPR-Cas12a system without altering the sensing principle. The Mode-1 USTB was therefore designated "Ho-to-Hi USTB" (hydrophobic-to-hydrophilic USTB). For Mode 2, the hydrophilic and hydrophobic domain of the three-segment probe are simply interchanged, such that its hydrophobic group oriented toward the surface (proximal) and the hydrophilic group extending outward (distal). This configuration enables target-specific triggering of surface wettability transition from a hydrophilic to a hydrophobic, which is subsequently transduced into a macroscopic change in liquid behavior under tube inversion (Fig. 12.34c). This sensing mode is referred to as the "Hi-to-Ho USTB."

Based on this design, both the Ho-to-Hi and Hi-to-Ho USTBs enable simple, ultrasensitive detection of SARS-CoV-2 RNA and African Swine Fever Virus DNA within 1 min, requiring only tube inversion and visual inspection. The Ho-to-Hi USTB achieved a detection limit as low as 0.25 aM (15 copies/110 μL), while the Hi-to-Ho USTB reached 1 aM (66 copies/110 μL). The sensitivity of both sensors surpasses that of commercial RT-PCR/PCR kits and far exceeds that of standalone CRISPR-Cas assays. The estimated laboratory cost per test for the USTBs is approximately $0.1. Through their groundbreaking innovations in sensing mechanism and signal readout, USTBs successfully break the longstanding "sensitivity-cost-convenience" trilemma in pathogen diagnostics.

References

1. Zhu N, Zhang D, Wang W, et al. A novel coronavirus from patients with pneumonia in China, 2019 [J]. N Engl J Med. 2020;382(8):727–33.
2. Bai Y, Yao L, Wei T, et al. Presumed asymptomatic carrier transmission of COVID-19 [J]. JAMA. 2020;323(14):1406–7.
3. Li R, Pei S, Chen B, et al. Substantial undocumented infection facilitates the rapid dissemination of novel coronavirus (SARS-CoV-2) [J]. Science. 2020;368(6490):489–93.
4. Wu F, Zhao S, Yu B, et al. A new coronavirus associated with human respiratory disease in China [J]. Nature. 2020;579(7798):265–9.
5. Niu P, Lu R, Zhao L, et al. Three novel real-time RT-PCR assays for detection of COVID-19 virus [J]. China CDC Wkly. 2020;2(25):453–7.
6. 2019-novel coronavirus (2019-nCoV) real-time rRT-PCR panel primers and probes[EB/OL].
7. Kudo E, Israelow B, Vogels CB, et al. Detection of SARS-CoV-2 RNA by multiplex RT-qPCR [J]. PLoS Biol. 2020;18(10):e3000867.
8. Jung Y, Park G-S, Moon JH, et al. Comparative analysis of primer–probe sets for RT-qPCR of COVID-19 causative virus (SARS-CoV-2) [J]. ACS Infect Dis. 2020;6(9):2513–23.
9. Shen S, Fu AY, Jamba M, et al. A rapid SARS-CoV-2 variant detection by molecular-clamping based RT-qPCR [J]. MedRxiv. 2021:2021.04. 01.21254484.
10. Wang Q, Chen L, Long Y, et al. Molecular beacons of xeno-nucleic acid for detecting nucleic acid [J]. Theranostics. 2013;3(6):395.
11. Nagura-Ikeda M, Imai K, Tabata S, et al. Clinical evaluation of self-collected saliva by quantitative reverse transcription-PCR (RT-qPCR), direct RT-qPCR, reverse transcription–loop-mediated isothermal amplification, and a rapid antigen test to diagnose COVID-19 [J]. J Clin Microbiol. 2020;58(9):e01438-20. https://doi.org/10.1128/jcm.01438-20.
12. Parida M, Sannarangaiah S, Dash PK, et al. Loop mediated isothermal amplification (LAMP): a new generation of innovative gene amplification technique; perspectives in clinical diagnosis of infectious diseases [J]. Rev Med Virol. 2008;18(6):407–21.
13. Nagamine K, Hase T, Notomi T. Accelerated reaction by loop-mediated isothermal amplification using loop primers [J]. Mol Cell Probes. 2002;16(3):223–9.
14. Lu R, Wu X, Wan Z, et al. Development of a novel reverse transcription loop-mediated isothermal amplification method for rapid detection of SARS-CoV-2 [J]. Virol Sin. 2020;35: 344–7.
15. Dong Y, Wu X, Li S, et al. Comparative evaluation of 19 reverse transcription loop-mediated isothermal amplification assays for detection of SARS-CoV-2 [J]. Sci Rep. 2021;11(1):2936.
16. Lu R, Wu X, Wan Z, et al. A novel reverse transcription loop-mediated isothermal amplification method for rapid detection of SARS-CoV-2 [J]. Int J Mol Sci. 2020;21(8):2826.
17. Yu L, Wu S, Hao X, et al. Rapid detection of COVID-19 coronavirus using a reverse transcriptional loop-mediated isothermal amplification (RT-LAMP) diagnostic platform [J]. Clin Chem. 2020;66(7):975–7.
18. Yan C, Cui J, Huang L, et al. Rapid and visual detection of 2019 novel coronavirus (SARS-CoV-2) by a reverse transcription loop-mediated isothermal amplification assay [J]. Clin Microbiol Infect. 2020;26(6):773–9.
19. El-Tholoth M, Bau HH, Song J. A single and two-stage, closed-tube, molecular test for the 2019 novel coronavirus (COVID-19) at home, clinic, and points of entry [J]. ChemRxiv. 2020; https://doi.org/10.26434/chemrxiv.11860137.v1.
20. Park G. Development of reverse transcription loop-mediated isothermal amplification assays targeting severe acute respiratory syndrome coronavirus 2 [J]. J Mol Diagn. 2020;1525:30090.
21. Broughton JP, Deng X, Yu G, et al. CRISPR–Cas12-based detection of SARS-CoV-2 [J]. Nat Biotechnol. 2020;38(7):870–4.
22. Yang W, Dang X, Wang Q, et al. Rapid detection of SARS-CoV-2 using reverse transcription RT-LAMP method [J]. MedRxiv. 2020:2020.03. 02.20030130.

23. Jiang M, Pan W, Arasthfer A, et al. Development and validation of a rapid, single-step reverse transcriptase loop-mediated isothermal amplification (RT-LAMP) system potentially to be used for reliable and high-throughput screening of COVID-19 [J]. Front Cell Infect Microbiol. 2020;10:331.
24. Lamb LE, Bartolone SN, Ward E, et al. Rapid detection of novel coronavirus/Severe Acute Respiratory Syndrome Coronavirus 2 (SARS-CoV-2) by reverse transcription-loop-mediated isothermal amplification [J]. PLoS One. 2020;15(6):e0234682.
25. Li Y, Zhou Y, Ma Y, et al. A mismatch-tolerant RT-LAMP method for molecular diagnosis of highly variable viruses [J]. Bio-protocol. 2019;9(21):–e3415.
26. Zhou Y, Wan Z, Yang S, et al. A mismatch-tolerant reverse transcription loop-mediated isothermal amplification method and its application on simultaneous detection of all four serotype of dengue viruses [J]. Front Microbiol. 2019;10:1056.
27. Zhu X, Wang X, Han L, et al. Multiplex reverse transcription loop-mediated isothermal amplification combined with nanoparticle-based lateral flow biosensor for the diagnosis of COVID-19 [J]. Biosens Bioelectron. 2020;166:112437.
28. Li H, Lu H, Tang Y, et al. A rebuilding-free nucleic acid detection strategy enables ultrasensitive genotyping, N-in-1 logic screening and accurate multiplex assay of dangerous pathogens [J]. Angew Chem Int Ed. 2022;61(40):e202209496.
29. Bokelmann L, Nickel O, Maricic T, et al. Point-of-care bulk testing for SARS-CoV-2 by combining hybridization capture with improved colorimetric LAMP [J]. Nat Commun. 2021;12(1):1467.
30. Ganguli A, Mostafa A, Berger J, et al. Rapid isothermal amplification and portable detection system for SARS-CoV-2 [J]. Proc Natl Acad Sci. 2020;117(37):22727–35.
31. Ganguli A, Mostafa A, Berger J, et al. Reverse transcription loop-mediated isothermal amplification assay for ultrasensitive detection of SARS-CoV-2 in saliva and viral transport medium clinical samples [J]. Anal Chem. 2021;93(22):7797–807.
32. Wei S, Kohl E, Djandji A, et al. Direct diagnostic testing of SARS-CoV-2 without the need for prior RNA extraction [J]. Sci Rep. 2021;11(1):2402.
33. Makarova KS, Zhang F, Koonin EV. SnapShot: class 2 CRISPR-Cas systems [J]. Cell. 2017;168(1):328–328.e1.
34. Huang Z, Tian D, Liu Y, et al. Ultra-sensitive and high-throughput CRISPR-powered COVID-19 diagnosis [J]. Biosens Bioelectron. 2020;164:112316.
35. Pang B, Xu J, Liu Y, et al. Isothermal amplification and ambient visualization in a single tube for the detection of SARS-CoV-2 using loop-mediated amplification and CRISPR technology [J]. Anal Chem. 2020;92(24):16204–12.
36. Gootenberg JS, Abudayyeh OO, Lee JW, et al. Nucleic acid detection with CRISPR-Cas13a/C2c2 [J]. Science. 2017;356(6336):438–42.
37. Zhang F, Abudayyeh OO, Gootenberg JS. A protocol for detection of COVID-19 using CRISPR diagnostics[J]. A protocol for detection of COVID-19 using CRISPR diagnostics, 2020;8.
38. Metsky HC, Freije CA, Kosoko-Thoroddsen T-SF, et al. CRISPR-based surveillance for COVID-19 using genomically-comprehensive machine learning design [J]. BioRxiv. 2020: 2020.02. 26.967026.
39. Arizti-Sanz J, Freije CA, Stanton AC, et al. Streamlined inactivation, amplification, and Cas13-based detection of SARS-CoV-2 [J]. Nat Commun. 2020;11(1):5921.
40. Myhrvold C, Freije CA, Gootenberg JS, et al. Field-deployable viral diagnostics using CRISPR-Cas13 [J]. Science. 2018;360(6387):444–8.
41. Zhang WS, Pan J, Li F, et al. Reverse transcription recombinase polymerase amplification coupled with CRISPR-Cas12a for facile and highly sensitive colorimetric SARS-CoV-2 detection [J]. Anal Chem. 2021;93(8):4126–33.
42 Ding X, Yin K, Li Z, et al. Ultrasensitive and visual detection of SARS-CoV-2 using all-in-one dual CRISPR-Cas12a assay [J]. Nat Commun. 2020;11(1):4711.

43. Jeon Y, Choi YH, Jang Y, et al. Direct observation of DNA target searching and cleavage by CRISPR-Cas12a [J]. Nat Commun. 2018;9(1):2777.
44. Wang Y, Zhang Y, Chen J, et al. Detection of SARS-CoV-2 and its mutated variants via CRISPR-Cas13-based transcription amplification [J]. Anal Chem. 2021;93(7):3393–402.
45. Paige JS, Wu KY, Jaffrey SR. RNA mimics of green fluorescent protein [J]. Science. 2011;333 (6042):642–6.
46. Fozouni P, Son S, De León Derby MD, et al. Amplification-free detection of SARS-CoV-2 with CRISPR-Cas13a and mobile phone microscopy [J]. Cell. 2021;184(2):323–333.e9.
47. Liu TY, Knott GJ, Smock DC, et al. Accelerated RNA detection using tandem CRISPR nucleases [J]. Nat Chem Biol. 2021;17(9):982–8.
48. Niewoehner O, Garcia-Doval C, Rostøl JT, et al. Type III CRISPR–Cas systems produce cyclic oligoadenylate second messengers [J]. Nature. 2017;548(7669):543–8.
49. Santiago-Frangos A, Hall LN, Nemudraia A, et al. Intrinsic signal amplification by type III CRISPR-Cas systems provides a sequence-specific SARS-CoV-2 diagnostic [J]. Cell Rep Med. 2021;2(6):100319.
50. Yángüez E, White G, Kreutzer S, et al. HiDRA-seq: high-throughput SARS-CoV-2 detection by RNA barcoding and amplicon sequencing[J]. BioRxiv. 2020:2020.06. 02.130484.
51. Bloom JS, Sathe L, Munugala C, et al. Massively scaled-up testing for SARS-CoV-2 RNA via next-generation sequencing of pooled and barcoded nasal and saliva samples [J]. Nat Biomed Eng. 2021;5(7):657–65.
52. Wu Q, Suo C, Brown T, et al. INSIGHT: a population-scale COVID-19 testing strategy combining point-of-care diagnosis with centralized high-throughput sequencing [J]. Sci Adv. 2021;7(7):eabe5054.
53. Aynaud M-M, Hernandez JJ, Barutcu S, et al. A multiplexed, next generation sequencing platform for high-throughput detection of SARS-CoV-2 [J]. Nat Commun. 2021;12(1):1405.
54. Yelagandula R, Bykov A, Vogt A, et al. Multiplexed detection of SARS-CoV-2 and other respiratory infections in high throughput by SARSeq [J]. Nat Commun. 2021;12(1):3132.
55. Ludwig KU, Schmithausen RM, Li D, et al. LAMP-Seq enables sensitive, multiplexed COVID-19 diagnostics using molecular barcoding[J]. Nat Biotechnol. 2021;39(12):1556–62.
56. Warneford-Thomson R, Shah PP, Lundgren P, et al. COV-ID: a LAMP sequencing approach for high-throughput co-detection of SARS-CoV-2 and influenza virus in human saliva [J]. MedRxiv. 2021:2021.04. 23.21255523.
57. Hou YJ, Chiba S, Halfmann P, et al. SARS-CoV-2 D614G variant exhibits efficient replication ex vivo and transmission in vivo [J]. Science. 2020;370(6523):1464–8.
58. Korber B, Fischer WM, Gnanakaran S, et al. Tracking changes in SARS-CoV-2 spike: evidence that D614G increases infectivity of the COVID-19 virus [J]. Cell. 2020;182(4): 812–827.e19.
59. Plante JA, Liu Y, Liu J, et al. Spike mutation D614G alters SARS-CoV-2 fitness [J]. Nature. 2021;592(7852):116–21.
60. Zhou B, Thao TTN, Hoffmann D, et al. SARS-CoV-2 spike D614G change enhances replication and transmission [J]. Nature. 2021;592(7852):122–7.
61. Benton DJ, Wrobel AG, Roustan C, et al. The effect of the D614G substitution on the structure of the spike glycoprotein of SARS-CoV-2 [J]. Proc Natl Acad Sci. 2021;118(9):e2022586118.
62. Volz E, Mishra S, Chand M, et al. Transmission of SARS-CoV-2 Lineage B. 1.1. 7 in England: insights from linking epidemiological and genetic data [J]. MedRxiv. 2021:2020.12. 30.20249034.
63. Starr TN, Greaney AJ, Hilton SK, et al. Deep mutational scanning of SARS-CoV-2 receptor binding domain reveals constraints on folding and ACE2 binding [J]. Cell. 2020;182(5): 1295–310. e20
64. Tegally H, Wilkinson E, Giovanetti M, et al. Detection of a SARS-CoV-2 variant of concern in South Africa [J]. Nature. 2021;592(7854):438–43.
65. Faria NR, Mellan TA, Whittaker C, et al. Genomics and epidemiology of the P. 1 SARS-CoV-2 lineage in Manaus, Brazil [J]. Science. 2021;372(6544):815–21.

66. Lasek-Nesselquist E, Lapierre P, Schneider E, et al. The localized rise of a B. 1.526 SARS-CoV-2 variant containing an E484K mutation in New York State [J]. MedRxiv. 2021:2021.02. 26.21251868.
67. Annavajhala MK, Mohri H, Wang P, et al. A novel and expanding SARS-CoV-2 variant, B. 1.526, identified in New York [J]. MedRxiv. 2021:2021.02. 23.21252259.
68. Hoffmann M, Hofmann-Winkler H, Krüger N, et al. SARS-CoV-2 variant B. 1.617 is resistant to bamlanivimab and evades antibodies induced by infection and vaccination [J]. Cell Rep. 2021;36(3):109415.
69. Jain M, Olsen HE, Paten B, et al. The Oxford Nanopore MinION: delivery of nanopore sequencing to the genomics community [J]. Genome Biol. 2016;17:1–11.
70. Quick J, Loman NJ, Duraffour S, et al. Real-time, portable genome sequencing for Ebola surveillance [J]. Nature. 2016;530(7589):228–32.
71. Quick J, Grubaugh ND, Pullan ST, et al. Multiplex PCR method for MinION and Illumina sequencing of Zika and other virus genomes directly from clinical samples [J]. Nat Protoc. 2017;12(6):1261–76.
72. James P, Stoddart D, Harrington ED, et al. LamPORE: rapid, accurate and highly scalable molecular screening for SARS-CoV-2 infection, based on nanopore sequencing [J]. MedRxiv. 2020:2020.08. 07.20161737.
73. Li J, Quan W, Yan S, et al. Rapid detection of SARS-CoV-2 and other respiratory viruses by using LAMP method with Nanopore Flongle workflow [J]. BioRxiv. 2020:2020.06. 03.131474.
74. Cyranoski D. Alarming COVID variants show vital role of genomic surveillance [J]. Nature. 2021;589(7842):337–9.
75. Simonetti M, Zhang N, Harbers L, et al. COVseq is a cost-effective workflow for mass-scale SARS-CoV-2 genomic surveillance [J]. Nat Commun. 2021;12(1):3903.
76. Bull RA, Adikari TN, Ferguson JM, et al. Analytical validity of nanopore sequencing for rapid SARS-CoV-2 genome analysis [J]. Nat Commun. 2020;11(1):6272.
77. Park SY, Faraci G, Ward PM, et al. High-precision and cost-efficient sequencing for real-time COVID-19 surveillance [J]. Sci Rep. 2021;11(1):13669.
78. Yeo LY, Chang HC, Chan PP, et al. Microfluidic devices for bioapplications [J]. Small. 2011;7(1):12–48.
79. Antonelli G, Filippi J, D'orazio M, et al. Integrating machine learning and biosensors in microfluidic devices: a review [J]. Biosens Bioelectron. 2024;263:116632.
80. Goncalves M, Goncalves IM, Borges J, et al. Polydimethylsiloxane surface modification of microfluidic devices for blood plasma separation [J]. Polymers. 2024;16(10):1416.
81. Luppa PB, Muller C, Schlichtiger A, et al. Point-of-care testing (POCT): current techniques and future perspectives [J]. Trends Anal Chem. 2011;30(6):887–98.
82. Dighe K, Moitra P, Alafeef M, et al. A rapid RNA extraction-free lateral flow assay for molecular point-of-care detection of SARS-CoV-2 augmented by chemical probes [J]. Biosens Bioelectron. 2022;200:113900.
83. Lee S, Bi L, Chen H, et al. Recent advances in point-of-care testing of COVID-19 [J]. Chem Soc Rev. 2023;52(24):8500–30.
84. Bartosik M, Moranova L, Izadi N, et al. Advanced technologies towards improved HPV diagnostics [J]. J Med Virol. 2024;96(2):e29409.
85. Cheng Y, Wu A, Liu L, et al. Immunochromatographic strip for rapid and sensitive detection of bupirimate residues in peach, orange, and carrot [J]. Food Chem. 2024;459:140417.
86. Younes N, Yassine HM, Kourentzi K, et al. A review of rapid food safety testing: using lateral flow assay platform to detect foodborne pathogens [J]. Crit Rev Food Sci Nutr. 2024;64(27):9910–32.
87. Wang M, Feng J, Ding J, et al. Color-and background-free Raman-encoded lateral flow immunoassay for simultaneous detection of carbendazim and imidacloprid in a single test line [J]. Chem Eng J. 2024;487:150666.

88. Kim S, Hao Y, Miller EA, et al. Vertical flow cellulose-based assays for SARS-CoV-2 antibody detection in human serum [J]. ACS Sensors. 2021;6(5):1891–8.
89. Goncharov A, Joung HA, Ghosh R, et al. Deep learning-enabled multiplexed point-of-care sensor using a paper-based fluorescence vertical flow assay [J]. Small. 2023;19(51):2300617.
90. Jiao Y, Du C, Zong L, et al. 3D vertical-flow paper-based device for simultaneous detection of multiple cancer biomarkers by fluorescent immunoassay [J]. Sensors Actuators B Chem. 2020;306:127239.
91. Kropaneva M, Khramtsov P, Bochkova M, et al. Vertical flow immunoassay based on carbon black nanoparticles for the detection of IgG against SARS-CoV-2 spike protein in human serum: proof-of-concept [J]. Biosensors. 2023;13(9):857.
92. Budd J, Miller BS, Weckman NE, et al. Lateral flow test engineering and lessons learned from COVID-19 [J]. Nat Rev Bioeng. 2023;1(1):13–31.
93. Mannino RG, Nehl EJ, Farmer S, et al. The critical role of engineering in the rapid development of COVID-19 diagnostics: lessons from the RADx tech test verification core [J]. Sci Adv. 2023;9(14):eade4962.
94. Osman SEG, Ahmed EMEMM, Hussain NE. Sensitivity and specificity of gold chromatography immuno-assays IgM/IgG antibody test for COVID-19: review of the current literature [J]. Sudan J Med Sci. 2020;15(5):71–9.
95. Li G, Wang A, Chen Y, et al. Development of a colloidal gold-based immunochromatographic strip for rapid detection of severe acute respiratory syndrome coronavirus 2 spike protein [J]. Front Immunol. 2021;12:635677.
96. Ionescu RE. Updates on the biofunctionalization of gold nanoparticles for the rapid and sensitive multiplatform diagnosis of SARS-CoV-2 virus and its proteins: from computational models to validation in human samples [J]. Int J Mol Sci. 2023;24(11):9249.
97. Zou M, Su F, Zhang R, et al. Rapid point-of-care testing for SARS-CoV-2 virus nucleic acid detection by an isothermal and nonenzymatic signal amplification system coupled with a lateral flow immunoassay strip [J]. Sensors Actuators B Chem. 2021;342:129899.
98. He J, Zhu S, Zhou J, et al. Rapid detection of SARS-CoV-2: the gradual boom of lateral flow immunoassay [J]. Front Bioeng Biotechnol. 2023;10:1090281.
99. Zhu X, Wang X, Li S, et al. Rapid, ultrasensitive, and highly specific diagnosis of COVID-19 by CRISPR-based detection [J]. ACS Sensors. 2021;6(3):881–8.
100. Sun Y, Yu L, Liu C, et al. One-tube SARS-CoV-2 detection platform based on RT-RPA and CRISPR/Cas12a [J]. J Transl Med. 2021;19:1–10.
101. Zheng C, Wang K, Zheng W, et al. Rapid developments in lateral flow immunoassay for nucleic acid detection [J]. Analyst. 2021;146(5):1514–28.
102. Wang N, Zhang J, Xiao B, et al. Microfluidic-assisted integrated nucleic acid test strips for POCT [J]. Talanta. 2024;267:125150.
103. Tabatabaei MS, Islam R, Ahmed M. Applications of gold nanoparticles in ELISA, PCR, and immuno-PCR assays: a review [J]. Anal Chim Acta. 2021;1143:250–66.
104. Liu D, Shen H, Zhang Y, et al. A microfluidic-integrated lateral flow recombinase polymerase amplification (MI-IF-RPA) assay for rapid COVID-19 detection [J]. Lab Chip. 2021;21(10):2019–26.
105. Hull IT, Kline EC, Gulati GK, et al. Isothermal amplification with a target-mimicking internal control and quantitative lateral flow readout for rapid HIV viral load testing in low-resource settings [J]. Anal Chem. 2021;94(2):1011–21.
106. Li Z, Ding X, Yin K, et al. Instrument-free, CRISPR-based diagnostics of SARS-CoV-2 using self-contained microfluidic system [J]. Biosens Bioelectron. 2022;199:113865.
107. Zhang J, Tang L, Yu Q, et al. Gold-platinum nanoflowers as colored and catalytic labels for ultrasensitive lateral flow MicroRNA-21 assay [J]. Sensors Actuators B Chem. 2021;344:130325.
108. Hu J, Wang L, Li F, et al. Oligonucleotide-linked gold nanoparticle aggregates for enhanced sensitivity in lateral flow assays [J]. Lab Chip. 2013;13(22):4352–7.

109. Vindeirinho JM, Pinho E, Gomes L, et al. RT-RAA with a lateral flow assay readout based on ssDNA hybridization for detection of RNA viruses–the case of SARS-CoV-2 [J]. Sensors Actuators B Chem. 2025;426:136864.
110. Li M, Yin F, Song L, et al. Nucleic acid tests for clinical translation [J]. Chem Rev. 2021;121(17):10469–558.
111. Pengcheng W, Jiaren S, Caixia S, et al. Recent advances of lateral flow immunoassay for bacterial detection: capture-antibody-independent strategies and high-sensitivity detection technologies [J]. TrAC Trends Anal Chem. 2023;166:117203.
112. Islam MA, Karim A, Ethiraj B, et al. Antimicrobial peptides: promising alternatives over conventional capture ligands for biosensor-based detection of pathogenic bacteria [J]. Biotechnol Adv. 2022;55:107901.
113. Serafin B, Kamen A, De Crescenzo G, et al. Antibody-independent surface plasmon resonance assays for influenza vaccine quality control [J]. Appl Microbiol Biotechnol. 2024;108(1):307.
114. Zhang T, Deng R, Wang Y, et al. A paper-based assay for the colorimetric detection of SARS-CoV-2 variants at single-nucleotide resolution [J]. Nat Biomed Eng. 2022;6(8):957–67.
115. Chen L, Choo J. Recent advances in surface-enhanced Raman scattering detection technology for microfluidic chips [J]. Electrophoresis. 2008;29(9):1815–28.
116. Yager P, Edwards T, Fu E, et al. Microfluidic diagnostic technologies for global public health [J]. Nature. 2006;442(7101):412–8.
117. Bargahi N, Ghasemali S, Jahandar-Lashaki S, et al. Recent advances for cancer detection and treatment by microfluidic technology, review and update [J]. Biol Proced Online. 2022;24(1):5.
118. Han X, Zhang Y, Tian J, et al. Polymer-based microfluidic devices: a comprehensive review on preparation and applications [J]. Polym Eng Sci. 2022;62(1):3–24.
119. Borók A, Laboda K, Bonyár A. PDMS bonding technologies for microfluidic applications: a review [J]. Biosensors. 2021;11(8):292.
120. Funano S-I, Ota N, Tanaka Y. A simple and reversible glass–glass bonding method to construct a microfluidic device and its application for cell recovery [J]. Lab Chip. 2021;21(11):2244–54.
121. Lin W, Yuan L, Gao Z, et al. An integrated sample-to-answer SERS platform for multiplex phenotyping of extracellular vesicles [J]. Sensors Actuators B Chem. 2023;394:134355.
122. Basiri A, Heidari A, Nadi MF, et al. Microfluidic devices for detection of RNA viruses [J]. Rev Med Virol. 2021;31(1):1–11.
123. Horny M-C, Freisa M, Poujouly C, et al. Electrochemical nucleic acid biosensor on a microfluidic chip to understand the coupling between electrochemistry and microfluidics [J]. J Chem Educ. 2024;101(2):605–11.
124. Sun Z, Lin K-F, Zhao Z-H, et al. An automated nucleic acid detection platform using digital microfluidics with an optimized Cas12a system [J]. Sci China Chem. 2022;65(3):630–40.
125. Chen S, Sun Y, Fan F, et al. Present status of microfluidic PCR chip in nucleic acid detection and future perspective [J]. TrAC Trends Anal Chem. 2022;157:116737.
126. Su W, Liang D, Tan M. Nucleic acid-based detection for foodborne virus utilizing microfluidic systems [J]. Trends Food Sci Technol. 2021;113:97–109.
127. Tian F, Liu C, Deng J, et al. A fully automated centrifugal microfluidic system for sample-to-answer viral nucleic acid testing [J]. Sci China Chem. 2020;63:1498–506.
128. Sarabi MR, Yigci D, Alseed MM, et al. Disposable paper-based microfluidics for fertility testing [J]. IScience. 2022;25(9):104986.
129. Anushka BA, Das PK. Paper based microfluidic devices: a review of fabrication techniques and applications [J]. Eur Phys J Spec Top. 2023;232(6):781–815.
130. Li W, Ma X, Yong Y-C, et al. Review of paper-based microfluidic analytical devices for in-field testing of pathogens [J]. Anal Chim Acta. 2023;1278:341614.
131. Davidson JL, Wang J, Maruthamuthu MK, et al. A paper-based colorimetric molecular test for SARS-CoV-2 in saliva [J]. Biosens Bioelectron X. 2021;9:100076.

132. Yuan M, Li C, Zheng Y, et al. A portable multi-channel fluorescent paper-based microfluidic chip based on smartphone imaging for simultaneous detection of four heavy metals [J]. Talanta. 2024;266:125112.
133. Baranwal J, Barse B, Gatto G, et al. Electrochemical sensors and their applications: a review [J]. Chem. 2022;10(9):363.
134. Mceachern F, Harvey E, Merle G. Emerging technologies for the electrochemical detection of bacteria [J]. Biotechnol J. 2020;15(9):2000140.
135. Naveen MH, Gurudatt NG, Shim Y-B. Applications of conducting polymer composites to electrochemical sensors: a review [J]. Appl Mater Today. 2017;9:419–33.
136. Saputra HA. Electrochemical sensors: basic principles, engineering, and state of the art [J]. Monatsh Chem. 2023;154(10):1083–100.
137. Liu Y, Xue Q, Chang C, et al. Recent progress regarding electrochemical sensors for the detection of typical pollutants in water environments [J]. Anal Sci. 2022;38(1):55–70.
138. He J, Xu X, Li M, et al. Recent advances in perovskite oxides for non-enzymatic electrochemical sensors: a review [J]. Anal Chim Acta. 2023;1251:341007.
139. Zhu C, Yang G, Li H, et al. Electrochemical sensors and biosensors based on nanomaterials and nanostructures [J]. Anal Chem. 2015;87(1):230–49.
140. Ronkainen NJ, Halsall HB, Heineman WR. Electrochemical biosensors [J]. Chem Soc Rev. 2010;39(5):1747–63.
141. Hashem A, Hossain MM, Al Mamun M, et al. Nanomaterials based electrochemical nucleic acid biosensors for environmental monitoring: a review [J]. Appl Surf Sci Adv. 2021;4: 100064.
142. Amiri M, Bezaatpour A, Jafari H, et al. Electrochemical methodologies for the detection of pathogens [J]. ACS Sensors. 2018;3(6):1069–86.
143. Raziq A, Kidakova A, Boroznjak R, et al. Development of a portable MIP-based electrochemical sensor for detection of SARS-CoV-2 antigen [J]. Biosens Bioelectron. 2021;178:113029.
144. Ayankojo AG, Boroznjak R, Reut J, et al. Molecularly imprinted polymer based electrochemical sensor for quantitative detection of SARS-CoV-2 spike protein [J]. Sensors Actuators B Chem. 2022;353:131160.
145. Deng Y, Peng Y, Wang L, et al. Target-triggered cascade signal amplification for sensitive electrochemical detection of SARS-CoV-2 with clinical application [J]. Anal Chim Acta. 2022;1208:339846.
146. Ramírez-Chavarría RG, Castillo-Villanueva E, Alvarez-Serna BE, et al. Loop-mediated isothermal amplification-based electrochemical sensor for detecting SARS-CoV-2 in wastewater samples [J]. J Environ Chem Eng. 2022;10(3):107488.
147. Malla P, Liu C-H, Wu W-C, et al. Synthesis and characterization of Au-decorated graphene oxide nanocomposite for magneto-electrochemical detection of SARS-CoV-2 nucleocapsid gene [J]. Talanta. 2023;262:124701.
148. Shi K, Yi Z, Han Y, et al. PAM-free cascaded strand displacement coupled with CRISPR-Cas12a for amplified electrochemical detection of SARS-CoV-2 RNA [J]. Anal Biochem. 2023;664:115046.
149. Kashefi-Kheyrabadi L, Nguyen HV, Go A, et al. Ultrasensitive and amplification-free detection of SARS-CoV-2 RNA using an electrochemical biosensor powered by CRISPR/Cas13a [J]. Bioelectrochemistry. 2023;150:108364.
150. Yakoh A, Pimpitak U, Rengpipat S, et al. Based electrochemical biosensor for diagnosing COVID-19: detection of SARS-CoV-2 antibodies and antigen [J]. Biosens Bioelectron. 2021;176:112912.
151. Karakuş E, Erdemir E, Demirbilek N, et al. Colorimetric and electrochemical detection of SARS-CoV-2 spike antigen with a gold nanoparticle-based biosensor [J]. Anal Chim Acta. 2021;1182:338939.
152. Gomide JA, Moço AC, Tanaka LT, et al. Advancement in SARS-CoV-2 diagnosis: a new and stable electrochemical biosensor for genomic RNA detection [J]. Bioelectrochemistry. 2025;161:108798.

153. Kashefi-Kheyrabadi L, Nguyen HV, Go A, et al. Rapid, multiplexed, and nucleic acid amplification-free detection of SARS-CoV-2 RNA using an electrochemical biosensor [J]. Biosens Bioelectron. 2022;195:113649.
154. Patil TP, Parkhe VS, Kundale SS, et al. Antisense oligonucleotide conjugated gold nanoconstructs-based electrochemical biosensor for detection of SARS-CoV-2 [J]. Appl Surf Sci Adv. 2024;22:100618.
155. Alafeef M, Skrodzki D, Moitra P, et al. Binding-induced folding of DNA oligonucleotides targeted to the nucleocapsid gene enables electrochemical sensing of SARS-CoV-2 [J]. ACS Appl Bio Mater. 2023;6(3):1133–45.
156. Kumar N, Shetti NP, Jagannath S, et al. Electrochemical sensors for the detection of SARS-CoV-2 virus [J]. Chem Eng J. 2022;430:132966.
157. Antiochia R. Developments in biosensors for CoV detection and future trends [J]. Biosens Bioelectron. 2021;173:112777.
158. Soh JO, Park BC, Park HS, et al. Multifunctional nanoparticle platform for surface accumulative nucleic acid amplification and rapid electrochemical detection: an application to pathogenic coronavirus [J]. ACS Sensors. 2023;8(2):839–47.
159. Najjar D, Rainbow J, Sharma Timilsina S, et al. A lab-on-a-chip for the concurrent electrochemical detection of SARS-CoV-2 RNA and anti-SARS-CoV-2 antibodies in saliva and plasma [J]. Nat Biomed Eng. 2022;6(8):968–78.
160. Mei Y, Lin X, He C, et al. Recent progresses in electrochemical DNA biosensors for SARS-CoV-2 detection [J]. Front Bioeng Biotechnol. 2022;10:952510.
161. Zambry NS, Obande GA, Khalid MF, et al. Utilizing electrochemical-based sensing approaches for the detection of SARS-CoV-2 in clinical samples: a review [J]. Biosensors. 2022;12(7):473.
162. Bisht A, Mishra A, Bisht H, et al. Nanomaterial based biosensors for detection of viruses including SARS-CoV-2: a review [J]. J Anal Test. 2021;5(4):327–40.
163. Sengupta J, Hussain CM. Graphene-based electrochemical nano-biosensors for detection of SARS-CoV-2 [J]. Inorganics. 2023;11(5):197.
164. Pramanik A, Gao Y, Patibandla S, et al. Bioconjugated nanomaterial for targeted diagnosis of SARS-CoV-2 [J]. Acc Mater Res. 2022;3(2):134–48.
165. Bai H, Wang Y, Li X, et al. Electrochemical nucleic acid sensors: competent pathways for mobile molecular diagnostics [J]. Biosens Bioelectron. 2023;237:115407.
166. Huang Z, Wang W, Wang Y, et al. Electrochemical detection of viral nucleic acids by DNA nanolock-based porous electrode device [J]. Anal Chem. 2023;95(45):16668–76.
167. Naorungroj S, Srisomwat C, Khamcharoen W, et al. Sequential flow controllable microfluidic device for G-quadruplex DNAzyme-based electrochemical detection of SARS-CoV-2 using a pyrrolidinyl peptide nucleic acid [J]. Anal Chem. 2023;95(34):12794–801.
168. Farzin L, Sadjadi S, Sheini A, et al. A nanoscale genosensor for early detection of COVID-19 by voltammetric determination of RNA-dependent RNA polymerase (RdRP) sequence of SARS-CoV-2 virus [J]. Microchim Acta. 2021;188:1–12.
169. Heo W, Lee K, Park S, et al. Electrochemical biosensor for nucleic acid amplification-free and sensitive detection of severe acute respiratory syndrome coronavirus 2 (SARS-CoV-2) RNA via CRISPR/Cas13a trans-cleavage reaction [J]. Biosens Bioelectron. 2022;201:113960.
170. Chaibun T, Puenpa J, Ngamdee T, et al. Rapid electrochemical detection of coronavirus SARS-CoV-2 [J]. Nat Commun. 2021;12(1):802.
171. Ji D, Guo M, Wu Y, et al. Electrochemical detection of a few copies of unamplified SARS-CoV-2 nucleic acids by a self-actuated molecular system [J]. J Am Chem Soc. 2022;144(30): 13526–37.
172. Alafeef M, Dighe K, Moitra P, et al. Rapid, ultrasensitive, and quantitative detection of SARS-CoV-2 using antisense oligonucleotides directed electrochemical biosensor chip [J]. ACS Nano. 2020;14(12):17028–45.
173. Kong D, Wang X, Gu C, et al. Direct SARS-CoV-2 nucleic acid detection by Y-shaped DNA dual-probe transistor assay [J]. J Am Chem Soc. 2021;143(41):17004–14.

174. Wang L, Wang X, Wu Y, et al. Rapid and ultrasensitive electromechanical detection of ions, biomolecules and SARS-CoV-2 RNA in unamplified samples [J]. Nat Biomed Eng. 2022;6 (3):276–85.
175. Wu Y, Ji D, Dai C, et al. Triple-probe DNA framework-based transistor for SARS-CoV-2 10-in-1 pooled testing [J]. Nano Lett. 2022;22(8):3307–16.
176. Wang X, Kong D, Guo M, et al. Rapid SARS-CoV-2 nucleic acid testing and pooled assay by tetrahedral DNA nanostructure transistor [J]. Nano Lett. 2021;21(22):9450–7.
177. Sun Y, Yang C, Jiang X, et al. High-intensity vector signals for detecting SARS-CoV-2 RNA using CRISPR/Cas13a couple with stabilized graphene field-effect transistor [J]. Biosens Bioelectron. 2023;222:114979.
178. Chen S, Sun Y, Fan X, et al. Review on two-dimensional material-based field-effect transistor biosensors: accomplishments, mechanisms, and perspectives [J]. J Nanobiotechnol. 2023;21 (1):144.
179. Liang Q-H, Cao B-P, Xiao Q, et al. The application of graphene field-effect transistor biosensors in COVID-19 detection technology: a review [J]. Sensors. 2023;23(21):8764.
180. Fattahi Z, Hasanzadeh M. Nanotechnology-assisted microfluidic systems for chemical sensing, biosensing, and bioanalysis [J]. TrAC Trends Anal Chem. 2022;152:116637.
181. Sun Y, Liu K, Zhang H, et al. A tube-based biosensor for DNA and RNA detection [J]. Sci Adv. 2025;11(18):eadu2271.

MIX
Papier aus verantwortungsvollen Quellen
Paper from responsible sources
FSC® C105338

If you have any concerns about our products,
you can contact us on
ProductSafety@springernature.com

In case Publisher is established outside the EU,
the EU authorized representative is:
Springer Nature Customer Service Center GmbH
Europaplatz 3, 69115 Heidelberg, Germany

Printed by Libri Plureos GmbH
in Hamburg, Germany